MARINE ALGAE
OF THE EASTERN TROPICAL
AND SUBTROPICAL COASTS
OF THE
AMERICAS

MARINE ALGAE OF THE EASTERN TROPICAL AND SUBTROPICAL COASTS OF THE AMERICAS

BY

WILLIAM RANDOLPH TAYLOR

ANN ARBOR

THE UNIVERSITY OF MICHIGAN PRESS

Published in the United States of America by The University of Michigan Press and simultaneously in Toronto, Canada, by Ambassador Books Limited

Library of Congress Catalog Card No. 59-9736

University of Michigan Studies
Scientific Series

Volume XXI

MANUFACTURED IN THE UNITED STATES OF AMERICA
BY THE LORD BALTIMORE PRESS, INC.

In appreciation
of the aid given by my wife
Jean Grant Taylor
who assisted in compiling
the initial list
from which this manual has grown,
furthered it under tropical suns,
and encouraged it through the years
of laboratory and museum work
at home and abroad

PREFACE

THIS work is the direct, if slowly matured, result of the three stimulating summer periods I spent at the former Dry Tortugas Laboratory of the Carnegie Institution of Washington (1924–26). Under conditions almost spartan, but ideal for field and simple laboratory studies, an admiration for the marine algae of the tropics was acquired which has grown through the years with increase in familiarity with their elegance of form and the complex economy of their structure. I have had many occasions to be grateful to the late W. H. Longley, the kindly Director, who introduced me to what was essentially the tropical marine flora of eastern America.

The reports resulting from work done at the Laboratory, published by the Institution (1924–28), caused a continuous flow of collections to come to me for identification, because for the first time an illustrated manual dealing with many of the American tropical algae became readily available in this country, and it stimulated people to collect algae in Florida and the West Indies. It was evident that a local flora for part of Florida would not serve for identification of algae from all parts of the eastern American tropics and subtropics, and the collectors naturally turned for help to the author of the one familiar book. Unfortunately, the applicable literature was very dispersed and hard to correlate, the collections in herbaria, except in Cambridge and New York, quite inconsiderable, and so I often could not come to clear decisions as to the identities of the plants sent. I soon realized that I needed, and was in the best position to initiate, a more comprehensive manual. It was a project to be developed over several years, though I did not imagine that thirty would elapse before it was ready to be published. With the help of my wife a skeleton manuscript was prepared, showing by references to the literature the sources of the geographical records of each alga. This annotated list has been repeatedly revised over the years, typed again and again, and expanded into its present descriptive form.

To the succession of graduate student assistants and illustrators who have helped convert the first draft into final shape I am most grateful. Dr. Hannah T. Croasdale helped by translating the descriptions of scores of species out of or into Latin. The last two complete typings were the exacting responsibilities of Dr. Lewellya W. Hillis and Egon M. Gross, to whom hearty thanks are due. It is no more possible to list all who have dealt with this manuscript than it would be possible to list the many people who have contributed material for study. In any case, most of the latter have been appreciatively mentioned in the shorter papers which I have occasionally prepared.

For loans of reference materials or opportunities to study in the museums I am deeply indebted to the directors and curators of the Farlow Herbarium of Harvard University and of the herbaria of the New York Botanical Garden, the University of California at Berkeley, the United States National Museum, the British Museum (Natural History), the Royal Botanic Gardens at Kew, the universities of Lund, Uppsala, Copenhagen, and Leiden, and the Muséum National d'Histoire Naturelle in Paris. The personal encouragement and technical advice given by the late F. S. Collins, M. A. Howe, W. A. Setchell, and F. Børgesen, all most distinguished phycologists, supported me for many years.

Institutional aid in field studies has not been lacking; the whole project has been dependent on ample time for research, opportunities to go into the field, and financial support. The chairmen of the Botany Department of the University of Michigan, successively Professors H. H. Bartlett, W. C. Steere, and K. L. Jones, and the deans of the College of Literature, Science, and the Arts have provided suitable freedom for research; the deans of the Horace H. Rackham School of Graduate Studies have provided grants for assistants and incidental expenses from the Faculty Research Fund. Every summer since the work was begun I have had the facilities and superb library of the Marine Biological Laboratory at Woods Hole, Massachusetts, at my disposal, and often a good part of the summer was devoted to microscopic studies, comparison with local species, and bibliographic work on this project at the Laboratory.

The American Philosophical Society made a grant toward the initial Bermuda studies; to it, to the directors of the Bermuda Biological Station for Research, Professor Dugald E. S. Brown and Dr. William H. Sutcliffe, Jr., and to Dr. A. J. Bernatowicz, who was my direct assistant in the Bermuda work, I owe much. As the guest of Captain G. Allan Hancock, Director of the Hancock Foundation of the University of Southern California, I was able to visit under the happiest circumstances many most interesting places en route to Trinidad and the Galapagos. The Director of the Institute of Jamaica, C. Bernard Lewis, and his assistant in charge of the algal collections, Miss Lena Pierce, gave unstinted and much appreciated aid during my Jamaica visit. The National Science Foundation made a generous grant toward the expenses of the field studies in Jamaica and of the second Bermuda visit; it later made an additional grant to supply an assistant and underwrite the manufacturing costs of this book, without which it could not have been completed and for which I am correspondingly grateful.

In offering this manual to botanists it is well to remind them that this is in the aggregate a pioneer work. Without doubt, essentially all species likely to be encountered are described. Much has been done to amplify the descriptions and render them comparable, but the range of variation in many marine algae is extraordinary, and formal descriptions fail to cover all contingencies. Very many new locality records have been added from museum material. References are given to records of the several species under the various names which appear to have been used in earlier reports. However, the book cannot claim monographic completeness for all groups, and the residue of records I treat as doubtful will at some future time be confirmed, assigned to established species, or, more probably, be abandoned for lack of evidence. Many illustrations of characteristic species have been prepared. Illustrations of minute species and of structural details have had to be omitted for lack of space, a matter of great regret. All things considered, I feel that anyone who is willing to study his specimens thoroughly, doing the necessary microscopic work, may through this book reach satisfactory identifications of most of his plants.

CONTENTS

INTRODUCTION

HISTORICAL SURVEY

THE history of the study of marine botany found in warmer American seas is not a richly documented one. It would seem that this algal flora became relatively familiar to botanists without the publication of many elaborate or formal studies. These plants, appealing to inquisitive travelers, reached European herbaria quite often, but seldom in extensive collections; special collectors were not sent to get them, as so often happened for land plants, birds, and insects. Persons interested in marine algae gradually came to know the general characteristics of the flora and the more striking species without these having been the subject of notable local surveys.

At first there were rather casual references to various "fuci" as curiosities; later algae were treated as seriously as other plants (Sloane 1696b, 1707–25, Browne 1756). Then quite adequate references and illustrations of them began to appear singly in general works, such as Dawson Turner's beautiful *Historia Fucorum* (1808–19). References of this sort later became expanded into species lists based on particular explorations, such as we find, on a very elegant scale, in Martius (1826) for Brazil and in Montagne for Cuba (1842). The possibility of preparing such lists arose with the appearance of the first really encyclopedic books on algae, like C. A. Agardh's *Species Algarum* . . . . (1820–28), where in one work one could find a reasonable summary of the known species of the whole world, and of their characteristics.

The third quarter of the nineteenth century brought a fine group of publications, which may be accepted as completing the first stage in our knowledge of the algae of the region. Curiously, activity then stopped for twenty-five years or more, and we have little of note until local studies on modern standards began to appear well after the beginning of the twentieth century. At this point it will be best to shift our attention to particular localities and publications relating to them.

Studies at the northern end of the mainland range begin, for practical purposes, with Harvey's *Nereis Boreali-Americana* (1852–58a), many of the most striking genera being described and illustrated from Florida material. Bailey's (1848) and Hooper's (1850) few references were hardly influential, and progress was not continuous. Farlow (1875, 1876) took up Harvey's records and added a few; Mrs. Floretta A. Curtiss (A. H. Curtiss 1899) and Mrs. G. A. Hall collected assiduously, but left it to J. G. Agardh and others to describe the plants in their more general works. No comprehensive studies were attempted before Hoyt's paper (1920) on the transitional flora of North Carolina and mine (1928) on that of southern Florida. Nevertheless, the Chlorophyceae of the entire area were treated quite carefully in Collins' *Green Algae of North America* (1909–18a), and so our knowledge of this group early exceeded that of all the others.

A short-lived interest in the algae of Cuba developed, chiefly shown by the publications of Montagne (1842, 1863), but he was only acting as the European expert and not as the active observer in the field. Cuban studies then ceased for a long time; Howe (1915), Sánchez Alfonso (1930), and Castellanos (1945) published a few identifications, but the most numerous records, particularly of Isle of Pines algae collected by E. P. Killip, appear in Taylor (1941b, 1954, 1955).

For Bermuda the story is much better; the islands are accessible, transportation about them is easy, and botanists have long been particularly interested in them. Early but inaccurate lists were published by Kemp (1857) and Rein (1873). Farlow collected there extensively in 1881 and 1900, but his discoveries chiefly appeared in exsiccatae. While the famous Challenger Expedition visited the colony and published a list of the algae, this was mainly based on records of earlier workers. Next, having contributed many Bermudian specimens to the *Phycotheca Boreali-Americana* and having made notable contributions to the flora, Collins consolidated his earlier short papers in a joint work with a fellow-collector (Collins and Hervey 1917), and Howe did likewise in his chapter of a general Bermuda flora (Howe 1918a). Finally, I, with A. J. Bernatowicz, visited the islands in 1949. Dr. Bernatowicz continued his studies for about two years; I returned in 1956 for further work and for brief visits in 1958 and 1959. The numerous observations we have made have been utilized in the preparation of the present book.

The Bahamas begin to appear substantially in algal literature with the collections made by Edward Palmer (Eaton 1875). Howe collected there extensively in 1904–7. Numerous records of his findings appear in the *Phycotheca* . . . and his published papers (1904 *et seq.*), but culminate in his chapter in Britton and Millspaugh's *Bahama Flora* (1920).

Jamaica seems to enter phycological history significantly with the famous work of Sir Hans Sloane (1707): 43 marine plants were reported, but after removing marine phanerogams, corals, and other extraneous items, only four recognizable species remain. Browne (1756) also mentions algae, as does Lunan (1814), but with no substantial additions. Modern collections begin with the visits of Pease and Butler (1891 and later), Humphrey (1893), and Duerden (1901), who all contributed specimens to the *Phycotheca* . . . . Collins (1901), using plants from other sources and his own collections, summarized these records in an extended paper, bringing the total species for the island to the impressive number of 224 (including the Myxophyceae but not the coralline algae), undoubtedly the best-authenticated list from the area up to that date. Small additions were made by me on specimens sent for determination (1929, 1933), and in 1956 I examined the material in the Science Museum at Kingston and collected very extensively around the island for three months.

Here one must introduce the notable list of West Indian marine algae compiled by Murray (1889). Murray had access to the splendid herbaria of the British Museum (Natural History) and Kew, including the numerous if small collections of algae deposited there from time to time by botanical travelers to the West Indies. Jamaica is represented, though sparingly, and so are many other islands, but Murray was quite uncritical, and he incorporated many records based on misidentifications, the *nomina nuda* of Mazé and Schramm, and other misleading material. In spite of all this, his is still the only comprehensive survey of the West Indian flora, and many of the records from smaller islands rest solely on his report.

Hispaniola, the second largest of the Antilles, has been sadly neglected, partly because of the relative difficulty of coastal transportation in the past. Børgesen (1924) reported a few species from the satellite Beata Island. Arndt collected quite extensively in the southwestern area in 1925–29 and with me reported the findings

(1929) in a substantial list, to which a number of species were added through collections made by C. R. Orcutt, H. H. Bartlett, and others (Taylor 1933, 1940, 1943).

Puerto Rico early furnished a substantial quota of over 90 species reported by Hauck (1888), but very little has since appeared in the literature. Howe (1915) visited the island, and the resulting collections lie in the herbarium of the New York Botanical Garden; a few specimens were issued in the *Phycotheca* . . ., but the bulk was held for presentation in the great "Flora of Porto Rico," a volume he did not live to complete.

The marine algae of the Virgin Islands (while still the Danish West Indies) were thoroughly studied and described by Børgesen (1913–20), and his papers are the most informative of any dealing with West Indian algae. Though concrete descriptions of many species were not provided, numerous, most acute observations are recorded for many others.

Famous among the works on algae of the Lesser Antilles are the books based on the collections of H. Mazé and A. Schramm in Guadeloupe, the identifications having been made by the brothers Crouan, well-known phycologists of Brest. These are all rare books in the original issues: the first edition was set in type in two columns on only one side of the page; it was published in 1865 by "A. Schramm et H. Mazé" at Basse-Terre. Included are names of 462 algae, including 13 diatoms. The second issue with the same authorship is (except for the title page) an autographic one, lithographed on both sides of the paper, the lines running across the page, and was published in 1866 at Cayenne, French Guiana. Numerous changes appeared, and the algal list rose to 502 names, including diatoms. The third issue is mislabeled as the second edition on the title page, and, Schramm having died, is ascribed to "H. Mazé et A. Schramm." It was published at Basse-Terre and is printed, the lines of type running across the page; 940 algae, including diatoms, are listed. This edition is relatively well known, because it has been reissued in facsimile. The brief descriptions accompanying some of the species in the earlier editions were curtailed in later ones; a great many of the supposed new species were never described. Although Murray (1889) accepted most of these records for his general catalogue, the names of a large proportion are invalid. Fortunately in the later editions most specimens were cited by number as well as by name and can be located in the herbaria at Paris, London, Kew, and perhaps elsewhere. Most of the plants mentioned as new can be assigned to earlier-described entities; sometimes the specimens in different herbaria under a given number are not identical. The collections, for all the faults of publication, were made with great discrimination.

Perhaps the last Antillean study calling for individual discussion is that made by Vickers (1905) at Barbados. She collected very critically on the island, and her plants are available, well documented, in several herbaria. After her death two parts of a study illustrated with beautiful drawings (the Chlorophyceae and Phaeophyceae) were published in one volume (1908).

Beyond this, space does not permit elaboration of the Antillean literature. Through the citations after the various species and the bibliography it will not be difficult to locate the sources of records from the other islands, insofar as these are based on the literature. For some, like Martinique, there is quite a little information; for others, like Trinidad and Tobago, remarkably little, and for most of the smaller islands there is none at all.

The mainland west of Florida has had an even poorer coverage of the shore line by botanists. I have elsewhere discussed the algae of the botanically barren or at least monotonous coast along the north shore of the Gulf of Mexico (1954a, c) and the slightly more varied Texas coast flora as well (1941a, b). The Caribbean coast of Mexico is probably richer; a few Mexican algae were reported as early as 1847 by J. G. Agardh, but nothing more substantial appeared until I was able to study and publish on the W. C. Steere collection from Yucatan (1935); the same paper reported a number of species from British Honduras collected by C. S. Lundell.

Next, Costa Rica comes into view with a small collection by C. W. Dodge and his party (Taylor 1933), and then Panama, with further collections by Dodge (Taylor 1929), by Howe (1910), and by Taylor (1942). The long coast from Panama to the Guianas has received but the scantiest attention. The only early work on the Netherlands Antilles is the paper by Sluiter (1908). These islands lie close to the Venezuelan coast, where A. F. Blakeslee collected a few species (Taylor 1929). Both regions were visited by the Hancock Expedition in 1939, when I collected rather more extensively, though the list for Venezuela is still brief (Taylor 1942). The

Guianas, and especially French Guiana, figured quite importantly in the reports by Montagne (1850, and scattered references) of tropical algae collected by LePrieur, but nothing further has appeared.

Brazil has been more favored. The algae share, if somewhat inadequately, in the great reports on the flora prepared by Martius (1826 *et seq.*), and at about the same time Greville (1833a) and Montagne (1839a) reported on algae collected by A. Saint Hilaire. A new group of collectors, notably H. Schenck and A. Glaziou, appeared some thirty years later, and Martens (1870, 1871), Zeller (1876), and Möbius (1889 *et seq.*) reported on their specimens in several papers. After another span of nearly half a century several important collections had accumulated in America, and with Howe (1928, 1931) and Taylor (1930a) reporting on them, the list of species increased very greatly. Consequently, I attempted a synoptic catalogue in 1931, and at various other times have reported on smaller groups of specimens. However, the most important recent reports are those of Joly (1950 *et seq.*), introducing a new era of field study. It is not intended to include the algae of Uruguay in this volume in any comprehensive way, for the flora shows a marked change from that of the better-known parts of Brazil. However, as I have seen a little material from Uruguay (Taylor 1939), I have included the extensions of range into that country of plants known somewhat to the north.

The foregoing may seem a long and perhaps tedious recital of minor researches. Even so, this summary is only selective. The bibliography, however, attempts to be representative. The obscurity of many of the sources has been a main hindrance to the proper study and appreciation of the tropical American marine flora. One must remember that many species records are based on references buried in such general works as those of Turner, the Agardhs, Kützing, and De Toni, and it would be a mistake to assume that all records of species have been found, evaluated, and incorporated here. On the other hand, in the catalogue of species there are hundreds of records taken directly from herbarium material I have seen, which have not been published. It is believed that the total list, and even the detailed distribution of species, is as complete as could be expected of a pioneer work.

Within the territory covered by this book, namely from Bermuda and from North Carolina (respecting the more southern species) to southern Brazil, we may recognize 760 species as present in the flora, together with 140 varieties and forms of these, of greater or less distinctiveness.

GEOGRAPHICAL DISTRIBUTION

The floras in the various parts of the arctic and cooler north temperate zones are subject to a great enough range of conditions, seasonal and otherwise, to cause distinctive changes in the vegetation from place to place. This is less true in the region covered by the present book, partly because the seasonal changes are much reduced, partly because the major currents tend to spread tropical conditions north and south of their strict limits. Thus, the marine algae of the Bermudas at 32° 20′ N.L. and of Florida at 25° N.L. are almost solely those one would expect in Jamaica or along the north coast of South America, well within the tropics. This extreme northward spread of the tropical flora to the Bermudas, the result of their position relative to the Gulf Stream current, is less strongly shown on the mainland. Even so, the flora in North Carolina at 34° 45′ N.L. is more marked by the tropical element it contains than by the north temperate element, at least in the summer and in deeper water. The inhospitable shore line, of unstable stretches of sand or mud almost devoid of good natural attachment areas for algae, makes the flora so poor from New Jersey to the Carolinas—and to some extent thence to central Florida—that an analysis of it is unsatisfactory.

The southern limits of the tropical flora are also rather vague. Certainly so far as we know them the genera along the Brazilian coast are essentially tropical in character (Taylor 1931), but southward from the Ilha São Francisco neighborhood at 27° S.L. species begin to appear which are not known in the tropical area to the north, so that while the tropical flora dominates much farther south than Rio de Janeiro, it is reduced but perceptible even into Uruguay, though a south temperate flora has here in large part replaced it (Taylor 1939a).

If we attempt to analyze the Caribbean flora and its relationships more closely, we find that most of the region is so ill-explored respecting marine botany that any deductions based on all known species would be utterly misleading, for a large percentage of the algae is known from few places, but is progressively recorded as other regions are competently investigated. If, however, we consider only rather well-known, frequently reported species, we find that about 28 per cent of them are only known within the Caribbean, 11 per cent range to the north (but are not necessarily characteristically northern), and a far larger proportion, 33 per cent, range south along the Brazilian coast. The rest are more cosmopolitan forms.

I am not willing to venture real lists of species representative of different floristic areas and doubt if a valid series could be made up; however, I may venture examples of these categories (Taylor 1955). For instance, *Ulva lactuca* is an outstanding cosmopolite, though less abundant in the tropics than elsewhere. Plants which range well to the north of the Caribbean and not to the south (or somewhat doubtfully in the first case) would include *Sargassum filipendula* and *Dasya pedicellata*. By contrast, *Caulerpa sertularioides* and *Dictyota dentata* range southward (for I consider Bermuda and southern Florida within the characteristically tropical Caribbean area). As restricted to the Caribbean area our examples are *Cymopolia barbata* and *Herposiphonia secunda*. It is even harder to be sure of any real limitations of distribution within the Caribbean (other than possible endemics) because of the poor knowledge we have of most of the northern South American coast. As an example of a possible restriction one may consider Cymopolia, which seems to be limited to the Greater Antilles, Mexico, Florida, the Bahamas, and the Bermudas, and not to occur along the eastern islands or southern shores at all.

There are, of course, local conditions which modify the flora within our range. These are chiefly simple restrictions due to topography, where for instance, to a large extent the coast is sandy or muddy and supports few macroscopic species of algae. There has been a belief that Trinidad and Tobago have floras restricted by a surface layer of reduced salinity due to the discharge of the Rio Orinoco, but I did not see any evidence of a special restriction when there, and it would require a much more expert and prolonged study of their shores than has yet been given to prove the point. Of course, the shores near the estuaries of all the large rivers will have a poor flora of larger algae. The Mississippi in particular, discharging into a rather circumscribed Gulf, does affect the salt concentration for quite a distance, but probably the major causes of the barrenness of the north coast of the Gulf of Mexico are the unstable nature of the shores, which are often backed by swamps, and the low temperatures of the winter season, perceptibly lower than those of the tropical areas of the West Indies and even of the more northern islands with a tropical flora.

ALGAL HABITATS

It is difficult to discuss the relative values of marine and of terrestrial ecology and to adapt the same terminology for marine vegetation which has been developed, in some complexity, for land plants. For land plants the cyclic climatic changes are relatively slow, in fact annual, and the persisting chemical environmental factors are chiefly of the substratum and are localized in their effects on the root systems. On the other hand, with the marine algae the chemical nature of the substratum probably has its main effect on the vegetation in determining which sporelings can attach, and the physical nature in turn has its effect on which of these can resist dislodgement and come to maturity. Far outweighing these in range of effects are the fluctuating conditions in the aquatic environment directly surrounding the absorptive, photosynthetic, and reproductive parts of the plants. Such not only vary in some respects on a long seasonal cycle, but in other ways vary radically on diurnal or semidiurnal cycles, which themselves change monthly. Tidal conditions affecting the amount of light received are superimposed on the cycle of daylight, the direction and force of the tidal currents change, and in estuarine regions the osmotic pressure of the medium may change to a degree easily lethal to unadapted plants.

With these great differences in the nature of the problems involved, it is not difficult to realize that instrumentation designed for terrestrial study is ill-adapted to work along the borders of the sea. The force of the surf under even average conditions prevents the establishment of detailed plots and instrument stations in exposed areas. Observations are generally limited to times of favorable tidal and weather conditions, and the inevitable storms may cause such rapid destruction of a population that continuity of occupation from year to year cannot be expected. Microclimatic situations abound; the environment on the upper face of a rock at low tide in the tropics is vastly different from that a few centimeters distant on the under side, and the flora differs accordingly. Some of the difficulties of study could be overcome if there were many trained investigators, but it is evident that these are lacking; compared with land ecology, the littoral and sublittoral marine field is indeed poorly supported. Contrast with the ecological study of the marine plankton is equally great, for biological oceanography is well suited to precise, if costly, instrumentation and is relatively handsomely supported with ships and other equipment, though the number of investigators is not large.

Accepting the fact that ecological researches cannot be carried on along the shore as well as in the fields and woods, and in most cases cannot yet advance much beyond the descriptive phase, it is worth while to consider in what regions it is possible to operate with most scientific precision. Certainly this will not be in the arctic and subarctic; I have had an advanced student working in northernmost Labrador for two seasons, and the restriction of the vegetation by ice action limited intertidal observations to specially protected places and the ecological data to those available by simple inspection during a very brief season. Even in northwestern Europe and temperate northeastern North America it is difficult to plan studies that adequately span conditions throughout the year because of the storms and cold of winter. It is perhaps in the Mediterranean region and the southeastern Atlantic United States or Gulf of Mexico, where warm temperate and subtropical conditions prevail and there is access to scientific laboratories, that researches have in the past been most readily undertaken, though they are indeed all too few.

It is rapidly becoming possible, however, to work effectively in more characteristically tropical American places. Actually, the algal floras at Bermuda at 32° 20′ N.L. and at Key West in Florida differ very little from those of places technically in the tropics. There is an excellently equipped laboratory at Bermuda suitable for ecological work, although the Dry Tortugas, Florida, Laboratory closed some years ago. There are now good laboratories on the Gulf of Mexico coast, in the Bahamas, Puerto Rico, Barbados, and Curaçao, and yet others developing in Cuba, Jamaica, Venezuela, and elsewhere, so that tropical studies are entirely practicable.

TEXT FIGURE 1. Abrupt rocky shore margin above a submerged shelf, at low tide. Stunted plants of *Sargassum bermudense* form a zone toward low-water line. The dark zone extending to high-water mark consists of small Gelidiums and similar plants. Bermuda.

Even under such good conditions ecological observations cannot be made with complete efficiency by one man working alone under marine conditions. Boats and boatmen will be required, and the manual aid of assistants in placing equipment and transcribing data —in short, a small but trained crew. The continuous working season is a great advantage in planning experimental work; the minimal tidal amplitude makes placing and reading of instruments easier. On the other hand the relatively stable water temperatures and insolation do restrict the types of study possible. Fully detailed and methodical studies of the American tropics lie in the future.

Let us consider, however, what we may find out by simple observations regarding the ecological conditions in some of the typical habitats met with in America. On the open rocky shores one may consider that the physical conditions for algal growth are most like those of northern coasts. One may look in vain here for a general zonation, although there may be an evident local one. The narrow tidal range relative to the area covered by spray is part of the reason. Conditions being suitable for a good algal growth, the low-tide line is likely to be bordered by a narrow, dense growth of stunted Sargassum (Text fig. 1) or less often of Turbinaria, but on an abrupt shore or cliff this may be absent. Above the Sargassum may be a rather wider zone of Colpomenia (Text fig. 2) or of dwarfed Laurencias, sun-bleached and pale. The top zone is likely to be a narrow pilose strip of Gelidium, Wurdemannia, or Gelidiella. One should not expect a zone of green algae, or a high zone of Myxophyceae, although these plants may be present, especially endolithic species, discoloring the calcareous rocks.

If, however, the rocks form a terrace at about low-water mark, the Sargassum zone is likely to be prominent and to border the terrace where it drops to deep water. Behind this the rock shelf presents one of the most deceptive ecological situations to be met along any shore. Unless the shelf is subject to the scouring action of wave-borne sand, it may be expected to be completely covered by a dense vegetation, but the plants may be so dwarfed, so covered with calcareous powder, as almost to defy recognition. The constituents vary very much from place to place but in relatively simple associations, each covering large areas. In the shallowest areas small Gelidiaceae and Centroceras may form the bulk of the vegetation, often densely colonized with tiny Padina plants, and the whole per-

TEXT FIGURE 2. Sloping rocky shore margin, at low tide. The *Sargassum bermudense* zone has in its upper part been overgrown with Colpomenia. Bermuda.

haps somewhat brown with diatoms. In places somewhat submerged even at low tide the aspect changes a little, but the concealment by calcareous material may continue. The flora here almost always is built on a stratum of crustose lithothamnia, and these even when living may be closely covered over broad areas with mats of dwarfed Digenia, Acanthophora, or Jania plants, or even *Avrainvillea rawsoni,* all covered with the smaller things. In slightly deeper pools, as one passes toward the Sargassum zone, the more conspicuous Chlorophyceae, which do not so much retain the lime, become evident. Almost certain to be among these is *Caulerpa racemosa,* as are *Cladophoropsis membranacea, Cladophora fuliginosa,* and *Anadyomene stellata* (Text fig. 3). Near the Sargassum, where the algae are seldom and briefly laid bare by the tide, Dictyosphaeria and Dictyota become important, and the Padinas, Digenia, and Acanthophora begin to show more typical growth forms. All of these plants, it must be remembered, also grow well under quite other conditions.

When the shore is composed of much-broken or eroded rocks, the floristic conditions are altered very much. In the West Indies the aeolian limestones and beach rock when eroded become extremely jagged and difficult, even dangerous, to traverse. The multitude of crevices produced gives shelter to many species, and there is more tendency for plants to reach full size. Here the Sargassa form scattered plants of full development in deeper water as well as dwarfed ones in the shallows; fine plants of Dictyota, Dilophus, and Bryothamnion are common, and all the plants of the shelf rock appear mixed more or less together. Here, too, one is much more likely on the higher rocks to find *Polysiphonia howei,* Lophosiphonia, and *Bostrychia tenella.*

Broad shallow depressions in the beach rock naturally tend to become filled with sand, and this sand to be densely colonized with marine phanerogams, of which the chief is *Thalassia testudinum.* The rather short blades of this plant commonly bear small algae, especially Fosliella, but seldom in obscuring quantity. The water surface at spring tides commonly almost exposes these colonies, but they extend into water of several decimeters depth. In sandy shoals of clear water not exposed to heavy wave action and not immediately underlain by rock, they cover vast areas and are of great importance. Other phanerogam genera found are Cymodocea and Diplanthera, but they are less widespread, and more commonly

TEXT FIGURE 3. Submerged rock shelf at low tide. The general cover is of Centroceras and other minute species, with rounded dark clumps of *Cladophora fuliginosa*, lighter ones of Cladophoropsis and various less evident species. Jamaica.

grow in even more shoal, protected places. The Thalassia colonies tend to accumulate sand and grow upward with it; at the abrupt edge of such a sea-grass bed one may step off to a clear bottom nearly a meter lower. Between the rhizomes of the Thalassia one may expect scattered Penicillus, Udotea, and Avrainvillea plants rooted in the sand, with Amphiroa over the surface between the culms and Laurencia on the bases, or Acetabularia on coral fragments (Text fig. 4), but the only algal genus likely to be common enough to signify in an ecological sense is the epiphyte Fosliella. Epiphytes of secondary importance would include *Giffordia duchassaigniana*, *Chrysonephos lewisii*, small Chondrias, and Ceramiums. The Diplanthera beds are even denser and are likely to support little except minute epiphytic species; the Cymodocea beds, though not so dense, are also inhospitable. All three are, however, very major ecological features from Bermuda southward through the Caribbean.

Clean sandy shallows on open shores not occupied by Thalassia beds are seldom colonized by algae, for the material is too shifting for their attachment. There are a few exceptional situations. Great beds of purple-black *Lyngbya majuscula* may lie at 5–10 dm. depth on white shell sand. A little below the high-tide line *Caulerpa fastigiata* may form mushroom-like masses 7 dm. tall and 0.5–3.0 meters in diameter, the filaments having consolidated the sand into very firm, marginally deeply undercut mounds. On the top of the mounds the filament tips are only shortly exposed where they are laid bare at low tide, but the filaments may stream out into the water for 10–15 cm. around the margins. Rocks over the sandy flats bear the usual shallow-water Laurencias, Dictyotas, and Halimedas, depending on their size, and even in very shallow places they may be covered with Cladophora, Gracilaria, or Grateloupia.

When the sandy shallows lie in small bays and coves protected by narrowness of the mouth, or by a reef, their flora is much richer. Small coves and bays seldom have a rocky bottom throughout; they generally have a sandy bottom with scattered rocky reefs, though the margin is often rocky. These conditions favor a rich flora of Dictyotas along the shore, with good growths of Caulerpas (Text figs. 5, 6), Acanthophora, Padina (Text fig. 7), well-developed Sargassums, and spongelike clusters of Galaxaura. Here *Halimeda opuntia*, which can only form compact masses in crevices of the surf-beaten beach rock, may form great colonies a meter or much more in

TEXT FIGURE 4. Colony of *Acetabularia crenulata* growing on coral rock within reach of the waves. Jamaica.

TEXT FIGURE 5. Portion of a submerged colony of *Caulerpa racemosa* at low tide. The dark masses toward the left consist of streamers of the diatom Biddulphia. Jamaica.

TEXT FIGURE 6. Portion of a submerged colony of *Caulerpa verticillata* forming the dappled background over which are spreading a few branches of *Caulerpa sertularioides*. Jamaica.

diameter, apparently not definitely attached to the bottom. Occasionally the segments remaining when these plants break up accumulate in masses sufficient to become consolidated into rock, but their relatively soft consistency generally prevents this. Zonation may be fairly distinct: in Text figure 8 appears a sharp band of *Codium intertextum* just above low-tide line.

Muddy coves and bays are more likely to show characteristic plant associations. The soft, shifting material discourages plants fitted to attach directly to particular objects. Even the rhizomatous Caulerpas, which do well enough in the sand, are unsuited, whereas the sod-forming "sea grasses" may partially succeed and in turn support algae which could not attach to the mud. Characteristic genera for these places would include Cymopolia, Penicillus (Text fig. 9), Avrainvillea (Text fig. 10), Udotea, and Halimeda, growing scattered over the bottom, with their taproot-like portions deeply sunk in the mud. Intertidally, around the shores, *Vaucheria dichotoma* and *Boodleopsis pusilla* may be expected.

It has long been known that the presence of mangroves along the shore favors the growth of certain species of algae. Two trees, in particular, are able in much of this territory to grow where the soil about their roots is perpetually saturated and periodically flooded with water, more particularly sea water. One of these, the Black Mangrove (*Avicennia nitida* Jacq.), forms a narrow fringe along the shore and sends up simple slender pneumatophores in abundance from roots beneath the surface spreading far beyond the span of the crowns of the trees. These give attachment to various algae which are little shaded by the foliage; the genera chiefly concerned are Rhizoclonium near the high-water line and Bostrychia throughout the intertidal band; in winter Monostroma also may appear in abundance, and other genera in small degree at times share the space, but the assortment is rather limited.

More familiar and more important is the Red Mangrove (*Rhizophora mangle* L.). This tree may fringe the seashore or shores of lagoons, but produces denser and deeper woodlands than Avicennia. Its stilt roots branch out seawardly in repetitious arching fashion and extend the groves in shallow water of the ocean as well as in brackish lakes or tidal rivers. These stilt roots in the dark depths of the groves are essentially barren of algae, but around the sunlit margins they favor a particular algal flora, and the mangrove rela-

Text figure 7. Portion of a submerged colony of *Padina sanctae-crucis*. Jamaica.

TEXT FIGURE 8. Abrupt rocky shore margin above a submerged shelf, at low tide. A narrow zone of *Codium intertextum* extends down to low-water mark. Bermuda.

TEXT FIGURE 9. The muddy bottom of a sheltered cove with a submerged colony of *Penicillus capitatus*. The dark areas especially toward the right side consist of Hypnea. Bermuda.

tives in other countries have parallel floras. The principal genus in the region under discussion is Bostrychia, and *B. montagnei* may form a heavy cloak about the intertidal part of each root (Text figs. 11, 12). It is often partly displaced by *B. tenella, Caloglossa leprieurii,* or *Catenella repens,* but these are more able to range to other substrata. Above such plants Rhizoclonium may form a green band. Occasionally *Murrayella periclados, Caulerpa verticillata,* Polysiphonia, Centroceras, Wurdemannia, or other plants are prominent, but likewise are not specific. The roots tend to reduce water circulation, and much mud is deposited all over these algae, so that they present a very dingy, gray appearance.

It is necessary to remember, however, that these algae all require a fair amount of light, and do not grow in the depths of the groves. Where the mangroves extend in open fashion up onto the shore, Vaucherias, Cladophoropsis, and Boodleopsis appear on the mud at the edge of the water, and Enteromorphas may accompany them on dead wood. Where the mangroves surround brackish lakes, they show characteristic changes in flora. The Bostrychias and their companions may be absent; the Wurdemannia may form unexpectedly long wefts in the water. Even if the water seems quite fresh Batophora may cover submerged dead branches, as seen around a Chara-bottomed lake. Again, if the bottom is muddy but the water more salty there may be many Penicillus plants with exceedingly large, loose capitula, or great masses of Chaetomorpha or Cladophora. If the lakes are fully salt with adequate tidal flow they will support a flora like that of muddy bays and coves.

Reefs in tropical and subtropical areas, no matter whether their origin be of igneous or sedimentary rock, are likely to have become so covered with corals and lithothamnia that, from the standpoint of colonization, later arriving algae, including secondary lithothamnia, are forced to attach to a calcareous substratum. The encrusting corals and above all the encrusting lithothamnia may bury the original materials many centimeters deep, and indeed if the region gradually submerges, the lithothamnia may develop rock masses of geologic importance, and the original rock will come to lie deeply buried. While all the uncritical literature tends to designate these as coral reefs, the coral animals are by no means the only organisms concerned. In some cases, as at Bermuda, at least the surface layers may be completely dominated by the tubes of serpulid worms. In a

TEXT FIGURE 10. The muddy bottom of a sheltered cove with scattered plants of *Avrainvillea longicaulis* emerging at low tide. The boat in the background rests on a Thalassia bed. Bermuda.

TEXT FIGURE 11. The muddy bottom of a sheltered cove bordered with mangroves. The Rhizophora stilt roots are covered between tide marks with shaggy *Bostrychia montagnei* coated with white mud, while on the mud below are multitudinous small black snails. Bermuda.

TEXT FIGURE 12. Part of one of these Rhizophora roots with the mud washed off the *Bostrychia montagnei* plants, shown here at about natural size. Bermuda.

large proportion of cases the algal lithothamnia (*ante* "nullipores") cover the major part of the surface, overgrowing the corals as they die. Sometimes they themselves may be seen to have formed the major part, occasionally even the whole, of the rock.

The outlying reefs may be much less heavily vegetated than a northerner would assume to be the case. On the windward side, and especially if they are nearly covered with actively growing corals and gorgonians, they are quite likely to be botanically barren. If they have few living corals the reefs may be nearly completely covered with lithothamnia, and then the situation is different. On these and on the dead corals other algae may grow, and where there is a little shelter they may grow quite well. If exposed at low water there will be little growth on the reef top except in the clefts and pools, although if persistently wet the rocks may be covered with Pocockiella. The pools may have heavy colonies of *Halimeda opuntia*, Anadyomene, Dictyosphaeria, dwarfed Padina, Galaxaura, and probably rather well-developed Digenia or *Laurencia papillosa* (Text fig. 13). In the lee of the reef the variety of growth will be greater; such places suit Codium and *Valonia macrophysa;* the Sargassa, Padinas, iridescent Stypopodium, and *Dictyopteris justii* will be more scattered and larger. Under the ledges the Rhodophyceae in particular will reach their greatest luxuriance, with Liagora, Asparagopsis, and Lophocladia prominent, or with such rarities as Thuretia or Dictyurus.

In discussing algal distribution thus far we have considered primarily the intertidal flora, or at least that easily accessible from the shore. A very substantial flora extends far deeper than this. The notion common among land botanists that few seaweeds, and these Rhodophyceae, go down to more than a few meters depth is easily destroyed by examination of the records of algae in this volume. Many extend deeply, though they are commonest near the surface; few, it is true, seem to be limited to great depths, and for these the records are too scanty to give assurance that they may not occur in shallower water. If we take as our starting level a depth of about 30 meters, we find a flora of some 130 species recorded as having grown this deep, and this consists of 56 Chlorophyceae, 15 Phaeophyceae, and 60 Rhodophyceae. Ranging to 50 meters, 60 species are recorded: 34 Chlorophyceae, 10 Phaeophyceae, and 16 Rhodophyceae. To 70 meters, 31 species: 21 Chlorophyceae, 1 or 2 Phaeophyceae, and 9 Rhodophyceae. To 90 meters, 23 species: 12 Chlorophy-

TEXT FIGURE 13. Rocky tide pool at low-water mark with submerged plants of *Galaxaura squalida* (center), *Laurencia papillosa* (rear) and *Dictyosphaeria cavernosa* (foreground, center). Jamaica.

ceae, 1 or 2 Phaeophyceae, and 9 Rhodophyceae. From these data we see that under Bermudian, Floridian, and Caribbean conditions, Chlorophyceae are quite as well adapted to growth down to a depth of 90 meters as are Rhodophyceae, retaining undiminished their strong green color. At this depth the bottom is characteristically soft and the temperature quite low.

The reduced light at such a depth is doubtless a factor reducing the quantity of algae growing there, in fact eliminating many algae, but not one acting differentially between algal classes. It is notable that almost all of the Chlorophyceae growing at 90 meters or lower are Siphonales. There is no such complete concentration in one order of Rhodophyceae, though most of them are Ceramiales, likewise considered a highly evolved group. Most of these Chlorophyceae have delicately fibrous bases adapted to growth on a very soft bottom, but the same can hardly be said for the associated Rhodophyceae. As almost all of these Chlorophyceae and most of the Rhodophyceae are able to range into quite shoal water, like those reaching the 30-, 50-, and 70-meter levels, and as dredging has not shown any great tendency toward formation of extensive colonies of single species, it does not appear that there is any clear bathymetric zonation in deeper waters of this area.

When the general botanist starts to review the algae of this area, the first to come to mind are the "gulfweeds," and too often these are the only tropical algae about which he has any information. The gulfweeds are, to be sure, important, but not of primary moment. They are of two kinds, *Sargassum natans* and *S. fluitans,* with differential characters explained in the systematic section to follow. Portions of littoral species often drift with these pelagic ones, but they do not multiply. Significant amounts of drifting gulfweeds are not found south of a line extending from about 16° or 17° N.L. in the Lesser Antilles to 14° N.L. near the Central American coast (Parr 1939). We may accept it that there is a very rapid growth and fragmentation of the floating plants north of this line, and there is no evidence that the pelagic flora needs to be constantly replenished from a benthonic source. Both species, passing out through the Straits of Florida, continue in excellent health into the drift area called the "Sargasso Sea" and multiply vegetatively there. They are scattered as patches or windrows, but present no obstacle to navigation. It is not unusual from the Greater Antilles to Bermuda to find them drifted ashore in great quantities (Text fig. 14). Indeed, they

TEXT FIGURE 14. Sandy beach below an abrupt rocky shore with "Gulfweed" (chiefly *Sargassum natans*) being washed ashore. Jamaica.

may drift far northward and occasionally be blown ashore on the coasts of Newfoundland or the British Isles. The wind and the water currents also bring the Sargassums into the Gulf of Mexico. While there seems to be some growth here, particularly in the summer, it is clear that it is not very considerable. Of the two, *S. natans* maintains itself best, but all the plants pass into an unhealthy state in the northwesterly sector late in the year.

COLLECTION AND PRESERVATION

The simple equipment needed for the shore collection of marine algae is quite different from that required for collecting other plants. First in importance is a pail, flexible plastic, wooden, or fiber preferred, of 2 to 3 gallons' capacity. One should have also a few pint jars and some smaller wide-mouthed bottles. As a substitute for the pail one may prefer a waterproof bag carried by a strap over the shoulder. In this the algae are kept wet rather than afloat. Some tool for scraping or breaking rocks is next in order; a suitable instrument can be made from a heavy file or a broken car spring by sharpening one end at a forge.

Beyond these, the clothing of the collector is the only consideration. Heavy canvas shoes which come above the ankle are the best footwear for protection from barnacle cuts and the knifelike edges of volcanic rock or eroded coral and shell limestone. The soles should be of thick rubber and deeply corrugated. Shoes with tough felt soles a half-inch thick, or soles of coiled rope, though very hard to obtain, are better, being more efficient in preventing slipping. Denim trousers and shirt, for protection from cuts and sun, are better than the customary bathing suit. This is particularly true in the tropics, in spite of the heat, where extended exposure to intense sunlight may be acutely dangerous. To these should be added a large straw hat or its equivalent, and dark glasses, for protection of the eyes.

Knowledge of the character and the period of the tides is the next requisite. It is possible to do some collecting at any time, but certain tide stages favor certain types of work. No serious student collects material from the dried or decaying mass of seaweed lying above high-tide line, and in warm countries algae washed ashore quickly spoil. Living and healthy material alone is of any use. However, one should not be too scornful of what fate brings in the way of material recently detached and drifting ashore. Algae from deep water cannot be obtained otherwise, except by dredging. Much just criticism has been leveled at records based on drifting material, but if one uses discretion in this regard, specimens of the utmost value can be had. One should choose a rising tide for this collecting, since then the algae will be borne in closest to the water's edge. One should also inspect the chosen coast to see in what sheltered spots the algae are accumulating, since wind and current tend to localize the richest harvests. Though an onshore gale brings much fine material, it is easier to see what specimens are worth collecting after the roughness of the waves has subsided. One may then often stand quietly at waist depth and select undamaged specimens from the great variety of plants floating by.

This type of collecting is admirable within its limitations, but only the locating of plants actually attached, growing in place, will acquaint the student with their growth habits and ecological relations and give sure data on their distribution. In the intertidal zone and a little below, this can be done afoot. Deeper waters can be explored with underwater swimming or diving gear, or a dredge; the waters of most of the area under study greatly favor the use of those independent diving outfits which contain their own air supply. With respect to the intertidal zone a few precautions must be observed. Work here can best be done as the tide reaches its lowest point, so that a few words of advice may be offered to those unacquainted with the sea. On the coasts with which we are chiefly concerned the tide reaches high and low points of approximately equal levels twice each day, and on each succeeding day these extremes occur about fifty minutes later in the day. Extreme or spring tides come twice monthly when the forces of moon and sun are in conjunction at new and full moon, and alternate with minimal or neap tides when these forces are in opposition near the first and last quarters of the moon. Exact details for selected stations are available in annual tables published under government auspices, and from these the data for intermediate points may be calculated.

In general in the eastern American tropics and subtropics the tidal range is small, but nevertheless important to the collector of algae. One should sharply observe local conditions, for they may greatly influence the success of a collecting expedition. For instance, a strong onshore wind causes the water to stand higher than it otherwise would for any stage of the tide. Near the narrow openings of coves or bays the strength of the current may be as important to safe collecting as is the tide level to access to the plants, and so one must ascertain the time and the duration of slack water, which does not necessarily occur when the tide is lowest.

In practice one goes out over the tidal flats or down over the rocks as the water falls, keeping at a comfortable wading depth, examining tide pools, rock crevices, overhanging rocks and cliffs, and looking under the tangle of Sargassums for small species. For this purpose a glass-bottomed pail or box is of the greatest utility. Protected sandy or even muddy coves, the barely submerged meadows of marine phanerogams, or even areas of shattered rocks are easily explored on foot in warm seas. Crustose types may be chipped or scraped off and, like the more delicate species, placed in bottles of sea water. It is more important to keep the water cool than to have the specimens floating about in a large quantity of it; also, the pail is easier to carry so. Bits of strong paper bearing collection data written with a very soft black pencil should be placed with the specimens. The containers should not be crowded, and duplication of coarse species should be avoided, for algae decay quickly if unduly crowded or warmed. If necessary, several pails may be used, kept in the shade, and the water changed at intervals. Many coarse algae, such as Codiaceae, Sargassums, Turbinarias, Galaxauras, Gracilarias, and the like, can be stowed in a sack if kept moist and in the shade, and these need not crowd the collecting pails. Cliffs dropping into deep water and jetties and wharves which cannot be approached afoot may be inspected from a rowboat, with due respect for the severity of the waves. With discretion this kind of collecting is simple and safe. As the tide rises to make unprofitable the search for shoal-water types, one may well turn to collecting in the drift.

The most notable hazards one should be prepared to avoid when working in these warmer seas include the severely stinging jellyfish and "Portuguese men of war," or Physalias; the black sea urchin Diadema, whose long, sharp, and fragile spines can administer highly infectious and painful punctures; and the eel-like morays, which are capable of savagely biting one's hands. The jellyfish and Physalias must be watched for, especially the latter, with their long tentacles. Diademas frequent sandy shallows where they generally can be easily seen, but may also be partly concealed in holes between rocks or algae, though not as often as other, more harmless sea urchins. Morays are particularly likely to lurk in holes in coral reefs, and it is wise never to venture a hand into any obscure cleft or hole without first probing with a long, heavy steel scraping tool or a net handle. An alert collector can always avoid all of these hazards.

The algae which grow below reach at low spring tides are difficult to study in position; they comprise a large proportion of the drift-weed washed ashore and are ordinarily available only in this state. From a boat, however, if the water is very clear, some little of their habit can be seen, especially through a glass-bottomed bucket. It is possible to secure specimens to a depth of about six feet or so if a heavy-tined right-angled digging fork (such as clams are dug with) is used on a handle about ten feet long, though such a tool is very difficult to operate because of the resistance of the water. The extreme clarity of the water aids collecting in many Florida and West Indies stations. In tropical seas one may obtain a much more just idea of the relation of the underwater vegetation to its surroundings by the use of a diving helmet with air hose and pump. However, while such an outfit is excellent for inspection, I have always found by experience that it is poor for collecting, since one's movements are greatly retarded, and small specimens are hard to handle. Face masks or goggles, snorkels, swimming fins, and compressed air tanks are often a greater help, and the first items are suited to even the least athletically inclined.

On the whole, in deeper water a dredge must eventually be resorted to. Intelligently directed dredging is not easily accomplished, but when well conducted a dredging trip can be an exceedingly informative method of collection and study. A great deal depends on suitable equipment and the co-operation of the assisting boatmen. A box-type dredge, strongly built with angle iron, and the longer margins of the rectangular opening developed as two-foot cutting edges, will serve well on a hempen line of three-fourths-inch diameter; with it one may dredge from a small launch to a depth of 10 meters, if sufficient weight is used to keep the dredge on the bottom. Deeper dredging should hardly be attempted without hoisting machinery, such as is found on fishing or lobstering boats. In the clear waters of warm seas algae not infrequently grow to depths of 50-200 meters, and large dredges operated from specially equipped ships often secure very valuable material from such levels. The type of bottom involved largely determines the bulk of the plant haul. Although sandy or even muddy bottoms yield more algae in tropical than in northern seas, the presence of scattered shells or rocks greatly increases the chance of a substantial bottom vegetation.

If scattered stones are too large, or mixed with boulders or reefs of rock, the dredge will probably catch sooner or later and, unless the boatman can extricate it, will be lost. The speed of the boat must be reduced to the point where the dredge will stay on the bottom most of the time, and then an experienced hand on the line will be able to tell whether it is drawing over cobbles or among dangerous rocks. Dredging among submerged coral heads or reefs requires the utmost skill and is particularly hazardous. It should never be attempted at moderate depths where the same species can be studied by diving. There is no question but that this type of collecting is a bit of a gamble: the loss of a dredge pitted against the chance of some rare specimens. To go far afield with but one dredge is to invite its early loss, a disastrous end of the day's endeavor, as I myself have bitterly experienced.

A few minutes' haul is best in shallow water. Better to catch a few specimens, save them in good condition, and try again, rather than haul a long time, packing sand or rocks on the plants to their destruction, causing a heavy lift and perhaps a torn net. Also, each unnecessary minute below is an unjustified risk of getting the dredge jammed under or between rocks and lost. In deeper water, such as 200 meters, the time required to get the dredge on the bottom, the difficulty of keeping it there, and the slow uphaul to prevent damaging material may require up to two hours. At such depths two or three times as much cable should be overside as the depth by sounding at the dredging station. In shoal waters somewhat less line is suitable, particularly in rocky situations; here one should use only the amount that will keep the dredge in proper position. This will help prevent fouling the line among boulders; if it does get caught, it is usually most readily freed with the line taut and the boat nearly above the dredge. Then, strong and sharp pulls at various angles will eventually free the dredge—or break the line.

Immediately the dredge comes aboard it should be emptied into a tub of cool sea water for sorting and sent below, while preparation is made for another catch. All the while the collector's boat will lie and wallow in the troughs of the waves, to the displeasure of unseaworthy passengers. To lean over and sort a tub of algae and moribund animals while the gear slats back and forth is a good test of anyone's resistance. However, it is of some comfort that dredging is not practicable in really rough water. A representative series of specimens should be segregated in sea water in pails and bottles, labeled, and put aside in the shade. Records should be made of the character of the bottom, the depth of the water at the start and finish of the haul, the state of the tide if in very shallow water, and the location. Separate pails for each haul are necessary unless hauls are repeated over the same bottom with substantially the same type of catch.

The material collected will eventually reach the collector's base of operations, and the problem of disposal then appears. In general, the methods of preservation are two: in liquid or by drying. For the first a suitable supply of bottles, tins, or casks is requisite; for the second one needs herbarium equipment. Marine algae present a difficult problem of liquid preservation because the fluids usually affect the cell walls rather drastically. For the most part, coarse species can be preserved for a time in neutral 4 per cent formaldehyde in sea water, if they are kept away from the light. More delicate species do better if alcohol amounting to 30 per cent is added to the weak formaldehyde, but then they shrink somewhat. All solutions containing acid should be avoided because they soften the intercellular substances, allowing the plants to fall apart. For this reason borax should be added to formalin solutions until they give a pink reaction when tested with phenolphthalein. Simple alcoholic preservation is best except for the risk of shrinkage. When practicable, the material should first be placed in weak alcohol, perhaps 20 per cent, and then after a few hours in progressively stronger mixtures until a safe grade is reached. In weak alcohol the specimens fall apart; in strong, they shrink. The optimum for any given species is somewhere between 50 and 85 per cent. If the material is to be stored indefinitely, 5 per cent of glycerine may be added to 70 per cent alcohol to prevent complete drying by accident. Specimens with a calcareous matrix preserved in formaldehyde may be altered by solution of the lime if the preservative becomes acid, as it naturally does in the light, so that an excess of borax is to be advised. Material destined for critical morphological or cytological study should be prepared by methods to be found in books on microscopical technique.

On expeditions far afield, methods adopted will have to be more wholesale in type. Relatively few specimens can then be segregated in bottles. The home-canning devices whereby a lid is crimped onto a can body, precisely as is done for preserved vegetables and fruits, is excellently adapted to preserving algae during long trips (Taylor 1950b). The cans are far cheaper and lighter than glass jars and when filled may be shipped great distances without protective packing. Algae in alcohol or formaldehyde may remain five years or more in these cans without mishap, provided they are fully filled. The writer has received in excellent condition shipments from as far as Brazil, the South Falkland Islands, and Chile, where each individual group of specimens from a common station was loosely wrapped, with a label, in a bag of cheesecloth, tied, and packed with many others in five-gallon tins of 4 per cent formaldehyde. These were soldered shut, crated, and shipped north, to be opened, sorted, and shifted to glass or paper many months later. Disintegration in formaldehyde does not soon become serious unless the material is exposed to strong light or unless the plants are very delicate. Heavy shells, stones, and lithothamnia should not be packed in the same containers with soft specimens and should be well wrapped to prevent rubbing or breakage.

Every collector occasionally is faced with the impossibility of securing even such simple reagents as formaldehyde or alcohol. While it cannot be recommended for preservation of critical reproductive structures, or even of some morphological details, it is easily possible to preserve marine algae quite adequately for purposes of identification and preparation of herbarium specimens by the addition to the sea water of copious quantities of common salt, even of the rather crude quality obtainable in the provision stores of poor tropical villages. Experiments have shown that if the brine first formed is drained and the algae relayered with fresh salt, many species, and even many rather delicate ones, may be preserved in this way for several years. It is of prime importance that the salt be intimately and quickly mixed with the specimens. Then almost everything will remain in excellent condition even under the most severe tropical conditions. Specimens prepared with salt retain their color and may be mounted as described below, using fresh instead of sea water.

On the whole, except for special purposes, marine algae are best preserved for taxonomic reference as dried specimens. They are as permanent in that state as vascular plants, for, though some are more brittle and so liable to damage in handling, they are not subject to attacks of dermestid or herbarium beetles. If kept in the dark they retain their color for a long time, but they fade in the light. Some seem able for an indefinite period to reassume their natural form when soaked in water; others give more trouble. For specimens belonging to puzzling groups it is always well to retain small samples in formaldehyde until suitable studies have been completed.

Equipment for drying algae may be very simple. The massive coralline algae and smaller lithothamnia on stones and shells are simply momentarily rinsed in fresh water and dried in a shady place. When labeled they may be wrapped in newspapers and packed away for shipment. The phycologist should make every effort to secure whole, full-grown specimens, for a big adult plant is much more valuable and informative for taxonomic purposes than a young one or a few fragments cut from a large one—too often with the size not specified on the label! These if uncalcified can be bent around or folded to dry within the size limits of an herbarium sheet. It is often possible to soak large specimens again, to study them, measure them approximately, and remount them for the herbarium.

For smaller species some method of support is requisite. Paper is the first essential, preferably good medium-weight herbarium paper with a substantial rag content. Standard sheets (11.5 by 16.5 inches, or 29.5 by 42 cm.), halves, and quarters make a good selection. The larger looseleaf-notebook drawing-paper sheets serve well, though limiting the mounting of the larger species to small portions. A number of pieces of unbleached, unsized muslin of the full-sheet size will be needed, some cheap, thin-waxed paper, herbarium blotters (or, failing these, folded newspapers), a pan big enough to allow immersion of the mounting paper, and a few metal sheets to fit the pan (zinc preferred); a large camel's-hair brush, large-bulb pipettes, and dissecting needles complete the equipment. It is possible to mount the specimens in a shoreside tide pool, with a bit of slate or glass as a support for the paper, but it is much easier to work indoors. Mounting is difficult on a ship unless she is tied up, because even a slight roll will send the water and the specimens across the pan to the utter destruction of characteristic and orderly arrangement.

Although there is no excuse for careless mounting under ordinary circumstances, the traveler in out-of-the-way places should mount specimens whenever and however possible. Fresh-water species should be mounted in fresh water, and marine species when alive in sea water, but salted or otherwise preserved marine specimens may be mounted in fresh water. It is well to put the collection number and date on the paper with a lead pencil before mounting each specimen, so as to prevent later confusion. In practice, enough water is placed in the pan to cover the support and the paper, but not deeply. The chosen specimen is washed free of dirt, laid on the paper, and roughly arranged. If it forms too thick a mat, branches should be removed from inconspicuous places, with care not to obscure the characteristic habit, but with the aim of so reducing the bulk that the individual branches will show clearly when the plant is dried. Next comes a final arrangement aided by brush, needles, or water from the pipette, the support being gradually lifted and stood aside on end for a few minutes to drain off the water. After draining, the sheets of paper with their plants should be laid face up on blotters and covered with either muslin or waxed paper. Which is chosen depends on the character of the specimen: cloth should be used on all slippery and on most delicate things; waxed paper is better on wiry specimens. Other layers of blotter, paper, specimen, and cloth are superimposed until the harvest is disposed of, and a moderate weight is then placed on the pile. After a couple of hours the blotters should be replaced by a dry set, and this repeated twice daily until the specimens are dry. Then the cloth or the waxed paper should be stripped very cautiously from the specimens and the mounts kept in folders under slight pressure until incorporated in the permanent herbarium. The specimens will stick to the paper more or less completely by virtue of their own adhesiveness. Wiry or nonlubricous specimens may often be made to stick by adding glue to the water in the mounting pans. If the cloths and the waxed papers are stored flat and free from wrinkles, they may be used indefinitely. The blotters may also be used repeatedly, but after they become much worn they should be used for the first change only, and as they become saturated with salt should be thrown away.

To these general directions must be added special methods for certain types and conditions. If one is working on a considerable task of collecting and mounting, it is important to use corrugated cardboard ventilators between pairs of blotters; the strapped pack of material may be placed above a heater, and the air circulating through the corrugations will dry the specimens in a few hours. The best type of corrugated board is faced on one side. Pieces are cut 12 by 34 inches, with the corrugations crosswise; when folded back to back, with the corrugations exposed, these are most effective. Various devices for heating are familiar to field workers. A tube of fireproofed canvas to tie around the suspended press is a good portable tool, one or two kerosene lanterns being placed on the ground at the bottom of the tube. The writer prefers to use aluminum boxes about 30 inches high and 15 by 20 inches at the top, open above and below. A stiff frame or wire grid should be fastened below the top of the box to support the loaded press, and a suitable baffle above the heaters, if needed, to spread the warmth evenly. If kerosene lamps or electric heaters are placed in the bottom of such a box (with allowance for free entrance of the air) and the press with specimens is put on the top frame (any unoccupied areas being covered with metal sheets), the rising hot air will dry the whole in 12 to at most 36 hours.

Certain precautions pay in dealing with some of the algal types. Sargassaceae and many Codiaceae, Galaxauras, Gelidia, and similar plants do not stick well to paper or, if they do, shrink so much as to cause curling and breaking under pressure. These are best dried between two cloths in newspaper folders. After drying with heat, it is well not to pack bulky algae under pressure immediately; they should be left exposed to the air in a shady place for a short time to become more flexible and less brittle. However, they should never be left overnight except under pressure. It is very important for future critical study that selected portions of all delicate types be mounted on mica to be filed with the specimens. Each mica sheet should be split in two immediately before use and the samples mounted on the freshly exposed faces. Collection data may be scratched on the mica, and the specimens dried under cloth as if they were on paper. Only plants which adhere well should be mounted in this way.

When one is preparing the specimens for insertion in the herbarium, those which have been dried between cloths should be mounted on standard herbarium paper, but not until the climate of the repository is reached, for if mounted at the seashore they will shrink and crack at an inland station. If glue is used, it should be as thick as possible; waterproof cements are generally preferable, for watery adhesives cause swelling and curling. Rubber cement and adhesive cellophane tapes should never be used, nor any tapes with a flexible plastic adhesive. Good cloth tape with ordinary gum adhesive is quite safe. Mounts on small paper sheets should be trimmed and attached to the standard sheets; very small or fragile mounts and mica slips should be placed in envelopes, with the label on the flap. Small pebbles and shells with algae attached likewise go into envelopes on standard herbarium sheets, but bulky corallines or large stones must be stored, with their labels, in boxes. Segmented corallines, such as Corallina, Cheilosporum, and Amphiroa, often disarticulate on the slightest pressure after drying. These are best flattened on a card heavily coated with liquid glue and dried in the press under a sheet of heavily waxed paper. While the glue is disfiguring, it prevents displacement of the segments, and so far no synthetic plastic has appeared which will remain clear and resoluble through the years. Data for the labels should include the exact locality, habitat of the plant (such as type of rock or other support, exposure to the air and light, and depth), date, collector's name, field number, and the name of the plant if it is known.

So far as possible, microscopic observations for identification should be made on dissections or transverse sections of fresh material or material well preserved in liquid. Frequently this is impracticable because of its delicacy or gelatinous nature. Sometimes the freezing microtome is the best instrument to use. At other times it is better to section dried specimens freehand, directly. Although these may, if sufficiently tough, be sectioned after softening in water or 50 per cent alcohol, it is more generally satisfactory to hold a bit of the dry specimen on a microscope slide and thinly shave the end of it with a sharp razor blade or a scalpel of good steel, collecting the sections in a nearby drop of water. Thin membranes such as those of Monostroma are best thus sectioned together with the supporting paper, bits 2 to 3 mm. square being cut for the purpose from desirable parts of the specimen, especially where the membrane is folded in overlapping layers. Often well-teased-out or gently crushed preparations will suffice, and sectioning will prove unnecessary. Most observations for taxonomic purposes can be effected without special staining, but it is sometimes helpful to stain with dilute aqueous solutions of iodine (Chlorophyceae), with Methylene Blue (especially for Phaeophyceae), or with Congo Red followed by a little weak caustic potash (Rhodophycean reproductive organs).

DESCRIPTIVE CATALOGUE

CHLOROPHYCEAE

CHLOROCOCCALES

PLANTS vegetating in a motionless phase, free, less often attached, never filamentous but sometimes coenobic or endophytic; cell wall usually thin or at least not gelatinous, often of elaborate and distinctive form; protoplast with one to several chloroplasts, pyrenoids, and nuclei; cell divisions absent in the vegetative phase; asexual reproduction by zoöspores, aplanospores, autospores, daughter coenobia, or other means; sexual reproduction by iso- or anisogametes.

ENDOSPHAERACEAE

Plants generally unicellular endophytes or parasites, scattered or at times gregarious, the form irregular; chloroplasts radiating, with or without pyrenoids; reproduction by biflagellate isogametes (or zoöspores?).

Chlorochytrium Cohn, 1872

Plants endophytic; cells rounded, with an irregularly thickened lamellated wall; chloroplast at first parietal, later irregular and radiating; vegetative cells frequently persisting in a starch-packed akinete-like condition.

Chlorochytrium moorei Gardner

Vegetative cells nearly spherical, 16–26 μ diam., with a somewhat radiating chromatophore containing a single pyrenoid; any cell potentially developing into a sporangium containing motile quadriflagellate zoöids, which are reported to be dimorphic, spherical and 6–7 μ diam. or pyriform and 2.5–3.5 μ diam., and are discharged through a pore in the cell wall.

Bermuda.

REFERENCE: *Chlorocystis cohnii,* Collins and Hervey 1917.

TETRASPORALES

Plants generally with a more or less prolonged quiescent or attached, often jelly-invested, vegetative phase, readily passing into a flagellate, free-swimming state by simple emergence from the pectic investment; cells uninucleate, generally with a single chloroplast and pyrenoid; when motile, with two to occasionally four anterior flagella.

PALMELLACEAE

Plants gelatinous, the gelatin homogeneous or stratified; cells chiefly peripheral, each with a parietal chromatophore.

KEY TO GENERA

1. Plants epiphytic, of indefinite form, not zonate; without structural gelatinous strands or stalks................ PSEUDOTETRASPORA
1. Plants lithophilous, with concentrically zoned lobes; the cells terminating gelatinous strands.................... PALMOPHYLLUM

Pseudotetraspora Wille, 1906

Colonies aggregated, becoming macroscopic, gelatinous; cells 2–4 together, somewhat peripheral, spherical or oval after division; chromatophore parietal, lobed or radiating, with a central pyrenoid; reproduction by oval akinetes.

Pseudotetraspora antillarum Howe

Thallus subglobose or variously constricted and wrinkled, mostly 2–20 mm. diam., often more or less flattened and perforate with age; yellowish brown when living, brownish or dark brown on drying; cells subglobose or ellipsoid, mostly 3–7 μ in maximum diameter.

Bahamas. Enveloping the leaves of sea grasses and the older parts of coarse algae, in shallow water.

REFERENCE: Howe 1920.

Palmophyllum Kützing, 1845

Plants forming spreading, gelatinous, horizontally flattened, concentrically zonate lobes; cells only crowded near the margins of the thalli, each terminating a gelatinous stalk which forms part of the substance of the thallus; cells round or oval, each with a parietal chromatophore, but neither pyrenoids nor pseudocilia; reproduction by zoöspores.

Palmophyllum crassum (Naccari) Rabenhorst

Plants to 7 cm. diam., olive green, marginally lobed, the surface relatively smooth and the texture firm; jelly diffuse below, firmer and striate near the surface; cells dispersed, but more numerous toward the surface, spherical and about 8 μ diam., or more often elliptical, 6.5–7.5 μ diam., 8.5–9.5 μ long.

Bermuda. Growing on stony material dredged from a depth of 90 m.

REFERENCE: Bernatowicz *in herb.*

ULOTRICHALES

Plants filamentous to foliaceous, simple or branched, without or with differentiation of attached base and free apex; in a few cases condensed toward a unicellular state or by irregular divisions becoming somewhat parenchymatous; cells showing a distinct wall, usually a single parietal chromatophore, pyrenoids, and a single nucleus.

KEY TO FAMILIES

1. Plants essentially filamentous 2
1. Plants of solitary cells, or cells loosely associated into gelatinous filaments .... CHAETOPELTIDACEAE, p. 53
1. Plants essentially membranous .... ULVACEAE, p. 54

2. Distinctive sporangia present .... GOMONTIACEAE, p. 53
2. Sporangial cells often enlarged, but not distinctive in form 3

3. Filaments typically unbranched, uniseriate, without hairs ULOTRICHACEAE, p. 46
3. Filaments branched, crowded and disk-forming or free, often bearing simple hairs .... CHAETOPHORACEAE, p. 47

ULOTRICHACEAE

Plants filamentous; filaments with or without an attaching base, cylindrical, typically unbranched, uniseriate; cells with thin to gelatinous walls, lateral platelike or bandlike chromatophores with or without pyrenoids.

Ulothrix Kützing, 1833

Filaments at first attached by a sterile basal cell, later sometimes free, cylindrical, normally unbranched, with thin to subgelatinous walls; cells with one parietal bandlike chromatophore which more or less completely encircles the cell and shows one to several pyrenoids; quadriflagellate zoöspores formed two to many in each cell, discharging through a pore or forming aplanospores within the sporangium, or larger single aplanospores may be formed; sexual reproduction by the formation of eight to sixty-four biflagellate gametes.

Ulothrix flacca (Dillwyn) Thuret

Filaments entangled, often in considerable skeins; color bright to dark green; filaments 10–25 μ diam.; cells 0.25–0.75 diameter in length; chromatophore covering the whole side wall of the cell, with 1–3 pyrenoids; sporangial cells swollen to 50 μ diam.

North Carolina.

REFERENCES: Blomquist and Humm 1946, Williams 1948b.

CHAETOPHORACEAE

Plants filamentous, erect from a basal holdfast or decumbent, typically branched, generally uniseriate, but sometimes forming disks by lateral approximation of the filaments or by parenchymatous divisions; terminal or lateral hairs frequently present; cells with thin to gelatinous walls, usually single nuclei, and single more or less dissected platelike or ringlike chromatophores with pyrenoids.

KEY TO GENERA

1. Plants discoid, irregularly though truly parenchymatous in the center, but marginally subfilamentous 2
1. Plants truly filamentous throughout, though when discoid these filaments sometimes pseudoparenchymatously congested in the center 3

2. Disks more than one cell thick in the center, the cells plurinucleate, without pyrenoids .... ULVELLA, p. 52
2. Disks monostromatic throughout, the cells uninucleate, with a pyrenoid .... PROTODERMA, p. 51

3. Filaments not laterally united 4
3. Filaments more or less laterally united, the disks small, epiphytic; central cells enlarging to form the sporangia PRINGSHEIMIELLA, p. 50

4. Hairs present, formed from the sides of intercalary vegetative cells .... PHAEOPHILA, p. 48
4. Hairs absent; growing within the cell membranes of other algae ENTOCLADIA, p. 49

Phaeophila Hauck, 1876

Plants epi- or endophytic, of branching uniseriate filaments the cells of which bear one to three long, eseptate hairs neither separated by a wall from the supporting cell nor swollen at the base; cells with lobed parietal chromatophores and several pyrenoids; reproduction by quadriflagellate zoöspores.

KEY TO SPECIES

1. Vegetative cells 8 μ diam. or less .... P. filiforme
1. Vegetative cells 9 μ diam. or more .... P. dendroides

Phaeophila dendroides (Crouan) Batters Pl. 2, fig. 4

Plants frequently and widely branched, epi- or endophytic; the cells 9–40 μ diam., 15–50(–80) μ long, generally cylindrical, partly somewhat irregular, containing a parietal lobed and interrupted chromatophore with several pyrenoids; hairs 1–3 on each cell, the bases often somewhat spirally twisted; zoösporangia subcylindrical to irregularly swollen, intercalary or terminal on short branches, 16–40 μ diam., 30–85 μ long.

Bermuda, Florida, Hispaniola, Virgin Isls. Growing on larger algae.

REFERENCES: *P. floridearum*, Børgesen 1913–20, 1924, Collins and Hervey 1917, Taylor 1928.

Phaeophila filiforme (Collins and Hervey) n. comb.

Filaments dichotomously or laterally branched; vegetative cells terete, 2–6 μ diam., 2-several diameters long; bristles continuous with the supporting cell and sometimes extending beyond the host, 4 μ diam. at the base, soon reduced to 2 μ, but the base not bulbous; sporangia developed from any older cell of a plant, oblate sphaeroidal, 6–12 μ diam., discharging the few zoöspores by a protruding pore beyond the surface of the host.

Bermuda. In the membrane of Lyngbya and perhaps other filamentous algae.

REFERENCE: *Endocladia filiforme*, Collins and Hervey 1917, p. 39.

Entocladia Reinke, 1879

Plants forming small patches in the membranes of the supporting host; the filaments spreading, branched, in the center of the disk tending to become subparenchymatously congested; hairs and setae absent; cell structure showing a single simple platelike chromatophore with one to several pyrenoids; asexual reproduction by the formation of zoöspores; sexual reproduction by the formation of isogametes.

KEY TO SPECIES

1. Cells short, length usually less than 3 times the diameter
   E. viridis, p. 50
1. Cell length generally much more than 3 times the diameter 2

2. Cells 5–13 μ diam., length 2–4 or more times breadth . E. vagans, p. 49
2. Cells frequently swollen to 20 μ diam., otherwise about 4 μ diam.; to 70 μ long .... E. ventriculosum, p. 50

Entocladia vagans (Børgesen) n. comb.

Plants of wide-spreading, irregularly branched filaments; cells subcylindrical, 5–13 μ diam., 2–4 times as long or more, near the middle often bearing a small unilateral lenticular cell the contents of which may divide to several acicular bodies; sporangia formed from cells near the host's surface, discharging about 15 zoöspores through a small papillar pore.

Virgin Isls. Endophytic in the cell walls of Griffithsia.

REFERENCE: *Endoderma vagans*, Børgesen 1913–20, p. 418.

Entocladia viridis Reinke

Plants of filaments with arborescent branching; the cells 3–6 μ diam., variable in length and width, about 1–6 diameters long, sometimes cylindrical, oftener irregularly swollen and distorted, those terminating the branches blunt or tapering.

Bermuda, North Carolina, Florida, Bahamas, Virgin Isls., Brazil. In the cell walls of various algae, particularly Rhodophyceae.

REFERENCES: Möbius 1890; *Endoderma minuta*, Möbius 1889, 1890; *Endoderma viride*, Børgesen 1913–20, Collins and Hervey 1917, Howe 1920, Hoyt 1920.

Entocladia ventriculosum (Børgesen) n. comb.

Plants forming large spots composed of long irregularly wide-branching filaments; cells very irregular in form, cylindrical and about 4 μ diam., or much inflated in the middle, to 20 μ diam., often 70 μ long.

Virgin Isls. Endophytic in the cell walls of Chrysymenia.

REFERENCE: *Endoderma ventriculosum*, Børgesen 1913–20, p. 420.

Pringsheimiella v. Hoehnel, 1920

Plants consisting of small monostromatic epiphytic disks with marginal growth, the central cells somewhat taller than those at the edge, which are radially elongate; cells in young plants sometimes bearing long colorless hairs; protoplasts each showing a large plate-like chromatophore with a single pyrenoid; asexual plants compact, with external walls somewhat gelatinously thickened, the cells near the center of the disk producing numerous zoöspores; sexual individuals more diffuse, with intercellular spaces, walls of equal thickness, the cells near the center of the disk producing many small gametes.

KEY TO SPECIES

1. Zoösporangia oval to subpyriform, 15–22 μ diam., 28–38 μ high; vegetative cells to 12 μ in surface diameter . . . . P. scutata

1. Zoösporangia shorter than broad, about 45 μ high, discharging by a definite pore above; vegetative cells 5–25 μ diam. .. P. udoteae

Pringsheimiella scutata (Reinke) Schmidt and Petrak

Disklike plants 1–2 mm. diam.; cells variable in size and shape, to 12 μ in diam., the marginal ones radially elongate, the central ones distinctly taller than broad; zoösporangia oval to subpyriform, 15–22 μ diam., 28–38 μ high.

Bermuda, Jamaica, Virgin Isls.

REFERENCES: *Pringsheimia scutata,* Collins 1901, Børgesen 1913–20, Collins and Hervey 1917.

Pringsheimiella udoteae (Børgesen) n. comb.

Plants small, disciform, the cells in the center irregularly disposed, but those toward the margin in radial branching rows and about 5–25 μ broad by 17–20 μ long, and 10–12 μ high in sectional view; cells in the central region sometimes maturing into urceolate sporangia about 45 μ high, and subequal or broader in radial diameter, discharging through a pore on the top.

Jamaica, Hispaniola, Virgin Isls., Tobago. Epiphytic on various larger algae.

REFERENCES: *Pringsheimia udoteae,* Børgesen 1913–20, p. 11, Taylor and Arndt 1929, Taylor 1942.

Protoderma Kützing, 1843, *emend.* Borzi, 1895

Plants forming small patches on the substratum, one cell thick, most of the disk appearing quite parenchymatous, the margin showing its filamentous character in short series of little-differentiated cells; hairs and setae absent; cell structure showing a single simple platelike chromatophore with one pyrenoid; asexual reproduction by biflagellate zoöspores produced four to eight in a sporangial cell, or by aplanospores.

KEY TO SPECIES

1. Cells 6–12 μ diam. in surface view; plant usually thin ..... P. marinum
1. Cells 10–24 μ diam.; plants 2–10 cells thick, the interior cells to 25–45 μ tall .... P. polyrhizum

Protoderma marinum Reinke

Plants forming a considerable thin, light-green layer; structure subparenchymatous, the angular cells 6–12 μ diam., irregular in shape and arrangement except at the margin, where they are in rather indistinct series.

Bermuda, Florida, Hispaniola.

REFERENCES: Collins and Hervey 1917, Taylor 1928, 1933.

Protoderma? polyrhizum Howe

Plants forming substantial, often confluent patches, 2–10 cells thick in the central portions, one cell thick at the margins; cells of the surface 10–24 μ diam., the walls 1.5–4.0 μ thick; cells of the central and ventral portion somewhat elongate vertically, to 25–45 μ tall; ventral layer sending septate rhizoidal extensions of 6–20 μ diam. into crevices of the substratum; cells of the dorsal surface forming zoösporangia of aplanosporangia without change of shape.

Bahamas. In its more substantial thallus of several cells in thickness this plant disagrees with the current definition of Protoderma.

REFERENCE: Howe 1924.

Ulvella Crouan, 1859

Plants epiphytic, disklike, without hairs or bristles, the cell arrangement in the center rather irregular and in more than one layer, but the branched filamentous arrangement evident near the margins, often with precocious forking of the elongated marginal cells; cells plurinucleate, with a parietal chromatophore without pyrenoids; reproduction by zoöspores formed from the cells of the central area.

Ulvella lens Crouan Pl. 2, fig. 7

Plants 1–5 mm. diam.; irregularly placed cells of the central portion 5–10(–20) μ diam.; cells filamentously oriented toward the margin, 3–4(–15) μ broad by 15–30 μ long, the peripheral ones often forked.

North Carolina, Florida, Virgin Isls. Epiphytic on other algae or growing on stones or coral fragments.

REFERENCES: Børgesen 1913–20, Hoyt 1920, Taylor 1928.

CHAETOPELTIDACEAE

Plants solitary or more or less filamentous, seldom aggregated to form a disk, the cells usually bearing one or more prominent simple or basally sheathed setae; cells with one nucleus, and usually one parietal chromatophore with one or two pyrenoids.

Diplochaete Collins, 1901

Cells solitary or in vague gelatinous filaments, globose or nearly so, bearing two or more long sheathless setae.

Diplochaete solitaria Collins

Cells 25–30 μ in diameter, the walls 5–8 μ thick, hardly lamellose; setae 2, arising from near the lower margin of the cell, tapering from a base which may be 4–6 μ thick.

Bermuda, Jamaica. Epiphytic on larger algae in shallow water.

REFERENCES: Collins 1901, 1909b, Collins and Hervey 1917.

GOMONTIACEAE

Plants filamentous, penetrating the substratum; sporangia formed from large embedded cells near the matrix surface, ultimately becoming detached from the vegetative filaments.

Gomontia Bornet and Flahault, 1888

Plants perforating shells or wood, usually reported as having creeping, branched filaments, the cells irregular, often crowded, or where more deeply penetrating, more regular and slender; each protoplast with a parietal lobed or reticulate chromatophore and one to several nuclei; asexual reproduction by zoöspores formed in sporangia which are developed near the surface of the substrate, and which enlarge, developing thick walls with specially indurated projections reaching from the surface.

Gomontia polyrhiza (Lagerheim) Bornet and Flahault

Plants coloring the inhabited shell grass green, their filaments widely and irregularly branching, 4–8 μ diam.; sporangia 30–125 μ diam., 150–250 μ long; zoöspores of 2 sorts, one 10–12 μ long by 5–6 μ diam., the other about 3.5 μ diam., 5 μ long; aplanospores 4 μ diam.

Bermuda, North Carolina, Florida, Bahamas, Jamaica, Virgin Isls.

REFERENCES: Collins 1901, Børgesen 1913–20, Collins and Hervey 1917, Howe 1920, Hoyt 1920, Taylor 1928.

ULVACEAE

Plants usually tubular or membranaceous, capillary to broad, occasionally reduced to one or two rows of cells; attached or becoming free-floating; one or two cells in thickness, the cells showing one or two large lateral chromatophores with, as a rule, single pyrenoids and a single nucleus; sexual and asexual plants morphologically indistinguishable; reproductive cells unaltered or slightly enlarged.

KEY TO GENERA

1. Adult plants at least in part tubular .... ENTEROMORPHA, p. 55
1. Adult plants not tubular 2

2. Plants filamentous, the cells biseriate .... PERCURSARIA, p. 54
2. Plants foliaceous 3

3. Blades one cell thick .... MONOSTROMA, p. 63
3. Blades 2 cells thick .... ULVA, p. 64

Percursaria Bory, 1828

Plants filamentous, the filaments slender, generally of two rows of cells more or less symmetrically placed side by side; cells rectangular, the chromatophores single parietal plates.

Percursaria percursa (C. Agardh) J. Agardh

Plants mostly of simple filaments infrequently proliferating, to several centimeters long, flexuous and twisted, light green; filaments of 1–4, most generally of 2, rows of cells, each row 10–15 μ wide, the cells to 28 μ long, walls rather thick.

Guadeloupe. Forming floating masses with other algae in brackish salt marsh pools, or stranded on rocks.

REFERENCES: *Enteromorpha contorta* and *E. percursa,* Mazé and Schramm 1870–77, Murray 1889.

Enteromorpha Link, 1820

Plants capillary to ample, simple or chiefly alternately branched, tubular or with the branches showing uniseriate filamentous tips; at first attached, sometimes later free-floating; adult holdfasts formed by downgrowths from cells of the subsolid stalklike portion; cells usually close-placed; chloroplast single, a lateral plate more or less covering the outer or lateral face of the cell, generally with one pyrenoid; reproduction by zoöspores and gametes produced on similar plants.

Plants of several species in this genus are able to grow luxuriantly in water much heated by the sun, in brackish water, in estuaries where they are alternately bathed in sea water and fresh water as the tides change, and even to migrate to fresh-water ponds. Under any of these circumstances the growth habit of the plants may be greatly altered, and identification becomes very difficult. If the local flora in nearby strictly marine stations is known, the relatively stable structural characters of cell shape, size, and arrangement will usually give the necessary clues to enable identifications to be made. It is not possible to apply the studies of Bliding and others on northern races, summarized by Kylin (1949), to these tropical species at this time.

KEY TO SPECIES

1. Plants flat or crisped, hollow only at the margins and the tapering stalklike base .... E. linza, p. 63
1. Plants locally tubular, mostly of simple nontubular strands 1–3 cells in cross-section .... E. chaetomorphoides, p. 57
1. Plants tubular throughout, except near the base and in filamentous proliferations 2

2. Plants simple or subsimple 3
2. Plants usually branched below, the cells not in longitudinal rows E. compressa, p. 60
2. Plants usually branched, the cells in marked longitudinal rows ... 6

3. Strands or blades without conspicuously thickened cell walls; plants usually large, the blades commonly 5–10 mm. diam., though less in dwarf or young individuals 4
3. Strands as seen in section showing thickened inner and outer cell walls, the cells not radially elongate; plants very small and slender, primarily north-temperate ...... E. micrococca, p. 57

3. Blades as seen in section showing walls thick throughout, though especially so on inner and outer surfaces, the cells markedly radially elongate; plants larger, probably primarily south-temperate .... E. bulbosa, p. 61

4. Plants with cells in longitudinal series at least below, cells 10–28 μ diam.; when branched the branches all from near the base, similar and subequal .... E. flexuosa, p. 61
4. Plants with the cells not in distinct longitudinal series 5

5. Cells 9–15 μ diam.; plants often very large .... E. intestinalis, p. 62
5. Cells 5–7 μ diam.; plants 1–10 cm. tall .... E. minima, p. 62

6. Branches essentially similar to the main axis 7
6. Branches markedly different from the main axis 9

7. Plants compressed, sparingly branched, linear; cells 4–8 μ diam. E. marginata, p. 57
7. Plants inflated 8

8. Branching chiefly near the base, the axes generally 1–2 mm. broad, the rectangular cells in clear longitudinal series throughout .... E. lingulata, p. 60
8. Branching distributed, the axes to 10–15 mm. broad, the subangular cells in longitudinal series in the younger parts, but the rows obscure in the primary axis .... E. prolifera, p. 59

9. Branches abruptly smaller, often thornlike; cells 14–16 μ diam. E. salina, p. 56
9. Branches of successive orders, each in turn smaller 10

10. Ultimate branchlets short, spinelike; longitudinal cell series only evident in the ultimate divisions ...... E. ramulosa, p. 60
10. Ultimate branchlets elongate 11

11. Branches ending in long uniseriate tips .... E. plumosa, p. 58
11. Branches not having long uniseriate tips 12

12. Main axes bearing erect slender pluriseriate branches which are not regularly and repeatedly forked .... E. erecta, p. 59
12. Main axes repeatedly redivided .... E. clathrata, p. 58

Enteromorpha salina Kützing

Fronds filiform, tubular, with a few sometimes opposite branches of 2 or more rows of cells, or in the youngest of a single series; cells in longitudinal series throughout, quadrangular, 14–16 μ diam., or slightly longer than broad.

V. **polyclados** Kützing. Fronds beset with more numerous spreading spinelike branchlets; cells 15–23 μ diam., 1.0–1.5 diameters long.

Bermuda, Florida, Louisiana, Bahamas.

REFERENCES: Collins 1909b, Howe 1920, Taylor 1928.

Enteromorpha marginata J. Agardh

Plants gregarious, several blades often stranded together; rich green, very slender, compressed, simple or a little proliferously branched; length 2–3 cm., width usually 15–20 cells, the whole blade 12–100 μ thick; cells in longitudinal series, particularly along the margins, rounded-subquadrate, 4–8 μ diam., the chromatophore covering the face of the cell, the walls thick.

Bermuda, ?Cuba, ?Hispaniola, Guadeloupe. Plants usually growing on the roots of grasses or other objects near the water's edge; primarily a northern species.

REFERENCES: Mazé and Schramm 1870–77, Murray 1889 (both including v. *longior*), Collins and Hervey 1917.

Enteromorpha chaetomorphoides Børgesen

Plants filiform, terete, simple or occasionally with a chiefly uniseriate proliferation near the attenuate uniseriate axis tip; axes sometimes uniseriate or of 2, generally of 3, rows of cells and about 45 μ diam.; occasionally thicker, of several longitudinal rows, and in these thicker parts hollow; cells subquadrate to rectangular, about 16–18 μ diam.

Jamaica, Puerto Rico, Virgin Isls., Barbados, Brazil. Plants of shallow lagoons and sheltered places, particularly entangled with other algae among the roots of mangroves.

REFERENCES: Børgesen 1913–20, Joly 1957; *C. torta,* Vickers 1908.

Enteromorpha micrococca Kützing *f.?*

Plants small, gregarious, attached, to 1.5 cm. tall, the simple or subsimple tubular blades not much twisted, tapering into the slender stalk; cells in no definite order, or below in vaguely longitudinal series, somewhat angular, 5.5–9.5 μ wide, rather longer than broad; the whole membrane 18–22 μ thick. The lateral walls not notable, but the inner and outer conspicuously thick, especially the inner, which reaches 3 μ.

Uruguay. The species is essentially a northern one and these specimens differ in the lesser degree of twisting that they show, the slightly larger cells, and thinner lateral cell walls. Plants of rock surfaces, intertidal in very exposed situations.

REFERENCES: Taylor, 1939a, 1957.

Enteromorpha plumosa Kützing

Solitary or scattered, attached, 1–3 dm. tall, repeatedly delicately, irregularly, or sometimes in part oppositely branched, nearly capillary throughout, the branches cylindrical, soft, usually with long uniseriate tips; cells subrectangular, in the tips about 8–12 μ diam., below in longitudinal and often also in transverse series, 12–20 μ wide, 18–26(–40) μ long, with chromatophores not covering the whole cell face.

Bermuda, Florida, Texas, Bahamas, Jamaica, Hispaniola, ?Puerto Rico, Virgin Isls., Barbados. Plants growing on rocks, shells, or other solid objects in relatively quiet water.

REFERENCES: Børgesen 1913–20, 1924, Collins and Hervey 1917, Howe 1920, Taylor 1928, 1941a, 1954a, Taylor and Arndt 1929; *E. hopkirkii*, Vickers 1908; *E. percursa*, Rein 1873, Hemsley 1884.

Enteromorpha clathrata (Roth) J. Agardh

Plants at first attached, later free-floating, light green, to about 4 dm. long; slenderly cylindrical and repeatedly branched, the axes reaching 0.5–2.5 mm. diam., the upper branches virgate, all tending to taper from the base but not ending in a prolonged single series of cells; cells in distinct longitudinal rows, or somewhat irregularly disposed below in old plants, subrectangular, 10–28 μ diam., 13–38 μ long.

Bermuda, Florida, Louisiana, Cuba, Hispaniola, Puerto Rico, Virgin Isls., Guadeloupe, Grenada, Brazil. When young growing attached, later forming floating masses in warm and quiet, often brackish waters.

REFERENCES: Greville 1833, Montagne 1863, Martens 1870, Mazé and Schramm 1870–77, Hemsley 1884, Hauck 1888, Murray 1889, Möbius 1890, Collins 1909b (including *E. crinita*), Børgesen 1913–20, Taylor 1933, 1954a (including *E. crinita*).

Enteromorpha erecta (Lyngbye) J. Agardh

Plants attached, soft and very slender, 1–2 dm. tall; the elongate axes distinct, cylindrical, tubular, laterally beset with many smaller erect branches; cells in the smaller branches often in transverse series, and in the whole plant in fairly distinct longitudinal rows, subrectangular or somewhat rounded, the radial walls rather heavy; the chromatophore occupying most of the face of the cell, which in the main axis may be 13–24 μ diam., 15–28(–38) μ long; in section the cells are subequal or taller than broad, the whole membrane 18–28 μ thick.

North Carolina, Cuba, Jamaica, Barbados, Brazil. Growing on rocks, woodwork, and larger algae in rather open situations.

REFERENCES: Collins 1901, Vickers 1905, 1908, Schmidt 1923, 1924, Williams 1949a; *E. clathrata* v. *confervoidea*, Murray 1889.

Enteromorpha prolifera (Müller) J. Agardh

Plants solitary or tufted, generally remaining attached, to 6 dm. tall, subcylindrical, 1.0–1.5 cm. diam., more or less abundantly proliferously branched, with occasional branches to the second degree; cells rounded subangular, 10–19 μ diam., in the younger parts always in longitudinal series; the membrane 15–18 μ thick in section, but relatively firm.

Bermuda, North Carolina, South Carolina, Florida, Texas, Cuba, ?Jamaica, Guadeloupe, Barbados, Venezuela, Brazil. Plants growing attached to rocks, shells, or other solid objects in rather sheltered situations about low-tide level.

REFERENCES: Collins 1901, Vickers 1905, Collins and Hervey 1917, Hoyt 1920, Howe 1928, Taylor 1941a, 1954a, Sanchez A. 1930; *E. intestinalis* v. *prolifera* and v. *tubulosa*, Mazé and Schramm 1870–77; *E. prolifera* f. *tubulosa*, Vickers 1905.

Enteromorpha ramulosa (J. E. Smith) Hooker

Plants tubular, rather stiff, much branched; branches with short, spinelike branchlets; cells rather rounded, showing longitudinal series only in the ultimate divisions.

Florida, West Indies. A plant of exposed situations. A more detailed description based on material from the Strait of Magellan (Taylor 1939a) may not fully apply to the tropical specimens.

REFERENCES: Collins 1909b, Taylor 1954a.

Enteromorpha lingulata J. Agardh Pl. 1, fig. 3

Plants generally tufted, sometimes even crowded into a turf; to 7 cm., seldom to 15 cm., tall, the individuals rarely simple, more commonly sparingly to abundantly branched, the branches distributed along the lower part of the primary axis with the more developed ones higher than the filiform branch initials lower down; near the base subsolid; above very slender and gradually dilated, terete, generally not over 1–2 mm. diam.; cells in clear longitudinal rows, about 9–12 μ wide, 9–28 μ long, sharply rectangular, the lateral walls in surface view thin or but slightly thickened.

North Carolina, Florida, Mississippi, Louisiana, Texas, Mexico, Caicos Isls., Cuba, Jamaica, Hispaniola, Puerto Rico, Virgin Isls., St. Barthélemy, Barbuda, Aves I., Martinique, Guatemala, British Honduras, Costa Rica, I. las Aves, Venezuela, Trinidad, Brazil. Common in warm temperate and tropical America on solid objects near the low-tide level, often in dense growths.

REFERENCES: Børgesen 1913–20, Taylor 1930a, b, 1933, 1935, 1936, 1941a, b, 1942, 1943, 1954a, Sanchez A. 1930, Blomquist and Humm 1946, Quirós C. 1948, Joly 1951, 1957.

Enteromorpha compressa (Linnaeus) Greville

Plants generally gregarious, attached, bright to dark green; to 3 dm. tall, tubular, more or less compressed or collapsed; above expanded, 2–20 mm. wide; below long, tapering, and characteristically with several branches from the gradually contracted stalklike base which are similar to the principal blade; cells in the adult plants irregularly placed, 10–15 μ diam., rounded subquadrate, the walls not thickened; in section vertically elongate, the whole membrane 13–20 μ thick.

Bermuda, Florida, Puerto Rico, Guadeloupe, Barbados, Grenada, French Guiana, Brazil, Uruguay. A plant of rocks and woodwork in moderately exposed situations. Most of the records are to be doubted, although those from the extremes of the range north and south seem authenticated.

REFERENCES: Montagne 1850, Martens 1870, Mazé and Schramm 1870–77, Dickie 1874a, b, Hemsley 1884, Murray 1889, Möbius 1890, Taylor 1930b, 1939a; *E. compressa* f. *lingulata,* Schmidt 1923, 1924; *E. complanata,* Martens 1870, Mazé and Schramm 1870–77, Murray 1889; *E. complanata* v. *confervacea* and v. *crinita,* Mazé and Schramm 1870–77, Murray 1889.

Enteromorpha bulbosa (Suhr) Kützing

Plants to about a decimeter tall, the subsimple tubular axes to 2–4 mm. diam.; when dry rather firm in texture and dark in color; the whole membrane 28–42 μ thick, the cells irregularly placed or occasionally oriented in vague longitudinal rows, rather angular, 7–22 μ diam., radially much taller than broad; lateral cell walls rather thick, the inner and outer walls even thicker, to 2.5–3.0 μ.

Brazil, Uruguay.

REFERENCE: Taylor 1939a.

Enteromorpha flexuosa (Wulfen) J. Agardh

Plants gregarious, to 15 cm. tall, simple or occasionally divided at the base into 2–3 similar branches; cylindrical above or becoming intestiniform, commonly to 5 mm., rarely to 1 cm. or more, in diameter; gradually tapering downward, all divisions slender and subsolid at the base; cells in longitudinal rows, or, especially above, somewhat irregularly disposed, about 10–28 μ wide, 8.5–28.0 μ long, rectangular to polyhedral, the lateral walls in surface view moderate to thick.

F. **submarina** Collins and Hervey. Fronds inflated, contorted, floating.

Bermuda, North Carolina, Florida, Mississippi, Texas, Bahamas, Cuba, Jamaica, Hispaniola, Virgin Isls., St. Barthélemy, Barbuda,

Guadeloupe, Barbados, Grenada, Costa Rica, Netherlands Antilles, Venezuela, Tobago, Brazil, Uruguay. A common species in tropical American waters, found near and just below low-tide level on sticks, coral fragments, shells, stones, etc. The variety was found floating in brackish water at Bermuda.

REFERENCES: Collins 1901, Vickers 1908, Børgesen 1913–20, Collins and Hervey 1917, Howe 1920, Hoyt 1920, Taylor 1928, 1930a, 1940, 1941a, 1942, 1943, 1954a, Taylor and Arndt 1929, Quirós C. 1948, Joly 1951, 1957.

Enteromorpha minima Nägeli

Plants small, gregarious, attached, the base a minute disk of crowded, branched filaments, the blades 1–10 cm. tall, simple or slightly proliferously branched, yellowish-green and rather soft; blade linear, dilated sharply above the stalk, width to 1–2(–5) mm., tubular or compressed, the membrane 8–10 μ thick or somewhat more, cells 5–7 μ diam., angular, arranged in no definite order, the walls rather thin; in section the cells appearing nearly cubical, the walls on the inner and the outer faces about equally thick.

Bermuda, North Carolina. Plants growing on rocks and shells in the lower part of the intertidal zone, especially in exposed situations. Primarily a species of northern shores.

REFERENCES: Collins and Hervey 1917, Williams 1951.

Enteromorpha intestinalis (Linnaeus) Link

Solitary or gregarious, at first attached, often becoming free-floating, subintestiniform, light green; 1–20 dm. or even more in length, 1 mm. to 10 cm. wide, membranous; frond tapering below, above the stalk elongate-attenuate, tubular, cylindrical, clavate or generally inflated and bullate; simple, or rarely sparingly branched from the very base, or proliferous; cells not arranged in any order in the inflated portion, rounded-polyhedral, in surface view 9–15 μ diam., the whole membrane 20–40 μ thick, the cells in section radially rounded-oblong, about 14–17 μ tall, the lateral walls thin but rather thicker on the inner side.

F. **cylindracea** J. Agardh. Fronds attached, cylindrical above the stalk, to about 3–10 mm. diam., and very long, even to several decimeters.

F. **tenuis** Collins. Fronds attached, clavate, from a tapering stipe; membrane 20–30 μ thick.

Bermuda, North Carolina, Texas, Jamaica, Puerto Rico, Guadeloupe, Grenada, Brazil, Uruguay.

REFERENCES: Martens 1870, Mazé and Schramm 1870–77, Zeller 1876, Hauck 1888, Murray 1889, Collins 1901, Hoyt 1920, Taylor 1939a; *E. intestinalis* v. ?*capillaris*, v. *filiformis*, and v. *tubulosa*, Martens 1871. These reports undoubtedly should in large part be referred to *E. flexuosa* and *E. lingulata*, but some from the extreme north- and south-range limits seem to be valid.

Enteromorpha linza (Linnaeus) J. Agardh

Plants simple, usually gregarious, the yellowish-green blades with a short stalk and tapering base, the blade flat, linear to lanceolate, or with a crisped margin, to 37 cm. tall, 30 mm. wide; tapering base of the blade hollow, the flat portion of the blade of 2 layers united by their inner cuticula to a combined thickness of 35–50 μ, except near the margins, where they are separated by a definite tubular space; cells of the stalk longitudinally seriate; those in the blade angular, in vague linear series or in no definite order, 10–15(–20) μ diam., about 12 μ tall in section.

Bermuda, North Carolina, South Carolina, Cuba, Jamaica, Brazil. Plants growing attached to rocks or woodwork, chiefly in the intertidal zone of rather exposed places. Primarily a northern species.

REFERENCES: Greville 1833, Murray 1889, Hoyt 1920, Schmidt 1923, 1924, Taylor 1929b, Sanchez A. 1930, Joly 1951, 1957; *Phycoseris linza*, Martens 1870, Hemsley 1885.

UNCERTAIN RECORDS

Enteromorpha sp.?

Brazil. Printz 1927; *Kallonema obscurum*, Dickie 1870, 1875.

Monostroma Thuret, 1854

Plants at first attached, saccate, later usually splitting into a broad flattened blade or narrow segments, and in some species becoming free; walls in the saccate stage one cell in thickness, the expanded state likewise of one cell layer, the cell walls generally thin, but sometimes gelatinous; cells usually with a single platelike chromatophore which nearly surrounds the protoplast and contains one pyrenoid; asexual reproduction by the formation in each cell of several quadriflagellate zoöspores; sexual reproduction by the similar formation of many biflagellate gametes.

Monostroma oxyspermum (Kützing) Doty

Plants attached, ultimately in dense tufts, soft, light green, to about 3–10 cm. tall, or in protected pools detached and much larger, even to 60 cm.; the initial sac early split, the very broad segments of the membrane becoming plane or irregularly ruffled; cells in surface view irregularly placed or somewhat grouped 2–4 together, generally 7–18 μ, infrequently to 26 μ, diam., rounded-angular to round or oval, the intervening walls moderately thick to very thick and gelatinous in appearance; the thallus membrane 20–40 μ, occasionally to 60 μ, thick, with the cells in section rounded rectangular and about 15–21 μ tall.

Bermuda, South Carolina, Florida, Brazil. Plants of the lower intertidal zone and pools, attaching to various solid objects, often in brackish water and on mangrove roots.

REFERENCES: *M. latissimum,* Collins 1909b, Collins and Hervey 1917, Taylor 1936, 1954a, Joly 1957; *M. orbiculatum,* Hemsley 1884, Murray 1889, Collins and Hervey 1917.

UNCERTAIN RECORDS

?Monostroma obscurum (Kützing) J. Agardh

Brazil. Type from France. *Ulva obscura,* Martens 1870.

Ulva Linnaeus, 1753

Plants at first irregularly subfiliform, soon expanded, marginally attached or substipitate, plane or crispate, orbicular, lobed, or elongate-laciniate; of two cell layers, these in close approximation or separated by the gelatinous walls of the cells, which may be much thickened both internally and externally; cells showing single chromatophores, usually on the external faces, with one or two pyrenoids; asexual reproduction by the production of four to eight quadriflagellate zoöspores in each sporangial cell; sexual reproduction by the dioecious production of biflagellate anisogametes, usually eight in each cell.

KEY TO SPECIES

1. Plants normally divided into narrow segments; in transverse section cells along the paler central portions of the older segments much taller than those near the margins . . U. fasciata, p. 66
1. Plants simple or with broad lobes 2
2. Blades not naturally perforate, though often riddled by animal attack; cells in transverse sections nearly square or taller than broad .... U. lactuca, p. 65
2. Blades regularly perforate; cells much broader than tall in section .... U. profunda, p. 67

Ulva lactuca Linnaeus

Plants attached, foliaceous, bright green; the holdfast small, the stalk inconspicuous or apparently absent, the blade lanceolate to rounded, often somewhat lobed and undulate or folded, to 6 dm. long or more, and relatively broad; membrane thick near the base, the marginal portions somewhat thinner; cells usually about as tall as broad in section, closely placed in surface view.

V. **latissima** (Linnaeus) De Candolle. Plants of irregular outline, in sheets reaching 1–3 m. long and nearly as broad, pale green, often torn, not flat; thickness of the plant 35–40 μ, the cells in section nearly square; becoming detached and drifting.

V. **rigida** (C. Agardh) Le Jolis. Plants attached, with a distinct holdfast and stalk, firm and rather stiff, dark green when well developed, becoming somewhat cleft into broad lobes which are often plicate; thickness of the plant 60–110 μ, the cells in sections of the blade higher than broad, and cell walls thicker than in the other forms.

V. **lacinulata** (Kützing) n. comb. Plants to 4 dm. tall, of fragile texture, dark in color; moderately lobed, the lobes, especially below, more or less elaborately dentate-laciniate.

Bermuda, Florida, Alabama, Mississippi, Texas, Bahamas, Caicos Isls., Cuba, Jamaica, Hispaniola, Puerto Rico, Guadeloupe, Martinique, St. Vincent, Barbados, Grenada, Costa Rica, Panama, Colombia, Venezuela, Trinidad, Brazil. Of the varieties, v. *rigida* is generally found attached to rocks or other objects in the intertidal zone of more or less exposed shores, and where exposed to the surf is often greatly stunted. V. *latissima* is found in quiet coves and harbors, and salt-marsh pools, lightly attached or loose on the bottom. V. *lacinulata* is a very distinct thing, not to be confused with plants marginally eaten by animals, and within our range was found loose or lightly attached in a walled, shaded retaining pond for fish at Bermuda.

REFERENCES: Harvey 1858, Montagne 1853, Martens 1870, Mazé and Schramm *p.p.*? 1870–77, Hemsley 1884, Hauck 1888, Möbius 1889, 1890, 1892, 1895, Mohr 1901, Vickers 1905, 1908, Gepp 1905, Howe 1918a, 1920, 1928, Børgesen 1913–20, 1924, Hamel and Hamel-Joukov 1931, Taylor 1933, 1936, 1940, 1941a, b, 1942, 1954a, Quirós C. 1948, De Mattos 1952. *Phycoseris plicata,* as understood by Martens 1871, may be a small, deeply lobed form of this species. For v. *latissima:* Collins and Hervey 1917, Hoyt 1920, Taylor 1929a, 1930a, 1936, 1941a, b, 1954a; *U. capensis* and *U. laetevirens,* Mazé and Schramm 1870–77; *U. latissima,* Martens 1870, Dickie 1874a, Hemsley 1884, Murray *p.p.,* 1889. For v. *rigida:* Collins 1901, Børgesen 1913–20, Collins and Hervey 1917, Hoyt 1920, Schmidt 1923, Taylor 1928, 1930a, 1933, Taylor and Arndt 1929; *U. lactuca* f. *rigida,* Schmidt 1924; *U. lactuca p.p.,* Mazé and Schramm 1870–77; *U. rigida,* Mazé and Schramm 1870–77, Murray 1889, Howe 1928; *Phycoseris rigida,* Martens 1870. For v. *lacinulata: Ulva lactuca* f. *lacinulata,* Feldman 1937; *Phycoseris lacinulata* Kützing 1849, Species Algarum p. 476; Tab. Phyc. 1856, 6:9, pl. 21.

Ulva fasciata Delile Pl. 1, fig. 4

Plants 1–15 dm. tall, from a very small hapteron and stalk, the base of the blade cuneate, above expanding, irregularly lobed, generally irregularly or sometimes pinnately divided into ligulate or linear lobes which may become several decimeters long, 0.5–2.5 cm. broad, the margins entire to irregularly ruffled and crenate, with a somewhat paler central portion; in section the cells of the midline region much taller than those of the margin, and the thallus much thicker, to 100 μ or somewhat more.

Bermuda, North Carolina, Florida, Texas, Cuba, Jamaica, Hispaniola, Puerto Rico, Virgin Isls., St. Barthélemy, Guadeloupe, Dominica, Martinique, Barbados, British Honduras, Costa Rica, Colombia, Netherlands Antilles, Venezuela, Tobago, Brazil, Uruguay. Common and widespread in the tropics. Plants of moderately exposed shores, growing attached to various kinds of solid objects. In warm and quiet water the lobes may be very much broader and the middle thickened zone obscure; in particularly exposed places the plants may be much dwarfed and the lobes short.

REFERENCES: Harvey 1858, Dickie 1874a, Mazé and Schramm 1870–77, Murray 1889, Collins 1901, 1909b, Vickers 1905, 1908, Grieve 1909, Børgesen 1913–20, Collins and Hervey 1917, Hoyt 1920, Taylor 1930a, b, 1933, 1935, 1936, 1939a, 1940, 1941a, 1942, 1943, 1954a, Hamel and Hamel-Joukov 1931, Questel 1942, Joly 1951, 1957, De Mattos 1952; *U. fasciata* f. *lobata,* Piccone 1886; *U. fasciata* f. *lobata* and f. *taeniata,* Sluiter 1908; ?*U. lobata,* Mazé and Schramm 1870–77, Dickie 1874b, Murray 1889; *U. latissima* and v. *lobata,* Mazé and Schramm 1870–77, Murray 1889; *Phycoseris fasciata,* Martens 1870, 1871; *P. lobata,* Zeller 1876; *P. nematoidea,* Grunow 1870.

Ulva profunda Taylor Pl. 8, fig. 3

Plants at first ovate, attached by a cuneate base and a stalk about 1 mm. long; later widely expanded, irregular, to 4 dm. or more in diameter, delicate; regularly clathrate, the openings generally 0.5–1.5 cm. but reaching 6 cm. diam.; cells of the blade angular, 18–36 μ diam., containing one, seldom 2, chromatophores which nearly cover the face of the cell and contain one, rarely 2, pyrenoids; thickness of the blade 30–60 μ; in section cells subequal to broader than tall, thin-walled (0.3 μ) laterally, but the outer membranes 1.5–3.0 μ thick; cells bounding the openings transversely elongate, reaching surface diameters of 18×72 μ or more.

Florida. Dredged from water of 15–67 m. depth. Abundant at 33 m. where, apparently, it was attached to shell and coral fragments.

REFERENCE: Taylor 1928.

PRASIOLALES

Plants filamentous to foliaceous; cells often showing regular arrangements in groups in the broader forms; cells with firm walls, uninucleate, the radiating chromatophores with a central pyrenoid; multiplication by fragmentation; asexual reproduction by the production of akinetes either by direct conversion of vegetative cells or after one division across the plane of the blade.

PRASIOLACEAE

Characters of the Order.

Prasiola (C. Agardh) Meneghini, 1838

Plants small, foliaceous, firmly membranous, usually attached by a distinct short stalk or the edge of the membrane, or in some forms eventually becoming free; cells dividing in groups of fours and multiples, the groups remaining more or less distinct.

Prasiola stiptitata Suhr

Plants tufted, dark green, much curled and shrunken when dry; variable in size and form, 2–6(–8) mm. long, the base narrow and stalklike, above expanding more or less, becoming in well-developed plants lanceolate to fan- or kidney-shaped, the apex truncate, the margins often incurled; cells 5–7(–10) μ diam., seriate in the stalk, in the upper part crowded in regular blocks; akinetes formed at the truncate end of the plant, spherical, 10–12 μ diam.

North Carolina. Usually found on rocks near or above high-tide level, especially where sea birds are accustomed to roost.

REFERENCE: Williams 1949a.

CLADOPHORALES

Plants filamentous, usually with a distinct basal holdfast, simple or branched, the filaments uniseriate; the cells with a thin to thick firm wall, a large central vacuole, a generally multinucleate cytoplast with a greatly reticulate or perhaps ultimately fragmented chloroplast and one to many pyrenoids.

CLADOPHORACEAE

Characters of the Order.

KEY TO GENERA

1. Filaments unbranched, or with few short simple branchlets 2
1. Filaments progressively, often abundantly, branched CLADOPHORA, p. 77

2. Filaments attached by a basal end; cells very large, walls firm; plants not firmly adhering to paper .... CHAETOMORPHA, p. 69
2. Filaments eventually free, or attached by lateral holdfasts 3

3. Filaments coarse, symmetrical, a mass not completely collapsing on removal from water; unbranched .... CHAETOMORPHA, p. 69
3. Filaments more slender and irregular of contour, a mass collapsing on removal from water; unbranched or with a few lateral or rhizoidal spur branches .... RHIZOCLONIUM, p. 75

Chaetomorpha Kützing, 1845

Plants filamentous, unbranched, attached by a basal holdfast cell; cylindrical or usually somewhat broader above the base, or in the form of more or less entangled free filaments; cells highly multinucleate, usually with firm, often heavy, lamellose walls; asexual reproduction by zoöspores formed in slightly enlarged cells; sexual reproduction similar, by isogametes.

KEY TO SPECIES

1. Plants erect, the filaments clearly basally attached 2
1. Plants of flexuous entangled filaments without specific basal attachment 6

2. Small epiphytes, the filaments 10–20 μ diam ...... C. minima, p. 72
2. Much larger plants several centimeters tall, chiefly growing on rocks 3

3. Erect filaments much broader at the summit than below, reaching 1.5 mm. diam. .... C. clavata, p. 73
3. Filaments of nearly uniform diameter, except very near the holdfast 4

4. Filaments less than 100 μ diam. .... C. nodosa, p. 72
4. Filaments 100 μ diam. or more 5

5. Basal cell very elongate, 4–12 times as long as the next higher cell . . . . C. media, p. 73
5. Basal cell less distinctive, about 2.5–4.2 times as long as the suprabasal cell . . . . C. aerea, p. 72

6. Filaments under 100 μ diam. . . . . C. gracilis, p. 70
6. Filaments usually over 100 μ diam. 7

7. Filaments geniculate, 100–210 μ diam.; cells 1–2 diameters long C. geniculata, p. 71
7. Filaments without axial deflections 8

8. Filaments rather stiff, 100–375 μ diam., cells 0.75–5 diameters long . . . . C. linum, p. 71
8. Filaments usually rather softer, cells about as long as broad 9

9. Filaments 80–180 μ diam. . . . . C. brachygona, p. 70
9. Filaments 300–700 μ diam. . . . . C. crassa, p. 72

Chaetomorpha gracilis Kützing

Plants entangled, filaments cylindrical, about 40–70 μ diam., the cells 2–4 diameters long.

Bermuda, Florida, Jamaica, Hispaniola, Virgin Isls., Guadeloupe, Costa Rica, Venezuela, Brazil. Loosely floating in shallow water.

REFERENCES: Mazé and Schramm 1870–77, Murray 1889, Möbius 1890, Børgesen 1913–20, Collins and Hervey 1917, Taylor 1928, 1929b, 1931, 1933, 1936, Taylor and Arndt 1929; *C. implexa* v. *montagneana* and *C. javanica* f. *tenuis*, Mazé and Schramm 1870–77, Murray 1889.

Chaetomorpha brachygona Harvey Pl. 2, fig. 9

Filaments generally soft, flexuous-entangled, 80–180 μ diam.; the cells 60–420 μ long, but usually about as long as broad; cell walls generally thick.

Bermuda, North Carolina, Florida, Louisiana, Texas, Mexico, Bahamas, Cuba, Jamaica, Hispaniola, Puerto Rico, Virgin Isls., St. Barthélemy, Guadeloupe, Martinique, Barbados, British Honduras, Panama, Netherlands Antilles, Brazil. Plants of shallow, muddy coves, growing in loose masses among coarser algae and sea grasses, but occasionally to several meters' depth.

REFERENCES: Harvey 1858, Murray 1889, Collins 1901, Børgesen 1913–20, Collins and Hervey 1917, Howe 1918a, 1920, 1928, Hoyt 1920, Taylor 1928, 1929b, 1935, 1936, 1940, 1941a, 1942, 1943, 1954a, Taylor and Arndt 1929, Hamel and Hamel-Joukov 1931, Joly 1951, 1957; *C. tortuosa*, Mazé and Schramm 1870–77, Murray 1889 *p.p.; Rhizoclonium capillare*, Vickers 1905, 1908. See note under *C. linoides* Kützing.

Chaetomorpha geniculata Montagne

Filaments somewhat stiff, loosely entangled, flexuous, occasionally rather abruptly geniculate but otherwise cylindrical, 100–210 μ diam.; the cells (1.0–)1.5–2.0 diameters long.

Guadeloupe, French Guiana.

REFERENCES: Montagne 1850, Mazé and Schramm 1870–77, Murray 1889.

Chaetomorpha linum (Müller) Kützing Pl. 2, fig. 8

Plants composed of loosely entangled, unattached filaments, yellowish green, somewhat stiff and curled; cylindrical or the cells slightly swollen, 100–375 μ diam., 1–2(0.75–5.0) diameters long.

Bermuda, North Carolina, Florida, Bahamas, Cuba, Jamaica, Hispaniola, Puerto Rico, Virgin Isls., St. Barthélemy, Guadeloupe, Martinique, Barbados, Costa Rica, Panama, Netherlands Antilles, Brazil. Forming entangled masses among larger algae in shallow water.

REFERENCES: Hauck 1888, Murray 1889, Collins 1901, Sluiter 1908, Collins and Hervey 1917, Howe 1918a, 1920, Hoyt 1920, Taylor 1928, 1942, Taylor and Arndt 1929, Hamel and Hamel-Joukov 1931, Questel 1942; *C. dubyana*, Mazé and Schramm 1870–77, Murray 1889 *p.p.; C. geniculata*, Hemsley 1885; *C. linum* v. *brachyarthra*, Collins 1901; *C. chlorotica*, Mazé and Schramm 1870–77, Hauck 1888, Murray 1889, Möbius 1890; *Conferva linum*, Montagne 1863; *Rhizoclonium linum*, Vickers 1905, 1908; *R. subramosum*, Mazé and Schramm 1870–77, Murray 1889.

Chaetomorpha crassa (C. Agardh) Kützing

Plants entangled, filaments 500–550 (300–700) μ diam.; cells about as long as broad, but occasionally to 2 diameters in length; cell walls thick.

Bermuda, Virgin Isls., St. Barthélemy. Among other coarser algae in shallow water.

REFERENCES: Børgesen 1913, Collins and Hervey 1917, Howe 1918a, Questel 1942.

Chaetomorpha minima Collins and Hervey

Filaments erect from a disciform base, to about 5 mm. tall; cylindrical or a little enlarged above to 10–20 μ diam., the nodes sometimes a little constricted; cells 2–4 diameters long, with lamellate walls.

Bermuda. Growing epiphytically on Codium and Cladophora.

REFERENCES: Collins and Hervey 1917, Howe 1918a.

Chaetomorpha nodosa Kützing

Plants erect, filaments dull green, basally attached, about 12 cm. long, stiff and straight, 75–80 μ diam.; the cells as long as broad or a little longer, occasionally a little swollen.

French Guiana.

REFERENCE: Montagne 1850.

Chaetomorpha aerea (Dillwyn) Kützing

Plants gregarious, bright green, to 10–15, occasionally to 30 cm., tall; attached by a slender, subclavate basal cell which has a disklike base lobed or fimbriate at the margins; basal cell to 130–150 μ diam. at the top, 7.5–10.5 diameters long, and about 2.5–4.2 times as long as the suprabasal cell; filaments slender toward the base, above to 150–350(–500) μ diam., stiff and straight, the cells to 1–2 diameters long, little constricted at the septa; zoöspores formed in the upper cells of the filament, which become cask-shaped to subglobose, 600–700 μ diam.

Bermuda, North Carolina, South Carolina, Cuba, Jamaica, Hispaniola, Virgin Isls., Brazil. Primarily a northern species, growing on rocks, especially under ledges, in exposed, even surf-beaten locations. The older records, especially the more southern ones, are in need of critical confirmation.

REFERENCES: Dickie 1874a, b, Hemsley 1884, Murray 1889, Collins 1901, Børgesen 1913–20, Hoyt 1920, Taylor 1954a, Joly 1957; *C. dubyana*, Murray 1889 *p.p.?*, Piccone 1889; ?*C. vasta*, Mazé and Schramm 1870–77, Murray 1889.

Chaetomorpha media (C. Agardh) Kützing

Plants erect, tufted, dark green; filaments to 7–10, seldom to 20, cm. tall, attached by a distinctive, stout, clavate basal cell which has a radicular attachment supplemented in older filaments by external branches from the sides; basal cell to 400–525 μ diam. at the top, 8–50 diameters long, and (4–)8–12 times as long as the suprabasal cell; the filaments 450–550 μ diam., the cells here 2–4 diameters long.

Bermuda, Cuba, Jamaica, Hispaniola, Puerto Rico, Virgin Isls., St. Barthélemy, Redonda I., Guadeloupe, Aves I., Dominica, Martinique, Barbados, Costa Rica, Panama, Colombia, Netherlands Antilles, Venezuela, Trinidad, Brazil. Growing on rocks and reefs in shallow water, moderately exposed to the waves.

REFERENCES: Greville 1833, Mazé and Schramm 1870–77, Martens 1870, Möbius 1889, 1890, Murray 1889, Howe 1928, Taylor 1929b, 1933, 1936, 1941b, 1942, 1943, 1954a, Hamel and Hamel-Joukov 1931, Feldmann and Lami 1937, Questel 1942, Joly 1951, De Mattos 1952; *C. antennina*, Mazé and Schramm 1870–77, Martens 1871, Zeller 1876, Murray 1889, Vickers 1905, 1908, Grieve 1909, Børgesen 1913–20, Joly 1957.

Chaetomorpha clavata (C. Agardh) Kützing

Plants tufted, erect, the filaments to 30 cm. tall; the cells near the base 500–750 μ diam., gradually increasing to 1.5 mm. diam. or more at the tip; lower cells 3–4 diameters long, the upper ones subequal, in shape from subcylindrical below to cask-shaped, and in the upper portions, especially as fruiting approaches, submoniliform; cell walls relatively thin.

Florida, Bahamas, Jamaica, Hispaniola, Virgin Isls., St. Barthélemy, Barbados, Venezuela, Brazil. Growing attached to rocks or other solid objects.

REFERENCES: Martens 1870, 1871, Collins 1901, 1918a, Vickers 1905, 1908, Børgesen 1913–20, 1924, Howe 1920, Taylor 1929b, 1933, 1941a, 1943, Taylor and Arndt 1929; *C. intestinalis*, Mazé and Schramm 1870–77, Murray 1889.

UNCERTAIN RECORDS

Chaetomorpha atrovirens Taylor

Jamaica, Venezuela. A northern species. Type from Massachusetts. Taylor 1937; *C. melagonium*, Collins 1901; *C. piquotiana*, De Toni 1:271.

Chaetomorpha billardieri Kützing

Guadeloupe. Type from Australia. De Toni 1:275; Mazé and Schramm 1870–77, Murray 1889.

Chaetomorpha breviarticulata Hauck

Cuba, Guadeloupe. Type from the Adriatic. De Toni 1:266; Murray 1889.

F. *montagneana* De Toni

Cuba, the type locality. De Toni 1:267.

Chaetomorpha linoides Kützing

Virgin Isls., Antigua, Guadeloupe. Type from the Virgin Islands. De Toni 1:274; Kützing 1847, Mazé and Schramm 1870–77, Murray 1889. Specimens seen in Kützing's herbarium under this name and from the first two stations agree in general with *C. brachygona* as described here. If their identity with the type is established Kützing's name may be the valid one.

Chaetomorpha pachynema Montagne

Guadeloupe, Brazil. Type from the Canary Islands. De Toni 1:270; Mazé and Schramm 1870–77, Zeller 1876, Murray 1889.

Chaetomorpha saccata Kützing

West Indies and Brazil. Type from the "Antilles." De Toni 1:271; Martens 1870.

Rhizoclonium Kützing, 1843

Plants filamentous, simple or sparingly branched, initially basally attached but eventually free, or with lateral rhizoidal branches, these tapering, spreading, of few cells; main filaments usually entangled, of moderately long cells somewhat unevenly articulated; cells with moderately to quite thick walls; chromatophores with numerous pyrenoids; nuclei one to a few in each cell; reproduction by zoöspores or by akinetes formed from little-modified cells.

KEY TO SPECIES

1. Cells 25 μ diam. or more; wall thick 2
1. Cells usually 10–25 μ diam. 4

2. Cells 40–100 μ diam.; the cells 2–4 diameters long; branches few to many, sometimes pluricellular .... R. hookeri, p. 77
2. Filaments more slender 3

3. Filaments 68–74 μ diam., cells 1.25–2.0 diameters long, the wall to 13 μ thick; rhizoidal branching from basal cells only R. crassipellitum v. robustum, p. 76
3. Filaments 40–70 μ diam., cells 1–2 diameters long; branches absent .... R. tortuosum, p. 76

4. Filaments 20–30 μ diam., cells 1–2 diameters long; branches scattered, rhizoidal, few to many .... R. riparium, p. 76
4. Filaments 10–15 μ diam., cells 3–7 diameters long; branches absent .... R. kerneri, p. 75

Rhizoclonium kerneri Stockmayer

Plants of entangled yellowish-green filaments, without branches; cells 10–15(–19) μ diam., 3–7 diameters long.

Bermuda, Jamaica, Hispaniola, Virgin Isls., Guadeloupe, Netherlands Antilles. From the margins of quiet pools, and especially on the roots of mangroves. Plants of this area with shorter cells may be referable to *R. kochianum*, but do not seem clearly distinct.

REFERENCES: Børgesen 1913–20, 1924, Collins and Hervey 1917, Howe 1918a, Taylor 1942; *R. implexum?* and *Chaetomorpha lanosa*, Mazé and Schramm 1870–77.

Rhizoclonium riparium (Roth) Harvey

Plants entangled, yellowish-green, forming considerable expanses on the substrate, the rhizoidal branches usually numerous; the filaments curled, 20–30 μ diam., the cells 1–2 diameters long.

V. **implexum** (Dillwyn) Rosenvinge has few branches or more often lacks them.

Bermuda, North Carolina, Florida, Mississippi, Bahamas, Cuba, Jamaica, Guadeloupe, Tobago, Brazil. Plants of the intertidal zone, growing on rocks and woodwork and often found in tidepools.

REFERENCES: Collins 1909b, Collins and Hervey 1917, Howe 1920, Hoyt 1920, Taylor 1942, 1954a, Joly 1951, 1957; *Chaetomorpha gracilis* v. *tenuior* and *C. submarina,* Mazé and Schramm 1870–77, Murray 1889; *Conferva riparia,* Montagne 1863; *R. lanosum,* Mazé and Schramm 1870–77.

Rhizoclonium crassipellitum W. and G. S. West

Filaments entangled, curved and genuflexed; rhizoids sometimes produced from the basal cell, but otherwise unbranched; filaments 30–45 μ diam., cells 1–2 diameters long, the walls to 13 μ thick and lamellate.

V. **robustum** G. S. West. Filaments 68–74 μ diam., the cells 1.25–2.0 diameters long; otherwise as in the typical plant.

Bermuda, Bahamas, Barbados. The species is typically a plant of fresh-water localities where it forms mats on wet earth, but at least in the variety it is also found near high-tide mark and in brackish ponds.

REFERENCES: Collins 1909b, Collins and Hervey 1917, Howe 1920.

Rhizoclonium tortuosum Kützing

Plants forming dull, rather dark green, entangled masses; rhizoidal branches absent; filaments curled, rather stiff, 40–70 μ diam., cells 1–2 diameters long.

Bermuda, North Carolina, ?Barbados. Plants of the intertidal zone, growing on rocks and woodwork, and often found in tidepools.

REFERENCES: Howe 1918a; *Chaetomorpha tortuosa,* Dickie 1874a, Murray 1889 *p. p.*?

Rhizoclonium hookeri Kützing Pl. 2, fig. 5

Filaments rather stiff, entangled; branches few to many, short but often pluricellular and sometimes rebranched, of irregular thickness, 40–80(–100) μ diam.; the cells 2–4 diameters long, with a wall 4–10 μ thick.

Bermuda, Florida, Bahamas, Jamaica, Guadeloupe, Panama. Plants of the intertidal margins of protected coves and pools, especially in mangrove thickets, but also found on the walls of grottos and occasionally in fresh water.

REFERENCES: Collins 1909b, Collins and Hervey 1917, Howe 1918a, 1920, Taylor 1928, 1929b, Feldmann and Lami 1937; *R. tropicum,* Mazé and Schramm 1870–77.

UNCERTAIN RECORDS

Rhizoclonium arenosum (Carmichael) Kützing
V. *occidentalis* Kützing

Dominica. Type of the variety from the West Indies. De Toni 1: 281; Grieve 1909.

Rhizoclonium bolbogoneum Montagne

French Guiana, the type locality. De Toni 1:281; Montagne 1850.

Rhizoclonium kochianum Kützing

Virgin Isls., Barbados. Type from Europe. Vickers 1908, Collins 1909b, Børgesen 1913; *R. flavicans,* De Toni 1:279.

Rhizoclonium sargassicolum Crouan

Guadeloupe, the type locality. Mazé and Schramm 1870–77, Murray 1889 (including f. *spiralis* and f. *tenuis,* all essentially *nomina nuda*).

Cladophora Kützing, 1843

Plants filamentous, sparingly to repeatedly branched, attached by rhizoidal extensions from the lower cells, these filaments spreading over the substratum and often giving rise to new erect shoots; growth primarily apical; chromatophores reticulate, with many pyrenoids, or of separate disks, the nuclei numerous in each cell; asexual reproduction by numerous zoöspores formed in the little-modified outer cells of the branches; sexual reproduction similar, by isogametes.

It is difficult enough to do justice to this genus in well explored northern seas but quite impossible to deal adequately with representatives from the American tropics. A few species are distinctive and common, such as *C. fascicularis* and *C. fuliginosa*, a few are distinctive but rare, as *C. catenifera* and *C. prolifera*. The other records are confused by numerous misidentifications and inadequate descriptions. I have recognized as plausibly recorded about thirty-five species, but have relegated to a doubtful list over forty other species the records for which do not seem well substantiated, for which I have seen no specimens, or which are clearly *nomina nuda* or *seminuda*. Doubtless some of the accepted class may in time be discarded and some of the rejected re-established with adequate descriptions and record specimens. I have relied greatly on the work of Collins (1909) and the compilation of De Toni (1889 *et seq.*); where the available material seemed ample I have reviewed the species descriptions they have provided, but where it has been scanty or lacking I have not ventured to depart much from the characters as they have delimited them.

As the characterization of many of these species is incomplete, it follows that the key based on the descriptions and specimens available will be difficult to use. Since for an inexperienced observer more frustration than satisfaction may result from trying to use it, I have selected a very few of the more frequently reported, including perhaps the commonest, species and a few very distinctive ones, and have prepared in addition a short key (p. 81) which will be of assistance to the beginner in making a preliminary sorting of his material.

KEY TO SPECIES

1. Low, pulvinate or matted, with the main filaments more or less prostrate 2
1. Low or tall, but erect; or if unattached or much entangled at least loosely branched, without a prostrate filament system 5

2. Lower part of the main axis 300–350 μ diam. . . C. intertexta, p. 82
2. Lower part of the main axis less than 300 μ diam. 3

3. Branching mostly irregular, not dichotomous ...... *C. howei*, p. 82
3. Branching mostly di- or trichotomous 4

4. Lower filaments generally more than 100 μ diam., cells cylindrical or nearly so; color dark .... *C. repens*, p. 82
4. Lower filaments generally less than 100 μ diam., cells of the branchlets ovoid or pyriform .... *C. frascatii*, p. 82

5. Main filaments generally exceeding 150 μ diam. 6
5. Main filaments seldom or never reaching 150 μ diam. 15

6. Lower cells often exceeding 10 diameters in length 7
6. Lower cells generally less than 10 diameters long 11

7. Plants with notably fasciculate branchlets 8
7. Plants without notably fasciculate branchlets 9

8. Plants coarse and stiff, dark green and becoming blackish on drying .... *C. prolifera*, p. 91
8. Plants rather soft, bright green and not blackening *C. fascicularis*, p. 91

9. Plants becoming detached and growing to maturity in that state *C. heteronema*, p. 84
9. Plants remaining attached, the nodes strongly constricted 10

10. Main axes regularly exceeding 300 μ diam. .... *C. catenifera*, p. 92
10. Main axes seldom reaching 300 μ diam. ...... *C. catenata*, p. 83

11. Branchlets with strongly constricted nodes 12
11. Branchlets abundant, long and slender, the nodes not strongly constricted 13

12. Lower cells of the axes 2–4 diameters long .... *C. hutchinsiae*, p. 90
12. Lower cells of the axes 4–6 diameters long ... *C. brachyclona*, p. 87

13. Cell diameter nearly the same in all parts of the plant *C. fuliginosa*, p. 83
13. Terminal divisions markedly more slender than the main filaments 14

14. Branches spreading, except at the extreme base; axes often exceeding 250 μ diam. .... *C. crucigera*, p. 87
14. Branching generally erect; axes seldom reaching 250 μ diam. *C. utriculosa*, p. 89

15. Cells with more or less regular constrictions 16
15. Cells without regular constrictions 17

16. A single constriction to be found near the base of a cell *C. constricta*, p. 92

16. Multiple constrictions shown by the cells of the main axes; branchlets often strongly hooked or circinate . C. uncinata, p. 92

17. Main filaments distinctly angled or flexuous 18
17. Main filaments more or less straight 26

18. Frond floating, except at the earliest stages 19
18. Frond always growing attached 21

19. Main filaments to about 80 μ diam.; branchlets strongly curved, the plant of a spongy texture .... C. crispula, p. 85
19. Branchlets not markedly curved, plants not spongy, the main filaments usually coarser 20

20. Main filaments generally exceeding 100 μ diam.; a marine or brackish-water species .... C. expansa, p. 85
20. Main filaments hardly reaching 100 μ diam.; primarily a fresh-water species .... C. fracta, p. 86

21. Branchlets short, acute, spinelike .... C. polyacantha, p. 86
21. Branchlets not spinelike 22

22. Branches often decumbent; rhizoidal branchlets numerous C. corallicola, p. 83
22. Not decumbent; rhizoids few or none 23

23. Branchlets long, in pectinate series at the tips of the branches .. 24
23. Branchlets not in pectinate series 25

24. Main axes to 140–160 μ diam., the branchlets 40–60 μ diam. C. gracilis, p. 90
24. Main axes to 60 μ diam., the branchlets about 35 μ diam. C. luteola, p. 88

25. Main filaments 80–120 μ diam., the branchlets 40–80 μ diam.; plants of rocky shores in the North Atlantic .. C. flexuosa, p. 89
25. Main filaments to 65 μ diam., the branchlets 20–28 μ diam.; plants of subtropical to South Atlantic waters . C. brasiliana, p. 85

26. Branchlets curved, densely placed at the branch tips C. scitula, p. 90
26. Branchlets straight or nearly so 27

27. Main filaments not over 60 μ diam.; cells of the branchlets 1–2 diameters long .... C. delicatula, p. 87
27. Main filaments exceeding 60 μ diam. 28

28. Branchlets acute; main filaments to about 75 μ diam. C. glaucescens, p. 86
28. Branchlets not acute 29

29. Main filaments 40–75 μ diam., cells 5–20 diameters long C. crispata, p. 85
29. Main filaments thicker 30

30. Plants small, stiff; branchlets 55–75 μ diam. .. C. sertularina, p. 91
30. Branchlets not so coarse, or, plants larger 31

31. Main branches distinct, virgate 32
31. Main branching chiefly dichotomous 33

32. Branchlets 20–30 μ diam. .... C. nitida, p. 88
32. Branchlets 32–36 μ diam. .... C. virgatula, p. 88

33. Substance soft and silky; cells 4–12 diameters long; branchlets rather erect .... C. crystallina, p. 89
33. Substance crisp or harsh; cells usually less than 5 diameters long; branchlets widely spreading 34

34. Cells 3–5 diameters long; branchlets scattered and distant C. piscinae, p. 84
34. Cells mostly 1–2 diameters long; branchlets closer, often in unilateral series .... C. rigidula, p. 84

KEY TO SELECTED SPECIES OF CLADOPHORA

1. Plants attached, more or less erect 3
1. Plants matted, entangled, or somewhat detached, forming masses in rocks or tide pools 2

2. Filaments 80–150 μ diam., rather loosely entangled . C. expansa, p. 85
2. Filaments 60–80 μ diam., notably curled and more closely entangled .... C. crispula, p. 85

3. Plants coarse and rather stiff, with the main filaments exceeding 300 μ diam. 4
3. Plants softer and usually more delicate 6

4. Plants loosely branched, bright green .... C. catenifera, p. 92
4. Plants closely branched, blackening when dried 5

5. Branchlets fasciculate and plants bushy .... C. prolifera, p. 91
5. Branchlets not fasciculate, often curved to one side, the plants somewhat cushionlike .... C. fuliginosa, p. 83

6. Branchlets distinctly clustered, the main axes over 200 μ diam. C. fascicularis, p. 91
6. Branchlets not clustered 7

7. Main axes 40–60 μ diam. .... C. delicatula, p. 87
7. Main axes 80–140 μ diam. .... C. crystallina, p. 89

Cladophora frascatii Collins and Hervey

Plants small, in matted tufts 1–2 cm. tall; below irregularly and for the most part widely dichotomously branched, above in part similar, in part laterally branched more or less at right angles, or unilaterally on the outer side of recurved branches; main axes 70–100 μ diam., the cylindrical cells 2–5 diameters long; branchlets 60–80 μ diam., their somewhat turgid cells 2–3 diameters long, the terminal branchlet cells obtuse.

Bermuda. Mat-forming plants of tide pools.

REFERENCE: Collins and Hervey 1917.

Cladophora howei Collins

Plants densely matted, with a pronounced basal system of creeping filaments, the cells of these irregular in form but about 150 μ diam., 1–3 diameters long; from the basal system arise filaments which are about 50 μ diam. and their cells 5–6 diameters long near the base, and which fork sparingly, the branches erect, tapering to 20–25 μ diam. in the branchlets, the cells to 15–20 diameters long near the branchlet tips.

Bermuda, Hispaniola. Plants forming dense mats in tide pools.

REFERENCES: Collins 1909b, Collins and Hervey 1917, Howe 1918a, Taylor 1943.

Cladophora intertexta Collins

Plants forming matted tufts, the coarse filaments chiefly prostrate, 300–350 μ diam., cells 1.0–1.5 diameters long, rarely to 3 diameters; basal filaments bearing upright branches which are simple or very sparingly unilaterally branched, 200 μ diam., the tip cells blunt.

Jamaica. Plants forming mats on the bottoms of pools, the erect filaments conspicuous; Chaetomorpha-like in aspect.

REFERENCES: Collins 1901, 1909b.

Cladophora repens (J. Agardh) Harvey

Plants dull green, forming soft spongy cushions; filaments widely dichotomous, 100–150 μ diam.; cells 2–3 diameters long below, but in the branchlets 6–8(–10) diameters in length.

Bermuda, ?Florida, Guadeloupe, Grenada, Panama. On inshore rocks in sheltered places.

REFERENCES: Harvey 1858, Mazé and Schramm 1870–77, Murray 1889, Collins and Hervey 1917, Taylor 1928, 1929b.

Cladophora corallicola Børgesen

Plants pulvinate, tufted, attaching to the substratum by rhizoids from decumbent filaments; the main erect filaments 125–150 μ diam., with cells 5–10 diameters long, ditrichotomously branched below, more scantily branched above; where flexuous, the cells often curved; branchlets about 70 μ diam., the cells 5–6 diameters long.

Bermuda, Virgin Isls. From shallow or moderately deep water, growing on stones or coral fragments.

REFERENCE: Børgesen 1913–20.

Cladophora fuliginosa Kützing Pl. 2, fig. 3; pl. 3, fig. 4

Plants tufted, to 2–6 cm. tall, erect or the tips recurved; when crisp, bright green, but drying, blackish-brown; branching below dichotomous to alternate, above mostly alternate, or partly unilateral and subpectinate on rather curved branches; main axes to 380 μ diam., the cells slightly swollen and 3–6 diameters long, the branchlets 150–160(–290) μ diam., the cells nearly cylindrical and the more distal very long, to 10–13 diameters.

Bermuda, Florida, Bahamas, Caicos Isls., Turks Isls., Cuba, Cayman Isls., Jamaica, Hispaniola, Virgin Isls., St. Barthélemy, Guadeloupe, Martinique, British Honduras, Netherlands Antilles. A common plant forming firm tufts or mats near low-tide line on rather exposed shores; also dredged to a depth of 55 m.

REFERENCES: Collins 1901, 1907, 1909b, Børgesen 1913–20, 1924, Howe 1918a, b, 1920, Collins and Hervey 1917, Taylor 1928, 1933, 1935, 1940, 1942, 1943, 1954a, Taylor and Arndt 1929, Sanchez A. 1930, Hamel and Hamel-Joukov 1931, Feldmann and Lami 1937; *Blodgettia confervoides*, Harvey 1858, Farlow 1871, Rein 1873, Mazé and Schramm 1870–77.

Cladophora catenata (C. Agardh) Ardissone

Fronds densely tufted, dark green, stiff, to 8 cm. tall; filaments much branched, ditrichotomous below and 200–250 μ diam.; above the branchlets opposite, unilateral or irregularly placed, 80–150 μ diam.; cells in general 3–6 diameters long, turgid, the nodes constricted.

Jamaica, ?Brazil.

REFERENCES: Collins 1909b; *Conferva catenata,* Greville 1833a.

Cladophora heteronema (C. Agardh) Kützing

Plants in tufts several centimeters tall, later becoming free; filaments below rather sparsely divided but above very much and irregularly branched, 120–280 μ diam., at the forks sometimes adnate; cells cylindrical or somewhat swollen, generally 4–15 diameters long, but individual segments sometimes only 1–3 diameters long; branchlets 30–60 μ diam.

Virgin Isls., Brazil. Plants of sheltered shallows growing among sea grasses, also in brackish water.

REFERENCES: Børgesen 1913; *C. fracta* f. *marina,* Hauck 1885.

Cladophora piscinae Collins and Hervey

Plants large and bushy, light green but darkening on drying, a little stiff in texture; main filaments 100 μ diam., branching below by wide equidistant forkings, the cells 3–5 diameters long, not swollen; branchlets distant, spreading.

Bermuda. Plants growing in a walled fish pond, not attached.

REFERENCE: Collins and Hervey 1917.

Cladophora rigidula Collins and Hervey

Plants large, somewhat matted, light dull green, remaining spongy even when removed from the water; primary filaments freely dichotomously branching at wide angles below, more sparingly branched above and the branches more erect, 120 μ diam.; cells 1–2 diameters long, not swollen; above bearing branchlets about 80 μ diam. at right angles in somewhat unilateral series, the tip cells about 3 diameters long.

Bermuda. A plant of very protected shallow water, having been found in a walled fish pond.

REFERENCE: Collins and Hervey 1917.

Cladophora crispula Vickers

Plants forming spongy masses of contorted filaments; freely alternately or oppositely branched, the branchlets near the tips somewhat unilaterally placed and strongly curved; filaments not greatly tapering, 60–80 μ diam. below, 45–50 μ diam. above, cells about 8 diameters long.

Bermuda, Bahamas, Hispaniola, Virgin Isls., Barbados. From the shallow water in sheltered places, and a similar but perhaps slightly larger form from water of several meters depth.

REFERENCES: Vickers 1905, 1908, Børgesen 1913–20, 1924, Collins and Hervey 1917, Howe 1918a, 1920, Taylor and Arndt 1929.

Cladophora crispata (Roth) J. Agardh

Forming loose masses; main filaments 40–75 μ diam., sparingly laterally or dichotomously branched below, but branching above alternate and more dense; branchlets 20–35 μ diam.; cells cylindrical, 5–20 diameters long below, proportionally even longer in the branchlets, but the branchlets blunt, not tapering to the tips.

Hispaniola. A fresh-water species of northern countries, but here reported from a brackish lake.

REFERENCE: Phyc. Bor.-Am. 2290.

Cladophora expansa (Mertens) Kützing

Plants light green, forming loose cushions or soft entangled masses of considerable extent; filaments much and loosely branched, the main branches angled-flexuous, elongate, (80–)100–150 μ diam., divaricately or alternately divided, bearing smaller spreading secondary branches; cells in the main branches 3–6(–12?) diameters long; branchlets somewhat unilateral, blunt, to 40 μ diam.

Bermuda. A plant of warm lagoons, forming large unattached masses in shallow water.

REFERENCES: Collins and Hervey 1917, Taylor 1957.

Cladophora brasiliana Martens

Plants dark green below, paler at the tips, to 2.5 cm. tall; the filaments ditrichotomously divaricately branched; cells of the main axes 6–8 diameters long, 65 μ diam. below, but above only 22–28 μ diam.; the branchlets attenuate, flagelliform, their cells to 10 diameters long.

Guadeloupe, Brazil. Apparently a plant of warm, shallow, protected, and perhaps brackish waters.

REFERENCES: Martens 1870, Mazé and Schramm 1870–77, Möbius 1889, 1892, Murray 1889.

Cladophora glaucescens (Griffiths *ex* Harvey) Harvey

Plants 10–40 cm. tall, attached, soft, yellowish or grayish green; primary filaments elongate, plumose, and somewhat clustered, the lower cells 50–75 μ diam., 4–6 diameters or more in length; branchlets long, alternate or sometimes unilateral, 25–40 μ diam., cells proportionally nearly as long as those below.

Bermuda, South Carolina, Florida, Texas, Hispaniola. A plant of upper tide pools, but probably not exposed to full sunlight since this is primarily a plant of northern waters.

REFERENCES: Harvey 1858, Hemsley 1884, Murray 1889, Collins 1909b, Taylor 1928, 1933, 1941a.

Cladophora fracta (Vahl) Kützing

Plants dull green, bushy, rather stiff, with scattered spreading branches; lower axes 60–90 μ diam., the thick-walled cells 15–20 diameters long; above, the branchlets often recurved, to 35 μ diam., cell length 3–6 times breadth.

Bermuda, Bahamas. Chiefly a plant of fresh water, but occasionally reported from shallow, protected brackish or salt-water stations.

REFERENCES: Howe 1918a, 1920.

Cladophora polyacantha Montagne

Plants tufted, 10–20 cm. tall, dull green, rather stiff in texture; filaments dichotomously forked below, 50–80 μ diam., the branches long and flexuous, bearing long similar secondary branches 30–50 μ diam., and numerous short, acute, spinelike branchlets of 2–3 cells each, 25–35 μ diam., but the long branches usually naked toward their tips.

Florida, French Guiana. Growing on rocks exposed to the waves.

REFERENCES: Montagne 1850, Taylor 1928; *C. flexuosa* f. *floridana*, Collins 1906.

Cladophora crucigera Grunow

Plants coarse and pale green; sparingly and widely dichotomously branched, the branches slightly adnate at the forks; branchlets scattered, short, of few cells, opposite or sometimes alternate, very widespreading; cells below 280–320 μ diam., 6–8 diameters long, but in the ultimate branchlets 75–110 μ diam., 3–4 diameters long, a little constricted at the nodes.

Guadeloupe, Brazil. Growing along the shore.

REFERENCES: Grunow 1867, Mazé and Schramm 1870–77, Murray 1889.

Cladophora brachyclona Montagne

Plants tufted, to 15 cm. tall, lax, light or yellowish green; main filaments di- polychotomous, 160–250 μ diam., the cells cylindrical and 4–6 diameters long; branches rather distantly spaced below, branchlets above short, often of a single cell, more crowded, rather unilateral and pectinate, 50–75 μ diam., the cells ellipsoid, 1.5–2.0 diameters long.

Bermuda. The Mediterranean type is reported to grow on submerged stones.

REFERENCES: Collins 1909b, p. 341, Collins and Hervey 1917.

Cladophora delicatula Montagne

Plants loosely tufted and soft, dull green, to 10 cm. tall; main filaments loosely branching, the divisions erect, 40–60 μ diam., the cells 4–6 diameters long; branchlets in short unilateral series seldom over 8 cells long, the cells 20–30 μ diam., 1–2 diameters long, somewhat contracted at the septa.

Bermuda, Florida, Jamaica, Puerto Rico, ?Guadeloupe, Venezuela, French Guiana. A plant of quiet water in protected coves.

REFERENCES: Montagne 1850, Mazé and Schramm 1870–77, Murray 1889, Collins 1909b, Collins and Hervey 1917, Taylor 1928, 1942.

Cladophora luteola Harvey

Plants tufted, pale green, bushy, delicately branched; main axes about 60 μ diam., flexuous with the forking angles rounded, branching irregular, often trichotomous; branchlets unilateral or opposite, at the tips pectinate and somewhat crowded, 35(18–50) μ diam., the cells 6–8 diameters long, not swollen.

Bermuda, Florida, Bahamas, Cuba. A plant of tide pools and similar protected places.

REFERENCES: Harvey 1858, Farlow 1871, Dickie 1874b, Hemsley 1884, Murray 1889, Collins and Hervey 1917, Howe 1920, Taylor 1928.

Cladophora nitida Kützing

Plants bushy, often much exceeding 10 cm. in height, soft and slippery, light green; main filaments 50–100 μ diam., with scattered erect alternate or unilateral branches above; branchlets unilateral, 20–30 μ diam., cells (4–)6–12 diameters in length, not swollen.

Bahamas, Jamaica, Guadeloupe, Brazil.

REFERENCES: Howe 1920, Taylor 1931; *C. ruchingeri,* Mazé and Schramm 1870–77, Murray 1889; *C. trichotoma,* Greville 1833, Mazé and Schramm 1870–77, Murray 1889, Möbius 1890, Collins 1901.

Cladophora virgatula Grunow

Plants 7–10 cm. tall, pale green, soft; the main filaments subsimple, 75–110 μ diam., the cells 3–6 diameters long; where branched briefly adnate; throughout beset with erect spreading subsecund branchlets 12–13 mm. long, the filaments here 32–36 μ diam., the much swollen cells 2–3 diameters long.

Guadeloupe.

REFERENCES: Grunow 1867, Mazé and Schramm 1870–77, Murray 1889.

Cladophora crystallina (Roth) Kützing

Plants pale green, silky, and of a soft texture, 10–30 cm. tall; filaments becoming slightly entangled, distantly ditrichotomously branched, the branches erect or spreading; main axes 80–140 μ diam., tapering to 25–40 μ diam. in the branchlets, the cells 4–12 diameters long; upper branchlets alternately secund or sometimes whorled, the cells not constricted at the nodes.

Bermuda, North Carolina, Bahamas, Jamaica, Guadeloupe, Barbados. On the intertidal rocks.

REFERENCES: Mazé and Schramm 1870–77, Murray 1889, Collins and Hervey 1917, Howe 1918a, 1920, Hoyt 1920; *C. sericea*, Vickers 1905, 1908.

Cladophora flexuosa (Dillwyn) Harvey

Plants attached, 10–20 cm. tall, light green; main filaments somewhat stiff, irregularly flexuous, cells 80–120(–160) μ diam., about 6 diameters long, with alternate flexuous branches 40–80 μ diam.; branchlets alternate or unilateral, curved and sometimes recurved, the cells about 2 diameters long.

Bermuda, North Carolina, Florida, ?Barbados. Primarily a plant of rocky shores in the temperate North Atlantic.

REFERENCES: Vickers 1905, 1908, Collins and Hervey 1917, Hoyt 1920, Taylor 1928.

Cladophora utriculosa Kützing

Plants tufted, 10–20 cm. tall, light green, texture rather stiff; branching erect, di- polychotomous below, the filaments 100–250 μ diam. near the base, with cells 6–8 diameters long; above with laterally disposed branchlets 70–100 μ diam., the cells 2–4 diameters long.

Bermuda, Bahamas, Jamaica, Hispaniola, Puerto Rico, Brazil, Uruguay.

REFERENCES: Piccone 1886, Collins 1909b, Collins and Hervey 1917, Børgesen 1915–20, Howe 1918a, Taylor and Arndt 1929, Taylor 1939a, Joly 1957; *C. longiarticulata*, Mazé and Schramm 1870–77.

Cladophora gracilis (Griffiths *ex* Harvey) Kützing

Plants tufted, to 3 dm. long or more, bright grayish green, glossy, somewhat harsh in texture; filaments flexuose, irregularly and angularly bent, the alternate branches spreading, 140–160 μ diam., the cells 3–5 diameters long; branchlets unilateral, comblike, moderately elongate and slender, 40–60 μ diam., the cells 3–5 diameters long, the tips acute.

Bermuda, Florida, Guadeloupe, Netherlands Antilles. Plants of rocky shores, usually growing just below low-tide level in moderately sheltered places. Chiefly a plant of northern seas.

REFERENCES: Mazé and Schramm 1870–77, Murray 1889, Sluiter 1908, Taylor 1928.

Cladophora hutchinsiae (Dillywn) Kützing

Plants attached, to 4 dm. tall, grayish-green, stiff and coarse; main filaments flexuous, sparingly branched, 120–300(–400) μ diam., the cells 2–4 diameters long; branches elongate, flagelliform, 160–240 μ diam., branchlets mostly on the ultimate and penultimate branches, rather short, scattered or in unilateral series, constricted at the septa, 90–100 μ diam., the cells 1–2 diameters long, the tips blunt.

Florida, Jamaica, Guadeloupe, Barbados.

REFERENCES: Collins 1901, Vickers 1905, Taylor 1928; *C. alyssoidea* and v. *gracillima,* and *C. hormocladia,* Mazé and Schramm 1870–77, Murray 1889.

Cladophora scitula (Suhr) Kützing

Plants small, dull green, stiff; the primary filaments erect, 75–110 μ diam. below, the cells 2–4 diameters long, the somewhat recurving branches connate at the base; above densely glomerate, the cells in the branchlets somewhat turgid and the tips blunt.

West Indies. Only known from the original material, which was epiphytic on larger algae.

REFERENCES: Kützing 1854, Collins 1909b.

Cladophora sertularina (Montagne) Kützing

Plants 2–4 cm. tall, erect, rather stiff and very much branched, the branches clustered; main axes 125 μ diam., the cells not swollen, 3–5 diameters long; branchlets 55–75 μ diam., the cells 1.5–3.0 diameters long.

Guadeloupe, French Guiana. Plants of more or less sand-covered rocks in shallow water.

REFERENCES: Montagne 1850, Mazé and Schramm 1870–77, Murray 1889.

Cladophora fascicularis (Mertens) Kützing Pl. 3, fig. 3

Plants large, bushy, to 30–50 cm. tall; main axes stout, sparingly alternately branched, 200–360 μ diam., with cells 2–5 diameters long; branchlets 1.5–3.0 mm. long, somewhat pectinately arranged, but densely clustered near the ends of the lesser branches, the cells 70–120 μ diam., 1.0–2.5 diameters long, the tip cells tapering.

Bermuda, North Carolina, Florida, Louisiana, Texas, Mexico, Bahamas, Cuba, Cayman Isls., Jamaica, Hispaniola, Puerto Rico, Virgin Isls., St. Barthélemy, Guadeloupe, Barbados, Grenada, British Honduras, Colombia, Venezuela, Brazil, Uruguay. A plant of moderately protected coasts, often abundant on rocks or jetties near low-tide line.

REFERENCES: Montagne 1850, Martens 1870, 1871, Mazé and Schramm 1870–77 (including f. *denudata* Crouan and f. *glomerata* Crouan, *nomina nuda*), Rein 1873, Dickie 1874a, Hauck 1888, Murray 1889, Collins 1901, Vickers 1905, 1908, Børgesen 1913, Collins and Hervey 1917, Howe 1920, Hoyt 1920, Schmidt 1923, 1924, Taylor 1928, 1930a, 1933, 1935, 1936, 1939a, 1940, 1941a, b, 1942, 1943, 1954a, Taylor and Arndt 1929, Sanchez A. 1930, Joly 1951, De Mattos 1952.

Cladophora prolifera (Roth) Kützing Pl. 3, fig. 5

Plants tufted, to 20 cm. tall, dark green, becoming blackish when dried, coarse and stiff; main filaments to 300–475 μ diam., the cells to 20 diameters long, copiously ditrichotomously branched, the branches rather erect, clustered toward the tips; branchlets 130–200 μ diam., cells 4–6 diameters long, the tip cells blunt.

?Bermuda, North Carolina, ?Florida, Puerto Rico, Guadeloupe, Barbados, Trinidad, Tobago, Brazil. Plants of the lower littoral, growing on rocks. Earlier authors may have confused *C. fuliginosa* with this species.

REFERENCES: Hooper 1850, Mazé and Schramm 1870–77, Martens 1871, Dickie 1874a, b, Hemsley 1884, Murray 1889, Möbius 1890, Vickers 1905, 1908, Collins 1909b, Hoyt 1920, Taylor 1929b, Joly 1957; ?*C. catenata*, Martens 1870; *C. charoides*, Mazé and Schramm 1870–77, Murray 1889; *C. fruticulosa*, Martens 1871.

Cladophora constricta Collins

Plants densely tufted, to 10 cm. tall; main filaments to 65 μ diam., bearing chiefly opposite branches; upper branching opposite or unilateral, the short branchlets about 25 μ diam., somewhat secund, at first spreading, later curved upward, the tip cell rounded-conical; the cells 5–20 diameters long throughout, somewhat clavate and commonly with an annular constriction a little above the lower end.

Bermuda, Jamaica.

REFERENCES: Collins 1909b, Collins and Hervey 1917.

Cladophora uncinata Børgesen

Plants in dense tufts, to 4–5 cm. tall, attached by multicellular rhizoids; erect filaments stiff, ditrichotomously branched, about 110 μ diam., the thick cell walls annulate; above more slender, about 65 μ diam., the cylindrical cells with smooth walls and 4–6 diameters long; branchlets secund, 35 μ diam., often becoming hooked.

Virgin Isls. Growing in shallow, protected places.

REFERENCE: Børgesen 1915–30.

Cladophora catenifera Kützing Pl. 3, fig. 1

Plants large, particularly stiff and rigid, to 50 cm. tall; the main filaments 300–500 μ diam. below, the cells cylindrical or a little constricted at the nodes, 8–10 diameters long; branching alternate or opposite, the branches spreading, but those near the ends a little clustered, 100–250 μ diam., the cells unconstricted, 1.5–2.0 diameters in length.

Bermuda, Jamaica.

REFERENCES: Collins and Hervey 1917, Howe 1918a.

UNCERTAIN RECORDS[1]

Cladophora albida (Hudson) Kützing

Guadeloupe, Brazil. Type from England. De Toni 1:325; Mazé and Schramm 1870–77, Möbius 1889, Murray 1889.

Cladophora anisogona (Montagne) Kützing

Guadeloupe. Type from Torres Strait. De Toni 1:311; Mazé and Schramm 1870–77, Murray 1889.

Cladophora bicolor J. Agardh

Guadeloupe. Source of name unconfirmed. Mazé and Schramm 1870–77.

(*Cladophora*?) *Conferva bicolor* Mertens and Schwartz

Jamaica, the type locality. De Toni 1:348; Murray 1889. Reported to be a mixture lacking any chlorophycean alga.

Cladophora brachyclados (Montagne) Harvey

Mouth of the Rio Grande; also Cuba, the type locality. Harvey 1858, Collins 1909b; ?*C. montagneana*, De Toni 1:303; *Conferva brachyclados*, Montagne 1863, who does not definitely state that it had a marine origin.

Cladophora catenatoides Crouan

Guadeloupe, the type locality. De Toni 1:352; Mazé and Schramm 1870–77.

Cladophora comosa Kützing

Guadeloupe. Type from fresh water in Italy. De Toni 1:303; Mazé and Schramm 1870–77.

Cladophora conferta Crouan

Guadeloupe, the type locality. De Toni 1:351; Mazé and Schramm 1870–77, Murray 1889.

[1] Plants appearing in this list on the Crouans' authority and published in Schramm and Mazé 1865 or 1866, referred to through Mazé and Schramm 1870–77, are essentially *nomina nuda*.

Cladophora cornea Kützing

Brazil. Type from the Adriatic. De Toni 1:344; Möbius 1892.

Cladophora crassicaulis Crouan, and f. *denudata* Crouan

Guadeloupe, the type locality for both. Mazé and Schramm 1865, 1870–77, Murray 1889.

Cladophora crouanii Murray

Guadeloupe, the type locality. De Toni 1:352; Murray 1889; *C. luteola* Crouan *non* Harvey, Mazé and Schramm 1870–77.

Cladophora dalmatica Kützing

Guadeloupe, Barbados. Type from the Adriatic. De Toni 1:321; Vickers 1908; *C. glebifera* v. *occidentalis* Crouan (varietal type from Guadeloupe), Mazé and Schramm 1870–77, Murray 1889.

Cladophora dichotomo-divaricata Crouan

Guadeloupe, the type locality. De Toni 1:351; Mazé and Schramm 1870–77, Murray 1889.

Cladophora echinus (Biasoletto) Kützing

Brazil. Type from the Adriatic. De Toni 1:343; Möbius 1889.

Cladophora eckloni (Suhr) Kützing

Guadeloupe. Type from South Africa. De Toni 1:330; Mazé and Schramm 1870–77, Murray 1889.

Cladophora enormis (Montagne) Kützing

Puerto Rico. Type from the Canary Islands. De Toni 1:346, Hauck 1888.

Cladophora fascicularioides Crouan

Guadeloupe, the type locality. De Toni 1:352; Mazé and Schramm 1870–77, Murray 1889.

Cladophora gracillima Crouan

Guadeloupe, the type locality. De Toni 1:352; Mazé and Schramm 1870–77, Murray 1889.

Cladophora hilarii Greville

Brazil, the type locality. Greville 1833, Martens 1870.

Cladophora javanica Kützing

Guadeloupe. Type from fresh water in Java. De Toni 1:303; Mazé and Schramm 1870–77.

Cladophora kuetzingii Ardissone

Guadeloupe. Type from the Mediterranean. De Toni 1:314; *C. laxa,* Mazé and Schramm 1870–77, Murray 1889.

Cladophora laetevirens (Dillwyn) Kützing

Bermuda, Colombia. Type from England. De Toni 1:327; Harvey 1861, Hemsley 1884, Murray 1889.

Cladophora macallana Harvey

Guadeloupe. Type from Ireland. De Toni 1:313; Mazé and Schramm 1870–77, Murray 1889.

Cladophora mauritiana Kützing

Guadeloupe. Type from Mauritius. De Toni 1:328; Mazé and Schramm 1870–77, Murray 1889.

Cladophora mexica Crouan

Guadeloupe, the type locality. Mazé and Schramm 1865, 1866, 1870–77, Murray 1889; *C. mexicana*, De Toni 1:353.

Cladophora montagnei Kützing

Cuba, the type locality. De Toni 1: 347; *Conferva aegagropila,* Montagne 1842. De Toni quite reasonably suggests that this may be *C. fuliginosa* Kützing.

Cladophora obtusata Zanardini

Guadeloupe. Type locality not known. Mazé and Schramm 1870–77, Murray 1889.

Cladophora ovoidea Kützing

Guadeloupe. Type from northern seas. De Toni 1:313; Mazé and Schramm 1870–77 (including f. *crassicaulis* Crouan), Murray 1889.

Cladophora pellucida Kützing

Bermuda, Brazil. Type from England. De Toni 1:306; Greville 1833, Martens 1870, Murray 1889.

Cladophora penicillata Kützing

Bermuda, Guadeloupe. Type from Italy. De Toni 1:316; *C. flavescens*, Mazé and Schramm 1870–77; *C. lutescens*, Murray 1889 (including v. *longiarticulata*).

Cladophora rudolphiana (C. Agardh) Harvey

Guadeloupe. Type locality not known. De Toni 1:321; Mazé and Schramm 1870–77, Murray 1889.

Cladophora socialis Kützing

Guadeloupe. Type from Tahiti. De Toni 1:347; Mazé and Schramm 1870–77, Murray 1889.

Cladophora submarina Crouan

Guadeloupe, the type locality. De Toni 1:351; Mazé and Schramm 1870–77, Murray 1889.

Cladophora subtilis Kützing

Guadeloupe. Type from the Adriatic. Mazé and Schramm 1870–77, Murray 1889.

Cladophora tranquebarensis (C. Agardh) Kützing

Guadeloupe. Type from fresh water in the East Indies. De Toni 1:304; Mazé and Schramm 1870–77, Murray 1889.

Cladophora trinitatis Kützing

Trinidad, the type locality. De Toni 1:340; Murray 1889.

Cladophora variegata (C. Agardh) Zanardini

Texas. "O. S." 1931.

(?*Cladophora*) *Hormiscia viridi-fusca* (Montagne) Kützing

Brazil, the type locality. De Toni 1:170; *Hormotrichum viridifuscum,* Martens 1870. Branched, so probably not *Urospora* (*ante Hormiscia*).

Cladophora zostericola Crouan

Guadeloupe, the type locality. De Toni 1:351; Mazé and Schramm 1870–77, Murray 1889.

Spongomorpha Kützing 1843

UNCERTAIN RECORD

Spongomorpha arcta (Dillwyn) Kützing

Guadeloupe. Type from England. *Cladophora arcta,* De Toni 1:335; *C. stricta* (type from Jutland), Mazé and Schramm 1870–77, Murray 1889.

SIPHONOCLADIALES

Plants uni- or multicellular, simple or branched, the branching irregular, or lateral from a primary axis, or organized in two or three planes into specialized thallus structures; cells generally multinucleate, with a netlike chromatophore or many disklike chromatophores; pyrenoids usually present.

KEY TO FAMILIES

1. Plants consisting of an axis with whorled, determinate branchlets DASYCLADACEAE, p. 97
1. Plants unbranched, irregularly branched or branched in a plane, but not in whorls .... VALONIACEAE, p. 107

DASYCLADACEAE

Plants each composed of a long axial cell attached to the substratum at the base by rhizoidal outgrowths, and bearing regular whorls of simple or forked branchlets of limited growth; reproduction by aplanospores or cysts, which in turn produce gametes.

KEY TO GENERA

1. Plants not calcified 2
1. Plants calcified 3

2. Whorls of branchlets closely set, the plant spongy; sporangia single and terminal on the basal cell of each branchlet cluster, between the divisions of the next degree ..... DASYCLADUS, p. 99
2. Whorls of branchlets distant, the plant open and bushy; sporangia several, clustered around rather than between the divisions of the next degree .... BATOPHORA, p. 98

3. Branchlet formation in many whorls producing a more or less massive, though sometimes small, thallus 4
3. Branchlets forming a single major terminal whorl usually joined into a disk near the apex of the slender stalk, or forming 2–3 independent disks 5

4. Plants simple, very small, unbranched, the surface continuous NEOMERIS, p. 99
4. Plants abundantly branched, several centimeters tall, the calcified branches with flexible joints .... CYMOPOLIA, p. 102

5. Spores not calcified .... ACETABULARIA, p. 103
5. Spores encased in lime 6

6. Spores contained in a solid mass of lime within a lime-free membrane; corona inferior present .... ACICULARIA, p. 106

6. Spores calcified but free, the sporangial membrane also calcified; corona inferior lacking .... CHALMASIA, p. 103

Batophora J. Agardh, 1854

Plants of moderate size, gregarious, not at all calcified; composed of a long erect axial cell basally attached, bearing rather distant whorls of repeatedly ditrichotomously forked determinate branchlets; sporangia large, several at a node, clustered about the bases of the branchlets.

Batophora oerstedi J. Agardh
Pl. 4, figs. 3, 4; pl. 5, fig. 4; pl. 6, figs. 3, 9

Plants bright green, soft and delicate, often growing in considerable clusters; 3–10 cm. in height, the simple axes becoming naked below, but above, the loosely whorled branchlets spreading to form a very soft cylinder 4–10 mm. diam., each branchlet dividing 1–7 times, the terminal 2–3 divisions hairlike, but deciduous on old or fertile plants; sporangia borne on the inner 1–4 series of forkings of the branchlets, about the bases of the next smaller series, several at each node, 325–450 μ diam., 500–1000 μ long, each containing numerous small obovoid or oblong ellipsoidal aplanospores 40–45 μ diam.

V. **occidentalis** (Harvey) Howe. Plants smaller, rarely over 4 cm. tall, slender, the branchlets crowded, forming a cylinder 2–4 mm. diam.; the aplanospores nearly spherical.

Bermuda, Florida, Texas, Mexico, Bahamas, Caicos Isls., Cuba, Jamaica, Hispaniola, Virgin Isls., St. Martin, Barbuda, Guadeloupe, British Honduras. The typical plants are found in the warm quiet water of lagoons, especially the borders of mangrove thickets, growing below but close to low-tide level. They can prosper in brackish water, and have been found under seemingly quite fresh-water conditions. The variety is found in more open water of normal marine salinity. The aspect of the very handsome sterile plants contrasts sharply with that of the darker green fertile ones, which appear as long clusters of small green spherules.

REFERENCES: Børgesen 1913–20, Collins and Hervey 1917, Howe 1918b, 1920, Taylor 1928, 1935, 1936, 1954a; *Coccocladus occidentalis* v. *laxus*, Howe 1904b; *Dasycladus conqueranti*, Mazé and

Schramm 1870–77. For the variety, Collins and Hervey 1917, Howe 1918a, 1920, Taylor 1928, 1935, 1936, 1941a; *Botryophora occidentalis,* Murray 1889, Collins 1901, Sluiter 1908; *Dasycladus occidentalis,* Harvey 1858, Rein 1873, Hemsley 1884.

Dasycladus C. Agardh, 1827

Plants small, gregarious, olive green, not calcified, of a spongy consistency; composed of an elongate axial cell bearing closely-placed whorls of short, compound determinate branchlets; gametangia solitary, terminal on the basal cells of these branchlets.

Dasycladus vermicularis (Scopoli) Krasser
Pl. 4, fig. 2; pl. 6, figs. 2, 8

Plants to 2–6 cm. tall, 3–6 mm. diam., arising from a base with rhizoidal lobes, the axis surrounded by compact whorls each of about 12 branchlet clusters, the basal cell of each cluster bearing 3–4 successive ditrichotomous series of cells and terminating in filiform or in short spinelike terminal cells; gametangia solitary, spherical, between the cells of the secondary branchlet whorls.

Bermuda, Florida, Bahamas, Caicos Isls., Cuba, Jamaica, Hispaniola, Brazil, I. Trinidade. Plants of reefs or tide pools, growing in shallow water and rather exposed places, often in the surf zone near low-tide level; also dredged to a depth of 55 m.; sometimes nearly covered by drifted sand, only the growing tips remaining exposed.

REFERENCES: Howe 1918a, b, 1920, Taylor 1928, 1930, 1933, 1954a, Sanchez A. 1930, Joly 1953c; *D. clavaeformis,* Hooper 1850, Farlow 1871, Murray 1889, Collins 1901, Butler 1902, Collins and Hervey 1917.

Neomeris Lamouroux, 1816

Plants small, generally gregarious, spindle-shaped or subcylindrical, more or less calcified, with an erect elongate axis tipped with a small cluster of bright green filaments, in the mature portion bearing closely placed whorls of short compound branchlets, each branchlet consisting of a basal cell bearing one whorl of branchlet cells with greatly expanded distal ends, each of which bears one delicate uniseriate hair filament; sporangia solitary, terminal on the primary branchlet cells between the cells of the secondary whorl, each containing one large aplanospore.

KEY TO SPECIES

1. Cortex continuous, faceted, the branchlets of the second order with flat approximated ends 2
1. Cortex rough, not faceted, the branchlets fusiform, their tips neither flat nor closely approximated .... N. cokeri, p. 100
2. Sporangia laterally coherent, forming bands around the plant N. annulata, p. 101
2. Sporangia mutually free 3
3. Plants 1–2 mm. thick, 15–20 times as long ...... N. dumetosa, p. 101
3. Plants 1.5–2.5 mm. thick, 4–8 times as long .... N. mucosa, p. 100

Neomeris cokeri Howe Pl. 5, fig. 3

Plants subcylindrical to clavate, 7–37 mm. long, 1.5–3.5 mm. diam., dark green above, white below; the basal clavate cells of the branchlet whorls each bearing 2 spindle-shaped cells, and these in turn having single terminal appendages of 2 sorts zonately segregated on the thallus: the first sort a single rather thick, clavate, less often arcuate cell, the other type with a basal cell similar though somewhat more slender, bearing ditrichotomously forked hairs, but both sorts deciduous and leaving the thallus with rather a rough surface; sporangia strongly calcified, separate or united into short rows but not forming rings about the thallus, 180–260 μ long including the stalk; spores obovoid or oblong ellipsoid, 82–94 μ diam., 140–190 μ long.

Bahamas. Growing under shelving rocks near low-water line.

REFERENCES: Howe, 1904b, 1909b, 1920.

Neomeris mucosa Howe

Plants subcylindrical or fusiform, 8–20 mm. long, 1.5–2.5 mm. diam., green above, grayish white below, the apical tuft usually inconspicuous, of ditrichotomous deciduous hairs; branchlets of the primary whorls each bearing 2 capitate polyhedral branchlet cells of the second order, 100–200 μ diam. in surface view, forming the facets of the continuous cortex; sporangia strongly calcified but mutually free, 178–215 μ long including the stalk, the spores 104–121 μ diam., 140–160 μ long.

Bahamas, Cuba, Netherlands Antilles. Plants of moderately exposed rocks at and near low-water mark, often with other species.

REFERENCES: Howe 1909a, b, 1920, Taylor 1942.

Neomeris dumetosa Lamouroux

Plants subcylindrical, elongate, 20–40 mm. long, 1–2 mm. diam., acute or acuminate at the apex, the terminal hair tuft inconspicuous; whorled basal branchlet cells bearing 3–8 capitate cells of the second order which are easily shed when past maturity, but these cells when intact forming a continuous cortex of polyhedral facets 100–185 μ diam.; sporangia calcified but free, 150–200 μ long including stipe, above which they are nearly spherical, 135–160 μ diam., soon deciduous.

Cuba, Jamaica, Virgin Isls.

REFERENCES: Farlow 1871, Murray 1889, Collins 1901, Howe 1909b.

Neomeris annulata Dickie Pl. 5, fig. 5; pl. 6, figs. 4–6

Plants sometimes solitary, often densely gregarious, 5–25 mm. tall, 1–2 mm. diam., substantially calcified below, less heavily so toward the green apex, which is surmounted by a tuft of fine deciduous simple hairs; whorls of branchlets distinctly visible, the expanded capitate ends of the outer cells polyhedral, forming a fairly even faceted surface; 2 cells of the outer cortical layer to each basal branch cell, the ends 80–135 μ diam.; sporangia obovate, the simple large aplanospore oval, 46–80 μ diam., 115–175 μ long.

Bermuda, Florida, Bahamas, Jamaica, Hispaniola, Puerto Rico, Virgin Isls., Nevis, Guadeloupe, Martinique, Barbados, I. Trinidade. Plants of shallow sunny situations, commonly in tide pools or on coral or shell fragments over a sandy bottom; also dredged to an extreme depth of 50 m., where very rare.

REFERENCES: Howe 1909a, b, 1918a, 1920, Børgesen 1913-20, 1924, Collins and Hervey 1917, Taylor 1928, 1942, Hamel and Hamel-

Joukov 1931, Feldmann and Lami 1936, Joly 1953c; ?*N. eruca*, Hauck 1888; ?*N. kelleri*, Vickers 1905, 1908.

Cymopolia Lamouroux, 1816

Plants bushy, the axes branched, calcified below but flexibly articulated; the juvenile zone of each axis tip characterized by a conspicuous tuft of bright green trichotomous filaments which end in delicate hairs; the mature axis consisting of a long coenocyte constricted at rather regular intervals and especially at each forking, but commonly without crosswalls, each segment between constrictions bearing several whorls of corticating branchlets of limited growth, these branchlets usually consisting of a basal cell and a cluster of four to ten distal cells, expanded at the ends and forming the smooth surface of the segments which are calcified except at the axial nodes; sporangia terminal on the initial branchlet cells.

Cymopolia barbata (Linnaeus) Lamouroux

Pl. 4, fig. 1; pl. 6, fig. 1

Plants to 1.0–2.0 dm. tall, the branches terete, segmented, calcified except at the flexible nodes, each terminated by a conspicuous tuft of trichotomous filaments forking 5–6 times, the lower 3–4 series with abundant chloroplasts, the upper more hairlike; these filaments, except the basal cell, deciduous; segments of the adult branches 2–4 mm. diam., 3–11 mm. long, moderately calcified and white with lime, the surface cells in face view 115–125 μ diam.; sporangia solitary, lying between the cells of the outer cortical zone, spherical to nearly pyriform, 160–200 μ diam.

Bermuda, Florida, Mexico, Bahamas, Caicos Isls., Cuba, Jamaica, Hispaniola, Puerto Rico. Plants of warm shallow water, growing just below low-tide level on stones or large coral fragments. On drying, the conspicuous tufts turn dark brown and the segmented branches become dull.

REFERENCES: Harvey 1858, Rein 1873, Hemsley 1884, Hauck 1888, Murray 1889, Collins 1901, Howe 1915, 1918b, 1920, Taylor 1928, 1933, 1940, 1943, 1954a.

Chalmasia Solms-Laubach, 1895

Plants stalked, the stalk terminating in a funnel-shaped disk composed of a whorl of branchlet cells constituting the sporangial rays, which are attached laterally by the layer of calcification, bearing dorsal lobes constituting a corona superior but without a corona inferior; rays ultimately producing a number of spherical, calcified, but free aplanospores.

Chalmasia antillana Solms-Laubach

Plants small, the disks 6 mm. diam., composed of 25-32 lightly calcified rays which are contracted at the base, inflated beyond, with a somewhat pointed apex; projecting lobes of the corona superior laterally free, attached to the contracted base of the rays, bearing 2–3 hair scars; aplanospores after decalcification about 150 μ diam.

Florida, Martinique. The fragmentary original collection of these plants was dredged from deep water south of Florida.

REFERENCES: Solms-Laubach 1895, Weber-van Bosse 1899, Collins 1909b, Taylor 1928.

Acetabularia Lamouroux, 1816

Plants commonly gregarious, erect from a lobed rhizoidal base, the slender stalk cylindrical or with slight annulate thickenings from the scars of successive whorls of delicate colorless evanescent branched hairs; bearing on the summit a disk or cup (or sometimes a series of them superimposed), composed of a whorl of branchlet cells; these cells showing two or three superimposed lobes attached near the stalk, laterally associated, the small upper ones in a ring collectively termed the corona superior, the large median ones constituting the rays of the disk or cup proper, and the small lower ones when present collectively termed the corona inferior; delicate colorless evanescent branched hairs borne on the lobes of the corona superior and where discarded represented by scars; rays of the disk after vegetative maturity functioning as aplanosporangia, the spore walls individually calcified.

KEY TO SPECIES

1. Mature disks less than 3 mm. diam.; corona inferior absent 2
1. Mature disks more than 4 mm. diam.; coronae both present 3

2. Disks 1.0–2.5 mm. diam., commonly irregular; coronal processes 22–35 μ diam. (measured lengthwise of the ray), bearing 2–3 hairs or rudiments; aplanospores 68–82 μ diam. .. A. pusilla, p. 104
2. Disks 1.5–5.0 mm. diam.; coronal processes 75–150 μ in radial diam., bearing 5–15 hairs or rudiments; aplanospores 88–190 μ diam. .... A. polyphysoides, p. 104

3. Rays lightly calcified, commonly free; ray ends not apiculate; outer margin of the lobes composing the corona inferior entire to emarginate 4
3. Rays usually firmly calcified; plant to 7 cm. tall; ray ends rounded, apiculate; outer margin of the lobes composing the corona inferior deeply indented .... A. crenulata, p. 105

4. Disks cup-shaped; ray ends rounded, emarginate, or with age markedly notched; lobes of the corona superior with 2–4 hair scars .... A. calyculus, p. 105
4. Disks nearly flat, the ray ends obtuse or truncate; each lobe of the corona superior with 2 hair scars only .... A. farlowii, p. 105

Acetabularia pusilla (Howe) Collins Pl. 6, fig. 13

Plants exceedingly small, the stipe 1–3 mm. tall; disk nearly flat, 1.0–2.5 mm. diam., with 6–17 lightly connected rays, these obovoid-clavate to clavate-subfusiform, blunt or slightly tapered at the tip; corona superior of small subcylindrical processes 22–35 μ diam., with 2, rarely 3, hair scars; corona inferior absent; 15–60 aplanosporangia in a sporangial ray, 68–82 μ diam.

Florida, Bahamas, Jamaica. Growing attached to coral fragments in very protected situations, but very difficult to find among larger algae because of its small size.

REFERENCES: Collins 1909b; *Acetabulum pusillum*, Howe 1909b, 1920, Taylor 1928.

Acetabularia polyphysoides Crouan

Plants very small, the stalk short, 5–10 mm. tall; the disk flat or cup-shaped, 1.5–5.0 mm. diam., composed of 11–25 rays connected by light and localized lime deposits, these rays obovoid to fusiform, contracted at the base, tapered and rounded at the apex; corona superior of projecting knobs with an oval upper surface bearing 3–13 hair scars elliptically disposed; corona inferior absent; sporangial rays containing 6–50 aplanospores, these 90–190 μ diam.

F. **deltoideum** (Howe) Collins. Rays usually 7 in number, much inflated, inversely deltoid or obovoid-deltoid; hair scars 6–8 on each coronal lobe.

Bermuda, Bahamas, Jamaica, Guadeloupe, Barbados.

REFERENCES: Mazé and Schramm 1870–77, Murray 1889, Solms-Laubach 1895, Vickers 1908, Collins 1909b; *Acetabulum polyphysoides,* Howe 1909a, b, 1920.

Acetabularia farlowii Solms-Laubach

Plants small, short-stalked, reaching a height of 1–2 cm.; the disk 4.0–7.5 mm. diam., composed of about 20–30 lightly coherent or laterally free rays; rays rounded or slightly emarginate at the blunt tips; coronal lobes rather crowded together, slightly emarginate at the distal end; 2 hair scars arranged in a radial row on each lobe of the corona superior; aplanosporangia large, 40–120 in each sporangial ray.

Florida. Growing on broken coral and shells near the low-tide level in sheltered localities.

REFERENCES: Solms-Laubach 1895, Howe 1905b, Taylor 1928.

Acetabularia calyculus Quoy and Gaimard

Stipes slender, 1.5–3.0 cm. tall; disks cup-shaped, 6–7 mm. diam.; rays 22–30, laterally lightly connected or essentially free, the distal end of each ray emarginate or ultimately with a broad square notch; coronal lobes remote, blunt, those of the corona superior with 2 hair scars in a radial line, or 3 triangularly placed, or seldom 4 scars; aplanospores about 80 in each sporangial ray, 160 μ diam.

Jamaica, Virgin Isls. Usually found in shallow water, attached to shells and coral fragments. It is able to grow well both in the shelter of mangroves and on rather exposed reefs.

REFERENCES: Collins 1909b, Børgesen 1913–20; *A. suhrii,* Solms-Laubach 1895.

Acetabularia crenulata Lamouroux Pl. 4, fig. 5; pl. 6, fig. 12

Plants to 7 cm. tall; disks at first funnel-shaped, later nearly flat, often 2–4 of successive ages superimposed, becoming 12–20 mm.

diam.; rays 30–80, closely united by moderate lime encrustation; the apices at first rounded, later truncate, the distal wall sometimes incrassate with an evident median apiculus, at least while young; lobes constituting the corona superior crowded, elongate, rounded or slightly indented peripherally, carrying 2 hair scars in a radial line; lobes of the corona inferior also crowded, elongate, very deeply indented peripherally; aplanospores to 500 in each sporangial ray, 65–80 μ diam.

Bermuda, Florida, Texas, Bahamas, Cuba, Jamaica, Hispaniola, Puerto Rico, Virgin Isls., St. Martin, Barbuda, Guadeloupe, Barbados, British Honduras, Colombia, Netherlands Antilles, Venezuela. Plants of shallow, protected lagoons and the borders of mangrove swamps, growing on shells, coral fragments, stones, mangrove roots and on coarse algae, commonly in large colonies. Occasionally found on rocks or jetties where exposed to mild wave action. When in exceedingly sheltered, warm pools the calcification may be much reduced. A notably attractive species, frequently encountered.

REFERENCES: Hooper 1850, Harvey 1858, Mazé and Schramm 1870–77, Farlow 1871, Rein 1873, Hemsley 1884, Hauck 1888, Murray 1889, Gardiner 1890, Solms-Laubach 1895, Collins 1901, Vickers 1905, 1908, Sluiter 1908, Børgesen 1913–20, 1924, Collins and Hervey 1917, Feldmann and Lami 1936, Taylor 1941a, b, 1954a; *A. caraïbica*, Kützing 1856, Mazé and Schramm 1870–77, Murray 1889, Sluiter 1908; *Acetabulum caule simplici*, Browne 1756; *A. crenulatum*, Howe 1903b, 1918a, b, 1920, Taylor 1928, 1933, 1935, 1936, Taylor and Arndt 1929, Sanchez A. 1930.

UNCERTAIN RECORD

Acetabularia peniculus (R. Brown) Solms-Laubach

Colombia. The type from Australia. De Toni 1: 421; Weber-van Bosse 1899.

Acicularia D'Archiac, 1843

Plants composed of an erect calcified stalk attached to the substratum by rhizoidal lobes, bearing at the top a whorl of equal branchlets or rays united by calcification into a disk; these rays near the center bearing both dorsal and ventral lobes constituting respectively the coronae superior and inferior; fertile rays at maturity filled with a mass of lime in which the spherical aplanospores are embedded.

Acicularia shenckii (Möbius) Solms-Laubach Pl. 6, fig. 11

Plants small, the stipes 1.0–1.5 cm., seldom to 3 cm., tall, rather stout; disks about 3–8 mm. diam., nearly flat, the margin crenulate; rays 30–50, wedge shaped, well joined side by side by the coating of lime, generally apiculate at the distal end; lobes constituting the coronae closely joined together, radially elongate and deeply cleft at the outer margin; aplanospores 100–200 to each sporangial ray, 60–80 μ diam.

Bermuda, Florida, Bahamas, Jamaica, Hispaniola, Puerto Rico, Virgin Isls., Guadeloupe, Martinique, Barbados, Brazil. Widely distributed, attached to shells, coral fragments, or stones in shallow, protected situations, but also dredged to a depth of about 73 m.

REFERENCES: Solms-Laubach 1895, Howe 1907, 1918a, 1920, Collins 1909b, Børgesen 1913–20, 1924, Collins and Hervey 1917, Taylor 1928, Feldmann and Lami 1937; *Acetabularia schenkii*, Möbius 1889; *A. caraibica*, Vickers 1905, 1908.

VALONIACEAE

Plants vesicular to filamentous, occasionally saccate or even solid, originating from a persisting initial coenocyte of limited growth, remaining vesicular with only hapteral outgrowths, or branching from the initial cell, the branch system coarse to delicate, irregular or organized into more or less regular plane blades, or developing into saccate or solid multicellular thalli, the cells still coenocytic; asexual reproduction by aplanospores or zoöspores; sexual reproduction where known anisogamous.

KEY TO GENERA

1. Plants truly unicellular, or at most the penetrating basal filamentous portion with occasional crosswalls ... HALICYSTIS, p. 109
1. Plants actually multicellular, because even if macroscopically of one or a few large coenocytes also possessing minute hapteral cells 2

2. Plants of obviously filamentous construction 3
2. Plants of apparently simple vesicular cells, or such cells in clusters, forming crusts or hollow sacs 12

3. Original cell constituting the greater part of a stalk different in character from the rest of the plant 4
3. Original cell not forming a distinctive stalk, usually similar to the other cells of the plant 5

4. Stalk normally simple, terminated by a dense tuft of branching filaments .... CHAMAEDORIS, p. 115
4. Stalk simple or sparingly branched, terminated by one or a few more or less plane netlike blades .... STRUVEA, p. 122

5. Plants foliar in aspect 6
5. Plants not foliar 7

6. Blades continuous, the association of radiating groups of filaments completed by intercalary cells to fill all spaces ANADYOMENE, p. 124
6. Blades in part continuous, but perforated by many openings of macroscopic size .... CYSTODICTYON, p. 124
6. Blades without any continuous part, netlike throughout MICRODICTYON, p. 120

7. Plants small, crustose, of evident adherent radial filaments PETROSIPHON, p. 117
7. Plants repent, in massive crusts, or erect, bushy 8

8. Plants bushy, the large primary cells at least locally divided in parenchymatous fashion .... SIPHONOCLADUS, p. 114
8. Plants bushy, entangled or crustose, the cells not parenchymatously divided 9

9. Cells in general microscopic 10
9. Cells macroscopic, even large 11

10. Plants with more or less distinct main and secondary axes, the slender filamentous branches joined into a spongy mass by haptera at the branchlet tips .... BOODLEA, p. 118
10. Plants without a distinct main axis, forming cushionlike masses CLADOPHOROPSIS, p. 117

11. Plants bushy, the cells of each order bearing conspicuous whorls of similar cells near each apex .... ERNODESMIS, p. 113
11. Plants not branching in definite successive whorls 12

12. Plants filamentous, the filaments less than 1 mm. diam. VALONIOPSIS, p. 112

12. Plants of separate or massed round or ovoid coenocytes, or if these cylindrical, exceeding 1 mm. diam. ...... VALONIA, p. 109
12. Plants of coenocytes which form solid parenchymatous structures, or of hollow sacs which may rupture and become explanate .... DICTYOSPHAERIA, p. 115

Halicystis Areschoug, 1850

Plants solitary or clustered, consisting of single large coenocytes arising terminally on a somewhat attenuate filamentous basal portion; protoplasm peripheral, with many nuclei, the lenticular chromatophores with pyrenoids; reproduction by macro- and microgametes produced by segregation of special areas of cytoplasm, subsequent cell division, maturation of uninucleate swarmers, and forcible discharge through pores in the cell wall.

Halicystis osterhoutii L. R. and A. H. Blinks Pl. 7, fig. 3

Plants unicellular, spherical to pyriform, commonly 4–7 mm. diam., but to over 3 cm. diam., attached by a slender stalklike portion; megagametes 10–15 μ long, 9–12 μ broad or more usually subspherical, 11 μ diam., biflagellate; microgametes about 4 μ diam., the swarmers dispersed every 2 weeks.

Bermuda, Florida, Bahamas, Jamaica, Grenadines. Plants of deep water, extending to at least 18 m. depth, or of shaded rock cliffs and caves, or under ledges, the attenuate portion of the plant endophytic in lithothamnioid algae, the inflated part extending beyond the surface.

REFERENCES: L. R. and A. H. Blinks 1930, Hollenberg 1935; *Halicystis* sp., L. R. Blinks 1927.

Valonia Ginnani, 1757

Plants coenocytic, with large vegetative cells and smaller hapteral or rhizoidal cells; conspicuous portion sometimes limited to the original vegetative cell, or this divided and redivided into similar coenocytes; asexual reproduction by zoöspores.

KEY TO SPECIES

1. Plants solitary or a few together, each limited to one large coenocyte attached by minute hapteral cells .. *V. ventricosa*, p. 110

1. Plants of few to many vegetative cells 2
2. Plants crowded, concrescent, small, the coenocytes 2–4 mm. diam., connected in obscure filaments, those lower in the mass elongate and often somewhat rhizoidal . . . . V. ocellata, p. 111
2. Plants of larger cells, the branches irregularly placed 3
3. Cell diameter 5–15 mm.; plants often forming large masses V. macrophysa, p. 110
3. Cell diameter less than 5 mm. 4
4. Cells 1–3 mm. diam., 5–10 mm. long, the plants tending to form dense tufts or balls . . . . V. aegagropila, p. 111
4. Cells 1.0–2.5 mm. diam., 5–20 mm. long, the branches chiefly repent . . . . V. utricularis, p. 112

Valonia ventricosa J. Agardh Pl. 9, figs. 4, 5

Plants solitary or few together, not forming a compact mass; individuals normally composed of one large oval to spherical coenocyte 1.5–3.0 cm. diam. and may reach 4 cm. in length; principal vegetative cell attached to the substratum by several minute hapteral cells at the slightly contracted lower end.

Bermuda, Florida, Mexico, Bahamas, Cuba, Cayman Isls., Jamaica, Hispaniola, Puerto Rico, Virgin Isls., Guadeloupe, Martinique, Barbados, Grenadines, Grenada, Old Providence I., Panama, Netherlands Antilles, Tobago, Brazil. Chiefly a plant of shallow water, the individuals scattered among colonies of Porites coral, Halimeda plants, and the like, or attached under ledges of reef rock, but also dredged to depths of about 30 m. A plant was measured which had reached 4 cm. diam. and 5.5 cm. in length, but this was quite unusually large.

REFERENCES: Murray 1889, 1893a, Collins 1901, 1909b, Howe 1903b, 1918a, b, 1920, Vickers 1905, 1908, Børgesen 1913–20, 1924, Collins and Hervey 1917, Taylor 1928, 1930a, 1933, 1939b, Taylor and Arndt 1929, Sanchez A. 1930, Hamel and Hamel-Joukov 1931, Feldmann and Lami 1936; *V. ovalis*, Mazé and Schramm 1870–77 *p.p.*, Murray 1889 *p.p.*

Valonia macrophysa Kützing Pl. 2, fig. 6; pl. 7, fig. 4

Plants seldom isolated, generally forming dark olive-green masses which may become 2–5 cm. or more in thickness and cover a considerable area; cells often crowded together and adherent, ovoid to obovate, clavate or irregular, branching from near the base or from exposed sides; vegetative cells 0.5–1.5 cm. diam., generally about 2.0(1.0–3.5) cm. long, near the base associated with a few smaller lenticular cells from which attaching rhizoids may arise.

Bermuda, Florida, Bahamas, Cuba, Cayman Isls., Jamaica, Hispaniola, Virgin Isls. A conspicuous plant of shallow water, probably much more widely ranging than the list above indicates. It is to be found on the shaded side of jetties and walls, in rock crevices, and under ledges, a little above or below low-tide line. It is often covered and concealed by epiphytes, such as Fosliella and Polysiphonia.

REFERENCES: Børgesen 1913–20, Collins and Hervey 1917, Howe 1918a, b, 1920, Taylor 1928, 1943, 1954a, Taylor and Arndt 1929.

Valonia ocellata Howe Pl. 9, figs. 6, 7

Plants densely caespitose or apparently crustose, forming cushions to 10–20 cm. broad, obscurely filamentous; cells of the upper surface angular, crowded, hardly a millimeter in diameter; below the cells elongate, stiltlike or rhizoidal; minute lentiform hapteral cells along the angles of contact of the upper vegetative cells.

Bermuda, Florida, Bahamas, Caicos Isls., Cuba, Cayman Isls., Hispaniola, Puerto Rico, Virgin Isls., Guadeloupe, Netherlands Antilles. Plants of rock surfaces and crevices on reefs between tide marks, and also found on mangrove roots in lagoons.

Young, crowded, and ill-formed plants of Dictyosphaeria are exceedingly hard to distinguish from this, unless the tenacular cells of Dictyosphaeria can be seen. Structurally it seems intermediate between Valonia and Dictyosphaeria, with more resemblance to solid forms of the latter genus, and it may prove not to be an independent species at all.

REFERENCES: Howe 1920, Taylor 1928, 1940, 1942, 1943, 1954a, Taylor and Arndt 1929, Feldmann and Lami 1936; *V. utricularis* f. *crustacea*, Børgesen 1913–20.

Valonia aegagropila C. Agardh Pl. 7, fig. 6

Plants at first attached, later free, eventually forming masses 4–20 cm. diam., composed of short branching filaments of rather large subcylindrical straight or arcuate coenocytes 1–3 mm. diam., 5–10 mm. long; branching from the sides or more usually the ends of the cells.

Florida, Bahamas, Caicos Isls., Cuba, Jamaica, Hispaniola, Puerto Rico, Virgin Isls., St. Barthélemy, British Honduras, Brazil. Primarily plants of shallow water, such as protected lagoons, attached to stones or later free; occasionally dredged to about 7 m. depth.

REFERENCES: Farlow 1871, Murray 1889, Collins 1901, Børgesen 1913–20, 1924, Howe 1920, Schmidt 1923, 1924, Taylor 1928, 1935, Taylor and Arndt 1929.

Valonia utricularis C. Agardh Pl. 9, fig. 10

Plants attached, spreading or creeping among coarser algae, in part later becoming erect and to 5 cm. tall; composed of short branching filaments of rather large cylindric-clavate, sometimes arcuate, cells 1.0–2.5 mm. diam., 5–20 mm. long.

Bermuda, North Carolina, Florida, Bahamas, Guadeloupe, Panama, Trinidad, Brazil. Most often found attached under ledges or large rocks in shallow water, on reefs or in slightly protected situations, and dredged to depths of about 2 m.

REFERENCES: Hemsley 1884, Murray 1889, Børgesen 1913–20, Collins and Hervey 1917, Taylor 1928a, 1929b, 1942, Williams 1948b; *V. syphunculus*, Mazé and Schramm 1870–77.

UNCERTAIN RECORDS

Valonia caespitosa Crouan

Guadeloupe, the type locality. De Toni 1:380; Mazé and Schramm 1870–77, Murray 1889. Essentially a *nomen nudum*.

Valonia tenuis Crouan

Guadeloupe, the type locality. De Toni 1:380; Mazé and Schramm 1870–77, Murray 1889. Essentially a *nomen nudum*, but the plant is probably a sterile Vaucheria.

Valoniopsis Børgesen, 1934

Plants cushionlike, of filaments somewhat repent and entangled below, where attached to the substrate by lobed or branched haptera near which crosswalls often appear, the ascending filaments somewhat stiff and arcuate; oppositely to subcorymbosely or unilaterally branched, often attaching to the substratum at the tips.

Valoniopsis pachynema (Martens) Børgesen

Plants forming wide cushions to 2.5 cm. thick, 5–7 cm. diam., of interlaced ascending filaments, repeatedly and often subcorymbosely branched, the younger branchlets erect, the older branches spreading and often arcuate; cells cylindrical, reaching 0.3–1.0 mm. diam. and several times as long.

Bermuda, Hispaniola.

REFERENCES: *Valonia pachynema,* Collins and Hervey 1917; *V. confervoides,* Murray 1889.

Ernodesmis Børgesen, 1912

Plants becoming bushy, starting as a single erect clavate cell, basally somewhat annulate, which bears a terminal cluster of similar branch cells, this pattern being repeated several times.

Ernodesmis verticillata (Kützing) Børgesen
Pl. 1, fig. 2; pl. 6, fig. 10

Plants to about 5 cm. tall, very bushy, borne on a single initial cell 10–20 mm. long and 1.5–2.5 mm. diam., which bears near the top a cluster of similar but somewhat smaller branch cells, these each in turn bearing similar branches to a total of 6 degrees, or perhaps more.

Bermuda, Florida, Mexico, Caicos Isls., Jamaica, Virgin Isls., St. Eustatius, ?Montserrat, Barbados, Venezuela, Tobago, Brazil. Plants of shallow, very quiet yet somewhat shaded pools, but reaching a depth of 9 m.

REFERENCES: Børgesen 1913–20, Howe 1918a, Schmidt 1923, 1924, Taylor 1928, 1929b, Hamel and Hamel-Joukov 1931; *Valonia caespitula* and *V. subverticillata,* Mazé and Schramm 1870–77, Murray 1889; *V. verticillata,* Mazé and Schramm 1870–77, Martens 1870 (including ?v. *major*), 1871, Dickie 1874a, Murray 1889, Collins 1901, Vickers 1905, 1908, Sluiter 1908.

Siphonocladus Schmitz, 1878

Plants originating as a single elongate cell of limited growth, attached by multicellular rhizoids at the base, the original cell becoming subdivided by irregular walls, especially above, the cells thus produced then growing out radially into filamentous appendages.

KEY TO SPECIES

1. Primary cell short, primary branching irregular forking, lateral branches few and irregular, plants stiff .... S. rigidus
1. Primary cell tall, lateral branches abundant, and plants ultimately brushlike, soft .... S. tropicus

Siphonocladus rigidus Howe Pl. 6, fig. 7

Plants tufted, pale green, small, the main axis short or indistinct; branching at first subdichotomous, later in part lateral or unilateral and irregular, the ultimate divisions 350–1100 μ diam., the tips sometimes attached by short tenacula; filaments mostly uniseriate, but in part 2–3 cells wide, the cells about as long as broad, except in the proliferations, where they are much longer.

Bermuda, Florida, Bahamas, Cuba, Jamaica. Plants of shallow water, growing at about 0.3–1.0 m. depth below low tide.

REFERENCES: Howe 1905, 1918a, 1920, Taylor 1928.

Siphonocladus tropicus (Crouan) J. Agardh Pl. 7, fig. 1

Plants solitary or clustered, erect, becoming brushlike, of a persistent primary cell which may reach a length of 5 cm., and a reported diameter of 1 cm., though usually less than 1 mm., the lower portion of the primary cell annulate, the upper part soon divided into many cells and on the surface bearing numerous even longer, much more slender branches similar to the primary cell in form and structure, which may even divide to a third degree, though these branchlets remain small.

Bermuda, Florida, Jamaica, Virgin Isls., Antigua, Guadeloupe, Barbados, Venezuela, Tobago. Plants of sheltered locations, epiphytic on coarser algae or lying loose on the bottom, and dredged from a depth of 18 m.

REFERENCES: Murray 1889, Collins 1901, Børgesen 1905, 1913–20, Vickers 1905, 1908, Howe 1909, 1918a, Taylor 1918, 1929b, 1942; *Apjohnia tropica*, Mazé and Schramm 1870–77.

Chamaedoris Montagne, 1842

Plants erect, brushlike, the elongated stalk monosiphonous, of large cells, annulate, somewhat calcified, attaching by rhizoidal extensions of the lower cell; terminal tuft not calcified, of branched filaments, the branches without a basal wall, but a wall formed across the filament just above each fork.

Chamaedoris peniculum (Ellis and Solander) Kuntze Pl. 5, fig. 2

Plants to 1–2 dm. tall, the stalk to 2 mm. diam., simple, or rarely branched, clearly annulate; the terminal uncalcified tuft commonly transversely flattened, reaching a diameter of 3.0–6.5 cm.

Florida, Bahamas, Cuba, Jamaica, Hispaniola, Puerto Rico, Virgin Isls., Guadeloupe, Martinique, Barbados, Grenada, Colombia, Brazil, I. Trinidade. Occasionally found in shallow water, growing under rock ledges; more generally washed ashore from deep water and dredged to a depth of 55 m.

REFERENCES: Børgesen 1913–20, Howe 1915, 1918b, 1920, Schmidt 1924, Taylor 1928, 1930a, 1933, 1941b, Taylor and Arndt 1929, Hamel and Hamel-Joukov 1931, Joly 1953c; *C. annulata*, Harvey 1858, Martens 1870, Mazé and Schramm 1870–77, Dickie 1874a, b, Zeller 1876, Murray 1889, Collins 1901, Vickers 1905, 1908.

Dictyosphaeria Decaisne, 1842

Plants rounded, multicellular, solid or more often hollow, and even continuing to grow after rupturing, attached by rhizoids contributed by several cells near the base; vegetative cells usually angular, macroscopic and coenocytic, increasing in number by internal division; attached to each other by minute tenacular cells along the lines of contact.

KEY TO SPECIES

1. Plants large, to 12 cm. diam.; cells usually to 1 mm. diam., often larger; the hollow thallus monostromatic . . . . . . D. cavernosa

1. Plants and individual cells smaller; thallus solid; cells often with spinelike projections extending into the vacuole
D. vanbosseae

Dictyosphaeria vanbosseae Børgesen

Plants sessile, small, nearly spherical, solidly parenchymatous, composed of angular cells averaging 500 μ diam., occasionally 700–800 μ diam., the cell wall frequently showing spinelike cellulose outgrowths 70–100 μ long, projecting toward the center of the cells; intercellular tenacula often proportionally larger and longer than in the next species.

Jamaica, Virgin Isls.

REFERENCE: Børgesen 1913–20.

Dictyosphaeria cavernosa (Forsskål) Børgesen Pl. 7, fig. 5

Plants sessile, green, hollow, 2–5 cm. diam., nearly spherical, becoming as much as 12 cm. diam. and irregularly lobed or partly collapsed, sometimes concrescent; sometimes rupturing without cessation of growth, irregularly saucer-shaped, and in deep water reaching a diameter of 30 cm.; thallus wall of a single layer of large angular vegetative cells 0.1–1.0 mm. diam., occasionally to 3 mm. diam.; connecting tenacular cells numerous, very minute, lenticular, often with a fimbriate distal margin.

Bermuda, Florida, Mexico, Bahamas, Caicos Isls., Turks Isls., Cuba, Cayman Isls., Jamaica, Hispaniola, Puerto Rico, Virgin Isls., St. Barthélemy, Nevis, Guadeloupe, Martinique, Barbados, Netherlands Antilles, Panama, I. las Aves, Brazil, I. Trinidade. Plants commonly found on rocks and reefs in the intertidal zone, where often ill developed, a little lower reaching full size, and dredged to a depth of 55 m.

REFERENCES: Børgesen 1932, Taylor 1940, 1942, 1943, 1954a; *D. favulosa*, Hooper 1850, Harvey 1858, Mazé and Schramm 1870–77, Farlow 1871, Dickie 1874a, b, Murray 1889, Collins 1901, Vickers 1905, 1908, Sluiter 1908, Børgesen 1913–20, Collins and Hervey 1917, Howe 1918a, b, 1920, Taylor 1928, 1930a, 1933, Taylor and Arndt 1929, Sanchez A. 1930, Hamel and Hamel-Joukov 1931, Joly 1950, 1953c; *D. valonioides*, Mazé and Schramm 1870–77, Murray 1889.

Petrosiphon Howe, 1905

Plants filamentous, chiefly decumbent and aggregated into a disk, closely attached to the substrate, into which numerous rhizoidal outgrowths may penetrate; marginally of one layer of uniseriate filaments of coenocytic units, but elsewhere pluristromatic and even producing short erect filaments.

Petrosiphon adhaerens Howe

Plants light green and a little calcified, to 2–6 cm. diam., round or a little lobed, striate near the margin; in old disks to 5 mm. thick in the center; marginal filaments 300–850 μ diam., straight or geniculate, dichotomously forked; cells 0.5–20.0 diameters long; reproduction by aplanospores.

Bermuda, Florida, Bahamas, Cuba, Jamaica. A plant of the underside of rocks and ledges in shallow water, on surf-beaten rocks and in tide pools.

REFERENCES: Howe 1905, 1918a, 1920, Taylor 1928.

Cladophoropsis Børgesen, 1905

Plants free-floating, or more generally tufted or turflike, forming a basal layer of pale or colorless entangled filaments attached by multicellular haptera, from which arise crowded erect green filaments without a distinct axis, but abundantly laterally branched; growth chiefly apical, the branches usually without a basal wall at the point of divergence; filaments multicellular, uniseriate, the cells multinucleate, the chromatophores at first clathrate, later fragmented, with many small pyrenoids.

KEY TO SPECIES

1. Plants attached; main filaments rarely to 250 μ diam. C. membranacea
1. Plants loose; main filaments usually over 350 μ diam. . . . C. macromeres

Cladophoropsis membranacea (C. Agardh) Børgesen

Pl. 2, fig. 1, pl. 3, fig. 2

Plants tufted or turflike, light green, glossy and often whitish when dried; 2–5 cm., occasionally to 10 cm., tall; erect filaments 170–270 μ diam., alternately, or above unilaterally, branched; branchlets 100–145 μ diam., the wall formation in the branchlets often long delayed and the first walls not basal to the branchlets; cells of irregular length, but long, not constricted at the nodes.

Bermuda, Georgia, Mexico, Bahamas, Caicos Isls., Cuba, Jamaica, Hispaniola, Puerto Rico, Virgin Isls., St. Barthélemy, Guadeloupe, Martinique, Barbados, Grenada, British Honduras, Panama, Netherlands Antilles, Venezuela, Tobago, Brazil. Growing on stones or woodwork in the intertidal zone, especially in quiet water. Commonly forming extensive mats, which may become infiltrated with fine sand.

REFERENCES: Børgesen 1905, 1913–20, Sluiter 1908, Collins and Hervey 1917, Howe 1918a, 1920, Taylor 1928, 1929b, 1933, 1935, 1940, 1941b, 1942, 1943, 1954a, 1955a, Taylor and Arndt 1929, Sanchez A. 1930, Hamel and Hamel-Joukov 1931, Feldmann and Lami 1937, Joly 1951, 1957; *Aegagropila membranacea* v. *caespitosa,* Martens 1870; *Blodgettia confervoides,* Hemsley 1884; *Cladophora membranacea,* Harvey 1858, Mazé and Schramm 1870–77, Rein 1873, Dickie 1874a, Hauck 1888, Murray 1889, Vickers 1908; *Cladophora bryoides, C. enormis,* and *C. membranacea* v. *caespitosa,* Mazé and Schramm 1870–77, Murray 1889; *Siphonocladus membranaceus,* Vickers 1905, Collins 1901; *Valonia confervacea* and *V. intricata,* Mazé and Schramm 1870–77, Murray 1889.

Cladophoropsis macromeres Taylor Pl. 2, fig. 2

Plants in loose masses, or entangled among other algae; filaments 15 cm. or more in length, the branching irregular below, irregular or unilateral above; diameter of the main filaments 375–460 μ, of the branchlets 210–295 μ.

Bermuda, Florida, Jamaica. Plants of very warm and quiet shallow pools.

REFERENCES: Taylor 1928, 1933, 1935.

Boodlea Murray and De Toni, 1890

Plants spongy, often concrescent; filamentous, the filaments arising from a basal attachment, branching freely in all directions, the branches commonly attaching by tenacular cells at the tips to other branches with which they come into contact.

KEY TO SPECIES

1. Branching throughout rather irregular; main filaments generally 200–350 μ diam., the ultimate branchlets 70–100 μ diam. B. composita
1. Branching in part irregular, but often opposite and developing in a single plane, with the blades concrescent; stipe distinct; main filaments 150–320 μ diam., the ultimate branchlets 60–80 μ diam. .... B. struveoides

Boodlea composita (Harvey and Hooker *fil.*) Brand

Plants bushy, forming spongy masses, the basal attachment soon diffuse; main filaments 200–350 μ, rarely to 450 μ, diam., the cells to 20 diameters long; branching on the main filaments at first unilateral or opposite, but soon becoming whorled or irregular, and in the lesser divisions altogether irregular, not flabellate; branchlets 70–100(–200) μ diam., when first formed without basal cross walls, which are acquired later.

Bermuda, Puerto Rico, Virgin Isls., Guadeloupe, Martinique, Barbados. A plant of shallow water growing in rock crevices and under ledges.

REFERENCES: Mazé and Schramm 1870–77, Murray 1889; *B. siamensis,* Børgesen 1913–20, Hamel and Hamel-Joukov 1931.

Boodlea struveoides Howe

Plants with a weak, simple or dichotomous stalk 5–30 mm. tall, the cells 200–450 μ diam., 4–40 diameters in length; main branch system sometimes irregularly but often oppositely branched; when oppositely branched it produces flat blades which cohere with others and may, like the irregular parts, form a spongy mass; cells of the main blade filaments 150–320 μ diam., the branchlets 60–80 μ diam. and the cells 2–4 diameters long, at the tips often adhering to other filaments by tenacula.

Bermuda. A plant of shallow water, growing on stones.

REFERENCE: Howe 1918a.

Microdictyon Decaisne, 1839

Plants sessile, expanded, membranous, crisp, the base attached by unicellular or sparingly divided rhizoids extending from the initial cell of the plant and from the larger neighboring cells; the membrane composed of monosiphonous filaments which branch more or less closely in one plane, and whose branchlet tips attach themselves to an opposed cell with a terminal wall thickening, forming an irregular open angular network.

KEY TO SPECIES

1. Branchlet tips unmodified, attaching directly to cells with which they come into contact 2
1. Branchlet tips developing a special hapteral cell by which they attach on contact .... M. curtissiae, p. 121

2. Main filaments 145–170 μ diam.; ultimate segments attaching by a smooth ring at the margin of contact .. M. boergesenii, p. 120
2. Main filaments to 600–800 μ diam.; ultimate segments attaching by a minutely crenellate ring at the margin of contact M. marinum, p. 121

Microdictyon boergesenii Setchell — Pl. 8, fig. 1

Plants foliaceous, pale green, the blades reaching at least 3.0 cm. diam. but usually crowded and somewhat concrescent, then of indefinite size and shape; adhering fairly well to paper on drying; meshes moderately large compared to the cell diameter; primary filaments obscure, 145–200 μ diam., the cells a little swollen, about 2.5–3.5 diameters long; ultimate ramifications with cells about 45–110 μ diam., 2–3 diameters long; special hapteral cells not present, the tips of unmodified branchlets abutting on another filament and attaching with a smooth, slightly thickened ring at the line of contact.

Bermuda, Florida, Jamaica, Virgin Isls., Guadeloupe, Brazil. Plants of moderately deep water, growing on larger algae, on stones, and on heavy shells; dredged to a depth of nearly 160 m.

REFERENCES: Setchell 1925, 1929, Taylor 1928, 1955a; *M. umbilicatum,* Collins 1901, 1909b, Gepp 1905b, Børgesen 1913–20.

Microdictyon marinum (Bory) Silva

Plants foliaceous, dull green, to 6–8 cm. tall or perhaps more, in crowded masses, not adhering to paper on drying; meshes commonly quite small, but sometimes of moderate size, to 0.6–1.3 mm. diam.; primary filaments distinct but not exceedingly pronounced, 600–800 μ diam. or somewhat more, the cells somewhat longer than broad; ultimate segments 300–400 μ diam., 1.5–2.0 diameters long; special haptera absent, the ultimate segments attaching directly to an opposed cell, with a thickening moderate terminal wall and with a variably crenellate thickened ring at the line of contact.

Bahamas, Cuba. Plants growing from just below low-tide line to a depth of 8 m. or more on rocks or stones and commonly associated with sponges.

REFERENCES: Silva 1955; *M. crassum*, Howe 1918b, 1920, Taylor 1955a.

Microdictyon curtissiae Taylor

Plants foliaceous, to 7.0 cm. long by 3.5 cm. broad, or perhaps larger, irregularly oval; not adhering well to paper on drying; meshes rather small in proportion to the filament diameters; primary filaments obscure but demonstrable, the cells 500–590 μ diam., with moderately swollen cells 2.0–2.5 diameters long; branching opposite or to 5–6 radiate from a node; lesser ultimate ramifications with cells 250–340 μ diam., the cells 2–3 diameters long; special hapteral cells markedly smaller than the segments bearing them, 43–114 μ diam., 31–82 μ long, the wall a little thickened, especially along the distal attaching margin where it is irregular to almost fibrocrenellate.

Florida.

REFERENCE: Taylor 1955a.

UNCERTAIN RECORD

Microdictyon calodictyon (Montagne) Kützing

Brazil. Type from the Canary Islands. De Toni 1:362; Dickie 1874b, Setchell 1929.

Struvea Sonder, 1845

Plants based on a simple or branched monosiphonous stalk, which attaches by multicellular branched rhizoids and above produces one or more broad blades of regular netlike filamentous construction, the filament tips attaching by definite hapteral cells.

KEY TO SPECIES

1. Stipes unbranched, or very rarely with 1–2 branches 2
1. Stipes more commonly branched 3
2. Fronds 3–5 cm. tall; midrib filaments with 4–6 pairs of branchlets S. anastomosans, p. 122
2. Fronds 7–10 cm. tall; midrib filaments with 10–15 pairs of branchlets .... S. elegans, p. 123
3. Each branch of the stipe with a separate network, which may reach a length of 7 cm. .... S. ramosa, p. 123
3. Stipe carrying a single netlike blade, which may reach a length of 3 dm. .... S. pulcherrima, p. 123

Struvea anastomosans (Harvey) Piccone Pl. 5, fig. 1; pl. 9, fig. 2

Plants densely entangled or attached to each other, to 3–5 cm. tall; the main filamentous axes 200–900 μ diam., unsegmented in the lower part of the stalk, segmented above, bearing in a plane 4–6 pairs of opposite branchlet filaments which divide and redivide with decreasing regularity to form the network, the ultimate filaments of the meshes 100–140 μ diam.; not infrequently the branchlet filaments developing irregular meshes out of the plane of the main blade; branchlet filaments of all orders at first continuous with the parent cell, later cut off by a crosswall.

V. **caracasana** (Grunow *ex* Murray and Boodle) Collins. Frond regularly bipinnate, the branchlet tips seldom attached.

Florida, Jamaica, Virgin Isls., Guadeloupe, Barbados, Venezuela, Trinidad, Tobago. Plants of reefs or other rather exposed places, growing in crevices of the rock or among coarse algae.

REFERENCES: Collins 1909b, Børgesen 1913–20, Taylor 1928, 1929b; *S. delicatula,* Mazé and Schramm 1870–77, Murray 1889, Vickers 1905, 1908.

Struvea ramosa Dickie Pl. 9, fig. 3

Plants arising from a lobed and often penetrating holdfast, to 10 cm. tall, stalked; stalks occasionally branched, the divisions 1–3 cm. long, 1.0–1.25 mm. diam., of 1–3 cells, the lower of which is 1–3 cm. long; above on each division the stalks bearing a flat blade to 7 cm. long, 5 cm. broad; midrib filaments bearing opposite lateral branchlets to 1–2 degrees, beyond which the branching is very irregular; the ultimate networks of small meshes about 0.25–1.0 mm. diam.

Bermuda, Netherlands Antilles. Plants of deep water, dredged from 90 m.

REFERENCES: Dickie 1874b, Hemsley 1884, Murray 1889, Howe 1918a, Taylor 1942.

Struvea elegans Børgesen Pl. 9, figs. 1, 8, 9

Plants with a creeping base attached by branched rhizoids; erect blades to 10–12 cm. tall, stalked, the stalk to 4.5 cm. long, its lower cell very large, slightly annulate at the bottom, about 1 mm. diam., 2.0–2.3 cm. long; the upper part of the stalk consisting of but a few cells; blade when young regular and symmetrical, lanceolate, to 4.0–7.5 cm. long, 3.0 cm. broad, with a pronounced midrib and apical growth, the midrib bearing opposite lateral branch filaments to the second or even occasionally to a fourth degree; when old the meshes large, bounded by long slender cells, often broken and the blade becoming irregular.

Florida, Virgin Isls. This is a plant of deep water only, and has been dredged from 40 m.

REFERENCES: Børgesen 1913–20, Taylor 1928.

Struvea pulcherrima (J. E. Gray) Murray and Boodle

Plants large, each with a long stalk bearing 2 branches below the blade itself; blades cordate, oval, to 30 cm. long, 20 cm. broad, with a distinct midrib filament and opposite lateral veins, the 2 basally free branches joining with and contributing to the formation of the network above.

Florida, Gulf of Mexico, Barbados. A handsome, very rare plant of deep water.

REFERENCES: Murray 1889, Collins 1909b; *Phyllodictyon pulcherrimum,* Dickie 1874a.

Cystodictyon Gray, 1866

Plants sessile, expanded, forming a clathrate membrane, the base attached by rhizoids from the initial cell and the lower cells of the main axes; the membrane composed of multicellular monosiphonous filaments which branch and rebranch in fanlike fashion in one plane, and attach where they abut on other filaments, the intervals in part being filled by smaller cells, in part left open as conspicuous lacunae.

Cystodictyon pavonium J. Agardh Pl. 1, fig. 1; pl. 8, fig. 4

Plants foliaceous, measuring to 14 cm. diam., or 10 cm. radially from the point of attachment; the stout, tapering main filaments radiating as veins from centers in various parts of the blade, composed of cells 180–300 μ diam., 2–10 diameters long; the openings in the blade rounded, to 0.5–10.0 mm. diam., or more; occasionally the main veins but little interconnected by membranous portions and the plants then very irregularly shaped.

Florida. Plants of deep water, rarely collected. Ordinarily only secured as fragments washed ashore, but dredged from 48 m. depth. Symmetrically developed plants if found would probably exceed Anadyomene in beauty of structure.

REFERENCES: Collins 1909b, Taylor 1928.

Anadyomene Lamouroux, 1812

Plants foliaceous, inconspicuously stalked, attached by rhizoids from the lower stalk and rib cells; blades formed by polychotomously branching flabellate series of cells forming the ribs of an open system which is closed by successively smaller fans and intercalary cells.

KEY TO SPECIES

1. Ribs of single cells in branching series .... A. stellata
1. Ribs of several cells in parallel series, less closely branched A. menziesii

Anadyomene stellata (Wulfen) C. Agardh Pl. 7, fig. 2; pl. 8, fig. 2

Plants erect, often crowded, the blades rounded and crisped, often a little concrescent, to 2.5–10.0 cm. tall and broad; the fanlike ribs barely visible to the unaided eye, to 0.25–0.33 mm. diam., the ultimate interstices finally filled by cells in lateral series along the ultimate veinlets.

F. **prototypa** Howe. Constituent filaments mostly free.

Bermuda, Florida, Mexico, Bahamas, Cuba, Jamaica, Hispaniola, Puerto Rico, Virgin Isls., Guadeloupe, Martinique, Barbados, British Honduras, Panama, Brazil, I. Trinidade. Plants rather frequent in the American tropics, found at about low-tide level and especially in crevices or attached under overhanging rocks, where if sufficiently sheltered they may withstand exposure between tides; dredged to a depth of 57 m. When living specimens are examined with a hand lens they are seen to be among the most beautifully constructed of marine plants.

REFERENCES: Hooper 1850, Montagne 1863, Murray 1889, Collins 1901, 1909b, Vickers 1905, 1908, Børgesen 1913–20, 1924, Collins and Hervey 1917, Howe 1918a, b, 1920, 1928, Schmidt 1923, 1924, Taylor 1928, 1929b, 1930a, 1933, 1935, 1942, 1943, 1954a, Sanchez A. 1930, Hamel and Hamel-Joukov 1931, Feldmann and Lami 1937, Joly 1950, 1953c; *A. flabellata*, Hooper 1850, Harvey 1858, Martens 1870, Rein 1873, Mazé and Schramm 1870–77, Dickie 1874b, Hemsley 1884. For the f. *prototypa*, Howe 1920.

Anadyomene menziesii Harvey

Plants to 25 cm. diam., the blade margin lobed, the stout ribs terminating in these lobes; ribs formed of short cells in parallel series, the marginal ones supporting fan-shaped groups of smaller cells which form the continuous membrane.

Gulf of Mexico. A very rare plant, known only from the original material dredged from 40 m. depth.

REFERENCE: Murray 1889.

SIPHONALES

Plants coenocytic; simply filamentous, or these filaments branched or aggregated to more or less elaborate shapes, free or calcified, in the less simple forms usually showing differentiation of rhizoidal and assimilative portions, and often of stoloniferous portions as well; coenocyte walls sometimes braced by extensions of the membrane across the cavity, and in some species at certain points with ringlike thickenings which almost divide the cavity into separate chambers; nuclei and small chromatophores very numerous, and pyrenoids absent or present.

KEY TO FAMILIES

1. Plants microscopic 2
1. Plants easily visible with the unaided eye; often very large 3

2. Cells bearing long colorless hairs; growing in the tissue of larger algae .... CHAETOSIPHONACEAE, p. 126
2. Cells without hairs; growing in the matrix of shells PHYLLOSIPHONACEAE, p. 189

3. Filaments free or casually intertangled, or enlarged to elaborate, relatively massive branched coenocytes 4
3. Filaments interwoven to form more or less complex thalli of characteristic shapes .... CODIACEAE, p. 156

4. Filamentous or massive, generally organizing rhizoidal, stoloniferous and foliar elements; walls internally supported by trabeculae .... CAULERPACEAE, p. 133
4. Filaments forming simpler plant types; walls not supported by trabeculae 5

5. Branching plumose, usually with distinct primary axes BRYOPSIDACEAE, p. 130
5. Branching not plumose 6

6. Reproductive organs distinct, lateral, often segregated by a pair of crosswalls, the zoöspores with a crown of cilia DERBESIACEAE, p. 127
6. Reproductive organs ill-defined, apparently aplanospores or akinetes; filaments constricted above the forks CODIACEAE, p. 156

CHAETOSIPHONACEAE

Plants consisting of branched microscopic coenocytes, vegetative crosswalls seldom present, although the filamentous types are often deeply constricted; long, unseptate hairs present; cytoplasm peripheral, with many disklike chromatophores containing pyrenoids and with many nuclei; sporangia segregated by a crosswall or involving entire coenocytes, the zoöspores with two or four flagella.

Blastophysa Reinke, 1888

Plants microscopic, epi- or endophytic, the coenocytic cells rounded cushion-shaped or lobed; cell wall thick, often locally particularly heavy, lamellose, and bearing long colorless unseptate hairs; multiplication by the formation of long outgrowths from the cells which swell at the tips, and from each of these portions a new cell is developed, separated from the then empty outgrowth by a wall; cell division, possibly by constriction, also sometimes present; sporangia little modified from the form of vegetative cells.

Blastophysa rhizopus Reinke, *f.*

Plants of oval or oval-acuminate cells which are scattered, or joined by slender colorless tubes, and each bearing 1–3 long colorless hairs very slightly enlarged at the base; coenocytes 25–52(–120) μ diam., 55–90(–130) μ long, with numerous rounded or angular chromatophores with single pyrenoids.

Bermuda, Virgin Isls., Saba Bank. Growing in the tissues of larger algae. Probably very much more common and widespread than the records indicate.

REFERENCES: Børgesen 1913–20, Collins and Hervey 1917.

DERBESIACEAE

Plants matted, with a creeping rhizomatous portion bearing attaching holdfasts and erect, simple or branched, assimilative filaments; asexual reproduction by zoöspores with an anterior circle of flagella, borne several together in sporangia which are usually lateral on the erect filaments.

KEY TO GENERA

1. Sporangia terminating a filament, or forming one side of a dichotomy .... BRYOBESIA, p. 129
1. Sporangia lateral on the filaments .... DERBESIA, p. 128

Derbesia Solier, 1847

Plants of unicellular filaments, primarily erect and simple or sparingly branched, stoloniferous below and attached by lobed holdfasts; zoösporangia lateral on the erect filaments, segregated by a pair of transverse walls.

KEY TO SPECIES

1. Filaments 100–600 μ diam. .... D. lamourouxii, p. 129
1. Filaments less than 100 μ diam. 2

2. Branching mainly dichotomous; filaments mostly under 50 μ diam.; sporangial pedicel about 15 μ diam. D. vaucheriaeformis, p. 128
2. Branching mainly lateral; filaments mostly over 50 μ diam.; sporangial pedicels 25 μ diam. or more .... D. marina, p. 128

Derbesia vaucheriaeformis (Harvey) J. Agardh

Plants forming erect, dense, brushlike tufts, the filaments dichotomously forked, 4–5 cm. tall, 25–50 μ diam.; walls sometimes present in pairs across the segments above a fork, about 30–35 μ apart; sporangia replacing branches, ovoid or broadly pyriform, 100–130 μ diam., 190–300 μ long, pedicellate, the pedicels about 15 μ diam., 50–100 μ long with 2 partitions near the base enclosing a cell 2–4 diameters long.

Bermuda, ?North Carolina, Florida, Bahamas. Probably chiefly growing on walls or stones near low-tide line in well protected situations, but also growing on coarser algae.

REFERENCES: Murray 1889, Collins and Hervey 1917, Howe 1920, Taylor 1928, Williams 1948b; *Chlorodesmis vaucheriaeformis*, Harvey 1858, Hemsley 1884.

Derbesia marina (Lyngbye) Kjellman

Erect filaments 50–70 μ diam., simple or sparingly laterally branched; intercalary cells occasional in the filaments near a branch, about as long as broad; sporangia lateral, obovoid to subspherical, 90–200 μ diam., 150–250 μ long, pedicellate, the stalk cells 30–35 μ diam., 30–70 μ long.

Bermuda, North Carolina. Reported as growing on coarse algae in shallow water.

REFERENCES: Collins 1909b, Collins and Hervey 1917.

Derbesia lamourouxii (J. Agardh) Solier

Erect filaments rather stiff, simple or subdichotomously forked, 100–600 μ diam.; sporangia lateral, solitary or a few unilaterally placed, globose, 300–550 μ diam., sessile or on a short, slender pedicel.

Bermuda, North Carolina, Guadeloupe. Plants of warm, quiet water.

REFERENCES: Collins and Hervey 1917, Williams 1948b; *Bryopsis balbisiana,* Mazé and Schramm 1870–77, *p. p.?*, Murray 1889.

UNCERTAIN RECORDS

Derbesia? fastigiata Taylor

Florida, the type locality. Taylor 1928.

Derbesia tenuissima J. Agardh

Bermuda. Type from the Mediterranean. De Toni 1:424; Murray 1889.

Bryobesia Weber-van Bosse, 1911

Plants coenocytic, with few crosswalls, of short repent branching filaments from which arise longer erect filaments the apices of which are transformed into sporangia cut off by a wall.

Bryobesia cylindrocarpa Howe

Filaments sparingly subdichotomous, 5–15 mm. long, 75–156 μ diam., the walls 3–10 μ thick, rarely septate, the branches occasionally with a wall at the base; sporangia constituting one arm of a dichotomy or pseudodichotomy, or appearing to be terminal, and the axis continuing by formation of a lateral innovation from below the sporangium; sporangia sessile, short-cylindric to obovoid, 90–180 μ diam., 150–450 μ long, the zoöspores 200–500 to a sporangium, ellipsoid or ovoid, 20–25 μ diam., 20–40 μ long.

Bahamas. Once dredged from about 7 m. depth.

REFERENCE: Howe 1920.

BRYOPSIDACEAE

Plants generally erect from a more or less extended rhizoidal holdfast, branching, the branchlets alternate, either continuing like the axis or of limited growth as lateral branchlets with more or less constricted bases.

Bryopsis Lamouroux, 1809

Plants erect, tufted or spreading, bushy, with more or less evident main axes and more or less pyramidally or uni- or bilaterally arranged branchlets formed successively from the apex; plants diploid, reproducing after the segregation of entire branchlets by a basal membrane, the contents undergoing meiotic divisions so that these branchlets eventually function as heterogametic gametangia, the smaller gametes brownish, the larger yellowish-green.

KEY TO SPECIES

1. Ultimate branchlets surrounding the lesser branches 2
1. Ultimate branchlets in 1–2 rows 3

2. Main axes coarse, sharply differentiated; the short ultimate branchlets generally 20–50 μ diam.; plants to 2 dm. tall B. duchassaignii, p. 133
2. Main axes more slender, not very distinct from the secondary branches; branchlets 40–80 μ diam. or more; plants rarely over 1 dm. tall .... B. hypnoides, p. 130

3. Branchlets very short and inconspicuous, mostly unilateral or in pairs near the middle of the axis .... B. ramulosa, p. 131
3. Branchlets longer, conspicuous 4

4. Frond triangular to lanceolate, light or somewhat olive green in color, with branchlets mostly in 2 rows ...... B. plumosa, p. 131
4. Frond linear-lanceolate, dark green, with 1–2 rows of branchlets B. pennata, p. 132

Bryopsis hypnoides Lamouroux

Plants more or less erect, tufted, to 10 cm. high, dull and rather dark green, the axis usually much branched; branches irregularly placed, progressively smaller, without very marked difference between the lesser branches and the rather slender branchlets.

F. **prolongata** J. Agardh. Loosely branched, the branches and axes also little differentiated.

Bermuda, North Carolina, Florida, Bahamas, Cayman Isls. Plants of shallow, warm and quiet, but not stagnant water.

REFERENCES: Harvey 1858, Murray 1889, Vickers 1905, 1908, Collins and Hervey 1917, Howe 1918, 1920, Taylor 1928, Blomquist and Humm 1946.

Bryopsis ramulosa Montagne

Plants densely tufted, dark green; the coarse erect axes sparingly branched; branchlets short, uniform, in scattered unilateral series or pairs, chiefly near the middle of the axis, leaving the tip and stalk bare.

Florida, Cuba, Virgin Isls., Barbados, Brazil.

REFERENCES: Montagne 1863, Dickie 1875, Murray 1889, Collins 1909b.

Bryopsis plumosa (Hudson) C. Agardh Pl. 9, fig. 11

Plants erect, tufted, to 10 cm. tall, light to olive green, the axis often naked below except for a few rhizoids, but above sparingly to much branched, branching generally pyramidal; branchlets distinct from the axis in size, roughly bilaterally attached to the lesser branches, sharply constricted at the base, obtuse at the apex, simple or very infrequently forked, not very slender, being 65–100 μ diam., grading rather evenly from the longest at the base of the branch to the initials at the apex so that the whole appears as a somewhat flat triangular blade, which may be 1.0–1.5(–3.0) cm. broad.

Bermuda, North Carolina, South Carolina, Florida, Mexico, Cuba, Virgin Isls., St. Barthélemy, Guadeloupe, Barbados, Grenada, British Honduras, Brazil, Uruguay. Plants typically of tide pools and sheltered locations though often exposed to moderate surf, growing near low-tide line.

REFERENCES: Harvey 1858, Mazé and Schramm 1870–77, Farlow 1871, Dickie 1874a, Hemsley 1884, Murray 1889, Möbius 1890, Børgesen 1913–20, Hoyt 1920, Schmidt 1923, Taylor 1928, 1935, 1939a, 1941b; *B. pennata* var., Mazé and Schramm 1870–77; *B. arbuscula*, Murray 1889.

Bryopsis pennata Lamouroux Pl. 9, fig. 12

Plants often in large tufts, dark green and sometimes iridescent, attached by rhizoidal holdfasts; the main erect filaments to 7 cm. tall, sparingly divided, often arcuate at the tips; branchlets distichous, of rather uniform length, giving a linear-lanceolate or oblong aspect to the narrow frond, which is about 5–8 mm. broad; main axes about 240–360 μ diam., the branchlets 75–150 μ diam.

V. **leprieurii** (Kützing) Collins and Hervey. Branchlets few, short, sometimes unilateral, in discontinuous series, the axis naked between.

V. **secunda** (Harvey) Collins and Hervey. Branchlets mostly unilateral, the frond tending to be one-sided.

Bermuda, North Carolina, Florida, Mexico, Bahamas, Cuba, Jamaica, Hispaniola, Puerto Rico, Virgin Isls., Guadeloupe, Martinique, Barbados, British Honduras, Colombia, Netherlands Antilles, Trinidad, French Guiana, Brazil. Plants of shallow, rather warm and quiet water, growing on woodwork or rocks.

REFERENCES: Mazé and Schramm 1870–77 *p. p.*, Murray 1889, Collins 1901, 1909b, Gepp 1905b, Vickers 1905, 1908, Collins and Hervey 1917, Howe 1918a, 1920, Taylor 1928, 1935, 1940, 1942, 1943, 1954a, Blomquist and Humm 1946, Joly 1957; *B. plumosa* v. *pennata*, Børgesen 1913–20, Hamel and Hamel-Joukov 1931; *B. plumosa* v. *densa*, *B. pennulata*, and *B. thujoides*, Mazé and Schramm 1870–77, Murray 1889. For the v. *leprieurii*, Collins and Hervey 1917; *B. leprieurii*, Montagne 1850, Mazé and Schramm 1870–77, Zeller 1876, Murray 1889, Vickers 1905, 1908; *B. plumosa* v. *leprieurii*, Børgesen 1913–20, Schmidt 1923, 1924; *B. plumosa* v. *ramulosa*, Mazé and Schramm 1870–77, Murray 1889. For the v. *secunda*, Collins and Hervey 1917, Taylor 1928; *B. harveyana*, Murray 1889, Collins 1901, Vickers 1905, 1908, Howe 1918a, 1928; *B. plumosa* v. *secunda*, Harvey 1858, Børgesen 1913–20.

Bryopsis duchassaignii J. Agardh

Plant light yellowish green, exceedingly soft, even lubricous, to 1–2 dm. tall, bushy, without distinction between blade and stalk; several virgate main branches of indefinite growth arising near the base of the plant, sparingly divided above, 0.75–1.25 mm. thick, these divisions bearing very many short, slender branches of limited growth which are closely surrounded by the very delicate ultimate branchlets 1.0–2.5 mm. long, 20–50(–75) μ diam.

V. **filicina** Collins and Hervey. Plants with a broad to narrowly lanceolate pinnate frond on a simple axis.

Bermuda, Florida, Bahamas, Cayman Isls., Jamaica, Hispaniola, Virgin Isls., Guadeloupe, ?Barbados. Plants of warm quiet shallow waters such as lagoons and the edges of mangrove thickets.

REFERENCES: Mazé and Schramm 1870–77, Murray 1889, Howe 1909, 1920, Børgesen 1913–20, Collins and Hervey 1917 (including the variety), Taylor 1928, 1933; *Trichosolen antillarum,* Montagne 1860, Mazé and Schramm 1870–77, Murray 1889.

UNCERTAIN RECORDS

Bryopsis caespitosa Suhr

Brazil. Type from South Africa. De Toni 1:428; Zeller 1876.

Bryopsis foliosa Suhr

Guadeloupe. Type from Australia. De Toni 1:432; Murray 1889, Collins 1909.

Bryopsis spinescens Zeller

Brazil, the type locality. De Toni 1:437; Zeller 1876.

CAULERPACEAE

Plants coenocytic, branched, slenderly filamentous, or quite large and then differentiated into rhizoidal, stoloniferous and erect portions, the latter assuming a great variety of forms; wall firm, braced internally by a system of trabeculae; reproduction by segregation of portions of the contents of a branch to form dimorphic swarmers, probably gametes, which are discharged through elevated papillae.

Caulerpa Lamouroux, 1809

With the characters of the family.

The highly differentiated erect branches of these often very beautiful plants enable us to divide them into quite distinct species without much danger of confusion. However, the variation within each species is very great, and although some of the forms are quite constant others are by no means so well defined. Under the circumstances it is not surprising that a great many varieties and forms have been described, and that they are of unequal technical worth. They are of very great value in a descriptive way, however, and should be retained for this reason. One must bear in mind that the apparent intergrades between species or major varieties usually represent the appearance on a portion of a plant of less specialized branches or branchlets, such as simple nearly cylindrical branchlets on a plant otherwise typical *C. racemosa;* this may well represent merely an inhibition of the full development potential of the plant. In another variety a similar simple branch form may be the fullest differentiation of which the variety is capable. The presence of a simple branch form in the first instance does not necessarily prove varietal identity with the second form. Presence on one plant of two highly differentiated branchlet forms, one of which could not be a morphological precursor of the other, is not to be expected. The taxonomic complexities here rival those of the Charophyceae, with a lesser range of structure on which to base the varietal nomenclature.

KEY TO SPECIES

1. Plants without distinction of stoloniferous and erect portions, or if these are distinguishable they are similar in form C. fastigiata, p. 136
1. Plants with contrasting stoloniferous and rhizoidal erect portions . 2

2. Erect blades flat, entire, sparingly proliferous from the stalk or face of the blades .... C. prolifera, p. 140
2. Erect portion of the plant either filiform or more massive, variously branched, lobed, or cleft......................... 3

3. Branchlets or lobes of the erect blades with minutely aculeate tips 4
3. Branchlets or lobes with smooth, rounded, or flattened tips 16

4. Branchlets normally and regularly forked 5
4. Branchlets, lobes, or teeth simple, or only occasionally forked ... 9

5. Plants large and bushy, with an obvious branched stalk 1–3 mm. diam., bearing slender pinnately divided branchlets on the upper portion only .... C. paspaloides, p. 149
5. Plants more delicate, the stalks of the erect portions not notably differentiated 6

6. Branchlets short, the divisions less than 10 diameters long 7
6. Branchlets with longer and verticillate divisions 8

7. Branchlets constricted sharply at each forking ... C. murrayi, p. 138
7. Branchlets unconstricted .... C. webbiana, p. 139

8. Plants small, to about 2.5 cm. tall, the verticils small and crowded, the stolons notably hairy, the branchlets constricted at the forks, not mucronate .... C. pusilla, p. 138
8. Plants to 7.0 cm. tall, the verticils conspicuous, well separated, the stolons naked or nearly so, the branchlets not constricted, bimucronate .... C. verticillata, p. 138

9. Branchlets filiform, in many ranks or randomly distributed 10
9. Branchlets or teeth 2-few ranked 11

10. Stolons naked .... C. cupressoides, p. 146
10. Stolons tomentose .... C. lanuginosa, p. 145

11. Erect branches bearing flat blades with broad flat marginal pinnae .... C. mexicana, p. 141
11. Branchlets or teeth terete or compressed 12

12. Branchlets or teeth usually short and not constricted at the base 15
12. Branchlets usually elongate and constricted at the base 13

13. Branchlets strongly compressed, tapering to the apex C. taxifolia, p. 142
13. Branchlets terete or somewhat compressed, cylindrical or expanded toward the apex 14

14. Branchlets nearly cylindrical, 0.3–0.5 mm. diam. C. sertularioides, p. 144
14. Branchlets somewhat compressed, gradually expanded from a slender base to the rounded, apiculate tip, 1.5–2.0 mm. diam. C. floridana, p. 143

15. Branches compressed above, even band-shaped, often notably spirally twisted, marginally dentate; the teeth generally in a single row on the outer edge of the spiral ..... C. serrulata, p. 145

15. Branches terete or compressed, not notably twisted; teeth generally 2 or more ranked, short, but in some varieties 2 or more diameters long and interpretable as branchlets C. cupressoides, p. 146

16. Branchlets nearly or quite cylindrical and clearly pinnately 2-ranked 17
16. Branchlets spherical, clavate or otherwise distally enlarged; if nearly cylindrical not regularly pinnate 18

17. Very small plants, 1–2 cm. tall, the cylindrical branchlets on repeatedly forked axes, the pinnules occasionally forked, occasionally divided in coralloid fashion .... C. vickersiae, p. 137
17. Quite large plants, reaching 1 dm. or more in height; the branchlets very slightly clavate near the tips, on undivided axes .... C. ashmeadii, p. 142

18. Branchlets subsessile, the very short stalks with a definite constriction at the summit .... C. microphysa, p. 155
18. Branchlets with a more obvious stalk, which is not at all constricted 19

19. Ends of the branchlets generally sharply swollen, yet varying from nearly cylindrical or clavate to subspherical, or frontally slightly compressed .... C. racemosa, p. 151
19. Ends of the branchlets terminating abruptly in a peltate disk C. peltata, p. 155

Caulerpa fastigiata Montagne Pl. 10, fig. 12

Plants forming considerable matlike colonies about 1.5–3.0 cm. high, delicate, filamentous, very little differentiated; the rhizoidal branches etiolate, the stoloniferous ones wide-ranging, to 150–210 μ diam., but hardly different in form from the more erect branches which are 90–120 μ diam. and fork dichotomously, irregularly, or oppositely; no specialized branchlets present, those of the last order at first a little clavate but later quite like the erect axes which bear them, and often redividing.

V. **confervoides** Crouan *ex* Weber-van Bosse. Plants becoming detached, floating free in the water, with stolons some decimeters long and branches even less differentiated than in the typical plant.

Bermuda, Florida, Bahamas, Cuba, Hispaniola, Virgin Isls., Guadeloupe, Barbados, British Honduras, Panama, Brazil. Plants of warm, quiet lagoons with sandy or muddy bottoms, and of the edges of mangrove swamps. By accumulating sand the close growth of this species may form firm tussocks 2–3 dm. tall and 1–2 m. in diameter; the surface may be exposed at extremely low tides; it is areolate, with small patches of dark green filament tips separated by lines of clear sand, and the margins of the tussocks are ringed with streaming filaments approaching v. *confervoides*. Transitions to the variety are not uncommon.

REFERENCES: Montagne 1863, Mazé and Schramm 1870–77, Murray 1889, Vickers 1905, 1908, Børgesen 1913–20, Collins and Hervey 1917, Howe 1920, 1928, Taylor 1928, 1929b, 1933, 1935, 1954a, Joly 1951, 1957; *Herpochaete fastigiata*, Martens 1870. For the v. *confervoides*, Mazé and Schramm 1870–77, Murray 1889, Collins 1909b, Taylor 1928.

Caulerpa vickersiae Børgesen Pl. 10, figs. 3–9

Plants very small, the short stolons inconspicuous but bearing a few scattered rhizoids, occasional pinnules, and the erect foliar branches on the upper side; erect branches hardly 1 cm. tall, about 150 μ diam., more or less stalked, the stalks naked or with a few short branchlets, above simple or occasionally branched, the blades flat, pinnate, the pinnules about 100 μ diam., up to 5 times as long, occasionally forked, the tips blunt.

V. **furcifolia** Taylor. Stolon 125–215 μ diam.; erect branches 60–85 μ diam., 4–7(–10?) mm. tall, 1–4 times forked, naked below, simply pinnate in small part above, the axes mostly beset with pinnate, irregularly placed, clustered, or subverticillate 1–3 times closely subdichotomously forked branchlets 50–85 μ diam., 170–300 μ long.

V. **luxurians** Taylor. Stolon inconspicuous, erect blades solitary or clustered, sessile or stalked, to 5–15 mm. tall, simple or more typically successively 1–5 times forked, 1–2 mm. broad, pinnate, the pinnules 60–85 μ diam., 0.5–1.0 mm. long.

Florida, Jamaica, Virgin Isls., Barbados, Costa Rica. Very inconspicuous little plants, to be sought among coarser algae and gorgonians on rocky reefs in shallow water, but also dredged from a depth of 20 m.

REFERENCES: Børgesen 1913–20, Taylor (including the varieties) 1928, 1933; *C. ambigua*, Vickers 1908, Collins 1909b.

Caulerpa murrayi Weber-van Bosse

Plants very small, the stolons relatively long, to 4–7 cm., irregularly branched, tomentose, the rhizoidal hairs simple or forked; rhizophorous branchlets poorly developed; erect branches formed at intervals of about 2–6 mm., 2–4 mm. tall, simple, chiefly toward the upper ends bearing opposite branchlets on various radii and simulating verticils, the branchlets successively 2–3 times dichotomous, sharply constricted at each fork, particularly toward the ends; terminal segments 2–3-mucronate.

Brazil. Only known from 2 dredged collections, and probably a deep-water species.

REFERENCE: Gepp 1905.

Caulerpa pusilla (Kützing) J. Agardh Pl. 10, fig. 11

Plants very small, the stolons entangled, tomentose with rhizoids, the erect branches to about 10–25 mm. tall, rhizoidiferous at the base, above with a few close-placed glomerulate whorls of spreading determinate branches which are closely dichotomous, markedly constricted at the forks, the terminal segments rounded or truncate, not mucronate.

Bermuda, Guadeloupe, Barbados, Brazil. Forming compact colonies on the sand in shallow water. Very young plants of *C. verticillata* may be confused with this, but the stolons of that species will have but scattered rhizoids.

REFERENCES: Vickers 1908, Collins 1909b, Collins and Hervey 1917; *C. verticillata* v. *pusilla*, Vickers 1905; *Stephanocoelium pusillum*, Martens 1870.

Caulerpa verticillata J. Agardh Pl. 10, figs. 1, 2

Plants tufted, the stolons slender and much-branched, naked or bearing a few rhizoids, with rhizoid-bearing branches below and foliar branches above; erect branches simple or abundantly irregularly forked, rising to a height of 7 cm., more or less naked below, but toward the top of each branch bearing conspicuous whorls of determinate branchlets, the whorls 5–8 mm. diam., often in short verticillate groups separated by 3–5 mm. on the naked axis; determinate branchlets about 100–210 μ diam. at the base, dichotomously branching 5–7 times, but not constricted at the forks, the ultimate divisions about 30–40 μ diam., with mucronate tips.

F. **charoides** (Harvey) Weber-van Bosse. Plants with the close-placed determinate branchlets alternate or opposite, scattered, not whorled.

Bermuda, Florida, Bahamas, Caicos Isls., Cuba, Jamaica, Hispaniola, Puerto Rico, Virgin Isls., Antigua, Guadeloupe, Martinique, Grenada, Netherlands Antilles, Venezuela, Brazil, I. Trinidade. Plants often forming very large colonies in shallow protected situations such as lagoons or the borders of mangrove swamps, growing on mud, roots, or stones, and dredged to a depth of over 30 m.

REFERENCES: Murray 1889, Collins 1901, Sluiter 1908, Børgesen 1913–20, Howe 1915, 1918a, 1920, Collins and Hervey 1917, Taylor 1928, 1954a, Taylor and Arndt 1929, Hamel and Hamel-Joukov 1931, Feldmann and Lami 1936; *C. pusilla,* Mazé and Schramm 1870–77, Murray 1889, Williams and Blomquist 1947. For the f. *charoides,* Collins 1901, 1909b, Børgesen 1913–20.

Caulerpa webbiana Montagne Pl. 10, fig. 10

Plants very small, creeping or assurgent, forming mats to 2 dm. diam., the branching stolons becoming copiously rhizoid-bearing in the older parts and with erect branches up to 15 mm. long on the upper side; erect branches bearing whorls, each of 2–6 determinate branchlets about 250 μ long, these terete at the base, flattened beyond, and closely 3–5 times dichotomous, the divisions of decreasing diameter, the tips furcate or mucronate.

F. **disticha** Weber-van Bosse. Erect branches flat, the determinate branchlets distichous, 2–3 times forked in the plane of the blade.

F. **tomentella** (Harvey) Weber-van Bosse. Plant more repent, determinate branchlets crowded among the rhizoids on the decumbent as well as on the erect portions, the divisions less flattened, at times in the creeping portions elongated and filamentous.

Florida, Jamaica, Puerto Rico, Virgin Isls., Guadeloupe, Brazil. Reported as found from the littoral down to a depth of 50 m., growing on coral fragments and sand.

REFERENCES: Mazé and Schramm 1870–77, Dickie 1874b, Murray 1889, Collins, Holden, and Setchell, Phyc. Bor.-Am. 1333, Børgesen 1913–20 (including f. *disticha*). For the f. *tomentella*, Vickers 1905, 1908, Taylor 1928, Williams and Blomquist 1947.

Caulerpa prolifera (Forsskål) Lamouroux Pl. 11, figs. 1–3

Plants often forming large colonies, the sparingly branched stolons wide-spreading, very slender but wiry, forming rhizoid-bearing branches below at intervals of 0.5–2.0 cm., and ascending light green foliar branches above at intervals of 1–5 cm.; erect branches with slender stalks 0.5–1.0 cm. long, bearing a single flat oval to linear-oblong entire blade 3–13 mm. wide, 3–15 cm. long, the base somewhat tapering, the apex narrowed but also obtuse; blade sometimes bearing similar proliferations from the face or the margin.

F. **obovata** J. Agardh. Plants thicker and coarser in all parts, the blades dark green, leathery, oval to obovate, 15–25 mm. broad, 25–55(–100) mm. long, commonly proliferous.

F. **zosterifolia** Børgesen. Blades narrow, hardly 6 mm. wide, linear-lanceolate, interrupted and richly proliferous.

Bermuda, North Carolina, Florida, Mexico, Bahamas, Cuba, Jamaica, Puerto Rico, Virgin Isls., Antigua, Guadeloupe, Martinique, Barbados, Grenada, Colombia, Venezuela, Brazil, I. Trinidade. The typical plant, with moderately thin, seldom-proliferating blades, grows on the sandy or muddy bottoms of quiet lagoons but little below low-tide level, and on rocks exposed to moderate wave action as well. However, it frequently appears in dredge hauls at depths to as much as 110 m. and is but a little more delicate at lower depths. F. *obovata* is known from shallow water and moderate depths to 20–40 m. also, but f. *zosterifolia* only from shallow water.

REFERENCES: Harvey 1858, Mazé and Schramm 1870–77, Farlow 1871, Rein 1873, Dickie 1874b, Hemsley 1884, Murray 1889, Piccone 1889, Collins 1901, Børgesen 1913–20, Collins and Hervey 1917,

Howe 1905, 1909, 1918a, b, 1920, Hoyt 1920, Schmidt 1923, 1924, Taylor 1928, 1936, 1941b, 1942, Hamel and Hamel-Joukov 1931, Humm 1952b, Joly 1953c. For the f. *obovata,* Børgesen 1913–20, Collins and Hervey 1917, Howe 1920, Taylor 1928; *C. prolifera* v. *firma,* Mazé and Schramm 1870–77, Murray 1889. For the f. *zosterifolia,* Børgesen 1913–20, Collins and Hervey 1917.

Caulerpa mexicana (Sonder) J. Agardh Pl. 12, figs. 2–5

Plants stoloniferous, the branching stolons slender, 0.5–1.25 mm. diam., probably seldom over 2–3 dm. in length, with slender descending rhizoid-bearing branches at intervals of a few millimeters, and erect foliar branches at intervals of 1–4 cm.; foliar branches briefly stalked, oblong or broadly lanceolate, 1–25 cm. tall, 5–16 mm. broad, simple or occasionally forked, pinnately divided, with a flat midrib 1.0–3.0 mm. broad and with closely placed, sometimes overlapping, opposite, ascending, flat, oval to oblong or somewhat arcuate and basally narrowed, acuminate and apiculate pinnules 2–10 mm. long.

F. **laxior** (Weber-van Bosse) n. comb. Plants generally smaller, the blades relatively narrow, linear or narrowly lanceolate, the pinnules more separated, spaced at intervals of 2–5 mm., often opposite, irregularly distributed or alternate, linear or tapering somewhat from the base, about 3–5 mm. long.

F. **pectinata** (Kützing) n. comb. Plants with somewhat narrower blades than usual, the pinnules separated by about their own width, not contracted at the base, a little upcurved.

Bermuda, Florida, Mexico, Bahamas, Caicos Isls., Cuba, Cayman Isls., Jamaica, Hispaniola, Puerto Rico, Virgin Isls., Guadeloupe, Barbados, British Honduras, Colombia, Venezuela, Tobago, French Guiana, Brazil, I. Trinidade. Plants attaching to coral fragments or small stones over a sandy or muddy bottom in shallow water, in such places as lagoons and mangrove thickets; dredged certainly to a depth of 73 m. and probably to 110 m. Very variable, in shallow water commonly dwarfed with close-branching rhizomes and close-set blades little over a centimeter long. The v. *laxior* occurs in deep, very sheltered inland lagoons and lakes with very active, often underground, flow of sea water.

REFERENCES: Harvey 1858, Rein 1873, Dickie 1874a, 1884, Murray 1889; *C. mexicana* v. *harveyana,* Mazé and Schramm 1870–77;

C. crassifolia, Howe 1905, 1915, 1918a, 1920, Børgesen 1913–20, Collins and Hervey 1917, Taylor 1928, 1936, 1940, 1942, 1948, 1954a, Feldmann and Lami 1937; *C. crassifolia* f. *mexicana,* Hauck 1888, Murray 1889, Vickers 1908, Børgesen 1913–20, Collins and Hervey 1917, Taylor 1928, 1929b, 1930a, 1935, 1941b, 1942; *C. pinnata,* Weber-van Bosse 1898; *C. pinnata* f. *mexicana,* Collins 1901, Vickers 1905, Schmidt 1924; *C. taxifolia* v. *crassifolia,* Mazé and Schramm 1870–77. For f. *laxior, C. crassifolia* f. *laxior,* Collins 1909b, Collins and Hervey 1917, *C. pinnata* f. *laxior,* Weber-van Bosse, 1898, p. 291. For f. *pectinata, C. crassifolia* f. *pectinata,* Collins 1909b; *C. pectinata,* Kützing 1849, p. 495; *C. pinnata* f. *pectinata,* Schmidt 1924.

Caulerpa taxifolia (Vahl) C. Agardh Pl. 12, fig. 1

Plants with widespreading naked stolons, giving off rhizoid-bearing branches below and foliar branches above; erect branches often quite close together, with stalks 1–3 cm. long ending in flat linear-oblong to linear, not infrequently locally contracted, blades, which reach a length of 5–15 cm. and a width of 6–12 mm.; blades simple or sparingly branched, regularly oppositely pinnate, the pinnules usually markedly upcurved, very strongly compressed, contracted at the base, tapering toward the tip and mucronate.

Bermuda, Mexico, Jamaica, Puerto Rico, Virgin Isls., Barbuda, Guadeloupe, Martinique, Barbados, Grenada, British Honduras, Colombia, Netherlands Antilles, I. las Aves. Plants of sheltered shores and shallow water, growing on sand, mud, or rocks, sometimes found in moderately exposed situations and dredged to 30 m. depth on sand and gravel.

REFERENCES: Mazé and Schramm 1870–77, Hauck 1888, Murray 1889, Collins 1901, Vickers 1905, 1908, Børgesen 1913–20, Howe 1915, Collins and Hervey 1917, Taylor 1935, 1941b, 1942, Feldmann and Lami 1936.

Caulerpa ashmeadii Harvey Pl. 11, fig. 4; pl. 18, fig. 9

Plants with a sparingly forking wide-spreading stolon at least 6.5 dm. long, 2.0–2.5 mm. diam.; at intervals of 2–6 cm. giving off stout descending branches which divide progressively into numerous slender rhizoids, and at similar intervals bearing ascending foliar branches consisting of stalks 1.75–2.0 mm. diam., 5–30 mm. long, and pinnate blades 2–10 cm. long, 13–25 mm. broad, the pinnules moderately close to rather widely (2–6 mm.) separated, moderately to sharply ascending, cylindrical below, becoming clavate toward the rounded truncate tip, 0.75–1.25 mm. diam., 7–18 mm. long.

Florida, Mexico, Virgin Isls., Brazil. Plants have been found on sandy or muddy bottoms at depths of from 3–36 m., and may occur in still shallower situations.

REFERENCES: Harvey 1858, Dickie 1874b, Murray 1889, Collins 1909b, Howe 1909, Børgesen 1913–20, Taylor 1928 *p. p.*, 1930a, 1936, 1941b.

Caulerpa floridana n. sp. Pl. 11, figs. 5, 6; pl. 18, fig. 10

Plants with sparingly forking, wide-spreading stolons to over 7 dm. long, usually 2.0–2.5 mm. diam., giving off stout, descending, rhizoid-bearing branches at intervals of 2–4 cm., and at slightly greater intervals bearing ascending foliar branches consisting of stalks 1–3 cm. long terminating in oval to oblong pinnate blades 3–13 cm. long, 1.7–2.7 cm. broad, the pinnules closely approximated, spreading to moderately ascending, contracted and terete at the base, compressed and gradually broadening above, usually 1.5–2.0 mm. wide, not inflated toward the tip, which is rounded and sharply apiculate.

Florida. Plants seemingly only known to grow in deep water, several collections having been secured at depths from 18 to 109 m. off a sandy bottom.

This handsome plant was considered to be a perfect and elegant form of *C. ashmeadii* in my report on Florida algae, but the late M. A. Howe considered it distinct and signified his intention to publish it as a new species, but had not done so at the time of his death. It is to be distinguished by the slightly compressed, gradually tapering pinnules which are contracted at the base and apiculate at the tip, whereas the pinnules of the typical *C. ashmeadii* in Howe's opinion were terete, not definitely contracted at the base, more locally swollen at the tip, and lacking any apiculus. I earlier considered plants with these features as old and decadent examples which had lost the terminal spine, but I have since seen delicate, attenuate plants and old, almost corneus plants of both forms and am inclined to accept Howe's opinion that the plants are distinct.

REFERENCE: *C. ashmeadii*, Taylor 1928 *p. p.*

Caulerpa sertularioides (Gmelin) Howe Pl. 13, figs. 1–7

Plants forming large colonies 1–2 m. diam. by an extensive system of coarse branching stolons which give off stout rhizoid-bearing branches below and foliar branches above; foliar branches flat, short-stalked, reaching a length of 10–15 cm., and a width of 13–22 mm., simple or occasionally branched, generally regularly and closely pinnate, the pinnules cylindrical, a little upcurved, about 180–330 μ diam., 3–11 mm. long, the tips rounded-conical and mucronate.

F. **brevipes** (J. Agardh) Svedelius. Blades nearly or quite sessile, 1–5 cm. long, rarely forked.

F. **corymbosa** Taylor. Blades briefly stalked, 2–8 cm. tall, becoming abundantly subcorymbosely branched and occasionally slightly rebranched to a second and third degree, yielding 10–20 blades on a single base; pinnules usually distichous, about 7 mm. long below, 3 mm. long on the upper blade divisions.

F. **farlowii** (Weber-van Bosse) Børgesen. Blades with pinnules sometimes in part nearly as in the typical plant, but most characteristically radially disposed, crowded on the axis to produce a terete rather than a flat frond.

F. **longiseta** (Bory) Svedelius. Blades with a naked stalk 1–3 cm. long, the whole reaching 18 cm. in length, not infrequently forked, distichous or occasionally tristichous.

Bermuda, Florida, Mexico, Bahamas, Caicos Isls., Cuba, Jamaica, Hispaniola, Puerto Rico, Virgin Isls., St. Barthélemy, Guadeloupe, Dominica, Martinique, Barbados, Grenada, British Honduras, Old Providence I., Costa Rica, Panama, Colombia, Netherlands Antilles, Venezuela, Trinidad, Tobago, Brazil. Rather a common species, though less so than *C. racemosa*. It forms large colonies in sandy shoals, particularly of f. *longiseta* at a few decimeters depth or where partially shaded, and small, dense colonies of f. *brevipes* on rocks more exposed to the light or the waves. It has been dredged to a depth of 110 m., but all the collections seen from beyond 3–4 m. depth are of very depauperate, attenuate individuals.

REFERENCES: Howe 1905b, 1915, 1918a, b, 1920, Vickers 1908, Sluiter 1908, Børgesen 1913–20, 1924, Collins and Hervey 1917, Taylor 1928, 1929b, 1930a, 1933, 1935, 1936, 1942, 1954a, 1955a, Taylor and Arndt 1929, Feldmann and Lami 1936, 1937, Questel 1942; *C. plumaris*, Harvey 1858, 1861, Mazé and Schramm 1870–77, Dickie 1874a, Hemsley 1884, Hauck 1888, Murray 1889, Vickers 1905, Schmidt 1923, 1924. For the f. *brevipes*, Sluiter 1908, Børgesen 1913–20, Collins and Hervey 1917, Taylor 1928, 1929b, 1935, 1939b, 1940, 1941b, 1942, 1943, Taylor and Arndt 1929; *C. plumaris* f. *brevipes*, Collins 1901. For the f. *corymbosa*, Taylor 1942. For the f. *farlowii*, Børgesen 1913–20, Taylor 1928, 1940, 1942; *C. selago*, Mazé and Schramm 1870–77, Murray 1889 *p. p.?* For the f. *longiseta*, Collins 1909b, Børgesen 1913–20, Sluiter 1908, Taylor 1941b; *C. sertularioides* f. *longipes*, Collins 1909b, Taylor 1928, Taylor and Arndt 1929; *C. plumaris* v. *boryana*, Martens 1870; *C. plumaris* f. *longiseta*, Collins 1901.

Caulerpa lanuginosa J. Agardh Pl. 14, figs. 1–2

Plants of moderate size, the densely tomentose stolons to 1–3 dm. long, emitting distant rhizoid-bearing branches below and erect assimilatory branches above; erect branches about 3–13 cm. tall, simple or with 1–4 branches, at the base briefly tomentose like the stolons, above with densely imbricate, slightly curved filiform branchlets distributed at random about the axis; branchlets about 3–5 mm. long, 60–180 μ diam., mucronate at the apex.

Florida, Bahamas, Jamaica, Martinique, Brazil. Plants preferring a sandy bottom, ranging from a little below low-tide line to a depth of 110 m. The branched rhizoid-like filaments on the stolons attach to and ensnare sand grains in a conspicuous fashion. The plants from deep water are small and delicate, the branchlets hardly over 60 μ in diameter, but those from shallow water are much coarser.

REFERENCES: Murray 1889, Howe 1909, 1920, Taylor 1928, 1930a; *C. lycopodium*, Harvey 1858, Murray 1889, Collins 1909b (incl. v. *delicatula*); *?C. selago*, Greville 1833a.

Caulerpa serrulata (Forsskål) J. Agardh *emend.* Børgesen Pl. 14, fig. 5

Plants with a strong, naked, branching stolon forming rhizoid-bearing branches below and erect assimilatory branches above; erect branches with a terete or compressed stipe and above a narrow, flattened, dichotomously forked lamina which is characteristically spirally twisted, sometimes locally constricted or expanded, or with proliferations, the margin entire or more usually serrate, especially on the outer side of the spiral, the mucronate teeth about as long as broad.

F. **angusta** (Weber-van Bosse) n. comb. Fronds narrow, the inner margin of the spiral entire but the outer dentate; constrictions few.

F. **lata** (Weber-van Bosse) Tseng. Fronds broader, with both margins dentate; constrictions numerous.

V. **boryana** (J. Agardh) Gilbert

F. **occidentalis** (Weber-van Bosse) Gilbert. Erect branches long-stalked, slender, terete or compressed, dichotomous above, the blades linear, flat, the marginal teeth distant, small, shorter than broad.

V. **pectinata** (Kützing) n. comb. Blades rarely twisted; proliferations often frequent; margin dentate or pectinate except where the blades are constricted.

Bahamas, St. Barthélemy, Guadeloupe, Barbados, Grenada, Venezuela.

REFERENCES: Børgesen 1932; *C. freycinetti,* Mazé and Schramm 1870–77, Murray 1889, Collins 1909b (including f. *angusta* and f. *lata*), Howe 1920. For the f. *angusta,* Weber-van Bosse 1898, p. 313; for the v. *boryana, C. freycinetti* v. *deboryana,* Collins 1909b (including f. *occidentalis*). For the v. *pectinata,* C. *freycinetti* v. *pectinata,* Weber-van Bosse 1898, Vickers 1908, Collins 1909b; *C. pectinata* Kützing 1849, p. 316 *p. p.?*

Caulerpa cupressoides (West) C. Agardh

Pl. 14, figs. 3, 4, 6; pl. 15, figs. 1–4; pl. 18, figs. 11–13

Plants forming extensive colonies, with branching stolons several decimeters in length, 1.5–2.5 mm. diam., anchored by stout descending rhizoid-bearing branches at intervals of 2–5 cm. and bearing erect foliar branches above at intervals of about 1–3 cm.; the erect branches 2.5–7.0 cm. tall, naked or nearly naked for the first 5–15 mm., rarely simple, generally copiously forking, the tapering terete green divisions forming a close cluster, each branch normally with 3 rows of short, imbricate, generally terete, apiculate branchlets which are about twice as long as the diameter of the supporting axis.

An extremely variable species, so that it appears best to approach the varieties and forms by an analytical key.

1. Branchlets generally in 2 ranks 2
1. Branchlets in more than 2 ranks 3

2. Plants tall, the erect stalks naked below; above the branchlets short, sometimes hardly more than marginal teeth .... v. flabellata
2. Plants low, the erect branches with branchlets to the base, these branchlets cylindrical, twice as long as the diameter of the axis or more .... v. lycopodium
2. Plants small, the small branchlets in 2–3 irregular ranks, their length about equal to the diameter of the axis .... v. serrata

3. Branchlets not twice as long as the diameter of the axis 4
3. Branchlets twice as long as the diameter of the axis or more 5

4. Frond sparingly branched and slender, with 3–4 rows of branchlets .... v. turneri
4. Frond more stout and bushy, with 5 or more rows of branchlets v. mamillosa

5. Branchlets not more than twice as long as the diameter of the axis 6
5. Branchlets much more than twice as long as the diameter of the axis, in several ranks .... v. lycopodium

6. Branchlets in 3 ranks, rarely and partially 4–5 .... v. cupressoides
6. Branchlets in several ranks .... v. ericifolia

V. **ericifolia** (Crouan) Weber-van Bosse. Erect axes surrounded by appressed multiseriate cylindrical branchlets about equal in length to the diameter of the axis.

V. **flabellata** Børgesen. Erect axes with a naked stalk 3–8 cm. long, above branching 3–5 times; to 2.0 dm. tall, the branches flat, with marginal branchlets about as long as broad, or reduced to mere serrations.

V. **lycopodium** (J. Agardh) Weber-van Bosse. Erect axes to 2.0 dm. tall, sparingly branched, the branchlets erect, in several ranks, all except the lower ones cylindrical, 2–6 times as long as the axis diameter.

F. **alternifolia** Weber-van Bosse. Erect axes dichotomously forked, with branchlets nearly to the base, the branchlets chiefly subnavicular and in 3 ranks, but in part often distichous or multiseriate and cylindrical.

F. **disticha** (Weber-van Bosse) Collins. Erect axes naked at the simple base, much branched and ramellate above; branchlets cylindrical and biseriate, opposite, near the summit occasionally multiseriate; except the lowest, branchlets up to twice the axis diameter in length.

F. **elegans** (Crouan) Weber-van Bosse. Erect axes dichotomously forked, the branchlets nearly all distichous and opposite, terete, and 3–6 times as long as the axis diameter, sometimes a few near the base subconical and irregularly disposed.

F. **intermedia** Weber-van Bosse. Erect, to 30 cm. tall, with a marked naked basal portion, the branchlets in 2 or more series, all except the lowest cylindrical.

V. **mamillosa** (Montagne) Weber-van Bosse. Plants with bushy erect axes, to a height of about 7 cm., several times forked above the stalk, the branchlets in several ranks, erect-spreading, obovoid to subnavicular, mucronate, hardly equal to the axis diameter in length.

F. **nuda** Weber-van Bosse. The base of the erect axis naked, destitute of branchlets.

V. **serrata** (Kützing) Weber-van Bosse. Plants with slender elongate erect axes irregularly branched, in part naked, in part and especially on the upper portions of the branches with small, erect or spreading, opposite or alternate, navicular branchlets in 2–3 ranks, in length hardly equal to the axis diameter.

V. **turneri** Weber-van Bosse. Plants with slender erect axes, the branchlets in 3–4 ranks, small, conical or subnavicular, hardly equaling the axis diameter in length, erect and rather appressed.

Bermuda, Florida, Mexico, Bahamas, Caicos Isls., Cuba, Jamaica, Hispaniola, Virgin Isls., St. Barthélemy, St. Eustatius, Guadeloupe, Martinique, St. Vincent, Barbados, Grenada, British Honduras, Panama, Colombia, Tobago, Brazil. A widely distributed, rather common tropical species which forms large colonies in sandy shallows, attaching to shells, stones, and coral fragments and anchored in the sand itself. The v. *flabellata* has been found to a depth of 110 m. While quite able to luxuriate in hot shallow tide pools or lagoons, this species does not seem to be at all adapted to shaded situations.

REFERENCES: Harvey 1858, Mazé and Schramm 1870–77 *p. p.*, Dickie 1874b, Murray 1889, Weber-van Bosse 1898 (including most varieties and forms), Collins 1901, Vickers 1908, Børgesen 1913–20, Collins and Hervey 1917, Howe 1918a, b, 1920, Taylor 1928, 1941b, 1943, 1954a, 1955a, Taylor and Arndt 1929, Sanchez A. 1930, Hamel and Hamel-Joukov 1931; *C. indica*, Mazé and Schramm 1870–77, Murray 1889; *C. juniperoides*, Phyc. Bor.-Amer. 79. For v. *disticha*, Collins 1909b, Taylor 1928. For v. *ericifolia*, Mazé and Schramm 1870–77, Collins 1901, Collins and Hervey 1920, Taylor 1928; *C. ericifolia*, Harvey 1858, Murray 1889. For v. *flabellata*, Børgesen 1913–20, Taylor 1928, 1935; *C. distichophylla p. p.* and *C. triangularis*, Mazé and Schramm 1870–77, Murray 1889. For v. *lycopodium* and its forms, Vickers 1905, 1908, Collins 1909b, Howe 1920, Taylor 1928, 1942, Taylor and Arndt 1929; *?C. cupressoides* v. *plumarioides*, Børgesen 1913–20; *C. cupressoides* v. *alternifolia p. p.*, v. *distichophylla p. p.*, and *C. triangularis p. p.*, Mazé and Schramm 1870–77; *C. lycopodium*, Williams and Blomquist 1942; *C. plumaris* v. *elegans*, Mazé and Schramm 1870–77, Murray 1889. For v. *mamillosa* and its form, Collins 1901, 1909b, Collins and Hervey 1917, Børgesen 1913–20, Taylor 1928, 1935, 1941b, 1942, 1943; *C. ericifolia*, Mazé and Schramm 1870–77. For v. *serrata*, Collins 1909b; *C. cupressoides* v. *alternifolia p. p.*, Mazé and Schramm 1870–77. For v. *turneri*, Collins 1901, 1909b, Taylor 1928, 1935, 1936, 1941b.

Caulerpa paspaloides (Bory) Greville

Pl. 16, figs. 1–4; pl. 18, figs. 8, 14, 15

Plants with a robust stolon up to 4 mm. diam., forming stout rhizoid-bearing branches below and bushy branches above; erect branches with a pronounced naked stipe 1–5 cm. long, which is closely dichotomously or irregularly digitately forked 1–4 times at the summit; the ultimate substipitate foliar divisions about 5–7 cm. long, terete or prismatic, densely covered with small, imbricate, pinnate, determinate branchlets set in 3–4 ranks, often keeled in aspect, with flattened sides, the ultimate branchlets (pinnae) simple or forked.

The varieties and forms of this species are particularly hard to define. The key below may be of some service. Three aspects seem fairly clear: the typical form has the foliar branches flattened on 3 sides because of the arrangement of the keeled determinate branchlets in 3 rows. The varieties *compressa* and *phleoides* have little evidence of flattening, and their ultimate branches are very short and crowded. The v. *wurdemanni* is a very loose and open form with long ultimate pinnae. When one tries to assign plants more closely than this, uncertainties multiply.

1. Foliar branches terete 2
1. Foliar branches compressed 4
2. Stalks very short, pinnate branchlets crowded . . . . v. compressa
2. Stalks several centimeters long 3
3. Pinnate branchlets crowded . . . . v. phleoides
3. Pinnate branchlets distant . . . . v. laxa
4. Pinnate branchlets very small; plants relatively attenuate . . . . v. laxa
4. Pinnate branchlets large 5
5. Foliar branches compact or moderately loosely covered by the pinnate branchlets . . . . v. paspaloides
5. Foliar branches loosely covered, the pinnae conspicuously long v. wurdemanni

F. **flabellata** Weber-van Bosse. Stalks to 11 cm. tall; foliar branches simple or branched, flabellately arranged, compressed, the pinnate branchlets alternately in 2 rows, fairly closely placed, the pinnae simple or forked, turned to one side.

V. **compressa** (Weber-van Bosse) Howe. Stalks 1–2 cm. long; foliar branches digitately disposed, 2–3 cm. long, terete, surrounded by very crowded, small pinnate branchlets, the spreading pinnae simple or once forked.

V. **laxa** Weber-van Bosse. Stalks as much as 18 cm. long; foliar branches to 28 cm. long, the very small pinnate branchlets rather distant, subverticillate or 3-ranked, the pinnae simple, in one row.

V. **phleoides** (Bory) J. Agardh. Foliar branches terete; pinnate branchlets crowded, with 2 ranks of pinnae.

V. **wurdemanni** Weber-van Bosse. Stalks about 8 cm. long; foliar branches 10 cm. long or more, the pinnate branchlets biseriate, subopposite or scattered, bearing simple biseriate pinnae turned to one side.

F. **phylloplaston** (Murray) Weber-van Bosse. Pinnae uniseriate.

Florida, Mexico, Bahamas, Cuba, Jamaica, Brazil. Growing in shallow water and dredged from a depth of 20 m.

REFERENCES: Harvey 1858, Murray 1889, Collins 1909b (including f. *flabellata*), Howe 1918b, 1920, Taylor 1928; *Chauvinia paspaloides*, Martens 1870. For the v. *compressa*, Howe 1920; *Caulerpa paspaloides* f. *compressa*, Taylor 1928. For the v. *laxa*, Taylor 1928. For the v. *phleoides*, Howe 1920, Taylor 1921; *C. paspaloides* f. *phleoides*, Taylor 1928; *Chauvinia phleoides*, Martens 1870. For the v. *wurdemanni*, Howe 1918a, Taylor 1928, 1941b, 1954a; *Caulerpa wurdemannia*, Hooper 1850. For the f. *phylloplaston*, Collins 1909b; *C. phylloplaston*, Murray 1891b.

Caulerpa racemosa (Forsskål) J. Agardh
Pl. 17, figs. 1, 3, 4, 6, 7; pl. 18, figs. 2–5, 7

Plants widespreading, with long, coarse branching stolons, becoming very densely entangled in old colonies, which often become 1–2 m. in diameter; stout descending rhizoid-bearing branches common; erect foliar branches often much crowded on the stolons, sometimes more remote, one to several centimeters tall, simple or very sparingly forked, covered with short clavate to spherical stalked branchlets.

This famous, ubiquitous tropical species is among the most variable in its variable genus. Even the noted monographer Mme. Weber-van Bosse did not venture to designate a typical form, in this departing from her custom with other species, nor did Børgesen in his review of Forsskål's specimens tell which form he had in hand. However, in correspondence he has informed me that Forsskål's specimen closely resembles Mme. Weber's figure 1, plate 33, which she considered to represent v. *clavifera*. It is probable, therefore, that those plants here designated v. *clavifera* should be considered to be close to the nomenclaturally typical variety. It appears best to approach the varieties and forms by an analytical key.

1. Branchlets abruptly expanded from the stalklike lower portion ... 2
1. Branchlets cylindrical or gradually enlarged from base to apex 4

2. Ends of the branchlets nearly spherical 8
2. Ends of the branchlets transversely more or less flattened 3

3. [Branchlets relatively few on each erect axis, each branchlet when fully developed ending in a flat, stalked disk; plants small .... C. peltata, p. 155]
3. Branchlets several on an axis, the end convex; plants large v. macrophysa

4. Branchlets cylindrical to narrowly clavate 5
4. Branchlets broadly clavate to turbinate 6
4. Branchlets near the bases of the erect axes cylindrical; those higher up gradually expanded toward the rather truncate apex .... v. cheminitzia

5. Erect axes with many densely imbricate branchlets .... v. laetevirens
5. Erect axes with few rather scattered branchlets .... v. gracilis

6. Erect axes sometimes flattened, especially where branchlets are few or lacking .... v. lamourouxii
6. Erect axes terete 7

7. Branchlets terete .... v. clavifera
7. Branchlets distally swollen, somewhat flattened on the exposed (tangential) side .... v. uvifera

8. [Branchlet stalks very short or lacking, somewhat constricted C. microphysa, p. 155]
8. Stalks equal to the swollen extremities of the branchlets v. occidentalis

V. **chemnitzia** (Esper) Weber-van Bosse. Plants usually slender, the erect branches covered with imbricate branchlets rather variable in form, often cylindrical near the base of the branch, but those above gradually enlarged upwardly to the truncate apex.

V. **clavifera** (Turner) Weber-van Bosse. Plants often crowded, the erect branches 1–11 cm. tall, the branchlets on them generally not crowded, irregularly disposed or opposite, or somewhat bilateral, generally briefly clavate on a short pedicel, the globular summit 1.0–2.5 mm. diam.

F. **reducta** Børgesen. Plants very small, the erect branches often forked, bearing a few irregularly placed cylindrical to turbinate branchlets.

V. **gracilis** (Zanardini) Weber-van Bosse. Plants with slender, elongate stolons, the few branchlets cylindrical, scattered or in clusters of small, slightly clavate branchlets.

V. **laetevirens** (Montagne) Weber-van Bosse. Plants robust, the erect branches to 12–30 cm. tall, the branchlets densely imbricate to very widely spaced, subcylindrical to clavate with a rounded apex, sometimes a little turned to one side.

F. **cylindracea** (Sonder) Weber-van Bosse. Branchlets cylindrical, the apex rounded.

V. **lamourouxii** (Turner) Weber-van Bosse. Plants tall, the erect branches to 16 cm. high, the branchlets alternate, subopposite, or few and widely scattered, pyriform or clavate; when branchlets are few the axis may be markedly compressed.

V. **macrophysa** (Kützing) Taylor. Erect axes to 3.0–6.5 cm. tall, the branchlets attached to them at intervals of 2–4 mm., obconical with the broad end hemispherical to slightly convex, 3–5 mm. diam.

V. **occidentalis** (J. Agardh) Børgesen. Erect axes 2–10 cm. tall, the branchlets not crowded, radially or distichously disposed, with stalks 1.0–2.5 mm. long, each abruptly expanded to a subspherical top 1.5–2.5 mm. diam.

V. **uvifera** (Turner) Weber-van Bosse. Erect axes 1.5–2.5 cm. tall, the branchlets crowded and imbricate, sharply ascending, the branchlet stalks 2–3 mm. long, gradually expanded to a rounded top 1.5–2.5 mm. diam., which is generally a little compressed tangentially with respect to the surface of the frond.

F. **condensata** (Kützing) Weber-van Bosse. Branchlets much crowded at the base of the erect axes, less numerous above.

Bermuda, Florida, Mexico, Bahamas, Caicos Isls., Cuba, Jamaica, Hispaniola, Puerto Rico, Virgin Isls., St. Barthélemy, St. Eustatius, Antigua, Guadeloupe, Martinique, Barbados, Grenadines, Grenada, British Honduras, Costa Rica, Canal Zone, Panama, Colombia, Netherlands Antilles, I. las Aves, Venezuela, Trinidad, Tobago, Brazil, I. Trinidade. The most ubiquitous plant of the genus, chiefly to be expected in shallow water. The species is common enough at a few decimeters below low-tide level, but has been reported from depths as great as 55 m. On sunny surf-beaten reefs and rocky shores the plants characteristically grow in compact colonies, the axes with short rounded branchlets. Dissimilar forms do grow, however, under seemingly identical conditions. In deep shady clefts one more often meets looser forms with longer erect axes, and they extend well up into the intertidal zone. In grottos and deep inland pools connected by limited, often underground channels the erect axes may be very long-attenuate, and the branchlets clavate-cylindrical. From deep water of 90–110 m. the v. *macrophysa* was dredged off the Florida coast, though similar forms have been reported from shallower stations.

REFERENCES: Howe 1907, 1920, Collins 1909b, Børgesen 1913–20, 1924, 1932, Collins and Hervey 1917, Taylor 1928, 1933, 1935, 1941b, 1942, 1943, 1954a, Taylor and Arndt 1929; *C. sedoides*, Mazé and Schramm 1870–77, Murray 1889 *p. p.* For the v. *chemnitzia, C. chemnitzia*, Piccone 1886, 1889; *Chauvinia chemnitzia*, Martens 1870. For *Caulerpa racemosa* v. *clavifera*, Vickers 1905, 1908, ?Sluiter 1908 (as f. *intermedia*), Børgesen 1913–20 (including f. *reducta*), Collins and Hervey 1917, Schmidt 1923, Taylor 1928, 1929b, 1930a, 1935, 1940, 1942, Taylor and Arndt 1929; *C. clavifera*, Harvey 1858, 1861, ?Mazé and Schramm 1870–77 (including v. *condensata*), Dickie 1874a, b, Hemsley 1884, Hauck 1888, Murray 1889 (including v. *condensata*), Möbius 1890, Howe 1907, 1920 (as f. *condensata abbreviata*), Schmidt 1924; *C. clavifera* v. *uvifera*, Mazé and Schramm 1870–77 *p. p.* For the v. *gracilis*, Collins 1909b, Taylor 1928. For the v. *laetevirens*, Howe 1903b, A. and E. S. Gepp 1905 (for f. *cylindracea*), Sluiter 1908, Collins 1909b, Børgesen 1913–20, Collins and Hervey 1917, Taylor 1928, 1929b, 1930a, 1942, Taylor and Arndt 1929. For the v. *lamourouxii*, Børgesen 1913–20, Collins, 1909b; *Chauvinia clavifera* v. *lamourouxii*, Martens 1890. For *Caulerpa racemosa* v. *macrophysa*, Taylor 1928; *C. racemosa* v.

clavifera f. *macrophysa*, Collins 1901,1909b. For the v. *occidentalis*, Børgesen 1913–20, Collins and Hervey 1917, Howe 1918a, b, Schmidt 1924, Taylor 1928, 1930a, Questel 1942, Joly 1953c; *C. chemnitzia* v. *occidentalis*, Murray 1889. For the v. *uvifera*, A. and E. S. Gepp 1905b, Sluiter 1908, Børgesen 1913–20, Collins and Hervey 1917, Schmidt 1923, 1924, Taylor 1928, 1929b, 1930a, 1933, 1942, Taylor and Arndt 1929, Quirós C. 1948, Joly 1957; *C. clavifera* v. *uvifera*, Mazé and Schramm 1870–77 *p. p.?*, Murray 1889, Vickers 1905; *C. uvifera*, Howe 1907.

Caulerpa peltata Lamouroux Pl. 17, fig. 2; pl. 18, fig. 1

Plants small, generally 1–3 dm. diam., the stolons freely forked, giving off rhizoid-bearing branches below and foliar branches above; erect branches 5–50 mm. tall, bearing one to several peltate branchlets consisting of a rather slender pedicel 1–2 mm. long, ending in a disk 1–2 mm. thick, 1.5–8.0 mm. broad.

F. **imbricata** (Kjellman) Weber-van Bosse. Erect axes with crowded branchlets.

Bermuda, Florida, Mexico, Barbuda, Brazil. Chiefly a plant of shallow water, growing on rocks, but dredged to a depth of over 30 m.

This species frequently does not fully develop the disklike branchlets on some of the branches. In such cases the branchlets vary from cylindrical to clavate or stalked-turbinate, and the plants closely resemble similarly depauperate forms of *C. racemosa*.

REFERENCES: Collins 1909 (including f. *imbricata*), Collins and Hervey 1917, Taylor 1928, Humm 1952b; *Chauvinia peltata*, Martens 1870, 1871.

Caulerpa microphysa (Weber-van Bosse) J. Feldmann
Pl. 17, fig. 5; pl. 18, fig. 6

Plants with slender creeping rhizomes bearing descending rhizoidal branches, which may attach to fragments of shell or may penetrate into the sandy sea bottom; erect branches 3–30 mm. tall, the branchlets crowded upon them, usually standing sharply at right angles to the axis, nearly sessile or on stalks to 0.5 mm. long, these generally with a definite constriction at the top, above which the branchlet is abruptly expanded and spherical, 0.9–2.4 mm. diam.

Bermuda, Florida, Jamaica, Puerto Rico, Nevis, St. Lucia, Netherlands Antilles, Venezuela. This delicate species is primarily a deep-water plant, having been dredged in water of depths from 9–110 m. However, it has also been found in shallower water attached to rocks in the shade. It may be related to *C. lentillifera* J. Ag., but it is a much smaller plant with far fewer vesicles which are not clearly in longitudinal series on the axis.

REFERENCES: Feldmann 1955; *C. racemosa* v. *microphysa,* Taylor 1928, 1940, 1942, Humm 1952; *C. racemosa* v. *clavifera* f. *microphysa,* Weber-van Bosse 1898, Vickers 1905.

UNCERTAIN RECORDS

Caulerpa coccinia Harvey

Florida. The source of this name has not been determined. Hooper 1850.

Caulerpa cylindrica Sonder

Barbados. The type from Australia. De Toni 1:474; Dickie 1874a, Murray 1889.

Caulerpa sedoides (R. Brown) C. Agardh

St. Eustatius, Grenada, Brazil. The type from Australia. De Toni 1:480; Murray 1889 *p. p.*, Sluiter 1908; *Chauvinia sedoides,* Martens 1870.

Caulerpa selago (Turner) C. Agardh

Grenada, Barbados, Brazil. The type from the Red Sea. De Toni 1:466; ?Murray 1889 *p. p.*, *Chauvinia selago,* Martens 1870.

CODIACEAE

Plants filamentous, the filaments branched, remaining separate or much more generally at maturity associated to form complexly organized coenocytic plants of characteristic forms, often with considerable specialization of parts of the filament system, but without crosswalls except to segregate reproductive organs, to repair damages, or in the simplest genera as fortuitous or irregular structures.

KEY TO GENERA

1. Plants persistently filamentous, not organizing massive thalli; filaments constricted above the forks .... BOODLEOPSIS, p. 157
1. Plants massive, organized into rhizoidal and foliar portions, and often with a compact stalk as well 2

2. Plants completely free of structural calcification (although often with attached or enmeshed calcareous sand) 3
2. Plants regularly calcified, at least in some parts 6

3. Erect plants, seldom cushion-shaped or decumbent; stalk when present not different in character from the branches above; structural filaments never torulose, the tips distinctly enlarged CODIUM, p. 184
3. Erect; stalk if present distinct from the blade; if absent the structural filaments more or less torulose and constricted above the forks 4

4. Foliar portion showing distinctive cortical filaments CLADOCEPHALUS, p. 163
4. Blades without differentiation of a cortex 5

5. Blades or subterete foliar lobes soft; without specialized structural filaments .... AVRAINVILLEA, p. 158
5. Blades thin, commonly porous; the ordinary filaments with adventitious diverticula which may show crosswalls and hapteral tips .... RHIPILIA, p. 162

6. Plants distinctly segmented .... HALIMEDA, p. 174
6. Plants without flexible uncalcified joints 7

7. Foliar portion consisting of a tuft of unconsolidated filaments PENICILLUS, p. 169
7. Foliar portion of one or more blades 8

8. Blades numerous, imbricate .... RHIPOCEPHALUS, p. 173
8. Blades single, or few, sometimes proliferous .... UDOTEA, p. 164

Boodleopsis A. and E. S. Gepp, 1911

Plants filamentous, forming uncalcified green cushions, the filaments loosely interwoven or marginally free, ditrichotomously or irregularly widely and repeatedly branched, the lower filaments thicker-walled and larger, occasionally emitting rhizoids, the upper more closely branched, with the branches sharply constricted at the base.

Boodleopsis pusilla (Collins) Taylor, Joly, and Bernatowicz
Pl. 27, figs. 1, 2

Plants green, forming a continuous turf, filamentous, matted, the filaments dichotomously or somewhat irregularly branched; the lower filaments to 90 μ diam., more or less buried in the mud, often colorless, of irregular contour, not sharply constricted, though frequently showing crosswalls, often attenuate into rhizoids; upper filaments more closely branched, 23–45 μ diam., strongly constricted at the base of each branch and frequently elsewhere; coralloid masses of short, unconstricted branches often found on the upper filaments; pyriform to subspherical sporangium-like structures sometimes present, 60–153 μ diam., 83–207 μ long; akinetes of very irregular form produced under laboratory conditions.

Bermuda, Bahamas, Cuba, Jamaica, Puerto Rico, Guadeloupe, Brazil. Forming green cushions on rocks and mud, and wide mats under mangroves near high-tide line.

REFERENCES: Taylor, Joly and Bernatowicz 1953; *Dichotomosiphon pusillus*, Collins 1909b, Collins and Hervey 1917, Howe 1918a, 1920, Feldmann and Lami 1936. There is a good possibility that this plant is a protonemal stage of some more elaborate member of the Codiaceae.

Avrainvillea Decaisne, 1842

Plants with a rather firm subterranean basal mass of rhizoids and enmeshed sand grains, the flabella sessile or stalked, green or brownish, often darkening when dried, the foliar portion terete to compressed or bladelike; uncalcified structural filaments often yellow or brown in color, dichotomously branched, constricted above the forks, often torulose or even moniliform; filaments of the cortical region not clearly differentiated; reproduction uncertain, perhaps by aplanosporangia terminal on short superficial filaments.

It is not unusual to find plants of *A. longicaulis*, *A. nigricans*, and perhaps other species with part or all of the blade exceedingly loose and open in texture, or even in lesser or greater part of completely unconsolidated, plumose filaments. This probably is the nature of the plant called *A. comosa* by Børgesen (1908).

KEY TO SPECIES

1. Plants digitately lobed above .... A. rawsoni, p. 159
1. Plants at maturity forming a plane blade above 2
2. Blade smooth, usually thin and relatively firm 3
2. Blade velvety, spongy, or hairy 6

3. Filaments of the blade of a generally uniform diameter of 20–30 μ, sometimes slightly torulose .... A. elliottii, p. 162
3. Filaments of the blade more slender at the surface than in the deeper layers 4

4. Filaments 30–40 μ diam. within, 14–17 μ near the surface, but the clavate filament tips to 17–27 μ diam. .... A. geppii, p. 161
4. Filaments more slender and the tips not clavate 5

5. Blades to 15 cm. broad, deeply cordate at the base; surface filaments markedly moniliform, 20–30 μ diam. within, 8–13 μ diam. at the surface .... A. asarifolia, p. 161
5. Blades to 7 cm. broad, shallowly cordate to cuneate at the base; filaments barely moniliform, or smooth, to 35 μ diam. within, 6–24 μ diam. at the surface .... A. levis, p. 162

6. Blades cuneate or oblong; interior filaments smooth but with basal constrictions in the branches, 28–70 μ diam., toward the surface tapering to 20 μ .... A. longicaulis, p. 160
6. Blades cuneate to suborbicular; interior filaments moniliform, 50–70 μ diam., tapering to 30 μ at the surface .. A. nigricans, p. 160

Avrainvillea rawsoni (Dickie) Howe Pl. 19, fig. 3

Plants caespitose, dull green, becoming brown or black in drying, arising from a massive base of rhizoids and sand which may become 2–3 dm. diam.; the foliar portions without stalks, consisting of more or less crowded, digitate, irregularly conical to clavate simple or cleft lobes 4–12 cm. long, 0.5–4.0 cm. diam., but these sometimes fused laterally and appearing rather flat; texture spongy but firm, the surface spongy to somewhat hairy; structural filaments smooth or submoniliform in the medulla, 28–68 μ diam., the walls thin; filaments toward the surface brown, moniliform, to 85–120 μ diam.

Bermuda, Bahamas, Cayman Isls., Jamaica, Hispaniola, Puerto Rico, Guadeloupe, Barbados, Panama. Plants of shallow water growing in the intertidal zone or just below low water, in sand or on sand-covered rocks.

REFERENCES: Howe 1907b, 1915, 1920, Taylor 1933, 1942, 1943, Hamel and Hamel-Joukov 1931; *A. nigricans*, Mazé and Schramm 1870–77, Murray 1889 *p. p.; Rhipilia rawsonii*, Dickie 1874a.

Avrainvillea longicaulis (Kützing) Murray and Boodle
Pl. 19, fig. 1

Plants 10–22 cm. tall; the stipe compressed, 3–10 cm. long, 5–10 mm. diam., sometimes gregarious from a rather massive base; the blade cuneate-rounded at the base, above round, obovate, or spatulate, about 6 cm. long, 5–7 cm. broad, moderately thick, not zonate, either smooth or hairy; structural filaments constricted at the base of each branch, smooth, 28–70 μ diam. in the medulla, somewhat less and sometimes slightly moniliform at the surface.

Bermuda, North Carolina, Bahamas, Salt Key Bank, Jamaica, Virgin Isls., Antigua, Guadeloupe, Barbados, Grenada, British Honduras, I. las Aves. Growing in the sand and mud of protected situations at a few decimeters depth below high-tide level, and thence to a depth of 30 m.

REFERENCES: Murray and Boodle 1889 (as to name only), Murray 1889, Collins 1901, Butler 1902, Vickers 1905, 1908, Howe 1907b, 1918a, 1920, Collins and Hervey 1917, Taylor 1935; *A. mazei,* Murray 1889, Børgesen 1913–20; *A. sordida* v. *longipes* and *Flabellaria fimbriata,* Mazé and Schramm 1870–77; *Rhipilia longicaulis,* Dickie 1874b, Hemsley 1884.

Avrainvillea nigricans Decaisne Pl. 19, fig. 2; pl. 25, figs. 11, 12

Plants to 1.5 dm. tall, dark brown or blackish in color, sessile or stalked, in which case the stalk 10–13 mm. diam., terete below, flattened above and transitional into the base of the blade; blade cuneate to rounded, somewhat irregularly lobed, to 6 cm. long, 8 cm. broad, rather softly leathery; structural filaments dull brown, notably moniliform, those of the medulla 50–70 μ diam., those of the surface about 30 μ; reputed sporangia superficial, 200–350 μ diam., 350–800 μ long.

F. **fulva** Howe. Stipe flatter and less distinct, blade narrower, thick and spongy, the filaments yellow or yellow-brown, coarser and more closely irregularly and widely branched than in typical plants.

Bermuda, Florida, Mexico, Bahamas, Caicos Isls., Cuba, Cayman Isls., Jamaica, Virgin Isls., Guadeloupe, Martinique, Barbados,

Brazil, I. Trinidade. Rather common plants of shallow water in the shelter of reefs, or of sandy shallows near mangroves, but also dredged to a depth of over 30 m.

REFERENCES: Murray *p.p.?*, 1889, Collins 1901, Vickers 1905, 1908, Howe 1907b, 1918a, b, 1920, Børgesen 1908, 1913–20, Collins and Hervey 1917, Taylor 1928, 1954a, Hamel and Hamel-Joukov 1931, Joly 1953c; *A. sordida* v. *longipes*, Mazé and Schramm 1870–77 *p. p.*

Avrainvillea geppii Børgesen

Plants brownish green, about 12 cm. tall; the stalk about 3–5 mm. thick, tapering and flattened distally; blade with a truncate-rounded base, broadly oblong, 5.5 cm. long, 8 cm. broad, the margin lacerate or slightly lobed, of loose texture, thin and indistinctly zonate; filaments of the medulla of the blade smooth to decidedly moniliform, 30–40 μ diam., tapering near the surface to 14–17 μ diam., but the filament tips at the surface subclavate, 17–27 μ diam.

Virgin Isls. Only certainly known in the original collection, dredged from a depth of 16 m.

REFERENCE: Børgesen 1913–20.

Avrainvillea asarifolia Børgesen

Plants dull grayish green, about 1–3 dm. tall; stipe terete below, compressed above, 3–23 cm. long, 6–8 mm. diam.; blade cuneate to cordate at the base, oblong reniform above, to about 8–10 cm. long, 15 cm. broad, entire or slightly lobed, blade texture rather firm, the surface smooth and usually zonate; filaments of the medulla of the blade smooth or a little torulose, 20–30 μ diam., in the cortical region more closely branched and moniliform, 8–13 μ diam., with slightly thickened walls.

Florida, Jamaica, Virgin Isls. Apparently a rather deep-water species, this plant has been dredged from about 4.6 to 90.0 m. depth. In character it is very close to the earlier description of *A. levis.*

REFERENCES: Børgesen 1908, 1913–20, Taylor 1928.

Avrainvillea elliottii A. and E. S. Gepp

Plants perhaps brownish, 6–13 cm. tall; the stout, slightly compressed stalks 1–4 cm. long, 0.6–1.0 cm. diam.; the blade truncate below, broadly expanded above, 3.5–5.0 cm. long, 5–10 cm. wide, the distal margin lobed, eroded or split; of moderate thickness, zonate; structural filaments smooth or slightly moniliform, 20–30 μ diam., little tapered.

Grenada.

REFERENCES: Gepp 1911, Collins 1912.

Avrainvillea levis Howe

Plants olive green to ashy brown in color, to about 11 cm. tall, the stalk terete, or compressed above, 0.5–5.0 cm. long, 5–6 mm. diam.; blade cuneiform-obovate to reniform with a cordate base, to 6 cm. long, 7 cm. broad, entire or somewhat lobed, compact of texture, smooth, generally zonate; filaments of the medulla of the blade to 35 μ diam., smooth or slightly moniliform, toward the surface reduced to 6–24 μ diam. and closely interwoven.

Florida, Bahamas, Caicos Isls., Cuba, Jamaica, Hispaniola, Grenada. Plants usually collected in shallow water, but dredged to about 90 m.

REFERENCES: Howe 1905b, 1907b, 1918b, 1920, Taylor 1928, 1943; *A. sordida,* Mazé and Schramm 1870–77 *p. p.*, Murray and Boodle 1889 *p. p.*

Rhipilia Kützing, 1858

Plants erect, green, uncalcified, subsessile or stalked; stalk without a distinctive cortex; foliar portion peltate, funnel-shaped or fan-shaped, of laxly interwoven filaments, producing a porous blade without a cortex; primary dichotomous flabellar filaments commonly bearing pseudolateral branches at right angles which may end in hapteral connections with other filaments.

Rhipilia tomentosa Kützing   Pl. 22, fig. 3

Plants usually solitary, stipitate, the stalks 0.5–1.5 cm. long, 2–4 mm. diam., each bearing a cuneate or rounded flabellum which may be thin to moderately thick and spongy, the margin lobed, fringed, or erose; structural filaments of the blade 30–70 μ diam., the tips a little clavate, the frequent pseudolateral filaments 33–60 μ diam., 50–200 μ long, often showing crosswalls, occasionally branched, with 2–3 short stout aculei at the tips or on lateral spurs, or with disciform haptera.

F. **zonata** A. and E. S. Gepp. Stipe slender, blade round to reniform, very thin and sometimes translucently zonate.

Bahamas, Cuba, Puerto Rico, Virgin Isls., Antigua, Guadeloupe, Brazil. Plants seemingly restricted to moderately deep water in this area, having been dredged from about 30 m.

REFERENCES: A. and E. S. Gepp 1911, Børgesen 1913–20, Howe 1915, 1918b, 1920, Taylor 1930a; *Avrainvillea laetevirens*, Mazé and Schramm 1870–77; *Udotea tomentosa*, Murray 1889, Howe 1907b. For f. *zonata*, A. and E. S. Gepp 1911, Collins 1918.

Cladocephalus Howe, 1905

Plants uncalcified, erect, from a base of matted rhizoids, the stipes distinct, the blades simple or fimbriate; medullary filaments green, parallel, sparingly dichotomous, smooth; cortex distinct, of repeatedly divaricately dichotomous, intricate, slender, eventually chlorophyll-free filaments.

KEY TO SPECIES

1. Stalk topped by a broad, simple blade .... C. luteofuscus
1. Stalk topped by a brushlike capitulum of narrow segments C. scoparius

Cladocephalus luteofuscus (Crouan) Børgesen Pl. 25, figs. 8, 9

Plants to 10 cm. tall; stalk 2–4 mm. diam., seldom branched, flattened above, the blade with a cuneate-rounded or truncate base, transversely oval, 4.5 cm. long, 7.0 cm. broad, smooth, somewhat zonate; structural filaments of the stipe 50–80 μ diam., the ultimate divisions 4–10 μ diam., densely interwoven to form a distinct cortex.

Florida, Virgin Isls., St. Martin, Guadeloupe. Apparently a plant of deep water, having been dredged from depths of 14 to 72 m.

REFERENCES: Børgesen 1908, 1913–20, Taylor 1928, Feldmann and Lami 1937; *Flabellaria luteofusca,* Mazé and Schramm 1870–77; *Udotea luteofusca,* Murray 1889, Howe 1907b.

Cladocephalus scoparius Howe Pl. 20, fig. 1

Plants dark green to blackish, drying yellowish, stalked; stalks 2–10 cm. long, 5–7 mm. diam., channeled, sometimes sparingly forked; flabellum linear to obovoid or more or less flattened, 3–8 cm. long, consisting of a tuft of subterete or flattened segments which are about 2 mm. wide; structural filaments of the medulla of the stalk 30–75 μ diam., smooth, in the cortex 6–11 μ diam.

Bahamas. Very distinctive plants of quiet shallow pools.

REFERENCES: Howe, 1905b, 1920.

Udotea Lamouroux, 1812

Plants erect from a tangled rhizoidal base, substantially calcified, the simple stalks each bearing a terminal fanlike or funnel-shaped blade; stalk and blade both composed of dichotomously branched filaments constricted above the forks and occasionally elsewhere, those of the stalk and in some cases those of the blade as well bearing lateral branches or simpler projections which constitute a distinctive cortical layer; reproduction not certainly known, but protonemal stages reported for some species.

KEY TO SPECIES

1. Flabellum a flat or proliferous blade 3
1. Flabellum funnel-shaped, sometimes strongly asymmetrical, rarely flat 2

2. Blade spongy, the base slightly decurrent; filaments of the blade interwoven, not firmly cemented together, 50–85 μ diam. U. sublittoralis, p. 165
2. Blade papery, the base truncate; filaments of the blade 30–135 μ diam., subparallel, definitely cemented together U. cyathiformis, p. 166

3. Blade composed of filaments without specialized lateral projections, and so completely ecorticate ..... U. conglutinata, p. 165

3. Blade composed of filaments bearing close-placed lateral projections which form more or less of a cortex 4

4. Corticating filaments forming a continuous cortical layer under a dense, rather smooth, but flexible calcareous coating 5
4. Cortex of projections which are not usually closely approximated; surface often minutely spongiose; less calcified 6

5. Branchlets forming the cortex of the flabellum capitate U. occidentalis, p. 169
5. Branchlets forming the cortex of the flabellum truncate but not capitate .... U. flabellum, p. 168

6. Projections on the superficial filaments in partial whorls U. verticillosa, p. 168
6. Projections on the superficial filaments in longitudinal rows 7

7. Projections simple, or digitately branched, in a single row on each superficial filament .... U. spinulosa, p. 167
7. Projections wartlike, truncate, or widely branched, in 2–4 rows U. wilsoni, p. 167

Udotea sublittoralis Taylor Pl. 22, fig. 6

Plants lightly calcified, hardly zonate, to 5 cm. tall; the terete stalks 2–10 mm. long, 0.75–2.0 mm. diam., slightly rough, the corticating filaments irregularly branched, the apices irregular, tapering or bent and sometimes truncate; blades funnel-shaped, occasionally asymmetrical, rarely flat, briefly tapering into the top of the stalk, 2.5–4.0 cm. long, 2–3 cm. diam., decidedly thick and of a spongy texture; the structural filaments widely dichotomous and interwoven, individually encrusted, 50–85 μ diam., without corticating branchlets or projections.

Florida, Bahamas, Guadeloupe. In moderately shallow water only, on sand in the shelter of reefs or islands.

REFERENCE: Taylor 1928.

Udotea conglutinata (Ellis and Solander) Lamouroux
Pl. 20, fig. 3; pl. 25, fig. 5

Plants fan-shaped, to 10 cm. tall, from a base of rhizoidal strands; the stalks to 2.5 cm. long, terete and to 2 mm. diam. below, flattened and expanded above, smooth and firm of surface, the cortex of widely branched peripheral filaments with slender, tapering, thick-walled ultimate segments; blades deeply cordate at maturity, rounded or somewhat reniform, to 7.5 cm. long, 11 cm. broad, whitish, slightly zonate, thin and firm, finely longitudinally fibrillose, the filaments 25–60 μ diam., without corticating projections.

Bermuda, Florida, Mexico, Bahamas, Cuba, Jamaica, Hispaniola, Virgin Isls., St. Martin, St. Eustatius, Guadeloupe, Grenada, Old Providence I., Panama, Colombia, Brazil. Apparently seldom found in very shallow water, but dredged to depths of 57 m.

REFERENCES: Harvey 1858, Dickie 1874b, Hemsley 1884, Murray 1889, Collins 1901, Sluiter 1908, Howe 1909b, 1918a, b, 1920, Børgesen 1913–20, 1924, Collins and Hervey 1917, Taylor 1928, 1936, 1939b, 1942, 1954a, Sanchez A. 1930, Humm 1952b; *Flabellaria conglutinata,* Mazé and Schramm 1870–77.

Udotea cyathiformis Decaisne Pl. 22, fig. 4

Plants funnel-shaped, whitish green, to 6–13 cm. tall; the stalks terete, to 1.5 cm. long, 2–4 mm. diam., smooth of surface, the corticating filaments closely packed, the end segments truncate, not greatly thickened; blades cyathiform, symmetrical or sometimes open on one side, 3.0–4.5(–10) cm. long, 3–5 cm. diam., attaching to the stalk directly from a small flattened or even concave base; texture thin, papery, but a little stiff, readily split, the surface smooth; composed of straight, nearly parallel filaments running radially to the margin, 30–50(–135) μ diam., without corticating filaments or projections.

Bermuda, North Carolina, Florida, Bahamas, Cuba, Jamaica, Virgin Isls., Guadeloupe, Barbados, Panama, Brazil, I. Trinidade. Seldom secured from very shallow water; dredged off a sandy bottom from quite moderate depths to below 90 m. In a considerable Jamaican colony of small plants growing in shallow water, some had fairly well-cemented blades, but in others the filaments, which reached a diameter of 190 μ, were mostly laterally free.

REFERENCES: Dickie 1874b, Murray 1889, Howe 1909b, 1918b, 1920, Børgesen 1913–20, Hoyt 1920, Taylor 1928, 1930a, Feldmann and Lami 1937, Joly 1953c; *Udotea conglutinata,* Vickers 1905, 1908.

Udotea spinulosa Howe Pl. 20, fig. 2; pl. 25, figs. 6, 16, 17

Plants whitish, well calcified, hardly zonate, to 8 cm. tall; stalks terete below, 1–2 cm. tall, 2 mm. diam., smooth or somewhat velvety in appearance, the corticating filaments closely branched, the ultimate segments tapering, thickened; blades simple, obtuse to transverse at the insertion, obovate, 5–6 cm. long, 4–8 cm. broad, the distal margin very thin and often laciniate; medullary filaments of the blade radiately subparallel, 46–84 μ diam., evident near the distal blade margin, gradually covered by the corticating projections so that the surface becomes spongiose and gradually more continuously covered toward the base of the blade; corticating projections 55–160 μ long, set in a single row on the superficial filaments, these processes crowned with 2–8 acuminate prongs.

Florida, Bahamas, Cuba, Virgin Isls. Described from material growing in shallow water, but chiefly a plant of moderately deep water, growing in sand, dredged from 9 to over 90 m. depth.

REFERENCES: Howe 1909b, 1920, Børgesen 1913–20, Taylor 1928, 1954a.

Udotea wilsoni Gepp and Howe Pl. 25, fig. 18

Plants to 13 cm. tall, green and lightly calcified, sometimes distinctly zonate; stalks 1.0–4.0 cm. long, simple or occasionally branched, terete and about 1 mm. diam. at the base, flat or channeled and to 3.4 mm. broad above; the stipe cortex of lateral, once-or-twice forked filaments with obtuse tips; blades single on each stalk or stalk branch, or 2–3 formed face to face attaching near each other toward the top of a stalk, 4–9 cm. long, 4–11 cm. wide, rounded triangular, the lateral margins tapering into the stalk, the distal margin rounded, crenate-lobed, very thin and often laciniate; structural filaments of the blade 40–50 μ diam., radially approximated, evident near the margin, the surface minutely granulose or spongy, the filaments near the base obscured by the development of corticating projections, these usually in 2 double rows directed toward the surfaces of the blade, or one pair of rows on the thicker parts, these being briefly pedicellate; projections 25–120 μ long, simple or forked, and very obtuse.

Bermuda, Florida, Bahamas, Anguila Isls., Cuba. A shallow-water plant, growing in sand or sandy mud among sea grasses.

REFERENCES: Howe 1918b, 1920, Taylor 1928.

Udotea verticillosa A. and E. S. Gepp

Plants to about 10 cm. tall, well calcified, grayish green, zonate, stalks 1.5–4.0 cm. tall, terete below and 1 mm. diam., a little flattened above, the corticating filaments 2–4 times dichotomous, the terminal segments sharply tapering; blades cuneate-subcordate at the base, broadly rounded, deeply lobed, to about 10 cm. wide, proliferous on the margin; structural filaments of the blade subparallel, sparingly dichotomous, not markedly constricted, progressively developing corticating appendages in an irregular fashion or in a clear arrangement in semiverticils on the superficial filaments, these appendages pedicellate, 30–180 μ long, the ends closely forked and acute.

Virgin Isls. Known only as dredged from depths of 9–40 m.

REFERENCE: Børgesen 1913–20.

Udotea flabellum (Ellis and Solander) Lamouroux
Pl. 20, figs. 4, 5; pl. 25, fig. 3

Plants broadly fan-shaped, gray-green and strongly zonate, flexible, although extensively calcified, to over 2 dm. tall above the dense basal mass of rhizoids and sand; stalks 1–2 cm. long, to 1 cm. wide and somewhat flattened above, terete and more slender below, with a smooth firm surface, the corticating filaments densely branched, with truncate, thickened, closely approximated ends; blades 0.5–2.1 dm. long, to 0.5–3.0 dm. broad, cordate below, rounded above, often reniform, often proliferous, the free margin irregularly crenate and the surface commonly becoming plicate-ribbed, sometimes cleft to several very narrow cuneate segments; surface continuous, smooth under a lens, corticated in the same manner as the stalk but the corticating branchlets shorter.

Bermuda, North Carolina, Florida, Bahamas, Caicos Isls., Salt Key Bank, Cuba, Cayman Isls., Jamaica, Hispaniola, Puerto Rico, Virgin Isls., St. Barthélemy, Guadeloupe, Martinique, Grenada,

British Honduras, Panama, Colombia, Brazil, I. Trinidade. Common plants in shallow water, growing in the sand, often found among sea grasses, but also dredged to a depth of more than 70 m.

REFERENCES: Harvey 1858, Børgesen 1913–20, 1924, Collins and Hervey 1917, Howe 1918a, b, 1920, Schmidt 1924, Taylor 1928, 1933, 1935, 1936, 1940, 1942, 1943, 1954a, Taylor and Arndt 1929, Sanchez A. 1930, Hamel and Hamel-Joukov 1931, Joly 1950, 1953c, Williams 1951; *Eudotea flabellata*, Gardiner 1890; *U. flabellata*, Dickie 1874b, Hemsley 1884, Hauck 1888, Murray 1889, Collins 1901; *U. halimeda*, Kützing 1849, Dickie 1874b; *Flabellaria incrustata*, Mazé and Schramm 1870–77.

Udotea occidentalis A. and E. S. Gepp

Plants gray below, greener near the margin, zonate, to 8 cm. tall, well calcified, stalks terete, to 1 cm. long, 1.5 mm. thick, the corticating branch filaments closely forked near the tips, the ends crowded together, obtusely rounded; blades with a cuneate base, obovate above, more or less proliferous; interior filaments of the blade 30–45 μ diam., bearing capitate, closely forked, lateral branchlets, their ends lobed and with 6–20 low, rounded prominences which may project through the layer of calcification.

Florida, Virgin Isls., Brazil. Apparently a plant of moderately deep water, growing at depths of 9–40 m. on a sandy bottom.

REFERENCES: Børgesen 1913–20, Taylor 1931; *U. argentea*, Collins 1909b; *U. halimeda*?, Martens 1870, Dickie 1874a, Murray 1889.

Penicillus Lamarck, 1813

Plants erect, superficially calcified, arising from a tangled basal cluster of rhizoidal strands; stalked, the stalks simple, each terminating in a capitular tuft of free dichotomously branched filaments which are generally constricted above the forks and not infrequently elsewhere; filaments of the capitula bright green at the growing tips but generally whitened by calcification as they mature; filaments of the stalks longitudinally disposed in the medulla, the lateral branches from these radially disposed and of specialized form, constituting a cortex; reproduction not certainly known, but a protonemal stage reported for some species.

KEY TO SPECIES

1. Filaments of the capitulum branching at wide angles, considerably tangled .... P. pyriformis, p. 170
1. Filaments branching at narrow angles, often lying nearly parallel and not entangled 2
2. Filaments slender, less than 0.3 mm. diam.; capitulum as long as broad, or a little longer .... P. capitatus, p. 171
2. Filaments more than 0.3 mm. diam. 3
3. Filaments conspicuously calcified; capitulum whitish-green P. lamourouxii, p. 172
3. Filaments very lightly calcified at the base, but for the most part bright green .... P. dumetosus, p. 172

Penicillus pyriformis A. and E. S. Gepp
Pl. 21, figs. 3, 5; pl. 25, fig. 1

Plants to 12 cm. tall, rather substantially calcified; stalks terete, or compressed above, 3.0–3.5 cm. long, 5–7 mm. diam., hardly penetrating the capitula, superficially calcified but the surface appearing minutely spongy; corticating filaments of the stalk loosely branched, the terminal segments acutely tapering and thickened; capitular tuft gray-green, compact and rather firm, oval to pyriform, sometimes crateriform at the top, tapering somewhat into the stalk, 5–7 cm. long, 3.0–4.5 cm. diam.; filaments of the capitulum 2–3 cm. long, 150–250 μ diam., rather much entangled, relatively strongly calcified.

F. **explanatus** Børgesen. Stalk to 6 cm. long, capitulum funnel-shaped to almost flat, very lax, to 10 cm. diam.

Bermuda, Florida, Bahamas, Caicos Isls., Cuba, Jamaica, Hispaniola, Virgin Isls., Anguilla I., St. Eustatius, Guadeloupe, Panama. Plants of sandy, protected shoals and shallow bays; the form from deeper water, having been dredged to a depth of 40 m.

REFERENCES: A. and E. S. Gepp 1905a, Sluiter 1908, Børgesen 1913–20, Howe 1918a, b, 1920, Collins and Hervey 1917, Taylor 1928, 1942, 1943, Taylor and Arndt 1929. For the f. *explanatus*, Børgesen 1913–20, Collins 1918.

Penicillus capitatus Lamarck Pl. 21, fig. 2; pl. 25, fig. 4

Plants to 1.5 dm. tall, rather substantially calcified, the stalks 3–10 cm. long, 2.5–3.0 mm. diam. below, 4–7(–10) mm. diam. at the summit, penetrating a little into the capitula, with a smooth, firm calcified surface; cortical filaments of the stalk closely branched, truncate and thickened at the slightly capitate tips; capitulum oblate-spherical to somewhat pyriform and then tapering a little into the stalk, rather compact, the capitular filaments mostly 2–3 cm. long, 125–200 μ diam.

F. **elongatus** (Decaisne) Gepp. Stalk to 15 cm. long, penetrating more than halfway through the capitulum, which is narrow and oblong in shape, with filaments reaching 300 μ diam.

F. **laxus** Børgesen. Stalk to 10 cm. or more in length, capitulum lax, loosely constituted of lightly calcified filaments seldom over 150 μ diam.

Bermuda, Florida, Mexico, Bahamas, Caicos Isls., Cuba, Cayman Isls., Jamaica, Hispaniola, Puerto Rico, Virgin Isls., St. Martin, St. Barthélemy, St. Eustatius, Guadeloupe, Dominica, Martinique, Grenada, British Honduras, Old Providence I., Canal Zone, Panama, Colombia, Netherlands Antilles, I. las Aves, I. Trinidade. In its range a common and characteristic plant of warm, muddy or sandy bays, where it may form considerable colonies or grow intermixed with sea grasses at a depth of 1–10 dm. It also grows well in more open shallows if protected by reefs. It is seldom met with in dredge hauls beyond 4–5 m., though reported from a depth of 40 m. In inland salt lakes of reduced salinity one usually finds the f. *elongatus,* while the f. *laxus* is usual in dredged material, though both forms may appear elsewhere.

REFERENCES: Harvey 1858, Mazé and Schramm 1870–77, Farlow 1871, Rein 1873, Hemsley 1884, Hauck 1888, Murray 1889, Collins 1901, Sluiter 1908, Børgesen 1913–20, Collins and Hervey 1917, Howe 1918a, b, 1920, Taylor 1928, 1933, 1935, 1936, 1939b, 1940, 1942, 1943, 1954a, Sanchez A. 1930, Hamel and Hamel-Joukov 1931, Joly 1953c. For the f. *elongatus,* Collins and Hervey 1917; *P. elongatus,* Mazé and Schramm 1870–77, Murray 1889, Sluiter 1908. For the f. *laxus,* Børgesen 1913–20, Collins and Hervey 1917.

Penicillus lamourouxii Decaisne Pl. 21, fig. 1; pl. 25, fig. 2

Plants to about 7 cm. tall, throughout conspicuously calcified; stalks 1–4 cm. long, soft, often hollow and compressed, not penetrating into the capitula, the surface rather smooth; the corticating filaments of the stalk sharply inflated above the basal constriction, forming here a considerable sac, above which they are closely branched, with truncate but not especially thickened tips; the rather loose capitulum to 5 cm. long, 2–4, seldom 5, cm. diam., oval to round, the filaments 300–500 μ diam.

V. **gracilis** A. and E. S. Gepp. Capitulum larger and denser, much more compactly filamentous, the filaments mostly 300–400 μ diam., little entangled.

Florida, Mexico, Bahamas, Caicos Isls., Cuba, Jamaica, Puerto Rico, Virgin Isls., St. Eustatius, Guadeloupe, Panama. Plants of protected sandy shoals and dredged to a depth of 73 m.

REFERENCES: Mazé and Schramm 1870–77, Murray 1889, A. and E. S. Gepp 1905a (including v. *gracilis*), Sluiter 1908, Børgesen 1913–20, Collins 1909b (including v. *gracilis*), Howe 1920, Taylor 1928, 1941b, 1942, 1954a.

Penicillus dumetosus (Lamouroux) Blainville
Pl. 21, fig. 4; pl. 25, fig. 15

Plants large, to 1.5 dm. tall or even taller, the stalks 2.0–8.0 cm. long, stout, 5–25 mm. diam. at the top, tapering a little toward the base, often compressed, the surface lightly calcified, soft and minutely spongy; cortical filaments of the stalk loosely branched, the terminal segments acutely tapering, thick-walled; the capitula soft, bright green, drying brownish, rounded or obovoid, tapering a little onto the stalks, to 10–15 cm. diam.; capitular filaments very lightly calcified only near the base of the tuft, 3.5–8.0 cm. long, loosely spreading, 400–800 μ diam.

F. **expansus** Børgesen. Stalk 2–4 cm. long; capitulum loose, 14–17 cm. diam.

Bermuda, Florida, Bahamas, Cuba, Cayman Isls., Jamaica, Hispaniola, Puerto Rico, Virgin Isls., Anguilla I., St. Barthélemy, Guadeloupe, Grenada, British Honduras, Old Providence I., Colombia. Plants of shallow, well-protected situations, growing in sand, but probably not adapted to muddy lagoons; dredged to 3.7 m. depth, but not reported from deep water except as the f. *expansus*, which came from about 30 m.

REFERENCES: Harvey 1858, Dickie 1874b, Hemsley 1884, Hauck 1888, Murray 1889, Collins 1901, 1909b, Børgesen 1913–20, Howe 1918b, 1920, Taylor 1928, 1935, 1939b, 1954a, Taylor and Arndt 1929; *P. clavatus* and *P. longiarticulatus*, Mazé and Schramm 1870–77. For the f. *expansus*, Børgesen 1913–20, Collins 1918.

Rhipocephalus Kützing, 1843

Plants erect, calcified, from a loose basal mass of rhizoids and sand; stipitate, the simple stalks stout and naked below, somewhat produced above, each bearing a capitulum of numerous small imbricate cuneate blades, which are monostromatic, composed of laterally approximated branched filaments constricted above the forks.

KEY TO SPECIES

1. Filaments of the young capitular blades quickly separating .... R. oblongus
1. Filaments of the capitular tuft permanently associated into cuneate blades .... R. phoenix

Rhipocephalus oblongus (Decaisne) Kützing

Pl. 22, fig. 1; pl. 25, fig. 7

Stalks 4.0–6.5 cm. long, to 6 mm. diam., subterete, the surface somewhat smooth but porous, continued into the terminal tuft for 0.75 of its length; medullary filaments of the stalk bearing lateral appendages 5–7 times dichotomously divided and terminating in short, obtuse digitate tips without especially thick end walls; capitula of green or bleached, lightly calcified filaments, to 5.5 cm. long, 1–3 cm. diam., the tip crateriform with growth initials at the pit bottom; the filaments 200–350 μ diam. at the base, 160–200 μ diam. above, below closely branched, above less so, when young associated laterally (but not firmly cemented) into small blades or flabellula.

Florida, Bahamas, Cuba. Plants of shallow water, growing on a sandy or muddy bottom. Readily confused when dried or in poor condition with species of Penicillus, but the thin end walls of the stipe cortex filaments will help distinguish it when the flabellula are obscured.

REFERENCES: Howe 1918b, 1920, Taylor 1928.

Rhipocephalus phoenix (Ellis and Solander) Kützing

Pl. 22, figs. 2, 5

Plants 7–12 cm. tall, dull green, lightly calcified above; stalks about 2–10 cm. long, to 3–8 mm. diam., terete, smooth and compactly calcified, the corticating filaments ultimately radially disposed, with somewhat generally thickened walls; capitular tuft about 3 cm. diam., 5 cm. long, the flabellula composing it retaining their identity until, near the base of the tuft, they became shredded; flabellar filaments tapering from 210–240 μ diam. at the short basal segments to 50–100 μ diam. near the tips.

F. **brevifolius** A. and E. S. Gepp. Capitulum cylindrical or elongate-conical, about 1.5 cm. diam., 4 cm. long, the flabellula about 6–8 mm. long, rather erect and closely imbricate.

F. **longifolius** A. and E. S. Gepp. Capitulum 3–4 cm. long and about as broad, the flabellula 10–15 mm. long, rather loosely divergent.

Florida, Bahamas, Cuba, Jamaica, Hispaniola, Puerto Rico, Guadeloupe, British Honduras, Old Providence I., Panama, Colombia. Plants of shallow water, often growing among marine phanerogams, but also dredged to 73 m. depth.

REFERENCES: Mazé and Schramm 1870–77, Farlow 1871, Murray 1889, Collins 1901, 1909b (including f. *longifolius*), A. and E. S. Gepp 1905a (including both forms), Howe 1918b, 1920, Børgesen 1924, Taylor 1928 (including f. *brevifolius*), 1933, 1935, 1939b, 1942, 1954a; *R. phoenix* v. *elatior,* Mazé and Schramm 1870–77, Murray 1889; *Penicillus phoenix,* Harvey 1858.

Halimeda Lamouroux, 1812 [1]

Plants erect from a fibrous holdfast, branched, with flexible joints alternating with calcified segments; segments moniliform, cylindrical

[1] The account of this genus has been prepared from a monograph written by Dr. Llewellya W. Hillis, which had not been published at the time this portion was sent to the printer.

or discoid, simple or lobed; structure filamentous, the coenocytic filaments closely parallel and indurated in the joints, loosely branched in the medulla of the segments and there laterally bearing special fascicles of branchlets the terminal cells of which are approximated to form a continuous cortex; filaments sometimes closed by wall thickenings at forkings and at the bases of the gametophores; reproduction by swarmers produced in large, dark green gametangia clustered on branched stalks borne on the surface of the calcified segments.

KEY TO SPECIES

1. Branches originating in a single plane, becoming displaced with age; a single basal attachment for each plant usually evident 2
1. Branches originating at random, not in one plane; with age, plants becoming attached at many places, the original holdfast becoming obscure .... H. opuntia, p. 176

2. Surface utricles nearly plane on the outer face, not over 85 μ diam.; the segment surface smooth 3
2. Surface utricles to about 50 μ diam.; each with a central spine, the segment surface thereby microscopically roughened H. scabra, p. 180
2. Surface utricles over 150 μ diam.; collapsing when dried, giving the segments a pitted appearance .... H. favulosa, p. 183

3. A small and rare species; segments obovoid, pyriform, or globose, smoothly calcified; branches not clearly originating in one plane .... H. lacrimosa, p. 179
3. Moderate to large species commonly over 5 cm. in height 4

4. Segments cylindrical; becoming forked in the lower parts where bearing branches .... H. monile, p. 182
4. Segments at least in great part flat 5

5. Segments essentially entire 6
5. Segments irregular to crenate or even deeply lobed 7

6. Base of the plant showing one or more subterete to narrowed stalklike segments; upper segments to 1.5 cm. diam.; well calcified; surface utricles when seen in section clearly in lateral contact for only 0.04–0.10 of their length ...... H. tuna, p. 178
6. Base of the plant without a stalk, or at most one segment a little differentiated; upper segments to 2–4 cm. diam.; little calcified; surface utricles in lateral contact for 0.20–0.66 of their length .... H. discoidea, p. 179

7. Segments distinctly ribbed, generally trilobed above
H. incrassata, p. 181
7. Segments indistinctly ribbed or smooth 8

8. Filaments of the central strand cohering at the nodes, connecting by pits or short tubes; surface utricles 27–45 μ diam., in lateral contact for 0.1–0.33 of their length ... H. simulans, p. 180
8. Filaments of the central strand fusing in pairs and above each node branching trichotomously .... H. gracilis, p. 177

Halimeda opuntia (Linnaeus) Lamouroux
Pl. 23, fig. 3; pl. 24, fig. 1

Plants whitish green, well calcified, of moderate height (about 1.0–2.5 dm.), spreading laterally to form large colonies attaching at various points but with no persisting primary base; densely branching in all planes, successive branches and successive segments often at right angles with each other; segments with a truncate lower margin or becoming slightly pedicellate, transversely oval to reniform or a little trilobed, 3 radiating ribs being distinctly visible on the surface; segments about 7–20 mm. wide, 4.5–12.0 mm. long; surface utricles 20–50(12–63) μ diam. in surface view, in section rarely 70 μ long, the side walls in contact for about 0.08 of their length; medullary filaments incompletely fused, generally in pairs, at the nodes, an opening developing at the point of contact, but each member of the pair continuing upward and dividing ditrichotomously independently of the other.

F. **cordata** (J. Agardh) Barton. Segments rounded, prolonged below into 2 well-marked auricles which overlap the next lower joint.

F. **minor** Vickers. Cushions small and dense, 3–4 cm. thick, successive segments often following in the same plane; lower segments triangular or triradiate with terete arms 2.5–3.0 mm. long, 2–3 mm. broad, or simple and cylindrical, but above the segments transversely somewhat oval, 4–5 mm. broad, 3.0–3.5 mm. long, truncate and pedicellate on the lower margin, slightly crenate on the upper; surface utricles 18–27 μ diam.

F. **triloba** (Decaisne) Barton. Cushions to 5 dm. or much more in thickness, upper segments about 10 mm. broad, 6 mm. long, transversely oval to reniform, cordate below, distinctly crenate above; lower segments about 10 mm. long and broad, deeply trilobed and 3–5-costate, the lobes terete.

Florida, Mexico, Bahamas, Caicos Isls., Anguila Isls., Cuba, Cayman Isls., Jamaica, Hispaniola, Puerto Rico, Virgin Isls., Tortola I., St. Barthélemy, Guadeloupe, Martinique, Barbados, Grenada, British Honduras, Old Providence I., Costa Rica, Canal Zone, Panama, Colombia, Netherlands Antilles, I. las Aves, Venezuela, Trinidad, Tobago, Brazil. A large species common in shallow water near low-tide line, but also encountered in dredge hauls to a depth of 55 m. (over 80 m. in Pacific stations). It grows on rocks, gravel, or sand in all except the most wave-swept situations. The f. *triloba* is the form most often encountered, except in relatively exposed places, where the typical plant prevails.

REFERENCES: Hooper 1850, Harvey 1858, Martens 1870, Mazé and Schramm 1870–77 *p. p.*, Farlow 1871, Dickie 1874b, Hemsley 1884, Zeller 1876, Hauck 1888, Möbius 1889, Murray 1889, Piccone 1889, Gardiner 1890, Collins 1901, Vickers 1905, 1908, Sluiter 1908 (including f. *cordata*), Howe 1903, 1909, 1918a, b, 1920, 1928, Børgesen 1913–20, 1924, Taylor 1928, 1929b, 1933, 1935, 1939b, 1940, 1942, 1943, 1954a, Taylor and Arndt 1929, Hamel and Hamel-Joukov 1931, Quirós C. 1948; *H. incrassata,* Mazé and Schramm 1870–77 *p. p.* For the f. *minor*, Vickers 1905, Taylor 1928. For the f. *triloba*, Børgesen 1913–20, Schmidt 1924, Taylor 1928, 1933, 1935, 1942, Taylor and Arndt 1929; *H. triloba* and *H. tridens,* Mazé and Schramm 1870–77 *p. p.*

Halimeda gracilis Harvey

Plants tall, lax, reaching 40 cm., more substantially calcified below than in the upper parts; segments subterete below, cuneate to subreniform above, not ribbed, the margins slightly undulate, 1.5–11.0 mm. broad, 0.5–9.0 mm. long, 1.0–2.0 mm. thick; cortical utricles 30–45(23–70) μ diam. in surface view, 50–90 μ long, lightly connected laterally for about 0.08 of their length; medullary filaments completely fused in pairs, sometimes in threes, at the nodes, not entangled, above the nodes branching trichotomously.

V. **opuntioides** Børgesen. Plants with larger, reniform, often crenate or even trilobed segments to 14 mm. broad, 9 mm. long, rather thick but fragile.

Jamaica, Puerto Rico, Virgin Isls., Barbados. Apparently a rather deep-water species, dredged to 60 m.

REFERENCES: Vickers 1908. For the v. *opuntioides,* Børgesen 1913–20.

Halimeda tuna (Ellis and Solander) Lamouroux Pl. 24, fig. 5

Plants about 1.0, seldom to 2.5, dm. tall, generally tufted, lightly calcified but dark green, branching in one plane; 1–2 of the lowermost segments thick, subterete, the others flat, cuneate to more generally round, transversely oval, or particularly reniform, 10–15 mm. wide, 7–11 μ long; when dried very slightly glossy, not ribbed, with entire margins; subcortical utricles in section turbinate to clavate, 30–110 μ diam.; cortical utricles 40–75(25–125) μ diam. in surface view, 60–130 μ long, the side walls in contact for 0.04–0.10 of their length; medullary filaments somewhat entangled, fusing 2–3 together at the nodes, above which they branch ditrichotomously.

F. **platydisca** (Decaisne) Barton. Less freely branched, more lax, less calcified and more flexible, the segments larger, to 2(–4) cm. diam., and rather more glossy.

Bermuda, Florida, Mexico, Bahamas, Caicos Isls., Anguila Isls., Cuba, Jamaica, Hispaniola, Puerto Rico, Virgin Isls., St. Martin, Saba Bank, Guadeloupe, Barbados, British Honduras, Panama, Colombia, Brazil. Common on rocks and reefs in shallow water; seldom dredged from deep water, but found to a depth of 32–80 m., especially the f. *platydisca.*

REFERENCES: Harvey 1858, 1861, Mazé and Schramm 1870–77, Dickie 1874a, b, 1884, Hemsley 1884, Hauck 1888, Murray 1889, Collins 1901, Vickers 1905, 1908, Sluiter 1908, Howe 1907b, 1909, 1918a, b, 1920, Børgesen 1913–20, 1924, Collins and Hervey 1917, Taylor 1928, 1930a, 1931, 1941b, 1942, 1945a, Taylor and Arndt 1929; *H. opuntia* and *H. platydisca,* Mazé and Schramm 1870–77 *p. p.*, Murray 1889 *p. p.* For the f. *platydisca,* Børgesen 1913–20, Taylor 1928, 1941b; *H. platydisca,* Mazé and Schramm 1870–77, *p. p.*

Halimeda discoidea Decaisne Pl. 24, fig. 2

Plants to 1.5–2.0 dm. tall, tufted or loose in habit, with a rather small rhizoidal mass, grayish green or whitish, sparingly branched, so lightly calcified as to adhere on drying, when their texture is almost papyraceous and the surface nitent; the basal segments subterete, those above flat, large, reaching 2–3 cm.[1] in width, cuneate when young or ill-developed, becoming transversely oval to reniform, entire, but sometimes with slightly truncated margins accentuated by heavy shrinkage during drying; subcortical utricles inflated and very large, to 135–260 μ diam.; surface utricles truncate on the ends, 37–85 μ diam., simple or fused laterally in pairs; medullary filaments more or less completely fused in twos or threes at the nodes and somewhat entangled.

V. **platyloba** Børgesen. With segments even larger and less calcified than in the typical plant, to 4 cm. broad.

Florida, Bahamas, Cuba, Jamaica, Puerto Rico, Virgin Isls., Guadeloupe, Martinique, Panama, Colombia, Venezuela, Brazil, I. Trinidade. This species is sublittoral, most often occurring at moderate depths, and has been dredged to 50 m., the v. *platyloba* in particular being a deep-water form.

REFERENCES: Howe 1907b, 1909, 1918b, 1920, Børgesen 1911, 1913–20 (including v. *platyloba*), Taylor 1928, 1942, Hamel and Hamel-Joukov 1931, Williams and Blomquist 1947, Joly 1953c.

Halimeda lacrimosa Howe Pl. 23, figs. 5, 6

Plants 2–5 cm. tall, fragile, lax, thinly but firmly calcified, gray-green, becoming whitish with age; without a distinct stalk, sparingly irregularly ditrichotomously branched, largely in a single plane, the lax branches sometimes attaching by rhizoids; segments obovoid, pyriform, or subglobose, 1–5 mm. broad and long; subcortical utricles clavate-capitate, 66–110 μ diam., each bearing 6–18 surface utricles; surface utricles obconical or somewhat flaring, 31–42 μ diam. in surface view, 40–110 μ long, the outer wall a little thickened, the lateral walls in contact for but 0.03–0.10 of their length but remaining attached on decalcification; medullary fila-

[1] The breadth of 40 cm. given by Børgesen in both his 1911 and 1913 discussions of this plant must surely be in error!

ments closely fusing 2–4 together at the nodes, these groups not entangled but sometimes slightly adherent.

Bahamas, Cuba. Growing from near low-water mark to a depth of 10–20 m.

REFERENCES: Howe 1909b, 1918b, 1920, Taylor 1954a.

Halimeda scabra Howe Pl. 25, figs. 10, 13, 14

Plants tufted and erect when to a height of 1 dm., or somewhat lax when to 2.5 dm. long, gray-green and moderately calcified, typically freely branched in one plane; segments flat, broadly ovate, straight or obtusely angled below, rounded and entire above, entirely dull of surface and harsh to the touch, 10–20 mm. broad, 7–13 mm. long; cortical utricles 27–50(–66) μ diam. in surface view with a prominent, often indurated central spine, laterally in but slight contact with each other and easily separated on decalcification; medullary filaments more or less fused in twos or threes at the nodes and often entangled; sporangia chiefly along the margins of the disks, pyriform, distichously arranged on simple or forked pedicels, 160–320 μ diam.

Florida, Bahamas, Caicos Isls., Salt Key Bank, Anguila Isls., Cuba. Plants of shallow water, perhaps not widespread but sometimes abundant, dredged to a depth of about 20 m.

Commonly mistaken for *H. tuna,* but with experience this species can be distinguished by its harsh texture, and under favorable conditions, with a good ×15 hand lens, the spines can be seen. When dry the extremely dull surface is quite characteristic, but its nature can easily be confirmed by decalcification and examination under a microscope.

REFERENCES: Howe 1905, 1907b, 1909, 1920, Taylor 1928, 1954a.

Halimeda simulans Howe Pl. 24, fig. 4

Plants to 1.5 dm. tall, green but well calcified, rather flabellate in appearance, freely branching in one plane above a very short stalk of 1–3 segments, the lower segments terete to cuneate or ovate, entire or somewhat trilobed, to 15 mm. wide, 10 mm. long, somewhat nitent, slightly ribbed; above more sparingly and elongate-branched, the segments broadly oval to reniform, the lower margin straight or concave, the upper entire or somewhat crenate; the upper segments 4–15 mm. broad, 2–11 mm. long; subcortical utricles in 2–3, rarely 4, series, the outermost turbinate to subglobose, 30–72 μ diam., the innermost obovoid to clavate, 41–110 μ diam.; surface utricles 27–45 (–60) μ diam. in surface view, 27–90 μ long, turbinate, firmly laterally attached, the walls in contact for 0.1–0.33 of their length; medullary filaments strongly coherent in a single group at the nodes, with thick and colored walls, all intercommunicating by open pits or short processes.

Bermuda, Florida, Bahamas, Caicos Isls., Cuba, Jamaica, Hispaniola, Puerto Rico, Virgin Isls., St. Barthélemy, Dominica, Martinique, British Honduras, Old Providence I., Panama, Colombia, Brazil. Generally a plant of moderately shallow water, growing in sand in sheltered places; occasionally found near low-tide level, but on the other hand dredged to a depth of 73 m.

REFERENCES: Howe 1907b, 1909, 1915, 1918a, b, 1920, Collins and Hervey 1917, Taylor 1928, 1935, 1939b, 1940, 1942, Taylor and Arndt 1929, Hamel and Hamel-Joukov 1931, Questel 1942, Williams and Blomquist 1947; *H. incrassata* v. *simulans*, Børgesen 1913–20.

Halimeda incrassata (Ellis) Lamouroux Pl. 23, figs. 1, 4

Plants erect, to 2.4 dm. tall from a massive base of rhizoids and sand, the upper parts dully green and moderately calcified, the lower segments forming an extensive, more heavily calcified stalk which may fork 2–3 times, the segments more or less fused together; above the free segments in the lower portions terete and 8.0 mm. diam., or more often in the distal parts compressed to flat, cuneate to transversely ovate, ribbed and generally somewhat 3–5-lobed, to 14 mm. broad; subcortical utricles globose or subglobose; surface utricles 42–84(34–105) μ diam., in section seen to be in contact for 0.05–0.12 of their length; medullary filaments rather closely connected in a single group at each node, communicating with each other by pits or, rarely, free; gametangia obovoid or pyriform, 200–380 μ diam., on rather long dichotomously forked pedicels, densely clustered, especially on the distal margins and free lobes of the fertile segments.

F. **gracilis** Børgesen. Plants with small, relatively thick segments, not distinctly ribbed or lobed.

F. **tripartita** Barton. Plants with the segments in the middle and upper parts of the branches deeply trilobed and the lobes nearly terete, the entire segments of the lower part of the plant likewise nearly terete.

Bermuda, Florida, Mexico, Bahamas, Caicos Isls., Anguila Isls., Cuba, Jamaica, Hispaniola, Puerto Rico, Virgin Isls., St. Martin, St. Barthélemy, Guadeloupe, Dominica, Martinique, Barbados, Grenada, British Honduras, Panama, Netherlands Antilles. Plants common in shallow water, growing on sheltered sandy flats among sea grasses, and occasionally at moderate depths down to 40 m.

REFERENCES: Harvey 1858 *p. p.*, Mazé and Schramm 1870–77 *p. p.*, Dickie 1874b, Hemsley 1884, Murray 1889, Sluiter 1908, Børgesen 1913–20, Hamel and Hamel-Joukov 1931. For the f. *gracilis*, Børgesen 1913–20 (as *H. tridens* f. *gracilis*), Collins and Hervey 1917. For the f. *tripartita*, *H. tridens* f. *tripartita*, Collins and Hervey 1917, Taylor 1928, 1935, 1936, 1942, 1943; *H. monile*, Mazé and Schramm 1870–77; *H. tridens*, Harvey 1858 *p. p.*, Mazé and Schramm 1870–77, Farlow 1871, Dickie 1874a, b, Hauck 1888, Murray 1889, Gardiner 1890, Collins 1901, Howe 1907b, 1909, 1915, 1918a, b, 1920, Collins and Hervey 1917, Taylor 1928, 1935, 1941b, 1942, 1943, 1954a, 1955a, Taylor and Arndt 1929, Sanchez A. 1930.

Halimeda monile (Ellis and Solander) Lamouroux Pl. 23, fig. 2

Plants to 1.0–2.5 dm. tall, dark green, firmly calcified, bushy; from a substantial base of rhizoids and sand, the lower segments somewhat stalklike and often flattened, above closely branched, the segments trilobed when supporting branches, the lobes terete, elsewhere the segments cylindrical, 1.5–2.0 mm. diam., 3–8 mm. long; subcortical utricles obovoid to clavate, 35–90 μ diam.; surface utricles 30–60(23–74) μ diam., seen laterally to be in firm contact for 0.10–0.33 of their length; medullary filaments as in *H. incrassata*.

F. **cylindrica** (Børgesen) Collins and Hervey. Plants slender, scantily branched, the segments below terete and seldom fused, slender throughout.

F. **robusta** (Børgesen) Collins and Hervey. Plants robust and densely branched, the lower segments cuneate and sometimes fused laterally.

Bermuda, Florida, Bahamas, Caicos Isls., Anguila Isls., Cuba, Cayman Isls., Jamaica, Hispaniola, Puerto Rico, Virgin Isls., Guadeloupe, British Honduras, Panama, Netherlands Antilles, Venezuela. Typical plants are common in shallow water at low-tide line. Those approaching f. *robusta* come from more exposed situations, those near f. *cylindrica* from exceedingly sheltered muddy lagoons and deeper water, but the species is seldom dredged from depths of more than 3–5 m.

REFERENCES: Murray 1889, Howe 1909, 1915, 1918a, b, 1920, Collins and Hervey 1917, Taylor 1928, 1935, 1942, 1943, Taylor and Arndt 1929, Sanchez A. 1930; *H. cylindrica*, Mazé and Schramm 1870–77, Murray 1889; *H. incrassata*, Harvey 1858 *p. p.*; ?*H. incrassata* f. *lamourouxii*, Sluiter 1908; *H. incrassata* v. *monilis*, Børgesen 1913–20. For f. *cylindrica* and f. *robusta*, Collins and Hervey 1917, Taylor 1928; *H. incrassata* v. *monilis* f. *cylindrica* and f. *robusta*, Børgesen 1913–20.

Halimeda favulosa Howe Pl. 24, fig. 3

Plants 9–22 cm. tall, the conspicuous base bulbous, of rhizoids and sand, the somewhat flattened stalk of 1–4 more or less fused segments, above which the congested but flaccid branching chiefly occurs; segments only moderately calcified, very irregular in shape, discoid, subsimple to trilobed, or to cylindrical, 4–9 mm. long, 2–9 mm. broad, 0.5–2.0 mm. thick; surface utricles 110–260 μ diam. in surface view, collapsing when dried, producing a pitted appearance, turbinate or obovoid in sections, 150–400 μ long, only slightly laterally coherent; axial filaments adhering at the nodes in a single group, closely connected by pores or by very short communicating processes.

Bahamas. Growing under overhanging rocks in shallow water near low-tide line.

REFERENCES: Howe 1905b, 1920.

Codium Stackhouse, 1797 [1]

Plants massive, from a matted, sometimes inconspicuous rhizoidal holdfast, of filamentous construction, cushion-shaped or branched, repent or erect, green, uncalcified; the structural filaments sometimes nearly divided by constricting thickenings of the walls, dichotomously branched, differentiated into axial strands and corticating branches, the latter enlarged at the ends into conspicuous cylindrical to turbinate utricles radially placed, which in turn often bear, around the distal end, delicate simple hairs, the utricles and the hairs both often nearly cut off by closure of the basal wall thickenings; reproduction by gametangia developed laterally on the peripheral utricles and cut off by a septum; gametes anisogamous, produced on the same or different plants, the macrogametes green, the microgametes more yellowish.

KEY TO SPECIES

1. Plants pulvinate or applanate, simple or lobed, attached generally over the lower surface 2
1. Plants decumbent, narrowly branched, with local secondary attachments .... C. repens, p. 186
1. Plants erect 3
2. Plants strongly attached and commonly in shallow water, firm; utricles generally 70–110 μ diam., less than 1 mm. long C. intertextum, p. 185
2. Plants lightly attached, perhaps only in deep water, spongy; utricles generally 130–390 μ diam. at the apex, more than 1 mm. long .... C. spongiosum, p. 185
3. Branches typically terete; utricles generally 120–350 μ diam., 460–850 μ long, the end walls usually markedly thickened C. isthmocladum, p. 186
3. Branches distinctly flattened, at least at the forks 4
4. Branching notably irregular, often appearing cervicorn; flattening moderate and often only evident at the forks; utricles generally 110–260 μ diam., 650–1150 μ or about 5 diameters long, the end walls often somewhat thickened ... C. taylori, p. 188

[1] For this account I have relied upon a manuscript revision of these species prepared by Prof. Paul C. Silva, and for this privilege am very much indebted. However, since the material has been rewritten to conform to the style of the rest of the volume, I must be responsible for errors which may have been introduced.

4. Branching rather regular, the flattening more general, often very conspicuous at the forks; utricles generally 220–500 μ diam., 1100–1750 μ long, or in the larger ones about 3 diameters long, the end walls thin .... C. decorticatum, p. 188

Codium intertextum Collins and Hervey

Plants scattered or more commonly in extensive colonies, applanate, firm, dull green, smoothly expanded or forming overlapping lobes several millimeters broad which are tightly adherent except along the margins; utricles cylindrical, more or less constricted below the apex, 70–110(45–215) μ diam., 575–720(400–880) μ long; apices rounded or truncate, sometimes indented; the end walls 3–20 μ thick, internally often alveolate to cribrose; hairs or scars common on older utricles in a band (60–)75–145 μ below the apex; gametangia pedicellate, fusiform or ellipsoidal, 56–108 μ diam., 220–330 μ below the apex.

Bermuda, Florida, Bahamas, Jamaica, Hispaniola, Puerto Rico, Virgin Isls., St. Kitts, Guadeloupe, Barbados, Netherlands Antilles, Venezuela, Brazil. Characteristically forming a distinct zone on rocks at low-tide line, but also at times on gorgonians or other firm objects and reported dredged from 20 m. depth.

REFERENCES: Collins and Hervey 1917, 1918, Howe 1918a, 1920 (including v. *cribrosum*), 1928, Taylor 1925, 1928, 1931, 1933, 1942, Feldmann and Lami 1937, Silva 1960; *C. adhaerens,* Montagne 1846, Schramm and Mazé 1865, 1866, Mazé and Schramm 1870–77 (including v. *arabicum*), Martens 1870, Hemsley 1884, J. Agardh 1888 *p. p.,* Hauck 1887, Murray 1889 (including v. *arabicum*), Möbius 1890, Collins 1901, Schmidt 1923a *p. p.,* 1924 *p. p.; C. difforme,* Dickie 1877, Vickers 1905, 1908, Collins 1909b, Børgesen 1913–20; *C. tomentosum β coralloides,* Martens 1871.

Codium spongiosum Harvey

Plants pulvinate or applanate, loosely adherent to the substratum, to 15 cm. thick and to 20(–50) cm. diam., spongy, somewhat lobed or ridged but not with erect branches; utricles subcylindrical, 130–390 μ diam., becoming with age 520 μ diam. at the apex but rather larger below, where to 850 μ diam., 1.5–4.0 (–6.0) mm. long; apices subtruncate to rounded, the wall somewhat thickened to 32 μ; hairs or scars abundant in a band below the apex of each utricle; gametangia borne on short pedicels about 360–660 μ below the apices of the utricles, lance-ovoid, 50–175 μ diam., 215–360 μ long.

Brazil. Only known in the area as dredged; reported from 22 m. depth.

REFERENCES: Silva 1960; *C.* sp., Taylor 1930a; *C. bursa?*, Joly 1953c.

Codium repens Crouan *in* Vickers

Plants procumbent, sometimes in extensive mats, irregularly to dichotomously branched; branches divaricate, sometimes entangled and anastomosing, often secondarily attached to the substratum, terete or slightly flattened, 1–5 mm. diam.; utricles subcylindrical to pyriform, narrower in thicker rather than in slender branches, 105–275 μ diam., 330–550 μ long in slender branches, 700–875 μ long in thicker ones, their apices rounded or slightly flattened with the end wall 4–6 μ thick; hairs or scars common, one or several on each utricle in a zone about 65–130 μ below the apex; gametangia rare, reported 1–3 on a fertile utricle, pedicellate, ovoid to stoutly fusiform, 90 μ diam., 220 μ long, in a zone 290–350 μ below the utricle apex.

Bermuda, Florida, Guadeloupe, Barbados, Netherlands Antilles, Brazil. Reported from shallow water, but even more often dredged, ranging from 3–55 m. depth.

REFERENCES: Vickers 1908, Collins 1909b, Schmidt 1923a, Taylor 1925, 1928, Feldmann and Lami 1937, Silva 1960; *C. tenue* v. *repens*, Mazé and Schramm 1870–77; *C. tomentosum*, Dickie 1874c; *C. tomentosum* v. *reptans*, Schramm and Mazé 1866; *C. tomentosum* v. *subsimplex*, Schramm and Mazé 1865.

Codium isthmocladum Vickers   Pl. 26, fig. 3

Plants erect, to 20 cm. tall, bushy, light green, dichotomously branched to 12 orders, rarely proliferous; branches terete, occasionally locally constricted, usually 2.5–8.0 mm. diam.; peripheral utricles subcylindrical or more distinctively clavate to pyriform, 120–350(–475) μ diam., 460–850 μ long, often constricted about one-third below the rounded or truncate apex; terminal utricle wall usually incrassate, 4–110 μ thick; hairs or scars few to several on each utricle, borne just below the apex; gametangia 2–several on a fertile utricle, pedicellate, lance-ovoid to oblance-ovoid, frequently asymmetrical, (48–)65–130 μ diam., 180–280 μ long.

Subspec. **clavatum** (Collins and Hervey) Silva. Lower branches to 5–10 mm. diam., often somewhat flattened; end walls of the peripheral utricles thin.

Bermuda, North Carolina, South Carolina, Florida, Mexico, Bahamas, Caicos Isls., Cuba, Jamaica, Hispaniola, Puerto Rico, Virgin Isls., St. Martin, Antigua, Guadeloupe, Martinique, Barbados, Guatemala, Honduras, Costa Rica, Canal Zone, Panama, Colombia, Venezuela, Trinidad, Brazil. A rather variable species as here interpreted; in shallow water found on reefs and under ledges below low-tide line or in pools as rather small, darker green, densely branched plants of firm texture, which commonly have slender utricles, but also dredged to depths as great as 73 m., when usually tall, paler green and lax, with broader utricles. The shallow-water form has generally been called *C. dichotomum* (=*C. tomentosum*), but Dr. Silva finds it unrelated to that European species and is not willing to separate it specifically from *C. isthmocladum*.

REFERENCES: Vickers 1908, Collins 1909b, Børgesen 1913–20, Collins and Hervey 1917, Howe 1918a, 1920, Schmidt 1923a, Taylor 1925, 1928, 1933, 1941, 1942 *p.p.*, 1954, Williams and Blomquist 1947, Williams 1948, 1949a, Pearse and Williams 1951, Silva 1960; *C. decorticatum*, Taylor 1930 *p.p.*, 1931 *p.p.*; *C. dichotomum*, Taylor 1943, Williams and Blomquist 1947; *C. elongatum*, Vickers 1908; *C. lineare*, J. Agardh, 1887 *p.p.*, Murray 1889 *p.p.*, De Toni 1889 *p.p.*, Taylor 1931 *p.p.*; *C. tomentosum*, Ashmead 1857, Harvey 1858 *p.p.*, Schramm and Mazé 1865, 1866, Mazé and Schramm 1870–77 *p.p.*, Martens 1870 *p.p.*, Farlow 1871, Dickie 1874a *p.p.*, 1874b *p.p.*, Farlow 1875 *p.p.*, 1876 *p.p.*, J. Agardh 1887 *p.p.*, Hauck 1888, Murray 1889 *p.p.*, De Toni 1889 *p.p.*, Möbius 1890 *p.p.*, Collins 1901, 1909b *p.p.*, Gepp and Gepp 1905, Vickers 1905 *p.p.*, 1908, Børgesen 1913–20 *p.p.*, Collins and Hervey 1917 *p.p.*, Howe 1918a *p.p.*, 1918b, 1920 *p.p.*, Schmidt 1923a *p.p.*, 1924 *p.p.*, Taylor 1931 *p.p.*; *C. tomentosum β tenue*, Martens 1871. For the subspecies *clavatum*, Silva 1960; *C. decorticatum* v. *clavatum*, Collins and Hervey 1917.

Codium decorticatum (Woodward) Howe Plate 26, figs. 1, 2

Plants erect, becoming large, reaching 1.0–1.5 m. in length, sparingly dichotomously branched or very bushy, seldom proliferous; branches terete, 6–25 mm. diam. and flattened only at the forks, or flattened throughout and at the forks to 6 cm. broad; utricles cylindrical or clavate, 220–500(115–850) μ diam., 1100–1750(790–2000) μ long; apices rounded, truncate or depressed, the terminal wall 4–8 μ thick; hairs or scars variable, numerous when present, in a zone 145–330 μ below the apex; gametangia lance-ovoid (58–) 72–125 μ diam., 185–300(144–390) μ long, several (to 7) on a fertile utricle, pedicellate, the stalks 10–15 μ long, borne 430–650 μ below the apices of the fertile utricles.

Bermuda, North Carolina, South Carolina, Florida, Virgin Isls., Guadeloupe, Costa Rica, Venezuela, Tobago, Brazil, Uruguay. Found in shallow water on rocks.

REFERENCES: Collins and Hervey 1917, Howe 1918a, 1928, 1931, Hoyt 1920, Taylor 1928, 1930 *p.p.*, 1931 *p.p.*, 1942 *p.p.*, Williams 1948, 1949a, Pearse and Williams 1951; Silva 1960; *C. decumbens*, Martius 1833; *C. elongatum* Montagne 1846, De Toni 1883, J. Agardh 1887, Möbius 1890, Collins 1909b, Børgesen 1913–20; *C. isthmocladum*, Taylor 1942 *p.p.*, Joly 1953 *p.p.*; *C. tomentosum*, Schramm and Mazé 1865, 1866, Mazé and Schramm 1870–77 *p.p.*, Martens 1870 *p.p.*, 1871, Rein 1873, Hoyt 1920, Taylor 1931 *p.p.*

Codium taylori Silva Plate 26, fig. 4

Plants erect, bushy, to 15 cm. tall, divaricately but unequally branched to 7–9 orders, falsely appearing cervicorn; branches usually flattened, especially at the dichotomies, 4–8(3–25) mm. broad, 3–4 mm. thick, obtuse at the tips; utricles cylindrical or clavate, 110–260(57–380) μ diam., 650–1150(550–1450) μ long; end walls slightly to moderately (23 μ) thickened; hairs or scars abundant, 2–several on a utricle in a zone 50–105 μ below the apex; gametangia 1–2 on a fertile utricle, pedicellate, the stalks to 8 μ long, attached to a slight swelling 275–430 μ below the apex of each utricle, ellipsoidal or cylindrical, 45–86 μ diam., 200–300(–350) μ long.

Bermuda, Florida, Mexico, Bahamas, Jamaica, Puerto Rico, Virgin Isls., Saba I., Guadeloupe, Barbados, Colombia, Netherlands

Antilles, Venezuela, Tobago, Brazil. Frequent in shallow water on rocks and reefs below tide levels.

REFERENCES: Silva 1960; *C. abreviatum,* Schramm and Mazé 1865, 1866, Mazé and Schramm 1870–77; *C. adhaerens,* Schmidt 1924, *p.p.; C. decorticatum,* Taylor 1930 *p.p.*, 1931 *p.p.*, 1942 *p.p.*, Joly 1957; *C. lineare,* J. Agardh 1887 *p.p.*, Murray 1889 *p.p.*, De Toni 1889 *p.p.*, Taylor 1931 *p.p.; C. dichotomum,* Taylor 1942; *C. pilgeri,* Taylor 1925, 1928, 1930, 1931, 1942; *C. repens,* Taylor 1942; *C. tomentosum,* Harvey 1858 *p.p.*, Schramm and Mazé 1865, 1866 *p.p.*, Mazé and Schramm 1870–77 *p.p.*, Martens 1870 *p.p.*, Dickie 1874a *p.p.*, 1874b *p.p.*, Farlow 1875 *p.p.*, 1876 *p.p.*, J. Agardh 1887 *p.p.*, Murray 1889 *p.p.*, De Toni 1889 *p.p.*, Vickers 1905 *p.p.*, 1908 *p.p.*, Børgesen 1913–20 *p.p.*, Collins and Hervey 1917 *p.p.*, Howe 1918a *p.p.*, 1920 *p.p.*, Schmidt 1923a *p.p.*, 1924 *p.p.*, Taylor 1925, 1928, 1930, 1931, 1942 *p.p.; C. tomentosum* v. *divaricatum,* Martens 1871; *Lamarckia tomentosa,* Kuntze 1898.

PHYLLOSIPHONIACEAE

Plants endophytic or endozoöic, of large oval cells or branched filaments without crosswalls, coenocytic, with many nuclei and many disklike chromatophores lacking pyrenoids.

Ostreobium Bornet and Flahault, 1889

Plants filamentous, of much-branched and apparently anastomosing coenocytic filaments very irregular in form and variable in diameter; asexual reproduction by aplanospores formed in the swollen ends of branches.

Ostreobium quekettii Bornet and Flahault

Plants of slender filaments, mostly 4–5 μ diam., much twisted in the tissues, with considerable local inflations to 40 μ diam., or more moderate vermiform enlargements; at the tips of the branches reduced to 2 μ diam.

Hispaniola, Colombia. Inhabiting the matrix of old, especially dead, shells.

REFERENCES: Børgesen 1924, Taylor 1942.

XANTHOPHYCEAE

HETEROSIPHONALES

Plants coenocytic, balloon-shaped to filamentous, with differentiation of rhizoidal and assimilative portions; chromatophores peripheral, small, sometimes with pyrenoid-like structures; food reserves of oil; asexual reproduction by zoöspores or aplanospores; sexual reproduction by iso- or anisogametes, or oögamous.

VAUCHERIACEAE

Plants filamentous, irregularly or dichotomously branched; filaments without regular wall formation except to segregate the reproductive organs; asexual reproduction by zoöspores produced singly in sporangia formed at the tips of branches, multiflagellate, the flagella in pairs over the whole surface; sexual reproduction oögamous, dioecious, or monoecious when fruiting organs of both sexes are sometimes borne together on a single stalk; the antheridia cylindrical, generally curved, producing many small biflagellate antherozoids; oögonia single or in groups, producing single eggs, which are fertilized in place and develop thick walls.

Vaucheria De Candolle, 1801

Plants of branched coenocytic filaments somewhat tufted or loosely entangled, or forming considerable mats, particularly in shallow water, the filaments without normal vegetative crosswalls or constrictions; antheridia more or less cylindrical, sessile or stalked, usually near the oögonia; oögonia single or grouped, sessile or stalked; oöspore when mature often with much stored oil.

KEY TO SPECIES

1. Antheridia and oögonia sessile or nearly so; filaments generally more than 80 μ diam. . . . . V. dichotoma, p. 191
1. Antheridia or oögonia stalked; filaments generally less than 80 μ diam. 2

2. Oögonia lateral, nearly sessile, nearly parallel to the filament V. nasuta, p. 191
2. Oögonia terminal on branchlets, without a hyaline stalk cell 3
3. Antheridia attached to the oögonia .... V. sphaerospora, p. 193
3. Antheridia attached below the oögonia 4
4. Oögonia turbinate to subclavate, 116–153 μ diam.; oöspores oblate-spheroidal, 90–136 μ in axial length .. V. bermudensis, p. 192
4. Oögonia clavate, the end nearly spherical, 140–210 μ diam.; oöspores strongly lenticular, 75–110 μ in length V. piloboloides, p. 192

Vaucheria dichotoma (Linnaeus) C. Agardh

Plants matted or in free-floating dark green masses, filamentous, the very coarse filaments dichotomously branched, 80–225 μ diam., occasionally with a brownish discoloration of the wall; dioecious, the antheridia usually more or less unilaterally disposed in a series along the filaments, sessile, erect or a little oblique, oval to ovoid with a slightly pointed apex, 100–144 μ diam., 137–225 μ long; oögonia solitary or a few together, subseriate and unilateral, sessile, erect, subspherical to obovoid, 260–360 μ diam., 280–380 μ long; oöspores nearly spherical, free from the wall and nearly filling the oögonia, 216–340 μ diam., the smooth brown wall slightly thickened.

Bermuda, Jamaica, Virgin Isls., ?Guadeloupe, Grenada. Forming wide pilose expanses on muddy or sandy protected shores about low-tide line; of coarse, partly erect filaments, attached by the filaments buried in the substratum.

REFERENCES: Collins 1909b, 1912, Børgesen 1913–20 (including f. *marina*), Taylor and Bernatowicz 1953; *?Chlorodesmis comosa*, Mazé and Schramm 1870–77. The f. *marina* Hauck does not seem worth segregating, in view of the measurements of the Bermuda material studied.

Vaucheria nasuta Taylor and Bernatowicz Pl. 27, figs. 6–9

Filaments forming a mat on the substratum, 28–56 μ diam., widely dichotomously branched; monoecious, the antheridia scattered or several together near the oögonia, stipitate, the stalk usually consisting of a straight or curved thin-walled hyaline supporting cell about 22 μ diam., 25–65 μ long, borne on a papillar enlargement of the filament; antheridia cylindrical to obovate, generally curved and often closely so, tapering a little at the ends, 17–28 μ diam. at the base, 50–115 μ long, emptying subterminally or by 1–4 lateral simple or tubular discharge pores; oögonia sometimes 2–3 together, sessile or subsessile, oblique to nearly parallel to the filament, subspherical to oval, asymmetrically attached, with a deflexed beak about 15–25 μ diam. and a little longer than wide; oöspores nearly filling the oögonia, subspherical to oval, 80–140 μ diam., 80–156 μ long, with a brown, slightly roughened wall about 3 μ thick.

Bermuda. Growing on muddy shores near low-tide level in very sheltered places, particularly under mangroves.

REFERENCE: Taylor and Bernatowicz 1953.

Vaucheria bermudensis Taylor and Bernatowicz Pl. 27, figs. 3–5

Plants forming diffuse colonies; basal filaments in the sandy substrate, colorless, sometimes septate; free green filaments 18–51 μ diam., irregularly widely branched; sporangial branches clavate, the aplanosporangia clavate to obovoid, 80–110 μ diam., 212–240 μ long; monoecious, the antheridia terminal on a branchlet, becoming displaced by an upgrowth from below and appearing lateral, borne on 1–2 empty supporting cells, subcylindrical to subfusiform, 26–50 μ diam., 219–464 μ long, with 1–2 lateral discharge pores; oögonia stalked, arising from below the antheridia and overtopping them, turbinate to clavate, separated by a crosswall from the filament, 116–153 μ diam., 285–510 μ long, slightly expanded near the tip; oöspores adherent to the distal end of the oögonium, generally oblate-spheroidal or somewhat compressed on the lower side, 105–141 μ diam., 90–136 μ long.

Bermuda. Forming very diffuse colonies, appearing as light-green clouds of filaments through which the sand below is visible. Found in very sheltered places in shallow water on a clean sandy bottom.

REFERENCE: Taylor and Bernatowicz 1952.

Vaucheria piloboloides Thuret Pl. 27, figs. 10, 11

Filaments sparingly branched, 40–60(–80) μ diam.; aplanosporangia sometimes formed at the ends of branches, 80 μ diam., 250 μ long; monoecious, the antheridia usually terminal on short branches, separated from the filament by empty cells, 30–45 μ diam., 150–200 μ long; cylindrical to spindle-shaped with a terminal and 1–2 lateral conical projections which become discharge pores; oögonia terminal on short branchlets near the antheridia, clavate, distally hemispherical, 140–210 μ diam., 320–500 μ long; oöspores lenticular, 105–145 μ diam., 75–110 μ thick, with a somewhat incrassate membrane, attached at the distal end of the oögonia.

Bermuda. Growing on very sheltered, sandy shores in shallow water.

REFERENCE: Taylor and Bernatowicz 1952.

Vaucheria sphaerospora Nordstedt

Filaments 25–60 μ diam.; plants monoecious, one oögonium and one antheridium occurring on each fruiting branchlet; antheridia curved toward and closely adjacent or appressed to the oögonia, each supported on a hyaline cell 54–56 μ long, which is attached to one side of an oögonium; antheridia fusiform, 38–57 μ diam., 128–157 μ long, with 3–6 pores through papillae 14–33(–45) μ long; oögonia 214–386 μ long, cylindrical below, swollen above, where 89–178 μ diam.; oöspores spherical, free from the oögonial wall, 86–141 μ diam.

Bermuda.

REFERENCE: Collins and Hervey 1917.

UNCERTAIN RECORDS

Vaucheria sp.

Guadeloupe. *Valonia tenuis,* De Toni 1:380; Mazé and Schramm 1870–77, Murray 1889.

CRYPTOPHYCEAE

FAMILY UNCERTAIN

Chrysophaeum Lewis and Bryan, 1941

Colonies much branched, yellow, flocculent, composed of very numerous cells; cells on long, very delicate, intertwined stalks, subcylindric, with pellicle and numerous chromatophores, lacking flagella and stigmata; with characteristic convulsive motion preceding reproduction; reproduction by aplanospores which produce zoöspores with two equal flagella laterally attached, one active, the other trailing.

Chrysophaeum taylori Lewis and Bryan

Colonies 1–6 cm. high, sulphur- to mustard-yellow, mucous, forming tufts of lacerate streamers; cells 49–118 μ long, 24–38 μ broad, ovate-cylindric, narrowed toward the base, with an anterior tubular invagination; chromatophores numerous, oval or irregular in outline; aplanospores about 25 μ diam.; zoöspores 16–32 from each aplanospore, about 9 μ long, the flagella 1.5 times longer.

Florida, Bermuda. Forming clumps of gelatinous streamers on coral rocks in quiet, shallow water.

REFERENCE: Lewis and Bryan 1941.

CHRYSOPHYCEAE

PHAEOTHAMNIACEAE

Chrysonephos Taylor, 1952

Plants filamentous; filaments dichotomously forked, multicellular, the cells with the lateral walls especially gelatinous, the two to four chromatophores yellowish-brown; zoöids produced after further gelatinization and swelling of the filament walls with rounding-up of the cells, the zoöids pyriform, with two chromatophores and two unequal flagella.

Chrysonephos lewisii (Taylor) Taylor — Pl. 28, figs. 2–7

Plants as in the genus; 1–3 cm. tall, sulphur-yellow, bushy, exceedingly delicate; filaments uniseriate, or locally, especially at a fork, pluriseriate, 30–38 μ diam. below, 4.5–7.7 μ diam. near the tips; the cells subcylindrical, a little turgid, with very thin walls between; zoöids pyriform, 7–8 μ diam., 8–9 μ long.

Florida, Bermuda, Jamaica. Forming exceedingly tenuous clouds of filaments over clean coral sand and sea grasses in the shallow water of very sheltered places.

REFERENCES: Taylor 1952; *Chrysophaeum lewisii*, Taylor 1951.

PHAEOPHYCEAE

ECTOCARPALES

Plants generally filamentous and branched, uniseriate with apical, trichothallic, or dominantly intercalary growth; sporophyte and gametophyte phases of similar appearance, reproducing by zoöspores and by iso- or anisogametes.

ECTOCARPACEAE

Plants filiform and more or less branched, or subsimple from a creeping, penetrating, or disciform base; generally uniseriate but occasionally some segments in the lower part with one or two longitudinal septa; reproductive organs lateral, replacing branchlets, or intercalary from transformed vegetative cells, and the plants in general producing zoöspores in unilocular sporangia, followed by a generation producing gametes in plurilocular gametangia; sporangia and gametangia occasionally found on the same plant.

KEY TO GENERA

1. Plants large and bushy, seldom less than 1 cm. tall at maturity ... 2
1. Plants very small, usually forming little colonies on larger algae 6

2. Reproductive organs intercalary in the vegetative branches 3
2. Reproductive organs terminal or lateral on the branches, though sometimes hair-tipped 4

3. Chromatophores discoid .... PYLAIELLA, p. 197
3. Chromatophores radiating .... BACHELOTIA, p. 198

4. Gametangia in dense racemose clusters .... SOROCARPUS, p. 208
4. Gametangia scattered, seriate or fasciculate 5

5. Chromatophores numerous, discoid or occasionally in the form of several short bands; plurilocular gametangia sometimes of 2–3 types in the same species .... GIFFORDIA, p. 205
5. Chromatophores few, elongate simple or forked bands ECTOCARPUS, p. 198

6. Special erect filaments absent .... PHAEOSTROMA, p. 205

6. Erect free filaments present; base of creeping filaments; chromatophores band-shaped .... HERPONEMA, p. 204
6. Erect free filaments present, base discoid; chromatophores discoid .... ECTOCARPUS, p. 198

Pylaiella Bory, 1823

Plant filiform, becoming large, attached by rhizoids more or less coalesced to a holdfast; filaments primarily uniseriate, widely oppositely or irregularly branched; cells with numerous discoid chromatophores; unilocular sporangia cask-shaped, formed in intercalary catenate series from cells of the vegetative branches, opening laterally; plurilocular gametangia usually intercalary, oblong to irregularly cylindrical, formed by repeated longitudinal and transverse division of several successive vegetative cells.

Pylaiella antillarum (Grunow) De Toni

Filaments densely caespitose, 1.5–3.5 cm. tall; branches short, patent, alternate, penicillate above; cells 11–25(–40) μ diam., shorter than broad or to 2–3 diameters long; sporangia usually formed in the middle of the branchlets, short, sometimes to 30 in a series, 15–18 μ diam. (perhaps immature?).

Bermuda, Bahamas, Guadeloupe, Barbados, ?Tobago, ?French Guiana. Growing on rocks and other objects in tide pools and in the surf zone. Lacking evidence of the character of the chromatophores one cannot be certain, but it is possible that some of these plants were actually *Bachelotia fulvescens*, *q. v.*, or that but one entity is involved.

REFERENCES: Howe 1920, Taylor 1929b; ?*Ectocarpus hooperi*, Mazé 1868; *E. antillarum*, Grunow 1867; ?*P. hooperi*, Vickers 1905, 1908.

UNCERTAIN RECORD

Pylaiella littoralis (Linnaeus) Kjellman

Brazil. Type from Europe. De Toni 3:531; ?*Ectocarpus littoralis* v. *brasiliensis* Grunow 1867, 1870.

Bachelotia (Bornet) Kuckuck *ex* Hamel 1939

Plants filamentous, the tufts with sparingly branched erect filaments arising from a creeping system; cells of the free filaments with radiating chromatophores; unilocular sporangia intercalary, seriate.

Bachelotia fulvescens (Bornet) Kuckuck

Plants tufted, composed of creeping primary and erect free portions; creeping filaments widely spreading and irregularly branched; erect filaments to 1.5–3.0 cm. tall, simple or sparingly branched, the cells frequently forming spreading spurlike branchlets, the occasional long branches similar to the axis, erect; cells 30–47 μ diam., 23–108 μ or 0.75–3.50 diameters long, length of those in the center of the branches equal to or less than their diameter, constituting a growth zone; young and senescent cells with small irregular chromatophores; mature healthy cells with chromatophores in 1–2 groups, radiating from the center; sporangia in long intercalary series, infrequently opposite in pairs, 57–65 μ diam., 26–31 μ long.

Bermuda, North Carolina, Florida, Bahamas, Cuba, Puerto Rico, Virgin Isls., Brazil. Growing on rocks, sometimes partially covered with sand; also found in lagoons or sometimes exposed to the surf.

REFERENCES: *Pylaiella fulvescens*, Børgesen 1913–20, Collins and Hervey 1917, Schmidt 1923, 1924, Blomquist and Humm 1946, Humm 1952a, Taylor 1954a, Joly 1957.

Ectocarpus Lyngbye, 1819

Plants yellowish to dark brown, usually freely branched, from a decumbent, rhizoidal, or, infrequently, a penetrating base; the uniseriate branches often terminating in colorless hairs, the axis sometimes corticated by rhizoids near the base; cells with parietal band-shaped, simple or forked, perhaps sometimes discoid chromatophores; reproduction by sessile or short-stalked unilocular sporangia which are scattered, replacing lateral branchlets, or by plurilocular gametangia which are stalked or sessile, and generally abundantly longitudinally and transversely divided, dehiscing by a pore; gametes isogamous.

KEY TO SPECIES

1. Plants generally to 2 cm. or more in height 2
1. Plants minute, microscopic, tufted, but rarely approaching 2 cm. .. 5

2. Branches often hooked; plants in spongy masses; gametangia ovoid to spherical .... E. breviarticulatus, p. 201
2. Branches not hooked at the tips 3

3. Gametangia acute to often hair-tipped .... E. siliculosus, p. 199
3. Gametangia not hair-tipped, usually obtuse 4

4. Gametangia usually terminal on the branchlets, very slender, 10–15 μ diam., to 250 μ long .... E. dasycarpus, p. 200
4. Gametangia usually lateral, shorter, 20–35 μ diam., commonly under 150 μ long .... E. confervoides, p. 200

5. Base a scutate disk .... E. subcorymbosus, p. 201
5. Base of free or centrally somewhat crowded branching filaments .. 6

6. Erect filaments simple, or only branched at the extreme base 7
6. Erect filaments with scattered branching, to 11 μ diam.; gametangia oval to truncate-cylindrical or clavate, 22–27 μ diam. E. rhodochortonoides, p. 202

7. Erect filaments 10–18 μ diam., their apices generally piliferous; gametangia basal, lanceolate to conical, 15–25 μ diam., 63–80 μ long .... E. elachistaeformis, p. 202
7. Erect filaments 7–12 μ diam., their apices blunt; gametangia suprabasal, lanceolate to oblong, about 27 μ diam., 60 μ long E. variabilis, p. 202

Ectocarpus siliculosus (Dillwyn) Lyngbye

Plants tufted, yellowish to brownish olive, at first attached and up to 3 dm. tall, later often free-floating, more or less entangled in the center; branching below at times pseudodichotomous, but above the branches alternate or unilateral, ascending; cells in the lower part of the main axes 40–60 μ diam., 4–5 diameters long, but above the cell length same as breadth or a little less; chromatophores irregularly band-shaped, often oblique; sporangia sessile or on short stalks, ellipsoid, 20–27 μ diam., 30–65 μ long, often on the same plant as the gametangia; gametangia sessile or usually very short-stalked, typically conico-subulate, 12–25 μ diam., 50–600 μ long, sometimes slightly curved, usually ending in a hair.

F. **arctus** (Kützing) Kuckuck. Rather stiff, the filaments twisted together and beset with rhizoids, the branches opposite or alternate; unilocular sporangia ovate-globose, minute; gametangia rather small and thick, ovate or ovate-oblong but rather irregular in shape, 20–30 μ diam., 40–50 μ long.

Bermuda, North Carolina, South Carolina, Florida, Mississippi, Texas, ?Barbados, French Guiana, Brazil. Growing on rocks, coarse algae, or other objects in rather shallow water. The more southerly records are open to question.

REFERENCES: Martens 1870, Mazé and Schramm 1870–77, Murray 1889, Hoyt 1920, Taylor 1936, 1941a, 1954a; *E. confervoides* v. *siliculosus,* Murray 1880; *E. spinulosus,* Montagne 1850; *E. viridis,* Harvey 1852. For the f. *arctus,* Hamel 1931–39, Collins and Hervey 1917, Howe 1918a; *?E. acanthoides,* Vickers 1905, 1908.

Ectocarpus dasycarpus Kuckuck

Plants tufted, not entangled, 5–7 cm. tall; main branching loosely pseudodichotomous, with small lateral branchlets; cells of the main filaments 20–40 μ diam., 2–3 diameters long, little constricted at the septa; gametangia sometimes sessile, generally terminal on one- to several-celled lateral branchlets, subcylindrical, 10–15 μ diam., the length variable, but to 250 μ, the tip not prolonged into a hair.

Mississippi. Epiphytic on coarse algae.

REFERENCE: Taylor 1954a.

Ectocarpus confervoides (Roth) Le Jolis

Tufts attached, more or less loosely entangled below, dark brown, to 5–30(–60) cm. tall; branching irregular, pseudodichotomous or chiefly alternate but not opposite, the branches tapering, often forming numerous rhizoids, particularly near the base of the plant, and rather sparingly hair-tipped; lower cells of the main branches 20–50 (–65) μ diam., 1–2(–3) diameters long; chromatophores ribbon-shaped, often forked; sporangia ovoid, sessile or on short stalks, 20–40 μ diam., 35–50 μ long; gametangia sessile or with short stalks, scattered, not ending in hairs, short-subulate to fusiform, 20–35 μ diam., 60–150(–250) μ long.

Bermuda, North Carolina, Florida, Texas, Brazil. Growing on woodwork, rocks, or coarser algae from low-tide level to moderate depths.

REFERENCES: Collins and Hervey 1917, Howe 1918a, Hoyt 1920, Schmidt 1923, 1924, Taylor 1928, 1936, 1941a; *E. confervoides* f. *halliae,* Collins 1906.

Ectocarpus breviarticulatus J. Agardh

Plants tufted, 2–4 cm. tall, of spongy ropelike strands of twisted filaments; filaments repent below, erect above; primary branching irregular, the filaments developing numerous short hooked branchlets, or these sometimes extending as rhizoids; the cells 27–54 μ diam., 1–2 diameters long, with numerous discoid chromatophores; gametangia ovoid to spherical, borne on a stalk cell and nearly at right angles to the filaments, to 57 μ diam., 62 μ long.

Jamaica, Virgin Isls., Redonda I., Guadeloupe, Barbados, Grenada, Costa Rica, Colombia, Netherlands Antilles, I. las Aves, Venezuela, Brazil. Growing on reefs and rocks in exposed situations.

REFERENCES: Dickie 1874b, Børgesen 1913–20, Taylor 1933, 1941b, 1942, Feldmann and Lami 1937, Joly 1957; *E. hamatus,* Mazé and Schramm 1870–77, Murray 1889, Vickers 1905, 1908; *E. spongioides,* Mazé and Schramm 1870–77, Murray 1889.

Ectocarpus subcorymbosus Farlow, *emend.* Holden

Plants small, tufted, to 1–3 mm. tall; bases crowded, disciform, of radiating filaments laterally conjoined, bearing 1–few-branched erect filaments; branches alternate or more rarely opposite, in the latter case the main filament or one of the branches may be represented by a long hair, or hairs may terminate the branches; the cells 12–15 μ diam., 2–4 diameters long, chromatophores disk-shaped; plurilocular gametangia terminal or when lateral, usually on a short stalk of 1–3 cells, sometimes crowded near the ends of branches, cylindrical, the apices bluntly rounded, 30–140 μ long, 18–25 μ diam., the segments 8–10 μ long, transversely 1–3-celled.

North Carolina. Epiphytic on marine phanerogams in shallow water.

REFERENCES: Taylor 1957, Blomquist 1958a; *Myriotrichia scutata,* Blomquist 1954.

Ectocarpus rhodochortonoides Børgesen

Plants very small, the basal portion of creeping filaments attached by rhizoids, bearing erect, sparsely branched filaments; the branches more or less equaling the primary axis, and the tips attenuate, with few chromatophores, but not true hairs with basal meristems; cells about 11 μ diam., 2–3 diameters long, but longer toward the filament tips, with bandlike, branched chromatophores; gametangia widely scattered, sessile, ascending, oval, truncate-cylindrical to clavate, very blunt, 22–27 μ diam., 27–33 μ long, often constricted at the septa, with cells to 10 μ diam.

Virgin Isls. Epiphytic on Padina in shallow water.

REFERENCE: Børgesen 1913–20.

Ectocarpus variabilis Vickers

Plants minute, the basal portion of creeping filaments, bearing erect filaments up to 2 mm. tall; subsimple or sparsely flabellately forked, the filament tips rounded, the cells 9–12 μ diam., to 3 diameters long; chromatophores band-shaped, several in each cell; gametangia on the lower parts of the filaments, sessile or stalked, lanceolate-oblong, obtuse, 20–30 μ diam., 45–75 μ long.

Jamaica, Virgin Isls., Barbados. Epiphytic on coarse algae in shallow water.

REFERENCES: Vickers 1905, 1908, Børgesen 1913–20.

Ectocarpus elachistaeformis Heydrich Pl. 29, fig. 9

Plants minute, epiphytic, the basal portion of wide-spreading ramifying decumbent or sometimes penetrating filaments 9.0–12.5 μ diam., attaching to the host by simple or branched haptera and giving rise to elongate erect, simple or rarely basally branched filaments; erect filaments to 1.0–1.5 mm. long, below 10–18 μ diam.; the cells 1–3 diameters long, but more slender and longer toward the hair tips; cells with several small (? or band-shaped) chromatophores; gametangia fusiform to narrowly conical, acute, borne near the bases of the erect filaments, seldom on the upper portion, 15–25 μ diam., 63–80(–?140) μ long.

Bermuda, Florida, ?Virgin Isls., Tobago. Epiphytic on various coarse algae in shallow water, the basal layer less distinct when the filaments are able to penetrate the surface of softer host plants.

REFERENCES: ?Børgesen 1913–20, Collins and Hervey 1917, Taylor 1928, 1942.

UNCERTAIN RECORDS[1]

Ectocarpus denudatus Crouan

Guadeloupe, the type locality. De Toni 3 : 565.

Ectocarpus fasciculatus (Griffiths) Harvey, v. *refractus* (Kützing) Ardissone

Brazil. Type from the North Sea. De Toni 3 : 554; Schmidt 1924.

Ectocarpus fenestroides Crouan

Guadeloupe, the type locality. De Toni 3 : 565; Mazé and Schramm 1870–77, Murray 1889.

Ectocarpus glaziovii Zeller

Brazil, the type locality. De Toni 3 : 547; Zeller 1876.

Ectocarpus guadeloupensis Crouan

Guadeloupe, the type locality, and Barbados. De Toni 3 : 565; Mazé and Schramm 1870–77, Murray 1889, Vickers 1905, 1908.

Ectocarpus heterocarpus Crouan

Guadeloupe, the type locality. De Toni 3 : 565; Mazé and Schramm 1870–77, Murray 1889.

Ectocarpus macrocarpus Crouan

Guadeloupe, the type locality. De Toni 3 : 565; Mazé and Schramm 1870–77, Murray 1889.

Ectocarpus minutulus Montagne

Cuba, the type locality, and Brazil. Montagne 1863, Zeller 1876, Murray 1889.

Ectocarpus moniliformis Vickers

Barbados, the type locality. Vickers 1905, 1908.

[1] The species listed here, proposed by the Crouans on the basis of Guadeloupe types, are all essentially *nomina nuda*.

Ectocarpus obtusocarpus Crouan

Guadeloupe, the type locality. Mazé and Schramm 1870–77, Murray 1889

Ectocarpus zonariae Taylor

Florida, the type locality. Taylor 1928.

Herponema J. Agardh *emend.* Hamel, 1939

Plants minute, filamentous, creeping or endophytic below, with short erect filaments on the host surface; growth intercalary, the filaments often attenuate and hairlike but without true basal hair meristems; cells with band-shaped chromatophores; unilocular sporangia and plurilocular gametangia terminal on short filaments from the basal layer or from the bases of the erect filaments.

KEY TO SPECIES

1. Erect filaments 5.0–8.0 μ diam. .... H. luteolum
1. Erect filaments 7.5–13.0 μ diam. .... H. tortugensis

Herponema tortugensis (Taylor) n. comb.

Plant a minute epiphyte; decumbent filaments forming a close mat or ramifying loosely over the host, bearing simple or sparsely branched erect filaments with gametangia; cells of basal filaments irregularly cylindrical, 7.5–12.5 μ diam., 12–27 μ long; vertical filaments about 10 cells long, reaching a height of about 0.2 mm., 7.5–13.0 μ diam.; cells 10 μ diam., 15–25 μ long; gametangia usually 1–3 on a filament, alternate or opposite, sessile, 7.5–15.0 μ diam., 25–50 μ long, oval to broadly cylindrical, obtuse, about 3–5 cells broad at the widest part.

Florida. Epiphytic on Zonaria in shallow water.

REFERENCE: *Ectocarpus tortugensis,* Taylor 1928, p. 108.

Herponema luteolum (Sauvageau) Hamel

Plants minute, epiphytic, the basal filaments forming a pseudoparenchymatous cushion on the host; short, simple or basally branched erect filaments crowded, hair-tipped, 100–400 μ long, 5–8 μ diam., the cells 1–3 diameters long; gametangia terminal on short basal filaments or on short branchlets from the bases of the erect filaments, 7–13 μ diam., 30–80 μ long, somewhat constricted at the septa and usually showing 2 series of cells in lateral view.

Bermuda. Epiphytic on *Dictyopteris justii* in shallow water.

REFERENCE: *Ectocarpus luteolus*, Collins and Hervey 1917.

Phaeostroma Kuckuck, 1893

Plants filamentous, epiphytic, forming small discoid colonies; the filaments near the center prostrate, sometimes crowded, becoming two- or three-layered and pseudoparenchymatous, but at least near the margin monostromatic, distinct and radially branching; cells with several platelike chromatophores in each cell; hairs with basal meristems present; unilocular sporangia formed by direct conversion of vegetative cells; plurilocular gametangia formed by subdivision of adjacent vegetative cells.

Phaeostroma pusillum Howe and Hoyt

Plants forming irregular discoid colonies 0.3–0.8 mm. diam.; vegetative cells often irregular, 5–10 μ diam., 10–16 μ long; hairs occasional, 8–10 μ diam.; sporangia obovoid or somewhat globose, 8–16 μ diam., or by previous division of a sporangial initial forming submoriform sori 16–48 μ diam., the individual sporangia then 5–8 μ diam.; gametangia sessile, ellipsoid to subconic, 15–18 μ diam., 22–27 μ long.

North Carolina, Bahamas. Epiphytic on Dictyota and other algae, and from these growing over the stolons of hydroids.

REFERENCES: Howe and Hoyt 1916, 1920, Howe 1920.

Giffordia Batters, 1893

Plants erect and bushy of habit, usually irregularly laterally branched, growth being generalized and intercalary; hairs with a basal meristem lacking, though branchlets may become distally attenuated and colorless; chromatophores numerous, discoid, occasionally in the form of very short bands; plurilocular gametangia and unilocular sporangia present, usually sessile, often seriate on the branches, the plurilocular structures sometimes of two or three types in the same species, with loculi of contrasting sizes, producing anisogametes.

KEY TO SPECIES

1. Plants minute, about 5 mm. tall .... G. rallsiae, p. 208
1. Plants larger, often several centimeters tall 2

2. Branches often arcuate near their tips, the branchlets pectinate, secund .... G. sandriana, p. 207
2. Branchlets not pectinate 3

3. Plants about 2 cm. tall, the gametangia scattered, usually more than 5 diameters long .... G. duchassaigniana, p. 207
3. Plants much larger 4

4. Gametangia to about 5 diameters long, more or less cylindrical, the tips blunt .... G. mitchellae, p. 206
4. Gametangia about 2 diameters long, conical ..... G. conifera, p. 207

Giffordia mitchellae (Harvey) Hamel Pl. 29, figs. 1, 2

Plants bushy or in large diffuse masses, from 2–3 cm. in length to a few decimeters, attached by rhizoids below, the longer filaments sometimes stranded together; branching abundant, alternate or secund; main axes to 35–50 μ diam., the cells 2–3 diameters long, in the branches to 15–20 μ diam.; older cells with discoid chromatophores; branches often with hairlike tips; sporangia rare, reported to be sessile, oblong, 25–45 μ diam., 60–80 μ long; gametangia polymorphic, sessile in series on the upper side of the branches, elliptic-oblong to linear-cylindrical, markedly obtuse at the apex, 20–30 μ diam., 50–150 μ long, the meiosporangia with cells about 6–7 μ wide, those in the megasporangia 10–17 μ wide.

Bermuda, North Carolina, Florida, Texas, Bahamas, Jamaica, Virgin Isls., Guadeloupe, Brazil. Common in shallow water, growing on coarse algae, sea grasses, shells, and rocks. Extensive, detached masses of sterile ectocarpoid algae often found in warm, quiet water may consist of this species.

REFERENCES: Hamel 1931–39; *Ectocarpus mitchellae,* Mazé and Schramm 1870–77, Murray 1889, Collins 1901, Børgesen 1913–20, Collins and Hervey 1917, Howe 1918a, 1920, Hoyt 1920, Taylor 1928, 1936, 1941a, 1954a, Joly 1957.

Giffordia conifera (Børgesen) n. comb.

Plants of moderate size, attached by rhizoids, with distinct ascending axes; branches above in part short and patent, alternate, but irregular in length and terminating in long hairs; cells about 40 μ diam., 0.5–4.0 diameters long, with numerous discoid chromatophores; sporangia ovate, usually solitary, erect on the branchlets in their axils; gametangia elongate-conical, solitary, sessile and erect on the branchlets in their axils, or 2–4, rarely many, in a row along the branchlet, 24–40 μ diam., 40–110 μ long.

Bermuda, North Carolina, Virgin Isls. In shallow water near the shore growing epiphytically or on stones, in either sheltered or exposed situations.

REFERENCES: *Ectocarpus coniferus*, Børgesen 1913–20 (1914, p. 8), Collins and Hervey 1917, Blomquist and Humm 1946.

Giffordia sandriana (Zanardini) Hamel

Plants tufted, 4–12 cm. tall; densely alternately and corymbosely branched below, the filaments somewhat twisted together and consolidated by rhizoids, in the lesser divisions the branches arcuate, pectinate, secund; primary filaments 30–100 μ diam., the branchlets 10–20 μ diam., the cells a little longer or shorter than broad; gametangia sessile, ovoid to conical, in series on the upper sides of the branchlets near the base, 15–35 μ diam., 50–60 μ long.

Bermuda. Growing on rocks or other objects in shallow water of sheltered places.

REFERENCES: Hamel 1931–39; *Ectocarpus sandrianus*, Collins and Hervey 1917.

Giffordia duchassaigniana (Grunow) n. comb. Pl. 29, fig. 10

Plants tufted, to 2 cm. tall, very soft, of entangled rhizoid-like filaments below, the tufts above of many erect irregularly branched filaments with hairlike tips; filaments 20–34 μ diam., the cells about 0.5 diameter long below, 1.5 diameters long above; sporangia sessile on the filaments, ascending, obovate or oval, to 70 μ diam., 110 μ long; gametangia sessile or with a single stalk cell, chiefly scattered, cylindrical to clavate or fusiform, bluntly rounded at the apex, 19–50 μ diam., 112–250 μ long.

Bermuda, North Carolina, Florida, Texas, Jamaica, Puerto Rico, Virgin Isls., Guadeloupe, Barbados, Netherlands Antilles, Tobago. Growing on coarse algae or rocks in shallow water, chiefly of sheltered places, and dredged to a depth of 20 m.

REFERENCES: *Ectocarpus duchassaignianus*, Grunow 1867, p. 45, Mazé and Schramm 1870–77, Hauck 1888, Murray 1889, Vickers 1905, 1908, Børgesen 1913–20, Collins and Hervey 1917, Hoyt 1920, Taylor 1928, 1933, 1936, 1941a, 1942, 1954a, Hamel and Hamel-Joukov 1931.

Giffordia rallsiae (Vickers) n. comb.

Plants minute, the basal portion of creeping filaments attached by rhizoids, the erect filaments about 5 mm. tall; sparsely and irregularly branched, the branches flexuous, irregular as to length, sometimes hair-tipped, of cells about 27–40 μ diam. which are very variable in length, reaching 5 diameters; sporangia oval or ovate, to 45 μ diam., 70 μ long; gametangia scattered, or on the bases of the longer branches, sessile or stalked, fusiform, the apex acute or a little attenuate, 27–40 μ diam., 80–120 μ long.

Bermuda, Texas, Jamaica, Virgin Isls., Barbados. Epiphytic on coarse Rhodophyceae, from shallow water.

REFERENCES: *Ectocarpus rallsiae*, Vickers 1905, p. 59, 1908, Børgesen 1913–20, Collins and Hervey 1917, Taylor 1936, 1941a.

Sorocarpus Pringsheim, 1862

Plants of erect, bushy habit; repeatedly and widely branched, the branches uniseriate, bearing lateral and terminal hairs with basal growth; plurilocular gametangia in small, densely crowded racemose clusters, producing anisogametes which fuse or germinate parthenogenetically.

Sorocarpus uvaeformis (Lyngbye) Pringsheim

Plants tufted, to 20 cm. tall; branching irregularly alternate, the filaments about 50 μ diam. below and 20 μ diam. above, the cells with a few discoid chromatophores; plurilocular gametangia oval, 20–25 μ long, 12–15 μ diam., on 1–2-celled stalks or subsessile, clustered near the bases of the branchlets, branches, or hairs.

North Carolina.

REFERENCE: Williams 1949a.

SPHACELARIALES

Plants generally filamentous, branched and polysiphonous; growth from large apical cells, the segments from which generally divide longitudinally in a regular and characteristic fashion; sporophyte and gametophyte phases of similar appearance, reproducing by zoöspores and by iso- or anisogametes.

SPHACELARIACEAE

Plants with a stoloniferous base, or erect from the holdfast; branches sometimes similar to the axis, naked or corticated by rhizoids, but in other cases the axis and its main divisions clothed with short branchlets derived from the vegetative axis or from the cortex.

KEY TO GENERA

1. Plants 0.5–3.0 cm. tall, usually without distinctive heavier axes
   SPHACELARIA, p. 209
1. Plants larger, the corticated main axes distinctly thicker than the whorled determinate branchlets .... CLADOSTEPHUS, p. 212

Sphacelaria Lyngbye, 1819

Plants small, forming subglobose tufts or spreading mats, attached by basal disks or stolons ramifying on or penetrating the support; erect axes subsimple to bushy, generally abundantly laterally determinately branched, the ultimate divisions irregular or two-ranked, but indeterminate branches infrequent, like the main axis; hairs present in some cases; segments from the apical cell divided transversely once before longitudinal segmentation; secondary transverse divisions present or absent; pedicellate, stalked, bi- or triradiate propagula of various shapes occur on many species; unilocular sporangia or plurilocular gametangia, alike or heteromorphic, occur on simple or forked lateral stalks.

KEY TO SPECIES

1. Propagula slender; arms many times longer than broad 2
1. Propagula stout; arms little if at all extended 3

2. Propagula triradiate at the summit .... S. fusca, p. 210
2. Propagula biradiate at the summit .... S. furcigera, p. 210

3. Filaments 45–80 μ diam.; propagula obconic or somewhat flattened; arms not extended, the cells at the tip of each arm transversely divided .... S. novae-hollandiae, p. 211
3. Propagula with evident arms 4

4. Filaments 25–60 μ diam.; propagula about as wide as long; arms 1–2 diameters long, obliquely divergent, the end cells undivided .... S. tribuloides, p. 211
4. Filaments 55–80 μ diam.; propagula much wider than long, the arms transverse with very large tip cells; stalks each 2-celled, articulated directly with the broad body of the propagulum .... S. brachygonia, p. 212

Sphacelaria fusca (Hudson) C. Agardh

Plants arising from a compact one-layered disk of radiating filaments, the erect filaments 1–3 cm. tall, 60–80 μ diam.; sparingly branched, the branches often long; propagula pedicellate at the base, triradiate, the stalk increasing in girth from base to apex, the straight arms not constricted at their bases, projecting at about 90° from each other, cylindrical or tapering somewhat to the tips; stalk with a buttonlike cell at the summit, but no terminal hair between the arms.

Bermuda. Reported to be saxicolous or epiphytic.

REFERENCES: Sauvageau 1900–1904, Collins and Hervey 1917.

Sphacelaria furcigera Kützing Pl. 29, fig. 5

Plants stoloniferous or endophytic below, the erect axes 0.5–3.0 cm. tall, 16–45 μ diam., sparsely branched; segments as long as, or somewhat longer than, broad, sparingly longitudinally divided, and without secondary transverse divisions; hairs present, 12–16 μ diam.; propagula slender, biradiate, the stalk to 24 μ diam., cylindrical or tapering to the base, at the summit with 2 cylindrical or somewhat tapering arms about equaling it in length, with a span reaching

450 μ, the segments undivided or with a single crosswall; sporangia spherical, 50–70 μ diam.; gametangia dimorphic: the one type cylindrical, 45–65 μ long, 24–28 μ diam. with cells about 3 μ diam., the other more irregularly shaped, 30–60 μ long, 28–40 μ diam., with cells about 6 μ diam.

Bermuda, North Carolina, Florida, Jamaica, Virgin Isls., Martinique, Barbados, Costa Rica, Venezuela. Epiphytic on various coarse bushy algae, or growing on lithothamnia or rocks.

REFERENCES: Sauvageau 1900–1904, Vickers 1905, 1909, Sluiter 1908, Børgesen 1913–20, Collins and Hervey 1917, Taylor 1928, 1954a, Blomquist and Humm 1946.

Sphacelaria tribuloides Meneghini Pl. 29, fig. 6

Plants tufted or scattered, attached by short stolons, the erect axes 4–5 mm. tall, bearing radially disposed erect branches; segments 25–60 μ diam., as long as broad or a little longer, sparingly longitudinally divided; hairs often present, 10–15 μ diam.; propagula pedicellate, biradiate, broadly triangular in side view, the arms thick, with a span of about 140–165 μ, without a terminal hair between the arms.

Bermuda, North Carolina, Florida, Mexico, Bahamas, Jamaica, Hispaniola, Puerto Rico, Virgin Isls., Guadeloupe, Martinique, Barbados, Brazil. Growing on rocks and coralline algae in shallow water, the tufts often infiltrated with sand; sometimes also on coarse, bushy algal species.

REFERENCES: Mazé and Schramm 1870–77, Murray 1889, Möbius 1892, Sauvageau 1900–1904, Vickers 1905, 1908, Børgesen 1913–20, Collins and Hervey 1917, Howe 1918a, 1920, Schmidt 1924, Taylor 1928, 1954a, Taylor and Arndt 1929, Hamel and Hamel-Joukov 1931, Blomquist and Humm 1946, Humm 1952a; *S. brachygonia,* Murray 1889 *p. p.*

Sphacelaria novae-hollandiae Sonder Pl. 28, fig. 8

Plants stoloniferous below, tufted, to 2 cm. in height; erect axes 45–80 μ diam., with irregularly disposed branches somewhat corymbosely clustered; segments of the axes somewhat shorter than broad; hairs often present, 15–20 μ diam.; propagula biradiate, in side view somewhat triangular, rather longer than broad, thick, almost round in midsection; original tip cells of the arms of the propagula divided by a wall at right angles to the lateral margin.

Bermuda, Jamaica, Martinique, Barbados. Growing on rocks, often with *S. tribuloides*.

REFERENCES: Sauvageau 1900–1904, Collins and Hervey 1917.

Sphacelaria brachygonia Montagne

Plants 1–2 cm. tall; filaments 55–80 μ diam., radially branched, length of the segments 0.3–0.6 the breadth; propagula generally borne in pairs one above the other on the same segment, with 2-celled stalks, from side view appearing transversely oval or spindle-shaped, the stalk attaching in the middle, the arms transverse or nearly so.

Brazil.

REFERENCES: Martens 1870, Murray 1889 *p. p.*, Sauvageau 1900–1904.

Cladostephus C. Agardh, 1817

UNCERTAIN RECORD

Cladostephus verticillatus (Lightfoot) Lyngbye

Brazil. Type from Scotland. De Toni 3:513; *C. myriophyllum,* Martens 1870.

TILOPTERIDALES

Plants filamentous, branched; growth trichothallic or localized intercalary, the segments usually ultimately undergoing limited longitudinal division; sporophyte and gametophyte phases of similar appearance, the asexual generation apparently reproducing by monospores and the sexual generation involving microgametes and motionless cells which may represent oöspheres.

TILOPTERIDACEAE

Plants attached by rhizoids, filamentous, repeatedly branched, elongating by intercalary growth zones; cells with numerous small disklike chromatophores; asexual reproduction in some cases by uninucleate spores, in others by quadrinucleate spores, both formed singly in sporangia; sexual reproduction probably by small biciliate microgametes produced in tubular "microgametangia," in some cases perhaps acting as antherozöids and associated with large motionless macrogametes.

KEY TO GENERA

1. Plurilocular "microgametangia" hollow, producing many motile swarmers . . . . HAPLOSPORA
1. Plurilocular structures not evidently hollow, producing a few nonmotile spores . . . . ACINETOSPORA

Haplospora Kjellman, 1872

Plants filamentous, monosiphonous above, more or less polysiphonous below; branching progressive, arising irregularly from all sides of the successive axes; asexual reproduction by nonmotile quadrinucleate spores formed singly in stalked, sessile, or intercalary sporangia; intercalary tubular plurilocular microgametangia known; uninucleate sporangia (oögonia?) partly immersed in the branches of sexual plants, each producing a nonmotile spore (or egg?).

Haplospora vidovichii (Meneghini) Bornet

Plants tufted, to 2 cm. tall, or in ropelike twists to 30 cm. in length; filaments 36–60 μ diam., primary branching distant, subdichotomous, becoming unilateral above, with numerous additional divaricate branches, some terminated by long hairlike tips and others by short, blunt spur branches; quadrinucleate sporangia divaricate, on one-celled pedicels, attached near the middle of the supporting cell, solitary, or rarely to 4 in a verticil; microgametangia often on the upper parts of the filaments, on 1–2-celled stalks, ovoid-lanceolate, 80–100 μ long; uninucleate sporangia (oögonia?) pedicellate, divaricate, ovoid, about 30 μ diam., 50 μ long.

Bermuda. On rocks or other objects in rather shallow water.

REFERENCES: Phyc. Bor.-Am. 2026; *Heterospora vidovichii*, Collins and Hervey 1917, Howe 1918a.

Acinetospora Bornet, 1891

Plants filamentous, entangled, with intercalary growth zones; sporangia dimorphic, lateral, in one case developing several zoöspores as in ordinary unilocular sporangia, in the other developing a single motionless monospore; plurilocular sporangia also lateral, terminal on short branchlets, not evidently hollow, with few and large cells, each of which produces an aplanospore-like body.

Acinetospora pusilla (Griffiths) Bornet

Plants usually epiphytic, tufted, to 4–8 cm. tall; filaments forming a mat below, becoming more erect above, flexuous, commonly bearing lateral rhizoids; cells 15–56 μ diam., the length about 1–6 diameters; chromatophores numerous, discoid or slightly elongated; growth localized, intercalary; lateral branches scattered or often in pairs, spreading at right angles or curved; plurilocular gametangia small, ovoid-lanceolate to fusiform, with large cells; monosporangia often 2–3 together, at the bases of the branches, sessile or on one-celled pedicels, subspherical, 31–42 μ diam., to 45–53 μ long; unilocular zoösporangia also reported.

North Carolina, Barbados. Plants of relatively shallow water and a rocky or muddy bottom; dredged from a depth of 5 m.

REFERENCES: Blomquist 1955; *Ectocarpus breviarticulatus,* Williams 1948; *Ectocarpus pusillus,* Dickie 1874a, Murray 1889.

DICTYOTALES

Plants thalloid, with alternation of equal generations; growth apical; asexual reproduction by aplanospores; sexual reproduction by nonmotile eggs and motile antherozoïds.

DICTYOTACEAE

Plants of moderate to large size, simple or branched, the divisions growing from apical cells or marginal rows of apical cells, forming fanlike, foliaceous, strap-shaped branching thalli, usually composed of cortical and medullary layers of cells; sexual and asexual generations distinct, the plants of the two phases similar in form; asexual reproductive elements consisting of sporangia in which four or eight nonmotile aplanospores (commonly called tetraspores) are produced; sexual reproductive elements, eggs, borne singly in the superficial oögonia, and probably uniflagellate antherozoïds, borne in great numbers in the superficial plurilocular antheridia.

KEY TO GENERA

1. Thalli branched, the branches strap-shaped to foliaceous 2
1. Thalli entire, or lobed or split, the segments essentially fan-shaped (sometimes with a rhizomorphic phase) 5

2. Branches without midribs 3
2. Branches with midribs .... DICTYOPTERIS, p. 226

3. Branches narrow, generally less than 1 cm. broad, and richly subdivided, each growing from an apical cell 4
3. Branches more than 1 cm. wide, less frequently forked, growing from an apical row of cells .... SPATOGLOSSUM, p. 225

4. Thallus with the medulla of a single layer of cells .. DICTYOTA, p. 217
4. Thallus with the medulla at least marginally 2 cells thick DILOPHUS, p. 215

5. Apical margin of the thallus rolled inward; a coarse, fibrous growth form known, capable of developing fan-shaped blades directly .... PADINA, p. 233
5. Apical margin of the thallus not inrolled 6

6. Sori with paraphyses among the tetrasporangia; blade ecostate STYPOPODIUM, p. 232
6. Sori lacking paraphyses 7

7. Blades ecostate, sporangia with 4 or 8 spores .. POCOCKIELLA, p. 231
7. Blades costate below, sporangia producing 8 spores .. ZONARIA, p. 230

Dilophus J. Agardh, 1882

Plants bushy, erect from an irregular holdfast, subdichotomously or in part somewhat alternately branched, the branches strap-shaped; structurally showing a medulla which may be one cell thick in the young portions, but along the margins of the blades and more extensively near the base it becomes two or more cells in thickness, the cortex remaining of one layer; otherwise as in Dictyota.

KEY TO SPECIES

1. Branches 0.3–1.0 mm. broad, often rather strongly flabellately disposed .... D. guineensis
1. Branches 0.5–3.0 mm. broad, rather erect and not notably in a plane .... D. alternans

Dilophus guineensis (Kützing) J. Agardh Pl. 30, fig. 1

Plants generally of small size, attached by rootlike holdfasts, 0.5–1.2 dm. tall, the habit dense, the plants often irregular, the main axis 0.5–1.5 mm. broad, bearing alternate primary branches which terminate in densely dichotomous or partially alternate terminal flabellate branch systems, the segments 0.3–1.0 mm. broad, not much narrower than the axis; the tips from somewhat obtuse to acute.

Bermuda, Florida, Bahamas, Caicos Isls., Cuba, Jamaica, Hispaniola, Virgin Isls., Antigua, Guadeloupe, Martinique, Barbados, British Honduras, Netherlands Antilles, Venezuela, ?Brazil. Generally growing on rocks in exposed situations, close above or below low-tide line, but also dredged to about 7 m. depth.

REFERENCES: Mazé and Schramm 1870–77, Murray 1889, Collins 1901, Vickers 1905, 1908, Børgesen 1913–20, Collins and Hervey 1917, Howe 1918a, 1920, 1928, Taylor 1928, 1933, 1935, 1940, 1942, 1954a, Taylor and Arndt 1929, Hamel and Hamel-Joukov 1931, Feldmann and Lami 1937; *Dictyota antiguae,* Dickie 1874a, Zeller 1876; *D. acutiloba, D. antiguae, D. bartayresiana p. p., D. bipennata, D. paniculata, D. radicans,* Mazé and Schramm 1870–77, Murray 1889; *D. guineense,* Mazé and Schramm 1870–77.

Dilophus alternans J. Agardh Pl. 30, fig. 3

Plants of moderate size, 1.0–1.5 dm. tall, densely bushy, rather opaque yellowish brown and slightly iridescent; primary branching rather obscurely alternate from axes 2–4 mm. broad, the lateral divisions subdichotomous to eventually regularly so, 0.5–3.0 mm. broad, suberect or somewhat flabellate; generally quite obtuse, or retuse at the apices.

Florida, Bahamas, Caicos Isls., Cuba, Cayman Isls., Jamaica, Hispaniola, Virgin Isls., Barbados, Brazil. This plant is infrequently reported, but occurs in shallow water growing on rocks. It has been dredged from depths of as much as 26 m.

REFERENCES: Murray 1889, Collins 1901, Vickers 1905, 1908, Howe 1920, Børgesen 1924, Taylor 1928, 1930a, 1954a; *Dictyota dentata p. p., D. dichotoma* v., *D. pinnatifida p. p., D. prolifera,* Mazé and Schramm 1870–77, Murray 1889; *?D. suhrii,* Murray 1889.

UNCERTAIN RECORD

Dilophus spiralis (Montagne) Hamel

Bermuda, Barbados, Brazil. Type from France. *Dictyota dichotoma p. p.*, De Toni 3:263; *D. spiralis*, Grunow 1867; ?*D. ligulata*, Murray 1889, Vickers 1908.

Dictyota Lamouroux, 1809

Plants erect, bushy, brown or sometimes iridescent, attached by a small irregular fibrous holdfast and often by secondary holdfasts on branches in contact with the substratum; generally dichotomously, less often pinnately, branched, each branch developing by regular segmentation from a conspicuous, shallow, convex apical cell; medulla of one layer of regularly placed cells, covered on each surface by a one-celled cortical layer; hairs developed in inconspicuous tufts scattered over the thallus; reproductive organs as typical for the family, the tetrasporangia scattered, the gametangia in sori.

KEY TO SPECIES

1. Ultimate branching alternate .... D. dentata, p. 224
1. Ultimate branching dichotomous or cervicorn, but sometimes also proliferous 2

2. Thallus margin aculeate-dentate .... D. ciliolata, p. 223
2. Thallus margin shallowly erose-dentate .... D. jamaicensis, p. 223
2. Thallus margin entire 3

3. Branching essentially dichotomous throughout, but the divisions sometimes a little unequal in size 4
3. Branching ultimately cervicorn to irregular, though when plant is young almost completely dichotomous .... D. cervicornis, p. 222

4. Branches regularly spirally twisted, sinuses broad, wide-angled D. volubilis, p. 220
4. Branches not consistently twisted 5

5. Branches to 1–2 mm. broad or more below, but ultimately filiform at least above, often with an abrupt transition; sinuses wide-angled .... D. divaricata, p. 221
5. Branches of nearly uniform width throughout, about 0.5–1.0 mm. broad or rarely more; branch sinuses narrow D. linearis, p. 219
5. Branches broader, not narrowed above 6

6. Internodal segments less than 4 breadths long; upper sinuses broad, with an angle of about 80° .... D. bartayresii, p. 219
6. Internodes longer; upper sinuses usually narrower 7

7. Internodes 5–10 breadths long .... D. dichotoma, p. 218
7. Internodes 15–20 breadths long .... D. indica, p. 221

Dictyota dichotoma (Hudson) Lamouroux Pl. 31, fig. 5

Plants to about 1 dm. tall, occasionally to 3.5 dm., bushy, usually regularly dichotomous, forking at angles of 15°–45°, usually with narrow sinuses, with little general decrease in width from the base to the upper branches, the lower segments a little broader below each fork than just above it, the terminal divisions somewhat tapered, but hardly acute; segments 2–5 mm. broad, the internodes 1.0–2.5 cm. long; rarely proliferous except at the base, which in color and texture is much like the rest of the plant.

V. **menstrualis** Hoyt. Plants quite large, nearly 3 dm. tall, the segments to 16 mm. wide; otherwise like the species in form and structure, but different in that it produces its gametes at monthly intervals during the spring tides of the full moon instead of every 2 weeks.

Bermuda, North Carolina, South Carolina, Florida, Bahamas, Caicos Isls., Cuba, Cayman Isls., Jamaica, Puerto Rico, Guadeloupe, Martinique, Barbados, Grenada, Colombia, Netherlands Antilles, Venezuela, Brazil. Frequent and widely distributed, but in the area probably less common than *D. cervicornis, D. divaricata,* and *D. bartayresii.* Generally found in shallow water, but reported from a depth of 55 m.

Dictyota dichotoma is notable for the regularity with which it may release its crops of gametes at fortnightly intervals, within about an hour before daybreak shortly after the maximum of spring tides. Where the tidal range is slight this phenomenon may not be well defined.

This species is variously interpreted by different authors and ranges widely in the dimensions of the segments. Even European material is not always strictly dichotomous. The early Caribbean records are all open to question, since this name may be found in herbaria applied to almost any of the Dictyotas. There is considerable doubt in my mind regarding the distinctness of *D. indica* as Collins applied the name to plants of this area, for it seems difficult to set a sharp limit between these specimens and *D. dichotoma.* However, in order not to confuse the records this species has been kept separate in this account. On the other hand there seems no possibility of distinguishing *D. pardalis* Kützing as understood by Børgesen and others; probably most of the plants so labeled have been irregularly branched individuals of *D. dichotoma.*

REFERENCES: Hooper 1850, Harvey 1852, 1861, Martens 1870, Mazé and Schramm 1870–77, Dickie 1874b, Hemsley 1884, Murray 1889 *p. p.*, Möbius 1890, Collins 1901, Gepp 1905b, Collins and Hervey 1917, Howe 1918a, 1920, Schmidt 1924, Taylor 1928, 1954a, 1955a; *D. attenuata p. p.*, Mazé and Schramm 1870–77; *D. dichotoma* v. *intricata*, Murray 1889. Probably also including *D. pardalis*, Dickie 1874a, 1875, Murray 1889, Vickers 1908, Sluiter 1908, Børgesen 1913–20, Schmidt 1923, 1924, Taylor 1928, 1933. For the v. *menstrualis*, Hoyt 1927; *D. dichotoma*, Hoyt 1920; ?*D. dichotoma* f. *latifrons p. p.*, Mazé and Schramm 1870–77.

Dictyota linearis (C. Agardh) Greville

Plants to 5–12 cm. tall, abundantly dichotomously branched, or a little irregular, the segments 0.5–1.0 mm. broad, hardly tapered, sometimes a little twisted, 0.5–2.0 cm. between forks, the sinuses generally narrow and the branch tips acute.

Bermuda, Caicos Isls., Cayman Isls., Jamaica, Hispaniola, Virgin Isls., Guadeloupe, Netherlands Antilles. Locally abundant on rocks near low-tide line, and dredged to a depth of 20 m.

Caribbean plants placed under this name are commonly broader than those from Europe usually associated with it, and it is possible that slender forms of *D. cervicornis* and *D. divaricata* have been mistaken for it.

REFERENCES: Montagne 1863, Mazé and Schramm *p. p.* 1870–77, Murray 1889, Sluiter 1908, Børgesen 1913–20, 1924, Collins and Hervey 1917, Howe 1918a.

Dictyota bartayresii Lamouroux Pl. 30, fig. 2

Plants usually low, in masses 0.5–2.0 dm. diam., becoming densely intermeshed, light brown, in texture crisp and relatively fragile;

branches when young frequently complanate, later bent and a little twisted; the segments seldom 2 mm., more often 4–7 mm., broad, the internodes 1–4 times as long as broad, the dichotomies equal with a divergence of 45°–90°, usually about 80°, the tips usually broadly rounded, less often somewhat acute.

Bermuda, Florida, Mexico, Bahamas, Cuba, Jamaica, Hispaniola, Virgin Isls., Antigua, Guadeloupe, Dominica, Martinique, Barbados, Grenada, Panama, Colombia, Brazil. Plants of a rocky or shelly bottom, very common in partially sheltered locations, where it chiefly occurs in shallow water, but it has been dredged to a depth of about 33 m.

REFERENCES: Collins and Hervey 1917, Howe 1918a, b, 1920, 1928, Børgesen 1924, Taylor 1928, 1929b, 1930a, 1933, 1954a, Taylor and Arndt 1929, Sanchez A. 1930, Hamel and Hamel-Joukov 1931; *D. bartayresiana*, Harvey 1852, 1861, Martens 1870, Mazé and Schramm 1870–77 *p. p.*, Dickie 1874a, b, Hemsley 1885, Zeller 1876, Möbius 1889, Murray 1889, Collins 1901, Vickers 1905, 1908, *p. p.*, Grieve 1909, Børgesen 1914; *D. crenulata*, v. and *D. dichotoma p. p.*, Mazé and Schramm 1870–77, Murray 1889; *D. patens*, Murray 1889.

Dictyota volubilis Kützing *sensu* Vickers Pl. 31, fig. 6

Plants small, to about 1 dm. tall, entangled, very widely and rather closely branching at angles of 60°–120°, the branches above not much narrower above than in the lower parts of the plant, about 1.5–3.0 mm. wide, and strongly spirally twisted.

Virgin Isls., ?Guadeloupe, Martinique, Barbados, Panama, Venezuela. Plants of very shallow, sheltered situations; generally found loose upon the bottom, but also dredged to a depth of 12 m.

Moderate twisting of the blades of Dictyotas is a commonplace character without taxonomic value; *D. dichotoma* in particular often shows it, and this species as originally defined has by some been reduced to synonymy under that name. Certainly Kützing's figure (1859, vol. 9, pl. 13, fig. II) is not very distinctive. However, one not rarely finds plants of the extreme form figured by Vickers, and for the present it seems worth while to segregate them under this descriptive name.

REFERENCES: Vickers 1908, Børgesen 1913–20, Taylor 1929b, 1942, Hamel and Hamel-Joukov 1931; *?D. dichotoma* f. *curvula p. p., ?D. indica* f. *torta,* Mazé and Schramm 1870–77.

Dictyota divaricata Lamouroux Pl. 31, figs. 3, 4

Plants of small to moderate size, 3–7 cm. high, the branches at first complanate and closely spreading, iridescent, later more erect, of looser, more tangled habit and inconspicuous brown coloring; branching regularly dichotomous, the angles of forking wide, generally 90°–120°, the width of the segments variable, commonly to 3 mm. broad in the lower portions and characteristically abruptly reduced to 0.1–0.2 mm. in the upper portions, though sometimes narrow throughout; often copiously proliferous and branches coming into contact often attached by rhizoids.

Bermuda, Florida, Mexico, Bahamas, Caicos Isls., Cuba, Jamaica, Hispaniola, Puerto Rico, Virgin Isls., Guadeloupe, Martinique, British Honduras, Old Providence I., Colombia, Netherlands Antilles, Trinidad, Brazil. Common on rocks in shallow water; found in both sheltered and exposed places, and dredged to about 55 m. depth.

REFERENCES: Collins 1901, Collins and Hervey 1917, Howe 1920, Taylor 1928, 1930a, 1935, 1939b, 1940, 1942, 1943, 1954a, Taylor and Arndt 1929, Humm 1952b; *D. bartayresiana* v. *divaricata,* Martens 1870, Mazé and Schramm 1870–77, Murray 1889; *D. dichotoma p. p.,* Dickie 1874b; *?D. dichotoma* f. *curvula p. p., D. dichotoma* v. *implexa, D. furcellata, D. linearis p. p., D. sandvicensis,* Mazé and Schramm 1870–77, Murray 1889.

Dictyota indica Sonder *in* Kützing Pl. 31, fig. 1

Plants to 2 dm. tall, resembling *D. dichotoma,* but the segments 1.0–1.5 mm. broad throughout, the internodes 1.5–4.0 cm. long, but slightly tapering downwards, the sinuses wide and rounded.

Bermuda, Florida, Cuba, Virgin Isls., Guadeloupe, Barbados. Found in very quiet sheltered places, and dredged to a depth of 27 m.

This plant is with difficulty to be distinguished from *D. dichotoma,* of which it may be a lax, narrow, widely-branched form of warm or quiet water.

REFERENCES: Kützing 1859, Martens 1870, Mazé and Schramm 1870–77, Dickie 1874a, Murray 1889, Vickers 1905, 1908, Børgesen 1913–20, Collins and Hervey 1917, Taylor 1928; *D. abyssinica, D. aequalis, D. cervicornis* v., *D. dichotoma* v., *D. linearis, p. p.,* and f. *major,* Mazé and Schramm 1870–77, Murray 1889.

Dictyota cervicornis Kützing Pl. 31, fig. 2

Plants bushy, erect, to 2 dm. tall, but little entangled; branching subdichotomous to notably asymmetrical, the reduced members of each fork frequently short and spurlike; in the typical plant branches quite slender, tapering little from the base of the plant upwards, the branch tips acute; the segments generally 1.0–2.5 mm. broad, with internodes 1.0–3.5 cm. long; very often proliferous from the faces of the segments.

Bermuda, North Carolina, Florida, Mexico, Bahamas, Caicos Isls., Cuba, Cayman Isls., Jamaica, Hispaniola, Virgin Isls., St. Barthélemy, Barbuda, Nevis, Antigua, Guadeloupe, Martinique, Barbados, Grenadines, British Honduras, Panama, Colombia, Netherlands Antilles, Venezuela, Tobago, Brazil, I. Trinidade. A common plant in shallow water, growing on rocks, shells, or larger plants, and dredged from a depth of 26 m.

The forms *spiralis, pseudodichotoma,* and *pseudobartayresii* which I proposed (1928) are of doubtful value. The plants differ little from their species counterparts: *D. volubilis, D. dichotoma,* and *D. bartayresii,* except for the presence of a distinct proportion of branching where one member is short and simple or nearly so, while the other continues to grow forward and divide.

REFERENCES: Mazé and Schramm 1870–77, Zeller 1876, Murray 1889, Collins 1901, Collins and Hervey 1917, Howe 1920, Taylor, 1928, 1929b, 1930a, 1933, 1935, 1936, 1940, 1942, 1943, 1954a, 1955a, Taylor and Arndt 1929, Sanchez A. 1930, Hamel and Hamel-Joukov 1931, Williams 1951, Joly 1953c; *D. acutiloba, D. bartayresiana p. p., D. cuspidata, D. dichotoma* v. and v. *curvula p. p., ?D. indica* f. *torta, D. linearis,* and *D. liturata,* Mazé and Schramm 1870–77, Murray 1889; *D. bartayresiana,* Vickers 1908 *p. p.; D. cirrosa,* Hohenacker Meeresalgen n. 429; *D. dichotoma p. p.,* Dickie 1875; *D. fasciola,* Harvey 1852, Martens 1870, Mazé and Schramm 1870–77, Hemsley 1884, Murray 1889 (including v. *abyssinica*), Collins 1901.

Dictyota ciliolata Kützing Pl. 32, fig. 3, Pl. 59, fig. 1

Plants to 1.5 dm. tall or probably more, rather regularly dichotomously branched, the angles acute or rounded but the branches erect, frequently a little spirally twisted, the segments to 12 mm. wide below a fork and 7 mm. wide just above it, 1.0–4.5 cm. long, the terminal divisions generally tapering and subacute; margins sparingly to closely and regularly aculeate-dentate, the projections typically ascending, each with a broad base, becoming terete toward the tip.

V. **bermudensis** n. var. Plants smaller throughout, in general 5–8 cm. tall, closely subdichotomously to clearly alternately branched, dark in color below, the segments 1.0–2.0 mm. broad, 3–10 mm. long, the margins sparingly and minutely dentate.

Bermuda, Florida, Mexico, Jamaica, Hispaniola, Virgin Isls., Guadeloupe, Martinique, Barbados, Grenada, British Honduras, Panama, Venezuela, Tobago, Brazil. Occasional on rocks in rather shallow, somewhat sheltered situations, and dredged to a depth of 24 m.

Marginal teeth are sometimes nearly completely absent from plants of this species; contrariwise, plants of other species with rudimentary proliferations are sometimes mistaken for it, so particularly close observation of obscurely characterized individuals is necessary.

REFERENCES: Howe 1918a, 1928, Taylor 1928, 1933, 1935, 1942, 1943, Taylor and Arndt 1929, Hamel and Hamel-Joukov 1931; *D. attenuata p. p.*, *D. dichotoma* f. *latifrons p. p.*, *D. naevosa,* Mazé and Schramm 1870–77, Murray 1889; *D. ciliata,* Hooper 1850, Harvey 1852, Mazé and Schramm 1870–77, Dickie 1874a, b, Hemsley 1884, Murray 1889, Möbius 1890, Collins 1901, Vickers 1905, 1908, Børgesen 1913–20, Collins and Hervey 1917; *D. dichotoma, D. linearis,* Hooper 1850; *Asperococcus intricatus,* Murray 1889, *p. p.*

Dictyota jamaicensis n. sp. Pl. 32, figs. 4, 5

Plants very bushy, to 12 cm. tall from a small fibrous holdfast which may show a few short terete flagellar proliferations; axes somewhat stupose below, often markedly twisted, dichotomously branched at intervals of 4–15, generally about 7 mm., the branches very widely spreading below but at narrow angles or subparallel in the uppermost divisions where the sinuses are minutely rounded; segments 1–2 mm. broad, the margins minutely erose-dentate at irregular intervals of 0.5–5.0 mm., generally about 1.0 mm.; tetrasporangia scattered or in small groups near the center of the segments, individually to about 80 μ diam.; oögonia in sori to about 500 μ long, 250 μ wide, scattered throughout the fertile segments.

Jamaica, Grenada, Venezuela. Growing attached to rocks in shallow water.

These plants seem rather large for *D. crenulata* as found on the western coast of Mexico, though not dissimilar in some respects; the habit is quite different from that of *D. ciliolata.*

Dictyota dentata Lamouroux Pl. 30, figs. 4, 5

Plants robust, bushy and erect, 1–2 dm. tall, the well-defined main axes with lateral branches repeatedly alternately branched; main axes to 6 mm. broad, the branches from them arising at intervals of 1–2 cm.; segments of the ultimate indeterminate divisions 2–3 mm. broad, bearing alternate spurlike branchlets 1–2 mm. long at intervals of 3–7 mm.; branch tips broadly rounded on young shoots, but with maturity generally becoming dentate or aculeate.

Bermuda, Florida, Bahamas, Caicos Isls., Cuba, Cayman Isls., Jamaica, Hispaniola, Puerto Rico, Virgin Isls., St. Barthélemy, Antigua, Guadeloupe, Martinique, Barbados, Grenada, Panama, Netherlands Antilles, Trinidad, Tobago, Brazil, I. Trinidade. A handsome species, locally abundant, growing on rocks in shallow, moderately exposed situations from the intertidal zone to a depth of 15 m., but most common at about 2 m. depth.

REFERENCES: Montagne 1863, Martens 1870, Mazé and Schramm 1870–77, Dickie 1874a, Hauck 1888, Murray 1889, Collins 1901, Vickers 1905, 1908, Sluiter 1908, Børgesen 1913–20, 1924, Collins and Hervey 1917, Howe 1918a, b, 1920, 1928, Schmidt 1923, 1924, Taylor 1928, 1929b, 1930a, 1933, 1940, 1954a, Hamel and Hamel-Joukov 1931, Feldmann and Lami 1937; *D. dentata* f. *mertensii,* Schmidt 1923; *D. mertensii,* Martens 1870, Vickers 1905, 1908, Sluiter 1908; *D. brongniartii,* Mazé and Schramm 1870–77, Martens 1870, Murray 1880; *?D. pinnatifida,* Schmidt 1924.

UNCERTAIN RECORDS

Dictyota crenulata J. Agardh

Jamaica, Virgin Isls., Barbados. Type locality the Pacific coast of Mexico. Mazé and Schramm 1870–77, Martens 1871, Hemsley 1885, Murray 1889, Vickers 1905, 1908, Børgesen 1913–20.

Dictyota cuspidata Kützing

Mexico, the type locality. Kützing 1859.

Dictyota pinnatifida Kützing

Antigua, the type locality, and Guadeloupe. Kützing 1859.

Spatoglossum Kützing, 1843

Plants erect, foliaceous, the blades flat, somewhat stalked below, not zonate and generally ecostate, irregularly, subpalmately to pinnately lobed; growth originating from groups of initial cells at the tips of the lobes, producing a plane margin, the blade of several cells in thickness with the median layers composed of larger cells than the single cortical layers; tetrasporangia scattered over both surfaces of the blade; dioecious, the oögonia dispersed over the surface, rarely a few together; antheridia in small, scattered, slightly raised sori.

Spatoglossum schroederi (Mertens) Kützing Pl. 33, fig. 5

Plants 1–3 dm. tall, often iridescent when growing, medium to dark brown on drying, moderately adherent to paper, deeply and repeatedly alternately, subdichotomously, or palmately lobed or branched, the irregular divisions 0.5–2.5 cm. broad, 0.5–7.0 cm. long between the forks, sometimes narrowed to stalklike bases, the margins undulate or irregularly dentate, the teeth in part acute; often marginally proliferous.

Bermuda, North Carolina, Florida, Mexico, Jamaica, Guadeloupe, Martinique, Barbados, Costa Rica, Panama, Tobago, Brazil. Uncommon; growing on shaded rocks in sheltered places.

REFERENCES: Martens 1870, 1886, Collins 1901, Vickers 1905, 1908, Collins and Hervey 1917, Howe 1918a, Hoyt 1920, Schmidt 1924, Taylor 1928, 1929b, 1933, 1942, Hamel and Hamel-Joukov 1931, Joly 1951, 1957, De Mattos 1952; *Taonia schroederi*, Harvey 1852, Mazé and Schramm 1870–77, Hemsley 1884, Murray 1889.

Dictyopteris Lamouroux, 1809

Plants small to large and bushy, from a substantial rhizoidal holdfast, branched, the branches strap-shaped, with a pronounced midrib and sometimes other veins; branches growing from an apical group of cells; colorless hairs in superficial sori, scattered or near the midrib; reproductive organs also in sori.

KEY TO SPECIES

1. Pinnate veinlets present 2
1. Pinnate veinlets absent 3

2. Blade margins entire; membrane between the veinlets one cell thick; branching irregularly alternate .... D. plagiogramma, p. 229
2. Blade margins macroscopically aculeate-dentate to ciliate; membrane between the veinlets 2–4 cells thick; branching dichotomous .... D. hoytii, p. 229

3. Blades dark, firm, 4 cells thick, 1.5–8.0 cm. wide .... D. justii, p. 226
3. Blades thin, almost transparent, narrower 4

4. Branching dichotomous 5
4. Branching alternate .... D. jamaicensis, p. 228

5. Plants small, erect, or indefinitely spreading and tangled; branches irregularly dichotomous, 0.5–5.0 mm. wide; in older plants often with marginal veins .... D. delicatula, p. 227
5. Plants erect, larger; the branches 5–15 mm. wide D. membranacea, p. 227

Dictyopteris justii Lamouroux — Pl. 33, fig. 1

Plants erect, forming large tufts which reach a height of 4 dm. or more; the dark brown, sometimes iridescent blades dichotomously branched, the segments strap-shaped, entire or irregularly crenulate or ruffled, obtuse or emarginate at the tips, 1.5–8.0 cm. broad; the midrib prominent and persistent whereas the older parts of the lamina disintegrate leaving a stalklike, often stupose base; laminar membrane becoming opaque and about 170 μ thick, of 2 medullary layers of larger cells with a corticating layer of smaller cells on each face; hairs in sori scattered over the blades; fertile sori large, appearing as patches on the upper parts of the blades.

Bermuda, Florida, Mexico, Bahamas, Cuba, Jamaica, Hispaniola, Puerto Rico, Virgin Isls., St. Barthélemy, Guadeloupe, Martinique, Barbados, Netherlands Antilles, Venezuela, Trinidad, Tobago, Brazil. Plants intertidal in habit, but ranging to about 40 m. depth, growing on rocks in moderately exposed situations.

REFERENCES: Hauck 1888, Collins 1901, Vickers 1905, 1908, Børgesen 1913–20, Collins and Hervey 1917, Schmidt 1924, Hamel and Hamel-Joukov 1931, Feldmann and Lami 1937, Humm 1952b; *?Haliseris areschougii,* J. Agardh 1848; *H. justii,* Montagne 1863, Martens 1870, Mazé and Schramm 1870–77, Dickie 1874a, b, Hemsley 1884, Murray 1889, Sluiter 1924; *Neurocarpus justii,* Howe 1915, 1918a, b, 1920, Taylor 1928, 1929b, 1930a, 1933, 1942.

Dictyopteris membranacea (Stackhouse) Batters

Plants with a distinct and characteristic odor when removed from the water, erect, bushy, 10–30 cm. tall, freely but irregularly dichotomously branched at intervals of 1–5 cm. and sometimes proliferating from stipe and midribs; the segments strap-shaped, entire, 5–15 mm. wide, briefly tapered at the tips, obliquely lacerate below; midrib evident, percurrent, eventually denuded of lamina below and thickened by secondary wall formation to form the stipe; lateral veinlets absent; membrane 2 cells thick, except near the margins and the midrib where somewhat thicker; sporangia in irregular bandlike sori on each side of the midrib; gametangia in larger but less compact groups occupying the greater part of the lamina.

North Carolina, St. Barthélemy, Guadeloupe, Brazil. Ordinarily growing on rocks at low-water level or a little above, but this plant has been dredged to a depth of 35 m.

REFERENCES: *Dictyopteris polypodioides,* J. Agardh 1848, Hoyt 1920; *Halyseris polypodioides,* Hemsley 1884, Murray 1889; *Neurocarpus membranaceus,* Taylor 1931.

Dictyopteris delicatula Lamouroux Pl. 33, fig. 3

Plants erect but small, 2–8 cm. tall, or indefinitely spreading and assurgent, the branches attaching to the substratum at several points or to each other at points of contact; branching dichotomous to irregular, the segments 0.5–5.0 mm. broad, the membrane 2 cells thick, the cells in distinct divaricate rows but not forming pinnate veinlets, the margins with longitudinally elongate cells and, in older states, with inconspicuous marginal ribs which may contain indurated cells; sori in single rows on each side of the midrib.

Bermuda, Florida, Mexico, Cuba, Jamaica, Hispaniola, Puerto Rico, Virgin Isls., Redonda I., Guadeloupe, Aves I., Dominica, Martinique, Barbados, Grenadines, Grenada, Costa Rica, Panama, Colombia, Netherlands Antilles, Venezuela, Trinidad, Tobago, Brazil, I. Trinidade. Very frequent, particularly as an epiphyte growing on coarser algae in rather shallow water, but seldom abundant; dredged to a depth of over 30 m.

REFERENCES: Hauck 1888, Collins 1901, Vickers 1905, 1908, Børgesen 1913–20, 1924, Collins and Hervey 1917, Schmidt 1923, 1924, Taylor 1939b, 1940, 1941b, 1942, 1943, 1954a, Hamel and Hamel-Joukov 1931, Feldmann and Lami 1937, Quirós C. 1948, Joly 1953c, 1957; *D. hauckiana*, Möbius 1889; *Dictyota cervicornis p. p.*, Mazé and Schramm 1870–77; *Haliseris delicatula*, Harvey 1852, Montagne 1863, Martens 1870, Mazé and Schramm 1870–77, Dickie 1874a, b, Piccone 1886, 1889, Murray 1889, Sluiter 1908, Grieve 1909; *Neurocarpus delicatulus*, Howe 1918a, Taylor 1928, 1929b, 1930a, b, 1933, Taylor and Arndt 1929; *N. hauckianus*, Howe 1928, Taylor 1929b, Taylor and Arndt 1929.

Dictyopteris jamaicensis n. sp. Pl. 32, fig. 2

Plants to 15 cm. tall or more, the lower axes denuded and stalk-like; at first subdichotomous, soon alternately branched in 1–3 degrees at intervals of 1–2 cm., the strap-shaped blades light brown, 7–11 mm. broad, the margins entire to minutely erose-dentate, somewhat undulate; membrane of one cell layer in the younger parts, of 2 layers in the older, where 42–60 μ thick; the cells in surface view somewhat rectangular, about 24–33 $\mu \times$ 33–42 μ; midribs conspicuous, lateral and marginal veins absent; sori of hairs scattered over the blades or somewhat restricted to 2 rows; reproductive organs not seen.

Jamaica. Dredged from depths of 33–73 m., where growing attached to shells and fragments of corals.

These plants probably had not attained maximum stature, but the membrane was fully matured below and the midribs developed into stalks some centimeters in length. They do not agree with *D. plagiogramma* because of the absence of veinlets, nor with *D. membranacea* because of the alternate branching, so they are considered to represent a hitherto undescribed species.

Dictyopteris plagiogramma (Montagne) Vickers Pl. 33, fig. 2

Plants erect, to a height of about 2.5 dm., profusely branched, pale and translucent; branching alternate to somewhat irregular, at intervals of 1.0–2.5 cm., the sinuses narrow but rounded, the segments 3–7 mm. broad, with a prominent midrib and pinnate veinlets running obliquely to the margin, the membrane otherwise in general one cell in thickness; margin entire, the cells not greatly elongated; sori of hairs in irregular rows beside the midribs; sporangia irregularly scattered near the midribs, 80–120 μ diam.

Bermuda, Florida, Mexico, Cuba, Jamaica, Hispaniola, Puerto Rico, Virgin Isls., Guadeloupe, Martinique, Barbados, Netherlands Antilles, Venezuela, Brazil, I. Trinidade. Probably growing on rocks, old corals, or shells as dredged from the moderate depths of 9–18 to as much as 55 m.

REFERENCES: Hauck 1888, Möbius 1889, Collins 1901, Vickers 1905, 1908, Børgesen 1913–20, Collins and Hervey 1917, Hamel and Hamel-Joukov 1931, Taylor 1942, 1943, Humm 1952b, Joly 1953c; *Haliseris plagiogramma*, Montagne 1863, Martens 1870, Mazé and Schramm 1870–77, Rein 1873, Dickie 1874a, b, Hemsley 1884, Murray 1880; *Neurocarpus plagiogrammus*, Howe 1928, Taylor 1928.

Dictyopteris hoytii n. sp. Pl. 32, fig. 1

Plants bushy, yellowish-brown, to 30 cm. tall, arising from a small cushion-shaped holdfast, the simple to frequently and irregularly branched lower axes denuded and stupose for 0.5–6.0 cm.; blades above usually dichotomous to 2–6 degrees, the sinuses narrow, the apices obtuse to retuse; blades strap-shaped, 1.4–3.0 cm. wide, 90–150 μ thick, undulate, the margins straight or obscurely crenate, conspicuously aculeate-dentate, the teeth 0.5–1.0 mm. long, at intervals of 0.75–2.0 mm.; structurally showing 2–4 medullary layers of large cells and a cortical layer of much smaller ones on each face; midribs conspicuous, extending to the branch apices; lateral veins delicate but evident, alternate or opposite at intervals of about 2–3 mm., extending obliquely to the margins, composed of rather small, thick-walled cells in irregular tiers of 5–6, the thickness to 185 μ; hairs in small groups scattered over the blades; tetrasporangia reported in small sori in lines parallel to the lateral veins; oögonia 65–130 μ diam., scattered over the blades singly or 2–3 together, except for a narrow marginal zone.

North Carolina, Venezuela. Growing on rocks and other hard objects from about low-tide line to a depth of 25 m.

This is a more closely branched plant than the South African *D. serrata,* of a lighter color and with broader, more abundantly aculeate segments which, though paler, seem to have more medullary layers than that species.

REFERENCE: *D. serrata,* Hoyt 1920.

Zonaria J. Agardh, 1841

Plants of moderate size, bushy, growing from a substantial felted rhizoidal holdfast, the branches plane, cleft into segments which become stalklike below and often proliferate; the stipes continued for a distance into the blades as midribs; blades growing from the distal margin by a row of cell initials, transversely banded by rows of hairs, structurally showing a medulla of regularly placed cells, each of which is covered at the surface by two cortical cells; sporangia in sori devoid of paraphyses, each producing eight aplanospores.

Zonaria tournefortii (Lamouroux) Montagne

Fronds erect, 3.5–17.0 cm. tall, stipitate, attached by a cushion at the base, parchmentlike, substance firm, color reddish brown; stipe subterete, elongated, branched, densely brown-tomentose; branches going off into cuneate, flattened blades flabellately incised, marked by vague lines radiating from the base and by distant, rather indistinct, concentric zonations parallel with the distal margin; stipes briefly continued as midribs onto the flattened segments of the lamina; sori forming irregular, spotlike patches scattered over the surface of the fertile blades.

North Carolina, Brazil. In this region only known as a plant of rather deep water; dredged from 25 m. and washed ashore.

REFERENCES: *Padina flava,* Greville 1833; *Stypopodium flavum,* Martens 1870; *Zonaria flava,* Hoyt 1920, Joly 1953c.

UNCERTAIN RECORD

Zonaria parvula C. Agardh

Bermuda. Type obscure. Murray 1889. A plant named *Z. parvula* by Greville is recognized as the sporophyte of *Cutleria multifida* (Smith) Greville, but Murray's reference to C. Agardh is not understood.

Pocockiella Papenfuss, 1943

Thallus flat, simple at first, later becoming lobed, growing from the distal end and often the lateral margins by a row of cell initials; frond composed of three tissues: a one-layered medulla, a subcortex of two or more layers of cells, and a single-layered cortex; in transverse section, the cells of the three tissues occurring in vertical rows; reproductive organs formed in scattered sori on both surfaces of the frond, the asexual sori lacking paraphyses, the sporangia producing four or eight aplanospores.

Pocockiella variegata (Lamouroux) Papenfuss Pl. 33, fig. 4

Plants to 3–8 cm. long, light brown, not much changed in color by drying; decumbent to erect, attached by rhizoids from the bases of the blades or in decumbent forms also from the lower surface; the thallus little cleft, becoming broadly deltoid to suborbicular or kidney-shaped, the lobes 1–7 cm. broad; the marginal row of apical cells developing a single layer of large medullary cells covered by 1–2 subcortical layers, and by a cortical layer of smaller cells; the number of subcortical layers usually equal on both sides of the blade, which is 100–300 μ thick; cortical cells overlying the medullary cells about 8 on one face and only 4 on the other.

Bermuda, North Carolina, Florida, Bahamas, Caicos Isls., Cuba, Jamaica, Hispaniola, Virgin Isls., St. Barthélemy, St. Eustatius, Nevis, Guadeloupe, Martinique, Barbados, Old Providence I.,

Panama, Colombia, Brazil. Common on intertidal rocks and shells, especially in exposed situations, and dredged to 55 m. depth.

REFERENCES: Joly 1953c, Taylor 1954a; *Aglaozonia canariensis,* Børgesen 1912, 1913–20, Howe 1920; *Gymnosorus collaris,* Martens 1870, Taylor 1931; *G. variegatus,* Collins 1901, Gepp 1905b, Sluiter 1908; *Zonaria lobata,* Mazé and Schramm 1870–77 *p. p.; Z. variegata,* Martens 1870, Mazé and Schramm 1870–77, Dickie 1874b, 1884 Hemsley 1884, Möbius 1889, Murray 1889, Piccone 1889, Vickers 1905, 1908, Børgesen 1913–20, 1924, Collins and Hervey 1917, Howe 1918a, b, 1920, 1928, Hoyt 1920, Schmidt 1923, 1924, Taylor 1928, 1930a, 1933, 1939b, 1943, Taylor and Arndt 1929, Sanchez A. 1930, Feldmann and Lami 1937; *Z. variegata* v. *discolor,* Martens 1870.

Stypopodium Kützing, 1843

Plants of considerable size, bushy, growing from a massive rhizoidal holdfast, at first plane, eventually cleft to numerous narrow segments and becoming bushy; growing from the distal margin by a row of cell initials, transversely banded by rows of hairs; structurally showing a medulla of irregularly placed cells over each of which lie several cortical cells; tetrasporangia in sori near the hair zones, associated with paraphyses.

Stypopodium zonale (Lamouroux) Papenfuss Pl. 28, fig. 1

Plants to 3.6 dm. or somewhat more in height, often highly iridescent in water, becoming blackish brown on drying, attached by a large holdfast of compacted rhizoids; the blade at first broadly fan-shaped and even to 15 cm. wide, with an irregularly growing margin and an attenuate base, soon variously cleft into cuneate or even strap-shaped segments 1–5 cm. broad, the narrow bases of which become invested with rhizoids; blades transversely zonate at irregular intervals of about 3–15 mm., because of bands of colorless hairs; sporangia in irregular sori bordering the hair zones.

Bermuda, Florida, Bahamas, Caicos Isls., Cuba, Cayman Isls., Jamaica, Hispaniola, Puerto Rico, Virgin Isls., St. Barthélemy, Guadeloupe, Dominica, Barbados, Panama, Colombia, Venezuela, Tobago, Brazil. Rather common and very handsome plants of shallow water, even from the intertidal zone, in areas moderately exposed to wave action, and dredged to a depth of 55 m.

REFERENCES: *Stypopodium fuliginosum,* Martens 1870; *S. lobatum,* Collins 1901, Gepp 1905; *Spatoglossum versicolor,* Martens 1870; *Taonia atomaria,* Murray 1889; *Zonaria fuliginosa,* Mazé and Schramm 1870–77; *Z. lobata,* Harvey 1852, 1861, Mazé and Schramm 1870–77 *p. p.,* Dickie 1874a, b, Hemsley 1884, Hauck 1888, Murray 1889, Vickers 1905, 1908, Børgesen 1913–20, Collins and Hervey 1917, Schmidt 1924; *Z. lobata* f. *padinoides,* Mazé and Schramm 1870–77; *Z. zonalis,* Howe 1918a, b, 1920, Taylor 1928, 1929b, 1930a, 1933, 1943, 1954a, Sanchez A. 1930, Hamel and Hamel–Joukov 1931, Joly 1953c.

Padina Adanson, 1763

Plants brown, or whitened by a thin deposit of lime, often conspicuously zonate, of moderate size, tufted, attached by a compact felted rhizoidal holdfast, the stipe often invested by rhizoids; consisting of one or more generally fan-shaped blades which may become split into narrow spatulate segments, growing from a row of initial cells bordering the distal, curved, involute margin; occasionally developing an extensive coarse, fibrous, ramified growth phase,[1] the branches of which develop from single prominent apical cells, and from these branches fanlike blades may ultimately arise at various points; blades two to several cells thick, zonate, the zones marked by concentric rows of hairs; blades commonly whitened with lime on one face or both, and bearing tetrasporangia, scattered or in sori between the hair bands; dioecious (except in one species), the gametangia also in sori between the hair bands, the oögonial sori indusiate, the antheridial sometimes naked.

KEY TO SPECIES

1. Plants with blades less than 6 cells thick throughout, excluding the stipe; sterile and potentially fertile glabrous zones alternating 2
1. Plants with blades 6–8 cells thick in the basal region above the stipe, 2 cells thick near the revolute margin; piliferous zones

[1] Misinterpreted by Collins (1901, p. 251) and made the type of the "genus" Dictyerpa.

distinct, the glabrous zones to 7 mm. wide; the blade surfaces little if at all lime-encrusted .... P. vickersiae, p. 236

2. Species with completely bistratose blades; sporangia sori indusiate, appearing in every second glabrous zone 3
2. Blades bistratose behind the involute margin, 3–4 layered in the basal region 5

3. Indusium persistent, conspicuous and visible with the hand lens or the unaided eye 4
3. Indusium evanescent, its base visible but only with a compound microscope; sporangia to 120 μ diam.; surface cells of the blade 25–35 μ wide .... P. sanctae-crucis, p. 237

4. Cells of the blade surface 25–38 μ wide; mature sporangia to about 80–110 μ diam.; a band of tetrasporangial sori occurring in the middle of each fertile zone .... P. haitiensis, p. 235
4. Cells of either layer of blade 30–50 μ wide; mature sporangia about 170 μ diam.; sori occurring in 1–3 equally spaced bands in each fertile zone .... P. perindusiata, p. 235

5. Sori generally forming a band on either side of each piliferous zone and placed close to it, but at times the lower band of sori not developed; sporangial sori indusiate, the indusia persistent; plants strongly encrusted on the upper surface, less on the lower; gametophytes monoecious ..... P. pavonica, p. 234
5. Sori generally forming a band above and usually at a distance from every second piliferous zone; sporangial sori naked; fertile glabrous zones narrower than sterile ones; usually the upper surface of the frond moderately encrusted; gametophytes dioecious .... P. gymnospora, p. 237

Padina pavonica (Linnaeus) Thivy [1] n. comb.

Blades more or less clustered, fanlike, usually partially curved into the shape of a trumpet, stipitate and stupose below, often substantially calcified on the upper surface and more lightly below; piliferous lines alternating on the 2 faces, but often obsolete on the lower side, 1.5–4.0(–6.0) mm. apart; blade above of 2 layers of cells and about 65 μ in thickness, below of 3–4 layers and about 130 μ thick; sporangial sori in more or less continuous bands on each

[1] In a doctoral dissertation completed in 1945 Mrs. Francesca Thivy proposed this combination and several new species of Padina; some of these have already been published for her by myself, and others appear here on succeeding pages.

side or at least on the upper side of the hair lines of the upper surface, 90–140 μ diam., covered by a persistent indusium; gametophytes monoecious, the sori of antheridia alternating with those of oögonia in bands on each side of the upper surface hair lines, the oögonia about 50 μ diam.

Bermuda; old and unconfirmed reports attribute this species to Florida, Cuba, Jamaica, Virgin Isls., Barbados, Grenada, Colombia, and Brazil, but little confidence can be placed in them. Generally found on rocks just below low-tide line in moderately protected situations, but in Europe this species has been dredged to a depth of 20 m.

REFERENCES: F. Thivy *ms.* n. comb.; *Fucus pavonicus* Linnaeus, Species Plantarum 2 : 1162, 1753; *P. pavonia,* Hooper 1850, Harvey 1852, 1861, Montagne 1863, Dickie 1874a, b, Hemsley 1884, Collins and Hervey 1917, Howe 1918a; *Zonaria pavonia,* Martens 1870; *Z. tenuis,* Martens 1870, Zeller 1876.

Padina haitiensis Thivy n. sp. Pl. 75, fig. 1

Plants stipitate, arising from a felted holdfast, 5–6 cm. tall, to 8 cm. broad, commonly split even to the stipe into broad lobes, ferruginous and stupose toward the base, brown below, hardly calcified on the lower surface, with the hair lines evident, somewhat more heavily calcified on the upper surface; zones between hair lines of the upper surface 1.5–2.0 mm. broad; blades of 2 cell layers from near the apical margin to near the stipe, the thickness 65 μ in the basal third but 105 μ immediately above the stipe; cells 25–38 μ wide, those of the lower layer about 0.66 as deep as those of the upper; zones alternately fertile or sterile, the indusiate sori lying in the middle or upper part of each fertile zone, 0.4–0.6 mm. broad, the sporangia 80–110 μ diam.

Turks Isls., Hispaniola.

REFERENCE: F. Thivy *ms.*

Padina perindusiata Thivy n. sp. Pl. 75, fig. 2

Plants exceeding 10 cm. in height, very lightly calcified on both sides, piliferous lines alternating on the opposite faces, the zones irregular in width, alternately fertile and 1.5–3.0 mm. broad, or sterile and 0.75–2.0 mm. wide; blades bistratose, 105 μ thick below, 90 μ in the middle portion, the cells 30–50 μ wide, the lower cell layer 0.33–0.5 as deep as the upper; sporangial sori 0.5–0.75 mm. wide, in one continuous or in 2–3 broken lines in the middle of each fertile zone, bordered by a conspicuous indusium; sporangia 170 μ diam.

Florida. Apparently a moderately deep-water species, as dredged from 9–14 m. depth.

REFERENCE: F. Thivy *ms.*

Padina vickersiae Hoyt — Pl. 34, fig. 1

Plants rather large, forming tufts 10–15 cm. tall, the originally suborbicular blades 5–15 cm. broad, becoming cleft and proliferous, split to narrow rhizoid-invested stalklike portions below; zonate, without calcification or very lightly calcified on the upper concave surface; hair lines about 2–7 mm., usually about 4 mm. apart; blades 2 cells and 50 μ thick near the growing margin, but 4 cells thick throughout most of the blade, and 6–8 cells and 150–220 μ thick in the lower portions; each interpilar zone potentially fertile, the reproductive organs in 1–2 irregular lines about the middle of the band; sporangial bands about 0.5 mm. broad, indusiate, occurring on both surfaces of the blade but chiefly on the lower, the sporangia 100–120 μ diam.; dioecious, the antheridial sori about 200 μ broad, the oögonia 30–65 μ diam.

Bermuda, North Carolina, Florida, Texas, Mexico, Bahamas, Cuba, Jamaica, Hispaniola, Virgin Isls., St. Barthélemy, Guadeloupe, Aves I., Martinique, Barbados, Grenada, British Honduras, Panama, Colombia, Netherlands Antilles, Venezuela, Trinidad, Brazil. Rather frequent, growing on the roots of mangroves, on rocks, old corals, or shells in very sheltered, perhaps occasionally more exposed places, and dredged to a depth of about 14 m.

REFERENCES: Hoyt *in* Howe 1920, Howe 1920, 1928, Taylor 1928, 1929b, 1930a, 1935, 1936, 1940, 1941a, 1942, 1943, 1954a, Sanchez A. 1930, Questel 1942, Joly 1951, 1957; *P. howeana,* Børgesen 1913–20, Schmidt 1923, 1924; *P. pavonia,* Mazé and Schramm *p. p.,* 1870–77; *P. variegata,* Montagne 1863, Möbius 1889, 1895, Vickers 1905, 1908, Collins and Hervey 1917.

Padina sanctae-crucis Børgesen Pl. 34, fig. 2

Plants clustered and often rather large, 5–15 cm. tall, the rather light brown blades 5–9 cm. broad, often split into narrower auricled segments, the stalklike portions invested by rhizoids; the blades markedly zonate, substantially calcified only on the upper concave side; bistratose, the cells 25–35 μ broad in surface view, the lower layer of cells 0.66–1.0 the height of those of the upper layer; hair lines alternating on the 2 surfaces, about 1.0–2.5 mm. apart, fertile in alternate zones, the sori in bands close above each second hair line; sporangia appearing in irregular bands, up to 120 μ diam. and when young showing an evanescent indusium; dioecious, the oögonia 35–50 μ diam., in 1–2 bands 0.1–0.25 mm. broad; antheridia in a continuous or broken band.

Bermuda, Florida, Bahamas, Caicos Isls., Anguila Isls., Cuba, Cayman Isls., Jamaica, Hispaniola, Puerto Rico, Virgin Isls., St. Barthélemy, Guadeloupe, Grenada, British Honduras, Netherlands Antilles, Brazil, I. Trinidade. Often abundant, growing on rocks, old corals, and shells in rather shallow water of moderately protected stations, but dredged to a depth of about 14 m.

REFERENCES: Børgesen 1913–20, Collins and Hervey 1917, Howe 1918a, 1920, 1928, Taylor 1928, 1929b, 1930a, 1935, 1940, 1943, 1954a, Taylor and Arndt 1929, Sanchez A. 1930, Feldmann and Lami 1937, Joly 1953c; *P. pavonia p. p.*, Mazé and Schramm 1870–77. *Dictyerpa jamaicensis*, Collins 1901 is a growth stage of a Padina, perhaps of various species, but certainly of this one.

Padina gymnospora (Kützing) Vickers

Plants tufted, 5–10 cm. tall, the blades 5–20 cm. broad, rounded, or split into narrower portions, the lower parts stalklike and stupose; usually rather moderately calcified on the upper surface; 2 cells and 50–60 μ thick near the growing margin, 3 cells and 75–110 μ thick below, where the cells of the middle layer may be taller than those of the surfaces, or even 4 cells thick near the stipe; hair lines alternating on the 2 faces, the sterile zones 2–4 mm. wide, alternating with fertile zones 1.5–3.0 mm. wide; sporangia in discontinuous bands 0.5–1.5 mm. wide, usually median between alternate hair lines and without indusia, 90–125 μ diam.; gametophytes dioecious; antheridia in 1–2 bands in alternate zones.

Bermuda, Florida, Cuba, Cayman Isls., Jamaica, Hispaniola, Puerto Rico, Virgin Isls., Guadeloupe, Barbados, Grenada, Costa Rica, Panama, Colombia, Netherlands Antilles, Venezuela, Trinidad, Brazil. Growing on rocks, old corals, or shells in rather shallow water in both sheltered and moderately exposed situations.

REFERENCES: Vickers 1905, 1908, Sluiter 1908, Børgesen 1913–20, 1924, Collins and Hervey 1917, Schmidt 1923, 1924, Taylor 1933, 1942, 1943, 1954a, Feldmann and Lami 1937, Joly 1957; *Zonaria gymnospora,* Martens 1870, Zeller 1876, Murray 1889.

UNCERTAIN RECORDS

Padina commersonii Bory

Florida, Puerto Rico, Virgin Isls., Antigua, Guadeloupe, Netherlands Antilles, Brazil. Type from Mauritius. De Toni 3:244; Mazé and Schramm 1870–77, Hauck 1888, Murray 1889, Sluiter 1908, Schmidt 1923, 1924; *P. tenuis,* Montagne 1863.

Padina durvillaei Bory

Jamaica, Brazil. Type from Chile. De Toni 3:245; Collins 1901; *P. durvillaei* f. *obscura,* Piccone 1886, 1889.

Padina fraseri (Greville) J. Agardh

Brazil. Type from Australia. De Toni 3:246; *Zonaria fraseri,* Martens 1870.

CHORDARIALES

Plants of filamentous construction, the filament strands single and becoming corticated, or more often grouped into branches of the multiaxial type, or, habit reduced and the plants more or less pulvinate or disciform; plants as ordinarily described reproducing by unilocular sporangia and often with plurilocular gametangia and isogametes, or gametangium-like structures, but also generally bearing functional gametangia on a small filamentous phase which may have developed from zoöspores originating in the unilocular sporangia.

KEY TO FAMILIES

1. Plants large, usually erect and branched, corticated 2
1. Plants small to minute, or if crustose sometimes a few centimeters in diameter 3

2. Plants with a single initial axial filament visible in young branches . . . . SPERMATOCHNACEAE, p. 249
2. Plants with four or five axial filaments . . . STILOPHORACEAE, p. 250
2. Plants with many initial axial filaments . . CHORDARIACEAE, p. 245

3. Plants disklike, crustose or pulvinate 4
3. Plants small, tufted from a spreading or penetrating base, the erect portion of free filaments with intercalary growth ELACHISTEACEAE, p. 244

4. Plants minute, pulvinate or punctiform, of spreading filaments sometimes bearing short erect branches MYRIONEMATACEAE, p. 239
4. Plants minute, or to several centimeters in diameter, disk- or crustlike 5

5. Sporangia of transformed surface cells LITHODERMATACEAE, p. 244
5. Sporangia lateral to paraphyses in sori . . . . . RALFSIACEAE, p. 243

MYRIONEMATACEAE

Small cushionlike plants, consisting of radiating branched filaments with apical growth closely laterally approximated, bearing short vertical filaments, unilocular sporangia, plurilocular gametangia, hairs, and paraphyses on the upper surface.

KEY TO GENERA

1. Basal filaments not coalescing into a disk . . . . . . ENTONEMA, p. 242
1. Basal filaments usually radial and uniting to form a disk 2

2. Paraphyses present, as well as uniseriate erect filaments ASCOCYCLUS, p. 241
2. Paraphyses absent 3

3. Basal layer simple . . . . MYRIONEMA, p. 239
3. Basal part usually of 2 cell layers . . . . HECATONEMA, p. 240

Myrionema Greville, 1827

Plants forming minute, flattened cushions or disks, a unistratose layer below of radiating crowded filaments, the filaments differing in habit and dimensions according to whether sporophyte or gametophyte individuals are concerned; forming above short erect assimilators and colorless hairs; unilocular sporangia on separate plants; plurilocular gametangia usually pluriseriate, more or less stalked.

Myrionema strangulans Greville

Plants forming small disks; basal filaments closely or sometimes a little loosely placed; cells 5.0–8.5 μ diam., 1–3 diameters long; hairs 8–13 μ diam.; erect assimilators rather densely crowded, their basal cells often in lateral contact, the filaments 50–100 μ long, somewhat clavate-moniliform, the cells cylindrical below, to subspherical and shorter above, 6–11 μ diam., with several discoid chromatophores in each cell; reproductive organs on the basal filaments or on the basal cells of the assimilators; sporangia ellipsoid or obovoid, 20–35 μ diam., 35–65 μ long; gametangia obtuse-cylindrical, sessile or on stalks of 1–2 cells, 7–11 μ diam., 15–50 μ long.

North Carolina, Florida, Virgin Isls. Epiphytic on larger algae.

REFERENCES: Hoyt 1920, Taylor 1928; *M. vulgare*, Børgesen 1913–20.

Hecatonema Sauvageau, 1897

Plants discoid below, the radial basal filaments sometimes once divided parallel to the substrate, producing short, simple or branched, erect filaments generally over the colony surface; specialized hairs with basal growth produced directly on the filaments of the disk or terminating the free filaments; chromatophores numerous, discoid or as very short bands; unilocular sporangia not certainly known; plurilocular gametangia formed on the disk or on the free filaments.

KEY TO SPECIES

1. Plants to about 1–2 mm. tall; gametangia lateral or terminal ..... H. terminalis
1. Plants with erect filaments usually less than 250 μ long; gametangia terminal ..... H. floridana

Hecatonema terminalis (Kützing) Kylin

Plants inconspicuous, tufted; bases disklike, of irregularly entangled primary basal filaments about 10–18 μ diam., cells 8–24 μ long, sometimes horizontally divided; erect filaments to 2 mm. tall, infrequently branched, cylindrical, or a little attenuate, 8–15 μ diam., cells 1–4 diameters long; sporangia little known, reported as ellipsoid or oval, 40–52 μ long, 24–30 μ diam., terminal on the branches; gametangia terminal or lateral on the erect filaments, on short stalks, ovoid to oblong, generally somewhat distorted, not rostrate, 48–120 μ long, 14–18 μ diam.

Puerto Rico. Growing on rocks, coarse algae, pilings, etc., in shallow water.

REFERENCES: Hamel 1931–39; *Ectocarpus terminalis,* Hauck 1888.

Hecatonema floridana (Taylor) n. comb.

Plant a minute epiphyte, with decumbent branches ramifying over the host or closely associated to form a disk; disk becoming 2 cells thick, giving rise to erect assimilative filaments and gametangia, and rarely to hairs; basal cells 7.2–10.8 μ diam. and 10.8–19.8 μ long; erect filaments 8.5–12.0 μ diam., cells 17.5–25.0 μ long, the filaments averaging about 12 cells in length; gametangia 12.5–30.0 μ diam. and 50–80 μ long, spindle-shaped, obtuse to acute at the apex.

Florida. Growing upon Zonaria in rather shallow water.

REFERENCE: *Phycocoelis floridana,* Taylor 1928, p. 109.

Ascocyclus Magnus, 1874

Plant minute, in the form of epiphytic colonies or disks of radiating branched filaments, from which arise colorless hairs with basal growth, short, clavate vegetative filaments, elongated hyaline unicellular paraphyses, and plurilocular gametangia.

KEY TO SPECIES

1. Base discoid; filaments 4–6 μ diam.; gametangia 5–12 μ diam. .... A. orbicularis
1. Base somewhat diffuse; filaments irregular, 7–8 μ diam.; gametangia 16–26 μ diam. .... A. hypneae

Ascocyclus hypneae Børgesen

Plants not discoid, the branching filaments creeping on and among the surface cells of the host; cells of these filaments irregular, often swollen in the middle, 7–8 μ diam.; paraphyses clavate, slender below, with a stalk of 1–4 cells, enlarging above to 10–16 μ diam., the length to 65 μ, the walls rather thick and the contents dark brown; gametangia stalked, oblong to fusiform, 16–26 μ diam., to 65 μ long.

Virgin Isls. Epiphytic on Hypnea in shallow water.

REFERENCE: Børgesen 1913–20.

Ascocyclus orbicularis Magnus

Plants disklike, 1–3(–10) mm. diam., formed of radiating filaments; erect vegetative filaments absent or rare, somewhat clavate; paraphyses sessile or on one-celled stalks, at first ovate and brown, later much elongate, truncate and saccate, hyaline, 8–25 μ diam., to 170 μ long; gametangia sessile or on one-celled stalks, cylindrical or slightly clavate, obtuse, uniseriate, about 20–75 μ long, 5–12 μ diam.

Bermuda. Epiphytic on marine algae and phanerogams in shallow water.

REFERENCES: Collins and Hervey 1917, Howe 1918a.

Entonema Reinsch, 1875

Plants partly endophytic, forming brown spots on larger algae, the filaments ramifying over and between the cells of coarser algae, with short erect free filaments several cells in length; specialized hairs with a basal growth zone present, extending outside the surface of the host.

Entonema parasiticum (Sauvageau) Hamel

Plants minute; basal portion of highly ramified filaments penetrating deeply between the cells of the host; basal filament cells 2–10 μ diam., 8–30 μ long; external filaments erect, crowded, simple, 60–90 μ long, the tips rounded or piliferous, the cells 6–8 μ diam., 6–12 μ long, each containing a single platelike chromatophore; hairs chiefly borne on the superficial basal filaments; gametangia sessile or on 1–2-celled stalks, subcylindrical, acute at the apex, 9–10 μ diam., to 50 μ long, showing 2 rows of cells in side view.

Bermuda. Partly endophytic, growing on various coarse Rhodophyceae.

REFERENCES: Hamel 1931–39; *Ectocarpus parasiticus*, Collins and Hervey 1917.

RALFSIACEAE

Crustose, mostly perennial plants, composed of radial filament systems laterally united to form a horizontal layer, from which series of cells arises to form a more or less pseudoparenchymatous thick crust, and also colorless hairs; plurilocular gametangia and unilocular sporangia are formed in superficial sori on different plants.

Ralfsia Berkeley, 1831

Plants rarely minute, often one to several centimeters in diameter, forming expanded, more or less attached, usually somewhat brittle crusts, which are at times multiple by the overgrowth of overlapping lobes; of two layers, the basal of radiating filaments, usually with rhizoids on the under surface, and the upper layer of parallel ascending assimilatory filaments, the erect cell series compact, laterally united, usually with dark-colored cell walls, and with the chromatophores concentrated in the cells near the surface of the thallus; hairs inconspicuous, in tufts; unilocular sporangia and plurilocular gametangia in superficial sori on different plants, the sporangia lateral at the bases of the free paraphysal filaments, the gametangia terminal on erect filaments.

Ralfsia expansa J. Agardh

Plants forming smooth, rounded, coriaceous dark brown crusts which become irregular, wrinkled, and concrescent with age, rather lightly attached to the substratum by rhizoids; basal layer of radiating filaments from which branches curve upward to form the bulk of the thallus, or sometimes the horizontal layer median and the filaments curving toward both surfaces; paraphyses clavate, about 5–14 cells and 100–170 μ long, with cells shorter than broad above, but about 4–7 diameters long below, the maximum diameter about 12 μ; sporangia pedicellate, oval-pyriform, 30 μ diam., 75–120 μ long; gametangia cylindrical, 1–2 on a surface cell of the thallus, 15–20 cells long, 5–6 μ diam., the lower cells sterile, the upper fertile and in part biseriate in side view by longitudinal division of the original cells.

?Florida, Mexico, Hispaniola, Virgin Isls., Martinique, Netherlands Antilles, Tobago, Brazil. Plants encrusting stones or other hard objects in shallow water.

REFERENCES: Harvey 1852, Murray 1889, Børgesen 1912, 1913–20, Taylor 1928?, 1929b, 1942, 1943, Hamel and Hamel-Joukov 1931, Feldmann and Lami 1932, Joly 1957; *Myrionema expansum* J. Agardh 1847.

UNCERTAIN RECORD

Ralfsia sp.

Florida. Taylor 1928, as figured.

LITHODERMATACEAE

Plant encrusting, strongly adhering by the under surface, growing peripherally; structure parenchymatous below; plurilocular gametangia originating as uniseriate lateral outgrowths of uniseriate filaments which grow up from the surface, or by transformation of outgrowths from the surface cells; unilocular sporangia formed by transformation of surface cells.

Lithoderma Areschoug, 1875

UNCERTAIN RECORD

Lithoderma sp.

Virgin Isls. Børgesen 1913–20.

Lithoderma extensum (Crouan) Hamel

Brazil. Type from France. De Toni 3:307 as *L. fatiscens;* Möbius 1890.

ELACHISTEACEAE

Tufted plants, showing a basal portion of densely intertwined filaments and long simple free filaments with intercalary growth as assimilators, the latter often bearing below short lateral branchlets; plurilocular gametangia and unilocular sporangia borne among the short branchlets, but also in some genera on the assimilators.

Elachistea Duby, 1830

Plants small brush- or cushionlike epiphytes, the basal portion of densely matted colorless filaments which often penetrate the host tissues, the vegetative filaments branched at the base only, uniseriate, of two sorts; in part long, free, straight assimilatory filaments with basal intercalary growth, the cells with numerous rounded chromatophores, but at their bases having crowded, often curved and moniliform short filaments or paraphyses of limited growth; unilocular sporangia pyriform, plurilocular gametangia slenderly cylindrical; both organs borne among the paraphysal filaments.

Elachistea minutissima Taylor Pl. 29, fig. 11

Plants minute, epiphytic; basal filaments ramifying throughout the cryptostomata of Sargassum, giving rise to erect, basally branched, assimilative filaments, paraphyses, and gametangia; assimilators reaching 425 μ in length, 9.4–13.0 μ diam., cell length equal or to twice as much as breadth; paraphyses at maturity 40–56 μ long, 4.5–5.0 μ diam. above, the lower cells subcylindrical, the uppermost ovoid; gametangia 3.76–5.60 μ diam., rarely with longitudinal walls, 37.6–81.0 μ long, narrowly cylindrical.

Florida. Epiphytic, nearly enclosed within the cryptostomata of Sargassum.

REFERENCE: Taylor 1928.

CHORDARIACEAE

Plants subspherical to filiform, simple or branched, sometimes hollow, of filamentous construction within a gelatinous matrix, a core of relatively colorless filaments giving off lateral assimilatory branches which constitute the cortex and contain most of the chromatophores; branch growth apical or trichothallic; colorless hairs with basal growth usually abundant; unilocular sporangia and plurilocular gametangia immersed, formed from branchlets of the cortical filaments, or sometimes functionally on the branches of a small filamentous gametophyte.

KEY TO GENERA

1. Assimilatory filaments free .... Levringia, p. 246
1. Assimilatory filaments embedded in jelly 2

2. Medulla of easily separable parallel filaments .... Eudesme, p. 246
2. Medulla pseudoparenchymatous, the constituent filaments firmly conjoined .... Cladosiphon, p. 248

Levringia Kylin, 1940

Plants slender, branched, tomentose, the axis of numerous united filaments each with a zone of trichothallic growth, proximal to which each develops lateral branches which subdivide and form the surface layer of free divergent assimilatory filaments, the elongate distal cells of which contain small discoid chromatophores; plurilocular gametangia usually short-stalked, pod-shaped, borne on the branching divisions at the base of the long assimilators.

Levringia brasiliensis (Montagne) Joly

Plants 6–14 cm. tall, greenish brown, sparingly branched below, usually with one dominant division, the branches 1.5–2.0 mm. diam., medulla colorless, pseudoparenchymatous, composed of the primary filaments and rhizoidal outgrowths from them; cortical assimilatory filaments closely branched at first, but distally simple for the outer millimeter or more, the cells to 16–22 μ diam., 40 μ long; gametangia on special short branchlets attached near the bases of the assimilators, irregularly cylindrical to slightly fusiform, blunt, about 25 μ diam. and to 125 μ long.

Brazil, Uruguay. Plants generally growing on the shells of Mytilus on rocky shores and in the intertidal zone.

references: Joly 1953a, 1957; *Mesogloia brasiliensis,* Montagne 1843, 1846.

Eudesme J. Agardh, 1880

Fronds branched, from a small disk holdfast, the branches terete, gelatinous; the axis solid or locally inflated, of longitudinal filaments laterally branched, large and slender ones intermixed, loosely conjoined; externally bearing a cortex of radiating peripheral assimilatory filaments borne in fascicles on the lateral divisions of the medullary filaments, the assimilators subsimple, cylindrical below, above of asymmetrical oval cells, often surrounding the unilocular sporangia, which are obovoid to spherical; plurilocular gametangia formed by the conversion of the upper portion of the assimilators, often formed on the same plants as the unilocular sporangia.

Eudesme zosterae (J. Agardh) Kylin

Plants relatively small and slender, 1 dm. to, rarely, 3 dm. tall, bright brown, gelatinous and slippery, the axis with a contracted base, the main portion usually 1–3 mm. diam., bearing few to numerous short, simple, or less often 1–3-times redivided, secondary branches, all flexuous to contorted, the surface often nodulose; peripheral filaments erect, rather stout and rigid, cylindrical below, moniliform above, where the cells are spheroidal, 20–40 μ diam.; sporangia formed in the tufts of assimilators, at their bases, ovoid to nearly globular, 50–100 μ long, 25–70 μ diam.; gametangia cylindrical to obtuse-conical, formed in rows by lateral outgrowth of the upper cells of the assimilators, sometimes bilaterally placed, each usually 3–6 cells or 15–45 μ long, 1–2 cells or 7.5–18.0 μ diam.

Bermuda, North Carolina, Florida, Bahamas, Cuba, Guadeloupe, Panama, Venezuela. Epiphytic on coarse algae and phanerogams in shallow, relatively quiet water.

REFERENCES: Taylor 1957; *Castagnea zosterae*, Howe 1918a, 1920, Hoyt 1920, Taylor 1928, *p. p.*, 1936, 1942, 1945a; *Cladosiphon zostericola*, Mazé and Schramm 1870–77, Murray 1889; ?*Mesogloia griffithsiana*, Hemsley 1884; *M. zostericola*, Hooper 1850; *Myriocladia capensis* and *M. gracilis*, Mazé and Schramm 1870–77, Murray 1889; *M. mediterranea*, Mazé and Schramm 1870–77. It is not clear how far the older references apply to *E. zosterae* as it is known in the north, or how far they should be referred to *Cladosiphon occidentalis* Kylin.

UNCERTAIN RECORD

Eudesme virescens (Carmichael) J. Agardh

Bermuda, Florida. Type from Scotland. De Toni 3:404 (as *Castagnea virescens*); Murray 1889; *Mesogloia affinis*, Hooper 1850; *M. virescens*, Hooper 1850, Harvey 1852, Dickie 1874b, Hemsley 1884.

Mesogloia C. Agardh, 1817

UNCERTAIN RECORD

Mesogloia vermiculata (J. E. Smith) Le Jolis

Bermuda. Type from England. De Toni 3:425; Hemsley 1884, Murray 1889.

Liebmannia J. Agardh, 1842

UNCERTAIN RECORD

Liebmannia leveillei J. Agardh

Mexico. Type from Corsica. De Toni 3:340; Murray 1889, Harvey 1852.

Cladosiphon Kützing, 1843

Axis spongy or hollow, the surrounding medulla of filaments more or less united into a pseudoparenchyma whose cells are four to eight diameters long; assimilatory filaments fairly sharply differentiated, simple or somewhat branched at the base; colorless hairs with basal growth present; plurilocular gametangia developed by lateral growth and subdivision of the distal cells of the assimilators.

Cladosiphon occidentalis Kylin

Plants erect from a small holdfast, to 1.5–2.0 dm. tall, paniculately much branched; branches firmly fleshy, from 0.75–2.50 mm. diam., the ultimate branchlets somewhat irregular; structure filamentous, the axial strands of filaments 50–112 μ diam., rather strongly cohering, of cells 200–400 μ long; these filaments giving off lateral divisions which form an assimilatory cortex, the strands of which at the base are composed of cells about 5.6–7.5 μ wide and 19–28 μ long, hardly constricted at the nodes, becoming more deeply and often unilaterally constricted above, where the cells are 11–15 μ diam., and 13–21 μ long; sporangia usually 1–3 in each fascicle of assimilators, of various ages, at maturity 30–38 μ diam., 56–80 μ long; gametangia normally on separate plants, formed by the transformation of assimilators, simple to much branched, the simple gametangia of 5–10 moniliform cells about 13 μ in diameter, or some with these cells longitudinally divided, gametangia of this type grading into short and broad pluriseriate organs which reach

19 μ in diameter and 48 μ in length and which may be oppositely, alternately, or secundly branched; colorless hairs present, 7.5–15.0 μ diam., the cells reaching 150 μ in length.

Bermuda, Florida, Virgin Isls. Plants epiphytic on sea grasses or coarse algae in shallow water, and dredged to 11 m. depth.

REFERENCES: Kylin 1940; *Castagnea zosterae*, Børgesen 1913–20, Collins and Hervey 1917, Taylor 1928 *p. p.*

Sauvageaugloia Hamel *in* Kylin, 1940

UNCERTAIN RECORD

Sauvageaugloia griffithsiana (Greville) Hamel

Bermuda. Type from England. De Toni 3: 407 (as *Castagnea griffithsiana*); Hamel 1939 (without a genus description); *C. griffithsiana,* Murray 1889.

SPERMATOCHNACEAE

Plants more or less widely branched, slender, the branches composed of one or more central filaments each of which is developed directly from an apical cell; medulla surrounded by assimilatory filaments.

Nemacystus Derbès and Solier, 1850

Thallus irregularly or more or less widely branched, solid or hollow, the initial axial filament at least in the young parts readily visible; the medulla pseudoparenchymatous, originating from downgrowths of the basal cells of the cortical assimilatory filaments; these assimilators simple, not markedly clavate; hairs with a basal meristem present; unilocular sporangia and plurilocular gametangia known, sometimes occurring on the same individual, the former developed on the bases of the assimilators, the latter on little special branch systems directly on the outer medullary layer.

Nemacystus howei (Taylor) Kylin Pl. 29, figs. 12–14

Plants very lubricous, axes 2–4 dm. long or probably longer, generally much tangled below; diameter of branches generally 0.3 mm., frequently less, becoming at most 1 mm. in the largest main axes; branching alternate, abundant, especially in younger portions;

medullary cells 58–166 μ diam., 132–1410 μ long; lateral assimilatory filaments 10.0–16.5 μ diam. toward their summits and 3.2–8.0 μ diam. toward their bases, the cells few to about 8 to a branch, cylindrical below to moniliform or reniform above, containing several small chromatophores; hairs 8–10 μ diam. at the base; sporangia located near the bases of the clusters of assimilators, 1–6, generally not over 2, in a fascicle, spherical to ovoid, straight or asymmetrical, 18–37 μ diam. by 20–45 μ long; gametangia borne near the base of the fascicle of assimilative filaments, replacing branches, up to 10 in a fascicle, 6.6–7.5 μ diam. and 36.5–66.0 μ long on a strictly gametangial plant, 1–2 cells in diameter, 10–20 or more cells in length, somewhat constricted at the transverse walls, never branching; on a plant bearing abundant sporangia there were found many gametangia 6.6–8.3 μ diam. and 50–78 μ long.

Bermuda, Florida. Dredged from 7–55 m. depth, the plants apparently entangled among coarser algae.

REFERENCES: Kylin 1940; *Castagnea howei,* Taylor 1928.

STILOPHORACEAE

Plants bushy, attached by a holdfast disk, stoutly filiform, freely dichotomously to irregularly branched, developing by the apical division of an axial group of filaments which produce a parenchymatous cortex ultimately bearing colorless hairs and sori of unilocular sporangia or plurilocular (?) gametangia associated with short assimilatory filaments; active plurilocular gametangia also produced on an alternate microscopic phase.

Stilophora J. Agardh, 1841

Characters of the family; filaments formed on the surface of the thallus, simple if sterile, branched if fertile, and then segregated into groups to which the reproductive organs are confined, these forming as lateral divisions of the sparingly branched fertile filaments.

Stilophora rhizodes (Ehrhart) J. Agardh

Plants more or less erect, but entangled, from a disklike holdfast, filiform, 1–3 dm. tall, pale brown, relatively stiff and brittle, freely and dichotomously to irregularly branched without persistent main axes; medulla of 4–5 longitudinal series of cells, or ultimately hollow, the cortex narrow, parenchymatous; branches nodulose with clumps of arcuate filaments 75–85 μ long which are slender and branched below, where about 3.5 μ diam., above undivided and asymmetrically moniliform, to 9.5–13.0 μ diam., the cells with lenticular chromatophores; sporangia and uniseriate gametangia usually on different plants, arising from the bases of the peripheral filaments; sporangia scattered, obovate or clavate, 36–56 μ long, 22–32 μ diam.; gametangia filiform, uniseriate, with 4–9 cells, 30–50 μ long, 9.5–11.5 μ diam.

Bermuda, North Carolina, Florida. Growing on rocks and other hard objects in sheltered, shallow water.

REFERENCES: Collins and Hervey 1917, Hoyt 1920, Taylor 1954a.

UNCERTAIN RECORDS

Stilophora antillarum Crouan

Guadeloupe, the type locality, but a *nomen nudum*. This is probably *Rosenvingea floridana* (Taylor) Taylor. Mazé and Schramm 1870–77, Murray 1889.

SPOROCHNALES

Plants bushy, the branches tipped with tufts of hairs; sporophytic, reproducing by zoöspores formed in unilocular sporangia; alternating with a microscopic, filamentous, oögamous gametophyte stage.

SPOROCHNACEAE

Plants often wiry, the tips of the branches and branchlets terminating in conspicuous tufts of chromatophore-rich hairs each of which has an independent growth zone just above the base; main axes composed of firmly united filaments growing from an apical meristem below the hair tuft; at maturity the axes with a firm pseudoparenchymatous medulla of large colorless cells covered with a small-celled chromatophore-rich cortex; unilocular sporangia borne on the cortex, sometimes associated with branched paraphyses; the zoöspores from these producing microscopic, monosiphonous, branched, filamentous gametophytes which are monoecious, and produce minute antheridia and oögonia.

KEY TO GENERA

1. Sporangia borne on branched paraphyses localized on special lateral branchlets .... SPOROCHNUS
1. Sporangia on fertile papillae scattered over the plant, sessile, associated on the same basal cells with paraphyses.......... NEREIA

Nereia Zanardini, 1845

Plants bushy, arising from a discoid holdfast, irregularly branched, without a dominant main axis, the long branches flexuous, bearing numerous papillae which may elongate into short spur branchlets; papillae, branchlets, and axes all terminating in conspicuous tufts of brown filaments; unilocular sporangia borne on the papillae below the hair tufts, sessile, on cortical cells which also bear short simple paraphyses.

Nereia tropica (Taylor) Taylor Pl. 29, figs. 3, 4; pl. 35, fig. 1

Plants large, to 35 cm. or more in height, 2–3 times alternately branched, the branches at times exceeding the axis in length, the larger divisions 1–2 mm. diam., the lesser slender, and all terminating in brown hair tufts, these hairs 17–26 μ diam., and 3–6 mm. long; younger branches beset with papillae and ultimately short spur branchlets which likewise terminate in hair tufts, but older parts of the plants depiliate; sori of sporangia and paraphyses crowded about the bases of the hair tufts; paraphyses of single clavate or obpyramidal cells 17–28 μ diam., 32–51 μ tall; sporangia subcylindrical to obovoid, 11–19 μ diam., 26–42 μ long.

Bermuda, Florida. Plants growing on a rocky bottom in deep water, dredged from depths of 9–55 m.

REFERENCES: Taylor 1955a; *Stilophora tropica,* Taylor 1928.

Sporochnus C. Agardh, 1817

Plants erect from a conspicuous fibrous holdfast, wiry, much branched, with a distinct primary axis and numerous lateral secondary axes, the divisions of the last degree bearing numerous determinate pedicellate branchlets which, like the axes, terminate in brown hair tufts; fertile areas restricted to the swollen distal parts of the branchlets, comprised of sori of branched paraphyses which terminate in enlarged rounded cells and which bear laterally several unilocular sporangia.

KEY TO SPECIES

1. Branchlet pedicels short, to 2 mm. long; fertile portions ovoid to short spindle-shaped .... S. pedunculatus
1. Branchlet pedicels longer, to 10 mm.; fertile portions ultimately long-cylindrical .... S. bolleanus

Sporochnus pedunculatus (Hudson) C. Agardh Pl. 35, figs. 4, 5

Plants in excess of 3 dm. tall, slender, freely branched, and the longer of these branches rebranched, the lateral branches reaching 2 dm. in length; axis and branches of all orders closely beset (at intervals of 2–5 mm.) with short branchlets, these generally 0.75–1.25 mm., rarely to 2 mm., long, 0.5 mm. diam., on pedicels usually 1 mm., rarely to 2 mm., long; when plants are young, the branchlets are tipped with a tuft of brown filaments 3–4 mm. long, which are shed from old plants; long branches also tipped with tufts of filaments, which reach 5 mm. in length.

Bermuda, North Carolina, Florida, Mexico, Brazil. Plants, usually of rocky substrata, growing in rather deep water, dredged from 5 to 55 m. and less certainly to a depth of 110 m.

REFERENCES: Hemsley 1884, Murray 1889, Hoyt 1920, Taylor 1928, 1930a, 1954a, Humm 1952b.

Sporochnus bolleanus Montagne Pl. 35, figs. 2, 3

Plants in excess of 8 dm. tall, coarse, branched, the branches beset with short branchlets which reach a length of 3.6 mm., a diameter of 0.5 mm., on pedicels 3–10 mm. long; when the plants are young, the branchlets are tipped with tufts of brown filaments reaching 8–10 mm. long, but they shed them when older; leading axes also tipped with a large tuft of brown filaments reaching a length of 12–15 mm.

Bermuda, Florida, Puerto Rico, Brazil. Plants of rocky shores, growing in exposed situations a little below low-water level, and dredged to a depth of 90 m.

REFERENCES: Hauck 1888, Collins and Hervey 1917, Howe 1918a, Taylor 1928, 1930a.

DESMARESTIALES

Plants of filamentous construction, the branches with subapical trichothallic growth at least at first, branches uniaxial, often with considerable cortication; plants as ordinarily described reproducing by unilocular sporangia, the zoöspores from which develop microscopic filamentous oögamous gametophytes.

DESMARESTIACEAE

Plants arising from a basal disk, filiform, or compressed and with a midrib, oppositely or alternately branched; in cross section showing an axial filament and a subparenchymatous cortex, sometimes also with short superficial assimilatory filaments; sporangia resulting from the conversion of superficial thallus cells, or of assimilatory filaments.

Arthrocladia Duby, 1830

Plants bushy, wandlike, the primary branches terete, bearing whorls of simple or alternately branched chromatophore-bearing filaments; structurally with a very large thick-walled axial filament, and a cortex which shows several layers of relatively large thin-walled cells within, and outwardly small cells with thick walls and brown chromatophores; unilocular sporangia seriate, moniliform, stalked, replacing lower branchlets of the whorled tufts of filaments.

Arthrocladia villosa (Hudson) Duby

Plants from small disks, to 4 dm. tall, loosely, slenderly, and widely branched, brown, drying quite greenish; axis to 1 mm. thick, rather stiff, irregularly branched below, alternately or characteristically oppositely 1–3-times branched above; on all but the oldest parts of the plant bearing tufts of filaments about 4 mm. long, which branch oppositely below, alternately above, to 21 μ diam. below, 7 μ diam. in the ultimate divisions; unilocular sporangia replacing branchlets of these filaments, alternately or unilaterally placed on the lower segments, to 15–20-seriate, the series 50–350 μ long, the individual sporangia 11–15 μ diam., 5.5–8.5 μ long, each producing several swarmers and developing a single lateral pore.

North Carolina.

REFERENCE: Williams 1949a.

PUNCTARIALES

Plants filiform or membranous-expanded, with intercalary growth and ultimately parenchymatous subdivision of the cells; plants as ordinarily described reproducing by unilocular sporangia and also with plurilocular gametangia or gametangium-like structures, but sometimes bearing functional gametangia and anisogametes on a very small filamentous phase.

STRIARIACEAE

Plants filiform, variously branched, sometimes tubular; of uniseriate or irregularly polysiphonous filaments, or parenchymatous and of more substantial diameter; growth intercalary, but cell division soon ceasing below; reproductive organs formed by the direct transformation of superficial cells or by division of them.

UNCERTAIN RECORDS

Stictyosiphon Kützing, 1843

Stictyosiphon charoides Zeller

Brazil, the type locality. De Toni 3:469; Zeller 1876.

Striaria Greville, 1828

Striaria fragilis J. Agardh

Guadeloupe. Type from the Baltic Sea. De Toni 3:472; Mazé and Schramm 1870–77, Murray 1889.

PUNCTARIACEAE

Plants of variable form, filiform, tubular, or phylloid, usually simple, generally stalked, from a basal disk; colorless hairs usually abundant; growth generally intercalary, structure parenchymatous; unilocular sporangia and plurilocular gametangia formed by the transformation of superficial cells of the erect plant, external or immersed, or on the creeping base, or on a filamentous phase of the plant.

KEY TO GENERA

1. Plants with elongate, simple or branched axes 3
1. Plants not elongate 2

2. Plants saccate, often deeply lobed .... COLPOMENIA, p. 260
2. Plants clathrate .... HYDROCLATHRUS, p. 260

3. Plants very small; erect axes simple or subsimple from a filamentous creeping base .... MYRIOTRICHIA, p. 256
3. Plants several centimeters tall at maturity 4

4. Plants unbranched 5
4. Plants branched 6

5. Axes flat, in the form of solid blades 7
5. Axes terete, hollow, sometimes locally constricted SCYTOSIPHON, p. 259

6. Axes and branches solid, often compressed ..... CHNOOSPORA, p. 263
6. Axes terete, hollow .... ROSENVINGEA, p. 261

7. Cells of the blade surface small, very different in size from those of the interior .... PETALONIA, p. 258
7. Cells of the blade surface almost as broad as those of the interior .... PUNCTARIA, p. 257

Myriotrichia Harvey, 1834

Plant filiform, the base sometimes creeping, of uniseriate filaments attached by rhizoids, the erect axis uniseriate below, sometimes parenchymatous above; beset with short radial branchlets; longitudinal axial growth intercalary, near the base; hairs colorless with basal growth; unilocular sporangia solitary, scattered, opposite, or verticillate, attached to the creeping base, the axis, or the branchlets; plurilocular gametangia lateral, solitary or aggregated into little clusters, uni- to pluriseriate; sporangia and gametangia sometimes occurring on the same plant.

KEY TO SPECIES

1. Gametangia in tufts on the ends of the erect filaments, about 8 μ diam., the cells uniseriate .... M. repens

1. Gametangia lateral or terminal on the erect filaments, single or in clusters, 12–30 μ diam., the cells in 2 or more rows . . M. occidentalis

Myriotrichia occidentalis Børgesen

Plants with basal filaments irregularly branched, the cells about 10 μ diam., 20 μ long; basal filaments bearing hairs or short erect filaments, rarely plurilocular gametangia; short erect filaments about 150 μ tall, 12 μ diam., consisting of about 6 cells, the tips obtuse; principal erect filaments hair-tipped, about 1 mm. tall, 18–24 μ diam., so far as known uniseriate, the cells shorter than broad or subequal, bearing occasional single or grouped, simple or sparingly branched, short lateral branchlets and hairs; gametangia sessile or pedicellate, fusiform to subconical, obtuse, 12–30 μ diam., 50–100 μ long.

Virgin Isls. Epiphytic on Dictyota at a depth of 10 m.

REFERENCE: Børgesen 1913–20.

Myriotrichia repens Hauck

Tufts 0.25–1.00 mm. tall, of branched creeping filaments bearing erect, simple, uniseriate axes 12–25 μ diam., the cells usually 2–4 diameters long; hairs sometimes formed on the basal layer and terminating the erect filaments; sporangia sessile or pedicellate, formed on the basal filaments, or 2–few together terminal or verticillate on the erect filaments, spherical or oval, about 25–50 μ diam.; gametangia clustered at the tops of the fertile filaments, uniseriate, about 8 μ diam.

Bermuda. Epiphytic on larger plants in shallow water.

REFERENCES: Collins and Hervey 1917, Hamel 1931–39.

Punctaria Greville, 1830

Thallus foliaceous, beset with tufts of hairs, tapering sharply to a short slender stalk and attached by a basal disk; when mature four to seven cells thick, the surface cells not conspicuously small; unilocular sporangia scattered; plurilocular gametangia somewhat clustered, angular, immersed below, the apex prominently projecting above the surface, pluriseriate, like the sporangia formed by the metamorphosis of the cortical cells; gametangia and sporangia occurring on separate plants, or on the same plant in that sequence, the swarmers developing to filamentous plantlets which may bear gametangia.

Punctaria latifolia Greville

Plants in the form of lanceolate blades arising from minute basal disks; texture soft and thinly membranous, subpellucid, yellowish or olive-brown, in drying becoming dull greenish and generally not adhering to paper; stalks 2–5 mm. long, the blades 10–45 cm. long, 2–8(–15) cm. broad, at the base abruptly cuneate, at the apex generally obtuse; 2–4 cells and 50–65(–85) μ thick, the cells in surface view 15–40 μ diam., in fairly clear rows.

North Carolina.

REFERENCE: Williams 1949a.

Petalonia Derbès and Solier, 1850

Fronds flat, each arising on a short filiform stalk from a holdfast disk; substance firm; structure internally of several layers of large cells intermixed with more slender filaments, externally of a distinctive layer of small chromatophore-bearing cells; fertile areas at first localized, later involving the whole surface of the frond, producing crowded plurilocular gametangia.

Petalonia fascia (O. F. Müller) Kuntze

Plants linear-lanceolate, 0.75–4.5 dm. long, often gregarious, arising from disciform bases, with slender stalks, the bases of the blades tapering, usually asymmetrical, the margins straight or irregularly curved, the apices tapering; colorless hairs numerous over the surface; substance membranous to leathery, dark brown, particularly when fertile; thickness of blade 145–200 μ in sterile parts and in fertile parts to 270 μ, structurally showing much larger cells within, those near the surface smaller; gametangia crowded, covering large areas of the blade, uniseriate, about 3.5–9.5 μ diam., 30–75 μ long.

V. **caespitosa** (J. Agardh) Taylor. Thinner and lighter brown, from a markedly asymmetrical base, the width 2.5–6.0 cm.

North Carolina, Florida, Brazil. Growing on rocks or other objects in shallow water. Primarily a northern species.

REFERENCES: Williams 1949a, Humm 1952a, Joly 1957.

Scytosiphon C. Agardh, 1811

Fronds from a small basal disk, simple and tubular, terete, firmly membranous; the wall of large cylindrical inner cells, the cortex of small rounded and angular ones; growth intercalary near the base; plurilocular gametangia forming a broad, continuous layer or localized spots, developed from the cortical cells, generally uniseriate, associated with paraphyses.

Scytosiphon lomentaria (Lyngbye) C. Agardh

Plants gregarious, terete, tapered, short-stalked; length 1.5–7.0 dm., diameter 5–10(–30) mm., characteristically locally constricted when mature; gametangia subcylindrical, crowded in large fruiting areas, 3.5–7.5(–9.5) μ diam., 45–65 μ long, uniseriate or frequently with oblique walls, less often biseriate with longitudinal walls, or forked; paraphyses 27–32 μ long (or more?), 8–13 μ diam., not exceeding the gametangia in height.

Bermuda, North Carolina, South Carolina, Florida. Growing on rocks in shallow water, just below low-water level, in moderately protected situations. In warmer waters these plants do not reach the large sizes often seen on northern shores.

REFERENCES: Harvey 1852, Collins and Hervey 1917, Howe 1918a, Blomquist and Humm 1946, Taylor 1954a.

Asperococcus Lamouroux, 1813

UNCERTAIN RECORDS

Asperococcus clavatus

Florida. Apparently a *nomen nudum*. Hooper 1850.

Asperococcus echinatus (Mertens) Greville

Florida. Type from the Atlantic Ocean. De Toni 3:494; Hooper 1850.

Colpomenia Derbès and Solier, 1856

Plants depressed spherical, hollow, simple, becoming constricted or lobed, sometimes with age expanded and more flat; the wall of two layers of cells, those within large and rounded, but those forming a one-celled surface layer small, angular, and containing chromatophores; hairs with basal meristems in scattered tufts; plurilocular gametangia crowded in small, scattered, superficial sori, associated with paraphyses.

Colpomenia sinuosa (Roth) Derbès and Solier Pl. 36, fig. 1

Plants light yellowish brown, solitary or clustered, sessile, spherical, becoming lobed, or irregularly expanded, individually reaching a diameter of 3–12 cm.; the wall 0.3–0.4 mm. thick, the large interior cells to 180 μ diam., nearly colorless, the intermediate ones smaller, the cells of the surface layer 3.7–7.5 μ diam.; gametangia cylindrical, 3.7–7.5 μ diam., 18.8–30.0 μ long; paraphyses obovate, to 11 μ diam., 47 μ tall.

Bermuda, North Carolina, Florida, Mexico, Bahamas, Cuba, Jamaica, Puerto Rico, Virgin Isls., Guadeloupe, Martinique, Barbados, Panama, Netherlands Antilles, Venezuela, Brazil, I. Trinidade. Common on intertidal rocks in exposed situations, and dredged to a depth of 13.5 m.

REFERENCES: Collins 1901, Vickers 1905, 1908, Børgesen 1913–20, Collins and Hervey 1917, Howe 1918a, 1920, 1928, Taylor 1928, 1930a, 1940, 1942, 1954a, Hamel and Hamel-Joukov 1931, Feldmann and Lami 1937, Blomquist and Humm 1946, Joly 1951, 1953c, 1957; *Asperococcus sinuosus* Harvey 1852, Dickie 1874a, b, Hemsley 1884, Murray 1889; *Hydroclathrus sinuosus,* Hauck 1888, Möbius 1892; *Soranthera leathesiaeformis,* Mazé and Schramm 1870–77.

Hydroclathrus Bory, 1826

Plants sessile, at first spherical, hollow, later expanded and flattened, early clathrate, the strands becoming subterete, the margins inrolled and often attached by rhizoids, so that the cortex of small chromatophore-bearing cells tends to enclose the medulla of large, rather colorless cells; hairs with basal meristems formed in depressed sori; plurilocular gametangia forming a continuous layer over the surface.

Hydroclathrus clathratus (Bory) Howe Pl. 36, fig. 5

Plants at first subspherical, but more often irregularly explanate, to several centimeters, even a few decimeters, in diameter, or clustered and covering large areas; perforated, the apertures to 1–3 cm. in diameter, seldom more, the intervening strands about 1–3 mm. broad; cells of the central parenchyma 50–150 μ diam., of the cortical layer 5.5–9.0 μ diam. and polygonal; older plants with few or no reproductive structures, but in younger individuals the gametangia in a continuous layer; gametangia 9–17 μ diam., 13–23 μ tall.

Bermuda, Florida, Bahamas, Cuba, Cayman Isls., Jamaica, Virgin Isls., Guadeloupe, Martinique, Barbados, British Honduras, Netherlands Antilles, Venezuela, Brazil. Often abundant on intertidal rocks in exposed situations, less commonly dredged to depths of 7 m.

REFERENCES: Howe 1920, Taylor 1928, 1930a, 1935, 1940, 1942, Sanchez A. 1930, Hamel and Hamel-Joukov 1931, Feldmann and Lami 1932; *H. cancellatus,* Harvey 1852, Martens 1871, Möbius 1890, 1892, Mitchell 1893, Collins 1901, Vickers 1905, 1908, Sluiter 1908, Børgesen 1913–20, Collins and Hervey 1917, Howe 1918a; *Asperococcus clathratus,* Mazé and Schramm 1870–77, Murray 1889; *Encoelium clathratum,* Martens 1870.

Rosenvingea Børgesen, 1917

Plants subsimple to bushy, attached by a basal disk, branched alternately or subdichotomously, the branches terete or compressed, with apical and intercalary growth; hollow except near the base where the cavity becomes filled with rhizoid-like filaments, the thallus wall of a few parenchymatous layers of cells of which the inner are large and nearly colorless, the outer small and containing the chromatophores; hairs single or in sori; plurilocular gametangia in sori, formed directly by upgrowth of the surface cells.

KEY TO SPECIES

1. Plants tangled and subrepent, or if erect most abundantly branched in the upper part; branches often flattened and adherent .... R. intricata, p. 262
1. Plants erect, generally taller, the branching more evenly distributed 2

2. Diffusely branched; branching irregular to subdichotomous, the divisions similar to the axis .... R. sanctae-crucis, p. 263
2. Sparingly branched, chiefly near the base; the branches slender, flexuous, tapering .... R. floridana, p. 262

Rosenvingea floridana (Taylor) Taylor Pl. 29, figs. 7, 8

Plants bushy, to 3.2 dm. tall, the primary axis short, soon indistinguishable, the branching fairly abundant near the base, above more sparse, irregular, with occasional aculeate projections, the distal segments long, flexuous, tapering, acute; diameter of main axis near the base about 0.3 mm., but above as much as 1.5 mm.; terete or compressed, hollow except near the base; hairs present, to 12 μ diam.; gametangia inconspicuous, in small sori or covering large areas, initially 3–5 cells long, the outer 1–3 cells dividing longitudinally, forming a stalked structure as much as 14 μ diam., but not over twice as long as broad.

Florida. Growing on rocks in shallow water.

REFERENCES: Taylor 1955a, *Cladosiphon floridana,* Taylor 1928.

Rosenvingea intricata (J. Agardh) Børgesen Pl. 36, fig. 2

Plants bushy or low, densely matted and tangled, to 3–4 dm. or more in height, soft in texture, usually abundantly branched, especially above, the branches often interadherent, contorted, 1–10 mm. diam., markedly narrowed from the main trunk to the branchlets, the tips attenuate; cells of the inner layer to 28–37 μ broad by 56–131 μ long; of the cortex angular, 9–19 μ diam.

Bermuda, North Carolina, Florida, Mexico, Jamaica, Guadeloupe, Barbados, Netherlands Antilles, Venezuela, Trinidad, Brazil. Growing on rocks in shallow, rather exposed situations, and dredged to a depth of 35 m.

REFERENCES: Collins and Hervey 1917, Howe 1918a, Taylor 1928, 1942, Williams 1951, Humm 1952b; *Asperococcus intricatus,* J. Agardh 1847, Harvey 1852, Mazé and Schramm 1870–77, Dickie 1874b, Murray 1889 *p. p.; A. ramosissimus* and *A. schrammii,* Mazé and Schramm 1870–77, Murray 1889; *Encoelium intricatum,* Martens 1870; *Striaria attenuata,* Collins 1901 *p. p.,* Vickers 1905, 1908; *S. attenuata* v. *ramosissima,* Collins 1901; *S. intricata,* Vickers 1905, 1908.

Rosenvingea sanctae-crucis Børgesen

Plants tufted, very briefly stipitate, to 20 cm. tall, terete or somewhat compressed, the surface uneven; branching sparse and irregularly alternate or subdichotomous, the branches obtuse at the tips, about 2 mm. diam.; walls of 3–4 cell layers, the inner cells large; the outer cells small and polygonal, each with a single platelike chromatophore; hairs chiefly scattered, 8–9 μ diam.; gametangia in irregular sori, cylindrical to clavate, 5–12 μ diam., 20–40 μ long or more.

North Carolina, Florida, Jamaica, Virgin Isls., Guadeloupe, Martinique, Venezuela. Growing on stones in shallow water of somewhat sheltered places.

REFERENCES: Børgesen 1913–20, Hamel and Hamel-Joukov 1931, Taylor 1940, 1942; *Asperococcus orientalis*, Hoyt 1920; *Striaria attenuata* f. *ramosissima*, Phyc. Bor.-Am. 737, 1640.

UNCERTAIN RECORD

Rosenvingea orientalis (J. Agardh) Børgesen

Guadeloupe. Type from the Philippines. *Asperococcus orientalis*, De Toni 3:495; Mazé and Schramm 1870–77, Murray 1889.

Chnoospora J. Agardh, 1847

Plants of moderate size, the base discoid, bushy above and moderately firm, the branches successively tapering toward the acute tips, solid, subterete or commonly distinctly compressed; medulla colorless, of elongate large cells parenchymatously disposed; toward the surface the cells decrease in size, and constitute the chromatophorous cortex; fertile areas often contiguous or merging, composed of numerous short, crowded, moniliform filaments borne on the cortex, some sterile, some becoming siliquiform, composed of a series of unilocular sporangia.

Chnoospora minima (Hering) Papenfuss Pl. 36, figs. 3, 4

Plants dull greenish brown, not adhering to paper upon drying, rather stiff, to 15 cm. tall, the axis repeatedly fastigiately di-polychotomous, closely or loosely branched, the segments 7–40 mm. long, of about equal length in any one individual, subterete to compressed, to 1 mm. thick, 2 mm. broad, slightly expanded at the forks, gradually tapering toward the upper divisions.

Redonda I., Guadeloupe, Panama, Netherlands Antilles, Venezuela, Tobago, Brazil. Growing on rocks in the intertidal zone.

REFERENCES: *C. atlantica,* J. Agardh 1847, Dickie 1874b, Joly 1951; *C. fastigiata,* Feldmann and Lami 1937; *C. fastigiata* v. *atlantica,* Möbius 1889; *C. fastigiata* v. *pacifica,* Mazé and Schramm 1870–77, Murray 1889; *C. pacifica,* Taylor 1929b, 1942, Joly 1957.

UNCERTAIN RECORD

Chnoospora implexa (Hering) J. Agardh

Guadeloupe. Type from the Red Sea. De Toni 3:466; Mazé and Schramm 1870–77, Murray 1889.

FUCALES

Plants of sturdy construction, parenchymatous externally, but often with a filamentous medulla; growth from an apical cell group in the adult branches; sporophytic, the sporangia carried in superficial conceptacles with branched paraphyses; the potential mega- and microspores after meiosis continuing to divide and passing through a cytological gametophyte phase; when matured the sporangia discharging their now gametangial contents in coherent masses from the conceptacles; these, disintegrating in the water, liberate gametes, the antherozöids fertilizing the eggs in a free state.

KEY TO FAMILIES

1. Axes subterete, to alate with a midrib, but not foliar; vesicles if present intercalary .... FUCACEAE, p. 264
1. Axes terete, bearing distinct foliar organs; vesicles usually present, lateral or immersed in the terminal branchlets SARGASSACEAE, p. 266

FUCACEAE

Plants with dichotomous or pinnate strap-shaped simple or costate branches, in some cases with piliferous cryptostomata, often with buoyant air bladders; receptacular portions of the thallus terminating main branches or forming stalked lateral branchlets, heterosporous, becoming sexual after meiosis, the oögonia each forming several eggs.

KEY TO GENERA

1. Branching dichotomous, the branches with a midrib bordered by a thinner membrane .... Fucus
1. Branching irregular to lateral, branches narrow, compressed, without a marginal membrane .... Ascophyllum

Ascophyllum Stackhouse, 1809

Plants cartilaginous or fleshy, the main axis irregularly to pinnately branched with ultimate branchlets pinnately disposed; no midrib or cryptostomata present; deciduous receptacles likewise pinnately disposed on short pedicels; heterosporous and eventually dioecious; the oögonia forming four eggs.

Ascophyllum nodosum (Linnaeus) Le Jolis

Plants olive to yellowish, reaching a large size, usually 3–6 dm., but occasionally to 3 m., long; erect from a disciform holdfast; main axis and principal branches compressed, coriaceous, with large single float bladders often 1.5–2.0 cm. diam. and 2–3 cm. long, at times much larger, greatly expanding the branches at irregular intervals; forking dichotomous or largely irregular to laterally pinnate; axis and branches pinnately beset with short, simple to forked, somewhat clavate compressed branchlets 1–2 cm. long, singly or in groups, which become converted into or replaced, in reproduction, by the deciduous, dioecious, oval, yellowish receptacles, which are single or in clusters of 2–5 on stalks 0.5–2.0 cm. long, the fruiting portion about 1–2 cm. long, 5–12 mm. wide.

Bermuda, ?Brazil. A northern species, only known to occur in Bermuda as drifted ashore, but not infrequent. A few filaments of *Polysiphonia lanosa* have been noted on such plants, conformable with a northern origin.

REFERENCES: De Toni 1895, Collins and Hervey 1917, Howe 1918a, Prat 1934.

Fucus Linnaeus, 1753

Plants erect from a disciform or irregular holdfast, generally dichotomously, but sometimes subpinnately, branched, the branches strap-shaped with a more or less distinct midrib, to which by destruction of the margins of the blade the lower portions of the plant are often reduced; air-filled bladders of definite form regularly present in some species, absent in others, or exceptionally the frond in any species irregularly inflated; piliferous cryptostomata generally present on sterile portions of the blades; receptacles terminal on main or lateral branches, when young usually flat, later in some species inflated; heterosporous, becoming hermaphrodite or dioecious, the oögonia forming eight eggs.

Fucus vesiculosus Linnaeus

Plants brown, often large, generally 3–9 dm. tall; branching usually dichotomous or a little irregular, often proliferous below, the branches above strap-shaped, about 10–15 mm. wide, with a marked midrib throughout; scattered cryptostomata, and vesicles 5–10 mm. diam., usually prominent, the latter generally paired on each side of the midrib; receptacles terminal on the branches, single, paired, or forked, broadly lanceolate to obovate, usually 1.5–2.5 cm. long; sexes produced by different individuals, the conceptacles from which the antherozöids derive are orange when the receptacles are opened, while those which produce oögonia are olive green.

North Carolina, ?Bermuda, ?West Indies, ?Brazil. Growing on rocks in or just below the intertidal zone. Except for the North Carolina records all reports of this northern species from within our range are very questionable.

REFERENCES: De Toni 1895, Collins and Hervey 1917, Hoyt 1920; *?F. ceranoides* and *?F. distichus,* Kemp 1857, Rein 1873, Murray 1889.

SARGASSACEAE

Plants with flat costate branches or transitional stages to cylindrical branches bearing numerous macroscopic spinelike projections, turbinate foliar organs, or costate leaves, usually with cryptostomata and buoyant air bladders; conceptacles in ordinary branches or in special receptacular branchlet systems, heterosporous, becoming sexual after meiosis, the oögonia each producing a single egg.

KEY TO GENERA

1. Foliar organs when present spinelike; vesicles and fertile regions formed in ordinary branches .... CYSTOSEIRA, p. 267
1. Foliar organs very narrow and branched, or broader, leaflike, with a distinct costa; vesicles and fertile branches lateral and axillary, or the latter in terminal panicles ..... SARGASSUM, p. 268
1. Foliar organs turbinate or obpyramidal, often with a vesicle in the distal end and often with distal marginal and longitudinal spine-bordered ridges; fertile branches crowded, axillary TURBINARIA, p. 284

Cystoseira C. Agardh, 1820

Plants very bushy, light brown, often iridescent, to several decimeters in height, attached to a basal disk which bears one or more primary axes; stems smooth or tuberculate, the branching very irregular and the branches very diverse in form, usually terete, sometimes flattened, and then sometimes with a costa, in some cases without appendages and in others more or less abundantly beset with large spinelike projections serving as foliar organs; vesicles single or catenate, formed by inflation of the branches; cryptostomata usually present on the branches and foliar spines; fertile receptacular areas usually developed near the tips of the branchlets, the conceptacles usually bisexual.

Cystoseira myrica (Gmelin) J. Agardh

Primary stem short, muriculate, closely branched, the very slender terete leading branches spreading, to 3 dm. long or more, alternately divided to about 3 degrees, throughout abundantly beset with terete truncate cylindrical foliar projections 0.5–1.0 mm. long; vesicles usually solitary, on short spinulose stalks, ellipsoid, to about 2 mm. diam., 3 mm. long, each terminated by a short spinulose projection; receptacles slender, clavate-cylindrical, 2–9 mm. long, bearing scattered acute spines.

Florida, Bahamas. Growing on shells or rocks in shallow water; apparently rare.

REFERENCES: Murray 1889, Howe 1920, Taylor 1928.

UNCERTAIN RECORD

Cystoseira ericoides (Linnaeus) C. Agardh

Florida. Type from Europe. De Toni 3:167; Hooper 1850.

Sargassum C. Agardh, 1820

Plants light brown; when attached, with rather massive lobed holdfasts, bushy, with (in our species) slender branches and distinct broad to filiform, simple or occasionally forked, entire or serrate leaf organs; lateral stalked bladders usually present; receptacles axillary or paniculate, generally nearly cylindrical, less often prismatic-compressed, or flattened, forked.

It is commonly exceedingly difficult to distinguish sharply between the species of Sargassum growing within the area. The available descriptive characters would seem ample, but prove to be very variable. Respecting leaves, for instance, we find that those at the base of a rather young plant are often much larger than is characteristic of the species. Even typically simple-leaved species may show individuals with freely forked leaves about the base. Older plants, especially when fruiting, may show much narrower leaves in the upper portions than on the growing vegetative shoots. The habit may vary a great deal, plants exposed to the surf being short and very densely tufted with broader leaves than sheltered plants nearby, which will be open of habit. Respecting quite specialized characters also, such as shown by the receptacles of *S. platycarpum* and *S. hystrix,* one is constantly finding anomalous conditions. One may well suspect that part of the confusion is due to hybridization.

In order that maximum confidence may be placed in the sampling, a careful survey should first be made of the plants in any limited area, to determine the variation shown with regard to habitat and age. The best plants at the climax of the fruiting stage should be selected as typical, and then such variants as plants exposed to extreme surf, and vigorous juvenile specimens associated with the typical ones. If, as frequently happens, two or more species grow intermixed, special care will be necessary, and in such cases the collection of dwarfed or stunted specimens is particularly futile.

KEY TO SPECIES[1]

1. Floating and sterile; cryptostomata absent or inconspicuous 2
1. Normally attached; cryptostomata generally present 3

2. Leaves linear, teeth aculeate; vesicles aculeate or often with a long, even phylloid, projection .... S. natans, p. 281
2. Leaves lanceolate, teeth broad; vesicles at most muticous S. fluitans, p. 281

3. Leaves mostly lanceolate to linear, to 7–50 breadths long 4
3. Leaves from narrowly oblong or lanceolate to ovate 11

4. Leaves thin 5
4. Leaves thicker and somewhat rigid 9

5. Leaves narrowly linear, the upper often forked; plants often blackening when dried 6
5. Leaves linear to lanceolate, the upper seldom forked 7

6. Stems nearly smooth; leaves 1–2 mm. wide, nearly entire; stalks equal in length to the vesicles ..... S. ramifolium, p. 278
6. Stems generally muriculate; leaves 2–4 mm. wide, clearly serrate; stalks half the length of the vesicles .... S. furcatum, p. 277

7. Plants very delicate, very loosely branching; leaves often 20–30 times as long as broad ... S. filipendula v. montagnei, p. 271
7. Plants bushy; leaves not becoming so long 8

8. Stems smooth; leaves serrate; cryptostomata scattered S. filipendula, p. 270
8. Stems muriculate; leaves serrate to nearly entire; cryptostomata nearly restricted to one row on each side of the costa S. acinarium, p. 271

9. Stems smooth, or slightly roughened below; leaves subentire, sometimes glaucous when dried .... S. cymosum, p. 278
9. Stems usually muriculate; leaves dentate; stalks as long as, or shorter than, the vesicles 10

10. Leaves with a prominent and often dentate costa, especially those from the base of the plant; vesicles 3–8 mm. diam., subsessile or briefly stalked .... S. pteropleuron, p. 274

[1] Although of systematic value when present, spines on stems or leaves, teeth on leaf margins, wings on leaf costas, and apiculae on vesicles are all very often reduced or obsolete on individual plants; specimens from warm sheltered coves generally have thinner leaves than might be expected on the same species growing on an exposed shore.

10. Costa not strongly ridged or dentate; vesicles 3–6 mm. diam., on stalks about as long as the diameter .... S. bermudense, p. 274

11. Cryptostomata 0.6–0.9 mm. diam.; receptacles prismatic-compressed, serrate .... S. platycarpum, p. 280
11. Cryptostomata obvious, but smaller 12
11. Cryptostomata minute, few or none; leaves large, broadly oval or obtuse-oblong .... S. hystrix v. buxifolium, p. 279

12. Receptacles palmated with subterete or flattened lobes; stems smooth; leaves acutely serrate .... S. hystrix, p. 279
12. Receptacles not palmated or flattened, the divisions terete throughout 13

13. Leaves lanceolate to linear-oblong, sinuate to obtusely dentate; cryptostomata in subsingle series on each side of the midrib; stems smooth .... S. rigidulum, p. 272
13. Cryptostomata scattered 14

14. Leaves lanceolate, 4–8 breadths long, sharply serrate-dentate; stems smooth, or sometimes muriculate in the younger parts S. vulgare, p. 272
14. Leaves lanceolate to ovate, 2–4 breadths long, aculeate-dentate and crisped; stems muriculate, or becoming smooth below S. polyceratium, p. 276

Sargassum filipendula C. Agardh Pl. 37, fig. 3; pl. 40, fig. 2

Plants 3–6(–10) dm. long, erect from a conical, spreading, lobed holdfast, the main axes smooth, usually sparsely forked, the principal branches dominant, with the branchlets producing long pyramidal forms; branchlets bearing alternate stalked leaves 5–8 mm. wide, 3–8 cm. long, which are thin, linear-lanceolate, simple or, on the lower portions of the plant, not infrequently forked, serrate, with a clearly marked midrib and numerous scattered cryptostomata, which may also appear on the stem; axillary vesicles spherical at maturity, 3–5 mm. diam., stalked, the slender stalks about 5 mm. long; receptacles simple, forked, or sparsely racemosely branching, axillary in the upper parts of the plant, the raceme 0.3–0.5 as long as the subtending leaf, which may be considerably smaller than in the vegetative part of the plant.

V. **laxa** J. Agardh. Plants exceedingly slender, the stems smooth, sparingly branched, leaves below to 2 mm. broad, 4 cm. long, somewhat serrate, those above to 0.5–1.0 mm. broad, 0.5–

3.0 cm. long, nearly entire; vesicles 1–4 mm. diam., on stalks as long as their diameter or somewhat longer.

V. **montagnei** (Bailey) Grunow. Plants very long and slender, the delicate stems nearly smooth, very loosely branched; leaves linear, to 3–6 mm. broad, to 15 cm. long, tapering to a subacute tip, often once or twice forked, or alternate or subpinnately divided; midrib distinct, cryptostomata few, small and scattered, margin entire or very slightly dentate; vesicles spherical or subpyriform, 2–3 mm. diam., with a mucro or a leaflet on the tip, the pedicel 0.5–3.0 times as long as the diameter of the vesicle; fertile branches loosely alternately divided, torulose, spreading, the ultimate divisions to 3 cm. long.

V. **pinnata** Grunow. Stems smooth, leaves forked or sparingly pinnate, to 1.0 mm. wide, to 3.3 cm. long, the divisions narrowly linear; vesicles 2.0–4.5 mm. diam.

Bermuda, North Carolina, Florida, Mississippi, Texas, Mexico, Bahamas, Cuba, Jamaica, Hispaniola, Puerto Rico, Virgin Isls., St. Barthélemy, Nevis, Redonda I., Guadeloupe, Dominica, Venezuela, Brazil. Common on shells, old corals, and rocks from just below low-tide level to a depth of about 33 m.; the v. *montagnei* probably from deep water. Very variable, especially in the southern parts of its range.

REFERENCES: Harvey 1852, Murray 1889, Grieve 1909, Howe 1918a, 1920, 1928, Hoyt 1920, Taylor 1928, 1930a, 1933, 1941a, b, 1943, 1954a, Taylor and Arndt 1929; *S. affine*, J. Agardh 1847, Harvey 1852, Mazé and Schramm 1870–77, Zeller 1876, Hemsley 1884, Hauck 1888, Murray 1889; *S. affine* v. *angustifolia*, Murray 1889; *S. montagnei*, Mazé and Schramm 1870–77, Murray 1889. For the v. *montagnei*, Collins and Hervey 1917, Taylor 1942, 1954a; *S. montagnei*, Harvey 1852.

Sargassum acinarium (Linnaeus) C. Agardh

Plants with a short, stout main axis quickly divided into several slender branches which are usually muriculate; leaves on the main branches and on the scattered short lateral branchlets, not crowded, rather thin, lanceolate to linear, 3–8 mm. broad in the basal clusters, 1–3 mm. broad on the erect branches, 3–7 cm. long, acute, the margin irregularly serrate or the teeth obsolete; costa evident, the cryptostomata of moderate size, rather few, in one irregular row on each side of it; leaves on the basal parts of the plant sometimes pinnately divided; vesicles spherical, on terete or often alate stalks, in length 1.0–1.5 times the vesicle diameter; fertile branches axillary, verrucose, sparingly and loosely alternately branched, in length more or less equal to the subtending bract leaves, or in the upper parts of the plant subpaniculate by suppression of the bracts.

Bermuda, Florida, Louisiana, Colombia, Venezuela.

REFERENCES: Setchell 1933; *S. linifolium,* Rein 1873, Hemsley 1884, Murray 1889, Collins and Hervey 1917, Howe 1918a, Taylor 1942, 1954a.

Sargassum rigidulum Kützing

Plants rather small, dark brown when dried, 1–3 dm. tall, gregarious, the slender erect axes seldom forking above the base, somewhat loosely bearing short lateral branchlets; axes smooth except for the tuberculate remains of the branchlets on their lower portions; leaves few on each branchlet, firm in texture, 2–4 cm. long, 2.0–3.5 mm. wide, linear-oblong to lanceolate, not acute, the margins entire to irregularly and rather inconspicuously crenate-serrate; costa evident, subpercurrent, the cryptostomata moderately small, in an irregular row or somewhat scattered on each side of it; vesicles absent or rather few, spherical, not apiculate, 2.5–3.0 mm. diam., on terete stalks hardly as long; fertile branches axillary, cymose, rather closely forked, tuberculate, the clusters about 0.3–1.0 as long as the subtending leaf, but with age and by loss of leaves sometimes appearing paniculate.

Bermuda, Bahamas, Cuba, Jamaica, Hispaniola, Puerto Rico, Virgin Isls., Guadeloupe, Netherlands Antilles, Venezuela, Brazil.

REFERENCES: Setchell 1937; *S. lendigerum,* Greville 1833, Montagne 1863, Martens 1870, Mazé and Schramm 1870–77, Dickie 1884, Murray 1889, Collins 1901, Børgesen 1913–20, 1914b, Collins and Hervey 1917, Howe 1918a, 1920, Taylor 1943.

Sargassum vulgare C. Agardh Pl. 38, fig. 1; pl. 40, fig. 5

Growing attached, plants erect to a height of 4 dm. or more, the main axis with few to many long side branches; branches smooth, or sometimes muriculate in the younger parts; leaves firm, narrowly lanceolate, 1.5–3.0 cm. long, 2–4 mm. wide, sharply serrate or subentire below, tapering to a generally asymmetrical base and to the apex, the costa distinct, cryptostomata small and scattered; vesicles numerous, spherical, 2.5–4.5 mm. diameter, on pedicels 0.5–2.0 mm. long; receptacles much branched, axillary or growing out into fruiting branches bearing vesicles and small leaves, and reaching a length of 2–3 cm.

V. **foliosissimum** (Lamouroux) J. Agardh. Plants with leaves closely congested, generally smaller, broader and more obtuse, rather undulate and twisted; receptacles shorter, not exceeding the vesicles and rather concealed by the leaves.

Bermuda, Florida, Mexico, Bahamas, Cuba, Jamaica, Hispaniola, Puerto Rico, Virgin Isls., Anguilla I., St. Barthélemy, Antigua, Guadeloupe, Dominica, Martinique, St. Vincent, Barbados, Grenada, British Honduras, Costa Rica, Colombia, Netherlands Antilles, Venezuela, Tobago, Brazil. Reported from the splash zone on exposed coasts and from high-water mark down to 6 m. depth under less arduous conditions. Apparently dominant in the Virgin Islands, but in Florida and the Bermudas, as probably in most places, rather uncommon. This name has often been applied incorrectly, especially by earlier collectors, and this may account for a report of the species at a depth of 55 m.

REFERENCES: Harvey 1861, Montagne 1863, Martens 1870, 1871, Mazé and Schramm 1870–77, Rein 1873, Dickie 1874a, b, Hemsley 1884, Hauck 1888, Murray 1889, Collins 1901 *p. p.*, Gepp 1905b, Sluiter 1908, Børgesen 1913–20, 1914b, 1924, Howe 1920, Schmidt 1923, 1924, Taylor 1928, 1935, 1941b, 1942, 1943, 1954a, Sanchez A. 1930; *S. vulgare* v. *laxum* and v. *oxydon,* Martens 1870; *S. cymosum* v. *dichotomum, ?S. leptocarpum, S. polyceratium p. p., S. spinulosum,* Mazé and Schramm 1870–77, Murray 1889. For the v. *foliosissimum,* Dickie 1874a, Murray 1889, Collins 1901, Vickers 1908, Børgesen 1913–20, 1914b, Collins and Hervey 1917, Schmidt 1923, 1924, Taylor 1942, 1954a, Taylor and Arndt 1929; *S. bacciferum p. p., ?S. integrifolum p. p., S. polyceratium p. p., S. trachyphyllum, S. turneri p. p., S. vulgare p. p.,* Mazé and Schramm 1870–77, Murray 1889; *S. foliosissimum,* Vickers 1905, *S. brevipes,* Murray 1889; *S. pteropus,* Hauck 1888.

Sargassum pteropleuron Grunow Pl. 39, fig. 1; pl. 40, figs. 4, 9

Plants growing attached, branching abundantly from a large cushionlike holdfast, erect to a height of 6–9(–17.5) dm.; main axis distinct, but with long side branches, all orders typically slightly to abundantly muriculate; leaves on main branches or on short spur branches, generally set with one edge toward the axis, generally linear to narrowly lanceolate, but on vigorous basal branches sometimes broadly lanceolate, 2.5–9.5 cm. long and 2.5–4.0(–7.5) mm. wide, strongly but generally irregularly serrate-dentate, the base subsessile, asymmetrically deltoid, the leaf apex sharply tapering, or in the broadest leaves somewhat obtuse, the costa distinct, especially highly ridged below, in its fullest development becoming raised to a notably serrate wing; vesicles numerous, spherical, 3–8 mm. diameter, subsessile to very shortly stalked; receptacular branches in loose axillary clusters.

Bermuda, Florida, Texas, Bahamas, Caicos Isls., Cuba, Dominica, Venezuela. A plant apparently growing well below low-tide level on rocks and old corals, not extending into really deep water, but only to about 4 m. In rare cases plants have linear, nearly prismatic leaves.

REFERENCES: Grunow 1867, Murray 1889, Grieve 1909, Howe 1920, Taylor 1928, 1936, 1941a, 1942, 1954a; *S. dentifolium*, Murray 1889.

Sargassum bermudense Grunow

Plants scattered or clustered, bushy, usually with a single main axis from the holdfast, to about 50 cm. tall, but often with one or more leading subdivisions arising near the base; axes stout below, flexuous above, moderately to conspicuously spiny; secondary branches generally very numerous, to 7–12 cm. long; leaves rather persistent on the branches although lost from the main axes below, fairly crowded, firm, simple or occasionally forked, broadly to more generally narrowly lanceolate, 1–3 cm. long, 1.5–2.5 mm. wide, the bases broadly deltoid, the apices generally acute, the margins subentire to more generally irregularly serrate; costa evident, subpercurrent; cryptostomata evident, somewhat scattered, or in the narrower leaves in an irregular row on each side of the costa; vesicles seldom absent, often numerous, spherical, rarely minutely apiculate,

3–6 mm. diam., on stalks about as long as the diameter; receptacles at first axillary to foliage leaves and coming to exceed them in length, for the most part loosely alternately branched, the divisions slender, little tapering, fertile throughout, becoming densely crowded in terminal panicles as the primary leaves are lost.

V. **contracta** Grunow. Plants scattered to more commonly densely gregarious, 5–10, seldom 15, cm. tall, the stems particularly stout and nodulose below, spiny above, sparingly divided and with side branches 5 cm. long or even less, or inconspicuous; the leaves much crowded, lanceolate, 1.0–2.5 cm. long, subacute, inconspicuously serrate; vesicles few or none; receptacles generally absent, but when present often in dense elongate panicles, much exceeding the leaves in length.

V. **hellebrandtii** Grunow. Plants scattered or in clusters, 30–70 cm. tall, with 1–7 primary axes from the base, these relatively slender and flexuous, spiny throughout, very loosely branched, the side branches to 10–15 cm. long; leaves at the base of the plant to 5–8 cm. long, conspicuously pinnate or furcate, the divisions tapering from a width of 6–8 mm. to acute tips, the leaves on the erect stems 2–3 cm. long, 2–3 mm. wide, rather thin, simple or sparingly laterally branched, linear to linear-lanceolate, sparingly serrate or entire; vesicles often numerous, receptacles often very loose, the divisions sometimes pedicellate.

V. **pinnatifida** Grunow. Plants 15–40 cm. tall, the axes sparingly branched, spiny; leaves imbricate, linear, very slender, rather thin, occasionally laterally branched, to 2–5 cm. long, 1–2 mm. wide, the tips attenuate and sometimes a little serrate; receptacles short, moderately closely clustered, the divisions often pedicellate.

V. **stagnalis** n. var. Plants to 1–2 m. tall, perhaps more, the fragile stems slender, flexuous, and almost completely smooth, very loosely branched, the branches of irregular length and 5–25 cm. long; leaves well spaced, thin and soft, narrowly oblong, seldom forked, 2–6 cm. long, 2–7 mm. wide, the apices often obtuse, the margins entire to sparingly, very obtusely serrate; vesicles small and few, sometimes very minutely apiculate, or bearing a little leaf; plants very scantily fruiting, the receptacles when present loosely branched.

Bermuda, Cuba. The typical form of this plant is common in Bermuda in shallow water where there is a little protection. The v. *contracta* is exceedingly common on rocky, exposed shores, forming a conspicuous belt in the intertidal zone. The varieties *hellebrandtii* and *pinnatifida* are different adaptations to quiet waters of open harbors. The v. *stagnalis* is found loose or lightly attached in warm, landlocked, but clear rocky ponds. This is a very variable species, which is confusing on first acquaintance. However, its range of form is not matched by any other species in the Caribbean. It is probably much more widespread than the records indicate.

The more typical plants of *S. bermudense* may be confused with plants of *S. vulgare,* and those of the v. *pinnatifida* with *S. ramifolium.* Leaves of *S. vulgare* are in general broader and more tapered, the base more asymmetrical, the serrations deeper and more acute, the cryptostomata smaller and more scattered than those of typical *S. bermudense,* but these characters would have to be assessed on well matured plants. As the stems of *S. ramifolium* are at most hardly perceptibly roughened and then only near the tips, and as the leaves tend to be much more branched than those of *S. bermudense* v. *pinnatifida,* there should be no great difficulty in distinguishing between them. The v. *hellebrandtii* similarly resembles *S. furcatum* somewhat, but is more open of habit, with the upper leaves much narrower than the lower, the stems with longer aculei, the vesicles on longer stalks. To this species one probably should refer *S. cymosum* v. *farlowii* (Grunow 1916) as a variant somewhat intermediate between v. *hellebrandtii* and v. *stagnalis.*

REFERENCE: Grunow 1916.

Sargassum polyceratium Montagne — Pl. 40, fig. 1

Plants growing attached, erect from a rather irregular disklike holdfast, reaching lengths of 4.5–9.0 dm.; strong lateral branches numerous, finally producing short spur branches; branches of all degrees generally abundantly muriculate, less often nearly smooth, or smooth in the older portions; stem leaves at first on the main axes, later closely set on the spur branches, characteristically with one margin toward the axis, lanceolate to more typically broadly ovate, 1.5–3.5 cm. long, 5–10 mm. broad, sessile, the asymmetrical base very broadly rounded or even transverse, the apex broadly obtuse to rather acute, the margin densely and deeply dentate-serrate; vesicles numerous, spherical, nearly sessile to more seldom on pedicels 2–6 mm. long; receptacular branches axillary, short and forking, not densely massed or conspicuous.

V. **ovatum** (Collins) Taylor. Differing from the species in habit; the main axes very densely beset with short spur branches bearing closely crowded leaves; leaves smaller than in the species, 1.5–2.0 cm. long by 3–8 mm. wide, greatly crisped, frequently forked; vesicles frequently lacking, rather inconspicuous among the leaves, round, 3–5 mm. diameter, on pedicels sometimes alate, usually short, but reaching 2 mm.; receptacles densely branching, in abundant axillary masses.

Bermuda, Florida, Bahamas, Caicos Isls., Cuba, Cayman Isls., Jamaica, Hispaniola, Puerto Rico, Virgin Isls., Guadeloupe, St. Lucia, Barbados, Old Providence I., ?Panama, ?Colombia, Venezuela, Brazil. Growing on rocks, corals, and shells in deep water and dredged to a depth of about 14 m. The variety is most commonly found in very shallow water on the reefs, but seems also to grow in as deep water as the typical form. In extremely sheltered, warm, clear water this species may reach a length of 13 dm., with an open habit of growth and small leaves.

REFERENCES: Montagne 1863, Mazé and Schramm 1870–77, Murray 1889, Howe 1920, 1928, Taylor 1928, 1929b, 1935, 1939b, 1940, 1941b, 1943, 1954a, Taylor and Arndt 1929, Sanchez A. 1930, Feldmann and Lami 1937; *S. bahiense,* Martens 1870; ?*S. cheirifolium,* ?*S. incisifolium, S. lendigerum p. p., S. polyphyllum,* Mazé and Schramm 1870–77, Hauck 1888, Murray 1889; ?*S. polyphyllum,* Martens 1870; *S. spinulosum p. p.,* and v. *ciliata,* Mazé and Schramm 1870–77, Murray 1889; *S. spinulosum* v. *ciliata,* Piccone 1889; *S. vulgare,* Collins 1901 *p. p.* For the v. *ovatum,* Taylor 1928, 1929b, 1933, Taylor and Arndt 1929; *S. vulgare* f. *ovata,* Collins 1901.

Sargassum furcatum Kützing

Plants to 35 cm. tall, several stout branches arising near the base, becoming slender above, bearing numerous shorter branchlets to 5–7 cm. long, all axes being sparingly and minutely muriculate;

leaves chiefly on the younger divisions, not crowded, thin, linear-lanceolate to linear, simple or commonly 1–4-forked, 1.5–5.0 cm. long, 2–4(–5) mm. wide, nearly entire to shallowly but clearly serrate, the costa evident, subpercurrent; the cryptostomata small, irregularly placed but crowded nearly into one row; vesicles fairly numerous on the longer branches, spherical, not apiculate, 3–4 mm. diam., on stalks about half as long as the vesicles; receptacles axillary, racemose, becoming conspicuous and 2–3 cm. long, the ultimate divisions terete, nodulose, interspersed with very small leaves.

?Mexico, Virgin Isls., ?Guadeloupe, ?Netherlands Antilles, Brazil.

REFERENCES: Kützing 1861; *S. vulgare* v. *furcata,* Grunow 1916.

Sargassum ramifolium Kützing

Plants to 25 cm. tall, often with several slender leading axes from a very short primary stem and these sparingly laterally branched, almost completely smooth, leafy at first but later denuded; leaves to 3–5 cm. long, thin, simple or more often sparingly pinnate or the tips forked, all divisions linear, 1–2(–3) mm. wide, the margins entire with an occasional minute tooth; midribs distinct, percurrent; cryptostomata rather small, in a single series on each side of the leaf; vesicles commonly rather few, sometimes absent, spherical, to about 3 mm. diam., on terete stalks of length about equal to the vesicle diameter; receptacles formed axillary to ordinary foliage leaves, becoming conspicuous as they replace the normal foliage; receptacular branchlets alternate, the last divisions forked, often briefly pedicellate, somewhat spindle-shaped and nodulose.

Hispaniola, Virgin Isls., Venezuela, Brazil.

REFERENCES: Kützing 1861, Martens 1870, Zeller 1876.

Sargassum cymosum C. Agardh Pl. 38, fig. 4

Plants dark brown, when dried often glaucous, with long simple divisions from the short primary axis or with these more or less copiously laterally branched; elongate axes slender, smooth; upper leaves elliptical, oblong-lanceolate or linear, the margin entire or faintly crenulate, the costa evident, the cryptostomata small; basal leaves sometimes sparingly pinnately divided; vesicles few, spherical, often mucronate, on stalks about equal to the diameter; fertile branches repeatedly dichotomously branched, filiform, verrucose, about half as long as the subtending leaves.

Bermuda, Florida, Bahamas, Cuba, Hispaniola, Guadeloupe, Barbados, Trinidad, Brazil. Probably growing on rocks or coral fragments in the sublittoral zone and probably more common in the more southern parts of its range.

REFERENCES: Greville 1833, Montagne 1863, Martens 1870, 1871, Mazé and Schramm 1870–77, Dickie 1874a, Zeller 1876, Piccone 1886, 1889, Möbius 1889, 1890, 1892, Murray 1889, Howe 1920, 1928, Schmidt 1923, Taylor 1928, 1930a, 1942, Taylor and Arndt 1929; *?S. cymosum* v. *stenophyllum,* Schmidt 1923, 1924, Joly 1951, 1957; *S. cymosum* and *S. dichocarpum,* Mazé and Schramm 1870–77, Murray 1889; *S. cheirifolium,* Möbius 1889; *S. esperi,* Montagne 1863, Piccone 1886, Möbius 1889, Murray 1889; *S. integrifolium,* Martens 1870, Mazé and Schramm 1870–77 *p. p.,* Piccone 1886, Murray 1889; *S. rigidulum,* Martens 1870, 1871, Möbius 1889, 1890, Schmidt 1923, 1924.

Sargassum hystrix J. Agardh

Pl. 37, fig. 1; pl. 38, fig. 2; pl. 40, fig. 6

Plants not large, rather bushy, the branches slender, smooth, radially disposed, about 3 dm. long; leaves oblong-elliptical, becoming acuminate, the base strongly asymmetric, costate, the margin generally serrate and the teeth acute, with small scattered cryptostomata; vesicles particularly numerous in the lower parts of the plant, spherical, usually mucronate, on very short stalks which are sometimes compressed; fruiting branchlets about one-third as long as the subtending leaves, cymose and crowded, branched, subterete, becoming broadly palmated with subterete lobes or teeth, and roughened by the conceptacular pores.

V. **buxifolium** (Chauvin) J. Agardh. Plants of moderate size, to 5 dm. tall, with a few main branches but numerous short lateral branchlets on which the large leaves are chiefly crowded; leaves to 6.0 cm. long, 1.5 cm. broad, oval or obtuse-oblong, entire, or minutely serrate at the tip only, the cryptostomata few and obscure, mostly in the distal part of the leaf; generally sterile.

V. **spinulosum** (Kützing) Grunow. Plants to 4 dm. tall, bushy, freely branched, leaves scattered, oblong, to about 2.5 cm. long, 1.3 cm. wide, the tip obtuse, the base asymmetric, the margin strongly aculeate-serrate, with a rather inconspicuous costa and small scattered cryptostomata 0.10–0.15 mm. diam.; vesicles axillary, 3–5 mm. diam., not apiculate, the pedicels one-third to one-half as long as the diameter of the vesicles; fertile branchlets axillary, 0.5–1.0 cm. long, 1–2 times alternately divided, the divisions with terete pedicels, these chief divisions partially or completely divided once or twice, but when the bract leaves are suppressed, appearing to be 3–6 times branched; in general with age becoming strongly palmated, the lateral or digitate lobes subterete, but fertile throughout, the conceptacles evident and the surface nodulose.

Bermuda, Florida, Mexico, Bahamas, Cuba, Jamaica, Puerto Rico, Virgin Isls., Guadeloupe, Grenada, Costa Rica, Trinidad, Brazil. Perhaps a deep-water species, as most of the available specimens were cast ashore; reported dredged from a depth of 57 m. The v. *buxifolium* is relatively frequent about the Florida Keys, but otherwise the species seems rather rare.

REFERENCES: J. Agardh 1847, Harvey 1852, Murray 1889, Børgesen 1913–20, 1914b, Collins and Hervey 1917, Howe 1920, Taylor 1928, 1941b. For the v. *buxifolium*, Taylor 1928, 1954a. For the v. *spinulosum*, Taylor 1941b.

Sargassum platycarpum Montagne Pl. 38, fig. 3

Plants rather small, usually not over 4 dm. tall, the short, stout main axes sparingly branched, these branches slender, smooth; leaves scattered on the main branches or more often somewhat crowded on short lateral shoots, 2–6(–9) mm. broad, to 2–3(–7.5) cm. long, lanceolate, acute, tapering into a petiole, the margins generally acutely and coarsely serrate; costa evident, the cryptostomata very large, 0.6–0.9 mm. diam., in a single irregular row on each side of it; vesicles on short compressed stalks, usually muticous; fertile branchlets closely cymose, crowded, the clusters 6–7 mm. long, short-stalked below, above compressed or triangular in section, the margins usually strikingly serrate.

Bermuda, Bahamas, Caicos Isls., Cuba, Jamaica, Hispaniola, Puerto Rico, Virgin Isls., Anguilla I., St. Barthélemy, Antigua, Guadeloupe, Dominica, Martinique, St. Lucia, Barbados, Grenada, Venezuela, Brazil. Commonly growing on more or less exposed rocks in shallow water.

REFERENCES: Mazé and Schramm 1870–77, Hauck 1888, Möbius 1889, Murray 1889, Collins 1901, Vickers 1905, 1908, Sluiter 1908, Grieve 1909, Børgesen 1913–20, 1914b, Howe 1920, Taylor 1940, 1942, 1943, 1954a, Taylor and Arndt 1929, Hamel and Hamel-Joukov 1931, Feldmann and Lami 1937; *S. liebmannii, S. polyphyllum, S. polyceratium p. p.*, and *S. turneri p. p.*, Mazé and Schramm 1870–77; *Carpacanthus platycarpus,* Martens 1870.

Sargassum fluitans Børgesen Pl. 39, fig. 2; pl. 40, fig. 7

Plants widely branching, to 5–8 dm. diameter or more, without a dominant axis, entangled, pelagic; stem smooth or very sparingly spinulose; leaves short-stalked, thick and firm, 2–6 cm. long, 3–8 mm. wide, narrow to ordinarily lanceolate, the base asymmetric, the tips obtuse to acute, serrate, the sharp teeth broadly flattened at the base; cryptostomata absent or few and obscure; vesicles oval to subspherical, about 4–5 mm. diam., on stalks 2–3 mm. long, often winged, the vesicle tips not apiculate; receptacles unknown.

Bermuda, Florida, Louisiana, Texas, Bahamas, Cuba, Jamaica, Hispaniola, Virgin Isls., Guadeloupe, British Honduras, Costa Rica, Panama. Only known in the pelagic state, in which it is carried north as far as southern Massachusetts; common and sometimes more abundant than *S. natans,* with which it is ordinarily associated.

REFERENCES: Børgesen 1913–20, 1914b, Collins and Hervey 1917, Howe 1918a, 1920, Taylor 1928, 1933, 1935, 1936, 1941a, b, 1942, 1943, 1954a, Sanchez A. 1930, Prat 1935a; *?S. affine,* Gardiner 1890; *S. bacciferum* v., Mazé and Schramm 1870–77; *S. hystrix* v. *fluitans,* Børgesen 1914b.

Sargassum natans (Linnaeus) J. Meyen
Pl. 37, fig. 2; pl. 40, figs. 3, 8

Plants widely branching, to several decimeters diameter, without a dominant axis, tangled, pelagic; stem smooth; leaves firm, acutely linear, 2.5–7.0(–10) cm. long, 2.0–3.5 mm. wide, serrate, the teeth slender, terete, their length to 0.5–0.75 the width of the leaf, cryptostomata absent, midrib not prominent; vesicles 3–5 mm. diam., borne on stalks as long, sometimes smooth, but more typically apiculate, or tipped with a long spine or reduced leaf; receptacles unknown.

Bermuda, North Carolina, Florida, Alabama, Louisiana, Texas, Mexico, Bahamas, Caicos Isls., Cuba, Jamaica, Hispaniola, Guadeloupe, Barbados, Grenada, British Honduras, Brazil. Only known in the pelagic state, in which it has been carried north as far as Newfoundland and the Isle of Jersey; often very common and generally associated with *S. fluitans.* The Brazilian record is certain as to identification, but suspiciously out of the normal range for this plant.

REFERENCES: Børgesen 1913–20, 1914b (including v. *ciliata*), Collins and Hervey 1917, Howe 1918a, b, 1920, Hoyt 1920, Taylor 1928, 1935, 1936, 1941a, b, 1943, 1954a, Taylor and Arndt 1929, Sanchez A. 1930, Parr 1939, Prat 1953a; *Lenticula marina serratis foliis,* or Sargasso, Sloane 1696a, b; *S. angustifolium* and *S. bacciferum p. p.,* Mazé and Schramm 1870–77, Murray 1889; *S. bacciferum,* Harvey 1852, 1861, Grunow 1867, Martens 1870, Rein 1873, Dickie 1874a, b, Hemsley 1884, Gardiner 1890, Mohr 1901, Collins 1901, 1906 (including f. *angustum*), Jennings 1917.

UNCERTAIN RECORDS

Sargassum chamissonis Kützing

Brazil. Type from the Pacific Ocean. Zeller 1876.

Sargassum desfontainesii (Turner) C. Agardh

Guadeloupe. Type from the Canary Islands. De Toni 3:17; Mazé and Schramm 1870–77, Murray 1889.

Sargassum diversifolium (Turner) C. Agardh

Guadeloupe, Brazil. Type from Egypt? Montagne 1839b. Mazé and Schramm 1870–77, Murray 1889.

Sargassum latifolium (Turner) C. Agardh

Brazil. Type from the Red Sea. De Toni 4:78; Greville 1833a.

Sargassum lendigerum (Linnaeus) C. Agardh

V. *fissifolium* Harvey *non* Grunow. Bermuda. Type locality unknown. Rein 1873, Murray 1889.

V. *vesiculifera* Grunow
Brazil, the type locality. Grunow 1867.

Sargassum liebmannii J. Agardh

Brazil. Type from the western coast of Mexico. De Toni 3:52; Piccone 1886, Murray *p. p.* 1889.

Sargassum maximiliani (Schrader) Martius

Brazil, the type locality. De Toni 3:110; Martens 1870.

The names given in Grunow's monograph of Sargassum (1915-16) constitute a special problem because of the very large number of secondary entities recognized by him, and the impracticability of a thorough check on his reference materials. Where it has not been possible to assign his records to reasonably well-known species, they have been held in the uncertain category, for there is nothing in his descriptions which gives assurance that plants of the rank of otherwise unrecognized species are involved, and the possibilities for confusion among ill-known segregates of varietal and form rank in this genus are very great. The applicable names follow:

S. bacciferum (Turner) C. Agardh: v. ?*kuetzingiana* Grunow, Florida, Texas, Cuba, Mexico, St. Martin, Guadeloupe; f. *spinuligera* Kützing, Atlantic Ocean.

S. cymosum J. Agardh: f. *apiculata* Grunow, Martinique; v. ?*bacciferoides* Grunow, Florida, Cuba; v. *delicatula* Grunow, West Indies, Trinidad; f. *diversa* Grunow, Guadeloupe; v. *esperi* (Sieber) Grunow, Martinique, Netherlands Antilles, Venezuela, Brazil; v. *farlowii* Grunow, Bermuda; v. *lendigerum* (Turner) Grunow, Virgin Isls., Venezuela; v. *poeppigii* Grunow, Cuba; f. *subfurcata* Grunow, Martinique; v. *subpinnata* Grunow, Venezuela.

S. desfontainesii (Turner) J. Agardh: v. *schrammii* Grunow, Guadeloupe.

S. filipendula C. Agardh: v. *attenuata* Grunow, "Spanish Main"; f. *berteroi* Grunow, ?Gulf of Mexico; v. ?*cappanemae* Grunow, Brazil; v. *carpophylloides* Grunow, "Straits of Florida"; v. *contracta* J. Agardh, Bahamas, Mexico, Venezuela; f. *cuspidulata* Grunow, Trinidad; f. *magdalenae* Grunow, Colombia; v. *subcomosa* Grunow, Trinidad.

S. hystrix J. Agardh, West Indies, Brazil: v. *ciliata* Grunow, Brazil; v. *subcristata* Grunow, Mexico; f. *undulata* Grunow, Venezuela.

S. linifolium (Turner) J. Agardh: v. *schottii* Grunow, Atlantic Ocean.

S. peregrinum Grunow, Bermuda.

S. platycarpum Montagne, West Indies and Tropical America: v. *?bermudiensis* Grunow, Bermuda; f. *mazei* Grunow, Guadeloupe; f. *subciliata* Grunow, Hispaniola, Guadeloupe; f. *subpruinosa* Grunow, "In mari Americano."

S. polyceratium Montagne, Cuba: v. *?martinicensis* Grunow, Martinique.

S. rigidulum Kützing: v. *brachycarpa* (J. Agardh) Grunow, Venezuela; v. *melneri* Grunow, Brazil; v. *stenophylloides* Grunow, Jamaica.

S. vulgare C. Agardh: v. *aspera* Grunow, Virgin Isls., Marie Galante I. (Guadeloupe), Brazil; f. *clevei* Grunow, Tortola; v. *froehlichii* Grunow, Brazil; f. *lanceolata* J. Agardh, West Indies.

Turbinaria Lamouroux, 1828

Plants with a rhizomorphous base, sometimes very extensive and then probably representing many individuals; rarely devoid of erect shoots, more usually with more or less crowded, erect, subsimple or freely branched axes which are smooth except for tuberculate scars of old leaves or branchlets; bearing numerous obconic or obpyramidal peltate foliar organs, these sometimes with an embedded air vesicle, sometimes with one or two rings of teeth about the distal end, sometimes in the obpyramidal species with the longitudinal ridges alate or even serrate; cryptostomata numerous on the leaves; fruiting branches densely forking, crowded in the leaf axils.

KEY TO SPECIES

1. Petiole ridges usually nearly entire, the lamina ordinarily with a vesicle .... T. turbinata
1. Petiole ridges usually dentate, the lamina without a vesicle T. tricostata

Turbinaria tricostata Barton

Stem sparingly branched, 5–15 cm. tall, the branches short; leaves obpyramidal, rounded-triangular on the truncate distal end, without vesicles, the margins usually aculeate-dentate; the petiolar part rather long, even twice as long as the diameter of the distal end, in the young plants terete, but in mature individuals 3-angled, the ridges usually serrate.

Bermuda, Bahamas, Caicos Isls., Cuba, Cayman Isls., Jamaica, Puerto Rico, Virgin Isls., Guadeloupe, Netherlands Antilles. Plants of the intertidal rocks or old coral reefs, or growing a little below low-tide line.

REFERENCES: Barton 1891, Collins and Hervey 1917, Howe 1918a, 1920, Taylor 1940, 1942, 1954a; *T. vulgaris p. p.*, with v. *decurrens* and v. *trialata*, Mazé and Schramm 1870–77, Murray *p. p.*, 1889.

Turbinaria turbinata (Linnaeus) Kuntze Pl. 39, figs. 3–5

Plants erect, to 4 dm. or more in height from a branching holdfast, virgate or with long branches and beset with shorter spur branches bearing the leaves; leaves obpyramidal, reaching 1 cm. long and broad on fertile branches, or somewhat larger where the plants are sterile; petiolar part of the leaf with 3 nearly smooth-margined ridges; distal part of at least the upper leaves generally vesicular, the flattened top with a fairly broad, rounded-triangular to nearly circular membranous expansion having an entire or more commonly an aculeate-dentate margin; cryptostomata numerous, scattered; receptacles abundant, densely forking, axillary or between the leaves.

Florida, Bahamas, Caicos Isls., Anguila Isls., Cuba, Jamaica, Hispaniola, Puerto Rico, Virgin Isls., St. Barthélemy, Antigua, Guadeloupe, British Honduras, Panama, Colombia, Netherlands Antilles, Brazil. Plants of the intertidal rock and coral reefs and tide pools, or growing a little below low-tide level.

REFERENCES: Howe 1918b, 1920, Taylor 1928, 1935, 1940, 1941b, 1942, 1943, 1954a, Taylor and Arndt 1929, Sanchez A. 1930, Feldmann and Lami 1937, Questel 1942; *T. trialata*, Martens 1870, Hauck 1888, Barton 1891, Collins 1901, Børgesen 1913–20; *T. vulgaris p. p.*, Mazé and Schramm 1870–77, Murray *p. p.*, 1889; *Sargassum turbinata*, Montagne 1863.

RHODOPHYCEAE

BANGIOIDEAE

Plants unicellular, filamentous or membranous; the cells not connected by protoplasmic strands, with axial or parietal chromatophores and generally with pyrenoids; showing a presumably haploid vegetative phase which may reproduce asexually by monospores and may also show sexual reproduction; sexual organs if present consisting of spermatangia produced by subdivision of ordinary thallus cells into many small colorless units, and carpogonia of sexually modified cells which have developed a soft-walled receptive extension or trichogyne; the fertilized carpogonium producing a nonfilamentous carposporophyte (cystocarp) of a few carpospores.

BANGIALES

Plants filamentous or membranous, simple or branched; cells usually with axial chromatophores; asexual reproduction by naked monospores; sexual reproduction if present by spermatia and carpogonia; the fertilized carpogonium developing several carpospores.

KEY TO FAMILIES

1. Plants small, filamentous, or larger and parenchymatous, branched or plane, but in such cases not developed about a special axial cell row .... BANGIACEAE, p. 286
1. Plants fairly large, ultimately bushy, usually dull violet or grayish in color, the outer tissues developed about an axial cell row .... COMPSOPOGONACEAE, p. 296

BANGIACEAE

General characters of the Subclass; axis, if pluriseriate, without differentiation of medullary and cortical tissues; cells uninucleate, the chromatophores in our genera central, with radiating lobes.

KEY TO GENERA

1. Plants filamentous 3
1. Plants foliaceous 2

2. Blades only a few cells wide, the cells in longitudinal rows ERYTHROTRICHIA, p. 291

2. Blades very broad, caespitose or expanded, the cells not in longitudinal rows .... PORPHYRA, p. 294

3. Filaments unbranched 4
3. Filaments at least proliferously branched 5

4. Filaments uniseriate, or at most with a few longitudinal walls; monosporangia formed from portions of individual vegetative cells .... ERYTHROTRICHIA, p. 291
4. Filaments becoming highly pluriseriate; monosporangia formed by the conversion of vegetative cells .... BANGIA, p. 292

5. Filaments creeping, epiphytic or partly endophytic, often forming circumscribed disks .... ERYTHROCLADIA, p. 289
5. Filaments erect 6

6. Branching sparse or proliferous, or in young plants absent; the main branches of irregular thickness with the cells scattered near the surface .... BANGIOPSIS, p. 288
6. Branching rather abundant, pseudodichotomous 7

7. Cells elongate, greenish gray .... ASTEROCYTIS, p. 287
7. Cells short, rose-red .... GONIOTRICHUM, p. 288

Asterocytis Gobi, 1879

Plants bushy, filamentous, lubricous, grayish green, irregularly and widely branched; cells usually irregularly uniseriately placed, elongate; asexual propagation by motionless spores.

Asterocytis ramosa (Thwaites) Gobi

Plants in soft tufts, steel-green, to 1–10 mm. tall, the filaments 12–20(–25) μ diam., unilaterally or subdichotomously branched; cells uniseriate, 5–8 μ diam., 8–20 μ long, oval to elongate; in reproduction the filaments basipetally transformed into akinetes, which become subglobose or ellipsoid and when mature are about 14 μ long, with walls 2 μ thick, discharged laterally from the filaments.

North Carolina, Florida, Bahamas, Hispaniola, Virgin Isls., Grenada. Epiphytic on algae and on marine phanerogams in shallow water.

REFERENCES: Børgesen 1913–20, 1924, Phyc. Bor.-Am. 2296, Howe 1920, Taylor 1928, 1942, Taylor and Arndt 1929, Williams 1949a.

Goniotrichum Kützing, 1843

Thallus erect, filamentous, gelatinous, dull rose-red, pseudodichotomously branched, below attached by enlarged cells, above more delicately filamentous, terete or a little irregularly thickened or flattened; cells short, often disklike, with central radiating chromatophores, the pyrenoid central and the nucleus excentric; membrane thick, especially laterally; reproduction by the formation of monospores, which are liberated by dissolution of the thallus membrane.

Goniotrichum alsidii (Zanardini) Howe

Filaments isolated or in rosy tufts, 0.3–6.0 mm. long and reaching a diameter of 12–20 μ, except where thickened near the base to 30–50 μ; rarely simple, generally freely irregularly to pseudodichotomously branched, the branches a little attenuate; cells uni- to pluriseriate, discoid to ellipsoid or nearly cylindrical, 7–13 μ diam., 4–13 μ long, the chromatophores violet-red, tending toward green in some specimens.

Bermuda, North Carolina, Texas, Bahamas, Cuba, Jamaica, Virgin Isls., Guadeloupe, Barbados, Tobago, Brazil. Epiphytic on coarser algae and sea grasses in shallow, warm water.

REFERENCES: Howe 1920, Hoyt 1920; *G. elegans*, Collins 1901, Vickers 1905, Børgesen 1913–20, Collins and Hervey 1917, Hamel 1929, Taylor 1941a, 1942, 1954a, Joly 1956a; *Bangia elegans*, Mazé and Schramm 1870–77, Murray 1889.

Bangiopsis Schmitz, 1897

Plants gelatinous, the filaments simple, later sparingly or proliferously branched, the holdfast an expanded cell not reinforced by rhizoids from above, the young filaments uniseriate, the old ones thickened by repeated irregular division and peripheral displacement of the cells, each of which appears to have a radiating chromatophore.

KEY TO SPECIES

1. Plants seldom developing primary branches .... B. subsimplex
1. Plants ultimately developing a few primary branches .. B. humphreyi

Bangiopsis subsimplex (Montagne) Schmitz

Plants dull violet, the filaments subsimple, flexuous, verrucose, sometimes proliferous, the branchlets 10 μ diam.

French Guiana. Growing on rocks in marine as well as fresh-water situations.

REFERENCES: Børgesen 1913–20; *Compsogon subsimplex,* Montagne 1850.

Bangiopsis humphreyi (Collins) Hamel

Plants at first simple, uniseriate filaments about 35 μ diam., the cells discoid, about half as long as broad; filaments becoming locally very much thickened, the cells subdivided and now rounded and dispersed in the peripheral jelly; primary branches few, irregular, similar to the main axis; lateral branches short, proliferous, similar to the young filament in structure.

Bermuda, Cuba, Jamaica, Hispaniola, Virgin Isls., Brazil. Growing in shallow water, attached to rocks or other massive objects.

REFERENCES: Collins, Holden and Setchell Phyc. Bor.-Am. 421, Hamel 1929, Hamel and Hamel-Joukov 1931, Taylor 1933; *B. subsimplex,* but later *Goniotrichum humphreyi,* Børgesen 1913–20; *G. humphreyi,* Collins 1901, Collins and Hervey 1917, Joly 1956a.

Erythrocladia Rosenvinge, 1909

Plants of decumbent filaments with apical growth, the apical cells often deeply cleft before division; branching irregularly alternate and diffuse, or more commonly pseudodichotomous and the radial branches closely laterally approximated to form a disk, the central portion by crowding becoming pseudoparenchymatous; chromatophores single, parietal; reproduction by monospores produced from the older cells of the thallus.

KEY TO SPECIES

1. Filaments laterally approximated to form a compact disk
   E. subintegra, p. 290
1. Filaments loosely conjoined or entirely free, at least in the outer portions of the colony 2

2. Plants rather compact in the central, older part 3
2. Plants without a compact portion .... E. vagabunda, p. 291

3. Cells 2.0–3.8 μ diam. .... E. pinnata, p. 290
3. Cells 3.0–12.0 μ diam. .... E. recondita, p. 290

Erythrocladia pinnata Taylor Pl. 41, fig. 4

Plants creeping between the layers of the host coenocyte membrane; freely branched, the branches formed near the upper end or the middle of the supporting cell, little more slender toward the tips than in the older parts; the first branches alternate, or less often unilateral, but later formed opposite other branches on the same axis cells; filaments 2.0–3.8 μ diam., cells 2–4 diameters long, cylindrical or slightly swollen in the middle; monosporangia mostly on the older parts of the plant, protruding through the host membrane, sessile or terminating short 1–3-celled branches, subspherical, 4.5–5.5 μ diam.

Tobago. Epiphytic on old Bryopsis plants.

REFERENCE: Taylor 1942.

Erythrocladia recondita Howe and Hoyt

Thallus endophytic or pseudo-epiphytic, creeping in the superficial cell walls of other algae, reaching a diameter of 0.2–1.5 mm. and usually remaining single-layered; consisting at first of free, irregularly radiating, and irregularly branching filaments; branching lateral and spreading, or somewhat dichotomous; cells varied and irregular in form, 8–25 μ long, 3–12 μ broad; monoecious, the stalked spermatangia ovoid, 2–4 μ diam.; carpogonia each furnished with a beak or trichogyne exserted about 4–8 μ; cystocarps each usually forming a single ovoid or oblong carpospore, these mostly 8–19 μ max. diam.

North Carolina. Superficially endophytic in coarser algae.

REFERENCE: Hoyt 1920.

Erythrocladia subintegra Rosenvinge Pl. 41, fig. 1

Plants with radial filaments laterally approximated to form small epiphytic disks at least to 50 μ diam., the cell arrangement toward the center of the disk becoming irregular; filaments branching by a forking of the apical cell (marginal to the disk), one or the other of the arms being cut off by a transverse wall; cells 4–6 μ diam.; reproduction by monospores.

Bermuda, Florida, Jamaica, Hispaniola, Virgin Isls., British Honduras, Colombia, Trinidad, Tobago, Brazil. Epiphytic on various algae and sea grasses, especially in shallow water, but dredged to about 16 m. depth.

REFERENCES: Børgesen 1913–20, Collins and Hervey 1917, Taylor 1928, 1929b, 1930b, 1931, 1933, 1935, 1942, Taylor and Arndt 1929, Joly 1951.

Erythrocladia vagabunda Howe and Hoyt

Thalli endophytic or pseudo-epiphytic in the superficial cell walls of other algae, 0.75–2.25 mm. diam.; of irregularly branching, uniaxially elongate, or irregularly radiating filaments, often appearing to anastomose, and commonly forming irregular compact patches 2–6 cells broad; branching mostly lateral, often spreading or rectangular; cells for the most part irregularly oblong in surface view, often curved or 1–2-lobed, 9–40 μ long, 6.5–15.0 μ broad; cystocarps usually forming single ovoid or oblong carpospores 12–25 μ diam.

North Carolina. Superficially endophytic in coarser algae.

REFERENCE: Hoyt 1920.

Erythrotrichia Areschoug, 1850

Plants erect, major portion of simple filaments, sometimes pluriseriate or strap-shaped and monostromatic; the bases small, multicellular and disciform, or of a few rhizoidal cells, or simply of the expanded end of the basal cell; vegetative cells each with a radiating chromatophore and a central pyrenoid; monosporangia segregated from vegetative cells by oblique walls; spermatangia formed by subdivision of previously vegetative cells; carpogonia formed by direct transformation of similar cells; cystocarps of but few carpospores.

KEY TO SPECIES

1. Plant with a unicellular lobed base; erect filaments uniseriate or with a few longitudinal walls .... E. carnea

1. Plant with a base formed by rhizoidal extensions from several cells; erect portion at first uniseriate, later becoming strap-shaped .... E. vexillaris

Erythrotrichia carnea (Dillwyn) J. Agardh

Plants of simple terete filaments each attached by a basal cell which becomes lobed, the lobes extended to short ramified rhizoids; filaments 0.5–2.0(–8.0) cm. tall, 16–26 μ diam. above, the cells swollen, 16–32 μ long; the base gradually tapered, 9.5–13.0 μ diam., with cells 15–40 μ long; usually uniseriate, occasionally with a few longitudinal walls in the thicker parts; reproduction by monosporangia usually cut off by an oblique wall near the distal end of the cell; the spore becoming spherical on discharge and 13–15 μ diam.

Bermuda, North Carolina, Florida, Bahamas, Hispaniola, Virgin Isls., Netherlands Antilles. Epiphytic on larger algae in shallow water along the shore and over reefs.

REFERENCES: Børgesen 1913–20, 1924, Collins and Hervey 1917, Howe 1918a, 1920, Hoyt 1920, Taylor 1928, 1942; *Bangia ceramicola,* Sluiter 1908.

Erythrotrichia vexillaris (Montagne) Hamel Pl. 41, figs. 2, 3

Plants at first each a uniseriate filament with the basal cell attenuated to form the holdfast, which is later reinforced by rhizoidal downgrowths from neighboring cells; later, by longitudinal walls developing in parallel planes, becoming first biseriate above and then strap-shaped with up to 24 cell rows; thalli about 20 μ thick, 185 μ broad, and to over 675 μ long, remaining monostromatic; protoplasts about 13–15 μ diam. at maturity.

North Carolina, Guadeloupe, Martinique, Venezuela, Trinidad. Epiphytic on coarser algae, perhaps from moderate depths.

REFERENCES: Hamel 1929, Hamel and Hamel-Joukov 1931, Taylor 1942, Williams 1949a; *?Bangia dispersa,* Montagne 1856; *?B. grateloupicola,* Mazé and Schramm 1870–77, Murray 1889.

Bangia Lyngbye, 1819

Thalli erect, filiform, without branches; affixed by a dilated base, wherein the original attaching cell is supplemented by intramatrical filiform extensions from the nearby cells; above more or less thickened, terete or irregularly constricted, sometimes tubular; cells with single, radially lobed chromatophores; monospores formed by the direct transformation of superficial vegetative cells, frequently in the same plants that are reproducing sexually; monoecious or dioecious, the spermatangia formed from vegetative cells by repeated division; carpogonia formed by the direct transformation of superficial vegetative cells; cystocarps of about eight cells.

KEY TO SPECIES

1. In young filaments the cells longer than broad and rather separated; colonies notably yellowish .... B. lutea
1. Even in young filaments the cells no longer than broad, and closely connected; yellowish brown where excessively exposed to the sun, but otherwise dull purple ..... B. fuscopurpurea

Bangia fuscopurpurea (Dillwyn) Lyngbye

Plants gregarious, the filaments basally attached, softly slippery, and to 0.5–2.0 dm. long, 20–220 μ diam., or somewhat thicker; pale yellowish brown to generally brownish purple, the filaments at first slightly tapered toward the base, the cells with single, radially lobed chromatophores; uniseriate, the segments equal to the diameter or to 0.25–0.33 as long, becoming quaternately and then more extensively radially divided, resulting in thick, somewhat torulose and constricted filaments.

Bermuda, North Carolina, Texas, Brazil, Uruguay. Generally growing on intertidal rocks and woodwork.

REFERENCES: Collins and Hervey 1917, Howe 1918a, Hoyt 1920, Taylor 1936, 1941a, Joly 1956a; *?B. atropurpurea*, Murray 1889; *?B. compacta*, Howe 1918a. (Not *B. fuscopurpurea*, Mazé and Schramm 1870–77, for Hamel (1929) found specimen no. 859 to be a mixture dominated by Lyngbya and without Bangia; not *B. fuscopurpurea* v. *guayanensis* Kützing, Montagne 1850 from French Guiana, for Hamel (1929) reports that this material is ceramiaceous.)

Bangia lutea J. Agardh

Plants gregarious, forming conspicuous yellowish colonies; filaments at first erect, later elongate, irregular and somewhat tortuous, the initial cells of the filaments at first rather separated and even longer than broad, later much broader than long, becoming divided lengthwise into fours and ultimately into several cells; cell contents usually aggregated toward the center.

Bahamas, Guadeloupe, Brazil. Forming considerable patches on rocks in the intertidal zone.

REFERENCES: Mazé and Schramm 1870–77 (perhaps in part, but reported by Hamel (1929) in some collections to be *Bangiopsis humphreyi*), Murray 1889, Howe 1920.

UNCERTAIN RECORD

Bangia ciliaris Carmichael

North and South Carolina. Type from Great Britain. De Toni 4:7; Harvey 1858, Williams 1949a.

Porphyra C. Agardh, 1824

Plants membranous, often large, each attached by a small holdfast, expanding above into a soft slippery blade of one or two cells in thickness; cells with a roughly stellate chromatophore and a pyrenoid, alike except near the base where they are extended into intramatrical rhizoids to form the holdfast; asexual reproduction by monospores involving large continuous areas of the frond; sexual reproduction by spermatia, produced by cell division and conversion of portions of the frond into spermatangial cells, and scattered carpogonia, which are formed from vegetative cells which develop a short trichogyne extending to the surface of the thallus, and which produce small clusters of carpospores.

KEY TO SPECIES

1. Spermatangial development marginal, as a pale band 2
1. Spermatangial development in small, pale longitudinal flecks scattered over the outer portions of the blade .... P. leucosticta

2. Small, densely caespitose, the blades 20–30 μ thick .... P. roseana
2. Large, generally widespreading, easily laid flat on paper, blades 30–75 μ thick .... P. umbilicalis

Porphyra leucosticta Thuret

Plants more or less ample, dull pink to reddish purple, the blades 10–15(–70) cm. long, apparently sessile, umbilicate at the base, the general contour oblong, becoming undulate-plicate and sometimes sparingly laciniate above; soft, monostromatic, 25–50 μ thick; the cells 1.5–2.0 times as high as wide, from the surface averaging 12–15 μ diam. including the walls; monoecious, the spermatangia formed in elongated spots 5–10 mm. long, 1.0–1.5 mm. wide, parallel to each other on the outer portions of the blade.

Bermuda, North Carolina, Florida. Growing on mangroves and rocks at or above low-tide level.

REFERENCES: *P. atropurpurea,* Collins and Hervey 1917, Howe 1918a; *P. lacinata,* Murray 1889.

Porphyra roseana Howe

Plants densely caespitose, tawny-violet or olivaceous, shortly stipitate, usually with several laminae from a common base; the main divisions suborbicular to linear-oblong, mostly 1–3 cm. long, strongly crispate, the margins commonly lobulate, monostromatic except at the extreme base, 20–30 μ thick, rigid on drying and scarcely adhering to paper; cells irregularly angular, mostly 13–25 μ diam. in surface view, separated by walls 5–10 μ thick; cells in sections subquadrate, the superficial walls firm, 5–10 μ thick; monoecious, the spermatangial sori marginal, 32 spermatia in each spermatangium; cystocarps forming submarginal sori, 8 carpospores in each cystocarp.

Brazil.

REFERENCE: Howe 1928.

Porphyra umbilicalis (Linnaeus) J. Agardh

Plants becoming large, 1–3(–8) dm. long and 0.5–3.0 dm. wide, olivaceous to brownish purple; blades narrow when young but generally ultimately broadly umbilicate about a rounded base, above entire, or divided into broad lobes, nearly flat to plicate, texture membranous, soft, rubbery, the thickness 30–75 μ, usually about 50 μ; monostromatic, the cells 8–25 μ diam., averaging about 18 μ, in sections subquadrate to somewhat higher than wide, 20–30 μ tall; dioecious or monoecious, when the 2 sexes are generally segregated on the frond, the spermatangia forming a pale deliquescing marginal band, the carpospore groups scattered in clusters over the general surface of the blade.

South Carolina, Florida, ?Bermuda, ?Brazil. Growing on rocks at or above low-tide level.

REFERENCES: *P. laciniata,* Grunow 1867, Martens 1870, Zeller 1876, Piccone 1886, 1889; *P. laciniata* v. *umbilicata,* Martens 1870; *P. vulgaris,* Greville 1833, Harvey 1858, Melvill 1875, Hemsley 1884, Taylor 1928.

UNCERTAIN RECORD

Porphyra sp.

French Guiana, ?Uruguay. Hamel 1929; *P. vulgaris,* Mazé 1868 (not seen).

COMPSOPOGONACEAE

Plants bushy, often large, grayish, dull violet or a little greenish, filamentous, very soft but not gelatinous, slender, alternately branched, tapering to uniseriate tips; cells with numerous small peripheral chromatophores of irregular shapes, and several nuclei; older parts of the thallus corticated, the cortex segregated from the axial row of large cells by periclinal walls, the outer segments dividing in an irregular fashion to form a single superficial layer of small cells; reproduction by monospores, segregated by oblique walls from the vegetative cells.

Compsopogon Montagne, 1846

Characters of the Family.

Compsopogon caeruleus (Balbis) Montagne

Plants brownish violet, or steel-gray, virgate, to 1–2 dm. tall, about 2 mm. diam. below, the axial cells not visibly forming nodes in the older portions; abundantly branched, capillary in the ultimate divisions.

Florida, Louisiana. Often abundant in tidal creeks near the coast; the genus is primarily fresh-water, as in Jamaica and other West Indian islands, but it has been collected in quite brackish water along the mainland coast of the United States.

From various fresh-water localities species have been reported under other names and may be different from this one, as follows: *C. leptoclados* Mont. and *C. chalybaeus* Kütz. from French Guiana, the latter also from Puerto Rico and Florida; also *C. aeruginosus* (J. Ag.) Kütz. from Cuba. Identifications on a sound basis are hardly possible in this genus at the present time, since studies to date have not demonstrated well-defined differential characters.

REFERENCE: Taylor 1954a.

FLORIDEAE

Plants of many forms, their cells uni- or plurinucleate, often conspicuously connected by a single large protoplasmic bridge, rarely with axial or stellate chromatophores, generally with several small lateral chromatophores with or without pyrenoids; usually showing both diploid and haploid vegetative phases, the diploid phase sporophytic, reproducing through tetrasporangia which undergo meiosis; tetraspores formed by three parallel walls dividing the sporangium into four parts (linear or zonate type), by three intersecting walls with the spores lying parallel to each other or parallel in pairs (tetrapartite, cruciate), or by simultaneous wall formation as four pyramidal spores meeting in the center, with spherical-triangular external faces (tetrahedral, tripartite); polysporangia produced by continued division from tetrasporangia, and parasporangia developed as accessory structures also known; tetraspores germinating to develop a haploid gametophytic phase, which is sometimes also able to reproduce by monospores; sexual organs spermatia in spermatangia, usually produced terminally on miniature branch systems borne on vegetative cells, and carpogonia with elongate receptive trichogynes, borne terminally on special (carpogenic) branches, sometimes associated with special cells destined to incubate the zygote nucleus (auxiliary cells, the joint structure constituting the procarp); the fertilized carpogonium or the auxiliary cell without or with extensive cell fusions producing obscure to extended branched filaments (gonimoblasts), which in greater or lesser part become converted into carposporangia and with certain associated structures constitute the diploid carposporophyte; or in other instances with auxiliary cells not located near the carpogenic branches (procarps absent), so that the zygote nucleus in the carpogonium (usually after division) is transferred by special (oöblast or sporogenous) filaments to the auxiliaries, where the formation of gonimoblasts is initiated as before; carpospores on release developing the diploid vegetative sporophytic phase.

The ordinary three-phase life cycle is termed "diplobiontic"; that in which meiosis takes place in the carpogonium and the vegetative sporophyte is lacking is termed "haplobiontic" and is regularly found only in part of the Nemalionales. Recent reports indicate that some of the larger species may have microscopic filamentous phases, but these observations are as yet hardly complete enough to justify major revisions in algal classification.

NEMALIONALES

Plants filamentous, the free uniseriate filaments creeping or erect, or the plants much larger and corticated, when developing relatively soft branches with either the central filament or the multiaxial type of structure; cells uninucleate, with axial or lateral chromatophores; asexual reproduction by monosporangia, very exceptionally by bi- or tetrasporangia; sexual reproduction by spermatia in spermatangia, formed (in the larger species) from the surface vegetative cells, and carpogonia borne on few-celled carpogenic branches; the zygote itself producing the carpospores either directly or after establishing connections with nutritive cells on the carpogenic branch.

KEY TO FAMILIES

1. Plants of uniseriate, branched filaments
   ACROCHAETIACEAE, p. 299
1. Plants of more complex structure 2
2. Axes simply forking, exceptionally with a surface of free filaments; cystocarps immersed 3
2. Axes beset with small branchlets; pericarps external
   BONNEMAISONIACEAE, p. 347

3. Without a continuous cortical surface and without pericarps 4
3. Cortical cells approximated to form a continuous surface, the immersed cystocarps surrounded by pericarp filaments CHAETANGIACEAE, p. 331

4. Multiaxial, the terminal filament branchlets radial HELMINTHOCLADIACEAE, p. 319
4. Uniaxial, the short lateral determinate filaments supplemented by rhizoidal downgrowths .... NACCARIACEAE, p. 346

ACROCHAETIACEAE

Plants small, filamentous, the free filaments with apical growth and more or less evident erect or prostrate axes; cell arrangement uniseriate, the cells uninucleate, with one or more chromatophores; asexual reproduction by mono-, bi-, or tetrasporangia, formed laterally or terminally; sexual reproduction by spermatangia and by carpogenic branches of one to three cells including the carpogonium; cystocarps small, with the carposporangia formed on sparingly branched gonimoblast filaments.

KEY TO GENERA

1. Chromatophores single, or very few in each cell 2
1. Chromatophores numerous, small, peripheral .. RHODOCHORTON, p. 318

2. Chromatophores parietal 3
2. Chromatophores axial, stellate .... KYLINIA, p. 299

3. Chromatophores consisting of simple or lobed plates ACROCHAETIUM, p. 302
3. Chromatophores consisting of spiral bands .... AUDOUINELLA, p. 317

Kylinia Rosenvinge, 1909

Plants filamentous, erect and tufted or endophytic, solitary or gregarious; filaments branched, uniseriate, the cells each with one or occasionally more stellate chromatophores, often with pyrenoids; asexual reproduction generally by monospores, occasionally by tetraspores; sexual reproduction infrequent, the spermatangia clustered on lateral branchlets, the carpogonia sessile, lateral on erect branchlets, producing small gonimoblasts bearing terminal carposporangia.

KEY TO SPECIES

1. Plants endophytic .... K. liagoriae, p. 301
1. Plants at least in part external to the host 2

2. Base a single cell, the modified original spore; without basal filaments .... K. crassipes, p. 300
2. Basal part of the plant filamentous, creeping, the erect filaments very short 3

3. Original spore cell not recognizable .... K. infestans, p. 301
3. Original spore distinguishable, divided into 2 cells before forming the filaments .... K. pulchella, p. 300

Kylinia crassipes (Børgesen) Kylin

Plant very small, 50–70 μ or a little more in height; basal cell persistent, subglobose to cylindrical, thick-walled, about 12 μ diam., emitting 1–2 erect or widely spreading filaments, branching at the base, bearing short uniseriate branchlets of 1–6 cells; cells cask-shaped or subglobose, 5–7 μ diam., 5–9 μ long, or toward the apices only 4–5 μ diam.; hyaline hairs sometimes present, terminating the branches; sporangia uniseriate, on the upper sides of the branchlets, sessile or on one-celled stalks, 5–6 μ diam., 6–8 μ long.

Bermuda, Florida, Texas, Hispaniola, Virgin Isls. Epiphytic on Centroceras and Hypnea, the doubtfully distinct v. *longiseta* on *Chaetomorpha brachygona,* in shallow water of sheltered places.

REFERENCES: Papenfuss 1945b; *Acrochaetium crassipes,* Børgesen 1909, 1913–20 (including v. *longiseta*), 1924, Collins and Hervey 1917, Howe 1918a, Taylor 1928, 1941a.

Kylinia pulchella (Børgesen) Papenfuss

Plants small, pulvinate, arising from a base of creeping branched filaments crowded near the center; the original spore at first dividing into 2 equal persistent cells before forming the basal filaments; erect filaments about 24 μ tall, of 2–3 cells, seldom more, the cells 5–6 μ diam., 1.5–2.0 diameters long, each with a stellate chromatophore and a central pyrenoid; hyaline hairs numerous, terminal on the erect, occasionally lateral and sessile on the creeping filaments, 5–7 μ broad, 9–10 μ long.

Virgin Isls. Growing on *Chaetomorpha media* on an exposed rocky coast.

REFERENCES: Papenfuss 1945b; *Acrochaetium pulchellum,* Børgesen 1913–20.

Kylinia infestans (Howe and Hoyt) Papenfuss

Plants tufted, to 90 μ tall, arising from a basal portion formed by filaments ramifying in the membranes of the host, eventually crowded and subparenchymatous; erect filaments very short, from one to rather few cells 4.5–6.5 μ diam., 1–2 diameters long, terminating in very long and slender alternate hairlike prolongations which much exceed the colored filaments in length, or the erect filaments of several cells, once or twice branched; sporangia rarely sessile on the creeping filaments, more often terminal on one-celled stalks, or 2–3 together, or the sporangia sessile or pedicellate on the upper sides of the branchlets, ovoid or ellipsoid, 6.0–8.5 μ diam., 10–14 μ long.

Bermuda, North Carolina. Ectozoic on hydroids growing upon larger algae in rather deep water on a coral reef.

REFERENCES: Papenfuss 1945b; *Acrochaetium infestans,* Howe and Hoyt 1916, Howe 1918a, Hoyt 1920.

Kylinia liagoriae (Børgesen) Papenfuss

Plants endophytic, sparingly, irregularly, and widely branched, the cells irregular, subcylindrical to cask-shaped or with a median inflation, 6–14 μ diam., about 35 μ long, with a central radiating chromatophore and a pyrenoid; terminal hyaline hairs occasional; sporangia sessile, on stalklike projections of the supporting cell or on one-celled pedicels, solitary or 2 together, spherical or obovate, to 14 μ diam., 15–20 μ long.

Florida, Virgin Isls. Endophytic in *Liagora pinnata* from reefs at moderate depths.

REFERENCES: Papenfuss 1945b; *Acrochaetium liagoriae,* Børgesen 1913–20; *Chantransia liagoriae,* Børgesen 1913–20 (by error); *Chromastrum liagoriae,* Papenfuss 1945a; *Acrochaetium collinsianum,* Børgesen 1913–20, Taylor 1928.

UNCERTAIN RECORDS

Kylinia secundata (Lyngbye) Papenfuss

Puerto Rico, Brazil. Type from the Faeroe Islands. *Chantransia secundata,* De Toni 4:68; Piccone 1886, Hauck 1886.

Kylinia virgatula (Harvey) Papenfuss

Florida. Type from England. *Acrochaetium virgatulum* f. *luxurians,* Taylor 1929b; *Callithamnion luxurians,* Murray 1889; *Chantransia virgatula,* De Toni 4:69, 6:57.

Acrochaetium Nägeli, 1861

Filamentous, gregarious, rose-red; from a basal holdfast cell, disk, or filaments giving rise to simple or branching uniseriate and immersed, or erect and free filaments, which may be greatly attenuate and hairlike or may terminate in true hairs; chromatophores in the vegetative cells single, seldom divided to several, parietal, platelike or lobed, with or without pyrenoids; sporangia terminal or lateral, scattered, clustered, or in unilateral rows, usually monosporous, occasionally bi-, tetra-, or polysporous; sexual reproduction infrequent, by spermatangia borne variously but usually clustered on the lateral branchlets, and unicellular carpogonia which may be lateral or intercalary, or rarely on one-celled stalks; cystocarps formed directly from the carpogonia, the outer cells of short gonimoblast filaments producing the carposporangia.

KEY TO SPECIES

1. Plants entirely external to the host 2
1. Plants with the basal portions partly on the surface, partly penetrating into the tissues of the host 14
1. Plants without superficially creeping filaments; the base, or the entire vegetative portion, within the host tissue 22

2. Base consisting of a single cell, modified from the original germinating spore 3
2. Base consisting of creeping filaments sometimes crowded or fused into a subparenchymatous disk 4

3. Cells of the lower filaments 4–5 μ diam.; basal cell 5–8 μ diam. A. dufourii, p. 305
3. Cells of the lower filaments 5.5–8.0 μ diam.; basal cell disciform, to 20 μ diam. . . . . A. sargassi, p. 306

4. Persistent germinated spore cell evident, giving rise to the short, creeping basal filaments .... A. radians, p. 309
4. Original spore cell not discernible in the mature basal structure 5

5. Erect filaments 6 μ diam. or less 6
5. Erect filaments in the lower erect portions of the plants more than 6 μ diam. 10

6. Sporangia fusiform .... A. netrocarpum, p. 309
6. Sporangia oblong, oval, ovate, or at any rate not tapering to the tip 7

7. Sporangia about twice as long as broad 8
7. Sporangia much shorter 9

8. Erect filaments little-branched; sporangia chiefly on the main filaments .... A. gracile, p. 307
8. Erect filaments freely branched above; sporangia chiefly stipitate on branches, near their bases ...... A. caespitiforme, p. 306

9. Sporangia chiefly lateral and terminal on short, opposite branchlets, 7–8 μ diam., 8–10 μ long .... A. globosum, p. 308
9. Sporangia on the main branches, 5.5–6.5 μ diam., 10.0–12.5 μ long .... A. leptonema, p. 308

10. Cells short, mostly barrel-shaped; unicellular hairs present A. sancti-thomae, p. 310
10. Cells longer; such hairs absent 11

11. Base a subparenchymatous disk, not evidently filamentous A. thuretii, p. 310
11. Base evidently filamentous, but in the center often congested 12

12. Sporangia in long series on the branches, mostly pedicellate A. seriatum, p. 310
12. Sporangia on short lateral branchlets 13

13. Monosporangia 14–16 μ long .... A. flexuosum, p. 308
13. Monosporangia 18–22 μ long .... A. sagraeanum, p. 309

14. Germinated spore persistent, recognizable at the bases of the erect filaments 19
14. Spore not discernible 15

15. Basal structure a subparenchymatous disk with a thick-walled endophytic process .... A. robustum, p. 315
15. Base of short superficial and endophytic free branched filaments 16

16. Plants less than 1 mm. tall 17
16. Plants more than 1 mm. tall 18

17. Erect filaments of few cells, sparingly branched, the tips not hairlike .... A. antillarum, p. 311
17. Erect filaments much longer, freely branched, with hairlike tips .... A. ernothrix, p. 316

18. Erect filaments without hair tips; monosporangia about 6 μ diam.; bisporangia 9 μ diam. .... A. bisporum, p. 312
18. Erect filaments often hair-tipped; monosporangia 8–12 μ diam.; tetrasporangia 7–12 μ diam. .... A. daviesii, p. 307

19. Original germinated spore cell elongating slightly and penetrating between the surface cells of the host.................. 20
19. Original spore cell sometimes supplemented by short lateral filaments 21

20. Germinated spore cell simply elongating as it penetrates the host; erect filaments about 6 μ diam. .... A. hoytii, p. 306
20. Germinated spore cell narrowest in the middle, considerably expanded at the lower end below the superficial host cells; erect filaments about 9–11 μ diam. .... A. unipes, p. 316

21. Monosporangia about 11 μ diam., 27 μ long; sexual plant not known .... A. opetigenum, p. 314
21. Monosporangia 10–18 μ diam., 18–27 μ long; monoecious, the sexual plants common .... A. affine, p. 305

22. Germinated original spore persistent and recognizable 23
22. Original spore not defined in the basal structures 25

23. Original spore body undivided 24
23. Original spore bisected by a wall before forming filaments A. comptum, p. 312

24. Growing on Liagora; filaments to 11 μ diam.; bispores sometimes present, sexual plants unknown .... A. occidentale, p. 314
24. Growing on Dasya and Dudresnaya; filaments 8–16 μ diam.; bispores unknown, sexual plants reported .... A. bornetii, p. 312

25. Base of ramifying, widely spreading filaments 26
25. Base in part of thick filaments congested into vertical multicellular structures; sporangia often with an indurated tip A. phacelorhizum, p. 315

26. Plants growing in the interstices of the loose tissues of the host A. avrainvilleae, p. 311

26. Plants basally embedded in the cell walls and gelatinous matrix of the host 27

27. Endophytic filaments moniliform, near the surface of the outer host membrane .... A. homorhizum, p. 313
27. Endophytic filaments not moniliform, usually diffuse 28

28. Endophytic filaments with conspicuously irregular cell shapes and diameters .... A. repens, p. 315
28. Endophytic filaments not conspicuously irregular 29

29. Endophytic filaments rather short; sporangia sessile, in series on the branchlets .... A. hypneae, p. 313
29. Endophytic filaments widely ramifying; sporangia few, on stalks, on the lower branchlet cells .... A. nemalionis, p. 314

Acrochaetium affine Howe and Hoyt

Plants to 3.5 mm. tall; basal cell (original spore) subglobose or ellipsoid, 14–26 μ diam., becoming subpyriform, 20–33 μ long, sometimes developing short creeping filaments of 2–5 cells; primary erect filaments 1–4 from the basal cell, 6–14 μ diam., often immediately divided to 1–4 filaments 4–8 μ diam., all rather stiff below, flexuous above, where sparingly laterally or subdichotomously branched, the divisions elongate-virgate, 3.0–5.5 μ diam., often terminating in hairs; cells throughout 3–9 diameters long; monosporangia uncommon, lateral and sessile or pedicellate, 10–18 μ diam., 18–27 μ long; monoecious, the spermatangia near the carpogenic branches, solitary or 2–3 together, lateral or terminal; cystocarps common, 3–8 spored; carpospores about 8–18 μ diam., 13–26 μ long.

North Carolina. Epiphytic on coarser algae as dredged from a coral reef.

REFERENCES: Howe and Hoyt 1916, Hoyt 1920, Papenfuss 1945a.

Acrochaetium dufourii Collins

Plants 200–600 μ tall, from a basal cell (original spore), 5–8 μ diam., bearing 1–3 erect filaments; branching rather sparse, sometimes opposite or alternate, more commonly secund, erect, not very closely set; cells 4–5 μ diam., 2–5 diameters long; monosporangia 5–6 μ diam., 7–10 μ long, sessile or on a one-celled pedicel, on the main filaments or on the branches, usually in secund series.

Bermuda, North Carolina. Epiphytic on Dictyota and Sargassum in rather shallow water.

REFERENCES: Collins and Hervey 1917, Hoyt 1920, Papenfuss 1945b; *Chantransia dufourii,* Collins 1911.

Acrochaetium hoytii Collins

Plants arising from the original round spore, 12–25 μ diam., which, when embedded in the host, later becomes vertically elongate to 30 μ; without basal filaments, but emitting 1–3 erect filaments which are about 6 μ diam., the cells 2–4 diameters long; branching chiefly in the lower portions, the branches very long and subsimple, terminating in hairs; monosporangia lateral on the upper parts of the branches, on one-celled pedicels, rounded-oblong, 6 μ diam., 15 μ long.

North Carolina. Epiphytic on *Dictyota dichotoma.*

REFERENCES: Collins 1908, Hoyt 1920, Papenfuss 1945a.

Acrochaetium sargassi Børgesen

Plants reaching 0.7 mm. tall, the basal cell disciform, giving rise to an erect filament which branches from the base, the upper branching sparse, secund or opposite, the branchlets simple or ramifying, very attenuate toward the apices; cells of the lower portion 5.5 μ diam., 9–18 μ long, in the upper part 2–3 μ diam., 30–40 μ long; monosporangia not abundant, somewhat secund, sessile or pedicellate, obovate, 10 μ long, 7 μ broad; spermatangia opposite, about 2 μ diam., carpogonia lageniform, sessile.

Florida, Virgin Isls. Epiphytic on various algae from rather shallow water.

REFERENCES: Børgesen 1913–20, Taylor 1928, Papenfuss 1945b.

Acrochaetium caespitiforme Børgesen

Plants with a basal portion of superficially creeping branched filaments with cells about 5 μ diam., 8 μ long; erect filaments scattered along the basal ones, about 0.7 mm. tall, radially branched or occasionally secund, the branches scattered; cells cylindrical, in the lower parts of the main filaments about 5 μ diam., 12 μ long, above about 2.5 μ diam., but in the branches a little smaller, with a pyrenoid; monosporangia on 1–2-celled stalks, or rarely sessile, narrowly oval, 6 μ diam., 11–12 μ long.

Virgin Isls. Epiphytic on *Padina vickersiae* from shallow water.

REFERENCES: Børgesen 1913–20, Papenfuss 1945a.

Acrochaetium gracile Børgesen

Plants tufted, to 1 mm. tall or more, the base of more or less crowded creeping filaments; erect filaments numerous from a common base, simple or with a very few long terminally attenuate branches; filament diameter 5.5 μ below, 2.0 μ near the tips, the lower cells about 10 μ long but above to 20 μ, with one pyrenoid; fertile branchlets chiefly on the middle parts of the filaments, of 1–2 cells, rarely more, with monosporangia terminal and sessile on the upper sides, but also frequent on the main filaments, usually on one-celled pedicels; sporangia narrowly oval, 6–8 μ diam., 14–16 μ long.

Virgin Isls. Epiphytic on the leaves of *Sargassum vulgare* in a relatively sheltered location.

REFERENCES: Børgesen 1913–20, Papenfuss 1945a.

Acrochaetium daviesii (Dillwyn) Nägeli

Plants of decumbent ramifying filaments, which on firm hosts may be crowded into a felted mass, bearing erect filaments to 6 mm. tall, 7–13 μ diam. near the base, usually about 8 μ diam. near the summit; branches abundant, largely unilateral, rather erect, the cells 2–4 diameters long, containing a pyrenoid; slender multicellular hair-like branch tips more or less abundant; fertile branches short, 1–2 times redivided, usually attached near the bases of the longer branches, but sometimes scattered; monosporangia sessile or short-stalked, 8–12 μ diam., 11–20 μ long, sometimes replaced by tetrasporangia which are 7–12 μ diam., 13–16 μ long.

Hispaniola. Growing on coarser algae.

REFERENCES: Børgesen 1924, Papenfuss 1945b.

Acrochaetium flexuosum Vickers

Plants tufted, arising from a common base of crowded, creeping filaments, to about 0.7 mm. tall, the numerous erect filaments alternately and fastigiately branched, 9–10 μ diam. below, 6–7 μ diam. above, the cells 18–35 μ long below, 30–35 μ long in the upper parts of the plant, with a large pyrenoid; monosporangia usually on small branchlets secund along the upper side of the main branches, 1–2 on each branchlet, ovate-oblong, 9–10 μ diam., 14–16 μ long.

Virgin Isls., Barbados. Growing on *Chaetomorpha media* in shallow water.

REFERENCES: Vickers 1905, Børgesen 1913–20, Papenfuss 1945a.

Acrochaetium globosum Børgesen

Plants arising from a one-layered basal disk composed of creeping filaments; erect filaments several from a common base, sparingly alternately long-branched, tapering to attenuate tips, below the cells 5–6 μ diam., about 14–30 μ long, but above the cells 2–3 μ diam., about 70 μ long; cells containing several pyrenoids; fertile branchlets numerous, alternate or opposite, short, 1–3 cells long, with one monosporangium terminal and 1–3 others sessile along the upper side of the branchlet; sporangia ovate, 7–8 μ diam., 8–10 μ long.

Virgin Isls. Epiphytic on *Chaetomorpha media* in shallow water.

REFERENCES: Børgesen 1913–20, Papenfuss 1945a.

Acrochaetium leptonema (Rosenvinge) Børgesen

Plants of parenchymatous disks from which repent filaments are formed; these are irregularly laterally branched, with somewhat swollen cells, 3–4 μ diam. and 1.5–3.0 times as long; erect filaments simple to sparsely branched, reaching a length of 300 μ; the cells cylindrical to somewhat swollen, 2–3 μ diam., 2–5 times as long; mono- (and tetra-?) sporangia unilateral or terminal, rarely opposite, on branches of the first and second orders, sometimes solitary or paired on unicellular stalks, and also sessile on the repent filaments, 5.5–6.5 μ diam., 10.0–12.5 μ long.

Bermuda, Florida. Growing on larger algae.

REFERENCES: Collins and Hervey 1917, Taylor 1928, Papenfuss 1945a.

Acrochaetium netrocarpum Børgesen

Plants tufted, to 0.4 mm. tall; a common base of short branched crowded filaments creeping over the host, bearing several erect filaments sparsely radially alternately branched, 5–6 μ diam.; the cylindrical cells 3–4 diameters long, above more slender and 3–4 μ diam., with a single pyrenoid; monosporangia chiefly sessile, but also on one-celled stalks in series along the filaments and branches, fusiform, the apex a little truncate, 4–5 μ diam., 10–11 μ long.

Virgin Isls. Epiphytic on *Caulerpa taxifolia.*

REFERENCES: Børgesen 1913–20, Papenfuss 1945a.

Acrochaetium radians Hamel

Plants with a creeping base, arising from a persistent round spore 10 μ diam., the filaments radiating, seldom branched; erect filaments to 1 mm. tall, subsimple; cells cylindrical, 6–10 μ diam., 2–4 diameters long, near the tips 3 μ diam., to 35 μ long; ovate monosporangia in clusters or lateral series near the bases of the longer lateral branches or in series and terminal on short spur branches, generally stalked, 7–10 μ diam., 10–12 μ long.

Guadeloupe.

REFERENCES: Hamel 1928b, Papenfuss 1945a.

Acrochaetium sagraeanum (Montagne) Bornet

Plants tufted or stratose, to 2–4(–6) mm. tall; the basal layer composed of contorted and entangled filaments, from which the erect filaments arise, these about 12 μ diam. below, the cells 4–8 diameters long, the branching not very dense, the lowest branches longest, the upper gradually shorter, 7–9 μ diam., all rather erect and bearing short unilateral branchlets above the axils; 1–2 monosporangia borne on each articulation of the branchlets, about 18–22 μ long, 8–10 μ diam.; tetrasporangia sometimes replacing the monosporangia, 28–34 μ long, 17–27 μ diam.

Bermuda, Barbados. Epiphytic, as on *Dictyopteris justii* from shallow water in a sheltered locality.

REFERENCES: Vickers 1905, Collins and Hervey 1917, Papenfuss 1945a.

Acrochaetium sancti-thomae Børgesen

Plants to 0.2–0.3 mm. tall, from a unistratose basal disk of more or less confluent filaments; erect filaments simple or with 1–2 branches, below of cells 8–9 μ diam., 16–18 μ long, with an apparently central pyrenoid, but terminating in long hyaline deciduous hairs; monosporangia generally sessile and uniseriate in long rows on the axes and branches, sometimes opposite, sometimes on stalks bearing 1–2 sporangia, obovate, 7 μ diam., 10 μ long.

Virgin Isls. Epiphytic on the leaves of *Sargassum vulgare,* from a sheltered locality in shallow water.

REFERENCES: Børgesen 1913–20, Papenfuss 1945a.

Acrochaetium seriatum Børgesen

Plants tufted, to 1 mm. tall, of a one-layered basal disk of crowded filaments from which several erect filaments arise which may branch radially from near the base, but above more abundantly and in pectinate secund series; cells of the lower branches about 8–10 μ diam., about 22 μ long, above 6–7 μ diam. and 30 μ long; cells with a lateral pyrenoid; monosporangia oval, sessile or on one-celled stalks, in long secund series on the upper sides of the upper branches, rarely scattered, 6–9 μ diam., 9–13 μ long.

Virgin Isls. Growing on various coarser algae in sheltered places.

REFERENCES: Børgesen 1913–20, Papenfuss 1945a.

Acrochaetium thuretii (Bornet) Collins and Hervey

Basal portion an irregular disk, 60–120 μ diam.; erect filaments 2–3 mm. tall, 7.5–12.0 μ diam. below, slightly less above, branching from the base, the branches alternate or unilateral, often ending in a hair of the same diameter as the branch; cells below 3–5, above 8–12, diameters long, with a pyrenoid; monosporangia on one-celled stalks, crowded on the upper side of the branches near the base, 1–2 on each of the lower 1–3 cells, 14–17 μ long, 9–11 μ diam.; sexual plants monoecious, the organs borne on short branchlets near the bases of the branches; the spermatangia few, on the end and the upper segments, the carpogonium solitary on a lower segment; carpospores 18–21 μ long, 9–13 μ diam.

Bermuda. Epiphytic on Codium and various other coarser algae.

REFERENCES: Collins and Hervey 1917, Papenfuss 1945a.

Acrochaetium antillarum Taylor

Plants mostly creeping between the layers of the host membrane, freely irregularly alternately or seldom subdichotomously branched, the tapering branches forming short, erect filaments projecting beyond the host membrane; in the creeping portion the cells irregular in form, generally subcylindrical with ends slightly rounded, becoming irregularly curved or laterally distended, 4.5–5.2 μ diam., 8–15 μ long, chromatophores probably axial, showing a prominent pyrenoid in the larger cells; erect filaments very short, simple or very sparingly alternately branched, the cells more thick-walled than below, more regularly cylindrical, 4.0–5.7 μ diam., 5.7–8.0(–12.6) μ long; monosporangial filaments with 1–4 subcorymbiform branches bearing terminal oval sporangia about 7.5–8.5 μ diam., 10.5–12.5 μ long.

Tobago. Epiphytic on Bryopsis as drifted ashore.

REFERENCE: Taylor 1942.

Acrochaetium avrainvilleae Børgesen

Plants to about 1 mm. tall, the basal portion of short, penetrating filaments; erect filaments simple below, above sparingly alternately branched, the branches erect, occasionally with 1–2 branchlets; main axis to 9 μ diam., the cylindrical cells about 33 μ long, but in the branchlets the diameter about 5.5 μ diam.; cells with one large pyrenoid; monosporangia sessile, rarely pedicellate, a few or several in series, dorsal on the lower segments of the branchlets, oval, about 11 μ broad, 22 μ long.

Virgin Isls. Epiphytic on Avrainvillea; dredged from a depth of 20 m.

REFERENCES: Børgesen 1913–20, Papenfuss 1945a.

Acrochaetium bisporum Børgesen

Plants tufted, from a base of short, sparingly branched filaments creeping over and penetrating between the cells of the host; erect filaments generally sparingly radially branched from the middle upwards; cells cylindrical, about 8 μ diam., 20 μ long; cells with pyrenoid; hairs absent; mono- or bisporangia present, dorsally seriate on the branches or spur branchlets, generally sessile, the monosporangia narrowly ovate, 6 μ diam., 10 μ long, the bisporangia broadly ovate, 9 μ diam., 14 μ long.

Virgin Isls. Epiphytic on *Acanthophora spicifera* from rather shallow water.

REFERENCES: Børgesen 1910, 1913–20, Papenfuss 1945a.

Acrochaetium bornetii Papenfuss

Plants 2–3 mm. tall; each basal cell (original spore) 12–15 μ diam., sending down into the host an irregular and contorted filament; erect filaments arising from both the basal cell and the secondary internal filaments, the cells 8–16 μ diam., 3–10 diameters long; branches virgate, few below, more abundant above, alternate or somewhat secund, sparingly rebranched; monosporangia sessile or shortly pedicellate near the bases of the branches; cystocarps similarly placed, dense, hemispherical; antheridia forming small, dense, short-pedicellate clusters at various points on the branches; sporangia, cystocarps, and spermatangia produced on different plants.

Bermuda, North Carolina. Growing on Dasya, Dudresnaya, and other large algae from rather sheltered situations.

REFERENCES: Papenfuss 1945a; *A. corymbiferum,* Collins and Hervey 1917, Hoyt 1920.

Acrochaetium comptum Børgesen

Thallus to 1 mm. or more in height, the base a distinct persistent spore enlarged and transversely divided to 2 cells, from the lower of which a stout descending filament penetrates the host while from the other an erect filament initiates the superficial tuft; erect filament simple or subsimple below, but above with numerous long branches with secund branchlets; cells of the main filaments cylindrical, 8–11 μ diam. below, about 35 μ long, but above barely 8 μ diam., each with a prominent pyrenoid; hairs apparently absent; monosporangia seriate on the upper sides of the lower branchlet segments, usually single on one-celled stalks, oval to obovate, 11–14 μ diam., about 18 μ long.

Virgin Isls. Epiphytic on *Liagora pinnata* in a rather sheltered locality.

REFERENCES: Børgesen 1913–20, Papenfuss 1945a, Williams and Blomquist 1947.

Acrochaetium homorhizum Børgesen

Plants tufted, to 1 mm. tall or more, arising from short thick-walled moniliform basal filaments lying in the outer membrane of the host; the primary erect filaments bearing a few long scattered branches which are radially alternate below but unilateral above, and may bear several secondary branches; cells in the erect filaments 9–11 μ diam., 4–5 diameters long below, above 8–9 μ diam. and about 7 diameters long, each containing one pyrenoid; monosporangia secund on the upper side of the lowermost branchlet cells, either seriate or scattered, generally sessile, occasionally on one-celled stalks, oblong-elliptic, 10–11 μ diam., 20–22 μ long.

Virgin Isls. Growing on *Champia parvula* in the sheltered water of a lagoon.

REFERENCES: Børgesen 1913–20, Papenfuss 1945a.

Acrochaetium hypneae (Børgesen) Børgesen

Plants with a creeping, subsimple base, the erect filaments sparingly alternately and distantly branched; hairs absent; cells cylindrical, 7–10 μ diam., below 2–3 diameters long, above 3–4 diameters long, each with a pyrenoid; monosporangia lateral in series on the upper sides of the branchlets, sessile, ovate, 7–9 μ diam., 10–12 μ long.

Bermuda, Virgin Isls. Epiphytic on Hypnea in shallow protected localities.

REFERENCES: Børgesen 1913–20, Collins and Hervey 1917, Papenfuss 1945a; *Chantransia hypneae,* Børgesen 1909.

Acrochaetium nemalionis (De Notaris) Bornet

Plants forming tufts to 4–5 mm. tall, arising from filaments ramifying diffusely within the host, the erect filaments about 10 μ diam., the cells 3–5 diameters long; branches alternate, sparse below, more numerous and secund above, and their tips often attenuate and hair-like; monosporangia solitary, or 2–3 together on a 1–2-celled stalk on the lower segments of the branchlets, seldom sessile.

Bermuda. On submerged branches of Tamarix; probably also to be expected as an epiphyte on coarse algae.

REFERENCES: Collins and Hervey 1917, Hamel 1928, Papenfuss 1945a.

Acrochaetium occidentale Børgesen

Plants to 1–2 mm. tall, the base endophytic, of elongate filaments descending from the thick-walled persistent one-celled spore which becomes 16 μ in diameter; primary erect filament arising from the opposite side of the spore, simple below, sparingly and corymbosely branching outside the host, where 10–11 μ diam., the cells 27–40 μ long, but a little more slender within the host and attenuated to the branch tips; cells with a pyrenoid; monosporangia sessile, usually 2 on the lower cells of a branch, oval-ovate, 9–12 μ diam., 18–20 μ long, rarely becoming bisporic.

Virgin Isls. Growing on *Liagora elongata* in relatively shallow water. Børgesen suspects that this may be the same as *A. barbadense*.

REFERENCES: Børgesen 1913–20, Papenfuss 1945a.

Acrochaetium opetigenum Børgesen

Plants 1–2 mm. tall, rising from the persistent spore cell, 17 μ diam., which becomes elongated, cuneate, penetrating below the surface of the host, sometimes forming very short lateral outgrowths of 2–3 cells in length at the host surface but no penetrating filaments; erect filaments one to a few on each spore; branching radially alternate, sparser below, more abundant in the upper parts of the plant, cells 8–11 μ diam. below, 13–14 μ diam. in the median part and 6–7 μ diam. in the upper branches, the cells in general 80 μ or more in length; branch tips not hairlike; cells with a large lateral pyrenoid; monosporangia 1–3, sessile on the upper side of the lower cells of the branchlets, oblong-oval or subclavate, about 11 μ diam., 27 μ long.

Virgin Isls. Epiphytic on *Dasya pedicellata* at a depth of 20 m.

REFERENCES: Børgesen 1913–20, Papenfuss 1945a.

Acrochaetium phacelorhizum Børgesen

Plants to 1 mm. tall or more, from an endophytic base of more or less aggregated filaments between the cells of the host, the erect filaments branched from the base, but most abundantly above, and there somewhat unilateral; cells 11–15 μ diam. below and 6–7 μ diam. near the tips, about 40 μ long below, 54 μ long above; cells with a large pyrenoid; monosporangia few, scattered or in short series on the branchlets, sessile, oval, 12–14 μ diam., 22–25 μ long, at the apex often with a small protuberant wall thickening.

Virgin Isls. Growing on Codium *spp.* in shallow water.

REFERENCES: Børgesen 1913–20, Papenfuss 1945a.

Acrochaetium repens Børgesen

Plants with endophytic filaments of irregularly shaped cells loosely ramifying between the surface cells of the host; erect filaments scattered, arising from the endophytic ones, short, about 0.5 mm. tall, tapering to almost hyaline tips; branching not abundant, radial, alternate, or rarely opposite; the lower cells 7–8 μ diam., about 124 μ long, the upper ones 2–3 μ diam., 50 μ long; cells with a pyrenoid; monosporangia oval to obovate, 8 μ diam., 14 μ long, 1–2 together on one-celled pedicels on the upper sides of the basal cells of the branchlets, or on the main filaments.

Virgin Isls. Growing on a Gracilaria-like plant from shallow water.

REFERENCES: Børgesen 1913–20, Papenfuss 1945a.

Acrochaetium robustum Børgesen

Thallus densely tufted, to 1 mm. tall, at first attached by a thick-walled process (probably the original spore) 20 μ long, developing a basal disk of filaments spreading over the surface of the host; erect filaments numerous from each disk, sparingly radially alternately branched, the branches similar to the axis; cells 9–10 μ diam., 1.5–2.0 diameters long, or toward the tips 5 μ diam. and 40 μ long; cells with a large pyrenoid; fertile branchlets 1–3-celled, each bearing a terminal and 1–2-pedicellate or sessile, lateral, oval-ovate monosporangium, particularly thick-walled, 11 μ diam., 14–16 μ long.

North Carolina, Virgin Isls. Epiphytic on *Sargassum vulgare* from a sheltered locality.

REFERENCES: Børgesen 1913–20, Papenfuss 1945a, Williams 1951.

Acrochaetium unipes Børgesen

Plants to 2 mm. tall, attaching by a single enlarged cell derived from the germinated spore, which remains hemispherical above, narrows to pass between the surface cells of the host, and spreads out again below them, attaining a diameter of 20–22 μ and a length of 35 μ; erect filaments single on each basal cell, simple below, sparingly radially alternately branched and rebranched above, tapering toward the tips; below the cells 9–11 μ diam., to 50 μ long and rather thick-walled, but near the tips about 5 μ diam., 50–60 μ long; cells with a pyrenoid; monosporangia sessile, in short series on the upper side of the branches, oblong-clavate, 9–11 μ diam., 20–22 μ long (perhaps not mature).

Hispaniola, Virgin Isls. Epiphytic on Dictyota from about 10 m. depth.

REFERENCES: Børgesen 1913–20, 1924, Papenfuss 1945a.

Acrochaetium ernothrix Børgesen

Plants tufted, about 0.4 mm. tall; the basal portion obscure, probably endophytic, but erect filaments arising from it are sparingly divided below, more copiously above, the branches often redivided; the cylindrical cells 5–6 μ diam., 15–18 μ long, with a pyrenoid; branchlets generally of 2–3 short, stout cells, terminating in an attenuate, nearly colorless tip; monosporangia few on each branchlet, sessile, or on one-celled stalks, or 2–3 on a branching stalk near the base of a branchlet, ovate-oblong, 5–6 μ diam., 8–10 μ long.

Virgin Isls. Endophytic in Centroceras from a sheltered locality behind a reef.

REFERENCES: Børgesen 1913–20, Papenfuss 1945a.

UNCERTAIN RECORDS

Acrochaetium barbadense (Vickers) Børgesen

Bermuda, Barbados, Guadeloupe, Brazil. Type from Barbados, inadequately described. Børgesen 1913–20, Collins and Hervey 1917, Papenfuss 1945a; *Callithamnion pedunculatum,* ?Martens 1870, Mazé and Schramm 1870–77, Murray 1889; *Chantransia barbadensis,* Vickers 1905, De Toni 6:47.

(?*Acrochaetium*) *Chantransia chiloensis* Reinsch

Virgin Islands, the type locality. De Toni 6:61; Reinsch 1874–75, Papenfuss 1945a.

Acrochaetium dictyopterides Hamel and Hamel-Joukov

Martinique, the type locality, but apparently a *nomen nudum.* Hamel and Hamel-Joukov 1931.

Acrochaetium doumerguei Hamel and Hamel-Joukov

Martinique, the type locality, but apparently a *nomen nudum.* Hamel and Hamel-Joukov 1931.

Acrochaetium madininae Hamel and Hamel-Joukov

Martinique, the type locality, but apparently a *nomen nudum.* Hamel and Hamel-Joukov 1931.

Acrochaetium savianum (Meneghini) Nägeli

Bermuda, Florida, Jamaica, Guadeloupe, Brazil. Type from Italy. De Toni 6:66; Collins 1901; ?*Callithamnion byssaceum* and ?*C. pallens,* Mazé and Schramm 1870–77, Murray 1889; ?*C. posidoniae,* Grunow 1870.

Audouinella Bory, 1823

Plants small, filamentous, creeping and immersed in the substratum or erect and tufted; the filaments branched, uniseriate, the cells with one or a few spiral chromatophores lacking pyrenoids; asexual reproduction by monospores or tetraspores terminal or lateral on the branches; sexual reproduction by spermatangia clustered on the lateral branchlets and by carpogonia which may be lateral and sessile on erect branches or terminal on one- or two-celled branchlets, or intercalary; gonimoblasts forming carposporangia terminally or in series.

Audouinella membranacea (Magnus) Papenfuss

Plants of creeping, uniseriate, primary filaments 6–8(–10) μ diam.; often crowded, growing between the membranes of the host; the cells irregular, 1.5–8.0 diameters long, with a few short bandshaped chromatophores without pyrenoids; erect filaments penetrating above the surface of the host, but a few cells in length, simple or sparingly branched, the cells 7–8 μ diam., 1.5–2.0 diameters long; terminal cells on main axes and branches eventually transformed into tetrasporangia, or these lateral, 12–20 μ diam., 17–30 μ long.

Bermuda, Colombia. Growing in the chitinous matrix of sertularians and perhaps other animals, and less reliably reported in the cell walls of larger algae.

REFERENCES: *Rhodochorton membranaceum,* Collins and Hervey 1917, Taylor 1942.

Rhodochorton Nägeli, 1861

Plants slenderly filamentous, the lower portions consisting of ramifying decumbent filaments or cellular disks which support erect, branching, uniseriate, fertile filaments; cells with numerous small parietal chromatophores; tetrapartite sporangia terminating branches, or short-stalked on special subapical branchlets; sexual apparatus imperfectly known.

KEY TO SPECIES

1. Creeping base endophytic; erect filaments to 18 μ diam. R. galaxauriae
1. Creeping base epiphytic; erect filaments to 45 μ diam. R. corynosporoides

Rhodochorton corynosporoides Bornet *ex* Hamel

Plants rose-purple, about 1 cm. tall, arising from a base of creeping filaments; erect filaments about 45 μ diam. near the base, the cells about 2 diameters long, but above 25–35 μ diam. and the cells to 4 diameters long; branches irregularly disposed, long and similar to the axis; cells with numerous lacerate platelike chromatophores without pyrenoids; tetrasporangia near the branch tips, few or many in a series, often opposite, sessile, subspherical, tetrapartite, 30–40 μ diam.

Guadeloupe. Epiphytic on Centroceras filaments from shallow water.

REFERENCES: Hamel 1928b, Papenfuss 1945a; *Callithamnion corynosporoides,* Mazé and Schramm 1870–77 (a *nomen nudum*), Murray 1889.

Rhodochorton galaxauriae Vickers

Plants appearing as small tufts, but in part conspicuously endophytic, the lower filaments irregularly and intricately branched, the erect filaments from the surface about 0.5 mm. tall, 18 μ diam., with cells 30–60 μ long; tetrapartite sporangia terminal, rarely lateral, subglobose.

Barbados. Growing on Galaxaura. Not well described and of uncertain value.

REFERENCES: Vickers 1905, Papenfuss 1945a.

HELMINTHOCLADIACEAE

Plants of moderate size, erect and coarsely branched, very mucous, sometimes partly calcified; structurally with an axial row of cells bearing lateral assimilators, or multiaxial with many filaments in the center developing lateral assimilative branches of the "fountain" type; monosporangia present or absent; sexual reproduction by spermatangia borne in loose clusters on the ends of the assimilatory filaments; carpogenic branches borne variously on the assimilators, usually of three cells, the terminal being the carpogonium and auxiliary cells being absent; cystocarps immersed among the assimilative filaments, without definite pericarps, the gonimoblasts generally compactly branched with the outer cells becoming the carposporangia.

KEY TO GENERA

1. Carpogenic branches formed terminally on young assimilatory filaments 2

1. Carpogenic branches lateral on lower segments of the assimilatory filaments 3
2. Plants not calcified; lower cells of the carpogenic branch not developing involucral filaments .... NEMALION, p. 320
2. Plants with a little superficial calcification among the surface filaments; lower cells of the carpogenic branch developing many short lateral involucral filaments ..... TRICHOGLOEA, p. 322
3. Plants generally substantially calcified, even when very mucous, sometimes even rather stiff; medulla compact ..... LIAGORA, p. 324
3. Plants soft, medulla very loose; not substantially calcified HELMINTHOCLADIA, p. 324

Nemalion Duby, 1830

Frond terete or a little compressed, soft and slippery, simple or furcate-branched; structure multiaxial, the filaments of the axis with scattered anastomoses in a pale medullary cord, giving off dichotomous corymbose fascicles of peripheral chromatophore-bearing assimilatory filaments which form the cortex; spermatangia hyaline, near the apices of peripheral filaments; carpogenic branches terminal as divisions of the peripheral filaments, immersed, three- to six-celled; cystocarps compact, the end cells of the gonimoblasts producing the carpospores without special investment.

KEY TO SPECIES

1. Main axes and branches slender and smooth 2
1. Main branches rather compressed and convolute, strongly contracted at their bases, about 1.5 cm. diam., the plants rather sparingly branched from near the base ..... N. schrammi, p. 321
2. Axes seldom branched, usually 2–3 mm. diam. N. helminthoides, p. 320
2. Commonly branched, the main branches about 5–7 mm. diam. N. longicolle, p. 321

Nemalion helminthoides (Velley) Batters Pl. 42, fig. 2

Plants dull purplish, exceedingly slippery, with several subsimple axes from each disklike base, these only occasionally branching near the distal end; 8–20(–70) cm. tall, 2–3(–6) mm. diam.; monoecious, the spermatangia borne on the youngest parts of the plant; carpogenic branches usually 4-celled.

Brazil, Uruguay. Growing on rocks near high-tide line. Some individuals become horny when dried; others are soft and easily adhere to the mounting paper.

REFERENCES: Taylor 1939, Williams and Blomquist 1947, Joly 1956.

Nemalion longicolle Børgesen

Plants to 3.5 dm. tall, purplish, gelatinous, arising from a small disciform holdfast, the base slender, immediately above broadened, with several long branches at intervals of 1–2 cm., these and the main axis alternately or subdichotomously branched, the branches scattered, the axils notably rounded and the segments below somewhat broadened; diameter of branches to 5–7 mm., the upper segments somewhat less; medulla of longitudinally interwoven filaments 3–14 μ diam.; cortex of moniliform filaments, the cells about 13–14 μ diam.; carpogenic branches of 4–6 cells, 9 μ diam., on 6–9-celled stalks, the cystocarps being developed near the middle of the cortical fascicles, the carposporangia about 11 μ diam., 14 μ long.

Virgin Isls. From rocks in shallow water in sheltered places near shore.

REFERENCES: Børgesen 1909, 1913–20.

Nemalion schrammi Crouan *ex* Børgesen Pl. 42, fig. 1

Plants to 3.4 dm. tall, attached by a small holdfast disk, dark reddish brown, highly gelatinous but tough; irregularly radially branched, and this chiefly near the base, often proliferous, the branches long, subsimple, to 1.5 cm. diam., tapering gradually from below but each branch strongly contracted at the point of origin, terete in the slender distal parts, somewhat compressed and convolute below; structurally of a medulla of loosely interwoven, rather thick-walled, sparingly branched filaments 2–12 μ diam.; assimilators moniliform; the cells oval, about 14 μ diam., 28 μ long; carpogenic branches of 3–6, usually 4, cells about 9 μ diam., terminating and ill-differentiated from the 3–4-celled stalks which bear them; carposporangia about 11 μ diam.

Virgin Isls., Guadeloupe, Brazil. From rocks in shallow water in sheltered places near shore.

REFERENCES: Børgesen 1909, 1913–20, Williams and Blomquist 1947; *Helminthocladia schrammi* (*nomen nudum*) Mazé and Schramm 1870–77, Murray 1889.

Trichogloea Kützing, 1849

Plants bushy, very gelatinous or mucous, radially alternately branched, with some lime usually deposited in the outer portion of the medulla; axis filamentous, the filaments loosely intermeshed, the cortex of radially disposed fascicles of assimilatory filaments derived from the medullary strands, these filaments dichotomous, moniliform, clavate-enlarged towards their tips; cells with a lacerate chromatophore and a pyrenoid; spermatangia clustered on the assimilators; carpogenic branches terminal, replacing the outer portions of assimilators, the lower cells of each branch producing many short, crowded, involucral filaments; cystocarps stalked in the outer parts of the fascicles of assimilators.

KEY TO SPECIES

1. Plants coarse, essentially lime-free; involucre compact ... T. herveyi
1. Plants more slender, the outer medulla with a visible deposit of lime; involucre loose .... T. requienii

Trichogloea herveyi Taylor

Plants bushy, to 17 cm. tall, highly gelatinous and very soft, dull purplish red, uncalcified, or a little invisible entirely diffuse calcification present; subsessile from a small holdfast, repeatedly alternately branched, the branches to 8 mm. diam., more generally about 4 mm. diam. below, 1.5–2.0 mm. diam. in the lesser branches, which taper to more or less blunt apices, and are generally not crowded, but occasionally the branching is dense, short, and coralloid; medulla not perceptibly calcified, rarely slightly thicker in dried specimens; medullary filaments ditrichotomously or irregularly branching at wide angles, in young branches 5.5–8.5 μ diam.; cortical assimilatory filaments slender near the medulla, reaching 18–20 μ diam. in the distal portion, the outer cells oval to obovoid, truncate, each with a single lacerate chromatophore and a prominent pyrenoid; spermatangia delicate, 2.5–3.5 μ diam., usually 2 on each little stalk cell, whorled near the outer ends of 3–5 terminal or subterminal cells of the cortical filaments; carpogonia terminal on divisions equivalent to ultimate assimilatory filaments, about 3–5 of the lower cells of the carpogenic branch portion developing short 1–3 times divided involucral filaments, the outer cells of which below old cystocarps reach 4.5–5.5 μ diam.; cystocarps immersed in the outer portion of the cortex, small, 50–120 μ diam., the obovoid carposporangia formed from terminal cells of the gonimoblasts, 10.0–12.5 μ diam., 15.0–17.0 μ long.

Bermuda. Growing on rocks from the low-tide level to a few decimeters below, in rather exposed situations.

REFERENCES: Setchell 1915 in Phyc. Bor.-Am. 2034, Collins and Hervey 1917, Howe 1918a (in all as a *nomen nudum*), Taylor 1951a.

Trichogloea requienii (Montagne) Kützing

Plants reaching 20 cm. in height, repeatedly branched from a small holdfast, bushy, purplish red; main branches 2–4 mm. diam., the lesser branches 1–2 mm. diam.; medulla 1–2 mm. diam., markedly calcified, perceptibly thicker, raised, white and sharply delimited in dried specimens; cortex 0.5–1.0 mm. thick, gelatinous, the cortical filaments with outer cells cask-shaped to truncate-ovoid, 12–22 μ diam., 22–28 μ long, but more slender toward the medulla; monoecious, the spermatangia whorled on 1–5 cells of the cortical filaments, terminal or more usually 2–5 cells distant from the outer ends; carpogonia terminating cortical filaments, the lower 3–4 cells of the carpogenic branch sparingly developing once- or twice-branched involucral filaments, not forming a dense mass below the cystocarp, the whorls sparse and remaining well separate, the end cells 3.5–5.5 μ diam. on young axes but to 8–10 μ diam. on old ones; axis bearing the involucre about 8 μ diam. when young, becoming 15–25 μ diam. below the older cystocarps with complete resorption of at least the upper 2–3 cross-walls; cystocarps to 45–145 μ diam., the very numerous carposporangia to 5.5–9.3 μ diam., 10–12 μ long.

Puerto Rico, Guadeloupe, Barbados.

REFERENCES: Hamel and Hamel-Joukov 1931; *T. lubrica,* Vickers 1905; *Helminthocladia cassei,* Mazé and Schramm 1870–77, Murray 1829; ?*Nemalion barbadensis,* Vickers 1905.

Helminthora J. Agardh, 1852

UNCERTAIN RECORD

Helminthora guadeloupensis Crouan

Guadeloupe, the type locality; essentially a *nomen nudum.* De Toni 6: 85; Mazé and Schramm 1870–77, Murray 1889.

Helminthocladia J. Agardh, 1852

Plants bushy, the branching alternate, radial; structurally of numerous longitudinal axial filaments forming a very loose medulla, their radially placed branches forming a cortex of assimilatory filaments which are broadest toward the tips, the whole embedded in a soft jelly; spermatangia in compact clusters laterally placed on the outer segments of the assimilators; carpogenic branches lateral on inner segments of the assimilators, of three or four cells; cystocarps becoming involucrate, of numerous crowded gonimoblasts whose end cells become the carposporangia.

Helminthocladia calvadosii (Lamouroux) Setchell Pl. 43, fig. 5

Plants bushy, very gelatinous, dull purple, from a firm, small conical holdfast sharply expanding to a distinct stout primary axis 2–6 dm. tall, to 15 mm. diam.; the branches terete, similar to but more slender than the primary axis, somewhat thicker in the middle than near the base and greatly tapering to the tips, alternately radially disposed, dividing to 2–3 degrees; assimilatory filaments terminating in pyriform cells 20–25 μ diam., with single, large, distal, notably radially lobed chromatophores each with a prominent pyrenoid.

Bermuda, Florida. Growing on stones near or even above low-tide line in sheltered places. The American plants are reported to be dioecious, and Feldmann (1939b) suspects that these plants may represent a new species of Nemalion.

REFERENCES: Collins and Hervey 1917, Howe 1918a; *Helminthora divaricata,* Harvey 1853, Rein 1873, Melvill 1875, Hemsley 1884.

Liagora Lamouroux, 1812

Plants bushy, soft, sometimes mucous, generally with a dispersed soft calcification, in some cases becoming firm; of multiaxial filamentous construction, the compact medulla of colorless longitudinal filaments rather free of lime, the radial branching divisions of these being the active assimilators; spermatangia in clusters lateral or terminal on the assimilators; carpogonial branches arcuate, formed on segments of the middle parts of the assimilators, usually four- but sometimes three- to six-celled; cystocarps generally rather loose and somewhat encroached upon by the assimilators, of slender gonimoblasts, the outer cells becoming carposporangia; in one species the carposporangial initial dividing into four cells.

KEY TO SPECIES

1. Branching mainly dichotomous, but sometimes with abundant lateral proliferations 2
1. Branching mainly alternate 4

2. Plants rather loose in habit; assimilatory filaments 13–21 μ diam. near the surface; spermatangial clusters compact, ovoid to globose or hemispherical .... L. farinosa, p. 326
2. Plants more compact; assimilatory filaments usually less than 13 μ diam.; spermatangial clusters loose 3

3. Plants moderately calcified, the surface of the branches not smooth; cystocarps with well-defined involucral filaments L. ceranoides, p. 326
3. Plants more heavily calcified, especially below, where rather stiff, the surface smooth; cystocarps visible at the surface of fertile plants, with a poorly defined involucre ..... L. valida, p. 327

4. Plants with conspicuous virgate primary axes and shorter lateral branches, substantially calcified in the older parts L. decussata, p. 330
4. Plants more irregular, or paniculate, less calcified 5

5. Calcification absent; plants rather soft; reproduction unknown L. pectinata, p. 328
5. Calcification present but often slight 6

6. Dioecious, or spermatangia unknown 7
6. Monoecious 8

7. Plants paniculately branched; very mucous, with but a little and chiefly axial calcification; assimilators strongly moniliform, 13–19 μ diam.; cystocarps with a few involucral filaments .... L. mucosa, p. 328

7. Plants subdichotomously branched below, alternate in the distal portions; rather well calcified; assimilators somewhat moniliform, 19–20 μ diam.; spermatangia unknown; cystocarps each with a conspicuous involucre . . . . L. megagyna, p. 328

8. Calcification chiefly axial, the assimilatory filaments extending beyond it; spermatangial clusters not usually on the last segment but on the subterminal and third cells of the assimilators, 25–40 μ diam.; involucre of ascending filaments present L. pinnata, p. 329

8. Calcification chiefly in the superficial mucus; spermatangial clusters on the terminal and subterminal cells of the assimilators, 13–20 μ diam.; involucre, if present, of only a few usually pendant rhizoidal filaments . . . . L. pedicellata, p. 329

Liagora farinosa Lamouroux Pl. 43, fig. 3; pl. 45, fig. 2

Plants loosely and widely dichotomously branching, forming rather tangled reddish masses, to 12 cm. diam. or more; calcification moderate, especially light in the distal parts, assimilatory filaments extending beyond the calcification, their cells cylindrical or slightly swollen, 13–21 μ diam.; plants dioecious, the spermatangia forming dense masses toward the tip of the corticating filaments, 40–60 μ diam.

Bermuda, Florida, Bahamas, Turks Isls., Cuba, Cayman Isls., Jamaica, Hispaniola, Virgin Isls., Guadeloupe, Martinique, Barbados, Brazil. Plants of rather shallow water, growing on rocks or old corals in moderately sheltered locations.

REFERENCES: Howe 1920, Taylor 1928, 1933, Taylor and Arndt 1929, Sanchez A. 1930; *L. cayohuesonica,* Murray 1889; *L. cheyneyana,* Murray 1889, Collins 1901; *L. corymbosa,* Børgesen 1913–20; *L. elongata,* Collins 1901, Vickers 1905, Børgesen 1913–20, Collins and Hervey 1917; *L. farionicolor,* Murray 1889; *L. viscida* v. *laxa, Galaxaura tomentosa,* and *G. valida,* Mazé and Schramm 1870–77, Murray 1889.

Liagora ceranoides Lamouroux Pl. 43, fig. 1; pl. 45, fig. 1

Plants forming rather compact, soft, nearly white tufts 4–7 cm. diam., the branching rather close in the distal portions; calcification moderate, the surface becoming farinaceous below; assimilatory filaments hardly extending beyond the calcification, branching at wide angles, moniliform, 6–11 μ diam.; spermatangial clusters forming platelike disks on the terminal filaments, which intermesh so that the spermatangia rather completely cover the surface of the plant; cystocarps with well-developed filamentous involucres.

Bermuda, Florida, Mexico, Bahamas, Caicos Isls., Cayman Isls., Jamaica, Hispaniola, Puerto Rico, Virgin Isls., Saba Bank, Guadeloupe, Barbados, Grenada, Panama, Netherlands Antilles, Brazil. Growing on rocks, old corals, shells, etc., generally in rather shallow water, but reported from a depth of 20 m.

REFERENCES: Harvey 1853, Mazé and Schramm 1870–77, Howe 1918a, 1920, Taylor 1928, 1933, 1942, 1943, Taylor and Arndt 1929, Joly 1953c; *L. distenta, L. patens p. p., L. pulverulenta* f. *tenuior, L. turneri, L. viscida* v. *coarctata,* Mazé and Schramm 1870–77, Murray 1889; *L. leprosa,* J. Agardh 1847, Harvey 1853, Mazé and Schramm 1870–77, Murray 1889, Vickers 1905; *L. pulverulenta,* Harvey 1853, Mazé and Schramm 1870–77 *p. p.,* Dickie 1874a, Murray 1889, Collins 1901, Vickers 1905, Børgesen 1913–20, Collins and Hervey 1917.

Liagora valida Harvey Pl. 43, fig. 2

Plants rather small, not over 1 dm. diam., dichotomously and rather closely branched; except at the tips calcification moderate to heavy, especially in the lower parts, which are stiff and chalk white; branches about 1 mm. diam., smooth; axial filaments 20–40 μ diam., with rhizoidal filaments 8 μ diam. intermixed; assimilators erect, branching 4–5 times, the outer cells oval to pyriform, 10–15 μ diam.; spermatangial clusters platelike, borne on the end cells of the assimilators; carpogenic branches of 4–5 cells, somewhat curved; cystocarps visible as minute red spots on the surface of fertile plants, about 400 μ diam., the whole surrounded by a few involucral filaments of oval cells, and developing numerous descending slender rhizoidal filaments as well.

Bermuda, Florida, Bahamas, Caicos Isls., Cuba, Cayman Isls., Jamaica, Hispaniola, Puerto Rico, Virgin Isls., Guadeloupe, Barbados, British Honduras, Panama. Plants of shallow water growing on exposed rocks and old corals.

REFERENCES: Harvey 1853, Dickie 1874b, Hemsley 1884, Murray 1889, Collins 1901, Vickers 1905, Børgesen 1913–20, Collins and Hervey 1917, Howe 1918a, 1920, Taylor 1928, 1935, 1942; ?*L. annulata,* Hauck 1888; *L. brachyclada,* ?*L. fragilis,* Mazé and Schramm 1870–77, Murray 1889; *L. tenuis,* Phyc. Bor.-Am. 687.

Liagora pectinata Collins and Hervey

Plants rather soft, without incrustation, dull red-purple, up to 20 cm. high; main branches nearly naked below, above with numerous alternate or secund more or less sinuous branches, bearing similar branchlets, the lesser divisions spreading, recurved, often pectinately secund, all tapering to an acute point; central strand about 150 μ diam., of closely packed filaments, the cells about 20 μ diam., 4–8 diameters long; fascicles of assimilators 600–800 μ long, dense, even-topped; filaments many times dichotomous, the lower cells subcylindrical, 5–7 μ diam., 5–8 diameters long, the end cells ovoid, 8–10 μ diam., 2–3 diameters long; fructification unknown.

Bermuda.

REFERENCE: Collins and Hervey 1917.

Liagora mucosa Howe

Thallus very mucous, 5–20 cm. high, the main divisions somewhat paniculate, the branches 0.32–0.60 mm. broad toward the apex, lightly and irregularly calcified, especially near the axis; fascicles of assimilators 150–300 μ long, 3 or 4 times dichotomous, the filaments markedly moniliform, the distal cells 13–19 μ diam., 13–24 μ long; plants dioecious, the spermatangia forming compact rounded tufts 25–50 μ broad, on ultimate or penultimate cells of the assimilators; the cystocarps compact, 100–200 μ broad, with a few ascending involucral filaments.

Bermuda, Florida, Bahamas, Cayman Isls., Guadeloupe, Barbados.

REFERENCES: Howe 1920, Taylor 1928; *L. paniculata* Vickers 1905; *Helminthora antillarum,* Mazé and Schramm 1870–77, Murray 1889.

Liagora megagyna Børgesen

Plants 12–14 cm. tall, irregularly branched, in part dichotomous but especially in the upper parts alternate, with rather short side branches in 1–2 degrees; calcified, when drying becoming rough; axis of large (100–160 μ diam.) and slender (11 μ diam.) filaments intermixed; assimilatory filaments with a distinctively large basal cell, but beyond it slender, ditrichotomous, and gradually increasing in diameter toward the surface where the cells are rather oval and about 19–20 μ diam., 30–35 μ long; erect carpogenic branches of 4 short cells 20–27 μ diam.; cystocarps with basal involucres.

Virgin Isls. Plants of shallow water.

REFERENCE: Børgesen 1913–20.

Liagora pedicellata Howe

Plants generally very flaccid, 4–16 cm. tall, irregularly alternately branched, more or less paniculate; external mucus conspicuous, superficially containing calcareous granules in a somewhat alveolate pattern; branches successively shorter, in the lesser divisions 1.2–2.0 mm. diam. inclusive of the mucus; medulla below 1.5–2.5 mm. diam., sometimes denuded, somewhat calcified, but toward the branchlet apices represented by about 4 filaments 35–104 μ diam. intermixed with rhizoidal filaments 8–13 μ diam.; assimilators 4–7 times ditrichotomous, distally moniliform, the cells obovoid, 12–16(–20) μ diam., 15–26 μ long; monoecious, the spermatangia in small tufts 13–20 μ diam. on the ultimate and penultimate segments; carpogonial branches erect, 20–25 μ diam., of 4–6 cells including the pedicel, cystocarps compact, 90–215 μ diam., peripheral, each borne on a stout axis 300–450 μ long; involucres absent or of but a few decurrent filaments.

Florida, Bahamas, Caicos Isls., Jamaica, Guadeloupe. On rocks in shallow water, sometimes in moderately exposed situations.

REFERENCES: Howe 1920, Taylor 1928; *Helminthora dendroidea*, Mazé and Schramm 1870–77, Murray 1889.

Liagora pinnata Harvey

Plants moderately densely paniculately branched, 2–3 dm. diam., the branches frequently somewhat complanate; calcification moderate, chiefly axial, the plant soft but not mucous; the assimilatory filaments not moniliform, the cells slightly swollen, 17–28 μ diam.; plant monoecious, the spermatangia in tufts 25–40 μ broad; cystocarps with well developed ascending involucres.

Florida, Bahamas, Jamaica, Virgin Isls., Guadeloupe, Venezuela. From rocks and dead corals in shallow water and moderately exposed places.

REFERENCES: Harvey 1853, Mazé and Schramm 1870–77, Murray 1889, Børgesen 1913–20, Howe 1920, Taylor 1928, 1942; *L. bipinnata* and *L. pinnata* v. *arbuscula*, Mazé and Schramm 1870–77, Murray 1889.

Liagora decussata Montagne

Plants tall, to 10 dm. or more, virgate, the main axes slender, sparingly divided, copiously radially beset with shorter, often opposite, slender branchlets about 0.75–1.50 cm. long; branch tips soft, exposing the colored assimilatory filaments, but otherwise substantially superficially calcified, the calcification tending to flake off when dried; assimilatory filaments several times ditrichotomous, widely spreading; the cells of the lower divisions long, subcylindrical, the upper ones shorter and more swollen, the terminal ones clavate to pyriform 7–11(–15) μ diam.

Jamaica, Guadeloupe, St. Vincent.

REFERENCES: Mazé and Schramm 1870–77, Murray 1889, Butler 1902, Hamel and Hamel-Joukov 1931.

UNCERTAIN RECORDS

Liagora dendroidea Crouan

Guadeloupe, the type locality. De Toni 6:96; Mazé and Schramm 1870–77, Murray 1889.

Liagora distenta (Mertens) C. Agardh

Guadeloupe, Brazil. Type from Europe. De Toni 4:92; Zeller 1876; *L. distenta* v. *complanata,* Mazé and Schramm 1870–77, Murray 1889.

Liagora opposita J. Agardh

Florida, Guadeloupe. Type from Florida. De Toni 4:89; ?*L. albicans, L. patens p. p., L. pulverulenta p. p.,* and *L. rugosa,* Mazé and Schramm 1870–77, Murray 1889.

Liagora pilgeriana Zeh

Brazil, the type locality. De Toni 6:95; Zeh 1912.

Liagora prolifera Crouan

Guadeloupe, the type locality. De Toni 6:96; Mazé and Schramm 1870–77, Murray 1889 (including *L. viscida* v. *laxa*).

Liagora rosacea Zeh

Brazil, the type locality. De Toni 6:95; Zeh 1912.

Liagora viscida (Forsskål) C. Agardh

Jamaica, Virgin Isls., Grenada, Netherlands Antilles. Type from Europe. De Toni 4:90, 6:88; Murray 1889, Sluiter 1908; *L. viscida* v. *coarctata,* Sluiter 1908; *L. viscida* v. *gracilis,* Mazé and Schramm 1870–77, Murray 1889.

CHAETANGIACEAE

Plants of moderate size, erect and bushy of habit, soft in texture; sometimes partly calcified; structurally multiaxial, the filaments in the center developing lateral branches of the "fountain" type, the outer cells of which may be joined into a continuous epidermis or may be extended into chromatophore-bearing filaments; monosporangia or tetrasporangia sometimes present; spermatangia scattered over the surface of the plant as little cells in small groups, or in conceptacular sacs; carpogenic branches usually three-celled, borne on inner forks of the lateral filaments, the cell below the carpogonium capable of originating nutritive cells; cystocarps immersed, the gonimoblasts developed within a pericarp of slender crowded filaments, which form a sac like a conceptacle, eventually discharging through a pore at the surface.

KEY TO GENERA

1. Cystocarp basally attached, compact, centrally placed in the flask-shaped ostiolate pericarp; not calcified; tetrasporangia not present 2
1. Carpospores developed centripetally from gonimoblast filaments lining the walls of the ostiolate pericarp and associated with paraphyses; some lime usually present in the thallus; tetrasporangia known .... GALAXAURA, p. 334

2. Outermost cells of the radial filaments of the cortex crowded but not notably modified in the mature condition, and not forming a coherent epidermal layer .... GLOIOPHLAEA, p. 332

2. Outermost cells of radial filaments of the cortex enlarged, hyaline, laterally conjoined to form an epidermal layer
SCINAIA, p. 333

Gloiophlaea J. Agardh, 1870

Plants bushy, dark red, freely dichotomously branched; branches terete, fleshy to cartilaginous, the medulla of longitudinal filaments, bearing excurrent lateral fascicles, from which at first a loose cortex of hyaline utricles mixed with colored cells is formed, but this invaded with age by anticlinal rows of moniliform filaments, the inner cells somewhat larger than those at the surface; monoecious or dioecious; spermatangia forming a continuous layer over large portions of the frond; cystocarps more or less scattered, lying in the inner cortex, discharging through a small pore; pericarp of a few pseudoparenchymatous or filamentous layers of cells; gonimoblasts loose, attached to a few-celled placenta, forming the carposporangia terminally.

KEY TO SPECIES

1. Plants small, abundantly branched at intervals of 2–3 mm., the dried segments about 0.25 mm. diam.; axial strand evident; cystocarps 160–175 μ diam. .... G. caribaea
1. Plants tall, frequently branched at intervals of about 2 cm., the dried segments about 2 mm. diam.; axial strand obscure; cystocarps 66–88 μ diam. .... G. halliae

Gloiophlaea halliae Setchell

Plants 11–13 cm. tall, 5–7 times dichotomous, the branches cylindrical, 2 mm. diam. (dried); outer cortex becoming 22–40 μ thick, of anticlinal corymbose colored filaments alternating with ovate utricles which, being overgrown, may be succeeded by a second utricular zone; monoecious; cystocarps rounded-pyriform, 100–118 μ long, 66–88 μ diam.; pericarp of 5–6 layers of filaments.

Florida, Venezuela.

REFERENCE: Setchell 1914.

Gloiophlaea caribaea Taylor

Plants arising from a short conical base, 3–4 cm. tall, bushy, reddish brown near the tips of the branches, much darker below;

the primary stem short, the branching above dichotomous, the divisions to 10–13 stages at least; when dried the branches 0.25 mm. diam. near the base of the plant and 0.8 mm. diam. near the tips, when soaked, up to 0.5–0.7 mm. diam. below and up to 1.25 mm. diam. near the tips; axis consisting of stouter filaments 6–14 μ diam., with relatively thick walls, loosely disposed, and having among them filaments, about 2 μ diam., which continue out to form the loose inner cortex; outer cortex about 33 μ thick, with a layer of ovate-turbinate colorless utricles supported by a layer of large oval cells, which, in converging series, are connected with the slender filaments by small rounded cells; slender colored filaments penetrating between the utricles; monoecious, the spermatangia widespread, formed in the younger parts of the plant on the ends of the superficial filaments between the utricles; cystocarps abundant, subspherical or depressed, 160–175 μ diam., with 5–8 layers of coherent filamentous pericarp.

Hispaniola. From shallow water near shore.

REFERENCE: Taylor 1943.

Scinaia Bivona, 1822

Plants bushy, repeatedly dichotomously branched, the texture firm-gelatinous, a more or less indistinct axial strand being present in the branches; structurally showing three layers, the strand in the center giving off lax dichotomous filaments containing most of the chromatophores and constituting the assimilatory zone, these ramifications ending in large cells polyhedral in surface view, without conspicuous chromatophores, which are closely joined laterally in a smooth epidermis; monosporangia spherical, scattered, formed between the epidermal cells; the spermatangia spherical, in small sori, where two or three are pushed to the surface between the epidermal cells; carpogenic branches developed at forkings of the medullary filaments, consisting of three cells, the outer being the carpogonium, the second developing before fertilization four lateral nutritive cells, and the third or lowest developing the rudiments of the pericarp filaments; the terminal cells of the gonimoblasts producing the carpospores; cystocarps lying below the peripheral layer, enclosed in a dense, obscurely filamentous pericarp which eventually opens to the surface by a pore.

Scinaia complanata (Collins) Cotton Pl. 42, fig. 3

Plants pale rose-red, 5–8 cm. tall, 8–9 times dichotomous, complanate, the axils broad and a little flattened but not constricted; segments 1.5–6.0 mm. diam.; medullary strand rather obscure in intact specimens, of 6–8 filaments; subsurface layer of round or pyriform cells; epidermis of cells rectangular in section, 22 μ wide, 34–35 μ tall, these occasionally separated by slender colored cells; cystocarps about 165 μ long, 200 μ diam.; pericarps thin, of 2–3 pseudoparenchymatous layers.

V. **intermedia** Børgesen. Branches 1–2 mm. diam., terete, and with the axial strand of 20–30 filaments visible throughout. Probably not sharply distinct from the species.

Bermuda, Florida, Virgin Isls., Netherlands Antilles, Venezuela, Tobago. Plants from moderately to quite deep water, growing on shells or coral fragments to a depth of 55 m.

REFERENCES: Setchell 1914, Collins and Hervey 1917, Taylor 1942, 1953a. For the v. *intermedia,* Børgesen 1913–20, Taylor 1928.

UNCERTAIN RECORD

Scinaia furcellata (Turner) Bivona

Bermuda, Florida, Guadeloupe. Type from England. (All American plants probably are to be referred to *S. complanata* (Collins) Cotton.) De Toni 4:104; Harvey 1853, Mazé and Schramm 1870–77, Dickie 1874b, Hemsley 1884, Murray 1889; *Ginannia furcellata,* Hooper 1850.

Galaxaura Lamouroux, 1812

Plants bushy, of moderate size, fairly soft to firm and wiry, usually dichotomously branched; of two general types: either pilose, brownish red when young, the cortex inside of the zone of assimilatory filaments somewhat calcified, or, smooth and without free assimilators, light pink or translucent, becoming dull whitish with age, lightly to moderately calcified to the surface; structurally composed of a medulla of slender colorless filaments which give rise by lateral branches to an inner more or less filamentous, even pseudoparenchymatous cortex of large colorless cells which externally bear long assimilatory filaments or the outer cells larger and forming a more or less coherent superficial or epidermal membrane and in some species these epidermal cells bear one-celled subspinulose extensions or specialized filaments two or three cells long; unicellular colorless hairs sometimes present; sporangia tetrapartite, pedicellate on one–few-celled stalks, or lateral at the bases of the specialized filaments; spermatangia formed in conceptacles; carpogenic branches usually three-celled, formed on the inner cortical filaments, the cystocarps immersed, usually in the cortex, surrounded by pericarps of slender filaments, the gonimoblasts peripheral, developing carposporangia centripetally; each cystocarp discharging by a pore to the surface.

This genus presents peculiar taxonomic difficulty because it is fairly clear that certain species pairs represent the sporophyte and the gametophyte phases of the same plant. However, since the relationship has been brought out in but a few cases, and there are obviously many more needing study, and since the situation has not been experimentally confirmed for any species, the taxonomic readjustments have never been completed.

KEY TO SPECIES

1. Free assimilatory filaments generally persisting, the surface therefore pilose 2
1. Free assimilatory filaments absent, or at least soon deciduous and of but one type 6

2. All assimilatory filaments long and of the same type, sometimes branched .... G. comans, p. 336
2. Assimilatory filaments of 2 types, long and short 3

3. Assimilatory filaments of the 2 types arranged in more or less distinct alternating zones .... G. subverticillata, p. 339
3. Assimilatory filaments of the 2 types intermingled 4

4. Distance between dichotomies often more than 1 cm.; long assimilators 0.15–0.70 mm. long .... G. flagelliformis, p. 338
4. Dichotomies rather closer 5

5. Medullary filaments diverse in thickness and the assimilatory filaments commonly borne directly on them .. G. delabida, p. 338
5. Medullary filaments more uniform; assimilators always borne on a distinct supporting cell .... G. lapidescens, p. 337
G. liebmanni, p. 337

6. Thallus naturally nearly or quite terete, though sometimes a little compressed by drying 7
6. Thallus in large part clearly flattened 11
7. Assimilators of a single type present, mostly deciduous but sometimes irregularly persistent, especially on the lower parts of the plant; the upper, naked branches sometimes a little rugose .... G. squalida, p. 339
7. Free filamentous assimilators absent 8
8. Conspicuously annularly constricted or rugose, the segments between dichotomies quite short .... G. rugosa, p. 340
8. Not annularly rugose, or but slightly so, and then the segments longer 9
9. Calcification slight; segments generally over 2 mm. diam. G. obtusata, p. 342
9. Calcification substantial; segments seldom reaching 2 mm. diam. 10
10. Segments 0.45–1.0 mm. diam., cylindrical, the articulation at the forks quite marked and the plants fragile G. cylindrica, p. 341
10. Segments 1–2 mm. diam., somewhat tapering to the ends G. oblongata, p. 341
11. Branches commonly subarticulated at the forkings and here surrounded by conspicuous tufts of assimilatory filaments; segments short, holdfast and stalk inconspicuous G. stupocaulon, p. 344
11. Branches without a tendency toward articulation, normally without tufts of long assimilators 12
12. Stalk inconspicuous; segments usually several times as long as broad; above often cross-banded .... G. marginata, p. 343
12. Stalk and cushion-shaped holdfast conspicuous, stupose; segments short, not notably banded .... G. frutescens, p. 344

Galaxaura comans Kjellman

Plants to 12 cm. tall, irregularly alternately branched, the internodes to 3 cm. long, apparently without any limy deposit, densely covered with assimilators, the total width reaching 3–5 mm.; medulla of loosely intertwined filaments 8–25 μ diam. and at its surface giving rise to a layer of loosely associated polyhedral cells each bearing one, seldom 2, assimilatory filaments; assimilators of one kind only, with a large oval cell about 50 μ diam., 80 μ long at the base, the main part of the filament beyond this generally simple, sometimes branched, below of oval, but above quickly changing to cylindrical, cells, about 19 μ diam., 2 diameters long; tetrasporic, the sexual phase possibly *G. oblongata*.

Bahamas, Caicos Isls., Cuba, Jamaica, Puerto Rico, Virgin Isls., Guadeloupe, Netherlands Antilles. On shells and probably other objects to a depth of 15 m.

REFERENCES: Kjellman 1900, Sluiter 1908, Børgesen 1913–20, Howe 1918c, 1920.

Galaxaura lapidescens (Ellis and Solander) Lamouroux

Plants bushy, densely irregularly and rather closely branched at intervals of about 1 cm., the branch diameter including the assimilators about 3 mm.; medulla of filaments 11–13 μ diam. or sometimes more; cortex of polyhedral cells 40–50 μ diam., each supporting 1–2 assimilators; these of short and long types intermixed, both with large oval or obovoid basal cells about 45 μ diam., 70 μ long; the 1–2 distal cells of the short assimilators usually oval, about 28–30 μ diam.; the long assimilators with 1–2 smaller oval cells above the basal, thence cylindrical, about 20 μ diam., the cells 2–3 diameters long.

Bermuda, Florida, Mexico, Bahamas, Cuba, Jamaica, Puerto Rico, Virgin Isls., Guadeloupe, Martinique, Barbados, Grenada, Old Providence I., Panama, ?Colombia. Plants of shallow water, and also dredged from a depth of 12 m.

REFERENCES: Mazé and Schramm 1870–77, Dickie 1874a, b, Hemsley 1884, Murray 1889, Collins 1901, Vickers 1905, Børgesen 1913–20, Howe 1918b, 1920, Taylor 1928, 1929b, 1939b, 1954a.

Galaxaura liebmanni (Areschoug) Kjellman

Plants 7–8 cm. tall, much branched, the branches with abundant assimilators to 5–6 mm. diam., with hardly any lime; axial filaments near the surface developing polyhedral supporting cells which in turn each bear 1–2 assimilators; short assimilators consisting of an enlarged, densely granular oval basal cell 80 μ diam., and usually 2 smaller distal cells, the nodes contracted; long assimilators with a similar basal cell and a long, occasionally branched filament, the lower 2–3 cells oval, the rest of the filament nearly cylindrical, 20 μ diam., the cells 1.2–2.5 diameters long.

Mexico, Netherlands Antilles. It is probable that this is a later name for the plant usually called *G. lapidescens,* not recognized by Kjellman.

REFERENCES: Kjellman 1900, Sluiter 1908; *Holonema liebmanni,* Areschoug 1855.

Galaxaura delabida Kjellman

Plants densely tufted, about 3 cm. tall, the branches spreading, the segments below about 2–5 mm. long, 1.5 mm. diam., terete or a little compressed; assimilatory filaments deciduous above, and the branches becoming somewhat annulate; supporting cells scarce, about 30 μ diam.; assimilators of 2 classes, generally attached to the outermost medullary filaments; short assimilators with an ellipsoid or pyriform basal cell 30–33 μ diam., 35–45 μ long, the 1–2 distal cells hemispherical to obovoid; long assimilators about 0.7 mm. long, the base of 1–2 ellipsoid or pyriform cells 35–45 μ diam., 36–60 μ long, those beyond decreasingly tumid and the filaments for the most part cylindrical, about 15 μ diam., the cells 1.5–2.0 diameters long.

Virgin Isls.

REFERENCES: Kjellman 1910, Børgesen 1913–20.

Galaxaura flagelliformis Kjellman

Plants bushy, to 12 cm. tall or more, rather tough in consistency; the tomentose branches spreading, 1.0–2.5 cm. or more between the forkings, rather slender, 2.5–3.0 mm. diam. including the assimilators, often depiliate below; medullary filaments about 12 μ diam., diffusely calcified, bearing a cortical layer of rather small polyhedral cells about 35 μ diam., each supporting 1–2 assimilatory filaments; assimilators intermixed, often on the same support, of short (3-celled) and long (0.15–0.70 mm.) types, both with oval to obovoid basal cells 55–60 μ diam., 80–90 μ long; 2 outer cells of the short assimilators cask-shaped or depressed-spheroidal; lower 1–2 cells of the long assimilators turgid, the others cylindrical, about 17 μ diam., 1.5–2.0 diameters long, very slightly constricted at the septa; tetrasporic, the sexual phase probably being *G. squalida*.

Bermuda, Florida, Bahamas, Caicos Isls., Cuba, Jamaica, Hispaniola, Puerto Rico, Virgin Isls., Nevis, Guadeloupe. Reported from shallow water, and dredged from a depth of 18 m.

REFERENCES: Børgesen 1913–20, 1924, Collins and Hervey 1917, Howe 1917, 1918a, c, 1920, Taylor 1928, 1943, 1954a, Hamel and Hamel-Joukov 1931.

Galaxaura subverticillata Kjellman Pl. 44, fig. 6; pl. 45, fig. 9

Plants 4–7 cm. high; branches dichotomous, subverticillately villous; internodes 3–10 mm. long, 1.0–1.5 mm. diam.; medullary filaments 7–28 μ diam.; the peripheral tissue hardly constituting a coherent cortex, composed of the supporting and basal cells of the dimorphic assimilators; supporting cells relatively undifferentiated; tumid basal cells oval or elliptical, 25–46 μ diam., 40–76 μ tall; extended assimilatory filaments 1 mm. long, forming narrow bands at subregular intervals alternating with wider bands of short filaments; short assimilatory filaments commonly three-, rarely two-celled (including the tumid basal cell), the terminal and subterminal cells smaller than the basal ones; shafts of the extended assimilatory filaments usually simple, 15 μ diam. below, the upper part 20 μ diam., the cells 1–3 times as long as wide; plants asexual, the sexual phase probably being *G. rugosa*.

Bermuda, Florida, Bahamas, Caicos Isls., Turks Isls., Cuba, Cayman Isls., Jamaica, Hispaniola, Puerto Rico, Virgin Isls., St. Barthélemy, Guadeloupe, Martinique, Barbados, British Honduras. Common in shallow water on rocks and old corals.

REFERENCES: Vickers 1905, Børgesen 1913–20, Howe 1918a, 1920, Taylor 1928, 1935, 1940, 1943, 1953a, Taylor and Arndt 1929, Chou 1945.

Galaxaura squalida Kjellman Pl. 44, fig. 3; pl. 45, fig. 6

Plants bushy, to about 10 cm. tall, attached by a broad discoid holdfast; branching dichotomous, erect, closer below than above, sometimes proliferous; branches terete, villous below, eventually more or less glabrous above; the segments 1–2 mm. diam., 0.5–2.0 cm. long, smooth or annulate-rugose; extended assimilatory filaments evenly distributed or subverticillate toward the apex of the young branches, seldom 1.0 mm. long, 15–18 μ diam., composed of short cylindrical cells 1.5–2.0 times as long as broad; medullary filaments 7–18 μ diam., cortical tissue parenchymatous, tristromatic; cells of the innermost layer more or less globose, some 2–3-lobed, 22–40 μ diam., 1–2 times wider than high; cells of the intermediate layer globose or ovate, 18–25 μ diam.; cells of the epidermis closely arranged, of 2 kinds: either angular in surface view, 14–22 μ diam., in section semilunate, 10–15 μ tall by 18–25 μ broad at the outer wall, or occasionally circular and slightly elevated above the general level of the surface; sexual conceptacles spherical, embedded in the medullary region, with ostioles opening through the cortical tissue; carposporangia ovate or elliptical, about 50 μ long by 30 μ diam.; asexual plants probably *G. flagelliformis.*

Bermuda, Florida, Bahamas, Caicos Isls., Turks Isls., Cuba, Jamaica, Hispaniola, Puerto Rico, Virgin Isls., Guadeloupe, British Honduras, Old Providence I., Panama, ?Colombia, Venezuela. Common in shallow water on rocks and old corals.

REFERENCES: Børgesen 1913–20, Collins and Hervey 1917, Howe 1918a, b, c, 1920, Taylor 1928, 1929b, 1933, 1935, 1939b, 1940, 1942, 1943, 1954a, Taylor and Arndt 1929, Sanchez A. 1930, Chou 1947.

Galaxaura rugosa (Ellis and Solander) Lamouroux

Plants bushy, to 5–7 cm. tall, densely and widely dichotomously branched, the branches 1–2 mm. diam., the segments 2–10 mm. long, terete, clearly transversely rugose, but otherwise smooth except that multicellular hairs occasionally appear in the older parts of the plant; lightly and superficially calcified, the filamentous medulla not calcified; cortex of radially branched sequences of cells, the innermost depressed-spherical, or obovoid, sometimes lobed, 30–40 μ diam., the 1–2 intermediate cells much smaller, oval to obovoid, the outermost flattened, firmly joined laterally into an epidermis, polyhedral in surface view and 18–32 μ diam.; plants sexual, the tetrasporic phase probably *G. subverticillata.*

Bermuda, Florida, Mexico, Bahamas, Cuba, Jamaica, Hispaniola, Puerto Rico, Virgin Isls., St. Barthélemy, Guadeloupe, Barbados, Grenada, Panama, Tobago, Brazil. From rocks and old corals in shallow water of somewhat sheltered places and dredged to a depth of 18 m.

REFERENCES: Mazé and Schramm 1870–77, Dickie 1874b, Hemsley 1884, Hauck 1888, Murray 1889, Collins 1901, Vickers 1905, Børgesen 1913–20, Howe 1920, Taylor 1928, 1929b, 1933, 1941b, 1942, 1943, 1954a, Sanchez A. 1930, Hamel and Hamel-Joukov 1931; *G. annulata,* Martens 1870, Dickie 1874a, Murray 1889; *G. plicata,* Martens 1871, Dickie 1874a.

Galaxaura cylindrica (Ellis and Solander) Lamouroux Pl. 44, fig. 1

Plants densely bushy, to 10 cm. tall or more, the slender branches firm and smooth, well calcified in the cortical region and becoming brittle on drying, repeatedly dichotomous, erect, sometimes proliferous, of a rather uniform diameter of 0.45–0.75, seldom 1.0, mm., the segments 5–10 mm. long; medulla not calcified; cortex compact, 2–3 cells thick, the inner cells large, subglobose to ovoid, 18–24 μ diam., the surface cells small, obconical, about 8 μ tall, 10–15 μ diam.

Bermuda, Florida, Cuba, Jamaica, Hispaniola, Puerto Rico, Virgin Isls., St. Barthélemy, Guadeloupe, Martinique, Barbados, Grenada, Costa Rica, Panama, Venezuela, Brazil. From rocks and old corals in rather shallow protected situations. Material more properly to be referred to *G. oblongata* may well be recorded here. Though the branch diameters seem to converge, there is in practice little difficulty in distinguishing *G. cylindrica* from *G. oblongata.*

REFERENCES: Martens 1870, Mazé and Schramm 1870–77, Dickie 1874b, Hauck 1888, Murray 1889, Collins 1901, Vickers 1905, Børgesen 1913–20, Collins and Hervey 1917, Taylor 1928, 1929b, 1930a, 1940, 1943, 1954a, Sanchez A. 1930, Quirós C. 1948.

Galaxaura oblongata (Ellis and Solander) Lamouroux

Plants bushy, 5–12 cm. tall, abundantly dichotomously branched, the smooth, moderately calcified branches 1–2 mm. diam., the segments somewhat broader at the summit than below, 4–10 diameters long; medulla not calcified; cortex somewhat calcified, of 3–4 cell layers and about 70 μ thick, the inner cells loosely arranged, very large, irregularly oval, 35 μ diam., bearing 1–2 smaller intermediate broadly oval cells, and the small flattened surface cells which are polygonal in surface view, about 10–17 μ diam.; carposporangia 16–20 μ diam.; tetrasporic phase possibly being *G. comans.*

Bermuda, Florida, Bahamas, Caicos Isls., Cuba, Jamaica, Hispaniola, Puerto Rico, Virgin Isls., Guadeloupe, Martinique, Barbados, Costa Rica, Canal Zone, Panama, Trinidad, Brazil, I. Trinidade. Plants frequent in shallow protected situations on rocks or old corals; dredged to a depth of over 30 m.

REFERENCES: Mazé and Schramm 1870–77, Dickie 1874b, Børgesen 1913–20, Howe 1918b, c, 1920, Taylor 1928, 1929b, 1930a, 1933, 1941b, 1942, 1943, Sanchez A. 1930, Hamel and Hamel-Joukov 1931, Joly 1953c; *G. fragilis,* Mazé and Schramm 1870–77, Murray 1889, Vickers 1905, Børgesen 1913–20, 1924.

Galaxaura obtusata (Ellis and Solander) Lamouroux
Pl. 44, figs. 4, 5; pl. 45, fig. 5

Plants of coarse appearance, to about 7–10 cm. tall, copiously branched, the branches 1.5–4.0 mm. diam., generally jointed at the forks, the terete segments 1.0–2.5(–4.0) diameters long; smooth, lightly calcified in the cortex, opaque when dry; the cortex in the tetrasporic plants composed of one layer of greatly enlarged cells each outwardly supporting a slender stalk cell which bears 1–2 distal cells closely laterally approximated and polyhedral in surface view, 27–45 μ diam., forming the epidermis; in the sexual plants the cells of the middle layer instead of being columnar are swollen, and touch in their middle portions.

V. **major** n. var. Plants to 17 cm. tall, less closely branched, the branches 3–4 mm. diam., sometimes jointed at the forks, the mature segments 3–9 diameters long, lightly calcified below, often nearly free of lime above, and translucent pink in the living state.

Bermuda, Florida, Bahamas, Cuba, Jamaica, Hispaniola, Puerto Rico, St. Eustatius, Guadeloupe, Barbados, Tobago, Brazil. From about low-tide line on very sheltered rocks or old corals, and dredged to about 53 m. depth.

Plants currently accepted as *G. obtusata* readily fall into 2 classes respecting the proportions of the segments between forks, as was recognized by J. G. Agardh (De Toni 4, p. 11). Those with very short stout segments correspond to Ellis' (1786, p. 113) figure, and so are typical of the species; *G. moniliformis* Kjellman (1910) belongs here. Those with segments several times longer than broad exceed Ellis' (1786, p. 114) *Corallina oblongata,* a species name currently used for quite a different plant.

REFERENCES: Martens 1870, Mazé and Schramm 1870–77, Dickie 1874a, Hauck 1888, Murray 1889, Collins 1901, Vickers 1905, Collins and Hervey 1917, Howe 1917, 1918a, 1920, 1928, Taylor 1928, 1933, 1942, Hamel and Hamel-Joukov 1931, Feldmann and Lami 1937; *Corallina obtusata,* Ellis and Solander 1786; *G. moniliformis,* Kjellman 1910, Sluiter 1908, De Toni 1924.

Galaxaura marginata (Ellis and Solander) Lamouroux
Pl. 44, fig. 2; pl. 45, figs. 7, 8

Plants rather large, to 5–14 cm. tall, the branches smooth, dull, frequently transversely banded, flattened, about 1.0–2.5 mm. broad on the face, the segments several times as long; the branches occasionally tipped with a brush of deciduous hairs; the inner cortex composed of a pseudoparenchymatous layer 2–3 cells thick of large, nearly colorless cells, with, in the tetrasporic plants, on the outer side a pilose layer, each laterally free division consisting of a rather slender stalk cell once divided at the summit and bearing 2 large thick-walled, oval to obovoid, sometimes apiculate, assimilative cells 22–35 μ diam., 38–50 μ long; in sexual plants the cortex bearing on the outer side one layer of small depressed cells forming a continuous epidermis, on the outer face of which as a pilose layer each of the cells may bear one narrowly oval or columnar apiculate cell.

V. **linearis** (Kützing) J. Agardh. Segments more slender, locally constricted, with tufts of long filaments at the constrictions.

V. **dilatata** (Kützing) J. Agardh. Segments broader, to 5 mm. or even wider below a fork, 7–15 mm. long, without evident constrictions; epidermal cells polygonal, 30–46 μ max. diam., the lateral walls rather thick, in section truncate-obpyramidal, 18–20 μ tall.

Bermuda, Florida, Bahamas, Cuba, Jamaica, Hispaniola, Puerto Rico, Virgin Isls., Antigua, Guadeloupe, Martinique, Barbados, Costa Rica, Panama, Venezuela, Tobago, Brazil. Common in shallow water on reefs and in tide pools, attached to stones and old corals, and dredged from a depth of 55 m. The tetrasporic plants of *G. marginata, G. frutescens,* and *G. stupocaulon* lack distinctive microscopic characters, and it may not always be possible to distinguish them satisfactorily.

REFERENCES: Mazé and Schramm 1870–77, Dickie 1874a, b, 1884, Collins 1901, Vickers 1905, Børgesen 1913–20, Collins and Hervey 1917, Howe 1918a, c, 1920, 1928, Taylor, 1928, 1929b, 1930a, 1933, 1941b, 1942, 1943, Taylor and Arndt 1929; *G. canaliculata,* Grunow 1867, Martens 1870, Mazé and Schramm 1870–77, Zeller 1876; *G. occidentalis,* Børgesen 1913–20; ?*Liagora dichotoma,* Howe and Taylor 1931; *Zanardinia marginata,* Murray 1889; *G. schimperi, G. umbellata,* Mazé and Schramm 1870–77, Murray 1889. For the v. *dilatata,* De Toni 4:110. For the v. *linearis,* Möbius 1889.

Galaxaura frutescens Kjellman

Plants densely bushy, to 5–8 cm. tall, the holdfast cushion-shaped and like the short subterete stalklike base of the plant conspicuous and stupose; branching somewhat flabellate, close, the segments subcuneate, about 5–7 mm. long, 2–3 mm. broad, rather much flattened, hardly zonate except when young; assimilators short, simple or forked, the stalk cells cylindrical to clavate, 11–15 μ diam., 26–28 μ long, the terminal cells obovoid or ellipsoid, 30–34 μ diam., 35–45 μ long.

Panama, Brazil.

REFERENCES: Kjellman 1900, Howe and Taylor 1931; *Liagora dichotoma,* Greville 1833.

Galaxaura stupocaulon Kjellman

Plants bushy, 6–7 cm. high, attached by a rather small, stupose holdfast; stalk and branches terete and villous below, flattened above, continuous, or proliferous from subarticulate constrictions, where encircled by tufts of extended assimilatory filaments; segments 2–7 mm. long, 1–2 mm. wide, fleshy-membranous, the margins slightly thickened; the medullary filaments 7–15 μ diam.; cortex subparenchymatous, of 1–2 layers, 20–40(–60) μ thick (excluding the stalk and the terminal cells), the inner cells large, subglobose or compressed, 20–44 μ tall, 25–50 μ broad, the outer cells smaller, subglobose or pyriform, each bearing 1–2 cylindrical or cuneate-cylindrical stalk cells, these simple or once forked, generally unicellular, 30–36 μ tall; terminal cells subglobose to pyriform, 20–32 μ diam., (20–)40–50 μ tall; extended assimilatory filaments frequently locally intermixed with the club-shaped assimilators, short, or at the articulations to 1 mm. long, with one or sometimes 2 tumid basal cells, 24–34 μ diam., 58–76 μ long; the shaft 15–23 μ diam., with cells 2–5 times as long as the diameter.

Brazil.

REFERENCES: Kjellman 1900, Chou 1945.

UNCERTAIN RECORDS

Galaxaura apiculata Kjellman

Barbados. Type from Japan. De Toni 6: 133; Vickers 1905.

Galaxaura dichotoma Lamouroux

Brazil. Type from the West Indies. De Toni 4: 116; Martens 1870.

Galaxaura fastigiata (?)Harvey

Bermuda. Type of *G. fastigiata* Decaisne from the Moluccas, but this may not be the plant Rein and Murray ascribe (probably by error) to Harvey. De Toni 4: 116; Rein 1873, Murray 1889.

Galaxaura fruticulosa (Ellis and Solander) Lamouroux

Bahamas. Type from Japan. De Toni 4: 115; Murray 1889.

Galaxaura indurata (Ellis and Solander) Lamouroux

Bahamas, Virgin Isls. De Toni refers this name to Actinotrichia, a distinctive genus not reliably known from our area. De Toni 4: 115; Murray 1889.

Galaxaura lichenoides (Ellis and Solander) Lamouroux

Bahamas, the type locality. Ellis figures a short-segmented plant which De Toni suggests resembles *G. rugosa,* while Kützing figures one from the West Indies which Kjellman (1910, p. 55) rightly observes to be like

G. squalida. De Toni 4:115; ?Kützing 1858, Murray 1889; *Corallina lichenoides* Ellis and Solander 1786.

Galaxaura ramulosa Kjellman

Brazil, the type locality. De Toni 6:114.

Galaxaura stellifera J. Agardh

Florida, the type locality. Perhaps a form of *G. cylindrica.* De Toni 6:126.

Galaxaura umbellata (Esper) Lamouroux

Brazil, the type locality. De Toni 4:111; Martens 1870.

Galaxaura veprecula Kjellman

Barbados, Tobago. Type from the Indian Ocean. De Toni 6:137; Vickers 1905.

NACCARIACEAE

Bushy plants, gelatinous, the branches growing from apical cells, the axes monosiphonous, producing short lateral filaments and becoming corticated by downgrowths from the basal cells of these; carpogenic branches of two or three cells; gonimoblasts formed from the carpogonium, more or less diffuse, growing out among the neighboring assimilators, forming the carposporangia terminally.

Naccaria Endlicher, 1836

Plants bushy, from a discoid holdfast, the branching irregular; growth from an apical cell by oblique divisions, producing a monosiphonous axis; young cells of this cut off lateral segments which initiate formation of the spirally placed, short, lateral, deciduous, chromatophorous filaments, downgrowths from the persistent lower cells of which develop a pseudoparenchymatous thallus wall; with additional growth the axial cells elongate greatly and the wall partly separates, leaving a space about the axial filament, though it remains attached at intervals; wall of a few layers of large hyaline cells and a surface layer of very small chromatophore-bearing cells; long hyaline hairs abundant on the surface; dioecious, the spermatangia borne in clusters at the bases of the assimilators; carpogenic branches two-celled, borne on a supporting cell derived from the basal cell of a lateral filament; cystocarps generally found as swellings towards the ends of the branchlets, gonimoblasts branching among the neighboring lateral assimilatory filaments and forming carpospores terminally.

Naccaria corymbosa J. Agardh

Plants bushy, to 7 cm. tall or somewhat more, abundantly alternately paniculately branched; very soft but the branches somewhat firmer below, slightly nodose by virtue of the relatively large axial cells, the segments about 3–4 diameters long; the branchlets 30–40 μ diam., microscopically plumose through the presence of the very many assimilators, which are sparingly branched, ultimately to about 100–150 μ long, moniliform, with cells truncate-oval or cask-shaped, about as long as broad, 8 μ diam. in the branchlets to 13 μ diam. when mature, but these assimilators largely shed from all major branches; cystocarps causing slight swellings of the smaller branches.

Bermuda, Florida.

REFERENCES: Collins and Hervey 1917, Howe 1918a.

BONNEMAISONIACEAE

Plants of moderate size, slender and extensively branched, with an evident axis; growth apical, an axial row of cells being surrounded by short, compact, branched cell rows constituting a pseudoparenchymatous cortex with a continuous surface; sporangia unknown; spermatangia in dense masses covering small lateral branchlets; carpogenic branches lateral on the corticating cell rows, after fertilization forming gonimoblasts on a large fusion cell, constituting a cystocarp with a definite superficial pericarp, which is chiefly developed from adjacent tissue.

Asparagopsis Montagne, 1840

Plants bushy, above repeatedly slenderly branched to brushlike tufts, the alternate branches beset with short branchlets; the stem and larger branches becoming subtubular; spermatangia covering enlarged branchlets; each carpogenic branch on the fifth to seventh cell of a corticating cell series, on which three pericentral cells are formed, the oldest functioning as the supporting cell of the three-celled carpogenic branch; pericarps subpyriform, becoming pedicellate by transformation of a branchlet.

Asparagopsis taxiformis (Delile) Collins and Hervey
Pl. 71, fig. 4

Plants 5–20 cm. or more in height, purple-violaceous; with fibrous holdfasts, and spreading by rhizomatous axes; erect branches rather stout, sparingly divided, naked below or with the stubs of lateral branches, closely alternately branched above, becoming pyramidal, these determinate branches slender, to 2–5 cm. long, repeatedly divided, in the ultimate branchlets consisting of 3 rows of cells with no evident central filament, becoming very soft, delicate, and plumose.

Bermuda, Bahamas, Jamaica, Virgin Isls., Guadeloupe, Barbados, Panama, Venezuela, Brazil. Sometimes common in relatively shallow water, but also dredged from a depth of over 30 m. and these specimens rather laxly branched.

REFERENCES: Collins and Hervey 1917, Børgesen 1913–20, Howe 1920, Hamel and Hamel-Joukov 1931, Taylor 1942, Joly 1953c; *A. delilei*, Mazé and Schramm 1870–77, Martens 1871, ?Rein 1873, Dickie 1874a, Zeller 1876, Hemsley 1884, Möbius 1890, Murray 1889, Collins 1901, Vickers 1905.

GELIDIALES

Plants with slender, wiry, subsimple or redivided branches; from an apical cell or less often a small group of cells ultimately developing a multiaxial type of structure, and becoming corticated; asexual reproduction by tetrasporangia formed at or below the thallus surface; sexual reproduction by spermatangia formed from the surface cells and carpogonia on carpogenic branches loosely associated with chains of nutritive cells; the carpogonia producing filamentous gonimoblasts, on which carposporangia are borne among the nutritive cells.

KEY TO FAMILIES

1. Tetrasporangia tetrahedral, tetrapartite, or irregularly divided GELIDIACEAE, p. 349
1. Tetrasporangia zonate .... WURDEMANNIACEAE, p. 361

GELIDIACEAE

Plants small to moderately large, tough and wiry, branched; asexual plants reproducing by tetrasporangia, the sporangia generally embedded in the cortex of localized areas or branchlets of the plant; spermatangia borne on the surface of considerable areas or only on branchlets of the male plants; carpogenic branches borne on inner filaments of the female plants, associated with chains of nutritive cells in the same areas; the carposporophyte ultimately produced by elongated gonimoblast filaments which run out from the fertilized carpogonium through the surrounding tissue, producing the carposporangia laterally, but not making definite connections with the nutritive cells.

KEY TO GENERA

1. Rhizines absent 2
1. Rhizines present at least in the older parts of the plants 4

2. Branch tips with a single apical cell .... GELIDIELLA, p. 349
2. Branch tips multiaxial 3

3. Cells of the central medulla somewhat thick-walled; sporangia generally tetrahedral .... GELIDIOPSIS, p. 352
[3. Cells of the central medulla thin-walled; sporangia zonate WURDEMANNIA, p. 361]

4. Rhizines chiefly in the central medulla; tetrasporangia developed successively, often in rows; cystocarps unilocular PTEROCLADIA, p. 358
4. Rhizines chiefly in the subcortical region; tetrasporangia not developed in any special order; cystocarps bilocular GELIDIUM, p. 353

Gelidiella Feldmann and Hamel, 1934

Plants bushy, wiry, the slender branches terete; the segments from the unicellular branch apices immediately organizing a multiaxial medulla of long, thick-walled cells and a cortical zone of a few layers of rather short cells, the outermost rounded and with chromatophores; sporangia usually in stichidial branchlets or enlargements of the end of main branches, formed progressively in rather well-defined divergent rows in parallel series, tetrahedral, tetrapartite, or irregularly divided.

KEY TO SPECIES

1. Plants several centimeters tall, the long, coarse axes with many short determinate branchlets .... G. acerosa, p. 351
1. Plants smaller and more delicate, without conspicuously differentiated main axes 2

2. Plants about 1 cm. tall, the segments less than 100 μ diam. G. trinitatensis, p. 350
2. Plants taller and coarser 3

3. Erect branches usually exceeding 300 μ diam. ... G. taylori, p. 351
3. Erect branches less than 300 μ diam. 4

4. Cortical cells small and irregularly disposed ..... G. setacea, p. 350
4. Cortical cells larger, 10 μ diam., in rather distinct longitudinal rows .... G. sanctarum, p. 351

Gelidiella trinitatensis Taylor

Plants minute, the long basal stolons entangled, cylindrical, about 55–75 μ diam., with erect terete or compressed axes up to 2–15 mm. tall, 75–90 μ diam., simple or irregularly and sparingly branched, terminally or laterally bearing terete to generally flat, simple or pinnate to subpalmate sporangial branchlets; tetrasporangia originating near the apices in clear V-shaped parallel rows, later obscured, the spores being shed first from the lower portions of the branchlets.

Costa Rica, Trinidad, ?Brazil. From rocks near the shore in shallow water.

REFERENCE: Taylor 1943.

Gelidiella setacea (Feldmann) Feldmann and Hamel

Plants to 2–3 cm. tall, becoming black when dried, in elongated tufts, radially freely branched, the divaricate branches entangled, the segments setaceous, 150–200 μ diam., the tips acute.

Guadeloupe, French Guiana.

REFERENCES: Feldmann and Hamel 1934; *Acrocarpus gracilis*, Mazé 1868; *Echinocaulon setaceum*, Feldmann 1931; *Gelidum spinescens*, Mazé and Schramm 1870–77.

Gelidiella sanctarum Feldmann and Hamel

Plants tufted, dull purple and blackening on drying; slenderly setaceous, creeping below, the erect filaments to 2 cm. tall, near their bases to 200 μ diam., in the middle of the plant 60–100 μ diam.; branches terete or a little compressed; simple or rebranched, acute at the tips; cortical cells to 10 μ broad, in vertical rows; tetrasporangia in stichidia 130–150 μ diam., 450 μ long, terminating the branches; sporangia rather irregularly divided, spherical or ovate, 30–40 μ long.

Guadeloupe.

REFERENCES: Feldmann and Hamel 1934; *Echinocaulon sanctarum* (*nomen nudum*), Algues des Antilles Française (exsic.), no. 83, 1931.

Gelidiella taylori Joly

Plants small, the rather short basal strands creeping, affixed by haptera, about 125 μ diam., erect axes 2–3 cm. tall, branched, especially toward the top, compressed, 310–550 μ wide; tetrasporangial branchlets lateral, slightly inflated, the sporangia scattered near the middle of each branchlet.

Brazil. Growing on rocks in shallow water.

REFERENCE: Joly 1957.

Gelidiella acerosa (Forsskål) Feldmann and Hamel Pl. 46, fig. 5

Plants to 5–14 cm. tall, tough and wiry, greenish yellow to dull purplish; the basal part decumbent and attaching freely by haptera, bearing free elongate secondary branches which are erect or arcuate-recurved, sometimes attached at the tips, terete or a little compressed, sparingly alternately branched; terete filiform determinate branchlets 2–6 mm. long, radially, secundly, or sharply bilaterally disposed; tetrasporic branchlets irregularly disposed, slightly clavate, with the sporangia progressively developed from the apex; cystocarps(?) causing pronounced unilateral swellings on the fertile branchlets.

Bermuda, Florida, Bahamas, Caicos Isls., Salt Key Bank, Cuba, Cayman Isls., Jamaica, Hispaniola, Puerto Rico, Virgin Isls., Guadeloupe, Martinique, Barbados, Grenadines, Grenada, British Honduras, Costa Rica, Panama, Netherlands Antilles, Venezuela, Trinidad, Tobago, Brazil. Common and sometimes abundant in the intertidal zone on rocks and old corals, and dredged to 7.5 m. depth.

REFERENCES: Feldmann and Hamel 1934, Taylor 1940, 1942, 1943, 1954a; *Ahnfeldtia pinnulata,* Harvey 1853, Murray 1889; *Echinocaulon acerosum,* Børgesen 1932; *E. rigidum,* Hamel and Hamel-Joukov 1931, Feldmann 1931, Quirós C. 1948; *Gelidiopsis rigida,* Vickers 1905, Børgesen 1913–20, Collins and Hervey 1917, Schmidt 1923, 1924; *Gelidium rigidum,* Martens 1870, Mazé and Schramm 1870–77, Rein 1873, Dickie 1874a, Hemsley 1884, Piccone 1886, 1889, Murray 1889, Collins 1901, Howe 1918a, 1920, Taylor 1928, 1929b, 1930a, 1933, 1935, Taylor and Arndt 1929; *G. rigidum* v. *radicans,* Murray 1889; *Sphaerococcus rigidus,* Montagne 1863.

Gelidiopsis Schmitz, 1895

Plants erect, irregularly and slenderly branched; branches growing from a multiaxial apex but developing with a pseudoparenchymatous structure, in section the elongated cells of the central medullary area rather small and thick-walled, those of the subcortical region somewhat larger and less thick-walled; tetrasporangia in somewhat spatulate branch tips, generally tetrahedral; cystocarps prominently projecting, solitary or a few together on the upper branches.

KEY TO SPECIES

1. Axes in the middle parts of the plant commonly compressed ..... G. planicaulis, p. 353
1. No parts of the plant normally compressed 2
2. Plants several centimeters tall, subdichotomously branched ..... G. gracilis, p. 352
2. Plants less than 4 cm. tall, irregularly laterally branched, much entangled .... G. intricata, p. 353

Gelidiopsis gracilis (Kützing) Vickers

Plants tufted, to about 8 cm. tall, wiry; repeatedly subdichotomously branched, the divisions erect and fastigiate, very slender and rather elongate, toward the tips sometimes a little swollen.

Barbados, French Guiana.

REFERENCES: Vickers 1905, Feldmann 1931; *Acrocarpus gracilis,* Montagne 1850, Kützing 1868.

Gelidiopsis intricata (C. Agardh) Vickers

Plants gregarious, 2.0–3.5 cm. tall, dull greenish or purplish; much branched, the setaceous branches entangled, rather wiry; lower axes spreading; erect axes very sparingly laterally branched, the divisions similar to the axes, widely divergent, sparingly rebranched, both 200–300(–425) μ diam.

Bermuda, Florida, Cuba, Virgin Isls., Guadeloupe, Barbados, Costa Rica, Brazil. Common on rocks in the intertidal zone and somewhat lower, particularly in crevices or on overhanging cliffs, often in rather exposed places. Dredged from a depth of 10 m.

REFERENCES: Vickers 1905, Feldmann 1931, Hamel and Hamel-Joukov 1931; *Wurdemannia setacea,* Algae Exsic. Am. Bor. 14, Phyc. Bor.-Amer. nos. 250, 1887.

Gelidiopsis planicaulis (Taylor) n. comb.

Plants to 8 cm. tall, bushy, blackish purple when dried, quite tough and wiry in texture, the erect axes very sparingly and generally bilaterally branched, irregularly alternate, occasionally opposite; axes terete near the ends, clearly compressed in the larger portions near the middle of the plant, where the width may reach 1.1 mm., though usually but half as much, and the thickness 185 μ; sections of the older stems having a surface layer with radially somewhat elongated cells.

Jamaica, Hispaniola, Dominica, Martinique, Costa Rica, Trinidad, Brazil.

REFERENCE: *Wurdemannia miniata* v. *planicaulis,* Taylor 1943, p. 158.

Gelidium Lamouroux, 1813

Plants of moderate to small size, with the axis erect and usually laterally branched, the branches firm, cylindrical or flattened; structurally the branches each developed from an apical cell, the axial row of cells giving rise to pericentral cell rows which are later paralleled by many other secondary medullary filaments, the whole surrounded by short, compact, radially branched cell series constituting a firm assimilatory cortex; slender thick-walled filaments (rhizines) generally present in the axis between the larger cells, localized in the inner cortex or somewhat dispersed, but not localized in the medulla; tetrapartite sporangia formed on asexual plants in the cortex of restricted areas of the branches; spermatangia forming more or less extensive patches on the branches of male plants; female plants with three-celled carpogenic branches developed on the pericentral cell rows, whence there are also developed chains of nutritive cells; after fertilization the gonimoblast filaments spread among the nutritive cells and form a bilocular cystocarp embedded in the thallus, with a median septum, the carposporangia terminal on the gonimoblasts, the discharge pores not particularly elevated.

KEY TO SPECIES

1. Plants very small, 0.5–1.5 cm. tall, the base creeping, the erect flat branches spatulate, simple or sparingly pinnate G. pusillum, p. 354
1. Plants larger and freely branched 2
2. Branchlets flat 3
2. Branchlets almost filiform, terete or compressed .. G. crinale, p. 355
3. Margins of the branchlets, especially those bearing sporangial sori, minutely serrate .... G. serrulatum, p. 357
3. Margins of the branchlets entire 4
4. Fertile branchlets small and rather congested, especially on the lower parts of the plant .... G. floridanum, p. 357
4. Fertile branchlets not aggregated .... G. corneum, p. 356

Gelidium pusillum (Stackhouse) Le Jolis Pl. 45, fig. 4

Plants small, solitary or forming loose turfs, creeping below, giving rise to erect blades 5–15 mm. long, subcylindrical below, flattened to 0.5–0.75 mm. broad above, sparsely pinnately proliferating; central part composed of slender colorless filaments with exceedingly thick confluent walls, surrounded by the inner cortex of short, large cells and the epidermal layer of rounded angular cells slightly elongated lengthwise of the axis, about 4–10 μ surface diam.; rhizines in the stalklike portions subcortical, in the blade portions invading the medulla, often seemingly absent altogether.

V. **conchicola** Piccone and Grunow. Plants very small, hardly 5 mm. tall, forming dense turfs, the blades flattened throughout, otherwise as in the species; mature tetrasporangia 18–24 μ diam., arranged in somewhat irregularly pinnate rows in the broader blades.

Bermuda, Florida, Texas, Cuba, Guadeloupe, Barbados, Grenadines, Grenada, Old Providence I., Costa Rica, Colombia, Trinidad, Brazil. An important member of the flora of the upper intertidal zone, growing on exposed rocks.

REFERENCES: Collins and Hervey 1917, Taylor 1928, 1933, 1936, 1941a, b, 1954a, Howe 1928; *G. ligulatonervosum,* Mazé and Schramm 1870–77, Murray 1889; ?*G. repens,* Vickers 1905. For the v. *conchicola,* Collins and Hervey 1917, Howe 1918a, Taylor 1928, 1939b.

Gelidium crinale (Turner) Lamouroux

Plants gregarious, forming expanded tufts usually 2–5 cm. tall; dull purple to yellowish brown, rather wiry; primary branches rhizomatous, spreading over the substratum, about 0.5 mm. diam., firmly attached by holdfasts; above forming erect subterete branches 1.5–7.5 cm. tall which are irregularly alternate below, irregularly to distinctly pinnate above; superficial cells of the thallus 5–16 μ diam., 6–16 μ deep; in tetrasporic plants the branchlets more pinnately disposed and distinctly flattened, often spatulate, the sporangia 10–22 μ diam., 13–25 μ long; sexual plants dioecious; spermatangia on the surface of the branchlets; cystocarps in somewhat spindle-shaped enlargements of the branchlets, usually solitary, occasionally paired.

V. **platycladum** Taylor. Plants to 7–10 cm. tall, dark reddish purple, the base with a few flagelliform branches, mostly erect, the main axes often being twisted together, flat except at the extreme base, to 550 μ wide and 125 μ thick, but somewhat broader at the forkings; branching irregularly dichotomous below, seemingly alternate above, the ultimate divisions subcorymbosely crowded, acute-tipped; in section rhizines few, inconspicuous, and in the outer part of the central tissue; tetrasporangia in the acute, very slightly expanded branchlets of the last 2–3 divisions, irregularly placed, discharging from the apices first and only later from the lower parts of the sori, the tips sometimes decaying.

Bermuda, North Carolina, Florida, Texas, Jamaica, Hispaniola, Guadeloupe, Barbados, Trinidad, Brazil, Uruguay; the v. *platycladum* from Florida and Texas. Generally growing on rocks, and in the northern part of its range as an intertidal species in exposed places.

REFERENCES: Collins 1901, Vickers 1905, Collins and Hervey 1917, Howe 1918a, Hoyt 1920, Taylor 1929a, b, 1933, 1936, 1941a, Taylor and Arndt 1929; ?*G. spathulatum*, Vickers 1905, Collins and Hervey 1917; ?*G. delicatulum* and *G. corneum* v. *crinalis*, Mazé and Schramm 1870–77, Murray 1889; *Acrocarpus crinalis*, Martens 1870. For the v. *platycladum*, Taylor 1943, 1954a.

Gelidium corneum (Hudson) Lamouroux

Plants bushy, cartilaginous, 3–50 cm. tall, with prominent main axes; branches 1–3 degrees pinnate, cystocarpic plants being somewhat more consistently twice pinnate than the simpler tetrasporic specimens; the segments 0.5–2.0 mm. broad, thick but flattened, the terminal divisions somewhat constricted at the base but with a broadly rounded apex; in section the axes showing rhizines chiefly in the outer portions.

Bermuda, South Carolina, Florida, Texas, Cuba, Virgin Isls., Guadeloupe, Martinique, Grenada, Panama, Colombia, Trinidad, Brazil. It is extremely doubtful if many of these records will ever be confirmed. While small specimens from Martinique and Panama seem to me to belong in Gelidium and are best under this traditional name, most of those I have examined were referable to other species of Gelidium or to Pterocladia, and so except for the two mentioned (Taylor 1943) all of these records are to be considered untrustworthy.

REFERENCES: Greville 1833, Harvey 1853, 1861, Mazé and Schramm 1870–77, Zeller 1876, Möbius 1889, Murray 1889, Gepp and Gepp 1905, Collins and Hervey 1917, Schmidt 1923, 1924, Taylor 1928, 1929b, 1930, 1936, 1941a, 1943, Howe 1928, Hamel and

Hamel-Joukov 1931; *G. corneum* v. *nitidum,* Martens 1870; *G. corneum* v. *caespitosa,* v. *pristoides,* v. *setacea,* and *G. nanum,* Mazé and Schramm 1870–77, Murray 1889; *G. cartilagineum,* Grunow 1867, Murray 1889; *Sphaerococcus corneus,* Montagne 1863.

Gelidium floridanum Taylor

Plants with a ramified base, the entangled branches terete, reflexed, and somewhat flagelliform, giving rise to erect subsimple strap-shaped axes which at full development become about 1 mm. wide, 13 cm. tall, sparingly pinnately branched below, commonly regularly distichous in the upper portions of well-grown plants; sections of main axes showing numerous rhizines more or less in small groups scattered through the central and especially the peripheral tissue; sterile branchlets spreading, strap-shaped, contracted at the bases and obtuse at the apices; fertile branchlets in crowded pinnate clusters on the lower portions of the main branches; tetrasporic branchlets at first simple, obovate, becoming ligulate, then pinnately branched, with the sori occupying the central portions of the divisions, surrounded by a firm sterile border; cystocarpic branchlets with one, seldom 2, cystocarps in each division, becoming pinnate, the cystocarps bilocular, swollen, and with slightly projecting ostioles; after discharge of tetraspores or of carpospores the fertile segments often becoming perforate.

Florida, Trinidad.

REFERENCE: Taylor 1943.

Gelidium serrulatum J. Agardh

Plants bushy, purplish, to 15 cm. tall, the holdfast fibrous; stalked, the stalks denuded below, but to 4–5 mm. broad, above 3–4 times pinnate, the branchlets narrowed at the base, alternate or opposite, the lower segments sublinear, the upper shorter, when young the margin minutely serrate; fertile branchlets rather abundant on the upper parts of the plant, obovate, the tetrasporangia in a large sorus, the margins of the blade clearly serrate and crisped.

Venezuela, Trinidad.

REFERENCES: J. Agardh 1847, Harvey 1853, Murray 1889, Taylor 1929b.

UNCERTAIN RECORDS

Gelidium caerulescens Kützing

Bermuda, North Carolina, Jamaica, Guadeloupe. Type from New Caledonia. De Toni 6:161; Mazé and Schramm 1870–77, Murray 1889, Collins 1901, Howe 1918a, Hoyt 1920. All American records are probably referable to *Pterocladia americana,* or less plausibly to *G. corneum.*

Gelidium coarctatum Kützing

Brazil, the type locality. De Toni 6:161; Martens 1870.

Gelidium multifidum Greville

Brazil, the type locality. De Toni 6:163; Greville 1833a, Martens 1870.

Gelidium parvulum Greville

Brazil, the type locality. Type seen: perhaps a Gelidiopsis. De Toni 6:163; Greville 1833a, Martens 1870, Zeller 1876.

Gelidium pectinatum (Schousboe) Montagne

Barbados. Type from Morocco. Vickers 1905.

Gelidium radicans Montagne

Cuba, Guadeloupe. Type from Cuba. De Toni 6:163; Mazé and Schramm 1870–77; *Sphaerococcus radicans,* Montagne 1863.

Gelidium supradecompositum Kützing

Jamaica, Puerto Rico, Brazil. Type from Brazil. De Toni 6:162; Martens 1870, Hauck 1888, Collins 1901.

Gelidium torulosum Kützing

Brazil, the type locality. De Toni 6:162; Martens 1870, Piccone 1886, 1889.

Gelidium variabile (Greville) J. Agardh

Brazil. Type from India. *Gelidiopsis variabilis,* De Toni 4:410, 6:244; Piccone 1886.

Pterocladia J. Agardh, 1852

Plants of moderate to small size, of very various forms, with the axes usually erect and usually laterally branched, the branches terete or flattened; structurally developed from an apical cell, the axial row soon obscured, paralleled by many other medullary filaments, some polyhedral in section and of large diameter, others (rhizines) very slender, thick-walled, and refractive, and lying chiefly in the central medullary region; tetrasporangia usually localized in sori at first lying near the tip of a branchlet, but sometimes later displaced by growth, formed progressively from near the apex and commonly in divergent rows in parallel series; cystocarps unilocular, the carposporangia formed in short chains on the gonimoblasts, discharging through a somewhat elevated, even beaklike, pore.

KEY TO SPECIES

1. Branching pinnate, seldom bipinnate, the blades linear-lanceolate, the branchlets narrowly linear .... P. bartlettii, p. 359
1. Branching 2–4-pinnate, the blades broader, the branchlets more than 0.5 mm. wide 2
2. Dominant main axes evident .... P. pinnata, p. 361
2. Primary axis quickly dividing into several widely spreading equal divisions; when in exposed situations relatively little branched .... P. americana, p. 360

Pterocladia bartlettii Taylor Pl. 46, fig. 2

Plants dull purplish, bushy, somewhat entangled below, very slender in all parts, arising from minute flattened holdfasts, from which grow a few cylindrical or compressed rhizomatous branches a few millimeters in length, which bear secondary holdfasts and which at first give rise to erect linear-lanceolate simple blades up to 1 cm. in length; these blades soon replaced by mature axes to 6–8 cm. tall, tapering to each end, compressed below, flat above, to 0.5 mm. wide or a little more, and to 125–170 μ thick, bearing alternately and bilaterally numerous subsidiary axes of 1–2(–3) degrees, which become erect; medullary filaments involved by very abundant intercellular rhizines, which dominate the interior for two-thirds of the axis structure; main axis and subsidiary axes closely beset with generally divaricate, occasionally ascending, branchlets 2–3 (–5) mm. long in bilateral series along the margins, these sometimes very regular and comblike; branchlets linear to spatulate, usually flat, minutely apiculate, the margins entire or in broader examples obscurely and irregularly serrate, but in some specimens subterete and tapering; tetrasporangia in stichidium-like enlargements of the branchlets, near the base of the plant; spermatangia not seen; cystocarps unilocular, single or occasionally 2 together in the ultimate branchlets, which become strongly curved, the slightly elevated ostiole being formed on the dorsal side.

Texas, Cuba, Jamaica, Hispaniola, Guadeloupe. From warm, quiet water near shore.

REFERENCE: Taylor 1943.

Pterocladia americana Taylor Pl. 46, fig. 3

Plants gregarious, spreading, stoloniferous and subterete below, this feature being inconspicuous in luxuriant tufts; the basal portion giving rise to simple erect axes to 0.5–2.0 cm. tall, or in bushy plants of larger size up to 5–6 cm. tall and more or less divided; the axis sparingly pinnately branched and rebranched, becoming obscure in the tetrasporangial plant, where the branches are fastigiate, but in the sexual plants frequently more regularly pinnate or pseudo-palmate; all axes quite flat, often submembranous, with the rhizines chiefly in the central region; the branchlets tending to appear erect, their bases subcylindrical, somewhat contracted and at first up-curved, though later spreading, flat in the distal portion, to 500–600 μ wide and 75–150 μ thick; tetrasporangia in the simple blades occupying the distal portion, the sporangia originating in clear pinnate rows; in the smaller crowded branching plants the sporangia appearing in practically unmodified branchlets of the last 1–2 orders, again occupying the full width of the thin blades; the unilocular cystocarps causing a slight broadening of the branchlet bearing them, with a marked swelling on the dorsal surface, where the ostiole is somewhat raised.

Bermuda, North Carolina, Florida, Texas, Jamaica, Hispaniola, Barbados, Costa Rica, Venezuela, Trinidad, Tobago. On exposed rocks forming mosslike growths, but in sheltered places taller and more branched; a plant of shallow water. Very small plants of this species may be mistaken for *Gelidium pusillum*. In brackish water or very protected situations the branches tend to become long and flagelliform and less clearly pinnate.

REFERENCE: Taylor 1943.

Pterocladia pinnata (Hudson) Papenfuss Pl. 46, fig. 1

Plants erect, bushy, from a fibrous holdfast, 5–20 cm. tall, reddish brown, cartilaginous, the branches often stalklike below, above forming flat blades to 3–4 times pinnate, below showing a faint costa, the linear segments 1–2 mm. broad for the most part, but in the branchlets 130–600 μ broad; tetrasporangia in obtuse or subspatulate branchlets progressively developed from near the growing tip, but not in clear pinnate rows; spermatangia in linear sori on the branchlets; cystocarps formed in the middle of terminal branchlets, strongly unilaterally projecting.

Hispaniola, Virgin Isls., Costa Rica, Colombia, Venezuela, Trinidad, Brazil, Uruguay.

REFERENCES: Joly 1957; *Pterocladia capillacea,* Taylor 1943, Joly 1951, De Mattos 1952; *Gelidium capillaceum,* Möbius 1890; *G. corneum* v. *pinnatum,* Martens 1870, Børgesen 1913–20.

WURDEMANNIACEAE n. fam.

Plants gregarious, wiry, the branches slender; in structure basically filamentous and multiaxial, the filamentous character obscured with age, the cell walls thick except in the center of the medulla; reproduction by zonate tetrasporangia borne peripherally in essentially unaltered branches.

Wurdemannia Harvey, 1853

Plants small, bushy, gregarious and entangled below, attaching by haptera at various points; texture wiry; growth multiaxial, with several small apical cells; medulla of rather large cells, the central ones thin-walled, elongate and in longitudinal rows, those more peripheral shorter and less definitely arranged, the cortex of a single layer of cells nearly rectangular in section, slightly radially elongate, polyhedral in surface view; cell walls generally thick, especially toward the surface; tetrasporangia zonate, restricted to the slightly enlarged tips of the branches.

Wurdemannia miniata (Draparnaud) Feldmann and Hamel

Plants tufted or forming broad turfs to 3 cm. thick, dull red or often bleached; entangled and rhizomatous or flagelliform below, above seemingly sparingly branched, the terminal divisions rather erect, attenuate; diameter of the branches about 150–400 μ; cells of the central part of the medulla with relatively thin walls; tetrasporangia zonate, scattered over the distal parts of very slightly enlarged branchlets, radially placed with the outer end hardly covered by neighboring cortical cells, diam. 23–26 μ, length 48–55 μ.

Bermuda, North Carolina, Florida, Bahamas, Cuba, Jamaica, Virgin Isls., Guadeloupe, I. las Aves. From reefs and rocks in the intertidal zone exposed to the surf, but also in sheltered places.

REFERENCES: Feldmann and Hamel 1934, Taylor 1940; *W. setacea*, Harvey 1853, Rein 1873, Murray 1889, Børgesen 1913–20, 1924, Collins and Hervey 1917 *p. p.*, Howe 1918a, b, 1920, 1928, Taylor 1928, 1930a, 1941b, Sanchez A. 1930, Williams 1949a; *?Gelidium variable*, Mazé and Schramm 1870–77, Murray 1889.

CRYPTONEMIALES

Plants showing various shapes from filiform to fleshy-membranous or rocklike; with either the multiaxial or the central filament types of structure, becoming corticated; asexual reproduction by tetraspores formed in sporangia at the thallus surface, or in sunken pits or conceptacles; sexual reproduction by spermatangia borne on surface cells or on the lining of conceptacles, and by carpogonia on carpogenic branches sunken in the cortex or in conceptacles, the carpogenic branches provided with typical auxiliary cells formed before fertilization, usually on accessory axes which may be adjacent to the carpogenic axes or more commonly are remote; carpogonia after fertilization ordinarily directly or indirectly producing oöblast filaments which connect with the auxiliary cells, from which the cystocarps are produced.

KEY TO FAMILIES

1. Plants encrusted with lime 2
1. Plants free from lime 3

2. Sporangia scattered between erect superficial filaments or in crater-like depressions .... SQUAMARIACEAE, p. 369
2. Sporangia in conceptacles with a definite perithecium-like wall CORALLINACEAE, p. 375

3. Plants crustaceous .... SQUAMARIACEAE, p. 369
3. Plants erect, bushy or foliose 4

4. Plants with a single evident axial cell row
ENDOCLADIACEAE, p. 367
4. Plants biaxial .... RHIZOPHYLLIDACEAE, p. 367
4. Plants multiaxial, or without an evident cell row 5

5. Carpogenic branches of 3 cells; auxiliary cells the supporting cells of the carpogenic branches .. KALLYMENIACEAE, p. 429
5. Auxiliary cells in separate, unrelated axes 6

6. Carpogenic branches of 2 cells, the oöblast filaments going direct to the auxiliary cells ...... GRATELOUPIACEAE, p. 415
6. Carpogenic branches of 5 cells, the oöblast filaments first fusing with cells in the carpogenic branches before going to the auxiliary cells .... DUMONTIACEAE, p. 363

DUMONTIACEAE

Plants of medium size, more or less branched, or plane and entire, soft; the original axial filament and apical growth soon obscured, the plants then appearing to have a pseudoparenchymatous cortex and a filamentous, often hollow medulla; sporangia when present tetrapartite; spermatangia superficial on the male plants, widely distributed; carpogenic branches scattered, of several cells, the fertilized carpogonium fusing with an intermediate cell in the branch from which oöblast filaments go out to auxiliary cells in separate axes, which in turn give rise to the scattered cystocarps.

KEY TO GENERA

1. Auxiliary cells terminal on their filaments; each carpogenic branch developing short lateral branchlets with the end cells of which projections from the carpogonium fuse before the oöblasts run out to the auxiliary cells .. ACROSYMPHYTON, p. 365
1. Auxiliary cells intercalary in their filaments; each carpogonium forming a projection which fuses with 1–2 cells of the carpogenic branch before the oöblasts run out to the auxiliary cells .... DUDRESNAYA, p. 363

Dudresnaya Bonnemaison, 1822

Plants erect and bushy, very gelatinous or lubricous, the branches alternately radially disposed and repeatedly dividing; axes monosiphonous, forming whorls of ditrichotomous filaments to produce the assimilatory cortex and many longitudinal rhizoidal filaments which surround the axial strand; tetrasporangia where known terminal on these cortical filaments, zonate; spermatangia whorled on the distal ends of the last two or three cells of the cortical filaments; carpogenic branches formed laterally on one of the lower forks of the cortical filaments, curved at the distal end, the trichogyne very long and exserted; the carpogonium after fertilization communicating with one or two of the intermediate cells of the branch serving as nutritive cells, from which the oöblasts emerge; auxiliary cells intercalary in specially thickened, simple or occasionally branched cortical filaments; cystocarps embedded in the cortex, without any special pericarp.

KEY TO SPECIES

1. Distal portions of the cortical filaments moniliform, the cells subspherical .... D. bermudensis
1. Cortical filaments with all cells nearly cylindrical .... D. crassa

Dudresnaya crassa Howe Pl. 43, fig. 6

Plants very lubricous, briefly stalked, above densely branched, 5–18 cm. tall, rose-red; branching radial, the main divisions vermiform, about 5 mm. diam., the very fragile, tapering terminal branchlets about 1.5 mm. diam.; axial filaments of cells 135–165 μ diam., 200–430 μ long; cortical fascicles formed in whorls of 4, attaching well below the distal end of the axial cell bearing them, becoming 4–6 times dichotomously divided, the cells subcylindrical, toward the surface slightly swollen, with larger chromatophores and 3.0–4.6 μ diam., but the terminal ones tapering; tetrasporangia unknown; dioecious, the spermatangial plants rather small and pale; carpogenic branchlets of 6–10 cells, the third and fourth cells below the carpogonium acting as nutritive cells and initiating oöblasts; auxiliary filaments showing 6–13 seriate enlarged intercalary cells, of which a relatively small one near the middle is the auxiliary cell; mature cystocarps short-stalked, about 130–165 μ diam., the distal assimilatory tip of the auxiliary filament commonly persisting and even branching.

Bermuda. Occasional in shallow water to a depth of about 3 m., in somewhat sheltered situations, but with free circulation of the water.

REFERENCES: Howe 1905b, 1918a, Collins and Hervey 1917, Taylor 1950c.

Dudresnaya bermudensis Setchell

Plant solitary or somewhat gregarious, bushy, to 9 cm. tall, rose-red; base discoid, stipe short, branching radial or slightly bilateral, or secund, the branches terete or a little compressed; cortical filaments in the more central portion of cylindrical cells, becoming moniliform near the surface; tetrasporangia unknown; carpogenic branches of several cells; auxiliary branches simple or 2–3 forked, curved, the cells near the middle swollen; cystocarps kidney-shaped, 112–126 μ diam.

Bermuda, Bahamas.

REFERENCES: Setchell 1912, Collins and Hervey 1917, Howe 1918a, 1920.

UNCERTAIN RECORD

Dudresnaya canescens J. Agardh

Florida. De Toni 4:1627.

Acrosymphyton Sjostedt, 1926

Plants erect and bushy, very gelatinous, the branches alternately radially disposed and repeatedly dividing; axes monosiphonous, forming whorls of ditrichotomous filaments to produce the assimilatory cortex and many longitudinal rhizoidal filaments which loosely surround the axial strand; tetrasporangia unknown; spermatangia whorled on the distal ends of the last two or three cells of the cortical filaments; carpogenic branches formed laterally on the lower forks of the cortical filaments, sharply curved, the carpogonium bearing a very long exserted trichogyne, from the subcarpogonial cells bilaterally developing short branchlets whose enlarged terminal cells serve as initial nutritive cells, receiving a short process from the carpogonium; from these nutritive cells the true, wide-ranging oöblasts at first emerge, but later other cells of the carpogenic branch and its appendages may become involved in an irregular fusion structure from which many oöblasts may extend; auxiliary cells enlarged and terminal on short, simple, or occasionally branched divisions of the cortical filament system; cystocarps embedded in the cortex, without any special pericarp.

Acrosymphyton caribaeum (J. Agardh) Sjostedt

Plants solitary or perhaps a few growing together, light red, bushy, to 16 cm. tall, dividing from near the base, 5–6 times alternately branched, the main branches 5–8 mm. diam., the attenuate terminal divisions less than 1 mm. diam.; axial filaments 60–130 μ diam., cortical assimilatory filaments in the larger branches whorled, attached near the middle of the axial cell, 5–7 times branched, distally moniliform, the cells obovate and 6.5–10.0 μ diam.; dioecious or monoecious, spermatangia in irregular alternately or oppositely branching clusters on the outer divisions of the assimilators; carpogenic branches 7–9 cells long, strongly curved, developing lateral branchlets 1–4 cells long from most of the cells below the carpogonium; auxiliary cells 12–14 μ diam., terminal on separate filaments which are 5–12 cells long; cystocarps 45–90 μ diam.

Bermuda, Florida, Saba Bank. Occasional in shallow water on rocks and dredged from a depth of 12 meters.

REFERENCES: Sjostedt 1926, Taylor 1952b; *Dudresnaya caribaea*, Collins and Hervey 1917, Howe 1918a, Taylor 1928; *D. bermudensis*, Phyc. Bor.-Am. 2195, *p. p.*

GLOIOSIPHONIACEAE

Plants soft and bushy; structurally composed of a central filament with compact whorls of assimilatory filaments; sporangia tetrapartite; carpogenic branches of three cells, carried on a supporting cell common to the associated auxiliary axis of seven or eight cells, the fifth being the functional auxiliary; cystocarps attached to the bases of the assimilatory filaments, without a pericarp.

Gloiosiphonia Carmichael, 1833

UNCERTAIN RECORD

Gloiosiphonia capillaris (Hudson) Carmichael

Bermuda. Type from England. De Toni 4: 1530; Hemsley 1884, Murray 1889.

ENDOCLADIACEAE

Plants bushy, the terete branches developing from an apical cell by oblique walls, the segments differentiating into elements of the prominent uniseriate axial filament and of the oblique branching cell systems, forming a loose medulla of irregular cells with thick mucilaginous walls, and a thinner cortical layer radially filamentous in origin if not in aspect; tetrasporangia irregularly tetrapartite, formed laterally on the last forking of the corticating cell rows; spermatangia formed as sori on the younger branchlets; procarps developed on outer cells of the oblique filaments, consisting of two-celled carpogenic branches with basal and auxiliary cells below them; gonimoblasts ramifying among the nearby thallus cells, which act as nutritive cells, forming clusters of small carposporangia.

Endocladia J. Agardh, 1841

Characters of the Family.

Endocladia vernicata J. Agardh

Plants bushy, forming extensive cushions, to 5–6 cm. tall; widely dichotomously branched, more loosely below, more densely above, the segments about 1 mm. in diameter, terete below, above a little flattened, and the acute uppermost divisions minutely spinulose on the margins.

Brazil. Growing on rocks and shells.

REFERENCES: J. Agardh 1841, 1851; *Acanthobolus brasiliensis*, Martens 1870, Zeller 1876.

RHIZOPHYLLIDACEAE

Plants crustose or erect and when bushy, dichotomously branching and tough; structurally with a medulla of one or more longitudinal filaments and a compact assimilative cortex of branches turning outward, these represented in crustose genera by a hypothallus of radial branching filaments and a perithallus of erect filaments borne on them; tetrapartite sporangia scattered or in nemathecia in the cortex; spermatangia in nemathecia, lateral on short, crowded, superficial filaments; carpogenic branches associated with quite similar separate auxiliary axes, the carpogonium fusing with an intermediate cell in the carpogenic branch, from which oöblast filaments go out to auxiliary cells in separate axes, which give rise to the cystocarps crowded in swollen nemathecia.

Ochtodes J. Agardh, 1872

Plants bushy, dichotomously or irregularly branched, terete in all divisions; branches usually developed from twin apical cells; plant showing two conspicuous axial filaments surrounded by rhizoidal strands and bearing lateral fascicles of filaments, the inner cells large and somewhat loosely placed but in the cortex closely compacted, forming moniliform anticlinal cell rows; tetrasporangia unknown; spermatangial sori surrounding the axis; cystocarps crowded together into strongly unilaterally projecting nemathecia.

Ochtodes secundiramea (Montagne) Howe

Plants bushy, 4–10 cm. tall, reddish purple, rather firm in texture but the jelly easily dispersed; copiously alternately branched, subcorymbose toward the tips, the lower divisions about 1.0 mm. diam., the ultimate 0.1 mm.; axis sometimes with one, much more usually with 2 prominent central filaments; cystocarps 2–3 together, very prominent and subglobose, at the bases of the branchlets.

Bahamas, Cuba, Jamaica, Hispaniola, Puerto Rico, St. Barthélemy, Guadeloupe, Martinique, Barbados, Grenada, British Honduras, Costa Rica, Panama, Trinidad, Tobago. Growing on rocks or other large objects in shallow water along moderately exposed shores.

REFERENCES: Howe 1920, Taylor 1929b, 1933, 1935, 1940, 1943, 1954a, Taylor and Arndt 1929, Feldmann and Hamel 1937; *Acanthococcus adelphinus*, Mazé and Schramm 1870–77; *Gracilaria divaricata*, Phyc. Bor.-Am. 789; *Hypnea secundiramea*, Murray

1889; *Ochtodes filiformis*, Murray 1889, Vickers 1905; *Sphaerococcus filiformis*, Dickie 1875.

SQUAMARIACEAE

Plants spreading, crustaceous; usually with a basal medullary or hypothallial layer of radiating branched filaments, from which a compact upper cortical or perithallial layer of ascending or vertical cell rows representing erect filaments arises, the whole sometimes a little encrusted with lime; growth peripheral from the ends of the filaments of the basal layer; sporangia tetrapartite, scattered between the erect filaments, in nemathecial groups or in craterlike conceptacles; spermatangia in tufts on the ends of erect, superficial or paraphysal filaments; carpogenic branches scattered or in nemathecia, short, lateral on paraphysal filaments; after fertilization the carpogonium fuses with an intermediate cell in the carpogenic branch, from which oöblast filaments go out to auxiliaries formed laterally at the bases of other paraphyses; cystocarps small, scattered.

KEY TO GENERA

1. Tetrasporangia in superficial nemathecia 2
1. Tetrasporangia in sunken conceptacles, without paraphyses HILDENBRANDIA, p. 369

2. Paraphyses absent .... CONTARINIA, p. 374
2. Paraphyses present .... PEYSSONNELIA, p. 370

Hildenbrandia Nardo, 1834

Plants crustose, widely spreading, the lower side strongly adherent, the constituent filaments strongly united to a firm though thin uncalcified crust; tetrasporangia irregularly zonate, borne in sunken conceptacles which discharge by a wide pore.

Hildenbrandia prototypus Nardo

Plants forming a thin orange-red to brownish red, ultimately dark, purplish red, crust, widely expanded, 0.2–0.5 mm. thick; basal hypothallus of cells in rows, supporting a perithallus of vertical, crowded, infrequently branched rows of cells 4.0–6.5 μ diam.; conceptacles scattered over the whole surface of the plant, about 100 μ diam., with a relatively large opening; tetrasporangia attached to the bottom and walls of the conceptacles, 16–30 μ long, 9–14 μ diam., irregularly divided into 4 cells, usually obliquely; paraphyses absent, but old sporangial walls persisting and evident in the conceptacles.

North Carolina, Florida, Jamaica, Virgin Isls., Brazil. Attached closely to rocks, chiefly in the lower tide pools and about low-tide level, in moderately or very exposed situations. In the fast-flowing fresh-water streams of Jamaica thin, light red incrustations of *H. rivularis* (Liebm.) J. Ag. are frequently encountered, and they may be expected elsewhere in the West Indies and Central America.

REFERENCES: Collins 1901, Børgesen 1913–20, Taylor 1928, Williams 1946; *H. nardii*, Zeller 1876.

UNCERTAIN RECORD

Hildenbrandia rosea Kützing

Brazil. Type from Europe. Zeller 1876.

Peyssonnelia Decaisne, 1841

Plants crustose, little to considerably calcified, attached directly or by more or less scattered rhizoids, the margins of the plant often somewhat free; hypothallus of radiating filaments usually dichotomously branched parallel to the substratum; upper perithallus layer formed of filaments borne upon these, simple or forked once or twice, becoming erect and closely laterally united; reproductive organs in nemathecia of unconsolidated filaments or paraphyses on the surface of the thallus; tetrapartite sporangia associated with paraphyses; plants often monoecious, the spermatangia lateral on paraphysal filaments; carpogenic branches formed on the bases of paraphyses, four- or five-celled; after fertilization fusion occurs with the hypogynous cell of the carpogenic branch as a nutritive cell, from which oöblast filaments arise; auxiliary axes placed similarly to carpogenic branches, the auxiliary cell usually being the second lowest cell of the axis; cystocarps scattered, consisting of a few large carposporangia.

KEY TO SPECIES

1. Hypothallus of filaments forming small lanceolate fan-shaped groups 2
1. Hypothallus essentially of rather straight juxtaposed filaments 5

2. Thallus generally calcified 3
2. Thallus calcified on the under side only, commonly splitting horizontally and regenerating the missing layers P. nordstedtii, p. 372

3. Thallus small and very thin, easily detached after decalcification P. armorica, p. 373
3. Thallus much thicker 4

4. Sometimes with radial lines but no structural veins; perithallus of 2 distinct regions, the cells of the upper region small and those of the lower large .... P. rosenvingii, p. 373
4. With distinct radial veins of large cells; perithallus, except the upper- and lowermost layers, of cells of a uniform size P. boergesenii, p. 374

5. Thallus membranaceous, calcified on the lower side only, the long apical cells as tall as the cells of the hypothallus P. rubra, p. 371
5. Thallus hard, well calcified throughout; the short apical cells taller than those of the hypothallus 6

6. Thallus thin, easily detached; color bright pink .. P. simulans, p. 372
6. Thallus thicker, firmly adhering; color red-purple P. conchicola, p. 372

Peyssonnelia rubra (Greville) J. Agardh

Plants pink or light red, membranaceous, somewhat calcified on the lower side, adhering by the entire lower face by short rhizoids and with overlapping cuneate-rounded lobes, 12–14 μ diam.; hypothallus more or less monostromatic, of long radiating dichotomous rows of oblong cells; the apical (marginal) cells as tall as the hypothallus cells and radially elongate; perithallus of ascending filaments 6–10 cells long, the lower cells rather taller, the upper ones shorter than broad.

Bermuda, North Carolina, Florida, Bahamas, Cuba, Jamaica, Hispaniola, Puerto Rico, Virgin Isls., Costa Rica, Panama, Colombia. Dredged from depths of as much as 33 m.

REFERENCES: Hauck 1888, Collins 1901, Weber-van Bosse 1916–17, Howe 1918a, b, 1920, Taylor 1928, 1929b, 1933, 1942, Taylor and Arndt 1929, Williams 1951.

Peyssonnelia simulans Weber-van Bosse

Plants bright pink, thin, but substantially calcified, attached by short rhizoids below but easily detached from the substratum; round or irregular, without conspicuous radial or concentric lines; hypothallus of radial filaments, the marginal (apical) cells short and broad, in radial section appearing taller than the hypothallus cells, which are 16–20 μ broad, 20–40 μ long, and in section 20–24 μ tall; perithallus of filaments easily separated by pressure after decalcification, each basal cell (seen in radial section) bearing 2 tiers of 3–6 cells, the lower cells as wide (in tangential section) below as those of the hypothallus, and a little taller than broad, 20 μ wide, 24–28 μ tall, but the upper cells not so tall; cystocarpic nemathecia about 200 μ thick, the paraphyses slender, cylindrical except near the base, obtuse, the gonimoblasts transversely divided into 3 cells; sporangia also in sori, tetrapartite, about 60–80 μ tall.

Virgin Isls. Dredged from depths of 18–46 m.

REFERENCE: Weber-van Bosse 1916–17.

Peyssonnelia conchicola Piccone and Grunow

Thallus red-purple, fading when dried, round, somewhat lobed, 1–3 cm. diam., membranous-crustose; strongly adherent by numerous short rhizoids, the hypothallus of straight, radiating filaments; cells oblong, appearing distromatic in transverse section, the cell tiers of the perithallus arising obliquely, the lower cells 2–3 times as tall as broad, the upper ones half as tall as broad; paraphyses subclavate, with longer cells below than above; sporangia tetrapartite, not on special stalks, to about 80–100 μ long, 23–35 μ diam.

North Carolina, Virgin Isls. Collected in shallow water.

REFERENCES: Weber-van Bosse 1916–17, Williams 1948b.

Peyssonnelia nordstedtii Weber-van Bosse

Plants crustose, to about 4 cm. diam., adherent by the whole lower surface, lightly calcified; hypothallus of large primary filaments and of secondary branching filaments in fan-shaped groups, the apical (marginal) cells both tall and short; perithallus of tiers of cells 12–20 μ diam., the cells 12–36 μ tall; vegetative thickness reported to increase by horizontal splitting of the thallus and duplication of the missing tissues, resulting in a blade of several similar layers; immature tetrasporangial nemathecia with paraphyses about 80 μ long.

Virgin Isls. Dredged from about 22 m. depth.

REFERENCE: Weber-van Bosse 1916–17.

Peyssonnelia armorica (Crouan) Weber-van Bosse

Plants crustose, dark pink, hard, thin, adhering by the entire lower surface, to 1–2 cm. diam.; easily detached when decalcified; filaments of the basal layer in small fanlike groups; nemathecia numerous, scattered.

Jamaica, Virgin Isls. Dredged to a depth of 27 m.

REFERENCES: Weber-van Bosse 1916–17; *Cruoriella armorica*, Collins 1901.

Peyssonnelia rosenvingii Schmitz

Plants forming brownish purple, thin, rounded crusts, the margins somewhat undulate, strongly adherent to the substratum below by abundant rhizoids; becoming 0.5 mm. thick or more, and cracking when dried, on the lower face encrusted with lime except about the narrow margin, the upper surface minutely radially striate; the hypothallus monostromatic, of fan-shaped groups of filaments, the cells large and 3–4 diameters long; perithallial filaments sparingly dichotomously branched, more slender above than below, 20–30 (–38) μ diam. below and the cells taller than broad, but above the cells shorter, isodiametric except the small surface layer; cystocarpic nemathecia hardly elevated above the general surface.

Bermuda, Florida, Jamaica, Hispaniola, Virgin Isls., Guadeloupe, Grenada, Brazil. Encrusting stones and the like, and reported from depths to 57 m.

REFERENCES: *Peyssonnelia dubyi*, Harvey 1853, Mazé and Schramm 1870–77, Dickie 1874b, Hemsley 1884, Murray 1889, Collins 1901, Weber-van Bosse 1916–17, Taylor 1928, 1943.

Peyssonnelia boergesenii Weber-van Bosse

Thallus completely adherent to the substratum, purplish to greenish when dried, smooth, with delicate but conspicuous radiating veinlets; marginal cells small; hypothallus of crowded filaments arranged in small lanceolate groups, the principal axes in older thalli distinct and of larger cells; perithallus about 500 μ thick, below of large cells 20–40 μ diam., 30–40 μ tall, those of the surface much shorter, 20 μ diam., 10 μ tall; nemathecia about 240 μ thick, the paraphyses slender with elongate cells below, short cells above, 12–40 μ long, 6–12 μ diam.; gonimoblasts obovate, 160 μ long, usually of 4 cells, carried on 2–3-celled stalks among the paraphyses.

Virgin Isls. Reported from shallow water.

REFERENCE: Weber-van Bosse 1916–17.

UNCERTAIN RECORDS

Peyssonnelia atropurpurea Crouan

Florida. Type from France. De Toni 4:1700; Farlow 1876, Taylor 1928.

Peyssonnelia imbricata Kützing

Brazil. Type from Newfoundland. De Toni 4:1703; Zeller 1876.

Peyssonnelia polymorpha (Zanardini) Schmitz

Virgin Isls., Venezuela. Type from the Adriatic Sea. De Toni 4:1701; Weber-van Bosse 1916–17.

Peyssonnelia squamaria (Gmelin) Decaisne

Brazil. Type from Europe. De Toni 4:1697; Martens 1870.

Contarinia Zanardini, 1843

Plants crustose, attached by the lower surface, sometimes partly calcified, the margins entire or lobed; medullary hypothallus of cells in flabellately radiating, horizontally branching filaments; upper cortical or perithallial tissue of vertical dichotomous filaments in a soft jelly; sporangia tetrapartite, terminating the erect filaments, in verrucose sori; spermatangia lateral on short superficial filaments; carpogenic branches together with auxiliary axes in nemathecia, where the cystocarps are ultimately crowded.

Contarinia magdae Weber-van Bosse

Thalli firmly adherent to their substrata, the lower portion calcified; perithallial filaments 20 μ diam. near the basal layer, 12–16 μ diam. near the surface, the cells 1–2 diameters tall, rather closely united into a firm thallus, but the files rather easily separated after decalcification; obovate sporangia irregularly tetrapartite, 16–20 μ diam., 36–40 μ long.

North Carolina, Virgin Isls., Panama, Venezuela. Growing on old coral fragments and coralline algae in both shallow and deep water; dredged to a depth of 38–40 m. The Panama and Venezuela material had smaller surface cells (to 10–12 μ diam. and 6–10 μ long) than that from the Virgin Islands, but the tetrasporangia were occasionally as large as 42 μ diam., 73 μ long.

REFERENCES: Weber-van Bosse 1916–17, Taylor 1942, Williams 1946.

CORALLINACEAE

Plants with a thin basal layer which may constitute the whole thallus, or which may develop into a massive calcareous crust or a system of rigid branches, or from which may arise erect, slender-branched, jointed axes, throughout showing the multiaxial type of structure with chromatophores in the peripheral cells, calcified except the intervening flexible joints of the segmented species; reproductive organs generally in conceptacles with a definite perithecium-like wall, sunken in the crust, or terminal on lateral enlarged branches; tetrasporangia, however, sometimes in soriform conceptacles, transversely divided (zonate), often associated with sterile paraphysal filaments; spermatangia on short filaments crowded in conceptacles; carpogenic branches usually three-celled, in the central part of each cystocarpic conceptacle, one or two being formed on each basal auxiliary cell; after fertilization union of the carpogonium with the auxiliary cell of the carpogenic branch occurs directly or by an oöblast filament, and then similarly with auxiliary cells at the bottoms of other branches, after which general fusions occur, so that the carposporangia ultimately arise marginally from a large fusion cell.

The Corallinaceae is a family perpetually aggravating to phycologists who attempt to study specimens from within this area. The

Corallineae are not too bad; if one avoids fragmentary or ill-grown material from the intertidal zone, it is usually possible to make identifications with confidence. It is only occasionally that some of the lesser-known and ill-described species come into question.

The situation respecting the Melobesieae is far different. Little can be done without decalcification, sectioning, and a critical histological study of fertile, especially tetrasporangial, representatives of each sample. The surface characters are valuable, but they are variable and hard to define. Furthermore both surface and histological characters have been described by authors with such different standards of workmanship, with such different ideas of what characters should be described, using descriptive words with such different meanings and making such discordant statements regarding measurements of critical structures, that it is practically impossible to glean from the literature balanced comparative accounts of related species at the present. To make the necessary corrective studies would involve far too much time as a part of the present task.

A monograph of world-wide scope was projected by Foslie. Had he been able to complete it, we would have had a comprehensive picture of this huge group from one viewpoint. Unhappily, he died before it was far advanced, but his many superb photographs of authentic material, his notes, and the key to Lithothamnium, edited and published after his death (1929), give us an invaluable work of reference respecting the appearance of these plants. Beyond this, the monographs of Lemoine (1911) and Hamel and Lemoine (1952) and the many scattered papers of these and other writers must be consulted.

I have tried to synthesize from the literature and from specimens as good a set of descriptions as I could, but short of prolonged monographic study no great degree of completeness or accuracy can be achieved, and the keys reflect these inadequacies. However, I feel that a great deal of material, providing the specimens are generous in quantity and well selected, can now be accurately determined. The habit should be checked against Foslie's plates (1929), and the keys used only as suggestive guides; in many instances it will be necessary to section the material. In spite of everything, determination of specimens in this group will remain a laborious matter and for the present is fraught with uncertainty.

KEY TO SUBFAMILIES

1. Crustose or fruticose nonarticulated genera .... MELOBESIEAE, p. 377
1. Except for the basal attachment region, more or less erect, bushy genera with alternating calcified segments and flexible articulations .... CORALLINEAE, p. 402

MELOBESIEAE

Plants crustose to fruticose; a basal layer of radial branching filaments constituting a hypothallus which is continuous with the medulla of the erect branches, usually covered by a perithallus of cell rows which are ascending in the crustose forms, and radially disposed as a cortex in the branched species, and sometimes show large, specialized heterocyst cells singly or in groups.

KEY TO GENERA

1. Tetrasporangia borne in individual crypts which often have a zonate distribution .... ARCHAEOLITHOTHAMNIUM, p. 378
1. Tetrasporangia grouped in conceptacles 2
2. Tetrasporangial conceptacles with numerous pores 3
2. Tetrasporangial conceptacles with single pores 4
3. Delicate crustose plants with but one continuous cell layer in vegetative parts .... MELOBESIA, p. 379
3. More massive crustose to fruticose plants of several cell layers LITHOTHAMNIUM, p. 380
4. Delicate plants of 1–3 cell layers in the vegetative portions (except *F. chamaedoris*, where to 10 cells thick) .. FOSLIELLA, p. 386
4. More massive crustose to fruticose plants of numerous cell layers .. 5
5. Vegetative tissues homogeneous except for consistent differences between hypothallus and perithallus ..... LITHOPHYLLUM, p. 390
5. Some of the cells of the vegetative tissues as heterocysts conspicuously larger and often thicker-walled than the rest 6
6. Heterocyst cells in tangentially placed platelike groups seen as short rows of cells in section .... POROLITHON, p. 400
6. Heterocyst cells scattered as individuals in the tissue, often obscure; perithallus with cells somewhat irregularly disposed GONIOLITHON, p. 395

Archaeolithothamnium Rothpletz, 1891

Plants encrusting, massive, calcified; hypothallus of cells in horizontal rows; perithallus of rectangular cells in vertical series; tetrasporangia often arranged in zones, solitary in individual crypts in the thallus and discharged through individual pores; cystocarps in superficial conical conceptacles with single apical pores.

KEY TO SPECIES

1. Thallus thin, 0.5–1.0 mm., generally of but one layer; conceptacles prominently beaked, becoming deeply overgrown within the layer which bore them .... A. dimotum
1. Thallus thicker, generally of several superimposed layers, 0.25–1.0 mm. thick; conceptacles superficial, with almost no beak, not overgrown by the same layer which bore them A. episporum

Archaeolithothamnium dimotum Foslie and Howe

Thallus forming crusts 0.5–1.0 mm. thick, closely adherent to the substratum and not developing proper excrescences; perithallus cells sometimes quadrate, sometimes vertically elongated, 6–11 μ diam., to 15 μ long; minute intermediate cells rather numerous; sporangia 40–60 μ diam., 70–85 μ long, each conceptacular chamber with an extension toward the surface (the beak or "apiculus"), 15–25 μ long, the conceptacles becoming deeply overgrown after spore discharge.

Bahamas, Jamaica, Hispaniola, Puerto Rico. On rocks at low-water mark.

REFERENCES: Foslie and Howe 1906a, Howe 1920, Lemoine 1917, 1924.

Archaeolithothamnium episporum Howe Pl. 76, fig. 1

Plants brownish red, widely expanded, in sterile parts smooth to subnitent, repeatedly overgrown, the individual layers 0.25–1.0 mm., the total increment to 5 mm. or more; smooth or with irregular excrescences or nodules 2–5 mm. diam.; hypothallus 30–170 μ thick, the cells 8–11 μ diam., 17–28 μ long; cells of the perithallus in distinct layers sometimes of short and tall cells alternating, cells mostly 5–8 μ wide, 8–15 μ tall; tetrasporangia irregularly divided, in superficial conceptacles, 27–50 μ diam., 65–96 μ long including the "apiculus," in sori 0.1–1.0 mm. broad or confluent, the pores level with the surface or but slightly protruding, becoming 16–22 μ diam.; surface of fertile areas roughened by exfoliation after discharge of the spores.

Canal Zone, Panama. On rocks and dead corals from low-water mark to a depth of several meters.

REFERENCES: Howe 1919, Taylor 1929b.

Melobesia Lamouroux, 1812

Plants crustose, completely attached to the support, composed of but a single layer of cells in the vegetative parts, each cell cutting off obliquely a small superficial associated cell, but in the neighborhood of the conceptacles of several layers of cells; tetrasporangia associated with sterile filaments in soriform conceptacles with several pores; spermatangia lateral on two-celled filaments distributed over the bottom of the conceptacle cavities; cystocarpic conceptacles showing three-celled carpogenic branches arising from the basal layer in the center surrounded by incomplete branches without carpogonia, and the peripheral ones contributing to the wall of the conceptacle.

Melobesia membranacea (Esper) Lamouroux

Plants at first a thin membrane, subrugose, later becoming calcareous, sometimes overlapping, reddish to purple or white; thin, of one layer of cells except about the conceptacles, where 4–5 layers exist; cells of the monostromatic part of the thallus when seen from above 9–20 μ long by 5–9 μ broad, and in section 5–18 μ deep; conceptacles generally hemispherical; tetrasporangial conceptacles about 110–140(–200) μ (or by Foslie 160–350 μ) diam., scattered superficially over the crust, sometimes confluent, wartlike, with 8–27 distinct pores; the apertures of the sexual conceptacles contracted by filiform cells projecting from the surrounding walls.

Florida, Jamaica, Hispaniola, Puerto Rico, Virgin Isls., Guadeloupe, Old Providence I., Panama, Venezuela, Tobago, Brazil. Epiphytic on coarse algae and marine vascular plants in shallow water.

REFERENCES: Mazé and Schramm 1870–77, Piccone 1886, Hauck 1888, Möbius 1889, Murray 1889, Collins 1901, Lemoine 1917, Taylor 1939b, 1940, 1942, 1943.

Lithothamnium Philippi, 1837

Plants forming firm calcareous crusts, or branching from a crust-like base; of two cell layers, the basal spreading, the upper of erect cell rows with the cells in transverse superposed zones; sporangial conceptacles soriform, superficial or somewhat immersed, at first isolated, later the septa of adjoining units in part disappearing leaving a common cavity with the roof traversed by several pores; sporangia bisporic or tetrasporic, sometimes both types intermixed; cystocarpic conceptacles superficial or slightly immersed, conical or subconical, the fugacious tip short and thin, with an apical pore; carposporangia arising from the peripheral portion of the fusion cell, the central parts of which for a time support withered remnants of the trichogynes.

KEY TO SPECIES

1. Thallus crustose 2
1. Thallus attached or becoming free, with knoblike bosses, or conspicuously branching 5

2. Sporangial conceptacles small, less than 300 μ diam. 3
2. Sporangial conceptacles large, 400 μ diam. or more 4

3. Thallus conforming to the substratum, without independent excrescences; sporangial conceptacles 160–260 μ diam. L. sejunctum, p. 381
3. Thallus developing as more or less foliar, often overlapping lamellae; sporangial conceptacles 100–140 μ diam. L. mesomorphum, p. 382

4. Crusts closely conforming to the substratum and often superimposed, but not strongly adherent; sporangial conceptacles 400–600 μ diam. . . . . L. syntrophicum, p. 381
4. Crusts thin, loosely surrounding the supporting objects, with age becoming free; sporangial conceptacles 500–700 μ diam. L. ruptile, p. 382

5. Surface elevations knob- or wartlike, or very short thick branches little if any longer than broad 6
5. Obviously branching, though the branches often much fused 8

6. Very irregular wartlike bosses or very short branches 3 mm. diam. or less .... L. aemulans, p. 383
6. Very short branches generally more than 3 mm. diam. 7

7. Branches generally simple; medullary cells to 22 μ long L. floridanum, p. 383
7. Branches generally irregularly dichotomous; medullary cells to 32 μ long .... L. heteromorphum, p. 383

8. Primarily attached species 9
8. Generally dredged as individual plants or clusters loose on the bottom 11

9. Branches 1.0–2.0 mm. diam., sometimes a little flattened; conceptacles 300–400 μ diam. 10
9. Branches 2–3 mm. diam., conceptacles 450–600 μ diam. L. brasiliense, p. 383

10. Ends of the branches abruptly rounded-truncate L. erubescens, p. 384
10. The branches compressed, or slightly tapering to the rounded ends .... L. incertum, p. 384

11. Branches always irregular, contorted, becoming coalesced with age; cells of the perithallus 10–32 μ long .. L. occidentale, p. 385
11. Branches commonly symmetrical; cells of the perithallus 8–9 μ long .... L. calcareum, p. 384

Lithothamnium sejunctum Foslie

Plants of small, thin, rounded, strongly adherent crusts, concentrically zonate, the margins lobed and white; hypothallus of rectangular cells 3–7 μ diam., 10–15 μ long; perithallus of ovoid cells 3–7 μ diam., 5–7 μ tall; sporangial conceptacles 160–260 μ diam., the roof with about 40 pores; cystocarpic conceptacles 200–300 μ diam.

?North Carolina, Virgin Isls.

REFERENCES: Foslie 1906, 1929, Lemoine 1917, Hoyt 1920.

Lithothamnium syntrophicum Foslie

Plants crustlike or lamellate, to 2–3 cm. diam. or more, rather closely surrounding the supporting objects; new layers repeatedly, irregularly, and loosely formed over each other, each 200–800 μ thick, at length forming small, irregular, knotty nodules; hypothallus of cells 6–9 μ diam., 12–25 μ long; perithallus thick, the cells quadrate or somewhat rounded, 4–7 μ diam. or a little taller; sporangial conceptacles commonly somewhat immersed but at times convex, subprominent, 400–600 μ diam., tetrasporangia about 60 μ diam., 120 μ long.

Bermuda, Florida.

REFERENCES: Foslie 1901, 1929, Howe 1918a, Taylor 1928.

Lithothamnium ruptile (Foslie) Foslie

Plants forming thin, irregular, often loosely superimposed crusts, becoming free; hypothallus of rectangular-ovoid cells 7–14 μ diam., 18–36 μ long; perithallus of quadrate cells 9–11 μ diam., or somewhat elongated and 7–11 μ diam., 10–25 μ tall; sporangial conceptacles large, 500–700 μ diam.

Virgin Isls.

REFERENCES: Foslie 1907a, 1929, Lemoine 1917.

Lithothamnium mesomorphum Foslie

Plants crustose to lamellate, the crusts somewhat glossy when dry, easily separated from the substrate, the lamellae sometimes overlapping, sometimes erect, 300–500 μ thick; hypothallus of rectangular cells 4–7 μ diam., 10–14 μ long, or sometimes 20–25 μ long; perithallus of cells 7–10 μ diam., 10–12 μ long below, above smaller, nearly isodiametric, 5–8 μ diam.; sporangial conceptacles 100–140 μ diam., 60 μ tall.

V. **ornatum** Foslie and Howe. Thallus 150–200 μ thick, less proliferous, forming small lamellae attached by a limited marginal area; hypothallus of cells 7–11 μ diam., 11–20 μ long or near the lower surface even 3–6 μ diam., 15–25 μ long; perithallus of cells toward the upper surface rounded or subquadrate, oblong in section, 4–9 μ diam.

Bermuda, Florida, Bahamas, Virgin Isls., Old Providence I., Colombia.

REFERENCES: Foslie 1901, 1929, Lemoine 1917, Howe 1918a, Taylor 1939b. For the v. *ornatum,* Foslie and Howe 1906a, Howe 1920, Taylor 1928.

Lithothamnium aemulans (Foslie and Howe) Foslie and Howe

Plants forming an irregularly nodulose crust 10–15 mm. thick, or thicker by increments of overgrowth, bearing tuberculate bosses, or less evidently much anastomosed branches, mostly 1.5–3.0 mm. diam., 2–5 mm. long; hypothallus of subquadrate cells 6–10 μ diam., or vertically elongated to 15 μ; sporangial conceptacles 400–500 μ diam.

Puerto Rico. Forming heavy crusts around coral fragments in shallow water.

REFERENCES: Foslie 1907, 1929, Howe 1920; *L. fruticulosum* v. *aemulans,* Foslie and Howe 1906a.

Lithothamnium floridanum Foslie

Thallus forming a hard crust 1 mm. thick, bearing short, subsimple branches which are more or less concrescent, 4–7 mm. tall, 3–6 mm. diam., sometimes nodulose; medullary cells usually 9–11 μ diam., 14–22 μ long; perithallus of subquadrate or vertically elongated cells 7–14 μ diam., 9–22 μ tall; sporangial conceptacles 400–600 μ diam.

Florida, ?Cuba, Hispaniola.

REFERENCES: Foslie 1906, 1929, Howe 1918b, Taylor 1943.

Lithothamnium heteromorphum (Foslie) Foslie

Plants with a crustose base bearing very short, irregularly dichotomous branches 3–4 mm. diam., with rounded and thickened tips; cells of the medullary hypothallus 9–14 μ diam., 14–32 μ long; perithallus cells subquadrate, 9–11 μ broad, or vertically elongated and then 7–11 μ diam., 9–22 μ tall; sporangial conceptacles 350–600 μ diam.

Florida, Brazil.

REFERENCES: Foslie 1908a, 1929; *L. brasiliense* f. *heteromorpha,* Foslie 1900a.

Lithothamnium brasiliense Foslie

Plants at first crustose, the primary layer 0.5 mm. thick, bearing dichotomous branches and forming rounded angular masses about

4 cm. diam.; branches short, 2–3 mm. diam., their apices rounded, often concrescent below but free above; perithallus of cells 8–14 μ diam., 14–24 μ tall; sporangial conceptacles somewhat prominent, 450–600 μ diam., the center depressed and traversed by about 30 delicate pores; tetrasporangia about 30 μ diam., 100 μ long.

Brazil.

REFERENCES: Foslie 1900a, 1929.

Lithothamnium incertum Foslie Pl. 78, fig. 2

Plants forming subhemispheric cushions 2.5–5.0 cm. high, 7.5–15.0 cm. broad, with a closely adherent crustose base, soon developing erect, ramified, anastomosing branches; terminal segments of the branches subterete, 1.0–1.5 mm. diam., or usually decidedly flattened and 2 mm. or more broad; conceptacles formed near the ends of the branches, depressed-hemispherical, 300–400 μ diam.

Bermuda, Florida, Bahamas. From rocks 3–12 dm. below low-tide line in exposed situations.

REFERENCES: Howe 1918a, Taylor 1928, 1940, Foslie 1929.

Lithothamnium erubescens Foslie

Plants to 2.0–3.5 cm. tall, 3.5–4.5 cm. broad, on basal crusts to 0.7 mm. thick; above subdichotomously branched, the short branches fastigiate, very blunt, terete or a little compressed, 1.25–1.75 mm. diam.; cells in the branches 7–10 μ diam., 12–22 μ long; sporangial conceptacles subprominent, 300–400 μ diam., the flattened central portion traversed by about 20 pores.

F. **prostrata** Foslie. Plants less branched than in the typical form, the branches more or less compressed and decumbent.

Bermuda, Brazil.

REFERENCES: Foslie 1900a, 1901, 1929; *Goniolithon mamillare*, Dickie 1874b.

Lithothamnium calcareum (Pallas) Areschoug Pl. 79, fig. 1

Plants with a basal crust 150–300 μ thick, but soon forming free masses to 2–4(–10) cm. diam.; usually rather sparingly branched, the branches typically divergent, sometimes dichotomous, sometimes vermiform, loose, or in some forms more crowded, but always 1–3 mm. diam.; basal crust with hypothallus about 100 μ thick, of irregularly ovoid cells 4–16 μ diam., 10–20 μ long; the perithallus as in the branches; medulla in transverse section of cells 3–5(–10) μ diam., 6–12 μ long, the perithallus of cells about 6 μ diam., 8–9 μ long, in tiers of 5–6; sporangial conceptacles scarce, generally grouped at the ends of the branches, convex or concave, from the surface 200–550 μ diam., with 30–60 pores and a nonporose border; viewed in section the cavity 200–400 μ diam., 150–300 μ high.

Colombia, ?Brazil.

REFERENCES: Foslie 1929, Taylor 1942; ?*L. crassum*, Piccone 1886. The data given by Foslie (1929) do not seem to agree with those given by Lemoine (1911) and Hamel and Lemoine (1953).

Lithothamnium occidentale (Foslie) Foslie Pl. 79, fig. 2

Thallus early becoming free, much branched, not elegant; when small the irregular branches short, divergent, 1.5–3.0 mm. diam., but in large masses often almost completely coalesced, only the tips projecting; perithallus of rectangular or slightly swollen cells 6–10 μ diam., 10–32 μ long; sporangial conceptacles inconspicuous; sporangia 75 μ diam., 180 μ long.

V. **effusa** (Foslie) Foslie. Small loose colonies; branches fewer, more slender and longer, strongly divergent, 1.0–2.5 mm. diam.

Florida, Hispaniola, Virgin Isls., Netherlands Antilles.

REFERENCES: Foslie 1908c, 1929, Lemoine 1917, 1924, Taylor 1928; *L. fruticulosum* f. *occidentalis*, Foslie 1906. For the v. *effusa*, *L. soluta* v. *effusa*, Foslie 1906.

UNCERTAIN RECORDS

Lithothamnium amplexifrons (Harvey) Lemoine

Guadeloupe. Type from Africa. Lemoine 1917; *Lithophyllum amplexifrons*, De Toni 4:1788; *Melobesia amplexifrons*, Mazé and Schramm 1870–77, Murray 1889.

Lithothamnium fasciculatum (Lamarck) Areschoug

Brazil. Type from Europe. De Toni 4:1782; Möbius 1889.

Lithothamnium lenormandi (Areschoug) Foslie

Jamaica, Brazil. Type from Sweden. De Toni 4:1756; Collins 1901; *Lithophyllum lenormandi,* Möbius 1890.

Lithothamnium lichenoides (Ellis and Solander) Heydrich
F. *agariciformis* (Johnstone) Foslie

Bermuda, "West Indies." Type from Ireland. De Toni 4:1751; Foslie 1900a; *Melobesia lichenoides,* Dickie 1874b.

Lithothamnium polymorphum (Linnaeus) Foslie

Guadeloupe, Grenada, Brazil. Type from Europe. De Toni 4:1724; Mazé and Schramm 1870–77, Dickie 1874b, Murray 1889, Möbius 1889, 1890.

Lithothamnium scabiosum (Harvey) Foslie

Brazil, the type locality. De Toni 4:1748; *Melobesia scabiosa,* Martens 1870.

Lithothamnium ungeri Kjellman

Bermuda. Type from Norway. Bigelow 1905, Howe 1918a.

[?*Lithothamnium*] *Spongites verruculosa* Zeller

Brazil, the type locality. Zeller, 1876.

Fosliella Howe, 1920

Plants forming thin, rather lightly calcified crusts, the lower face adherent to the substratum; structurally showing one to few cell layers, the basal layer consisting of a radially disposed, closely united filament system; distinctively large hair-bearing (trichocyte) vegetative cells sometimes present; conceptacles of sporangia superficial or slightly immersed, conical or hemispheric-conical, with a single apical pore; sporangia with a short foot and associated with evanescent trabecular filaments; cystocarpic conceptacles smaller than those of sporangia but otherwise similar.

KEY TO SPECIES

1. Thallus strictly monostromatic .... F. atlantica, p. 387
1. Thallus 1–3 cells thick 2
1. Thallus 8–10 cells thick except near the unistratose margin F. chamaedoris, p. 389

2. Trichocytes present .... F. farinosa, p. 388
2. Trichocytes absent; thallus cells 6–7 μ broad ..... F. lejolisii, p. 387
2. Trichocytes presumably absent 3

3. Thallus cells 9–18 μ broad .... F. affinis, p. 388
3. Thallus cells 10–12 μ broad; thalli commonly superposed
F. bermudensis, p. 388

Fosliella atlantica (Foslie) n. comb.

Thallus an irregular crust, monostromatic, rarely segregating a second cell layer; cells 10–25 μ diam., 18–32 μ long (Lemoine) or 18–40 μ diam., 32–60 μ long (Foslie); sporangial (?) conceptacles conical, elevated, 500–800 μ diam.

Hispaniola, Virgin Isls. Growing on bryozoa and dead corals.

REFERENCES: *Mastophora atlantica,* Foslie 1906, p. 27; *Melobesia atlantica* Lemoine 1917.

Fosliella lejolisii (Rosanoff) Howe

Plants in the form of delicate dull pink or white crusts often closely crowded on the host, about 0.5–2.0 mm. diam., 15–30 μ thick, the cells of the radial rows as seen from the surface 6-10 μ long, 6-7 μ broad, or larger if bearing a fork and somewhat resembling trichocytes; in section the disk near the margin one cell thick, or with small, superficial cells cut off on the upper side near the distal end of each cell of the primary layer; toward the central part each cell of the primary layer is divided in the plane of the substratum so that the disk is 2–4 cell layers in thickness, the lowest layer being of cells 2–3 μ high, the second of cells 17–20 μ high, and the superficial layers like that at the base; sporangial and cystocarpic conceptacles convex or subhemispherical, frequently crowded, or almost confluent, 150–250(–300) μ diam., the apical pore apparently without filiform cells; tetrasporangia 50–80 μ long, 30–50 μ diam.; conceptacles of spermatangia 75–100 μ diam.

Florida, Mexico, Bahamas, Cuba, Jamaica, Hispaniola, Aves I., Barbados, Venezuela, Tobago, French Guiana, Brazil. From coarse algae and marine phanerogams in rather shallow, sheltered places. Lemoine (1917) considers that this species does not reach the tropics and that specimens so identified are probably *F. farinosa* with few or no trichocytes.

REFERENCES: Howe 1920, 1928, Sanchez A. 1930, Taylor 1942, 1954a, Humm 1952b, Joly 1957; *Melobesia lejolisii,* Möbius 1890, Collins 1901, Taylor 1928.

Fosliella affinis (Foslie) n. comb.

Thalli crustose, very thin, little calcified, suborbicular with lobed margins, becoming gregarious; cells of the thallus 9–18 μ broad, 14–22 μ long; conceptacles hemispheric-conical, 60–120 μ diam., or some 170–260 μ diam.

Virgin Isls. On rocks and shells.

REFERENCES: *Litholepis affinis,* Foslie 1906, p. 17; *Melobesia affinis,* Lemoine 1917.

Fosliella bermudensis (Foslie) n. comb.

Thalli delicate, irregular in form and spread, about 30 μ thick, commonly confluent and superposed even to 150 μ of aggregate thickness; hypothallus of cells which are quadrate or vertically elongated, 10–12 μ diam., 10–18 μ tall; cortical cells small; sporangial conceptacles crowded, superficial, hemispheric-conical, 150–180 μ diam., tetrasporangia about 30 μ diam., 55 μ long.

Bermuda, Hispaniola.

REFERENCES: *Litholepis bermudensis,* Foslie 1905a; *Melobesia bermudensis,* Foslie 1901, p. 22, Howe 1918a, Taylor 1933.

Fosliella farinosa (Lamouroux) Howe

Plants forming whitish fragile crusts on the support, usually to 2–5 mm. diam. or more, often crowded; of one continuous layer in the vegetative portions, the cells in surface view variable in size, sometimes about 12–20 μ long by 7–10 μ broad, sometimes 18–30 μ long by 15–18 μ broad, often bearing superficial cells cut off on the upper side near the distal ends of the cells of the primary layer; the colorless swollen cells or trichocytes each terminating a cell row, 22–40 μ long, 12–30 μ broad, commonly prolonged to a hair on the upper surface; sporangial and cystocarpic conceptacles 140–250 μ diam., those of spermatangia 60–80 μ; tetrasporangia (40–)50–90 μ long by (20–)30–50 μ diam.

V. **chalicodictya** Taylor. Plants small, forming minute, coalescent, lightly calcified crusts which individually may reach 1 mm., rarely 2 mm., diam.; constituent filaments sinuous to angularly bent, laterally cohering at the points of contact, or anastomosing, to form a definite network, the cells generally 10–12 μ diam., 15–20 μ long, with small superficial cortical cells superposed at the ends; occasional trichocytes, 18–22 μ diam., also present.

V. **solmsiana** (Falkenberg) Taylor. Plants of meandering filaments locally united at points of contact but chiefly laterally distinct, not forming a regular disk.

Bermuda, North Carolina, Florida, Bahamas, Cuba, Jamaica, Hispaniola, Puerto Rico, Virgin Isls., Guadeloupe, Barbados, British Honduras, Old Providence I., Costa Rica, Panama, Colombia, Venezuela, Netherlands Antilles, Tobago, Brazil. From coarse algae and marine phanerograms growing in rather shallow, sheltered places.

REFERENCES: Howe 1920, Taylor 1939b, 1942, 1943, 1954a, 1955a; *Melobesia confervicola,* Mazé and Schramm 1870–77, Lemoine 1917; *M. farinosa,* Mazé and Schramm 1870–77, Piccone 1886, Hauck 1888, Murray 1889, Collins 1901, Vickers 1905, Sluiter 1908, Lemoine 1917, 1924, Howe 1918a, Hoyt 1920, Taylor 1928, 1933, 1935, 1936, Taylor and Arndt 1929. For the v. *chalicodictya,* Taylor 1939b. For the v. *solmsiana,* Taylor 1939b, 1940; *Melobesia farinosa* v. *solmsiana,* Lemoine 1924, Taylor 1928, 1933.

Fosliella chamaedoris (Foslie and Howe) Howe

Thalli forming small, dull rose or purplish crusts 60–150 μ thick; hypothallus cells quadrate or horizontally elongated, 7–10 μ diam., to 12–13 μ long; perithallus several cells thick, the cells subquadrate or horizontally elongated, 7–10 μ diam., 10–14 μ broad; sporangial conceptacles convex or subconical, 150–200 μ diam.; bisporangia 30–40 μ diam., 50–60 μ long.

Bahamas, Virgin Isls. Encircling the stalks of Chamaedoris in shallow water.

REFERENCES: Howe 1920; *Lithophyllum chamaedoris,* Foslie and Howe 1906a; *Melobesia chamaedoris,* Foslie 1908b, Lemoine 1917.

Lithophyllum Philippi, 1837

Thallus wholly crustose or with erect branches from a crustose base, thin and brittle to thick and rocklike, the lower surface entirely adherent to the substratum or with free margins; hypothallus (or medulla in erect portions) several layers of cells in thickness, the cells elongated in the direction of growth; perithallus usually thinner than the hypothallus, the cells quadrate with secondary connections in the vertical walls; conceptacles partly immersed, opening by a single pore; sporangia bipartite or tetrapartite, restricted to the periphery of the floor of the conceptacle.

KEY TO SPECIES

1. Thallus essentially plane, conforming to the contours of the substratum, at most minutely roughened 3
1. Thallus roughened by verrucae or by branches a few millimeters in length 2
1. Thallus with numerous long, often anastomosing branches 9

2. Crusts about 1 mm. thick; conceptacles rarely to 200 μ diam. L. absimile, p. 393
2. Crusts to 3–6 cm. thick; conceptacles about 300 μ diam. L. munitum, p. 393

3. Conceptacles 120 μ diam. or less .... L. caribaeum, p. 391
3. Conceptacles larger 4

4. Very small plants epiphytic on segmented coralline algae L. corallinae, p. 391
4. Larger plants growing on other substrata 5

5. Conceptacles not over 300 μ diam. 6
5. Conceptacles generally exceeding 300 μ diam. 7

6. Hypothallus of cells 6–9 μ diam. .... L. erosum, p. 392
6. Hypothallus of cells about 4 μ diam. .... L. intermedium, p. 391

7. Thalli over 1 mm. thick .... L. bermudense, p. 393
7. Thalli less than 0.5 mm. thick 8

8. Thallus surface smooth .... L. pustulatum, p. 392
8. Thallus surface squamulose .... L. prototypum, p. 392

9. Branches closely anastomosing nearly to the ends, the colony presenting a highly convoluted aspect ...... L. congestum, p. 393
9. Branches anastomosed below, but this not conspicuous from the surface where the branch tips are free .. L. daedaleum, p. 394

9. Branches much flattened, thin, showing relatively broad plane or curved faces somewhat attached to each other on lines of contact .... L. platyphyllum, p. 394

Lithophyllum corallinae (Crouan) Heydrich

Plants forming irregularly rounded, convex, gray-lilac to rufescent crusts 1–5 mm. diam., the entire lower face fastened to the support, or the margins sometimes free; thickness 80–400 μ; the sporangial conceptacles deeply immersed, outwardly wartlike, 200 (–350) μ diam., the sporangia bisporic, 50–88 μ long, 18–32 μ diam.; spermatangial conceptacles not projecting, small; cystocarpic conceptacles of about the same size as the sporangial, the cells around the canal papilliform.

North Carolina. Epiphytic on Corallina.

REFERENCE: Williams 1948b.

Lithophyllum caribaeum (Foslie) Foslie

Plants crustose, gregarious, the crusts very thin, not concealing the substratum, the surface roughened; hypothallus of one layer of rectangular cells 4–7 μ wide, 5–12 μ long; perithallus very compact, the cells rectangular, 3–6 μ wide, 3–8 μ tall; conceptacles very numerous, 80–120 μ diam.

Bahamas, Jamaica, Hispaniola, Puerto Rico, Virgin Isls.

REFERENCES: Foslie 1907a, Lemoine 1917, 1924, Howe 1920, Taylor 1933.

Lithophyllum intermedium (Foslie) Foslie

Thalli at first of rounded crusts about 3 mm. thick, with lobed or crenulate margins, these crusts later confluent, forming a minutely mammillate surface; hypothallus of cells 4 μ diam., 10–12 μ tall; perithallus of cells 7–11 μ diam., 9–22 μ tall; conceptacles very numerous, 150–250(–300) μ diam.

Bermuda, North Carolina, Florida, Jamaica, Puerto Rico, Virgin Isls., Barbados.

REFERENCES: Foslie 1906, Lemoine 1917, Hoyt 1920, Taylor 1928; *Goniolithon intermedium*, Foslie 1901, Howe 1918a; *Lithothamnium incrustans*, Collins 1901.

Lithophyllum erosum Foslie

Thallus at first appearing as an irregularly crenate crust, later these crusts confluent, 0.1–0.2 mm. diam.; hypothallus of cells 6–9 μ diam., 9–18 μ long; perithallus of subquadrate cells 5–9 μ broad, or becoming elongated and to 7–11 μ tall; sporangial conceptacles convex, 140–240 μ diam., tetrasporangia about 40 μ diam., 70 μ long.

Virgin Isls.

REFERENCES: Foslie 1906, Lemoine 1917. Lemoine gives the thickness of the crusts as 40 μ.

Lithophyllum pustulatum (Lamouroux) Foslie

Plants at first flat, suborbicular, completely attached by the under face, becoming thick-convex or pulvinate, or ultimately often confluent and superimposed; 2–10 mm. diam., 2–3 cells thick, or rarely to 8 cells and 350 μ thick; conceptacles conspicuous, scattered over the thallus, 300–500 μ diam., the tetrasporangia 80–130 μ long, 30–70 μ diam.

Bermuda, North Carolina, Florida, Bahamas, Jamaica, Hispaniola, Virgin Isls., Guadeloupe, Martinique, Barbados, Grenada, Old Providence I., Costa Rica, Panama, Colombia, Brazil. Epiphytic on coarse algae in water of slight or very moderate depths.

REFERENCES: Howe 1918a, 1920, Taylor 1939b, 1940; *Dermatolithon pustulatum,* Hoyt 1920, Taylor 1928, 1929b, 1933, Taylor and Arndt 1929; *Melobesia pustulata,* Dickie 1874a, b, Hemsley 1884, Piccone 1886, Murray 1889, Collins 1901; *M. verrucata,* Mazé and Schramm 1870–77, Dickie 1874a, Murray 1889.

Lithophyllum prototypum Foslie

Crusts small, at first suborbicular, the margins thin and rather free, later irregularly confluent, about 300 μ thick, forming squamulose flakes over the surface; cystocarpic conceptacles scattered over the central parts of the crust, slightly conical, 450–550 μ diam.

Florida, Bahamas, Jamaica, Puerto Rico, Virgin Isls. On the shells of Pinna in shallow water.

REFERENCES: Foslie 1905, Lemoine 1917, 1924, Howe 1920, Taylor 1928; *Goniolithon udoteae,* Foslie 1901; *Lithothamnium prototypum,* Foslie 1897.

Lithophyllum bermudense Foslie and Howe

Thallus formed of crusts 1–2 mm. thick, the total increment of overlapping layers reaching 8–9 mm. in thickness; hypothallus of one, occasionally 2, layers of somewhat oblique cells 9–15 μ diam., 25–50 μ tall; perithallus of horizontal layers and vertical tiers of cells with conspicuous terminal and lateral interconnections; conceptacles (?sporangial) convex or subhemispherical, 400–600 μ diam. in surface view, but slightly elevated, with a single pore.

Bermuda, Florida. Growing on calcareous pebbles.

REFERENCES: Foslie and Howe 1906a, Howe 1918a.

Lithophyllum (?) absimile Foslie and Howe

Thallus crustose, the crusts about 1 mm. thick, bearing verrucae about 1–3 mm. wide and tall; hypothallus poorly developed, the cells 7–11 μ diam., 9–22 μ tall; perithallus of subquadrate cells about 5–7 μ diam.; conceptacles crowded, convex, 100–160(–200) μ diam.

Jamaica, Virgin Isls.

REFERENCES: Foslie 1907b, Lemoine 1917.

Lithophyllum (?) munitum Foslie and Howe

Thallus brownish red when living, forming irregularly lobed crusts to 3–6 cm. thick with densely crowded verrucae, or with sparingly and irregularly divided processes or branches 2–4 mm. diam., to 4 mm. tall, sometimes anastomosing; hypothallus cells 5–8 μ diam., 12–24 μ long; cells of the perithallus subquadrate or rounded in vertical section, 5–9 μ diam., or shorter or longer; conceptacles (? of sporangia) slightly convex, about 300 μ diam. in surface view.

Bahamas, Jamaica, Puerto Rico. Growing on dead corals near low-water mark.

REFERENCES: Foslie and Howe 1906a, Howe 1920.

Lithophyllum congestum (Foslie) Foslie

Plants 4–9 cm. tall, 8–16 cm. broad, isolated or gregarious, individually with a limited basal crust about 1 mm. thick, bearing densely crowded branch systems; branches terete or a little compressed below, 1–2 mm. diam., repeatedly irregularly divided, upwardly fastigiate, often more compressed, dilated or lobed, to 3–8 mm. broad, plane or folded, free or complexly anastomosed and highly convoluted; sporangial conceptacles crowded on the upper and outer branches, convex to subconical yet not prominent, about 300 μ diam., with about 20 minute pores; bisporangia 55–75 μ diam., 80–100 μ long.

St. Barthélemy.

REFERENCES: Foslie 1900, 1929, Lemoine 1917; *Goniolithon congestum,* Foslie 1898.

Lithophyllum daedaleum Foslie and Howe

Thalli with a thin adherent primary crust bearing abundant short, crowded, repeatedly subdichotomous, subfastigiate, and sometimes anastomosing branches about 2 mm. thick and to 6 cm. tall, their apices obtuse, truncate, or depressed, or when irregularly dilated and fused, 3–15 mm. diam.; cells of medulla and perithallus cuboidal or subglobose to cylindrical-oblong, 8–14 μ diam., 8–36 μ long, long and short cells often alternating; sporangial conceptacles convex, not prominent, about 300 μ diam.

> V. **pseudodentatum** Foslie and Howe. Branches commonly broadened upward, plano-compressed or subflabelliform, with anastomoses sometimes reaching 2 cm. in breadth.

Puerto Rico, Virgin Isls., Trinidad. On surf-beaten rocks near low-water mark.

REFERENCES: Foslie and Howe 1906a (with the variety), Lemoine 1917.

Lithophyllum platyphyllum (Foslie) Foslie

Plants attached, 5–7 cm. tall, 9–14 cm. broad, densely fastigiately branched, the branches irregularly divided, terete or subcompressed below, occasionally only a little expanded above but more characteristically dilated into broad, complanate, plane or folded anastomosing expanses 1–2 cm. broad, 1.25–2.0 mm. thick; conceptacles on the concave side of the outer branches, slightly convex, 200–300 μ diam.

St. Martin.

REFERENCES: Foslie 1900, 1929; Lemoine 1917; *Goniolithon platyphyllum*, Foslie 1898.

UNCERTAIN RECORDS

Lithophyllum (Dermatolithon) polyclonum Foslie
F. *flabelligera* Foslie

West Indies, the type locality. Foslie 1905b (including f. *flabelligera*), 1909, Lemoine 1917.

Lithophyllum zonale (Crouan) Foslie

Guadeloupe. Type from France. Foslie 1905; *Melobesia zonalis*, Foslie 1900b; *Hapalidium confervicola*, Mazé and Schramm 1870–77, Murray 1889.

Goniolithon Foslie, 1898 [1]

Thallus crustaceous, expanded, calcified, adnate to the substratum, of two layers, the hypothallus of several layers of cells, the perithallus of many layers, scattered heterocysts being present in the perithallus tissues, consisting of isolated, much enlarged cells; tetrasporangia in jug-shaped conceptacles with a prolonged apex, each at maturity with a single pore.

KEY TO SPECIES

1. Crustose throughout 2
1. Sometimes with a persistent crust but with numerous erect branches 6

2. Surface essentially smooth 3
2. Surface verrucose, warted, embossed, or mammillate 4

3. Smooth, subject to irregularities of the substratum; firmly attached conceptacles 300–400 μ diam. .... G. accretum, p. 396
3. Surface irregular, crusts lightly attached, often marginally free, conceptacles 0.6–1.2 mm. diam. .... G. solubile, p. 396

4. Projections, though low, segregated .... G. mamillare, p. 397
4. Projections involving the whole crust and generally coarse, contiguous 5

[1] This treatment does not conform to that given part of the genus by Setchell and Mason (1943), but they did not treat many of the Caribbean species, and it is not practicable for any but a specialist with the opportunity to section and study authentic material to make just disposal of them.

5. Surface minutely tesselate, becoming verrucose or mammillate, the elevations low in comparison with their widths; conceptacles 300–400 μ diam. .... G. boergesenii, p. 397
5. Surface with bosses 2–4 mm. diam., 5 mm. high; conceptacles 200–400 μ diam. .... G. affine, p. 397
5. Surface with bosses 4–10 mm. diam. and high; conceptacles 1.0–1.2 mm. diam. .... G. dispalatum, p. 398
6. Branches much anastomosed, in part terete but distally becoming flattened, forming a more or less chambered mass G. acropetum, p. 400
6. Branches essentially terete or at most compressed 7
7. Plants commonly free; the branches erect in the attached form but on free plants branching very loose, irregular and cervicorn .... G. strictum, p. 399
7. Plant more persistently remaining attached, or if free, the masses dense 8
8. The crustose portion most prominent, the branches simple or subsimple, 0.8–1.0 mm. diam., to 5–10 mm. long G. decutescens, p. 398
8. The crustose portion inconspicuous, the branches abundantly forking, the divisions much anastomosed, 1–5 mm. diam. G. spectabile, p. 399

Goniolithon accretum Foslie and Howe

Thallus light rose pink, forming a thin crust of indefinite shape closely adherent to the substratum and its irregularities, 80–340 μ thick, or to 1.3 mm. by overgrowth increments; cells of the hypothallus subquadrate, 8–14 μ diam., or transversely oblong and then 17–27 μ long; perithallus of roundish or subquadrate cells 4–9 μ diam., or vertically elongated to 10 μ; heterocysts present, 20–25 μ long; sporangial (?) conceptacles 300–400 μ diam.

Florida, Bahamas, Puerto Rico, Virgin Isls. Growing on surf-beaten rocks or pebbles in protected places.

REFERENCES: Foslie and Howe 1906a, Howe 1920, Taylor 1928; *Lithophyllum accretum*, Lemoine 1917.

Goniolithon solubile Foslie and Howe Pl. 76, fig. 3

Thalli purple, irregularly crustose, more or less closely adherent, the crusts 0.2–0.8 mm. thick, often becoming superposed, the surface irregular to nodulose; sporangial conceptacles superficial, conical, apiculate, 0.6–1.2 mm. diam.; tetrasporangia 30–40 μ diam., 70–80 μ long.

Florida, Bahamas, Jamaica, Hispaniola, Puerto Rico, Virgin Isls., Barbados.

REFERENCES: Foslie 1907b, Howe 1920, Taylor 1928, 1933, 1943; *G. notarisii,* Foslie 1900a; *G. propinquum* and f. *imbicilla,* Foslie 1908c; *Lithophyllum propinquum,* Lemoine 1917.

Goniolithon mamillare (Harvey) Foslie

Encrusting thallus suborbicular, lobed, thin; densely beset with mammilliform projections, or with age developing erect terete branches 0.75–1.50 mm. tall; the surface also sometimes showing small easily detached adventitious crusts; hypothallus ill-differentiated; perithallus of cells 7–15 μ diam., 10–20 μ tall; conceptacles large, conical, immersed in the ends of the projections, to 1.4 mm. diam.

Florida, Virgin Isls., Brazil. Growing on rocks.

REFERENCES: Taylor 1928, 1931, Foslie 1929; *Lithothamnion mamillare,* Dickie 1874b; *Melobesia mamillare,* Martens 1870; ?*Porolithon mamillare* v. *occidentalis,* Lemoine 1917.

Goniolithon boergesenii Foslie Pl. 76, fig. 2

Thallus crustose, firmly adherent, smooth or becoming verrucose or mammillate, the surface minutely and regularly tessellated; hypothallus of a single layer of cells about 8 μ diam., 20–25 μ long; perithallus of cells in ill-defined rows, 8–12(–20) μ diam., 7–18(–25) μ tall; conceptacles 300–400 μ diam.

Florida, Bahamas, Hispaniola, Virgin Isls., Barbados, Costa Rica.

REFERENCES: Foslie 1901, Howe 1920, Taylor 1928, 1933; *Porolithon boergesenii,* Lemoine 1917.

Goniolithon affine Foslie and Howe

Thallus crustose, 2–3 cm. diam., about 1 cm. thick, composed of layers 1–2 mm. thick, bearing verrucae or short, often coalescent bosses 2–4 mm. diam., 5 mm. high; hypothallus in section showing cells 9–11 μ diam., 14–25 μ long; perithallus of subquadrate, rounded, or vertically elongated cells 6–9 μ diam., 7–11 μ tall; heterocysts scattered; sporangial conceptacles subconical, 200–400 μ diam., the tetrasporangia 20–25 μ diam., 50–70 μ long.

Puerto Rico. On rocks.

REFERENCES: Foslie 1907, Lemoine 1917.

Goniolithon dispalatum Foslie and Howe

Plants crustose, compressed to subglobose or irregular, 1.5 cm. thick, the crusts becoming superimposed, on the surface forming tubercules and very short subtruncate branches 4–10 mm. diam.; the surface conspicuously exfoliating; hypothallus poorly developed, of cells 11–22 μ diam., 18–40 μ long; perithallus of subquadrate cells 11–18 μ diam., or of vertically elongated cells 9–18 μ diam., 11–25 μ tall; heterocysts 22–29 μ diam., or vertically elongate, 18–29 μ diam., 25–40 μ tall, or horizontally expanded, 12–25 μ diam., 22–36 μ long; sporangial conceptacles conical or depressed-conical, 1.0–1.2 mm. diam.

F. **subsimplex** Foslie and Howe. Thallus simple or nearly so, the branches obsolescent, barely elevated to subglobose, subellipsoid or irregular.

Bahamas.

REFERENCES: Foslie 1908c, Howe 1920.

Goniolithon decutescens (Heydrich) Foslie

Thalli cushion-shaped, forming primary crusts to 0.4–1.0 mm. thick, bearing simple or subsimple, terete or subterete, occasionally anastomosed branches 5–10 mm. long, about 0.8–1.0 mm. diam.; hypothallus of subquadrate cells 12–15 μ diam. or these somewhat elongated, 8–15 μ diam., 12–30 μ tall; cells of the perithallus partly subquadrate, 8–15 μ diam., partly vertically elongate to 24 μ tall; conceptacles 0.8–1.0 mm. diam. seen from the surface.

Bermuda, Florida, Bahamas, Caicos Isls., Hispaniola, Puerto Rico, Virgin Isls., Panama, Tobago. Growing on the roots of mangroves near low-water line.

REFERENCES: Howe 1918a, 1920, Taylor 1928, 1942, 1943. Howe (1920) suggests that *G. rhizophorae* and *G. strictum* v. *nanum*, both Foslie and Howe (1906), may be reducible to synonymy under the present species, but he retains *G. strictum* as clearly independent.

Goniolithon strictum Foslie Pl. 48, fig. 4; pl. 77, figs. 1, 2

Plants branched, generally becoming free, forming very fragile masses 10–14 cm. broad, 6 cm. thick; repeatedly loosely subdichotomous, the irregular branches spreading or often recurved, the divisions 1.0–1.5 mm. diam., the distance between the forks quite great; branches terete or compressed, in attached forms crowded, erect and fastigiate, straight or a little curved, terete, attenuate to the rounded tips; cells of the perithallus as tall as broad or somewhat taller, 12–20 μ diam., 12–25 μ tall; "heterocysts" frequent.

V. **nanum** Foslie and Howe. Branches crowded, subfastigiate, terete or subcompressed, 0.8–1.0 mm. diam., mostly 5–10 mm. tall.

Florida, Bahamas, Cuba, Jamaica, Hispaniola, Puerto Rico, Virgin Isls. Especially in the free state a very conspicuous, distinctive species, but the variety much less so.

REFERENCES: Foslie 1901, 1929, Howe 1918b, 1920, Taylor 1928. For v. *nanum*, Foslie and Howe 1906; *Lithophyllum strictum* v. *nana*, Lemoine 1917, 1924.

Goniolithon spectabile Foslie Pl. 78, fig. 1

Plants abundantly subdichotomously and irregularly branched, ultimately forming cushions 0.5–1.2 dm. diam., the branches tapering from the base, much anastomosed, 1.5–3.0(–5.0) mm. diam., the terminal divisions moderately erect, hardly over 2.0 mm. diam., generally obtuse, ultimately roughened with conceptacles.

F. **nana** Foslie. Cushions small, 2–3 cm. diam., the branches much more densely packed, 0.75–1.00 mm. diam.

F. **intermedia** Foslie. Intermediate between the species and f. *nana* in compactness and diameter of the branches.

Bermuda, Florida, Bahamas, Caicos Isls., Puerto Rico, Netherlands Antilles.

REFERENCES: Foslie 1929 (including the forms). Howe (1920) includes this species with a query under *G. decutescens,* a species Foslie (1929) does not recognize. Since the descriptions differ greatly they are kept separate for the present.

Goniolithon acropetum Foslie and Howe

Plants forming masses 4–12 cm. high, 6–15 cm. diam. but these often confluent; primary crust inconspicuous; erect branches repeatedly subdichotomous, much anastomosed, subterete or compressed, 1.25–3.10 mm. diam., subtruncate to capitate, of flattened divisions 10–25 mm. broad and 1–2 mm. thick above, producing by anastomosis tubular or infundibuliform chambers; young branches conspicuously decutescent, the exfoliations papyraceous, revolute, more or less zonate; medullary cells of the branches 11–20 μ diam., 14–40 μ long; perithallic cells 9–14 μ diam., 11–21 μ long; heterocysts numerous; sporangial conceptacles subconical to mammilliform, 1 mm. diam.; tetrasporangia 40–102 μ diam., 90–168 μ long.

Bahamas, Puerto Rico. Growing on corals and calcareous algae in about 1–4 dm. of water in a sheltered place.

REFERENCES: Foslie and Howe 1906b, Howe 1920, Foslie 1929; *Lithophyllum acropetum,* Lemoine 1917.

Porolithon Foslie, 1909

Plants crustose or branching, generally massive; hypothallus of several layers and perithallus thick; interspersed heterocysts grouped in transverse disks of distinctly large cells or scattered as individual large cells; sporangial conceptacles with paraphyses grouped in the center, the tetrasporangia lateral.

KEY TO SPECIES

1. Plants crustose 2
1. Plants branched .... P. antillarum

2. Perithallic cells 6–9 μ diam.; sporangial conceptacles 150–250 μ diam. .... P. pachydermum
2. Perithallic cells 7–14 μ diam.; sporangial conceptacles 300–400 μ diam. .... P. improcerum

Porolithon pachydermum (Foslie) Foslie

Plants of thin crusts; hypothallus poorly developed, the cells 4–10 μ diam., 8–17 μ long; perithallus compact, the cells 6–9 μ diam., 8–11(–14) μ tall; grouped heterocyst cells 8–16 μ diam., 17–29 μ tall; sporangial conceptacles convex but not prominent, 150–250 μ diam., the tetrasporangia 30–40 μ diam., 60–70 μ long; cystocarpic conceptacles convex, 200–300 μ diam.

F. **nexilis** Foslie and Howe. Cells broader than those of the typical plant, in vertical section 6–10 μ diam., 7–14 μ tall; heterocysts 11–20 μ wide, 22–43 μ tall.

Bahamas, Jamaica, Puerto Rico, Virgin Isls., Barbados. Growing on corallines and rocks.

REFERENCES: Foslie 1909, Lemoine 1917, Howe 1920; *P. oncodes,* Weber and Foslie 1904, Lemoine 1917; *Lithophyllum oncodes* f. *pachyderma,* Foslie 1904; *L. pachydermum,* Foslie 1906. For f. *nexilis,* Foslie 1909.

Porolithon improcerum (Foslie and Howe) Lemoine

Plants crustose, 1.0–1.5 cm. diam.; thin but irregular, 0.3–0.8 mm. thick; hypothallus of a single layer of cells 9–14 μ diam., 11–25 μ long; perithallus of rounded or vertically elongated cells 7–14 μ diam., 9–25 μ tall; heterocysts in small horizontal groups, the cells of which are 11–25 μ diam., 25–36 μ tall; sporangial conceptacles convex-subconical, 300–400 μ diam.; tetrasporangia 20–30 μ diam., 50–80 μ tall.

Bahamas, ?Jamaica.

REFERENCE: Howe 1920.

Porolithon antillarum (Foslie and Howe) Foslie

Plants forming somewhat columnar, rather flat-topped masses 10–30 cm. high, 8–20 cm. broad; the branches much fused, producing below a subsolid mass with lacunae, above somewhat more open, the divisions 3–12 mm. diam., 5–25 μ broad, subconical, compressed or subterete, the apices truncate or retuse, the interstices funnel-shaped or tubular; surface smooth, subpulverulent or marginally minutely corrugated or rugulose; medullary cells 7–10 μ diam., 7–18 μ long, sometimes alternating one short after 2 long cells; perithallic cells subquadrate or roundish, 7–10 μ diam.; heterocysts 14–20 μ diam., 20–33 μ long, grouped in small transverse plates and occurring in both medulla and perithallus; sporangial conceptacles somewhat convex, 150–300 μ diam., tetrasporangia 20–40 μ diam., 38–70 μ long.

Puerto Rico. Growing on a coral reef at low-water mark and evidently contributing to its structure.

REFERENCES: Foslie 1909; *Lithophyllum antillarum*, Foslie and Howe 1906b.

Mastophora Decaisne 1842

UNCERTAIN RECORD

Mastophora lamourouxii (Decaisne) Harvey

Guadeloupe. De Toni 4:1774; Mazé and Schramm 1870–77, Murray 1889, Lemoine 1917.

CORALLINEAE

Plants bushy, often gregarious from a basal crust similar in structure to Melobesieae, the erect axes branched, segmented, the calcified segments with both medulla and cortex or perithallus, but the articulations with little or no perithallus and no lime, although the medullary filaments are more or less thick-walled.

KEY TO GENERA

1. Conceptacles scattered over the surface of the segments AMPHIROA, p. 403
1. Conceptacles marginal or terminal 2
2. Conceptacles central in the tips of the terminal segments at the time of formation, though later often overpassed by the growth of lateral branches 3
2. Conceptacles formed at the extended upper angles, or along the upper margins of flat segments .... CHEILOSPORUM, p. 408
3. Branching dichotomous .... JANIA, p. 412
3. Branching lateral .... CORALLINA, p. 409
3. Branching lateral in sterile parts, dichotomous where fertile ARTHROCARDIA, p. 411

Amphiroa Lamouroux, 1812

Plants bushy, forming tufts or wide loose mats; arising from an inconspicuous basal crust which may early become obsolete; repeatedly dichotomously or irregularly branched, branches segmented, the articulations usually of one, or more often several, transverse series of cells; calcified segments cylindrical or flat, structurally with a medulla of small oval cells alternating with long cells in one or several transverse zones; cortex of a few layers of small rounded cells; conceptacles small, lateral along the faces of the segments, largely immersed yet somewhat projecting, discharging through a terminal pore.

KEY TO SPECIES

1. Segments of the thallus in great part broadly flattened; large, irregularly branched species 2
1. Segments throughout terete or compressed; rather regularly dichotomous species 3

2. Segments compressed to plane, linear to narrowly deltoid, rarely forked; commonly with a distinct midrib and irregular, thin margin .... A. tribulus, p. 406
2. Segments subterete to plane, linear to broadly triangular, when commonly forked; without any midrib, the margin smooth A. hancockii, p. 406

3. Segments terete throughout 4
3. Segments somewhat compressed 5

4. Segments seldom distally forked, slender, the ends commonly with padlike swellings .... A. fragilissima, p. 403
4. Segments often distally unequally forked, coarse, the ends without swellings .... A. rigida v. antillana, p. 404

5. Segments moderately compressed in the upper parts of the plant, hardly exceeding 0.5 mm. in width .... A. beauvoisii, p. 405
5. Segments strongly compressed throughout most of the plant, often exceeding 0.5 mm. in width .... A. brasiliana, p. 405

Amphiroa fragilissima (Linnaeus) Lamouroux Pl. 47, figs. 1, 2

Plants pulvinate-caespitose, or forming extensive mats, 2–4 cm. tall, rather regularly dichotomously branched, the angles rather acute, sometimes trichotomous or with adventitious branches, the segments 150–600 μ diam., forking at the joints, occasionally elsewhere as well, with segments commonly 8–20 times as long as broad, characteristically prominently swollen at the ends; medulla of the segments of 4–8 transverse rows of cells 55–90 μ long, alternating with 1–2 rows of short ones 15–30 μ long; structure of the flexible articulations similar but the cell walls thicker; conceptacles lateral, few to several to a segment, distinctly prominent, 300–340 μ diam., tetrasporangia about 25 μ diam., 50 μ long.

Bermuda, North Carolina, Florida, Bahamas, Cuba, Jamaica, Hispaniola, Puerto Rico, Virgin Isls., St. Barthélemy, Guadeloupe, Dominica, Martinique, Barbados, Grenadines, Grenada, British Honduras, Costa Rica, Panama, Colombia, Netherlands Antilles, Venezuela, Tobago, Brazil. Widespread and often abundant, growing in shallow water near low-tide line, forming isolated clumps in slightly exposed places, or great carpets in very sheltered coves; dredged to about 7 m. depth.

REFERENCES: Harvey 1853, Mazé and Schramm 1870–77 *p. p.*, Hemsley 1884, Hauck 1888, Murray 1889, Collins 1901, Vickers 1905, Sluiter 1908, Grieve 1909, Børgesen 1913–20, 1924, Howe 1918a, 1920, Hoyt 1920, Taylor 1928, 1929b, 1930a, 1933, 1935, 1939b, 1940, 1942, 1943, 1954a, Taylor and Arndt 1929, Sanchez A. 1930, Hamel and Hamel-Joukov 1931, Feldmann and Lami 1937, Quirós C. 1948; *A. cuspidata*, *A. cyathifera* as synonyms in Weber-van Bosse 1904; *A. debilis*, Harvey 1853, Hemsley 1884, Murray 1889, Collins 1901.

Amphiroa rigida Lamouroux v. **antillana** Børgesen

Pl. 47, fig. 3; pl. 48, fig. 1

Plants tufted, strongly calcified, regularly dichotomously branched with but occasional adventitious branchlets, not generally articulated directly at the points of forking, the intervals between the forks shorter in the upper portions than below, the angles generally very wide; branching segments commonly unequally forked at the top and each arm bearing a branch whose articulation may be as much as 1–2 mm. above the point of forking; segments long, cylindrical, very little tapered, 1.0–1.5(–2.0) mm. diam. below, perhaps only half of this in the branchlets, the tips rounded; structurally the medulla of the segments generally showing 2 transverse zones of cells about 100 μ long alternating with one of short cells 20 μ long; articulations of 2 transverse zones of long, thick-walled cells.

Florida, Bahamas, Jamaica, Hispaniola, Virgin Isls., Guadeloupe, Barbados, British Honduras, Panama. Plants of shallow water growing among marine phanerogams, and dredged to 11 m. depth.

REFERENCES: Børgesen 1913–20, Howe 1920, Taylor 1928, 1935, 1942, Taylor and Arndt 1929; *A. arthrocladia*, Mazé and Schramm 1870–77; *A. irregularis*, Mazé and Schramm 1870–77, Murray 1889; *A. rigida*, Vickers 1905, Taylor 1929b.

Amphiroa beauvoisii Lamouroux

Plants forming cushions 2–5 cm. tall, rose-red, the branching dichotomous, not solely at the flexible nodes, especially close in the upper parts, the branches often recurved; lower segments cylindrical, 600–650 μ diam., at least 4–5 times as long as broad; upper segments 400 μ diam., compressed and often cuneate-linear, and bifurcate at the tips, which become rounded and enlarged; medulla of the segments generally showing 2–3 arched bands of long thin-walled cells alternating with one of short cells; articulations almost entirely of medullary cells, with thick walls, in 4 bands.

Brazil. In tide pools, and dredged to a depth of 20 m.

REFERENCES: Piccone 1886, Schmidt 1923, 1924, Joly 1951; *A. exilis*, Martens 1870, 1871, Möbius 1889.

Amphiroa brasiliana Decaisne Pl. 48, fig. 2

Plants to about 5 cm. tall, closely to loosely dichotomously branched at wide angles in small, crowded plants or at narrow angles in tall ones; segments subterete below, strongly compressed above where with parallel sides, or cuneate to distally bifurcate, 0.3–1.5 mm. wide exclusive of the forks, 1–4 mm. long; distal ends of the segments below a node simple, or variously forked; the articulations nearly covered by the cortex of the segments along the lateral margins, usually with 3 medullary zones of successively shorter but relatively long cells topped by one of nearly isodiametric cells, substantially corticated with small cells in anticlinal series, but not calcified; the large calcified segments with medullary zones of (1–) 2–4 layers of long cells repeatedly alternating with one, seldom 2, layers of small ones; conceptacles prominent, often numerous and crowded on the faces of the segments, 270–420 μ diam.

North Carolina, Florida, Guadeloupe, Brazil, Uruguay. This species is very close to *A. beauvoisii* and may not be separable from it, but seems to be potentially a larger plant.

REFERENCES: Martens 1870, Zeller 1876, Hoyt 1920, Howe 1928, Taylor 1930a, b; *A. brasiliana* v. *ungulata*, Mazé and Schramm 1870–77, Murray 1889; *A. fragilissima*, Mazé and Schramm 1870–77 *p. p.; A. ungulata* v. *antillarum*, Schramm and Mazé 1865, 1866.

Amphiroa hancockii Taylor Pl. 47, figs. 6–8

Plants arising from an inconspicuous basal crust, to 8 cm. tall or more, irregularly and divaricately stoutly 2–4-chotomously branched, with numerous adventitious branches; branching complanate to less usually radial; segments smooth, sometimes almost terete, particularly in the lower segments and the adventitious branchlets, more generally uni- or bifacially flat, triangular, or even canaliculate; segments straight or sometimes arcuate, or at the upper end 2–4-fid, so that the arms may be even 2–4 times as long as the undivided part of the segments, and occasionally this division repeated without articulation; terminal branchlets tapering; diameter of segments to 5–8 mm. or more just below a fork, commonly 3–5 mm., length to 30 mm., and, where articulations are omitted between forkings, length to 50–60 mm.; conceptacles crowded, nearly covering the fertile segments, little elevated, 210–270 μ diam., with a single inconspicuous pore.

Hispaniola, St. Lucia, Panama. Plants of deep tide pools growing among corals. Terete flagellar branches may, exceptionally, have segments 70 mm. long.

REFERENCES: Taylor 1942, 1943; ?*A. tribulus*, Taylor 1929b.

Amphiroa tribulus (Ellis and Solander) Lamouroux Pl. 47 figs. 4, 5

Plants loosely bushy, 2.0–6.5 cm. diam., widely ditrichotomously or irregularly branching at the flexible articulations; the segments very irregular in shape, more or less terete throughout, or only short and subterete near the base, but intermediate and terminal segments at other times flattened, oblong to narrowly cuneate, 2–4 mm. broad, with a more or less distinct midrib at least on one face, the thinner margin sometimes minutely erose-dentate; conceptacles numerous, on the faces of the segments.

Florida, Mexico, Bahamas, Cuba, Jamaica, Hispaniola, Puerto Rico, Guadeloupe, Martinique, British Honduras, ?Panama. Occasional in the sublittoral and dredged to a depth of 18 m.

REFERENCES: Mazé and Schramm 1870–77, Murray 1889, Howe 1903c, 1915, 1918b, 1920, Taylor 1928, 1933, 1935, 1943, Taylor and Arndt 1929, Sanchez A. 1930; *A. anceps*, *A. brasiliana* v. *major*, *A. foliacea*, Mazé and Schramm 1870–77, Murray 1889; *A. dilatata*, Murray 1889; *A. galaxauroides*, Mazé and Schramm 1870–77.

UNCERTAIN RECORDS

Amphiroa breviarticulata Areschoug

Dominica. Type from the West Indies. De Toni 4:1810, 6:706 (as *Arthrocardia*); Murray 1889, Grieve 1909.

Amphiroa charoides Lamouroux

Jamaica, Guadeloupe. Type from Australasia. De Toni 4:1810, 6:704 (as *Metagoniolithon*); Collins 1901; *A. verrucosa*, Mazé and Schramm 1870–77, Murray 1889.

Amphiroa crassa Lamouroux

Guadeloupe. Type from Australia. De Toni 6:702; Mazé and Schramm 1870–77, Murray 1889.

Amphiroa dubia Kützing

Guadeloupe. Type from South Africa. De Toni 4:1820; Mazé and Schramm 1870–77, Murray 1889.

Amphiroa nodulosa Kützing

Guadeloupe, Barbados, Venezuela. Type from Colombia. De Toni 4:1821; Mazé and Schramm 1870–77, Murray 1889, Vickers 1905.

Amphiroa variabilis Harvey

Brazil, the type locality. De Toni 4:1817, 6:705 (as *Arthrocardia variabilis*); Martens 1870.

Cheilosporum Areschoug, 1852

Plants erect and bushy, fragile; branching dichotomous or lateral; segmented, the basal segments more or less terete and subcaulescent, those above compressed to flat, obcordate, obsagittate or lunate, simple or lobed; segments firmly calcified, the medulla with alternating transverse zones of small elliptical cells and elongated cells, the cortex tissue of small cells only; the uncalcified articulations consisting of a single series of long, thick-walled cells; conceptacles spherical, immersed in the margins and tips of the upper lobes of the segments, protuberant, with one or two pores.

Cheilosporum sagittatum (Lamouroux) Areschoug Pl. 48, fig. 3

Plants forming dense tufts, to 6–13 cm. tall, fairly regularly and loosely dichotomously branched; segments terete near the base but chiefly much flattened, reversely cuneate to sagittate, somewhat thickened along the axial line; anterior corners projecting at angles of 55°–70° measured from the middle of the segment, somewhat rounded-obtuse and subterete to acute and more spreading; axial length about 1.34 mm.; width across the anterior corners about 1.4 mm., the ratio of length to width about 3:4 in the more spreading examples; conceptacles single on each side below the dorsal margin of each fertile segment, causing a slight notch.

F. **minor** De Toni. Plants to 3 cm. tall.

Brazil.

REFERENCES: Möbius 1889, 1890.

UNCERTAIN RECORDS

Cheilosporum anceps (Kützing) Yendo

Brazil. Type from South Africa. De Toni 4:1823; *Corallina anceps,* Möbius 1889, 1890.

Cheilosporum cultratum (Harvey) Areschoug

Brazil. Type from South Africa. De Toni 4:1831; Martens 1870, Möbius 1890.

Cheilosporum planiusculum (Kützing) Yendo

Brazil. Type from Vancouver Island. De Toni 4:1825; *Corallina planiuscula,* Piccone 1886; *C. planiuscula* f. *antennifera,* Martens 1871, Möbius 1889; f. *normalis,* Möbius 1890; v. *polyphora,* Martens 1871.

Corallina Linnaeus, 1758

Plants with calcified, often confluent crustose basal disks spreading on the substratum; these bases giving rise to an indefinite number of erect axes which are terete to compressed, generally branching in a plane, the branching usually oppositely pinnate; articulated, the joints cylindrical to flattened, the articulations ecorticate, with one zone of long, thick-walled cells; conceptacles formed by the conversion of lateral or terminal pinnules, or occasionally lateral, or occupying branchlets of a single segment lateral to intercalary segments, each with a pore at the apex, simple or bearing hornlike projections.

KEY TO SPECIES

1. Branching irregularly and loosely pinnate, or subverticillate, the divisions attenuate; segments of the main axes terete or nearly so, 1.5–6.0 diameters long .... *C. cubensis*, p. 409
1. Branching regularly and rather closely pinnate, the short segments often in part distinctly flattened 2
2. Small, to 4 cm. tall, plumose, conceptacles horned; commonly epiphytic .... *C. subulata*, p. 410
2. Larger, to 10 cm. tall, complanate, conceptacles generally not horned; commonly epilithic .... *C. officinalis*, p. 410

Corallina cubensis (Montagne) Kützing Pl. 50, figs. 3, 4

Plants tufted, 1–3 cm. tall, densely crowded, the slender main axes irregularly, often oppositely branched; bearing at the upper ends of the segments irregularly pinnate, usually opposite, sometimes verticillate branchlets which may be simple or 1–2 times forked, commonly not lying strictly in a plane; adventitious branchlets also present; axis of segments to 180 μ diam. and 600 μ long, the branchlets of segments to 100 μ diam. below, tapering toward the apex; tetrasporic conceptacles urn-shaped, about 250 μ broad, 450 μ long, the tetrasporangia about 70 μ diam., 180 μ long.

Bermuda, North Carolina, Florida, Texas, Mexico, Bahamas, Caicos Isls., Cuba, Jamaica, Hispaniola, Virgin Isls., Guadeloupe, Martinique, Grenada, Brazil, I. Trinidade. Growing in shallow water between tide levels and to a depth of 26 m. on stones, lithothamnia and coarser algae.

REFERENCES: Harvey 1853, Collins 1901, Børgesen 1913–20, Howe 1918b, 1920, Hoyt 1920, Taylor 1928, 1933, 1940, 1941a, b, 1943, 1954a, 1955a, Taylor and Arndt 1929, Sanchez A. 1930, Hamel and Hamel-Joukov 1931, Humm 1952c, Joly 1953c; *C. ceratoides*, Kützing 1858, Dickie 1884, Murray 1889 *p. p.; Jania cubensis*, Mazé and Schramm 1870–77, Martens 1870, Dickie 1874b, Hemsley 1884, Murray 1889.

Corallina subulata Ellis and Solander Pl. 50, figs. 1, 2

Plants bushy, to 2–4 cm. tall, the main axes dichotomously branched, the divisions conspicuously flat, plumose, the width of each blade 0.9–3.5 mm.; segments at the base of the plant and of the main divisions terete, cylindrical to pyriform, to 240 μ diam.; median segments of the main branches broadly deltoid, costate, commonly with the distal angles truncate, the upper margins bearing 1-several branchlets; these segments to 0.7–1.0 mm. broad, 0.36–0.66 mm. long; lateral branchlets of 1-several segments, which may be simple or forked, the divisions spatulate or terete and terminally attenuate, about 60 μ diam.; conceptacles in the branchlets, ovate to elliptical, smooth or with a hornlike projection even longer than the conceptacle.

Mexico, Bahamas, Jamaica, Hispaniola, Puerto Rico, Guadeloupe, Barbados, Colombia, Brazil. Commonly epiphytic on coarser algae.

REFERENCES: Martens 1870, Dickie 1874a, b, Piccone 1886, Hauck 1888, Murray 1889, Collins 1901, Howe 1920, 1928, Taylor 1933, 1941b, 1943; *Amphiroa brasiliana*, Möbius 1889; *C. ceratoides, C. cuvieri, C. plumifera, C. trichocarpa*, Mazé and Schramm 1870–77, Murray 1889 *p. p.; C. cuvieri* v. *subulata*, Schmidt 1923, 1924.

Corallina officinalis Linnaeus

Plants tufted, to 10 cm. tall, relatively coarse, in the initial stages pinnately much-branched, the main branches in turn to tripinnate, the ultimate pinnae to some extent with cylindrical pinnules; segments in general more cylindrical in the lower parts of the plant, more flattened in the distal parts, to flat-cuneate; conceptacles ovate-subspherical, without horns, except in the spermatangial conceptacles.

?Bermuda, North Carolina, St. Lucia, Brazil, Uruguay. Growing on rocks or shells in the intertidal zone and dredged to moderate depths.

REFERENCES: Martens 1870, 1871, Murray 1889, Schmidt 1923, 1924, Taylor 1939a, Joly 1957; *C. officinalis* v. *fastigiata*, Zeller 1876; v. *mediterranea*, Grunow 1867.

UNCERTAIN RECORDS

Corallina carinata Kützing

Brazil. Type from South Africa. De Toni 4:1854; Zeller 1876.

Corallina granifera Ellis and Solander

Florida. Type from North Africa. De Toni 4:1845; Taylor 1928.

Arthrocardia Areschoug, 1852

Plants bushy, laterally branched in the sterile parts, dichotomously branched where fertile; articulations of one zone of uncalcified cells; segments often terete in the stalklike portions, rather thick and compressed or flat above, sometimes with a median ridge; structurally showing medullary zones of cells of equal length; conceptacles developed at the forward edge of otherwise ordinary segments, single or grouped, rather erect and each with an apical pore.

Arthrocardia stephensonii Manza

Plants 2–4 cm. tall, the branching alternate below, dichotomous in the fertile parts, the lower segments terete, the upper ones 1–2 mm. long and broad, compressed, sometimes with a rather obscure midrib, lobed, the lobes with somewhat acute angles; conceptacles apiculate, single in the middle of the forward margin of each fertile segment.

Brazil. Growing on rocks in exposed places near low-water mark.

REFERENCE: Joly 1957.

UNCERTAIN RECORD

Arthrocardia palmata (Ellis and Solander) Areschoug

Brazil. Type from South Africa. De Toni 4:1825 (as *Cheilosporum palmatum*); Martens 1870. For v. *filicula* (Lamarck) Areschoug, De Toni

4:1826 (as *Cheilosporum palmatum* v. *filicula*); *Corallina filicula,* Zeller 1876.

Jania Lamouroux, 1812

Thallus generally erect from a small basal disk, repeatedly dichotomously branched, the branches segmented, with flexible articulations between the calcified segments composed of one zone of long, eventually thick-walled cells; conceptacles of all types solitary, formed in inflated terminal segments, discharging by an apical pore, but commonly becoming horned, and these projections initiating new branches which continue the upward growth.

KEY TO SPECIES

1. Plants epiphytic, in minute, often depressed tufts hardly 3 mm. diam. .... J. pumila, p. 414
1. Plants much larger 2
2. Branching erect and the habit corymbose, the segments generally 125–200 μ diam., 3–6 diameters long, the branch tips acute .... J. rubens, p. 413
2. Branching wide-angled, spreading 3
3. Segments generally much less than 100 μ diam., 4–6 diameters long; entire habit exceedingly delicate .... J. capillacea, p. 412
3. Segments in the lower parts of the plant generally over 100 μ diam., 2–4 diameters long, somewhat more slender in the ultimate branchlets .... J. adherens, p. 413

Jania capillacea Harvey Pl. 49, fig. 4

Plants erect, capillary, 4–8 mm. tall, the branching regularly dichotomous at very wide angles, the branches sometimes recurved; branches 45–100(30–?150) μ diam., the segments 4–6(–10) times as long as broad; the conceptacles found as flattened swellings at or near the ends of the ultimate branches, opening by a distinct apical pore, frequently bearing 2 hornlike projections, which develop into branches and often in time form additional conceptacles.

Florida, Texas, Bahamas, Cuba, Jamaica, Hispaniola, Virgin Isls., Barbados, British Honduras, Costa Rica, Panama, Colombia, Brazil. Growing on coarser algae in shallow, sheltered locations and dredged to a depth of about 15 m.

REFERENCES: Harvey 1853, Murray 1889, Børgesen 1913–20, Howe 1920, 1928, Hoyt 1920, Taylor 1928, 1929b, 1930a, 1933, 1935, 1939b, 1940, 1941a, b, 1942, 1954a, Taylor and Arndt 1929, Feldmann and Lami 1937; *Corallina capillacea*, Collins 1901.

Jania adherens Lamouroux Pl. 49, figs. 1, 2

Plants erect, capillary, 1.0–3.5 cm. high, branching wide-angled (generally 45°–60° or more), the lower branches often arcuate; branches 100–200 μ diam. below, sometimes less in the uppermost branches, the segments 2–4(–6) diameters long; articulations always present at the base of each branch, and commonly present at intervals between each forking; segments bearing branches often slightly dilated and retuse at the upper end; branch apices conical, acute; conceptacles vasiform, about 240–300 μ broad, 300–340 μ long, laterally horned, eventually developing branches.

Bermuda, North Carolina, Florida, Bahamas, Jamaica, Hispaniola, Puerto Rico, Virgin Isls., St. Barthélemy, ?Guadeloupe, Aves I., Brazil, I. Trinidade. Growing on stones and coarse algae or phanerogams in shallow water, and to a depth of 18 m. or more.

REFERENCES: Martens 1870, Børgesen 1913–20, 1924, Howe 1920, Taylor 1928, 1930a, 1933, 1940, 1943, Taylor and Arndt 1929, Feldmann and Lami 1937, Williams 1949a, Joly 1953c; ?*Jania comosa*, Mazé and Schramm 1870–77, Murray 1889; ?*J. decussato-dichotoma*, Børgesen 1913–20; *J. tenella*, Kützing 1858, Mazé and Schramm 1870–77, Murray 1889.

Jania rubens (Linnaeus) Lamouroux Pl. 49, fig. 3

Plants tufted, erect, sometimes forming extended cushions, rose-red, attached by small discoid holdfasts, the height 2–3(–6) cm.; branching dichotomous, the angles very narrow and the habit very erect and corymbose, the branching tips acute; lower segments of the plants somewhat cask-shaped, often about 2 diameters long, the ordinary segments of the central portions of the plant cylindrical, 125–200(–240) μ diam., 3–6(–9) diameters long, several segments often intervening between points of branching; upper branchlets 65–100 μ diam.; branch-bearing segments broadened to 240–300 μ diam. at the top; tetrasporangial conceptacles at first terminal, vasiform, later horned, eventually intercalary below a fork, with a median protruding ostiole; monoecious, the cystocarpic conceptacles like the sporangial; the spermatangial, however, remaining terminal, lanceolate to fusiform, without marginal horns.

Bermuda, North Carolina, Florida, Bahamas, Caicos Isls., Cuba, Jamaica, Hispaniola, Puerto Rico, St. Barthélemy, Barbuda, Guadeloupe, Grenada, British Honduras, Colombia, Tobago, Brazil, Uruguay. Growing on reefs, large dead corals, coarse algae and other objects in quiet, shallow water, and dredged to about 30 m. depth. This species, like the other larger Janias, is occasionally incorporated in growing sponges and may then have a very confusing appearance.

It is doubtful whether the older records of this species from our area are to be trusted, but the more recent ones are confirmed and the plant is so widespread that its occurrence throughout the area is very probable. Slight annular constrictions of the segments may give the false impression of very short segments, but decalcification will quickly demonstrate the true articulations. Some Uruguayan specimens show dimensions in no way remarkable, but others range to segment diameters of 270–330 μ and lengths of 1.35–2.50 mm. This approaches the f. *longifurca* (Zanardini) Preda (1908, p. 13) which is credited with segments 300–400 μ diam.

REFERENCES: Harvey 1853, Mazé and Schramm 1870–77, Martens 1871, Piccone 1886, Murray 1889, Möbius 1890, 1892, Howe 1920, Schmidt 1923, 1924, Taylor 1928, 1929b, 1935, 1939a, 1942, 1943, 1955a, Taylor and Arndt 1929, Sanchez A. 1930, Joly 1951, 1957; ?*J. longifurca,* Mazé and Schramm 1870–77, Murray 1889; *Corallina rubens,* Hauck 1888, Möbius 1889, Collins 1901, Howe 1918a; *C. rubens* f. *intermedia,* Hauck 1888.

Jania pumila Lamouroux Pl. 49, fig. 5

Plants very small, to 2–3 mm. tall, often decumbent and forming a colony 0.5–1.0 cm. diam.; often with several branch systems from a common basal disk, occasionally simple and limited to a single conceptacle-bearing segment, but more usually dichotomously dividing 2–3 times; axes of short, stout segments which are subcylindrical, about 60–150 μ diam., and 200 μ long, with rounded ends when bearing branches; conceptacles in triangular to cuneate segments, the lateral angles acute to horned, the horns developing into tapering branchlets of 2–4 segments; conceptacular segments very variable, to 250 μ broad, 800 μ long in tetrasporangial and cystocarpic individuals; spermatangial conceptacles thick and spindle-shaped, about 140 μ broad, and 250 μ long.

Bermuda, Florida, Cuba, Jamaica, Hispaniola, Puerto Rico, Virgin Isls., Guadeloupe, British Honduras, Brazil. Occasional on Spatoglosum, Dictyopteris, and Turbinaria, and perhaps other coarse Phaeophyceae, in shallow water; a particularly distinctive species.

REFERENCES: Mazé and Schramm 1870–77, Murray 1889, Børgesen 1913–20, Taylor 1928, 1933, 1935, 1940; Williams and Blomquist 1947; *Corallina pumila*, Hauck 1888, Collins 1901, Howe 1918a; *Jania pygmaea*, Mazé and Schramm 1870–77, Murray 1889.

UNCERTAIN RECORD

Jania fastigiata Harvey

Guadeloupe, Brazil. Type from South Africa. De Toni 4:1854; Mazé and Schramm 1870–77, Murray 1889, Möbius 1890.

GRATELOUPIACEAE

Plants terete, compressed, or foliaceous, alternately, radially, or bilaterally branched; growth multiaxial, the medulla of more or less anastomosed primary filaments often associated with more slender rhizoidal filaments; cortex of branched filaments, the inner portion somewhat loose with cross connections between the cells, the outer part compact, of branched anticlinal cell rows, or the filamentous relation obscured; sporangia tetrapartite, scattered or in projecting nemathecia; carpogenic and auxiliary branches distinct, scattered or in localized areas, formed in the inner cortex on special accessory filament fascicles; carpogenic branches of two cells, the lower forming a vegetative side branchlet; auxiliary cells intercalary in the lower part of each fertile fascicle; cystocarps immersed, causing but little swelling at the surface, showing the enlarged auxiliary cell persistent at the base and a slight development of pericarpial filaments; spore discharge through a definite pore.

KEY TO GENERA

1. Sporangia scattered in the cortex of ordinary branches or at most in small scattered groups 2
1. Sporangia in special marginal leaflets .... CRYPTONEMIA, p. 426

2. Thallus clavate, simple, or occasionally proliferated CORYNOMORPHA, p. 429
2. Thallus branched or foliaceous 3

3. Terete but not complanate, compressed, or flat; soft, almost gelatinous; cortex of a few layers of cells rapidly decreasing in size toward the surface .... HALYMENIA, p. 416
3. Terete and radially branched, or pinnately branched and complanate, or flat; softly fleshy to almost cartilaginous; cortex outwardly of marked anticlinal cell rows .... GRATELOUPIA, p. 423

Halymenia C. Agardh, 1817

Plants of moderate to considerable size, foliaceous or bushy, generally of a gelatinous or softly fleshy consistency; entire or variously lobed or branched; structurally showing in the medulla slender filaments well separated in a soft jelly, often radiating from conspicuous ganglia; the cortex of large cells within, small cells without, not in evident filamentous arrangement; sporangia tetrapartite, scattered and immersed in the cortex; cystocarps immersed, with a pericarp of slender filaments, discharging through a definite pore.

KEY TO SPECIES

1. Plants branched, the branches terete except for a slight flattening below the narrow axils; 1–2 dm. tall, the forkings 2–3 cm. apart .... H. agardhii, p. 417
1. Plants entire or branched, but at least the main divisions flattened 2

2. Pinnately branched or lobed 3
2. Entire, irregularly lobed, or proliferous 4

3. Divisions in several degrees, at least the terminal filiform H. floresia, p. 418
3. Divisions less repeated, the broad ultimate segments strongly serrate .... H. pseudofloresia, p. 418

4. Blades entire, or rarely with a slight tendency to form lobes 5
4. Blades when fully mature markedly lobed, sometimes somewhat dentate or serrate 8

5. Stellate ganglia frequent in the medulla or subcortical tissue 6
5. Stellate ganglia absent or rare 7

6. Blades more gelatinous, nearly round; subcortical cells large, 40–65 μ diam. .... H. rosea, p. 422
6. Blades more membranous, elongated ovate; subcortical cells small, 10–20 μ diam. .... H. vinacea, p. 422

7. Blades thick, 125–300 μ .... H. integra, p. 422
7. Blades thin, 35–40 μ .... H. hancockii, p. 421

8. Surface smooth 9
8. Surface with prominent papillae .... H. duchassaignii, p. 419

9. Blades particularly thin, 60–200 μ thick 10
9. Blades in thicker parts much exceeding 200 μ; cortex usually of more than one layer 11

10. Blades 60–120 μ thick; stellate ganglia 15–65 μ diam. H. bermudensis, p. 419
10. Blades 125–200 μ thick; stellate ganglia 80–180 μ diam. H. echinophysa, p. 420

11. Blades rather small, 4–10 cm. wide; cystocarps protuberant on both sides of the blade; large medullary ganglia conspicuous; superficial jelly easily dispersed from dried specimens .... H. floridana, p. 420
11. Blades larger, to 5–6 dm. wide; cystocarps protuberant on one side of the blade; large medullary ganglia rare, but small subcortical stellate cells common; surface jelly firm H. gelinaria, p. 420

Halymenia agardhii De Toni Pl. 51, figs. 1, 2

Plants bushy, the small base disciform, the stalk slender or obsolete, the thallus quickly enlarging above it, 5–20 cm. tall, firmly gelatinous, 4–12 times dichotomously branched, complanate in habit, the sinuses broad below, narrow above; the segments (2–)4–7 mm. diam., terete above, cuneate or sometimes definitely compressed below, in general decreasing in width from base to apex and a little flattened at the forkings; tetrasporangia scattered; cystocarps small, not projecting.

Bermuda, North Carolina, Florida, Guadeloupe, Tobago. Occasionally found growing on stones in shallow water, but more often at depths of 4–14 m. With age the flattening increases so that in the lower portions the segments may be three times as broad as thick.

REFERENCES: Collins and Hervey 1917, Howe 1918a, Hoyt 1920, Taylor 1928, 1942; *Chrysemenia dichotomoflabellata*, Mazé and Schramm 1870–77, Murray 1889; *Halymenia decipiens*, ?Murray 1889; ?Vickers 1905; Phyc. Bor.-Am. 647.

Halymenia floresia (Clemente) C. Agardh

Pl. 45, fig. 12; pl. 51, fig. 3

Plants large, to 4 dm. long, 3 dm. or more across, very soft, gelatinous, pinnately to subdichotomously branched, the major divisions usually marginally beset with slender linear branchlets; width of main divisions usually 5–20 mm., occasionally to 35 mm., but often very slender and the branchlets abundant, producing a very bushy plant in contrast to individuals with broad main divisions and few branchlets; tetrasporangia scattered, immersed in the cortex, 25–30 μ long.

Bermuda, North Carolina, Florida, Mexico, Jamaica, Hispaniola, Virgin Isls., Guadeloupe, Barbados, Grenada, British Honduras, Colombia, Brazil. Frequently washed ashore, so probably common throughout the area; dredged from depths of 11–40 m.

REFERENCES: Hooper 1850, Harvey 1853, 1861, Mazé and Schramm 1870–77, Dickie 1874a, b, Murray 1889, Collins 1901, Børgesen 1913–20, Hoyt 1920, Taylor 1928, 1930a, 1933, 1935, 1936, 1941b, 1942, 1954a.

Halymenia pseudofloresia Collins and Howe

Plants foliaceous, the blades membranous-gelatinous, when dried dull red, the surface dull and smooth to minutely rugose; stipes 5–10 mm. long, blades cuneate at the base, obovate, suborbicular to very irregular in general shape, in length 7–36 cm., deeply and irregularly to subpalmately lobed, the divisions 1–8 cm. broad, often with stipitate marginal proliferations; lobes or proliferations lanceolate, serrate to subpinnately divided, the teeth acuminate-deltoid; medulla loosely filamentous, the filaments 10–16 μ diam., often obliquely transverse; stellate ganglia infrequent, 20–40 μ diam., cortex and subcortex relatively firm, the subcortical cells ellipsoid, 20–50 μ diam., obscurely anastomosed; cortex 2–6 cells thick, the cells as seen from the surface irregular in shape, 4–13 μ diam. and in section 2–5 times taller than broad; tetrasporangia 12–14 μ diam., 14–26 μ long.

Bermuda, St. Barthélemy. From sheltered places and rather shallow water.

REFERENCES: Collins and Howe 1916, Collins and Hervey 1917, Howe 1918a.

Halymenia duchassaignii (J. Agardh) Kylin Pl. 52, fig. 2

Plants foliaceous, to about 15 cm. tall, divided from near the base into a few irregularly cleft oblanceolate segments 1–4 cm. broad, the margins very irregularly lobed and erose-dentate, the surface strongly papillate.

Guadeloupe.

REFERENCES: Kylin, 1932; *Halymenia floresia* v., Schramm and Mazé 1866; *Kallymenia papulosa,* Mazé and Schramm 1870–77; *Meristotheca duchassaignii,* J. Agardh 1879.

Halymenia bermudensis Collins and Howe Pl. 53, fig. 1

Plants stipitate, the stalk 2–10 mm. long, 0.45–0.75 mm. diam., subterete, sometimes branched, sometimes subrhizomatous; frond foliaceous, with a dull surface, color violet-red when dried, membranous, from a little firm to quite gelatinous, suborbicular, cordate or obovate, simple or copiously lobed with lobes like the initial blade, the margins plane or ruffled, entire to obscurely and coarsely dentate, 4–30 cm. wide, 60–120 μ thick; subcortex 1–3 cells thick, the cells 13–25 μ diam.; stellate ganglia present at least in older portions, 15–65 μ diam., with 3–10 processes; cortex of the frond 1–3 cells thick, the cells subquadrate in section, rather firm-walled, in surface view angular, 5–10 μ diam.; tetrasporangia 8–10 μ diam., 12–16 μ long.

Bermuda, Florida. Plants of shallow water, growing under ledges, in rock clefts, pools, and caves and on Rhizophora roots, where there is shade and protection.

REFERENCES: Collins and Howe 1916, Collins and Hervey 1917, Howe 1918a; *Rhodymenia palmata,* Hemsley 1884, Murray 1889.

Halymenia echinophysa Collins and Howe

Plants foliaceous, membranous to gelatinous, surface dull, color when dry grayish purple; suborbicular, 10–18 cm. broad, 125–200 μ thick, deeply irregularly or subpalmately lobed, the lobes obovate or suborbicular, 2–6 cm. broad, their margins sinuate or erose-dentate; medulla loosely filamentous, the filaments (including the walls) 10–14 μ diam.; cortex submonostromatic, the surface cells rounded or somewhat vertically elongated, 4–8 μ diam., separated by 5–10 μ, the peripheral jelly 10–18 μ thick; subcortex of 2–4 layers of closely connected ovoid-ellipsoid or flattened thick-walled cells, the outer 6–10 μ diam., the inner 25–65 μ diam.; some of the inner subcortical cells 80–180 μ diam., echinate-stelliform, projecting into the medulla and showing 15–40 processes.

Bermuda.

REFERENCES: Collins and Howe 1917, Collins and Hervey 1917, Howe 1918a; *Kallymenia reniformis*, Hemsley 1884, Murray 1889.

Halymenia floridana J. Agardh Pl. 53, fig. 2

Plants foliaceous, texture membranaceous, rose or purplish; blades 5–20 cm. tall, 4–10 cm. wide, borne on a slender stipe a few millimeters long; at first ovate, entire, but later forming numerous ovate lobes, which are cuneate below and taper toward the obtuse apices, finally becoming laciniate, somewhat palmatifid; medulla traversed by many irregularly branched filaments of various sizes, frequently anastomosing and forming numerous conspicuous stellate ganglia; cortex and subcortex 1–4 cells thick, the cells usually of fairly uniform diameter; surface jelly conspicuous, easily dispersed from dried specimens; cystocarps scattered, forming swellings on both surfaces.

North Carolina, Florida, Colombia, Netherlands Antilles. Probably a deep-water plant; dredged from depths of 24–43 m.

REFERENCES: Hoyt 1920, Taylor 1928, 1942; *H. ligulata*, Collins 1906.

Halymenia gelinaria Collins and Howe

Plants foliaceous, fleshy-membranous to very gelatinous, a strong purplish red; attached by a small basal disk, the stipe 3–5 mm.

long; the blades cuneate below, orbicular to obovate, to 5–6 dm. broad, entire or proliferous, the margins lobed, dentate, erose-crenulate, or occasionally laciniate; thickness 60–600 μ; medulla of blade moderately compact to generally quite loose, the filaments mostly 8–14 μ diam., many being obliquely transverse, with a few inconspicuous stellate ganglia with nodes 20–65 μ diam.; the sub-cortex a close network of anastomosing filaments, the stellate nodal cells 13–20 μ diam.; cortex 1–4 cells thick, in surface view the cells 3–10 μ diam.; tetrasporangia scattered, 13 μ diam., 18–26 μ long; cystocarps numerous, 120–240 μ diam., slightly protuberant on one side of the blade.

North Carolina, Florida, Brazil.

REFERENCES: Collins and Howe 1916, Hoyt 1920, Taylor 1928, 1936; *H. floridana*, Phyc. Bor.-Am. 749A, Blomquist and Williams 1947; *H. floridana* f. *dentata*, Phyc. Bor.-Am. 750.

Halymenia hancockii Taylor

Plants with slender cylindrical stipes 1–3 mm. long; blades lanceolate or more generally oblanceolate, the base narrowly tapering, the apex obtusely rounded, the margin entire; generally simple, rarely bifurcate, occasionally with marginal or submarginal proliferations similar to the primary blade, especially after an injury; to about 10 cm. tall, 2 cm. broad, pinkish to dull reddish-purple; thickness 35–40 μ (soaked), with a cortex one cell layer in thickness, the cells about 8–10 μ diam. in surface view and 8 μ tall; the medulla loosely filamentous, conspicuous stellate cells absent, 3–4-branched subcortical cells present, inconspicuous; very slightly nitent when dry; sporangia numerous, scattered between the cortical cells, tetrapartite, 12–17 μ diam. in surface view; pericarps 200–280 μ diam., without an evident pore, the outer wall generally one cell thick, showing a considerable space loosely filled with slender filaments between it and the cystocarp.

Colombia. Apparently a deep-water species, known as dredged from 22–24 m. depth.

REFERENCE: Taylor 1942.

Halymenia integra Howe and Taylor

Plants membranous-gelatinous, dull red, slightly glossy, with a stipe 1–6 mm. long, the plane blade cuneate at the base, broadly lanceolate to obovate, the apex obtuse to acute, 3.5–18.5 cm. long, 1–6 cm. broad, 125–300 μ thick, rarely lobed or sparingly proliferous; medulla loosely filamentous, the walls obscure but the protoplasts 1.0–5.5 μ diam., sparingly anastomosing; subcortex of 2–3 layers of ellipsoid to subglobose cells connected by 8–12 slender processes, those of the outer layer 20–50 μ diam., of the inner 100–180 μ; cortex gelatinous, the outer cells ovoid or angular in surface view, 6–18 μ diam., widely spaced and as much as 5–10 μ apart; surface jelly 8–10 μ thick.

Brazil. Only known from the original dredged collection.

REFERENCE: Taylor 1930a, Howe and Taylor 1931.

Halymenia vinacea Howe and Taylor Pl. 52, fig. 1

Plants foliaceous, membranous or somewhat gelatinous, dull of surface and dull purple in color, very short-stalked, the base of the blade cuneate, the blade in general ovate, the apex somewhat narrowed but obtuse, simple or sparingly lobed, 1.6–8.0 cm. wide, 3.5–18.7 cm. long, 50–150 μ thick; medulla loosely filamentous, the filaments 3–8 μ diam.; numerous stellate ganglia present, 25–50 μ diam. with 5–12 long branches; subcortex of 1–2 layers of small cells 10–20 μ diam. with a few slender subjacent filaments; cortex of about one cell layer, the cells angular in surface view, 6–13 μ diam.; tetrasporangia 14–18 μ diam.

Brazil. Apparently a deep-water plant; only known from the original dredged collection.

REFERENCES: Taylor 1930a, Howe and Taylor 1931.

Halymenia rosea Howe and Taylor

Plants foliaceous, when dried the surface dull and the color dull red; the stalks very short, the blades moderately gelatinous, obovate or suborbicular, entire, 3–4 cm. wide, 4.5–6.0 cm. long, 50–180 μ thick (when young?); stellate ganglia frequent in the medulla or inner cortex, 20–55 μ diam. with 7–16 long projections; subcortex of about one cell layer, the cells ovoid to ellipsoid, 40–65 μ diam., connected by short radiating filaments and with smaller angular cells interspersed; cortex nearly monostromatic, the cells in surface view rounded or somewhat angular, 4–10 μ diam.

Brazil. Apparently a deep-water species.

REFERENCES: Taylor 1930a, Howe and Taylor 1931, Joly 1957.

UNCERTAIN RECORDS

Halymenia angusta (J. Agardh) De Toni

West Indies, the type locality. De Toni 4:1543.

(Halymenia) Gelinaria dentata Crouan

Guadeloupe, the type locality; essentially a *nomen nudum.* Mazé and Schramm 1870–77, Murray 1889. Not the older *H. dentata* Suhr; see Collins and Howe 1916.

Halymenia dichotoma J. Agardh

Bermuda, Guadeloupe. Type from Europe. De Toni 4:1540; Murray 1889; *Chrysymenia dichotoma,* Mazé and Schramm 1870–77.

Halymenia pennata Crouan

Guadeloupe, the type locality; essentially a *nomen nudum.* De Toni 6:540; Mazé and Schramm 1870–77, Murray 1889.

Grateloupia C. Agardh, 1822

Plants firmly gelatinous or fleshy to submembranous, foliaceous to branched, the branches terete to compressed or flat, solid or hollow; branching dichotomous to pinnate or proliferous; structure filamentous, the medulla of anastomosing filaments, the cortex of fascicles appearing as anticlinal rows of cells; sporangia tetrapartite, scattered; spermatangia forming patches over the surface; cystocarps scattered or somewhat grouped, small, immersed in the cortex.

KEY TO SPECIES

1. Thallus much and narrowly branched 2
1. Thallus simple or sparingly branched, the divisions more or less foliaceous and often proliferous 3

2. Branching dichotomous or subdigitate, the branches plane, the habit fastigiate .... G. dichotoma
2. Branching pinnate, bipinnate, or radial, the distal branches often terete, filiform; a very polymorphic species........ G. filicina

3. Blades simple or sparingly divided, generally linear, less often lanceolate, commonly with marginal lobes............. G. gibbesii
3. Blades at first linear-lanceolate, becoming lance-ovate and strongly asymmetrical, the margins undulate .... G. cuneifolia

Grateloupia filicina (Wulfen) C. Agardh Pl. 54, figs. 2, 3

Plants with one or several fronds from a scutate base, reaching a length of 75 cm.; linear, tapering to base and apex, compressed to flat, 2–5 mm. wide, firmly gelatinous, rarely subsimple, commonly marginally pinnate, sometimes proliferating from the face, sometimes apparently radially branched; the divisions linear, about 2 mm. wide, the lower often again pinnate, the upper shorter; tetrasporangia generally several together, immersed in the branches; monoecious, the spermatangia scattered over the surface, developed from the peripheral cells, 4–5 μ diam.; cystocarps also immersed and in groups in the branches, about 180 μ diam.

North Carolina, Florida, Bahamas, Cuba, Jamaica, Hispaniola, Puerto Rico, Virgin Isls., Guadeloupe, Dominica, Martinique, Grenada, British Honduras, Costa Rica, Panama, Colombia, Venezuela, Trinidad, Brazil. Plants of rather sheltered shores, growing in shallow water on rocks, old corals, etc.

REFERENCES: Greville 1833, Harvey 1853, Martens 1870, Mazé and Schramm 1870–77, Piccone 1886, Hauck 1888, Murray 1889, Collins 1901, Vickers 1905, Grieve 1909, Børgesen 1913–20, Hoyt 1920, Schmidt 1923, 1924, Taylor 1928, 1929b, 1933, 1935, 1936, 1942, 1943, 1954a, Humm 1952a, Joly 1957; *Gracilaria ramossissima, Grateloupia aucklandica, G. filicina* v. *bipinnata*, v. *congesta*, v. *elongata, G. lancifera*, and *G. prolongata*, Mazé and Schramm 1870–77, Murray 1889; *G. filicina* v. *filiformis,* Mazé and Schramm 1870–77, Grieve 1909; *Sporochnus pennatula,* Montagne 1863.

Grateloupia dichotoma J. Agardh

Plants gregarious from basal disks, 3–7 cm. tall, the blades simple below, strap-shaped and 3–5 mm. wide, dichotomously or irregularly dividing into flat linear segments about 1–3 mm. broad, often with small marginal proliferations, the terminal segments sometimes divergent, sometimes cervicorn or arcuate; tetrasporangia scattered, embedded in the cortex, about 14 μ diam., 27 μ long; cystocarps 4–6 together in upper segments, about 180 μ diam.

Jamaica, Virgin Isls., Guadeloupe, Costa Rica, Trinidad, Tobago.

REFERENCES: Mazé and Schramm 1870–77, Murray 1889, Collins 1901, Børgesen 1913–20, Taylor 1929b, 1933.

Grateloupia cuneifolia J. Agardh Pl. 54, fig. 4

Plants to 4 dm. tall, from a disciform base, briefly stipitate, the primary base commonly dividing very irregularly into a few similar blades asymmetrically expanded from the cuneate base of the frond, linear, becoming lanceolate to oval or broadly lunate, the margin irregularly undulate and sometimes lacerate, in width to 4–5 cm., often marginally proliferous, the proliferations at first linear-lanceolate; tetrasporangia scattered, formed in the cortical layer, about 19 μ diam., 30 μ long.

Jamaica, Virgin Isls., Redonda I., Guadeloupe, Martinique, Venezuela, Trinidad, Tobago, Brazil, Uruguay.

REFERENCES: Harvey 1853, Mazé and Schramm 1870–77, Martens 1871, Murray 1889, Möbius 1890, Børgesen 1913–20, Taylor 1929b, 1940, 1941b, 1942, Hamel and Hamel-Joukov 1931, Joly 1951, 1957, de Mattos 1952; *G. cutleriae, G. lanceola,* Mazé and Schramm 1870–77, Murray 1889.

Grateloupia gibbesii Harvey

Fronds membranous to fleshy, blackish purple, 15–50 cm. long, simple or scantily divided, the divisions linear to lanceolate, flat, 1–4 cm. broad, attenuate, the margins of the lobes entire to ligulate-proliferous, or the proliferations finely pinnate.

South Carolina, Florida, Guadeloupe, Venezuela. Grateloupias are very polymorphic, and one may suspect that *G. gibbesii* Harvey may be a variant of *G. cuneifolia,* while the *G. gibbesii* of Hoyt's report (1920, pl. 113, fig. 2) may be a broad, marginally proliferous form of *G. filicina.*

REFERENCES: Harvey 1853, Mazé and Schramm 1870–77, Murray 1889, ?Hoyt 1920, Taylor 1928.

UNCERTAIN RECORDS[1]

Grateloupia cutleriae (Binder) Kützing

Dominica. Type from Chile. De Toni 4:1569; Grieve 1909.

Grateloupia furcata Crouan

Guadeloupe, the type locality. De Toni 6:546; Mazé and Schramm 1870–77, Murray 1889.

Grateloupia prolongata J. Agardh

Jamaica, Guadeloupe, Dominica. Type from the Mexican Pacific coast. De Toni 4:1565; Collins 1901, Grieve 1909.

Grateloupia semibipinnata Crouan

Guadeloupe, the type locality. De Toni 6:547 (as *G. semipennata*); Mazé and Schramm 1870–77, Murray 1889.

Grateloupia spinulosa Crouan

Guadeloupe, the type locality. De Toni 6:546; Mazé and Schramm 1870–77, Murray 1889.

Grateloupia subverticillata Crouan

Guadeloupe, the type locality. De Toni 6:546; Mazé and Schramm 1870–77, Murray 1889.

Grateloupia versicolor J. Agardh

Dominica. Type from the Mexican Pacific coast. De Toni 4:1565; Grieve 1909.

Cryptonemia J. Agardh, 1842

Plants stipitate below, above flat, entire, lobed, or palmately or alternately branched with strap-shaped divisions, in the lower portion often with a definite midrib; firm in texture; medulla rather thick, of slender filaments and with many stellate cells; cortex compact, the inner cells larger, the outer small and in short anticlinal rows, the outermost of subrectangular close-placed cells; sporangia tetrapartite; carpogenic branches developed on inner cortical cells,

[1] These Grateloupia names originating with Crouan are essentially *nomina nuda*.

two-celled, usually with one short lateral sterile filament; auxiliary cells in different fasicles near their bases, the several lateral sterile filaments with long terminal cells; cystocarps immersed, each with a scanty pericarp of persisting filaments, and discharging through a pore.

KEY TO SPECIES

1. Midrib absent 2
1. Midrib present at least near the bases of the blades; margin irregular but nearly entire .... C. luxurians, p. 428
2. Blades hardly stalked; margins undulate to clearly dentate C. crenulata, p. 427
2. Blades stalked; margins entire .... C. bengryi, p. 427

Cryptonemia crenulata J. Agardh Pl. 58, fig. 4

Plants briefly stipitate, to 15 cm. tall; much and irregularly dichotomously to palmately divided, the linear to oblong branches without midribs, 0.3–1.5 cm. broad, their margins undulate to dentate, the teeth often forked, sometimes proliferous, the terminal segments indented at the tips.

Bermuda, North Carolina, Florida, Jamaica, Hispaniola, Puerto Rico, Virgin Isls., Guadeloupe, Martinique, Barbados, Colombia, Venezuela, Tobago, Brazil. Perhaps sometimes in shallow water, but also clearly a moderately deep-water species, as dredged from depths of 14–37 m.

REFERENCES: Harvey 1853, Rein 1873, Dickie 1874b, Hemsley 1884, Piccone 1886, Hauck 1888, Möbius 1889, Murray 1889, Collins 1901, Vickers 1905, Børgesen 1913–20, 1924, Collins and Hervey 1917, Howe 1918a, Hoyt 1920, Taylor 1928, 1931, 1941b, 1942, Hamel and Hamel-Joukov 1931, Williams and Blomquist 1947, Joly 1957; *Acrodiscus crenulatus,* Schmidt 1924; *Botryoglossum platycarpum,* Hemsley 1884; *Phyllophora crenulata,* Martens 1870; *Rhodymenia subdentata,* Mazé and Schramm 1870–77, Murray 1889.

Cryptonemia bengryi n. sp. Plate 80, fig. 1

Plants to 14 cm. tall, bushy, wine-red, dull to subnitent when dry; one or more axes arising from a small lobed holdfast, terete and stalklike below for 1–5 cm. and then alternately radially branched to 1–3 degrees, the lesser divisions becoming crowded, these bearing stipitate ecostate blades 2–8 cm. long, the divisions 10–15 mm. broad, 100–195 μ thick, firmly membranous, irregularly lanceolate to oblanceolate, rounded-cuneate at the base, commonly deeply cleft, with 1–5 alternate or terminal and subpalmate lobes, the margins often somewhat thickened, entire or appearing a little erose-subdentate near the tips; branching and proliferation from short terete projections developed near the ends of the smaller divisions of the stalk; cuticle firm, cortex of one cell layer, the cells subangular in surface view and 5.0–8.5 μ diam. with rather thick walls between, rounded-rectangular in section, about 8.0–8.5 μ broad, 10.0–11.5 μ tall; subcortex of transversely oval cells in 2 layers, the inner discontinuous; medulla moderately loose, of a network of filaments 2.5–6.6 μ diam., associated with refractive ganglia and their filamentous extensions, these ganglia not numerous, the bodies about 8.5 μ diam. and with 6–7 radii, or the refractive filaments showing junctions of 3–5 strands without gangliar enlargements; reproductive organs not seen.

Jamaica, ?Brazil. Plants were found growing on the rocks in the surf zone at a depth of somewhat less than a meter.

Cryptonemia luxurians (Mertens) J. Agardh Pl. 58, fig. 3

Caulescent, to 20 cm. tall, the stem denuded and subterete below, alate above, continuing into the blades as a midrib which may be inconspicuous or absent from the upper segments; 3–7 times subdichotomous, dividing into many linear-oblong branches 10–15 (–20) mm. wide, the margins undulate; sometimes abundantly proliferous from the margins and from the stems; tetrasporangia in small marginal leaflets.

Bermuda, Florida, Guadeloupe, Martinique, Barbados, Grenada, Colombia, Trinidad, Brazil.

REFERENCES: Martens 1870, 1871, Mazé and Schramm 1870–77, Dickie 1874a, Piccone 1886, Murray 1889, Gepp 1905b, Vickers 1905, Collins and Hervey 1917, Howe 1918a, Schmidt 1923, 1924, Taylor 1928, 1929b, 1941b, Feldmann and Lami 1937, Williams and Blomquist 1947; *C. lactuca,* Mazé and Schramm 1870–77, Murray

1889; *Nitophyllum platycarpum,* Murray 1889; *Phyllophora ?lactuca,* Greville 1833.

Aeodes J. Agardh, 1876

UNCERTAIN RECORD

Aeodes marginata (Roussel) Schmitz

Guadeloupe. Type from Algeria. De Toni 4:1580; *Schizymenia marginata,* Mazé 1870–77, Murray 1889.

Corynomorpha J. Agardh, 1872

Plants simple, fleshy, clavate from a discoid base, terete or prismatic, sometimes with similar proliferations; medulla loosely filamentous, the filaments anastomosing; cortex of anticlinal fascicles of dichotomous moniliform filaments; sporangia unknown; cystocarps small, grouped in spongy nemathecia, immersed in the swollen distal ends of the thalli.

Corynomorpha clavata (Harvey) J. Agardh Pl. 64, fig. 3

Plants solitary or gregarious, stipitate, dull purple-red, firmly fleshy, the slender stalks 2–5 mm. long, sometimes with a thickened collar, tapering from a single, or rarely forked, terete clavate thallus about 2.0–4.5 cm. long, the distal end obtuse, 3–7 mm. diam.; reproduction unknown.

Bermuda, Florida, Guadeloupe. Growing in the lower intertidal region on exposed rock ledges, and thence to a depth of 4.5 m.

REFERENCES: Taylor 1928, Blinks 1931; *Acrotylus clavatus,* Harvey 1850, Mazé and Schramm 1870–77, Murray 1889.

KALLYMENIACEAE

Plants foliaceous or branched, erect and rather soft; when mature with a filamentous or subparenchymatous medulla and thin, small-celled, assimilative cortex; sporangia scattered or in the branch tips, irregular to tetrapartite; procarps consisting of a supporting auxiliary cell which bears a three-celled carpogenic branch and a sterile cell, or a number of carpogenic branches; gonimoblasts formed from the auxiliary cells or the basal cells of the carpogenic branches, which then have wide connections with the auxiliaries; cystocarps invested and penetrated by a nutritive tissue of filaments developed from the cells near the procarp.

KEY TO GENERA

1. Medulla simply filamentous and the subcortex apparently parenchymatous .... KALLYMENIA, p. 431
1. Medulla of slender rhizoidal and coarse filaments intermixed; short rhizoidal filaments interspersed between the subcortical cells .... CALLOPHYLLIS, p. 430

Callophyllis Kützing, 1843

Plants compressed or foliaceous, more or less gelatinous; growing from an apical meristem; structurally filamentous, the medullary filaments chiefly longitudinal, the primary medullary filaments very coarse and thick-walled, with slender filaments of a rhizoidal type interspersed, the redividing lateral branches from these medullary filaments producing a tissue the inner portion of which consists of large, loosely associated cells interspersed with short-celled filaments of a rhizoidal nature and the outer portion, of anticlinal rows of small chromatophorous cells; sporangia tetrapartite, scattered in the cortex; cystocarps entirely immersed or protuberant, covered by a special thickening of the thallus wall, discharging by one or more pores.

KEY TO SPECIES

1. Segments between the forkings not very much broader near the base of the plant than in the mature upper parts .. C. divaricata
1. Segments below several times as broad as those above .. C. microdonta

Callophyllis divaricata (Greville) Howe and Taylor

Plants bushy, 4–7 cm. tall, firm in texture, repeatedly subdichotomously and rather divaricately branched, the segments 1–4 mm. broad, 0.2–0.5 mm. thick, oblong to linear, sometimes lobed, the tips obtuse.

Brazil.

REFERENCES: Howe and Taylor 1931; *Chondrus divaricatus,* Greville 1833, Martens 1870.

Callophyllis microdonta (Greville) Falkenberg

Plants 5–6 cm., perhaps more, in height, firmly membranous, dark red; arising from a small holdfast with a short, slender stalk, immediately broadening to a flat blade, this blade of branched, strap-shaped segments closely alternately to digitately divided, the divisions 3–12 mm. wide below, 1–3 mm. wide at the tips, the distances between the forks often less than the widths of the segments.

Brazil.

REFERENCES: Taylor 1930a, Taylor and Howe 1931; *Odonthalia microdonta,* Greville 1833, Martens 1870, Möbius 1890.

UNCERTAIN RECORDS

Callophyllis discigera J. Agardh

Guadeloupe, Brazil. Type from South Africa. De Toni 4:277; Martens 1871, Murray 1889; *Rhodymenia discigera,* Mazé and Schramm 1870–77.

Callophyllis laciniata (Hudson) Kützing

Bermuda, Brazil. Type from England. De Toni 4:278; Hemsley 1884, Murray 1889.

Kallymenia J. Agardh, 1842

Plants forming expanded blades, subsimple or broadly lobed, without veins; structurally with a medulla of branched, interlaced, and anastomosing filaments which support externally a subcortex of large polygonal cells and beyond these, smaller, rounded, chromatophore-bearing cells to the surface; sporangia scattered in the cortical layer, tetrapartite; cystocarps wartlike, immersed or projecting; carposporangia formed in large masses, invested by rhizoidal outgrowths from the neighboring cells.

KEY TO SPECIES

1. Plants very small, simple or lobed, or developing into very irregular, sparingly branched, bandlike blades ..... K. limminghii
1. Plants large, simple or lobed, the blades characteristically perforated .... K. perforata

Kallymenia limminghii Montagne Pl. 80, fig. 2

Plants at first with simple or 1–2–branched slender stipes 1–3 mm. long, bearing ovate to kidney-shaped firm, bright red membranous blades 0.5(–1.5) cm. long and nearly as broad, these eventually elongating or by proliferation producing dark red, strap-shaped blades 3–12 mm. wide and to 5 cm. long, which may be simple or bear a few similar branches; margins entire to very strongly and irregularly erose-dentate, sometimes with the teeth becoming secondary haptera or commonly proliferously extended into stipes and blades as in the juvenile phase; mature blades to 150 μ thick, with 1(–2) layers of red cortical cells 5–7 μ diam., 2(–3) layers of nearly colorless subcortical cells 25–30 μ diam., and a substantial medulla of strongly intertwined filaments 7.5–9.5 μ diam.; reproduction not seen.

Bermuda, Florida, Bahamas, Caicos Isls., Guadeloupe. In deep shade under overhanging rocks or in clefts in the rocks at or a little below low-water line, and dredged thence to considerable depths.

REFERENCES: Montagne 1860, Mazé and Schramm 1870–77, Murray 1889, Howe 1920, Taylor 1928.

Kallymenia perforata J. Agardh Pl. 60, fig. 3

Plants large, 1–2 dm. or more in diameter; texture soft, gelatinous, color very pale pink; form irregularly orbicular, reniform or lobed, rather regularly perforate with small holes in young blades, these in part irregular and larger in older plants, to 1.0–2.5 cm. diam.; blades moderately thick, the medulla sparse, of short, mostly transverse filaments widely spaced with scattered round cells and mucilage between them; the subcortex of large, rounded, thin-walled, close-placed cells, these often projecting where connected with medullary filaments, and sometimes stellate with several subcylindrical projections; cortex mostly of a single layer of very small cells; cystocarps scattered between the perforations.

Bermuda, Florida, Virgin Isls. Apparently strictly a deep-water plant; dredged from 30–90 m. Perhaps, as Børgesen suggests, nearer Chrysymenia than Kallymenia.

REFERENCES: Børgesen 1913–20, Collins and Hervey 1917, Taylor 1928.

UNCERTAIN RECORD

(Kallymenia) Meredithia microphylla J. Agardh

Barbados. Type from Europe. De Toni 4:308; Vickers 1905.

GIGARTINALES

Plants showing various forms from filiform to fleshy-membranous or crustose; corticated, with either the multiaxial or central filament types of structure; asexual reproduction by tetraspores formed in sporangia scattered over the plant just below the surface, or in restricted areas on branchlets; sexual reproduction by spermatangia borne on surface cells in more or less restricted areas, and by carpogenic branches originating in the cortex; typical auxiliary cells when present more or less remote from the carpogonia, established before fertilization, consisting of enlarged intercalary cells of the cortex filaments; carpogonium after fertilization producing oöblast filaments which transmit the zygote nuclei to the auxiliaries, from which the carpospore-bearing gonimoblasts are produced.

KEY TO FAMILIES

1. Texture softly fleshy or nearly gelatinous
NEMASTOMATACEAE, p. 434
1. Texture firmer 2
1. Small, parasitic plants .... HYPNEACEAE, p. 464

2. Branches essentially terete 3
2. Axes and main branches compressed or flat 5

3. Branch tips often hooked; axes near the tips showing a single central filament .... HYPNEACEAE, p. 464
3. Branch tips erect and growth multiaxial 4

4. Tetrasporangia zonate; inner medulla of longitudinal filaments
SOLIERIACEAE, p. 455
4. Tetrasporangia tetrapartite; medulla parenchymatous throughout .... GRACILARIACEAE, p. 438

5. Tetrasporangia zonate 6
5. Tetrasporangia tetrapartite 8

6. Plants small, the branches constricted into oval segments
RHABDONIACEAE, p. 461
6. Plants of moderate size, subfoliar or branched in a plane 7

6. Plants of moderate size, slenderly branched, the lesser divisions pinnate .... PLOCAMIACEAE, p. 452
7. Growth multiaxial .... SOLIERIACEAE, p. 455
7. Growth from scattered apical cells along the margins of the blades .... RHODOPHYLLIDACEAE, p. 463
8. In section the outer cells not radially seriate GRACILARIACEAE, p. 438
8. In section the outer cells definitely radially seriate 9
9. Axes terete below, above usually compressed; sporangia in superficial nemathecia .... PHYLLOPHORACEAE, p. 469
9. Axes and branches terete or but little flattened; sporangia in immersed sori .... GIGARTINACEAE, p. 472

CALOSIPHONIACEAE

Plants bushy, very mucous, radially alternately branched; growing from a transversely dividing apical cell, the axial row of cells persistent, more or less enveloped by rhizoids; corticating filaments much branched, whorled on the axis, the inner tissue loose, the outer branches more closely associated; sporangia unknown; carpogenic branch three-celled; cystocarps immersed, the gonimoblasts developed outwardly from the first or auxiliary cell, completely converted to carposporangia; pericarps absent.

Calosiphonia Crouan, 1852

UNCERTAIN RECORDS

Calosiphonia vermicularis (J. Agardh) Schmitz

Guadeloupe. Type from Spain. De Toni 6:569; *Lygistes vermicularis,* Murray 1889; *Nemastoma vermicularis,* Mazé and Schramm 1870–77.

Calosiphonia verticillifera (J. Agardh) Setchell

Bermuda, Florida. Type from Florida. De Toni 6:569; Setchell 1912, Collins and Hervey 1917, Taylor 1928, Kylin 1932.

NEMASTOMATACEAE

Foliaceous or branching plants, rather soft, with a longitudinally filamentous medulla and a compact cortex of radially fasciculate filaments; sporangia tetrapartite; spermatangia developed on the outermost cells of the cortex; carpogenic branches of three cells, developed laterally on the cortical filaments; auxiliary cells scattered, intercalary, developed in inner segments of corticating filaments; the gonimoblasts nearly completely maturing into carposporangia, the cystocarps without a distinct pericarp.

KEY TO GENERA

1. Medullary tissue somewhat calcified and the plants becoming grayish when dried .... TITANOPHORA, p. 437
1. Medullary tissue not at all calcified 2

2. Cystocarps completely immersed, without discharge pores NEMASTOMA, p. 435
2. Cystocarps immersed but sometimes slightly elevating the cortex, discharging through a pore .... PLATOMA, p. 436

Nemastoma J. Agardh, 1842

Plants erect, gelatinous to fleshy, foliaceous, lobed, or branched and the branches more or less terete; apical growth multiaxial, the medulla wide, of slender filaments; cortex more compact, of radial branched chromatophorous filaments, in the inner part intermixed with rhizoid-like threads; sporangia scattered, tetrapartite; cystocarps small and without pores, not projecting; the gonimoblasts arising from the oöblasts near the auxiliary cells, completely maturing into carposporangia.

Nemastoma gelatinosum Howe Pl. 42, fig. 4

Plants very soft and lubricous, light purplish-vinaceous, forming subhemispheric or somewhat flattened bushy tufts 6–16 cm. high, 5–9 times closely subdichotomous, the branching mostly in one plane, sometimes subpalmate, sometimes subpinnately distichous, the branches unequal, mostly 3–6 mm. in diameter, terete or flattened, often small and close near the base, above looser and to 15 mm. broad, slightly tapering, obtuse or subacute; medullary filaments 7–12 μ diam.; cortex of rather divaricate ditrichotomous submoniliform filaments 80–140 μ long, loosely imbedded in mucus and easily separable, the cells mostly obovoid, those of the surface usually 3–6 μ diam., 3–9 μ long.

Bermuda. From just below low-water mark to 3 m. depth in relatively sheltered situations.

REFERENCE: Howe 1918a.

UNCERTAIN RECORD

Nemastoma canariensis (Kützing) J. Agardh

Guadeloupe. Type from the Canary Islands. De Toni 4:1663; *Gymnophloea canariensis,* Mazé and Schramm 1870–77, Murray 1889.

Platoma Schmitz, 1897

Plants gelatinous to fleshy, foliaceous or compressed, lobed, or marginally or pinnately branched; medulla thick, generally of loosely arranged slender filaments more or less interwoven with rhizoids, embedded in a soft jelly; cortex of branched chromatophorous filaments, the inner part loose, penetrated by rhizoidal filaments, the outer more compact, of small cells in anticlinal rows and often with gland cells present; sporangia tetrapartite, scattered in the outer cortex; cystocarps small, scattered, immersed in the thallus and hardly projecting, each with an obscure discharge pore.

KEY TO SPECIES

1. Foliaceous, entire or somewhat lobed .... P. tenuis
1. Fronds plane, irregularly 2–3 times pinnately divided, at least the broader segments flattened .... P. cyclocolpa

Platoma tenuis Howe and Taylor

Blades subsessile, flat, extremely soft and gelatinous, after drying the color purplish-vinaceous and the surface dull; nearly orbicular or subreniform, subentire or sparingly lobed, here and there minutely undulate-crenulate, 1.5–6.5 cm. long or wide, 60–90 μ thick (when imperfectly soaked); medulla rather compactly filamentous, its filaments homogeneous, the rather loose cortex of mostly 3–4 times dichotomous, fastigiate filaments, their cells usually obovoid or pyriform, the superficial ones commonly 3–5 μ long, the inner ones 6–10 μ long; cystocarps embedded, about 50 μ diam., naked or with a few pericarpial filaments.

Netherlands Antilles, Brazil.

REFERENCES: Taylor 1930a, 1942, Howe and Taylor 1931, Williams and Blomquist 1947.

Platoma cyclocolpa (Montagne) Schmitz Pl. 43, fig. 4

Plants forming flat reddish-purple fronds, attached by a small disk, becoming 4–8 cm. tall, at first sessile, subreniform, flat, rather thick, becoming much lobed or most commonly several times branched, the branching irregularly dichotomous or palmate to pinnate, the axils rounded; the divisions when broad appearing crenate or deeply lobed, and when freely branched the compressed to terete segments becoming as long or longer than broad between each fork; axial filaments 8–11(–14) μ diam., cortex of short filaments 2–3 times forked, the outer cells obovoid, about 5–7 μ diam., 7.5–9.0 μ long.

Bermuda, Guadeloupe, Netherlands Antilles. Dredged from a depth of 43 m. on shells or coralline algae, and also found in shallow water.

REFERENCES: Collins and Hervey 1917, Howe 1918a, Taylor 1942; *Nemastoma multifida*, Mazé and Schramm 1870–77, Murray 1889.

Titanophora (J. Agardh) Feldmann, 1942

Plants attached by a basal disk, compressed or foliaceous, irregularly, laterally, or dichotomously branched, whitish red, dull, when dried subverrucose; somewhat calcified in the central tissues; medulla of thick filaments irregularly anastomosing and with scattered branches; cortex of short subdichotomous filaments; gland cells present; cystocarps surrounded by pericarpial filaments, immersed deeply in the cortex, which is hardly if at all elevated, discharging by a pore.

Titanophora incrustans (J. Agardh) Børgesen

Plants foliaceous, when dried the surface appearing roughened, the color becoming grayish green, about 14 cm. tall (incomplete); the primary axis subsimple, to about 2 cm. wide, marginally pinnately branched, the branches similarly foliaceous, with narrowed or even stalklike bases, the axils narrowly rounded; or the blades crowded, strap-shaped, subdichotomous, very variable in width; cells of the cortical filaments short.

Florida.

REFERENCE: Børgesen 1949.

GRACILARIACEAE

Plants branched, even bushy, the branches slender to coarse, terete to strap-shaped, firm and often cartilaginous; axes developing from apical cells, forming a parenchymatous medulla and a narrow small-celled assimilative cortex which may bear delicate colorless hairs; sporangia tetrapartite, scattered, just below the cortex surface; spermatangia scattered, cut off from the surface cells; carpogenic branches of two cells, the auxiliary cell not clearly differentiated, the carpogonium after fertilization merging with cells borne on the same supporting cell and probably equivalent to an auxiliary, and then producing a large fusion cell which gives rise to the gonimoblasts; cystocarp with a sterile basal placenta and a thick projecting pericarp opening by a pore.

KEY TO GENERA

1. Tetrasporangia scattered over the general branch system GRACILARIA, p. 438
1. Tetrasporangia in special branchlets .... CORDYLECLADIA, p. 452

Gracilaria Greville, 1830

Plants usually bushy from a small discoid base, terete or flattened, fleshy to cartilaginous, dichotomously, irregularly, or proliferously branched; sporangia formed just below the surface of the plants, tetrapartite; spermatangia generally cut off from the lining cells of vase-shaped crypts; cystocarps hemispherical, having a large cellular basal placenta tissue, within a prominent superficial pericarp composed of several layers, with the wall of which it may be connected by filaments, the outer cell rows radiating; discharging through a pore.

The species in this genus are notoriously variable. Some of those recorded in our area are hard to distinguish, the early descriptions not covering the same characters in comparable terms, and they are so scarce in collections that preparation of good descriptions is not yet possible. Consequently, an analytical key is difficult to formulate. Proper disposition of the many uncertain records from the descriptions accompanying them is seldom possible, particularly for those reported by Mazé and Schramm. Only in a few cases when authentic specimens could be studied was it possible to secure evidence on which to exercise considered judgment respecting these plants.

So far as possible one should secure a large series of specimens from a population from all variations of habitat and in the most mature stages of growth before identification is attempted. Identification of individual specimens is often impossible, and when there is any suspicion that the plants are fragmentary or that they were dwarfed or otherwise modified by exceptional circumstances under which they grew, it should not be attempted. Nevertheless, after extended experience with the Caribbean flora one should be able to recognize most Gracilarias from the area quite readily.

KEY TO SPECIES

1. Plants decumbent to assurgent, very thick and very irregular, subentire to marginally branched .... G. crassissima, p. 443
1. Plants essentially erect-spreading, foliaceous or bushy 2

2. Branches flat throughout, except the stipe or proliferations 3
2. At least the distal branches terete or compressed 7

3. Branched in several degrees 4
3. Main blades sparingly branched in one or a very few degrees 5

4. Repeatedly branched, the segments (except the basal ones) strap-shaped or oblong .... G. mammillaris, p. 447
4. Segments tending to taper strongly from the center toward both the bases and the tips of the branches .. G. foliifera, p. 446

5. Margins lacerate-dentate above and with many linear proliferations .... G. ornata, p. 446
5. Margins essentially entire 6

6. Segments relatively broad, the margins crisped .. G. cuneata, p. 448
6. Segments narrow, the margins plane .... G. curtissiae, p. 449

7. Terete throughout 12
7. Main axes and axillary regions usually compressed or flattened .. 8

8. Branching primarily dichotomous 9
8. Branching primarily pinnate 11

9. Divisions of all degrees widely divergent, the older with patent filiform proliferations; notably fragile .. G. venezuelensis, p. 448
9. Divisions more or less erect; not notably fragile 10

10. Rather sparingly branched, the divisions erect G. foliifera v. angustissima, p. 447
10. Repeatedly branched, the upper divisions sometimes secund, sometimes incurved; the flattening nearly limited to the regions of the forks .... G. compressa, p. 444

11. Main axes compressed, marginally branched, the branching irregular, loose .... G. cervicornis, p. 445
11. Main axes compressed, narrow; ultimate branching much congested, not remaining in a plane .... G. ferox, p. 444
11. Main axes strap-shaped, branching typically marginal and pinnate, lying in a plane .... G. domingensis, p. 446

12. Branches cylindrical, and at least in part abruptly constricted at the base 13
12. Branches broad at the base, or tapering toward it, but not locally constricted 14

13. Branches few, in one degree, the thick branches often arcuate G. cylindrica, p. 450
13. Branches many, of 2 or more degrees, the branches very slender .... G. blodgettii, p. 449

14. Coarse, moderately branched plants, the distal branching often arcuate and secund 15
14. Abundantly branched species, commonly slender 16

15. Plants cartilaginous, firm; horny when dried ..... G. debilis, p. 442
15. Plants softer, fleshy, shrinking greatly but becoming cartilaginous in drying .... G. usneoides, p. 442

16. Main branching alternate 17
16. Main branching dichotomous; smaller plants copiously branched, the branches congested, almost cartilaginous G. damaecornis, p. 443

17. Plants firm in texture, rather coarse, radially branched, the branchlets often secund on the outside of curved lesser branches .... G. armata, p. 441
17. Large soft or even fragile plants, slender, the branching radially alternate above, often somewhat dichotomous below 18

18. Cortex of 2–3 layers of small cells; spermatangia in crypts; cystocarps with radial trabeculae .... G. verrucosa, p. 441
18. Cortex of 4–6 layers of small cells; spermatangia in superficial sori; cystocarps without trabeculae .... G. sjoestedtii, p. 449

Gracilaria verrucosa (Hudson) Papenfuss Pl. 56, fig. 2

Plant bushy, 1–3 dm. tall, with age often becoming free; texture firmly fleshy, color dull purplish red to purplish, grayish, or greenish translucent; branches 0.5–2.0 mm. diam., repeatedly dividing, alternately or occasionally nearly dichotomously branched, with numerous lateral proliferations, terete throughout, tapering to the ultimate branchlets; cells of the medulla 300–450 μ diam., with rather thin walls; cortex of 2–3 layers of small cells; tetrasporangia numerous, scattered over the branchlets, oval, from the surface 22–30 μ diam., in section 30–33 μ long; cystocarps very prominent, often numerous.

Bermuda, North Carolina, Florida, Texas, Mexico, Cuba, Jamaica, Hispaniola, Puerto Rico, Virgin Isls., St. Barthélemy, Guadeloupe, Martinique, Barbados, Grenada, British Honduras, Venezuela, Trinidad, Brazil. (It is probable that some of these station records should be applied to *G. sjoestedtii* Kylin.) Common in the warm, shallow water of sheltered bays, attached to shells, stones, or other objects; often found in slightly brackish water.

REFERENCES: Papenfuss 1954; *G. confervoides*, Greville 1833, Harvey 1853, 1861, Mazé and Schramm 1870–77, Dickie 1874a, b, *p. p.*, Hemsley 1884, Hauck 1888, Möbius 1889, Murray 1889, Collins 1901, Vickers 1905, Børgesen 1913–20, Collins and Hervey 1917, Howe 1918b, Hoyt 1920, Schmidt 1923, 1924, Taylor 1928, 1929b, 1930, 1933, 1935, 1936, 1941a, b, 1943, 1954a, Sanchez A. 1930; *G. confervoides* v. *capillaris*, Vickers 1905; ?*G. acanthococcoides*, *G. apiculata p. p.*, ?*G. corticata*, ?*G. tuberculosa p. p.*, Mazé and Schramm 1870–77, Murray 1889; *Sphaerococcus confervoides*, Martens 1870; *S. confervoides* v. *setaceus*, Martens 1871; ?*S. divergens*, Zeller 1876.

Gracilaria armata (C. Agardh) J. Agardh

Plants about 3 dm. tall, fleshy-cartilaginous; terete throughout, the elongated main axes radially branched, the branches spreading, in the middle parts about 4–8 cm. long, beset with simple or forked aculeate branchlets, particularly secund on the outside of incurved branches.

Bermuda, Florida, Mexico, Jamaica, Virgin Isls., Grenada, Colombia, Brazil. (These records are to be viewed with suspicion.)

REFERENCES: Harvey 1853, Dickie 1874b, Hemsley 1884, Murray 1889 *p. p.*, Möbius 1890, Schmidt 1923, 1924, Taylor 1941b.

Gracilaria usneoides (Mertens) J. Agardh

Plants bushy, 4–8 cm. tall, fleshy, on drying cartilaginous and shrunken, reddish in color; subdichotomously widely branched, the upper divisions subsecund to divaricate or cervicorn, the young branchlets subdistichous, the tips obtuse.

Virgin Isls., Brazil. Perhaps plants of moderate depths, dredged to 9 m. These are reported to be very similar to *G. debilis*, but softer; perhaps they are not distinct.

REFERENCE: Børgesen 1913–20.

Gracilaria debilis (Forsskål) Børgesen Pl. 45, fig. 10; pl. 57, fig. 3

Plants usually bushy, coarse, reaching a height of 2 dm., pale straw-colored to pinkish in the living state, drying dull grayish brown and becoming horny, not adhering well to paper; abundantly branched, the lower branching irregularly alternate, the subterete segments to 3–6 mm. diam., cervicorn above, the ultimate segments varying from short and acute to flagelliform; in section showing a medulla of large, thick-walled cells, 240–320 μ diam., which gradually grade into the smaller-celled subcortex; surface cells about 5.4–7.5 μ diam., rounded or oval; cystocarps prominent.

Bermuda, Florida, Bahamas, Caicos Isls., Jamaica, Hispaniola, Puerto Rico, Virgin Isls., Guadeloupe, Barbados, Grenada, British Honduras, Venezuela, Brazil, I. Trinidade. Plants of shallow water and from moderately sheltered locations.

REFERENCES: Børgesen 1932; *G. arcuata, G. cartilaginea, ?G. dura, G. obtusa, ?G. secunda, ?G. secundata, G. sonderi, G. squarrosa, G. usneoides,* Mazé and Schramm 1870–77, Murray 1889; *G. cornea,* Mazé and Schramm 1870–77, Dickie 1874a, Piccone 1886, 1889, Murray 1889, Collins 1901, Vickers 1905, Howe 1920, Taylor 1928, 1929b, 1933, 1935, 1936, 1942, Taylor and Arndt 1929, Joly 1953c; *Gracilaria poitei*, Hemsley 1884; *G. wrightii p. p.*, Mazé and Schramm 1870–77, Hauck 1888, Murray 1889, Collins 1901, Vickers 1905, Børgesen 1913–20, Collins and Hervey 1917; *Plocaria complanata p. p.*, ?*P. compressa p. p.*, Mazé and Schramm 1870–77, Murray 1889.

Gracilaria damaecornis J. Agardh Pl. 55. fig. 2

Plant bushy, fleshy-cartilaginous, 7–15 cm. tall; the branches terete, 1.5–4.0 mm. diam., irregularly repeatedly and widely dichotomously forked below, more crowded and often subsecund above, the ultimate segments generally short, erect, blunt or tapering.

Bermuda, Florida, Bahamas, Caicos Isls., Cuba, Jamaica, Hispaniola, Virgin Isls., St. Barthélemy, Guadeloupe, St. Lucia, Barbados, Colombia, Venezuela, Tobago. Plants of shallow water, growing on old corals, stones, etc.

REFERENCES: Murray 1889, Collins 1901, Collins and Hervey 1917, Howe 1920, Taylor 1928, 1929b, 1941b, 1942, 1943, Sanchez A. 1930; *G. circinnata, G. divaricata, G. flabellata*, ?*Plocaria complanata p. p., P. damaecornis*, Mazé and Schramm 1870–77, Murray 1889; ?*G. circinnata*, De Toni 6:262; *G. divaricata*, Hemsley 1884.

Gracilaria crassissima Crouan *ex* J. Agardh

Pl. 55, fig. 4; pl. 57, fig. 4

Plants dull, pale reddish brown to almost colorless when living, shrinking enormously on drying, and losing altogether their habit of growth, 1.0–1.5 dm. diam.; at first irregularly palmate, the lobes extending into marginal branches which are cylindrical to flattened, to 5 mm. thick, to 2 cm. broad, becoming coarsely and closely irregularly alternately rebranched, or somewhat cervicorn, the branchlets tangled together, often recurved; structurally showing a medulla of large, thick-walled cells reaching a diameter of 460 μ, generally about 380 μ, somewhat smaller toward the cortex, which consists of radially compressed cells about 3.6 μ diam. viewed from the surface.

Bermuda, Florida, Bahamas, Jamaica, Hispaniola, Puerto Rico, Guadeloupe. Plants of shallow and of moderately deep water, growing attached on reefs and to rocks at depths of 3 to 11 m.

REFERENCES: Schramm and Mazé 1866, Mazé and Schramm 1870–77, Murray 1889, J. Agardh 1901, Howe 1920, Taylor 1928, 1929, 1933, 1942; *G. horizontalis,* Collins and Hervey 1917, Howe 1918a.

Gracilaria compressa (C. Agardh) Greville

Plants erect to a height of about 1.5–2.0 dm., fleshy rose-purple; branching irregularly dichotomous below, in the main divisions becoming radially alternate, spreading, above sometimes incurved and the divisions secund; in the central portions of the plant and especially below the forks somewhat compressed, the distal segments generally nearly terete, sometimes compressed, hardly tapered to their bases but attenuate to the tips; cells in the medulla thin-walled, to about 300 μ diam.; cystocarps scattered, ovate-hemispherical, the apex sometimes a little produced.

Florida, Texas, Mexico, Jamaica, Virgin Isls., Guadeloupe, Colombia, Brazil. Perhaps a plant of shallow water, but specimens have been dredged from a depth of 5 m.

This name has been applied to *G. verrucosa*-like specimens in which there seems to be a distinct flattening of the axes at the forks and perhaps a little elsewhere. It will be difficult to sharply distinguish between it, *G. verrucosa,* and *G. foliifera* v. *angustissima.*

REFERENCES: Harvey 1853, Dickie 1874a, Murray 1889, Collins 1901, Børgesen 1913–20, Schmidt 1923, 1924, Taylor 1928, 1941a, b; *G. caudata, Plocaria compressa p. p.,* Mazé and Schramm 1870–77, Murray 1889.

Gracilaria ferox J. Agardh Pl. 56, fig. 4

Plants erect, to 8–10(–35) cm. tall, densely branched and bushy, main branches 1.0–1.5 mm., occasionally 2 mm., in diameter, terete or flattened and to 5 mm. broad; branching irregular throughout, pinnate or alternate below to subdichotomous or cervicorn in the short, acute, terete, ultimate segments; medullary cells 250–350 μ diam. in the center, small outwardly; surface cells rounded-angular, 8–20 μ, generally about 14 μ, diam.; cystocarps very prominent, 0.75–1.25 mm. diam., appearing sessile upon, rather than immersed within, the branches.

Bermuda, Florida, Mexico, Caicos Isls., Cuba, Jamaica, Hispaniola, Puerto Rico, Virgin Isls., St. Barthélemy, Guadeloupe, Martinique, Grenada, Costa Rica, Panama, Colombia, Venezuela, Brazil. Plants of shallow water, chiefly about low-tide level, but descending to about 7 m.

REFERENCES: Grunow 1867, 1870, Mazé and Schramm 1870–77, Dickie 1874b, Hemsley 1884, Piccone 1886, 1889, Murray 1889, Collins 1901, Børgesen 1913–20, Collins and Hervey 1917, Howe 1918a, Taylor 1928, 1929b, 1930a, 1933, 1941b, 1942, 1943, 1954a, Taylor and Arndt 1929, Sanchez A. 1930; *G. acanthophora p. p.*, *G. armata* v. *gracilis* and f. *oceanica*, *G. bicuspidata*, *G. cervicornis p. p.*, *G. chondrioides p. p.*, *G. coronopifolia*, *G. curtiramea*, *G. dendroides*, *G. divaricata*, *G. grevillei p. p.*, *G. luteopallida*, *G. patens* v. *gracilis*, *G. poitei*, *G. ramulosa p. p.*, ?*G. secunda p. p.*, *Plocaria aculeata*, Mazé and Schramm 1870–77, Murray 1889.

Gracilaria cervicornis (Turner) J. Agardh

Plants to 1.5–2.5 dm. in length, fleshy-membranous, repeatedly pinnately branched, becoming complanate, the ultimate divisions subdichotomous; older segments somewhat compressed, to about 2–5 mm. broad, the branchlets more terete and but a third as broad; segments often marginally beset with short dentiform projections; cystocarps formed in the more slender segments, more or less marginal, hemispherical, apiculate.

Bermuda, Florida, Mexico, Bahamas, Caicos Isls., Cuba, Jamaica, Puerto Rico, Virgin Isls., St. Barthélemy, Guadeloupe, Martinique, Barbados, Panama, Colombia, Netherlands Antilles, Venezuela, Trinidad, Tobago, Brazil. Common in shallow water on shells, stones, and other objects.

REFERENCES: Harvey 1853, 1861, Mazé and Schramm 1870–77, Dickie 1874a, b, Hemsley 1884, Piccone 1886, 1889, Hauck 1888, Möbius 1889, 1890, 1892, Murray 1889, Collins 1901, Vickers 1905, Howe 1909, 1920, Børgesen 1913–20, Schmidt 1923, Taylor 1928, 1933, 1941b, 1942; *G. cervicornis* f. *acanthophora*, Schmidt 1924; *G. acanthophora p. p.*, *G. bipennata*, *G. cervicornis p. p.*, ?*G. divaricata p. p.*, *G. grevillei p. p.*, *G. ramulosa p. p.*, ?*G. spinescens*, *Plocaria oligacantha*, *P. vaga*, Mazé and Schramm 1870–77, Murray 1889;

Rhodymenia acanthophora, Möbius 1889; *Sphaerococcus acanthophorus*, Kützing 1849, Martens 1870; *S. cervicornis*, Martens 1870.

Gracilaria domingensis Sonder Pl. 57, figs. 1, 2

Plants to 3.5 dm. tall, the main axis sparingly divided, chiefly near the base, the leading axes strap-shaped, usually 4–12 mm. broad, generally bearing numerous lesser, narrow, marginal branches in pinnate fashion, which may redivide; cystocarps chiefly in lesser branchlets which, if small, may be markedly distorted.

Jamaica, Hispaniola, Guadeloupe, Martinique, Grenada, Costa Rica, Colombia, Tobago, Trinidad, Brazil.

These plants have been placed as a variety of *G. cervicornis*, but when typically developed they are so distinct that they are kept separate here. When the secondary branch system is extensive and irregular, the resemblance to *G. cervicornis* becomes close.

REFERENCES: Harvey 1861, Dickie 1874a, Murray 1889, Collins 1901, Butler 1902, Taylor 1941b; *G. cervicornis* v. *domingensis*, Joly 1951; *G. polymorpha*, *Plocaria bipennata*, *P. polymorpha* and vars. *arborescens* and *latifrons*, *P. squarrosa* p. p., Mazé and Schramm 1870–77, Murray 1889.

Gracilaria ornata Areschoug

Plants bushy, to 15 cm. tall, membranous-cartilaginous, dark red; at first dichotomously branched, but above tending to be somewhat pinnate, the flat segments cuneate below, entire, 7–12 mm. broad, above lacerate-dentate and with many proliferations from margins and faces of the blades, these much branched, the divisions linear; cystocarps hemispherical, scattered over the blades.

St. Kitts, Brazil.

REFERENCES: Areschoug 1855, De Toni 1900.

Gracilaria foliifera (Forsskål) Børgesen Pl. 55, fig. 1

Plants sparingly bushy, 10–30 cm. tall, dull purple or faded; the lower part relatively slender, above more coarse, thick, subterete to compressed or expanded and 2–15 mm. or somewhat more in width, sublinear or laciniate, the margin often proliferous; branching of one to several degrees, usually in the plane of the blade, ditrichotomous or alternate; tetrasporangia 20–35 μ diam., 30–45 μ long, formed in the upper mature branches just below the surface; pericarps projecting strongly on faces or margins of the blades.

V. **angustissima** (Harvey) Taylor. Branches subterete, the flattening evident only below the forkings, and usually unaccompanied by much proliferation, the habit being open and fastigiate.

Bermuda, North Carolina, South Carolina, Florida, Texas, Mexico, Bahamas, Caicos Isls., Cuba, Jamaica, Hispaniola, Grenada, British Honduras, Panama, Colombia, Venezuela, Trinidad, Brazil, Uruguay. Common on rocks and shells in quiet, shallow water.

REFERENCES: Børgesen 1932; *Chondrus multipartitus,* Greville 1833a, b; *G. chondroides, G. corallicola, ?G. corticata, ?G. damaecornis* f. *minor, ?G. dentata, G. patens, G. wrightii p.p., Plocaria chondroides, ?P. compressa p.p., P. corticata p.p.,* and v. *chondroides, P. disticha, P. divaricata p.p., ?P. flabelliforme, P. lacinulata, P. multipartita, ?P. tridactylites,* Mazé and Schramm 1870–77, Murray 1889; *Gracilaria lacinulata,* ?Piccone 1886, Børgesen 1913–20, Howe 1920, Schmidt 1923, 1924, Taylor 1928, 1929b, 1935, 1936, 1941a, b, 1943, 1954a, Sanchez A. 1930; ?*G. intermedia,* J. Agardh 1901; *G. multipartita,* Harvey 1853, 1861, Dickie 1874a, b, Hemsley 1884, Möbius 1889, Murray 1889, Piccone 1886, 1889, Collins 1901, Vickers 1905, Collins and Hervey 1917, Hoyt 1920; ?*G. tridactylites,* J. Agardh 1901; *Sphaerococcus multipartitus,* Montagne 1863, Martens 1870 (with v. *aeruginosus* and v. *elongatus*). For the v. *angustissima,* Taylor 1937b, 1941a; *G. mexicana p. p.* and ?*Plocaria durvillaei,* Mazé and Schramm 1870–77, Murray 1889; *G. multipartita* v. *angustissima,* Hauck 1888.

Gracilaria mammillaris (Montagne) Howe Pl. 59, fig. 4

Plants more or less gregarious, the fronds briefly stalked, 0.5–1.0 dm. tall, dull red, firmly fleshy, irregularly dichotomously branched, subflabellate, the segments cuneate below to oblong or linear above, 3–5(–10) mm. broad and quite flat, often rather thin, the sinuses rounded, the terminal divisions obtuse or emarginate; medulla of thick-walled cells 50–125 μ diam., cortex of 1–3 layers of small cells; cystocarps hemispherical, scattered over the blade.

Bermuda, North Carolina, Florida, Cuba, Jamaica, Hispaniola, Puerto Rico, Guadeloupe, Martinique, Panama, Colombia, Netherlands Antilles, Venezuela, Trinidad, Brazil. Plants of rather shallow water, growing on rocks in exposed situations and dredged to a depth of 18 m.

REFERENCES: Howe 1918a, Taylor 1928, 1929b, 1930a, 1935, 1940, 1941b, 1942, 1943, 1954a, Blomquist and Humm 1946; *G. dichotomo-flabellata,* Mazé and Schramm 1870–77, Murray 1889, Collins and Hervey 1917; *Rhodymenia mamillaris,* Mazé and Schramm 1870–77, Hauck 1888, Murray 1889.

Gracilaria venezuelensis Taylor Pl. 55, fig. 3

Plants bushy, involved, at least to 10–13 cm. tall, thin and fragile in texture, color rose to dark reddish purple; axis sparingly alternately or dichotomously branched below, more closely and more definitely dichotomously or polychotomously complanately branched above, locally with numerous subsimple marginal branchlets; the axis somewhat compressed-oval in section in the lower parts, thin and flat above, 2–3 mm. wide below, becoming wider in the middle portion and often particularly so at the forks, but much tapered to the narrow (0.5–1.0 mm.) tips; cystocarps prominent, marginal or facial, 0.75–1.5 mm. diam.

Venezuela. Apparently plants of moderate depths, dredged from 3.5–9.0 m.

REFERENCE: Taylor 1942.

Gracilaria cuneata Areschoug

Plants to 12–18 cm. tall, rosy purple; with a short subterete stalk, expanding into a membranous blade which is cuneate at the base, repeatedly dichotomously cleft above into lobes 5–12 mm. broad, generally with divaricate tips, and somewhat marginally crisped or proliferous; cystocarps scattered, hemispherical.

"West Indies," Venezuela, Brazil.

REFERENCE: Areschoug 1855.

Gracilaria curtissiae J. Agardh Pl. 54, fig. 1

Plants to 4 dm. tall, foliaceous, membranous, the branching dichotomous to polychotomous, the segments sometimes contracted at the base, strap-shaped or lanceolate, to 3.5 cm. broad, 5–22 cm. long, and to 0.75–1.0 mm. thick; structurally showing a medulla of colorless slightly compressed cells reaching 120 μ diam., the subcortical cells also depressed, 30 μ diam. or less; outer cortical cells cylindrical in tiers of 2(–3), 9–13 μ diam.; cystocarps scarcely apiculate, scattered irregularly over the frond.

Florida, Jamaica, Grenada. Perhaps only known from a moderate depth of water; dredged from 6 to 9 m.

REFERENCES: Collins 1901, Taylor 1928.

Gracilaria sjoestedtii Kylin

Thallus lax, 20–55 cm. tall, the terete divisions attenuate, about 1–2 mm. diam. below, but much less in the branchlets; radially freely branched, the main axes often percurrent; cortex of 4–6 layers of small cells; spermatangia in open, superficial sori; cystocarps without special trabecular filaments radiating between the gonimoblasts to the pericarp.

North Carolina, South Carolina, Florida, Jamaica, Hispaniola, St. Barthélemy, Martinique, Venezuela.

REFERENCES: Kylin 1930; *G. confervoides* v. *longissima,* Taylor 1941a, b; *Gracilariopsis sjoestedtii,* Dawson 1953.

Gracilaria blodgettii Harvey Pl. 56, fig. 1

Plants erect, bushy and abundantly branched, to about 2 dm. tall, light pink in color, adhering well to paper when dried; branching radial, alternate, spreading or ascending, repeated to 2–4 degrees, the main axes about 2 mm. diam., the branchlets about 0.5–1.0 mm. diam.; all divisions cylindrical, even the branchlets little tapered except at the tips, but all, or at least the younger, clearly constricted and pedicellate at the base; medulla of very large (to 930 μ diam.) colorless cells with walls little thickened, passing abruptly to an inner cortex of 1–2 layers of much smaller cells, and the outer cortex of a single layer of rounded-angular cells 12–20 μ diam.; tetrasporangia to 30 μ (perhaps more) diam.; cystocarps abundant, projecting considerably from the axis.

Florida, Texas, Jamaica, Hispaniola, Guadeloupe. Plants from deep water, dredged from 11–36 m. depth.

REFERENCES: Harvey 1853, Murray 1889, Collins 1901, Taylor 1928, 1933, 1936, 1941a, Taylor and Arndt 1929; *Cystoclonium difficile,* Mazé and Schramm 1870–77, Murray 1889.

Gracilaria cylindrica Børgesen Pl. 56, fig. 3

Plants erect, to 3 dm. tall, fleshy, rose-red in color, relatively very sparingly and simply branched, less often bushy; terete throughout; the primary axis tapered to a short and slender stipe, but in general cylindrical, 2–4 mm. diam.; branchlets simple, radially alternate, generally long, usually very sharply constricted or pedicellate at the base and occasionally elsewhere, cylindrical, often arcuate and sometimes blunt at the tips, about as thick as the primary axis; structurally showing a very broad medulla of large, rather thin-walled, colorless cells to 450–525 μ diam., with 1–2 somewhat smaller layers outside them, followed by a chromatophore-bearing cortex of about 2 layers of very small cells; tetrapartite sporangia scattered, immersed in the cortex; cystocarps scattered, somewhat projecting.

Florida, Virgin Isls., Guadeloupe, Colombia. A deep-water species, dredged from about 15 to 110 m. depth. From shallower water some specimens were secured perhaps intermediate between this and *G. blodgettii,* but usually the two species are easily distinguished.

REFERENCES: Børgesen 1913–20, Taylor 1928; *G. blodgettii,* Børgesen 1909; ?*G. secundiramea,* Mazé and Schramm 1870–77, Murray 1889; *Plocaria compressa p. p.,* Mazé and Schramm 1870–77.

UNCERTAIN RECORDS

Gracilaria arcuata J. Agardh (*non* Zanardini)

Florida, Virgin Isls. Type from Florida. De Toni 6:259.

Gracilaria caudata J. Agardh

Florida, Mexico, Virgin Isls., Barbados, ?Colombia. Type from the Virgin Islands. De Toni 4:443, 6:259; J. Agardh 1852, Harvey 1861 (with doubt).

Gracilaria chondroides (Kützing) Crouan

?Virgin Islands, Brazil. Type from Brazil. De Toni 4:456, 6:264.

(Gracilaria) Plocaria dactyloides (Sonder) Crouan

Guadeloupe. Type from Australia. Mazé and Schramm 1870–77.

Gracilaria dentata J. Agardh

Jamaica, Virgin Isls., Guadeloupe, Martinique, Barbados, Brazil, Colombia. Type materials from (a) northwestern Africa, and (b) Martinique. De Toni 4:450, 6:265; Vickers 1905, Børgesen 1913–20; *G. rangiferina*, Harvey 1861, Piccone 1886; *Sphaerococcus rangiferinus*, Martens 1870.

Gracilaria divaricata Harvey

Bermuda, Florida. Type from Florida. De Toni 4:455; Harvey 1853, Collins and Hervey 1917.

Gracilaria dura (C. Agardh) J. Agardh

?Florida, Puerto Rico. Type from Spain. De Toni 4:442, 6:256; Hooper 1850, Hauck 1888.

Gracilaria dura Harvey

Colombia, the type locality. A *nomen nudum*. Harvey 1861.

(Gracilaria) Plocaria flagelliformis (?Kützing) Crouan

Guadeloupe. Probably a *nomen nudum*. Mazé and Schramm 1870–77.

Gracilaria krugiana Hauck

Puerto Rico, the type locality. De Toni 4:453, 6:272; Hauck 1888.

Gracilaria latifrons Crouan

Guadeloupe, the type locality. Schramm and Mazé 1865; *Plocaria polymorpha*, De Toni 6:275; ?Mazé and Schramm 1870–77.

Gracilaria lichenoides (Linnaeus) Harvey

Netherlands Antilles. Type from Ceylon. De Toni 4:430, 6:252; Sluiter 1908.

Gracilaria mexicana (Kützing) Crouan

Guadeloupe, Mexico. Type from Mexico. De Toni 6:274; Mazé and Schramm 1870–77 *p.p.*, Murray 1889 *p.p.*

Gracilaria microdendron J. Agardh

Mexico, the type locality. De Toni 6:262; J. Agardh 1907.

Gracilaria prolifica (Kützing) Crouan

Texas, Guadeloupe. Type from Texas. De Toni 6:275; Mazé and Schramm 1870–77, Murray 1889; *Gigartina prolifica*, De Toni 4:228, 456.

Gracilaria ramulosa J. Agardh

Brazil. Type from Australia. De Toni 4:439; Greville 1833a.

Gracilaria salzmanni Bornet

Brazil, the type locality. De Toni 4:453, 6:272; Möbius 1889, Taylor 1931.

Cordylecladia J. Agardh, 1852

Plants filiform, widely branched, subcartilaginous, the medulla of large, elongated cells, the outer layer of small colored cells in anticlinal rows; sporangia tetrapartite, in the cortex of special branchlets; cystocarps subspherical, sessile on the branches.

Cordylecladia peasiae Collins

Plants arising from a more or less distinct crustaceous base, purplish brown, rigid, the branches very slender, dichotomously divided, with occasional scattered or secund, usually quite short, ramuli; sporangia in somewhat swollen and darkened tips of the branches and branchlets; cystocarps globular, sessile along the main branches.

Jamaica.

Examination of the Phycotheca specimen showed a medulla of rounded thick-walled cells and no recognizable cavity, in this differing from Coelothrix. However, there was a tiny cluster of delicate filaments between these cells near the center which, if not a fungal infection, would require that this plant be removed from the Gracilariaceae. A poorly-known species.

REFERENCES: Phyc. Bor.-Am. 791, Collins 1901.

PLOCAMIACEAE

Plants bushy, the branches compressed to flat and membranous, pinnately dividing from the margin to form a flat blade, the lateral branches bearing branchlets in small groups; apical cell and axial cell row evident, the cortex seemingly parenchymatous; tetrasporangia borne on special small branchlets (sporangiophylls), zonate; carpogenic branches, of three cells, borne on supporting auxiliary cells; cystocarps scattered along the margins of the thallus or in special branchlets, without evident pores.

Plocamium Lamouroux, 1813

Characters of the Family.

Plocamium brasiliense (Greville) Howe and Taylor Pl. 59, fig. 2

Plants purplish red, 5–9 cm. (or more?) tall; 2–3 times pinnate, ecostate or lightly costate at the extreme base, the acuminate pinnae in alternating pairs, the lower of each pair falcate, or long-deltoid, entire or lightly toothed on the outer margin, 2–4 mm. long, 0.5–1.5 mm. wide at the base, the other pair pinnate or pinnatifid in much the same fashion as the larger branches, or remaining subsimple; thallus 130–300 μ thick in median parts; superficial cells mostly 10–35 μ max. diam. in surface view, the large cells of medulla and subcortex usually showing through the cortex and 50–125 μ max. diam.; sporangiophylls (immature?) simple or once or twice furcate, ovoid or ellipsoid, obtuse, 150–230 μ long, the sporangia in a double series.

Netherlands Antilles, Brazil. Apparently a plant of deep water.

REFERENCES: Taylor 1930a, 1942; Howe and Taylor 1931, Joly 1951, 1957, de Mattos 1952; *Thamnophora brasiliensis*, Greville 1833, Martens 1870, 1871, Zeller 1876.

UNCERTAIN RECORDS

Plocamium coccineum (Hudson) Lyngbye

Jamaica, Brazil. Type from Europe. De Toni 4:490; Greville 1833, Martens 1870, Zeller 1876, Murray 1889. Möbius 1890, 1892, Taylor 1930a; *P. coccineum* f. *binderiana*, Möbius 1890. It is probable that the Brazilian records are based on slender specimens of *P. brasiliense*, but characteristic material does occur much farther south.

Plocamium membranaceum Suhr

Brazil. Type from South Africa. De Toni 4:593; *P. latiusculum*, Martens 1871.

SPHAEROCOCCACEAE

Plants bushy, divaricately or laterally branched; the branches terete or flattened; medulla with an evident central axis surrounded by a tissue of lesser filaments, or the axis obscure and the surrounding tissue parenchymatous; cortex compact, of larger cells within, smaller ones at the surface; the zonate tetrasporangia scattered in the branch tips or in special branchlets; cystocarps scattered over the thallus in special branchlets, protuberant; gonimoblasts developing outwardly from the large fusion cell, surrounded by a sterile region, only the outer two to four cell layers of the gonimoblast tissue forming carposporangia.

Caulacanthus Kützing, 1843

UNCERTAIN RECORDS

Caulacanthus rigidus Kützing

Brazil. Type from Africa. De Toni 6:153; Zeller 1876.

Caulacanthus ustulatus (Mertens) Kützing

Brazil. Type from Europe. De Toni 4:141, 6:153; *C. fastigiatus*, Martens 1870, Möbius 1889.

Sphaerococcus Stackhouse, 1797

UNCERTAIN RECORDS

Sphaerococcus aculeatus Kützing

Brazil, the type locality. De Toni 4:397, 6:275.

Sphaerococcus dendroides Kützing

Brazil. Martens 1870 (with a Kützing reference: Tab. Phyc. 18, pl. 82 a–c, which bears the name *S. chondroides,* so that the name *dendroides* may simply be an error).

Sphaerococcus dumosus Martius

Brazil, the type locality. Martius 1833, Martens 1870.

FURCELLARIACEAE

Plants bushy, the branches variously disposed, terete to compressed, or foliaceous; structurally showing a filamentous medulla without any central axis, the cortex compact, of larger cells within, smaller ones toward the surface; tetrasporangia zonate, formed in the cortex; spermatangia superficial; carpogonial branches formed on inner cortical cells, composed of three or four cells; auxiliary cells conspicuous, formed on other inner cortical cells before fertilization; cystocarps formed between the cortex and the medulla, most of the gonimoblast cells being matured as carposporangia; pericarps and discharge pores absent.

Halarachnion Kützing, 1843

UNCERTAIN RECORD

Halarachnion ligulatum (Woodward) Kützing

Florida, Barbados, Brazil. Type from England. De Toni 4:1653, 6:540; *Halymenia ligulata,* Hooper 1850, Harvey 1853, Dickie 1874a, Murray 1889, Möbius 1890, Taylor 1928. Probably most, if not all, of these reports were based on Halymenias.

SOLIERIACEAE

Plants plane or bushy, subsimple or branched; in structure the medulla clearly filamentous, the cortex obscurely so, appearing sub-parenchymatous with large cells within; tetrasporangia scattered at the surface, immersed, zonate; carpogenic branches of three or four cells borne on the inner cortex; auxiliaries more or less evident, scattered, borne similarly; cystocarps showing a central mass of sterile tissue, often connected by strands with a filamentous sheath, the carposporangia discharged through a pore.

KEY TO GENERA

1. Plants foliaceous, variously cleft or lobed ..... MERISTOTHECA, p. 460
1. Plants repeatedly branched, the branches terete or flattened 2

2. Plants fleshy, the branches terete or the larger compressed; the cystocarps immersed, although causing some swelling of the fertile branchlets .... AGARDHIELLA, p. 456
2. Plants generally cartilaginous, the branches terete or compressed, often papillose or spiny; the cystocarps generally in papillae .... EUCHEUMA, p. 457

Agardhiella Schmitz, 1896

Plants branching, the branches terete; the medulla filamentous, diffuse, the cortical tissue outwardly compact, small-celled, often with delicate unicellular hairs, the inner cells larger, developing downgrowing rhizoidal filaments which contribute to the medulla; spermatangia formed in patches on young branches; auxiliary cells intercalary in cortical cell series; cystocarps large, in the inner cortex, showing a sterile central tissue from the surface of which the carposporangia are produced, the whole invested by a partly filamentous tissue which is connected with the central mass by strands; the cystocarps discharging by a pore.

KEY TO SPECIES

1. Branching radial, the divisions terete .... A. tenera
1. Branching distichous, the larger divisions compressed A. ramosissima

Agardhiella tenera (J. Agardh) Schmitz

Plants bushy, to about 3 dm. tall, from deep rose-red to pinkish translucent, firm-fleshy, the attachment at first disklike, later fibrous; branches 1–3(–20) mm. diam., terete throughout, 1–3 times alternately radially or somewhat unilaterally branched, branchlets with constricted base and elongate-acuminate apex; tetrasporangia long-oval, zonate, often germinating and producing plantlets *in situ;* cystocarps scattered, evident to the unaided eye, yet little projecting.

North Carolina, South Carolina, Florida, Texas, Mexico, Bahamas, Cuba, Jamaica, Hispaniola, Puerto Rico, Virgin Isls., Guadeloupe, Barbados, Costa Rica, Colombia, Venezuela, Tobago, Brazil. Plants of shallow water and moderate depths, dredged to 30 m.

REFERENCES: Collins 1901, Vickers 1905, Børgesen 1913–20, Howe 1920, Hoyt 1920, Taylor 1928, 1933, 1936, 1941a,b, 1942, 1943, 1954a, Sanchez A. 1930, Feldmann and Lami 1937; ?*Gracilaria tuberculosa, Halymenia ramosissima, Rhabdonia dura* and v. *gracilis,* Mazé and Schramm 1870–77, Murray 1889; *Rhabdonia tenera,* Harvey 1861, Mazé and Schramm 1870–77, Hauck 1888, Murray 1889; *Solieria chordalis,* Harvey 1853, Dickie 1874a, Collins 1901.

Agardhiella ramosissima (Harvey) Kylin Pl. 58, fig. 5

Plants 20–30 cm. tall, fleshy to somewhat gelatinous; attached by a small disk, the axis repeatedly oppositely or sometimes irregularly distichously branched, all but the lesser divisions compressed, the ultimate segments attenuate and more or less terete.

V. **dilatata** J. Agardh. Branches short, to 1 cm. broad, beset with numerous short and slender branchlets.

V. **harveyana** J. Agardh. Axes terete near the base, above more compressed; the lower branches subdistichous, the upper less so and more terete; branchlets short, subfusiform, often with a few spinous projections.

Bermuda, North Carolina, Florida, Puerto Rico, Virgin Isls., Guadeloupe, Venezuela. Dredged from depths of 18–55 m.

REFERENCES: Kylin 1932; *Chrysymenia ramosissima,* Harvey 1853, Mazé and Schramm 1870–77; *Rhabdonia ramosissima,* Hauck 1888, Murray 1889, Børgesen 1913–20, Hoyt 1920, Taylor 1928, 1942. For the v. *dilatata,* Taylor 1928; *Chrysymenia ramosissima* v. *latifrons,* Mazé and Schramm 1870–77; *Rhabdonia ramosissima* v. *latifrons,* Murray 1889.

Eucheuma J. Agardh, 1847

Plants radially or bilaterally branched, terete or compressed, often abundantly papillose; medulla compactly filamentous, or of large cells interspersed with rhizoidal filaments; cortex of large cells within, small ones without, in radial series; cystocarps generally in papillae, laterally projecting, seldom in the smooth parts of a thallus, with a fusion cell or parenchymatous tissue in the center connected with the filamentous pericarp by strands of sterile tissue; carposporangia single or in series of two on the ends of the gonimoblasts.

KEY TO SPECIES

1. Plant cartilaginous, very coarse, relatively sparingly branched, the terete branches usually nodulose or spiny .. E. isiforme, p. 459
1. Plant repeatedly progressively branched 2

2. Branches chiefly strap-shaped, flat, except for marginal spur branchlets 3
2. Branches terete to strongly compressed 4

3. Branching chiefly pinnate, the marginal spur branchlets rather evenly disposed .... E. echinocarpum, p. 458
3. Branching very irregular throughout .... E. schrammii, p. 458

4. Plants relatively fleshy, adhering well to paper; main branches and divisions at most compressed, the terminal divisions elongate, terete .... E. acanthocladum, p. 458
4. Plants subcartilaginous, not adhering well to paper; the ultimate pinnules sometimes densely congested, especially along the margins of the flattened upper branches ... E. gelidium, p. 459

Eucheuma schrammii (Crouan) J. Agardh

Plants membranous when young, cartilaginous when old, complanate-spreading, subdichotomously or palmately branched with broadly rounded axils, the lower segments broad, the outer divisions narrower or even linear, but altogether irregular in all parts of the plant; cystocarps hardly projecting from the surface, but the overlying tissue often apiculate.

Virgin Isls., Guadeloupe.

REFERENCES: Kylin 1932; *Mychodea schrammii,* Mazé and Schramm 1870–77, Murray 1889.

Eucheuma echinocarpum Areschoug

Plants when young rather gelatinous, becoming thick and firmer toward maturity; flat throughout and complanate, the primary branching subpalmate, these divisions 1–4 dm. long, 5–15 mm. broad, pinnate, with rather few chief secondary branches, but ultimately many marginal dentate spur branches rather evenly disposed, 3–20 mm. long.

?Florida, ?Jamaica, Guadeloupe, Brazil.

REFERENCES: Möbius 1889, ?Collins 1901.

Eucheuma acanthocladum (Harvey) J. Agardh Pl. 50, fig. 5

Plants large, reaching a height of 3 dm. or more, and widely spreading, the color pale reddish, general texture firmly fleshy, but not cartilaginous; main axis and branches slightly flattened in section, or becoming oval or terete distally; branching usually abundant, but not crowded, to 5, 6, or higher orders; irregularly alternate to somewhat dichotomous in appearance, often distinctly secund, the ultimate divisions frequently cervicorn, sometimes flagelliform.

Florida, Texas, Barbados. Attached to rocks in shallow water and dredged to a depth of 9 m.

REFERENCES: Murray 1889, Taylor 1928, 1941a, b; *Chrysymenia acanthoclada,* Harvey 1853.

Eucheuma gelidium (J. Agardh) J. Agardh

Plants subcartilaginous, the erect complanate fronds to 10–12 cm. tall, with a main axis sparingly divided, terete to flat, the main divisions 3–5 mm. broad, repeatedly pinnate, the intermediate divisions especially flattened, marginally subentire to densely spinulose-proliferous, the proliferous terete branchlets forked and in typical material congested.

Bermuda, North Carolina, Florida, Puerto Rico, Guadeloupe, Barbados, Panama, Brazil.

REFERENCES: Mazé and Schramm 1870–77, Hauck 1888, Murray 1889, Vickers 1905, Collins and Hervey 1917, Howe 1918a, Hoyt 1920, Taylor 1928, 1930a, 1936, Kylin 1932; *Mychodea guadelupensis, M. pennata, M. polycantha,* and *Nemastoma jardini* v. *antillarum,* Mazé and Schramm 1870–77, Murray 1889.

Eucheuma isiforme (C. Agardh) J. Agardh

Pl. 45, fig. 11; pl. 50, figs. 6, 7

Plants large, to 3–5 dm. or more in height, harsh and cartilaginous in texture, color usually pale straw, yellowish brown, or reddish; branching sparse to ample, very irregular, apparently radial to complanate, of one to about 5 orders, the main branches similar to and often equaling the axis, commonly tapering, terete throughout, 3–8 mm. diam., smooth or more often nodulose or conspicuously spinose, the projections scattered or, especially when spinose, whorled; axis showing a slender medulla of close-placed thick-walled filaments, the inner cortex broad, of large thick-walled cells, the outer of smaller radially disposed and elongated cells; tetrasporangia in the outer cortex of the main branches and in the spinous projections; cystocarps in the spinous branches only, large and hemispherical, with a distal pore.

Bermuda, Florida, Mexico, Bahamas, Cuba, Hispaniola, Virgin Isls., St. Barthélemy. Generally met with in sheltered places and shallow water, but also dredged from a depth of 55 m.

REFERENCES: Harvey 1853, Dickie 1876b, Hemsley 1884, Murray 1889, Børgesen 1913–20, Howe 1918a, 1920, Taylor 1928, 1933, 1941b; *E. denticulatum*, Collins and Hervey 1917; ?*E. spinosum*, Mazé and Schramm 1870–77, Murray 1889; *Hypnea wurdmannia*, Hooper 1850.

UNCERTAIN RECORDS

Eucheuma nudum J. Agardh

Guadeloupe. Type locality not known. De Toni 4:368; Mazé and Schramm 1870–77, Murray 1889.

Eucheuma speciosum (Sonder) J. Agardh

Grenada. Type from Australia. De Toni 4:375; Murray 1889.

Meristotheca J. Agardh, 1872

Plants foliaceous, palmately cleft or irregularly lobed; medulla loose, filamentous; cortex pseudoparenchymatous, the inner of larger cells, the outer of smaller cells in anticlinal rows; cystocarps with a sterile central parenchyma, surrounded by the active gonimoblasts which in turn are surrounded and separated into segments by filamentous pericarp tissue.

Meristotheca floridana Kylin

Plants foliaceous, rose-red, fleshy, to 1.2–3.5 dm. tall, 4–5 cm. broad, usually somewhat lobed to irregularly or palmately cleft; the segments oblong to cuneate, 0.5–3.0 cm. broad, marginally conspicuously erose to repeatedly laciniate-dentate, the surface smooth or with many dentate or laciniate prominences; medulla of loosely anastomosing filaments; subcortex of larger cells, outer cortex of small cells in anticlinal, sparingly forking rows; sporangia zonate, scattered in the cortex of smooth plants; cystocarps elevated, even as stalked spherical papillae in some individuals, in others in thorn-shaped projections.

Bermuda, North Carolina, Florida, Bahamas, Barbados, Panama. Apparently a plant of deep water, as only known from specimens washed ashore.

REFERENCES: Kylin 1932; *M. duchassaignii*, J. Agardh 1879 *p.p.;* Vickers 1905, Howe 1910, 1915, 1920, Collins and Hervey 1917, Hoyt 1920, Taylor 1928.

RISSOËLLACEAE

Plants foliaceous, irregularly lobed or cleft; medulla loosely filamentous; radial cortical filaments repeatedly forked, the inner cortex loose, the outer more compact; tetrasporangia zonate, scattered and immersed; procarps consisting of an auxiliary cell and several two- or three-celled carpogenic branches on the same supporting cell; gonimoblasts primarily growing inwards, producing prominent cystocarps with large fusion cells in the middle surrounded by groups of seriate carposporangial cells; cystocarps discharging through an evident pore.

Rissoëlla J. Agardh, 1849

Characters of the Family.

UNCERTAIN RECORD

Rissoëlla denticulata (Montagne) J. Agardh

Barbados. Type from Peru. De Toni 6:228; Dickie 1874a, Murray 1889.

RHABDONIACEAE

Plants cylindrical or flattened, radially or bilaterally branched; medulla filamentous, cortex of close, radial rows of cells, the inner large, the outer small and rich in chromatophores; tetrasporangia zonate, scattered over the thallus in the outer cortex; carpogenic branches two- to five-celled, the auxiliary cell developed after fertilization; cystocarps immersed in the thallus, the carposporangia radiating from a large lobed fusion cell, the elevated overlying cortex acting in place of a specialized pericarp, the cystocarp discharging through a pore.

Catenella Greville, 1830

Plants creeping or assurgent, branched, the articulate branches of terete or compressed spindle-shaped or broadly oval segments, attaching to the substrate or other branches by haptera at intervals; medulla of a loose network of slender filaments around an obscure primary filament; cortex of anticlinal rows of cells, the inner larger, the outer smaller; tetrasporangia appearing in specialized branchlet segments; cystocarps usually solitary in shortened and much thickened terminal segments.

KEY TO SPECIES

1. Haptera formed from normal though somewhat attenuate segments of the thallus .... C. impudica
1. Haptera formed as more or less flagellar outgrowths from points of branching, but not as regular branch segments .. C. repens

Catenella impudica (Montagne) J. Agardh

Plants dull violet in color, creeping, complanate-expanded, 3–4 cm. tall; ditrichotomous, the younger segments slender, with the terminal ones more terete and acute; the older segments broader and more flattened, 2.2 mm. broad, 4–5 mm. long, elliptical-oblong; hapteral attachment by the flattened ends of elongate terminal thallus segments projecting in the axils of branch forks and in the same planes; sporangia in branches, each of a single obovate segment, forming a transverse zone in the middle.

Guadeloupe, Costa Rica, British Guiana, Surinam, French Guiana, Brazil. Growing on mangrove roots and shells in the intertidal zone.

REFERENCES: Montagne 1850, Möbius 1885, 1889, Murray 1889 *p. p.*, Post 1936.

Catenella repens (Lightfoot) Batters Pl. 66, fig. 13

Plants creeping or pulvinate, dull purple, to 3 cm. tall, with repent and assurgent branches, the haptera terminating uncorticated flagellar outgrowths chiefly formed at points of forking and not in the plane of branching of the thallus; branching ditrichotomous below but clearly pinnate above, the axis and branches divided into dorsiventrally compressed, ellipsoid to ovate segments 3–5 times longer than broad; sporangial segments pointed; cystocarpic segments obovate to subspherical.

Bermuda, Florida, Bahamas, Cuba, Jamaica, Virgin Isls., Tortola I., Guadeloupe, Martinique, British Honduras, Panama, Brazil. Growing on rocks and old corals, wharves, and mangrove or other tree roots within the intertidal zone and in rather sheltered places. When growing in brackish water the segments become nearly terete.

REFERENCES: Feldmann and Lami 1936, Taylor 1942, Joly 1951, 1957; *C. opuntia,* Børgesen 1913–20, Howe 1920, Hamel and Hamel-Joukov 1931, Post 1936; *C. opuntia* v. *pinnata,* Murray 1889, Collins 1901, Collins and Hervey 1917, Howe 1918a, Taylor 1928, 1935; *C. pinnata,* Harvey 1853, Hemsley 1884.

RHODOPHYLLIDACEAE

Plants plane or bushy, subsimple or branched; when mature the medulla either diffusely filamentous or apparently parenchymatous, the cortex of larger cells within, smaller ones without; procarps present, the carpogenic branches of three cells, each with an auxiliary borne on the same supporting cell; cystocarps developed from the enlarged auxiliaries within a pericarp formed by division of the neighboring cortex cells.

Cystoclonium Kützing, 1843

UNCERTAIN RECORDS

Cystoclonium difficile (C. Agardh) J. Agardh

Brazil, the type locality. De Toni 4:315; *Sphaerococcus difficilis,* Martens 1870.

Cystoclonium purpureum (Hudson) Batters

Cuba. Type from England. De Toni 4:314; *Sphaerococcus purpurascens,* Montagne 1863.

Rhodophyllis Kützing, 1847

Plants membranous, the dichotomous divisions abundantly proliferous at the margins; the medulla of branched filaments simulating veinlets, sometimes associated with large angular cells, the primary cortex of one layer of large cells on each face; tetrasporangia scattered, zonate, superficially immersed in the blade or in marginal proliferations; plants monoecious, the spermatangia scattered over the surface; procarps present, the three-celled carpogenic branches borne on a primary cortical cell series consisting of three cells, the second being the auxiliary which lies close to the carpogonium, the gonimoblasts producing carposporangia terminally; the cystocarps prominent, covered by the thickened overlying cortex acting as a pericarp, but lacking a pore.

Rhodophyllis gracilarioides Howe and Taylor

Plants to about 7 cm. tall, the stipitate blades reddish, cuneate from a subterete base, plane, palmate below, subpinnate above, 150–250 μ thick (when soaked), the main segments 3–6 mm. broad, the veins very obscure or invisible, the ultimate lobes lanceolate, deltoid, or subterete, acuminate, sometimes proliferous; the surface in older parts obviously reticulate, the meshes much larger than the obscure ones formed around the larger cortical cells; medulla of 2–4 rather irregularly disposed layers of large cells 50–260 μ diam. in cross section; cortex submonostromatic, its cells partly angular, chiefly 15–26 μ in longer diameter in surface view, the subcortical cells 35–55 μ diam., the outer walls firm, 8–20 μ thick.

Brazil. The type collection dredged and probably from deep water, but also reported from the littoral.

REFERENCES: Taylor 1930a, Howe and Taylor 1931, Williams and Blomquist 1947.

Calliblepharis Kützing, 1843

UNCERTAIN RECORD

Calliblepharis jubata (Goodenough and Woodward) Kützing

Brazil. Type from England. De Toni 4:466; Piccone 1886, 1889.

HYPNEACEAE

Plants bushy, laterally branched, branches terete; structurally showing a persistent central filament developed from an apical cell, and a cortex which matures into a pseudoparenchymatous tissue with large cells within, small cells without; tetrasporangia more or less localized in somewhat swollen branchlets, formed at the surface, zonate; procarps present, the carpogenic branches of three cells on a supporting cell which forms additional corticating series, an inner cell of these forming an auxiliary near the carpogonium; cystocarp swollen, showing a group of elongate cells anchoring the central mass to the pericarp, the carposporangia formed terminally on close-branched gonimoblasts.

KEY TO GENERA

1. Plants small to moderate in size, usually bushy ..... HYPNEA, p. 465
1. Plants very small parasites upon Hypnea ..... HYPNEOCOLAX, p. 469

Hypnea Lamouroux, 1813

Plants bushy, virgate or spreading; branches slender, terete, often with spinulose branchlets and crozier tips; tetrasporangia terminal on the corticating cell series, becoming immersed, zonate, grouped in the swollen parts of special short lateral branchlets which are often fusiform; cystocarps each enclosed in a hemispherical poreless pericarp.

KEY TO SPECIES

1. Some branches with evident thickened crozier tips H. musciformis, p. 467
1. Branches with erect or slightly curved but not hooked tips 2

2. Stems spinulose below, at least in part the spines stellate or cornute .... H. cornuta, p. 467
2. Stems smooth or if spiny, the spines simple 3

3. Plants forming extensive, rather dense mats ..... H. spinella, p. 465
3. Plants tufted or in very soft, loose cushions ... H. cervicornis, p. 466

Hypnea spinella (C. Agardh) Kützing

Sterile plants subcartilaginous in texture, forming extensive densely entangled mats to 2.5 cm. thick; branching in all directions, the branches spreading, the smaller ones spinelike; the slender branches concrescent, the ultimate divisions to 2.2 mm. long, subulate, tapering; sporangial plants similar, but branches rather thicker than those in the sterile specimens, the sporangia in an annular zone at about the middle of each of the ultimate divisions.

Bermuda, Cuba, Jamaica, Hispaniola, Virgin Isls., Grenada, Costa Rica, Colombia, Venezuela, Trinidad, Tobago, Brazil. A plant of rocks and reefs exposed to the surf. Børgesen (1913–20, p. 385) suspects that this is but a dwarfed form of *H. cervicornis.*

REFERENCES: Mazé and Schramm 1870–77, Martens 1870, Murray 1889, Børgesen 1913–20, Collins and Hervey 1917, Howe 1918a, 1929, Taylor 1930a, 1933, 1941b, 1942, 1943, Joly 1957; *H. spinella* f. *major,* Mazé and Schramm 1870–77, Murray 1889; *Sphaerococcus spinellus,* Montagne 1863.

Hypnea cervicornis J. Agardh — Pl. 73, fig. 2

Plants forming tangled tufts and extensive, rather fragile reddish or bleached mats; the chief axes to 2–15 cm. long, the lower branches decumbent, those above widely divaricate, subdichotomous, in the upper portions alternate and somewhat cervicorn, the ultimate divisions tapering to the tips; fertile branches 4–8 cm. long, exceeding the general tufts of the plant, freely radially branched and the crowded ultimate divisions fertile; tetrasporangia in rather crowded, linear or ovate, forking, subulate branchlets; cystocarps in similar branchlets on the carposporic plants.

Bermuda, Florida, Texas, Mexico, Cuba, Cayman Isls., Jamaica, Hispaniola, Virgin Isls., St. Barthélemy, Guadeloupe, Martinique, British Honduras, Panama, Colombia, Netherlands Antilles, Venezuela, Brazil. Common in quiet shallow water, redder and more compact in relatively exposed places, diffuse in warmer and quieter spots and then almost decolorized if exposed to full sunlight. Dredged to a depth of about 7 m.

REFERENCES: Harvey 1853, Grunow 1867, Martens 1870, Mazé and Schramm 1870–77, Murray 1889, Børgesen 1913–20, 1924, Collins and Hervey 1917, Taylor 1928, 1930a, 1933, 1935, 1936, 1940, 1941a, b, 1942, 1954a, Taylor and Arndt 1929, Hamel and Hamel-Joukov 1931; *Gracilaria confervoides,* Dickie 1875 *p. p.; H. acanthoclada, H. armata* v., *H. hamulosa p. p., H. rissoana p. p., H. spinella, ?H. valentiae p. p.,* Mazé and Schramm 1870–77, Murray 1889; *?H. divaricata,* Harvey 1853, Martens 1870, Murray 1889, Collins 1901; *H. pannosa,* Phyc. Bor.-Am. 590.

Hypnea cornuta (Lamouroux) J. Agardh

Plants bushy, about 2 dm. tall, repeatedly alternately branched, the branches tapering to the extremities, large and small divisions intermixed; simple spinuliform branchlets about 2.2 mm. long numerous; older branches more or less abundantly beset with acute, sessile or briefly pedicellate, stellate, peltate or reflexed branchlets of 3–6 points; cystocarps usually several together in simple branchlets.

Bermuda, North Carolina, Florida, Mexico, ?Cuba, Hispaniola, Virgin Isls., Guadeloupe, British Honduras, Colombia, ?Venezuela. Occasional in shallow water on shells or old corals. When cornute spines are scarce it may be hard to distinguish this species from fertile plants of *H. cervicornis.*

REFERENCES: Harvey 1853, 1861, Mazé and Schramm 1870–77, Dickie 1874b, Hemsley 1884, Murray 1889, Børgesen 1913–20, 1924, Taylor 1928, 1929b, 1933, 1935, 1954a, Williams 1951; *H. arborescens* v., *H. divaricata, H. gracilarioides, H. hamulosa p. p., H. rissoana p. p.,* Mazé and Schramm 1870–77, Murray 1889.

Hypnea musciformis (Wulfen) Lamouroux Pl. 73, fig. 1

Plants very bushy, often entangled, texture somewhat fragile, fleshy; color dull purplish red, or bleached; the bases disklike, ill-defined; erect branches to 10–20(–50) cm. tall, about 1–2 mm. diam., the leading branches dividing several times, beset with numerous, short, divaricate spur branchlets 1–5(–10) mm. long; ultimate branches usually with these in a secund series; tips of the branches often elongate, naked, typically swollen and crozier-hooked, bearing a crest of small branchlets on the back; tetrasporic branchlets somewhat siliquose or spindle-shaped and rostrate, with the numerous sporangia girdling the wider part; cystocarpic branchlets divaricate, often with spurlike divisions, the cystocarps strongly swollen.

Bermuda, North Carolina, South Carolina, Texas, Mexico, Bahamas, Caicos Isls., Cuba, Cayman Isls., Jamaica, Hispaniola, Puerto Rico, Virgin Isls., St. Barthélemy, Guadeloupe, Martinique, Barbados, Grenada, British Honduras, Costa Rica, Panama, Colombia, Netherlands Antilles, Venezuela, Trinidad, Tobago, French Guiana, Brazil, Uruguay. Common, growing on rocks, old corals, and shells in shallow, sheltered situations. In the tropics this species seems to be less tall than farther north, the branches more arcuate, the habit more entangled, and the crozier tips far more conspicuous and numerous. In color the plant appears to be a more reddish purple. The crozier tips are able to encircle objects with which they come to lie in contact, and act as clasping organs.

REFERENCES: Greville 1833, Hooper 1850, Montagne 1850, Harvey 1853, 1861, Martens 1870, 1871, Mazé and Schramm 1870–77, Dickie 1874a, Zeller 1876, Hemsley 1884, Piccone 1886, 1889, Hauck 1888, Möbius 1889, 1890, Murray 1889, Collins 1901, Vickers 1905, Børgesen 1913–20, 1924, Collins and Hervey 1917, Howe 1918a, b, 1920, 1928, Hoyt 1920, Schmidt 1923, 1924, Taylor 1928, 1929b, 1930a, 1933, 1935, 1936, 1939a, 1940, 1941a, b, 1942, 1943, 1954a, 1955, Sanchez A. 1930, Joly 1951, 1957, de Mattos 1952; *H. alopecuroides,* Kützing 1868; *H. arborescens, H. armata, H. harveyi, H. musciformis* v. *spinulosa, H. nigrescens, H. valentiae,* Mazé and Schramm 1870–77, Murray 1889; *H. musciformis* f. *tenuis,* Murray 1889; *H. musciformis* f. *spinella,* Hauck 1888; *H. esperi,* Martens 1871; *H. rissoana,* Martens 1870, Dickie 1874a, Zeller 1876, Murray 1889 *p. p.; Sphaerococcus musciformis,* Montagne 1863.

UNCERTAIN RECORDS

Hypnea corymbosa Crouan

Guadeloupe, the type locality; essentially a *nomen nudum.* De Toni 6:281; Mazé and Schramm 1870–77, Murray 1889.

Hypnea hamulosa (Turner) Montagne

Martinique, Brazil. Type from the Red Sea. De Toni 4:477, 6:279; ?Martens 1870.

Hypnea krugiana Hauck

Puerto Rico, the type locality. De Toni 4:482; Hauck 1888. Perhaps referable to *H. spinella.*

Hypnea pannosa J. Agardh

Florida, Brazil. Type from the Pacific coast of Mexico. De Toni 4:482, 6:280; Martens 1870, Taylor 1928.

Hypnea setacea Kützing

Guadeloupe, French Guiana. Type from French Guiana. De Toni 4:484; Montagne 1850, Mazé and Schramm 1870–77, Murray 1889.

Hypnea spicifera (Suhr) Harvey

Guadeloupe. Type from South Africa. De Toni 4:475, 6:279; Mazé and Schramm 1870–77, Murray 1889.

Hypnea valentiae (Turner) Montagne

Grenada. Type from the Red Sea. De Toni 4:479, 6:280; Murray 1889 *p. p.*

Hypneocolax Børgesen, 1920

Plants parasitic, minute, sessile, verruciform or nearly stellate; growth multiaxial; structure pseudoparenchymatous, of branching cell series in firm jelly, the medulla of larger irregular cells and the superficial tissues more compact, of smaller cells in more definitely anticlinal rows; asexual plants with bi- or tetrasporangia formed from the surface cells; spermatangia covering the surface of fertile plants, formed terminally on very slender cell series; carpogenic branches not known; cystocarps apparently without pores, immersed in hemispherical processes, with a large multilobed cell at the base, and terminal carposporangia.

Hypneocolax stellaris Børgesen

Plants about 0.75 mm. diam.; cortex covered with a cuticle 20–25 μ thick; bisporangia 16–22 μ diam., 30 μ long; cystocarps several in each plant; carposporangia 20–22 μ diam.

Virgin Isls. Growing on *Hypnea musciformis* in a shallow, sheltered locality.

REFERENCE: Børgesen 1913–20.

PHYLLOPHORACEAE

Plants bushy, dichotomously branched, the branches cylindrical to membranous, firm in texture; growth from a marginal or apical meristem producing a pseudoparenchymatous medulla and a close cellular cortex; sporangia seriate, tetrapartite, crowded in nemathecial sori; spermatangia developed on outgrowths from superficial cells; procarps present, the carpogenic branch of three cells carried on a supporting cell which serves as the auxiliary; gonimoblasts first growing inward, becoming interwoven with thin sterile filaments producing irregular masses of carposporangia without a definite sheath; sporangial or gametangial reproduction greatly modified in some species.

Phyllophora Greville, 1830

UNCERTAIN RECORDS

Phyllophora brodaei J. Agardh

Jamaica. Type from Europe. De Toni 4:231; Murray 1889.

Phyllophora gelidioides Crouan

Guadeloupe, Barbados. Type from the Canary Islands. De Toni 6:196; Vickers 1905; *Gelidium ligulatonervosum,* Mazé and Schramm 1870–77 (essentially a *nomen nudum*) is, so far as seen, *G. pusillum.*

Stenogramme Harvey, 1841

UNCERTAIN RECORD

Stenogramme interrupta (C. Agardh) Montagne

Florida. De Toni 4:239; Harvey 1853, Murray 1889.

Gymnogongrus Martius, 1833

Plants small, bushy, repeatedly forking; from moderately firm to horny, the branches cylindrical to flat, sometimes proliferated laterally; structurally showing a medulla of angular-rounded cells, the cortex of firmly coherent radial rows of several small cells; sporangia borne in the outer layer of small swellings or nemathecia on the branches, often imperfectly tetrapartite.

KEY TO SPECIES

1. Blades with segments narrowly strap-shaped, to about 2 mm. broad .... G. tenuis
1. Blades with subterete segments less than 1 mm. diam. .. G. griffithsiae

Gymnogongrus griffithsiae (Turner) Martius Pl. 60, fig. 1

Plants in pulvinate tufts, from a basal disk, to 2–5 cm. tall, dark purplish; repeatedly branching, fastigiate-dichotomous, sometimes polychotomous; branches slender, little tapering, to about 0.3–0.7 mm. diam., the tips pointed or a little flattened; nemathecia scattered, at first unilateral and pulvinate on the lower side of the branches, later more extensive, about 1 mm. diam. and enveloping them; sporangia generally imperfectly tetrapartite.

North Carolina, Guadeloupe, Tobago, Brazil, Uruguay.

REFERENCES: Martens 1870, Möbius 1890, 1892, Hoyt 1920, Taylor 1929b, 1930a, Joly 1951, 1957; *G. furcellatus* and v. *patens*, Mazé and Schramm 1870–77, Murray 1889.

Gymnogongrus tenuis (J. Agardh) J. Agardh Pl. 60, fig. 2

Plants bushy, 2–6 cm. tall, dull purple, membranous to cartilaginous in texture; subfastigiately dichotomous in branching, or near the apices appearing polychotomous, the divisions flat, linear, the lower segments to 2–3 mm. broad, the upper more slender; cystocarps formed in the median and upper segments, large for the size of the blade, about a millimeter broad, unilaterally projecting.

Mexico, Jamaica, Virgin Isls., Guadeloupe, Dominica, Martinique, Venezuela, Trinidad. Growing on rocks exposed to the surf, but reported able to grow in polluted water.

REFERENCES: Harvey 1853, Mazé and Schramm 1870–77, Murray 1889, Grieve 1909, Børgesen 1913–20, Taylor 1929b; *Gelidium fastigiatum, Gymnogongrus capensis, G. crenulatus, G. densus, G. dilatatus, G. linearis, G. pygmaeus, G. tenuis* v. *angusta,* Mazé and Schramm 1870–77, Murray 1889.

UNCERTAIN RECORDS

Gymnogongrus disciplinalis (Bory) J. Agardh

Guadeloupe. Type from Peru. De Toni 4:244; *Plocaria disciplinalis,* Mazé and Schramm 1870–77, Murray 1889.

Gymnogongrus tenuis (J. Agardh) J. Agardh, v. *angusta* J. Agardh

Dominica. Probable type of the variety from the west coast of Mexico. Grieve 1909.

V. *brevifolia* Holmes
Dominica, the type locality. Grieve 1909.

Ahnfeltia Fries, 1835

UNCERTAIN RECORD

Ahnfeltia durvillaei (Bory) J. Agardh

Guadeloupe. Type from Peru. De Toni 4:256, 6:201; Murray 1889; *Plocaria durvillaei,* Mazé and Schramm 1870–77, probably a Gracilaria.

GIGARTINACEAE

Plants plane or bushy, subsimple or branched, developing with an axis of several filamentous initials, the medulla ultimately pseudoparenchymatous, the cortex obscurely filamentous, the small outer cells in rows and containing chromatophores; sporangia in sori in the blade, developed from forking cell systems among the medullary filaments, tetrapartite; spermatangia formed in sori on the surface; procarps present, the three-celled carpogenic branch carried on a supporting auxiliary which, after fertilization, forms slender, branching gonimoblast filaments which ramify into the medulla and produce a diffuse cystocarp.

Chondrus Stackhouse, 1797

UNCERTAIN RECORD

Chondrus crispus (Linnaeus) Stackhouse

Bermuda. Type from Europe. De Toni 4:180; Murray 1889.

Iridaea Bory, 1826

UNCERTAIN RECORD

Iridaea littoralis Crouan

Guadeloupe, the type locality. Essentially a *nomen nudum.* Mazé and Schramm 1870–77, Murray 1889.

Gigartina Stackhouse, 1809

Plants foliar to bushy, firm-fleshy, dark reddish purple, simple to abundantly subdichotomously branched, the branches compressed to bladelike; developing a filamentous medulla surrounded by a cortex of anticlinal branched rows of successively smaller cells with the outermost containing the chromatophores; tetrapartite sporangia in immersed, rather spreading sori of indefinite form, the sporangia developed from cells of the inner cortex; spermatangia in groups developed terminally on close-branched filaments from superficial cells; cystocarps generally crowded in special fertile nodules or branchlets; loosely branched gonimoblast filaments consisting of larger nutritive cells supporting more distal divisions of smaller cells which form the carposporangia, eventually surrounded by a limiting layer of the medulla; cystocarps unilaterally projecting, the fleshy pericarp eventually rupturing.

KEY TO SPECIES

1. Entangled, the branches terete .... G. acicularis, p. 473
1. Bushy, the branches more or less flattened 2

2. Progressively alternately branched, complanate ..... G. teedii, p. 473
2. Relatively few main branches bearing crowded marginal branchlets .... G. elegans, p. 474

Gigartina acicularis (Wulfen) Lamouroux Pl. 60, fig. 6

Plants entangled, in purple or blackish-green cushions, 3–8 cm. tall, firm in texture, below the fibrous strands repent and attached by haptera, from these the axes turning up and sparingly irregularly dichotomous or pinnate, or near the tips subsecund, the segments 0.5–1.0 mm. diam., somewhat attenuate toward the base and in the uppermost segments; tetrasporangia in sori, causing slight unilateral swellings on the lower part of the branches; cystocarps subglobose, unilateral, single or in series of 2–4 near the middle of a branchlet, the tip of which is often reflexed.

Bermuda, North Carolina, Florida, Cuba, Jamaica, Hispaniola, Virgin Isls., Guadeloupe, Brazil, Uruguay.

REFERENCES: Greville 1833, Martens 1870, Mazé and Schramm 1870–77, Murray 1889, Børgesen 1913–20, Collins and Hervey 1917, Howe 1918a, Taylor 1928, 1943, 1954a, Blomquist and Humm 1946; ?*G. acicularis* v. *bipinnata*, Martens 1870; *Sphaerococcus acicularis*, Montagne 1863.

Gigartina teedii (Roth) Lamouroux Pl. 61, fig. 3

Plants with erect, subcartilaginous fronds, dull purplish, 10–30 cm. tall; repeatedly pinnately branched, compressed below, nearly terete in the lesser divisions, complanate, the ultimate pinnules spinuliform, subdistichous, and spreading; tetrasporangia in sori along the margins of the segments; cystocarps one or few together, causing marginal swellings on small pinnules.

Venezuela, Brazil.

REFERENCES: Hemsley 1884, Murray 1889, Möbius 1890, Taylor 1942, Joly 1951, 1957; *Chondroclonium teedii,* Martens 1870.

Gigartina elegans Greville Pl. 61, fig. 1

Plants bushy, subcartilaginous, dark purplish red; main axes to 5–12 cm. tall, sometimes with several equal branches, terete near the base, flat and to 2–3 mm. broad above, more or less naked below, but in the upper portion along the margins irregularly to closely distichously beset with short subterete to flattened branchlets 1–4 cm. long, which may bear crowded shorter branchlets in turn, especially on fertile plants; cystocarps strongly protuberant on the sides of small fertile branchlets.

?Texas, Brazil, Uruguay.

REFERENCES: Greville 1833, Martens 1870, Taylor 1941a.

RHODYMENIALES

Plants showing various shapes from filiform to fleshy-membranous, sometimes hollow; corticated, with a modified multiaxial type of structure, commonly appearing parenchymatous; asexual reproduction by tetrasporangia in sori or scattered over the plant just below the surface; sexual reproduction by spermatangia borne on surface cells in more or less restricted areas and by carpogonia in procarps sunken in the cortex; auxiliary cells established by segmentation indirectly from the cell supporting the carpogenic branch; cystocarps enveloped by a pericarp.

KEY TO FAMILIES

1. Foliaceous, or, if bushy, with a distinct main axis, seldom dichotomous; sporangia tetrapartite .. RHODYMENIACEAE, p. 475
1. Bushy, without a distinct main axis, or if with a stalk, the branchlets obviously septate; sporangia tetrahedral CHAMPIACEAE, p. 486

RHODYMENIACEAE

Plants plane or bushy, nearly simple or somewhat freely divided, the divisions flat and subdichotomous or subcylindrical and radially branched, solid or hollow, soft to tough-membranous; developing from an apical meristem, the innermost cells large, forming a parenchymatous medulla, or the medulla filamentous, or the medullary region merely a jelly-filled space; the cortex of larger cells within, the smaller ones without often in short anticlinal series and containing the chromatophores; sporangia sometimes in sori, tetrapartite, formed between these superficial cells; carpogenic branches three-celled, the supporting cells each also cutting off an auxiliary mother cell; gonimoblasts extensively branched, most of the cells forming rather small carposporangia, the whole eventually enveloped by a loose pericarp.

KEY TO GENERA

1. Cavity of the pericarp about the cystocarp filled with a filamentous net of tissue 2
1. Cavity outside the cystocarp mass devoid of a filamentous envelope 3

2. Cortex originally filamentous, the anticlinal filaments crowded, sometimes reduced to one or two cells .... FAUCHEA, p. 476
2. Cortex of a single cell layer .... LEPTOFAUCHEA, p. 477

3. Thallus with a strong, slender stem system supporting vesicular branchlets .... BOTRYOCLADIA, p. 483
3. Thallus without such differentiation 4

4. Thallus with a cellular medulla 5
4. Thallus with a jelly-filled medullary cavity 6

5. Medulla parenchymatous, of large firm-walled cells within, slightly smaller ones without .... RHODYMENIA, p. 484
5. Medulla a parenchyma of large cells with smaller ones in short filaments lying between them .... AGARDHINULA, p. 481

6. Thallus dichotomously branched, of short segments very strongly constricted, and septate at each forking COELARTHRUM, p. 482
6. Thallus alternately branched, or foliaceous 7

7. Thallus hollow throughout, terete or compressed, seldom flattened; medullary cavity without filaments or with very few of them .... CHRYSYMENIA, p. 478
7. Thallus locally hollow, often terete but also flattened; medullary cavity traversed by numerous filaments CRYPTARACHNE, p. 480

Fauchea Montagne, 1846

Plants foliaceous or alternately branched, flat; parenchymatous medullary tissue of large cells; cortical tissue of small cells arranged more or less in anticlinal rows; tetrasporangia in irregular superficial nemathecioid sori; pericarps without papillae or spines.

KEY TO SPECIES

1. Blades peltate, the stipe seldom marginal, entire or irregularly once lobed; cortex very thin, hardly recognizably filamentous F. peltata
1. Blades marginally stipitate, repent, usually 2–3 times branched; cortex thicker, the anticlinal filaments once branched, 2–4 cells long .... F. hassleri

Fauchea hassleri Howe and Taylor

Plants thin when dry, lubricous when moist, deep red; stipitate, radiately dichotomous, 1–4 times forked, 4–6 cm. broad (or more?), the axils mostly rounded, the segments oblong or cuneate, mostly 2–5 mm. broad, sometimes 10–12 mm. broad under the dichotomies, 150–280 μ thick (soaked); the apices usually subtruncate or obtuse, the margins entire; medullary layer of 1–3 series of large colorless cells, these mostly oval or elliptic and 60–130 μ by 100–260 μ in cross section, passing abruptly to a thin subcortical layer usually of a single series of cells 15–22 μ in long diameter; cortical layer of anticlinal or divaricate cell rows or filaments 2–4 cells long, the cells (protoplasts) minute, 3–6 μ long, immersed in a copious jelly; cystocarps marginal, pericarps sessile, verruciform-truncate or subhemispheric, 0.75–1.00 m. in diameter.

Brazil. Dredged from rather deep water.

REFERENCES: Taylor 1930a, Howe and Taylor 1931.

Fauchea peltata Taylor Pl. 54, figs. 5–7

Plants growing appressed to the substratum and attached by a short peltate or occasionally marginal stalk, 0.5–1.0 cm. tall, 1–9 cm. broad; disk or blade irregularly rounded, crenate, or digitately lobed, the lobes rarely secondarily divided; fleshy, increasing in thickness from the margin, which is about 200 μ thick, toward the base of the stipe; structurally showing a medulla of very large colorless cells about 175 μ diam., with about one layer of somewhat smaller cells on each side, and a cortex of 2 cell layers, the inner of fairly closely approximated rounded cells about 6–10 μ diam., the outer of smaller round cells 4.0–5.5 μ diam., separated in surface view by 1–5 times their own diameters; reproductive structures unknown.

Jamaica, Venezuela, Brazil. Growing on dead coral fragments at a depth of 20–40 m. Sterile, and so the generic disposition is tentative.

REFERENCES: Taylor 1942; *Chrysymenia planifrons*, Taylor 1930a.

Leptofauchea Kylin, 1931

Plants solid, flat, regularly dichotomous; medulla of large cells only; cortex of a single layer of small cells; pericarps without spines.

KEY TO SPECIES

1. Branching irregular, the segments about 2.5 mm. broad; medulla of one cell layer .... L. brasiliensis
1. Branching regularly dichotomous, the segments 4–8 mm. broad; the medulla of 2–3 cell layers .... L. rhodymenioides

Leptofauchea brasiliensis Joly

Plants 3–4 cm. tall, branching above a very short stipe, the ecostate divisions erect, sparingly and irregularly branched, often dichotomous, 2 mm., seldom to 5 mm. broad, subspatulate, briefly tapering to obtuse tips, 280–290 μ thick when mature and with but a single medullary layer; sporangial sori 80–90 μ thick, formed by proliferation of the cortex, producing rows of cells near the ends of which the sporangia occur; tetrahedral sporangia 12–14 μ diam.

Brazil. Growing on rocks in shallow water.

REFERENCE: Joly 1957.

Leptofauchea rhodymenioides Taylor Pl. 60, figs. 4, 5

Plants to 6 cm. tall (perhaps more), moderately branched, deep rose-pink in color, very slightly glossy when dried; holdfast somewhat lobed, stipe ill defined; axis rather evenly 3–5 times (or more?) dichotomously branched, the complanate divisions diverging 60°–90°, rarely more, 1.0–1.5 cm. apart, 4–8 mm. broad above the forks, not much wider below them, about 200 μ thick, the apices broadly rounded; the medulla of 2–3 layers of thin-walled cells 60–140 μ diam., with a layer of smaller cells on each side; the cortex of one complete cell layer, the cells 9–21 μ diam., with scattered cells outside it, these 3.5–6.0 μ diam.; sporangial sori appearing as irregular bands nearly as wide as the branch, to 4 mm. long, darker in color and duller than the vegetative portion, consisting of slender paraphysal filaments 6–8 cells or about 75–100 μ long, 4.5 μ diam., laterally bearing oval tetrapartite sporangia 51–77 μ long, 28–46 μ diam., attached to the second or third cell from the base; pericarps prominent, marginal, to 1.0–1.5 mm. diam.

Netherlands Antilles. Apparently a deep-water species; dredged from 43 m. depth.

REFERENCE: Taylor 1942.

Chrysymenia J. Agardh, 1842

Plants terete or a little compressed, but seldom flattened, hollow, alternately branched; medullary filaments few or wanting; inner cortical cells large, crowded, some bearing on the inner side sessile gland cells; outer cortex of one to three layers of small cells; tetrasporangia scattered over the thallus between the cortical cells; cystocarps scattered, partly immersed, the cortex elevated as a pericarp to form a more or less projecting pore.

KEY TO SPECIES

1. Thallus divided by constrictions into cylindrical segments C. enteromorpha
1. Thallus not segmented though freely branched 2

2. Branching primarily dichotomous, though with occasional adventitious branches .... C. halymenioides
2. Branching alternate and radial, though irregular....... C. ventricosa

Chrysymenia ventricosa (Lamouroux) J. Agardh Pl. 62, fig. 3

Plants purplish, delicately membranous, from a cuneate stipe 3–15 mm. long rising to height of as much as 30 cm., compressed below, 1–3 times radially alternately branched, the terete branches irregularly placed, sometimes opposite, alternate, or appearing almost dichotomous, crowded or 2–3 cm. apart, about 0.5–2.0 cm. diam.; when very young subconical; cystocarps hemispherical.

Bermuda, Jamaica, Virgin Isls., Venezuela. Probably a deep-water species; reported dredged from a depth of 90 m.

REFERENCES: Børgesen 1913–20, Taylor 1942.

Chrysymenia enteromorpha Harvey Pl. 62, fig. 2

Plants briefly stipitate, to 20–25 cm. tall or somewhat more, rose-red, delicately membranous, at first simple, saccate, cylindrical, the primary axis forming stipitate branches from the summit and sides, repeating to 3–4 degrees; medullary filaments absent; young branches of a single cell layer with a few small superficial cells along the angles of contact, the mature segments with a firmer layer of large inner cortical cells and a nearly complete outer cortical layer leaving only the central portion of each inner cortical cell exposed; gland cells crowded on certain of the inner cells facing the cavity; cystocarps projecting, with an apical pore.

Bermuda, North Carolina, Florida, Virgin Isls., Brazil. Apparently strictly a deep-water plant, growing on coral, rock fragments, or old shells; dredged from 15–90 m. depth.

REFERENCES: Harvey 1853, Dickie 1874b, Murray 1889, Børgesen 1913–20, Collins and Hervey 1917, Hoyt 1920, Taylor 1928, 1930a.

Chrysymenia halymenioides Harvey Pl. 62, fig. 1

Plants with subterete or compressed, hollow, rather thick gelatinous-membranous rose-red blades 7–10 cm. long; divisions slender at the base, but above 0.5–1.0 cm. wide, repeatedly dichotomously cleft, divided into broadly linear divergent segments with rounded axils, the branch tips obtuse; cystocarps prominent, rounded-conical.

Bermuda, Florida, Jamaica, Netherlands Antilles. Apparently a deep-water species; dredged from 43 m. depth.

REFERENCES: Harvey 1853, Murray 1889, Collins 1901, Collins and Hervey 1917, Taylor 1928, 1942.

UNCERTAIN RECORDS

Chrysymenia dichotoma J. Agardh

Bermuda. Type from Europe. Dickie 1874a. Perhaps this is *Halymenia dichotoma* (De Toni 4:1540), perhaps a misidentification.

Chrysymenia furcata Crouan

Guadeloupe, the type locality; essentially a *nomen nudum*. De Toni 6:304; Mazé and Schramm 1870–77, Murray 1889.

Cryptarachne Kylin, 1931

Plants foliaceous and broadly lobed or cleft, and with numerous secondary marginal blades, often locally hollow but not saccate or tubular; inner cortical cells large, closely placed, with occasional sessile gland cells facing the mucilage-filled center of the thallus, which is loosely traversed by slender colorless filaments; outer cortex of one to three layers of small cells.

KEY TO SPECIES

1. Thallus freely branched; medullary filaments numerous 2
1. Thallus broad, foliaceous, distally cleft into broad lobes; medullary filaments few .... C. planifrons

2. Branches rather narrow, irregular, contracted at the base C. dickieana
2. Basal part of the plant broad; with numerous distal branches about 2.5 cm. wide .... C. agardhii

Cryptarachne agardhii (Harvey) Kylin Pl. 63, fig. 2

Plants foliaceous, gelatinous-membranous, rose-red, from a short stipe expanding through a cuneate base to a blade 10–20 cm. long, dichotomously to palmately laciniate, the lobes narrowly cuneate, approximate above the narrow axils, 2.0–2.5 cm. broad, the margins commonly erose-dentate; occasional inner cortical cells bearing 2–4 small gland cells on the inner side; medullary area traversed by numerous rhizoidal filaments about 20 μ diam.

Bermuda, North Carolina, Florida, Virgin Isls.

REFERENCES: Kylin 1931; *Chrysymenia agardhii,* Harvey 1853, Dickie 1874a, Murray 1889, Børgesen 1913–20, Collins and Hervey 1917, Hoyt 1920, Taylor 1928.

Cryptarachne dickieana (J. Agardh) Kylin

Plants with rather flat blades, 6–10 cm. tall; when fresh, thick and gelatinous-membranous, drying subcartilaginous; very irregularly and chiefly alternately pinnate, narrow and of irregular width, the axes bearing small linear blades 2–5 mm. wide on the margins; rhizoidal filaments particularly abundant in the medullar region; numerous small cells with dense contents present between the cells of the middle cortical layer; cystocarps sessile, scattered; pericarps subspherical.

Brazil.

REFERENCES: Kylin 1931; *Chrysymenia dickieana,* J. Agardh 1892.

Cryptarachne planifrons (Melvill) Kylin Pl. 63, fig. 1

Plants foliaceous, to 4 dm. tall, from a very short stalk widely expanded and erect, rather thick, gelatinous, the margin with deltoid lobes, those below more obscure or lacking, those above short, broad, and widely separated; medullary region narrow, with few rhizoidal filaments.

Florida, Virgin Isls., Netherlands Antilles. Probably altogether a deep-water species; dredged from 30 m. depth.

REFERENCES: Kylin 1931, Taylor 1942; *Chrysymenia curtissiana,* J. Agardh 1885; *C. planifrons,* Murray 1889, Børgesen 1913–20, Taylor 1928.

Agardhinula De Toni, 1897

Plants fleshy, dichotomous, palmate or laciniate, flat throughout; medulla of several layers of large cells with small cells interspersed;

cortex of small cells in anticlinal rows of two or three cells; sporangia tetrapartite, in small or confluent, rather gelatinous sori; cystocarps prominent, within a hemispherical, ostiolate pericarp.

Agardhinula browneae (J. Agardh) De Toni

Plants 10–30 cm. tall, the bases cuneate, the blades plane, sparingly divided, the segments 1.0–5.5 cm. broad, the upper in general narrower than the lower ones, each often widening from the base upward, with rounded but rather narrow sinuses; in texture rather thick; cystocarps dispersed over the blade.

?North Carolina, Florida.

REFERENCES: Hoyt 1920, Taylor 1928.

Coelarthrum Børgesen, 1910

Plants erect, branched, articulated, the branching dichotomous, the segments hollow with a thin membranous wall, the cavity filled with a very soft jelly; wall of two cell layers, the outer of small cells loosely and irregularly associated; the inner layer of much larger, close-placed cells, some bearing on the side facing the cavity one to five glands on a small supporting cell; diaphragms at the ends of the segments of cells similar to the inner wall layer; sporangia tetrapartite, borne between the cells of the outer layer and projecting toward the surface; cystocarps scattered, pericarps hemispherical, discharging by an apical pore.

Coelarthrum albertisii (Piccone) Børgesen Pl. 61, fig. 6

Plants to 4–5 cm. tall, forming rather dense masses; the branches often attached to each other at points of contact, the segments nearly terete, to 6 mm. diam., 10 mm. long.

Bermuda, Virgin Isls., Guadeloupe, Venezuela. In sheltered angles of exposed reefs in shallow water and dredged from 18–27m. depth.

REFERENCES: Børgesen 1910, 1913–20, Collins and Hervey 1917, Howe 1918a, Taylor 1942.

Botryocladia Kylin, 1931

Plants with a more or less extended branch system, the axes slender, terete, strong, bearing stipitate bladderlike spherical or obovoid branchlets; inner cortex of one or several layers of large, usually colorless cells, some of them bearing sessile gland cells on the inner side facing the mucilage-filled cavity; the outer cortex sometimes discontinuous, of small cells with chromatophores; tetrapartite sporangia scattered on the branchlets; cystocarps also scattered on the branchlets, slightly projecting.

KEY TO SPECIES

1. Bladders usually 1–2(–4), rarely several together on a stalk, relatively large; gland cells in groups of 4–8 .... B. pyriformis
1. Bladders small and very numerous along the branches; 1–2 gland cells on each supporting cell .... B. occidentalis

Botryocladia pyriformis (Børgesen) Kylin Pl. 64, fig. 2

Plants 1–2 cm. tall, few-branched, the rather long pedicels or branches terminating in pyriform bladders 4–9 mm. long, generally 2–6 to a plant; bladders showing structurally an inner layer of large, nearly colorless and rather thick-walled cells 55–130 μ in diameter and angular from surface view, occasionally interspersed with smaller subspherical cells bearing 4–8 clavate gland cells; externally the bladders covered by an outer cortical layer of spherical-compressed cells 4–8 μ diam.; cells of intermediate size border the large interior cells between the inner and outer layers and form a nearly or quite complete intermediate layer across their faces.

Bermuda, Florida, Jamaica, Virgin Isls., Netherlands Antilles, Brazil. Occasionally found growing on intertidal rocks, especially in the shade of overhanging ledges and in rock crevices; dredged to a depth of 43 m. Plants are occasionally found with 2–3 branches, each bearing several bladders, and these can only be distinguished from *B. occidentalis* by ascertaining the gland-cell arrangement. Dredging has produced specimens with subsolitary bladders reaching the enormous size of 2.6–4.0 cm. diam., 5.0 cm. long.

REFERENCES: Kylin 1931, Taylor 1942; *B. skottsbergii*, Williams and Blomquist 1947; *Chrysymenia pyriformis*, Børgesen 1913–20, Collins and Hervey 1917, Howe 1918a, Taylor 1928, 1930a.

Botryocladia occidentalis (Børgesen) Kylin Pl. 64, fig. 1

Plants to 2.5 dm. tall, the axes frequently branched, wiry, bearing numerous shortly pedicellate, ovoid-pyriform to subspherical bladders to 4–5 mm. long, radially or bilaterally disposed; structurally the bladders showing an inner layer of large, nearly colorless and rather thin-walled cells 55–110 μ diam., angular from surface view, frequently bearing 1–2 small gland cells on the inner face; externally the bladders covered by an outer cortical layer of spherical-compressed cells 8–14 μ diam.; between the inner and outer layers cells of intermediate size border the large interior ones, but do not form a complete cell layer.

Bermuda, North Carolina, Florida, Bahamas, Puerto Rico, Virgin Isls., Guadeloupe, Martinique, Barbados, Grenada, Costa Rica, Colombia, Brazil. Apparently a plant of moderate to quite deep water; dredged to a depth of 55 m.

REFERENCES: Kylin 1931; *Chrysymenia obovata,* Mazé and Schramm 1870–77, Dickie 1875; *C. uvaria,* Harvey 1853, Mazé and Schramm 1870–77, Dickie 1874a, Hemsley 1884, Piccone 1886, 1889, Hauck 1888, Murray 1889, Vickers 1905, Børgesen 1913–20, Collins and Hervey 1917, Howe 1918a, 1920, Hoyt 1920, Schmidt 1923, 1924, Taylor 1928, 1933, 1936, Hamel and Hamel-Joukov 1931; *C. uvaria* v. *occidentalis,* Børgesen 1913–20; *C. uvifera,* Hooper 1850; *Gastroclonium uvaria,* Martens 1870.

Dendrymenia Skottsberg, 1923

UNCERTAIN RECORD

Dendrymenia flabellifolia (Bory) Skottsberg

Guadeloupe. Type from Chile. *Rhodymenia flabellifolia,* De Toni 4:517; Mazé and Schramm 1870–77, Murray 1889.

Rhodymenia Greville, 1830

Plants forming large membranous fronds, simple to dichotomously or palmately divided, often proliferous from the margins, in some cases the divisions narrow, strap-shaped; structurally of two layers, the internal of oblong colorless cells, the cortical of small cells with chromatophores, in the fertile portions becoming thicker by proliferation of the cortical cells in anticlinal series; tetrapartite sporangia formed between these series of superficial cells, in sori; cystocarps each with a somewhat swollen, loose pericarp, opening by a distinct pore.

KEY TO SPECIES

1. Plants of a lax, elongated habit, the branching erect .. R. occidentalis
1. Plants flabellar, the branches spreading .... R. pseudopalmata

Rhodymenia occidentalis Børgesen Pl. 59, fig. 3

Plants arising from disks which may send out supplementary processes in time; the blades narrowed at the base but hardly terete or stalklike, di- polychotomously irregularly branched with lateral proliferations, the flat membranous divisions erect, strap-shaped, about 4 mm. broad but narrower at each forking, the margins slightly sinuate, the apices obtuse; about 150 μ thick above, structurally showing 2–3 medullary layers of large thick-walled cells within, 2–3 layers of smaller ones on each side without, and a cortical layer of quite small cells; reproductive structures unknown.

Virgin Isls., where locally abundant. Plants of deep water, dredged from 18–27 m. depth.

REFERENCE: Børgesen 1913–20.

Rhodymenia pseudopalmata (Lamouroux) Silva

Plants reddish purple, fleshy-membranous, each with fibrous holdfasts, stoloniferous, showing slender, simple or sparingly branched, erect stems 1–4 cm. long, or these inconspicuous, but with flagellar outgrowths from the base; stems each bearing at the top a blade 1–8 cm. long and broad, which is cuneate at the base, at first simple, later generally 1–4 times dichotomously divided and complanate, the segments 2–10 mm. broad, cuneate to strap-shaped or somewhat distally tapering, the tips spatulate to subacute, the margins entire; tetrasporangia in sori just behind the tips of the terminal segments; cystocarps sessile, prominent, generally marginal but also on the faces of the distal segments.

V. **caroliniana** n. var. Plants becoming larger than the type of the species, the stipes very short, the spreading blades reaching a length of 12 cm., a width of 15 cm., being dichotomously branched at angles of 30°–110°, to as many as 7 degrees, the segments to 6 mm. wide below, 3.5 mm. wide above and very little tapered; sporangial sori transversely oval, very near the tips of the slightly spatulate terminal segments.

North Carolina, South Carolina, Brazil. Growing on rocks and reefs at low-tide level.

REFERENCES: *R. palmetta*, Möbius 1890, 1892, Hoyt 1920, Joly 1951, 1957, de Mattos 1952.

UNCERTAIN RECORD

Rhodymenia palmata (Linnaeus) Greville

Brazil. Type from Europe. De Toni 4:512; Greville 1833a, J. Agardh 1852.

Hymenocladia J. Agardh, 1852

UNCERTAIN RECORD

Hymenocladia divaricata (R. Brown) Harvey v. *tropica* Crouan

Guadeloupe, the type locality. De Toni 4:503; Mazé and Schramm 1870–77, Murray 1889.

CHAMPIACEAE

Plants usually bushy, branches terete or compressed, delicately membranous to quite soft; from an apical meristem developing superficial small assimilatory cells, an inner cortex of large cells and a medullary cavity traversed by more or less distinct longitudinal filaments bearing lateral secretory cells; tetrahedral sporangia formed from cortical cells and lying just below the surface; spermatangia borne on groups of surface cells; carpogenic branches three- or four-celled, each borne on an inner cortical cell, the auxiliary secondarily derived from the supporting cell; after fertilization the gonimoblasts, and in turn the large carposporangia, formed from a large fusion cell, the whole covered by a prominent ostiolate pericarp.

KEY TO GENERA

1. Thallus with a strong, slender stem system supporting septate vesicular branches .... GASTROCLONIUM, p. 489
1. Thallus without such differentiation 2
2. Thallus wiry throughout, not segmented, the cavity obscure COELOTHRIX, p. 488
2. Thallus obviously hollow 3

3. Thallus constricted and solid only at the bases of the branches, otherwise hollow .... LOMENTARIA, p. 487
3. Thallus divided by numerous transverse septa .... CHAMPIA, p. 490

Lomentaria Lyngbye, 1819

Plants tufted, repeatedly irregularly or unilaterally branched, the slender branches tapering near each end, terete or compressed, if hollow usually closed at the bases of the branches; tetrahedral sporangia somewhat grouped; spermatangia formed in patches on the surface of the plant; after fertilization the auxiliary cell associated with the fused cells of the carpogenic branch and other structures to form a nutritive region, the auxiliary itself bearing the gonimoblasts, the basal portions of these remaining sterile but the upper and larger part forming carposporangia, the cystocarp enclosed within a projecting ostiolate pericarp.

KEY TO SPECIES

1. Base creeping; erect branchlets simple .... L. rawitscheri
1. Base simple; plants bushy, well-branched .... L. baileyana

Lomentaria rawitscheri Joly

Plants rosy, tufted, to 5–7 mm. tall; lower axes irregularly branched, decumbent, 300–500 μ diam., bearing numerous erect simple branchlets 1–2 mm. tall, about 0.3–0.6 mm., occasionally to 1.0 mm., thick, oval, acute-tipped; sporangia oval, 37 μ diam., formed in short branchlets.

Brazil. Growing on rocks in shallow water.

REFERENCE: Joly 1957.

Lomentaria baileyana (Harvey) Farlow

Plants tufted, often widely expanded, 3–7(–20) cm. tall, densely branching; dull purplish red to pink, usually bleached below, drying much brighter; substance soft; branching irregularly alternate, the hollow branches terete, tapering near each end, curved, the ultimate segments 1–2 cm. long, often somewhat unilaterally placed, 0.5–1.5 mm. diam.; surface of the thallus showing both larger and smaller cells; tetrasporangia 30–55 μ diam., scattered under the cortical layer in the branches; pericarps usually scarce, sessile, external on the branches, somewhat produced at the ostiolate tip.

Bermuda, North Carolina, South Carolina, Florida, Brazil. Growing on marine plants, usually in shallow water, but dredged from a depth of 33 m.

REFERENCES: Kylin 1931, Taylor 1937, 1954a; *Chylocladia baileyana,* Harvey 1853; *Lomentaria uncinata,* Martens 1870, Collins and Hervey 1917, Howe 1918a, Hoyt 1920, Taylor 1928; *L. uncinata* v. *filiformis,* Collins and Hervey 1917; *L. reflexa* f. *uncinata,* Piccone 1889.

UNCERTAIN RECORDS

Lomentaria clavellosa (Turner) Gaillon

Brazil. Type from England. De Toni 4:573; Kylin 1931; *Chondrothamnium clavellosum,* Martens 1870.

Lomentaria orcadensis (Harvey) Collins

Bermuda, North Carolina. Type from the Orkney Islands. Taylor 1937; *Lomentaria rosea,* Williams 1949a; *Chylocladia rosea,* De Toni 4:575, 6:311; Murray 1889.

?*Lomentaria tenera* (Liebmann) Kützing

Mexico, the type locality. De Toni 4:556.

Coelothrix Børgesen, 1920

Plants forming firm cushions, the thallus much and irregularly branched, the slender terete branches rigid; medullary region hollow, the space usually small or even obscure, some of the large inner cells of the ill-defined filaments facing it bearing single sessile gland cells; wall composed of several layers of large firm-walled cells decreasing in size toward the surface, the outer cortex of a single definite cell layer; sporangia tetrahedral, aggregated in slightly swollen branch tips.

Coelothrix irregularis (Harvey) Børgesen

Pl. 45, fig. 3; pl. 46, fig. 4

Plants 2–3 cm. high, bushy or in cushions, loosely to densely entangled, notably bluish-iridescent; branches terete, somewhat tapering, frequently interadherent and fusing, 0.50–0.75 mm. diam.; somewhat hollow, showing a medulla of a few filaments of stout, elongated cells with very gelatinous walls loosely connected with the cortex, and ovoid-clavate mucilaginous gland cells projecting into the cavity; cortex of thick-walled cells reaching 45–80 μ diam., and 120–200 μ in length, surrounded by an epidermis of rather elongated cells, 18–27 μ diam. by 45–54 μ in length; reproduction by tetrasporangia carried in short, swollen, ovoid pedicellate branches.

Bermuda, Florida, Bahamas, Cuba, Jamaica, Hispaniola, Puerto Rico, Virgin Isls., Guadeloupe, British Honduras. Plants of the intertidal zone or to about 9 m. below it, growing in rock crevices and under ledges.

It is worthy of note that Børgesen observed no filaments in the cavity of this plant. Always obscure, they can often be detected in longitudinal section and sometimes are recognizable in transverse section. Their presence does, however, remove one of Kylin's objections to placing this genus in the Champiaceae, to which its tetrahedral sporangia relate it.

Care must be taken to distinguish this from Wurdemannia and from *Gigartina acicularis.* The restriction of the tetrasporangia to short branchlets is not in conformity with Børgesen's description, but does suggest Collins' *Cordylecladia peasiae,* with which on the other hand the medullary structure does not at all agree.

REFERENCES: Børgesen 1913–20, 1924, Taylor 1928, 1933, 1935, 1943, Taylor and Arndt 1929, Hamel and Hamel-Joukov 1931; *Cordylecladia irregularis,* Harvey 1853, Murray 1889, Collins 1901, Howe 1918a, b, 1920; *C. ringens,* Collins and Hervey 1917 *p. p.; Chylocladia ringens,* Mazé and Schramm 1870–77, Murray 1889; *Laurencia perforata,* Collins 1901, Collins and Hervey 1917.

Gastroclonium Kützing, 1843

Plants with a definite, terete, strong and solid stem below, irregularly dichotomously branched; above hollow, divided by thin septa; cystocarps consisting of carposporangia borne directly on a large fusion cell, branching gonimoblast filaments being absent; pericarps without a special pore.

Gastroclonium ovatum (Hudson) Papenfuss Pl. 61, fig. 2

Plants with a fibrous holdfast, supporting several erect stems to 15 cm. tall, 1–2 mm. diam.; these dichotomously or irregularly branched, naked below, above bearing branchlets in racemose fashion; branchlets obovate, becoming subtorulose, of several sections, generally simple, 3–20 mm. long, 1–3 mm. diam., rarely rebranched.

Colombia, Brazil.

REFERENCES: *Gastroclonium ovale,* Martens 1870, Taylor 1941b.

Champia Desvaux, 1808

Plants bushy, tufted from a small fibrous base, the branches repeatedly alternately divided, hollow, septate at intervals; growth from a group of apical cells; the adult axis composed of an interrupted peripheral layer of small cells, a deeper continuous layer of large thin-walled cells, and a series of longitudinal filaments which run from the apex of the plant down the outer part of the central cavity and are anchored at each septum and provided with a gland cell in each cavity; sporangia tetrahedral, scattered, numerous in the branches, formed from cortical cells so that they lie just below the cortical layer; spermatangia formed in patches over the surface of the branches of small plants; after fertilization the auxiliary mother cell enlarges greatly and fuses with other cells to form a nutritive cell, while the gonimoblast filaments grow out from the auxiliary, only the end cells forming carposporangia; cystocarps surrounded by delicate filaments within ovate, sessile, ostiolate pericarps.

KEY TO SPECIES

1. Branches of the plant rarely over 1.5 mm. diam., generally 0.5–1.0 mm. .... C. parvula
1. Thicker branches of the plant 2–4 mm. diam. .... C. salicornoides

Champia parvula (C. Agardh) Harvey Pl. 61, fig. 4

Plants tufted, the tufts dense, pale, dull red, pinkish brown or greenish, crisply membranous in texture, to 3–10 cm. tall; branching alternate, the branches 0.5–2.0 mm. diam., tapering, segmented, the hollow segments cask-shaped, 1.0–1.5(–5.0) diameters long, the tips of the branches obtuse; tetrasporangia 55–100 μ diam., numerous in the segments; cystocarps relatively few, in prominent superficial pericarps.

Bermuda, North Carolina, Florida, Bahamas, Caicos Isls., Jamaica, Hispaniola, Puerto Rico, Virgin Isls., Guadeloupe, Dominica, Barbados, Venezuela, Brazil. From Thalassia and other sea grasses, chiefly in shallow water but dredged from a depth of 37 m. These plants are in general much smaller, with more slender branches and less swollen segments, than those from farther north.

REFERENCES: Harvey 1853, Hauck 1888, Murray 1889, Collins 1901, Vickers 1905, Børgesen 1913–20, 1924, Collins and Hervey 1917, Howe 1918a, 1920, Hoyt 1920, Schmidt 1923, 1924, Taylor 1928, 1933, 1936, 1942, 1954a, Hamel and Hamel-Joukov 1931, Joly 1957; *Chrysymenia chylocladioides p. p.*, and *C. tenera*, Mazé and Schramm 1870–77, Murray 1889; *Chylocladia parvula*, Mazé and Schramm 1870–77; *Lomentaria parvula*, Zeller 1876.

Champia salicornoides Harvey Pl. 61, fig. 5

Plants from a basal disk and a short stipe to 4–12 cm. tall, subgelatinous to softly membranous, pale red in color; paniculately branched, the branches repeatedly irregularly opposite or occasionally verticillate, strongly constricted at the nodes, terete or a little compressed, 1.5–4.0 mm. diam., the segments about as long as broad; thallus wall of a single inner layer of large cells, and between these some smaller ones which may be secretory, like the cells midway along the interseptal filaments; cortex of very small cells, often discontinuous over the central portions of the large inner cells; general cuticle thick; tetrasporangia to 80 μ diam., scattered over the thallus, strongly projecting from between the cortical cells; the sessile pericarps broadly conical, scattered over the branches.

Florida, Bahamas, Caicos Isls., Puerto Rico, Virgin Isls., Guadeloupe, Barbados, Brazil. Seldom found in shallow water, this species has been dredged from a depth of 27 m.

REFERENCES: Harvey 1853, Børgesen 1913–20; *C. salicornioides*, Hauck 1888, Vickers 1905, Howe 1920, Taylor 1928; *Chylocladia salicornoides*, Murray 1889; *C. muelleri* and ?v. *cactoides*, *C. salicornia*, *Chrysymenia chylocladioides p. p.*, *C. subverticillata*, Mazé and Schramm 1870–77, Murray 1889.

UNCERTAIN RECORD

Champia compressa Harvey

Brazil. Type from South Africa. De Toni 4:561; *C. viellardi,* Zeller 1876.

CERAMIALES

Plants generally slenderly filamentous and branched, sometimes coarse, strap-shaped, or membranous; naked or corticated, with the central-filament type of structure developing from an apical cell; asexual reproduction by tetraspores formed in sporangia external or more or less covered by the cortex of ordinary branches, or grouped in specialized sporiferous branchlets or stichidia; sexual reproduction by spermatangia borne on the axial filaments, often in masses, or covering areas of the flat bladelike plants, or covering special colorless branchlets (spermatangial clusters or "antheridia"), and by carpogonia on carpogenic branches borne, together with sterile cells, on the axial or the pericentral cells; auxiliary cells formed after fertilization from the cell supporting the carpogenic branch; gonimoblasts developed from the auxiliaries to form a mass of carposporangia, which may be naked, partly enveloped by branchlets, or covered by a pericarp.

KEY TO FAMILIES

1. Foliaceous or, if bushy, with flat branches
   DELESSERIACEAE, p. 542
1. Not foliaceous and rarely with flat branches 2
2. Axes clothed with uniseriate, branched, chromatophorous filaments .... DASYACEAE, p. 555
2. Axes naked, corticated, or covered with branchlets 3
3. Axes uniseriate, but the nodes often corticated
   CERAMIACEAE, p. 492
3. Axes polysiphonous .... RHODOMELACEAE, p. 568

CERAMIACEAE

Plants usually bushy; branches usually terete, in some genera uniseriate, in others corticated; growth from the tip producing an axial row of cells, true colorless hairs often present, also hyaline hairlike extensions from the tips of branches; cortication if present usually developed first about the nodes, consisting of single ring of cells, or spreading over the internodes in more or less filamentous fashion, later appearing parenchymatous, or else consisting of an investment of rhizoidal filaments; sporangia superficial or stalked on the branches, single or whorled at the nodes, or in corticated species in the cortex, ultimately dividing in tetrapartite or tetrahedral fashion; spermatangia developed on special determinate branchlets, forming small colorless clusters, or covering the cortex of portions of larger species; carpogenic branches of four cells, borne on supporting cells, which may also give rise to sterile cells and after fertilization to one or two auxiliaries near the carpogonium; cystocarp composed of groups of gonimoblast filaments, the outer cells of which produce the carposporangia; cystocarps naked, enveloped in jelly, or partly enclosed by subtending filaments from below.

KEY TO GENERA

1. Filaments ecorticate, or with rhizoidal cortication, rarely the surface when heavily invested falsely appearing parenchymatous 2
1. Filaments regularly corticated at the nodes, the cortication essentially parenchymatous, often spreading over the internodes 13

2. Plants bushy, with the main axes at least distally more or less closely covered with determinate branched ramelli 3
2. Plants of open branching, commonly without sharp differentiation between axes and determinate branchlets 4

3. Plants with the divisions plumose, very soft; often iridescent WRANGELIA, p. 502
3. Plants very gelatinous, the ramelli very short; often whitened by a little calcareous deposit .... CROUANIA, p. 495

4. Branching alternate or seemingly dichotomous 5
4. Branching opposite, whorled or secund, at least in some parts of the plant 10
4. Branches associated in a network, without any main axis HALOPLEGMA, p. 513

5. Without distinction between axes and branches; the cells large, often moniliform; often with branched trichoblasts at the distal ends of the cells .... GRIFFITHSIA, p. 514

5. Branched trichoblasts absent; distinctions between axes and secondary branches generally evident 6

6. Branchlets usually potentially indeterminate 7
6. Branchlets of the last degree clearly determinate, with 1–3 papilliform cells at the nodes .... DOHRNIELLA, p. 501

7. Involucres absent 8
7. Cystocarps surrounded by slender involucral filaments MESOTHAMNION, p. 512

8. Asexual reproduction generally by radiating chains of seirospores; vegetative cells uninucleate .... SEIROSPORA, p. 510
8. Asexual reproduction by tetraspores; vegetative cells plurinucleate 9

9. Tetrasporangia sessile on the branchlets ... CALLITHAMNION, p. 504
9. Tetrasporangia terminating special lateral branchlets COMPSOTHAMNION, p. 511

10. Plants erect and bushy, one to several centimeters in height 11
10. Plants very minute, and chiefly creeping or matted 12

11. Gland cells generally present; sporangia tetrapartite ANTITHAMNION, p. 497
11. Gland cells absent; sporangia tetrahedral .. SPERMOTHAMNION, p. 518

12. Erect branches pinnate or bipinnate, the axes 20–25 μ diam. GYMNOTHAMNION, p. 522
12. Erect branches penicillate, the last divisions secund; main erect axes 34–46 μ diam. .... GRALLATORIA, p. 496

13. Ultimate branchlets narrowly corticated at the nodes only, the indeterminate branches fully corticated .... SPYRIDIA, p. 538
13. Cortication progressive and similar in branches of all degrees 14

14. Cortication complete and regular, the cells in longitudinal rows, the nodes generally with whorls of spines CENTROCERAS, p. 537
14. Cortication complete, but irregular .... CERAMIUM, p. 523
14. Cortication incomplete, or not very regular; spines absent from the nodes 15

15. Creeping, the pinnate branches flattened ... REINBOLDIELLA, p. 536
15. Erect, terete branches regularly present, although the base may be composed of terete creeping filaments .. CERAMIUM, p. 523

Crouania J. Agardh, 1842

Plants bushy, exceedingly mucilaginous, abundantly alternately branched, the lax divisions terete, tapering to the tips; axes of a single series of large cells with contiguous whorls of about four ramelli which are polychotomous below, alternately branched above, immersed in jelly; sporangia irregularly tetrahedral to tetrapartite, formed on the distal ends of the lower cells of the ramelli; cystocarps lateral or apparently terminal on fertile divisions of the ramellar clusters, usually single, more or less immersed, but without special involucral filaments.

KEY TO SPECIES

1. Tetrasporangia single in each branchlet fascicle .... C. attenuata
1. Tetrasporangia several in each fascicle .... C. pleonospora

Crouania attenuata (Bonnemaison) J. Agardh

Plants at first tufted, 1–3 cm. tall, becoming free and entangled among other algae, when the length may reach 5 cm.; rosy, or whitened by a light deposit of lime; branching irregularly alternate, the diameter of the branches usually about 0.2–0.5 mm.; axial cells to 170 μ diam., 300 μ long or somewhat more, ecorticate; ramuli spreading, the basal cell about 25–30 μ diam., 30 μ long, the upper ramelli with cells 6–8 μ diam. and 2–4 diameters long, sometimes blunt, sometimes continued into hair tips; tetrasporangia solitary on the distal ends of the basal cells of the ramuli, 40–50 μ diam., 60–65 μ long; spermatangia in groups of 2–3, terminating the divisions of the ramelli; cystocarps usually single, each in rather short branch.

Bermuda, Florida, Bahamas, Cuba, Jamaica, Hispaniola, Virgin Isls., Guadeloupe, Barbados. Not uncommon in shallow water, growing on various objects, and dredged to a depth of 20 m.

REFERENCES: Hooper 1850, Harvey 1853, Mazé and Schramm 1870–77, Murray 1889, Collins 1901, Vickers 1905, Børgesen 1913–20, Collins and Hervey 1917, Howe 1918a, 1920, Taylor 1928, 1954a, Taylor and Arndt 1929.

Crouania pleonospora Taylor

Plants to 6.5 cm. tall, tufted, very lightly calcified below, but in the upper parts rosy, free of lime; moderately branched, the branches markedly contracted at the base, 0.25–1.0 mm. diam.; axial cells to 250–400 μ diam., subequal to about twice as long as the diameter, in the smaller branches uncorticated, but later partly covered by a few rhizoidal filaments which first arise from the basal cells of the ramuli, and which in the oldest parts may loosely but thickly cover and obscure the axes; ramuli whorled, borne somewhat above the middle of each supporting cell, di- polychotomous, the basal cells about 30 μ diam., 50 μ long, the cells of the end segments about 8 μ diam., 24 μ long; tetrasporangia most common in later branchlets, single to several in a ramular fascicle at various forks, to 60–90 μ diam.; cystocarps solitary or 2 together, in small, spindle-shaped branches on small plants.

Bermuda, Florida. Growing in shallow water on various objects, particularly on coarse algae, and dredged to a depth of 9 m.

REFERENCE: Taylor 1928.

UNCERTAIN RECORDS

Crouania australis (Harvey) J. Agardh

Guadeloupe. Type from Tasmania or Australia. De Toni 4:1418; Murray 1889; *C. attenuata* v. *australis*, Mazé and Schramm 1870–77.

Grallatoria Howe, 1920

Plants creeping, filiform, ecorticate, dorsiventral, bearing two or three (or falsely five) short branchlets at a node, two being lateral, one dorsal; the lateral branchlets forking at the base, one division producing a short hapteral branchlet, the other an erect one; the dorsal branches repeatedly subdichotomous or trichotomous; chromatophores irregular or spiral; sporangia on branchlets from the dorsal branches, usually tetrahedral.

Grallatoria reptans Howe

Plants forming a velvety mat, dull purple brown, the repent filaments simple or sparingly dichotomous, 38–66 μ diam., the cells

1.5–2.0 diameters long; opposite determinate lateral branchlets from nearly every node, of which the deflexed hapteral divisions are 150–450 μ (4–6 cells) long, simple or once forked, 20–26 μ diam., the erect divisions simple, 200–400 μ (4–8 cells) long, about 20 μ diam. below, 8 μ diam. at the apex; dorsal erect branches 1–4 mm. high, usually borne at alternate nodes of the repent filaments, 34–46 μ diam. below, the basal cell 1.5–2.0 diameters long, but the cells above 3–6 diameters; upper branching penicillate, the branchlets subsecund along the inner sides of the penultimate divisions; sporangia tetrahedral, generally solitary on one-celled pedicels, obovoid or pyriform, 40–50 μ diam., 52–64 μ long; dense ovoid, globose, or pyriform cysts 60–150 μ diam. occasional near the ends of the deflexed filaments.

Bahamas. Forming a coating in rock crevices and on shells and algae near low-water mark, where exposed to the surf.

REFERENCES: Howe 1920.

Antithamnion Nägeli, 1847

Plants tufted, of uncorticated uniseriate filaments; branches often alternate below, but above repeatedly opposite or whorled, their attachment often considerably below the end of the supporting cell; sometimes one element in a branch pair suppressed, or secund, or in the ultimate branchlets truly alternate; cells uninucleate, with many small rounded or bandlike chromatophores; gland cells often present; sporangia tetrapartite, carried on the smaller branchlets, or often replacing branchlets of the last order; spermatangia in patches on ultimate branchlets; carpogenic branches formed on the lowest cells of such branchlets, the cystocarp consisting of a mass of carposporangia upon a stalk of a few sterile cells.

KEY TO SPECIES

1. Plants obviously flat-spreading; branchlets unilaterally seriate on the upper side of the supporting lateral branches A. plumula, p. 500
1. Plants usually bilaterally branched 2
2. Plants forming dense tufts 2–4 cm. tall, rarely more, the main axes sparingly divided 3

2. Plants repeatedly branched, even if small, and the tips not densely brushlike 4

3. Branchlets distichous, stout, strongly tapered; the "intercalary spores" (?gland cells) immersed, occupying most of the segment within which they lie . . . . A. cristatum, p. 499
3. Branchlets in 4 rows, slender, the branch tips densely brushlike; gland cells superficial, applied to the upper side of 3 cells A. cruciatum, p. 498

4. Determinate branchlets often ending in gland cells A. butleriae, p. 499
4. Determinate branchlets bearing gland cells laterally A. antillarum, p. 499

Antithamnion cruciatum (C. Agardh) Nägeli

Plants tufted, somewhat intricate below, alternately branched, but above rather sparingly divided; the branches long, at the tips the branchlets densely tufted or appearing ocellate-congested; height to 5 cm., color dull rose-red; the main axis below 50–90 μ diam., cells 90–300 μ long; lateral branches spreading, short, of rather even length, opposite in alternately placed pairs, or in fours, producing a 4-rowed aspect; branchlets alternate, in 2 rows, or somewhat unilateral, 9–14 μ diam., with cells 4–8 diameters long; gland cells numerous, near the bases of the ultimate branchlets, applied to the upper side of 3 cells; tetrasporangia 75–85(–100) μ long, 50–55 (–70) μ diam., replacing the branchlets of the last order, mostly on the upper side of the short branches, sessile or short-stalked; cystocarps to 400 μ diam.

V. **radicans** (J. Agardh) Collins and Hervey. Primary filaments decumbent and attached by rhizoids, forming a tangled mass, and these primary filaments producing short, erect branchlet systems above in secund fashion, and eventually a few lax, indefinite branches as well.

Bermuda, North Carolina, Barbados. Growing on coarser algae in shallow water, or at very moderate depths.

REFERENCES: Vickers 1905, Howe 1918a, 1920. For the v. *radicans*, Collins and Hervey 1917, Blomquist and Humm 1946.

Antithamnion antillarum Børgesen

Plants tufted, arising from decumbent filaments having cells 40–50 μ diam., 70–200 μ long; attached to the substratum by multicellular haptera with discoid ends; erect branches opposite on the rhizomatous axes, each with a distinctive short basal cell, the cells above it 20–40 μ diam., 50–150 μ long; erect branches alternately distichously branched, the subdeterminate secondary branches short, bearing one, rarely to 4 branchlets on the outer side, these of about 2–4 cells about 10 μ diam., 13 μ long, with a gland 13 μ wide, 20 μ long on the inner side in contact with 2 cells of the branchlet; tetrasporangia attached on the axillary side of the basal branchlet cells, sessile, subcylindrical, tetrapartite, 40 μ diam., 90 μ long.

Virgin Isls. Growing epiphytically on larger algae in a sheltered place in shallow water.

REFERENCE: Børgesen 1913–20.

Antithamnion butleriae Collins

Plants erect, the axes simple or with a few leading branches, which may be dichotomous, alternate, or occasionally opposite, the diameter near the base of the branch 30 μ, cells 3–6 diameters long, the walls thick; leading branches ecorticate, naked below, above each cell normally bearing a pair of determinate branches issuing at about two-thirds the height of the cell; the lowest simple, subulate, of 2–6 isodiametric cells; farther up the frond these branches are compounded with similar smaller branchlets, appearing first on the lower side of the ramulus, but usually dorsiventrally opposite; terminal cell of a determinate branch commonly glandular, short, touching the basal branchlet cells on each side; branchlet tip cells also sometimes glandular.

Bahamas, Caicos Isls., Jamaica, Virgin Isls., Barbados. Plants of very shallow water, growing on coarser algae.

REFERENCES: Collins 1901, Vickers 1905, Børgesen 1913–20, Howe 1920.

Antithamnion cristatum (Kützing) n. comb.

Plants to about 2 cm. tall, the axes sparingly alternately branched, chiefly near the base, the divisions long, percurrent, main axis otherwise naked below; the segments 2–3 diameters long, with thick walls; primary branches plumose, the axial cells short, about as long as broad, and about 80 μ diam.; when young, closely distichous, the branchlets opposite, short, stout, strongly tapering, acute, with cells 23–30 μ diam. and 0.5–1.0 diameter long; in old branches the fertile branchlets subverticillate; intercalary "spores" or gland cells conspicuous in the branchlets, solitary or seriate, segregated by longitudinal division of branchlet cells and occupying most of the segments concerned; tetrapartite sporangia sessile, secund, solitary or seriate, oval or obovoid, about 27 μ diam.

Brazil.

REFERENCES: *Sporacanthus cristatus,* Kützing 1855 (Tab. Phyc. 5, p. 24, pl. 82), Martens 1870. If the structures called "intercalary spores" by Kützing are actually monospores this generic assignment may not hold.

Antithamnion plumula (Ellis) Thuret

Plants bushy, bright rose, considering their delicacy rather stiff, the flattened habit of the fronds evident at a glance, to 5–12 cm. tall; alternately to irregularly branched, branches 100–200 μ diam., the cells 3–5 diameters long; all main branches beset with opposite (infrequently tetrastichous), short, more or less recurved branches about 0.5–0.7 mm. long, 50 μ diam., the cells 2–3 diameters long, these branches bearing on the upper side a series of ultimately spinuliform branchlets which taper from 20 μ at the base to their indurated tips; gland cells present on the upper sides of the branchlets, each attached to a single cell, sometimes seriate; tetrasporangia elliptical to globose, to 35–40 μ long, 25–30 μ diam., sessile or on short stalks, on the lower branchlets of the upper part of the plant; cystocarps large, usually 2–4 together, on the upper branches and more or less surrounding the supporting branchlet; spermatangia forming elliptical patches on the upper sides of the branchlets.

Bermuda, Florida. Apparently a plant of moderately deep water, growing on rocks, wharves, shells, etc.

REFERENCES: Murray 1889, Taylor 1929a.

UNCERTAIN RECORDS

Antithamnion floccosum (Müller) Kleen

Bermuda, Grenada. Type from northwestern Europe. De Toni 4:1411; Murray 1889.

Antithamnion terniramеum Hamel and Hamel-Joukov

Martinique, the type locality. Apparently a *nomen nudum*. Hamel and Hamel-Joukov 1931.

Ballia Harvey 1840

UNCERTAIN RECORDS

Ballia prieurii Kützing

French Guiana, the type locality, and apparently fresh-water. De Toni 4:1396; Montagne 1850.

Ballia pygmaea Montagne

French Guiana, the type locality, and apparently fresh-water. De Toni 4:1397; Montagne 1850.

Dohrniella Funk, 1922

Plants erect, the main axes sparingly branched, in the upper portion radially and alternately bearing a single simple cell; branchlets with one or more short basal cells, above the cells cylindrical and toward the tips at each node bearing two or three small lateral papilliform cells which occasionally in turn bear simple hairs; chromatophores numerous, in general linear, but short and nearly round in the papillar cells; lenticular gland cells sometimes cut off laterally from the branchlet cells near their distal ends; sporangia tetrahedral, pedicellate, single at the ends of branchlet cells; spermatangial clusters rounded, similarly situated; cystocarps unknown.

Dohrniella antillarum (Taylor) Feldmann-Mazoyer Pl. 65, fig. 1

Plants soft and delicate, 1–2 cm. tall, repent and attached by haptera below, the creeping portion 32–48 μ diam., the cells 95–420 μ long; ascending above, the erect axes seldom branched, about 40–47 μ diam., the cells cylindrical, below 380–420 μ long, in the upper part hardly 60 μ long; determinate branchlets borne on each segment of the upper parts of the plant, spirally disposed, single at each node, 5–15 cells long, the lowermost 1–3 cells short, those above 13–19 μ diam., 38–60 μ long; the distal two-thirds of the axial cells of each branchlet bearing at the outer end 1–3 radially disposed papilliform cells about 5.5–7.5 μ diam., 6–11 μ long; hairs and gland cells absent; cell walls rather thick and gelatinous, especially in the main axes; reproductive organs not seen.

Hispaniola, Guadeloupe. Growing in shallow water on dead Gorgonians and doubtless on other objects.

REFERENCES: Feldmann-Mazoyer 1940; *Actinothamnion antillarum,* Taylor 1929.

Wrangelia C. Agardh, 1828

Plants filamentous, bushy, the apical cells developing persistent central axes of rather large cells, with the indeterminate main branches from it usually two-ranked; main axes ecorticate, or corticated by filaments developed from the nodes and the bases of the determinate lateral branchlets, which cortex may become pseudoparenchymatous; reproductive organs borne on the branchlets and toward the apices of the main branches, more or less involucrate; sporangia tetrahedral, pedicellate; spermatangia in clusters which are terminal on short special branchlets developed from the nodes; cystocarps terminal on the ultimate branchlets, densely surrounded by slender involucral filaments.

KEY TO SPECIES

1. Plants solitary or clustered, not forming extensive colonies 2
1. Plants generally in considerable colonies, little more than 1.0–1.5 cm. tall .... W. argus
2. Plants large, regularly branched in 2 ranks, the axis becoming completely corticated by regular longitudinal filaments W. penicillata
2. Plants smaller, irregularly few-branched, the axis surrounded at the nodes by irregular tangled filaments .... W. bicuspidata

Wrangelia argus Montagne Pl. 66, figs. 7, 8

Plants in extensive, almost turflike colonies 1.0–1.5 cm. tall or little more, very soft and often iridescent, purple-red; plumose-branched in 2 indistinct ranks; axes not corticated or with a very few filaments about the nodes; determinate lateral branchlets terminating in attenuate cells, occasionally in single spinelike cells, or the terminal cells deciduous; tetrasporangia produced abundantly at the nodes, 60–70 μ diam., surrounded by slender, short-celled involucral filaments.

Florida, Mexico, Bahamas, Caicos Isls., Jamaica, Puerto Rico, Virgin Isls., Guadeloupe, Martinique, Barbados, Costa Rica, Panama, Colombia, Netherlands Antilles, Venezuela, Trinidad, Tobago. Common on rocks in shallow water and exposed situations.

REFERENCES: Collins 1901, Vickers 1905, Børgesen 1913–20, Howe 1920, Taylor 1928, 1929b, 1936, 1942, Hamel and Hamel-Joukov 1931, Feldmann and Lami 1937; *Callithamnion beuii,* Mazé and Schramm 1870–77, Murray 1889; *Wrangelia plebeja,* Harvey 1853, Mazé and Schramm 1870–77, Murray 1889, Sluiter 1908.

Wrangelia bicuspidata Børgesen Pl. 66, figs. 9, 10

Plants lax, sparsely and somewhat irregularly alternately branched, the axes to about 8 cm. in length; axes loosely surrounded at the nodes by tangled filaments; branchlets ending in attenuate cells, frequently tipped with 1–2 spines, or these deciduous, the tips then appearing blunt; tetrasporangia to 60–90 μ diam., surrounded by a few incurved, tapering involucral filaments about 25 μ diam.; sexual reproduction not seen.

Bermuda, Florida, Bahamas, Caicos Isls., Virgin Isls., Hispaniola. Occasional on marine phanerogams and other algae, perhaps other objects; collected in water of very moderate depth and dredged from as much as 40 m.

REFERENCES: Børgesen 1913–20, Howe 1920, Taylor 1928, Taylor and Arndt 1929; *Spondylothamnion multifidum,* Phyc. Bor.-Am. 891.

Wrangelia penicillata C. Agardh Pl. 66, figs. 5, 6; Pl. 74, fig. 5

Plants sturdy, widely and regularly alternately branched in 2 ranks, the axes to about 0.5–1.5 dm. in length, the branches with a spread of 1–2 dm.; axes corticated by the downgrowth of filaments from the nodes, these filaments becoming approximated and giving rise to a rather regular pseudoparenchymatous cortex; branchlets terminating in attenuate cells, frequently in single spinelike cells, or these deciduous, the tips then appearing blunt; tetrasporangia to 80 μ diam., the involucral filaments loosely arranged, the end cells 3–5 diameters long; spermatangial clusters terminal on short lateral branchlets developed at the nodes; clusters 55–60 μ diam., surrounded by a few stout involucral filaments; cystocarps terminal on short branchlets, surrounded by very many slender incurved involucral filaments.

Bermuda, Florida, Bahamas, Caicos Isls., Jamaica, Virgin Isls., Guadeloupe. Epiphytic on coarser algae, chiefly in water of moderate to considerable depth, and dredged to a depth of 30 m.

REFERENCES: Harvey 1853, Mazé and Schramm 1870–77, Hemsley 1884, Murray 1889, Børgesen 1913–20, Collins and Hervey 1917, Howe 1918a, 1920; Taylor 1928; *W. verticillata,* Mazé and Schramm 1870–77; *Dasya pellucida,* Mazé and Schramm 1870–77, Murray 1889.

Callithamnion Lyngbye, 1819

Plants tufted, erect from a disk, fibrous holdfast, or decumbent strands of monosiphonous branching filaments; ecorticate or with rhizoidal cortications, the branching seemingly dichotomous, or alternate; cells plurinucleate, with several to many small rounded to band-shaped chromatophores; sporangia borne on the upper side of the branches, tetrahedral, or rarely transversely bipartite; spermatangia forming small colorless tufts, borne near the bases of the branchlets on the upper side; in preparation for the carpogenic branch a fertile axial segment early forms two large opposite auxiliary mother cells, one serving as the support of the carpogenic branch, which is of four cells and lies partly across the fertile segment; after fertilization the auxiliary mother cells each cut off above large auxiliaries which jointly give rise to the paired gonimoblast masses, of which all cells except the lowest become carpospores.

KEY TO SPECIES

1. Uncorticated, or with a few rhizoidal filaments not significantly covering the axis 2
1. At least the base of the axis substantially covered by the rhizoidal filaments 3

2. Completely uncorticated 4

2. Main axes near the base with a few rhizoids descending from the bases of branches or as supplementary holdfasts, but these not covering the axis 5
3. Main axes near the base lightly corticated in fully developed plants 6
3. Main axes corticated heavily through much of their length 7
4. Upper branching conspicuously distichous, the branchlets 33–37 μ diam. near their bases, sharply tapering, becoming recurved, very acute .... C. uruguayense, p. 506
4. Upper branching subdichotomous, the branchlets 10–20 μ diam., the end cell rounded .... C. halliae, p. 505
5. Main filaments in the central parts of the plant about 160 μ diam.; tetrasporangia about 27 μ diam.; cystocarps geminate, trilobed-cordate; plants 2–4 cm. diam. ...... C. cordatum, p. 507
5. Main filaments 40(–140?) μ diam., tetrasporangia about 28–40 μ diam.; plants 3–12 cm. tall .... C. byssoides, p. 506
6. Upper branching subdichotomous, but becoming corymbose-crowded toward the apices of the branchlets, the ultimate branchlets conspicuously more slender .... C. corymbosum, p. 507
6. Upper branching alternate, bilateral, the branchlets not crowded toward the apices, not particularly slender C. roseum, p. 508
7. Habit exceedingly soft, seemingly subgelatinous ... C. herveyi, p. 508
7. Habit more spongy, the branchlets relatively rigid C. felipponei, p. 508

Callithamnion halliae Collins

Plants bright rose, to 5 cm. tall, usually with numerous percurrent axes, straight below, becoming flexuous near the top, with 1–2 series of similar alternate branches; short lateral divisions at first seemingly subdichotomously forked, but with growth the later ones evidently dichotomous; filaments not corticated, to 200 μ diam. below, the ultimate divisions 10–20 μ diam., in general the cells about 4 diameters long except where the filaments are densely branched; tip cells of the branches obtuse, without terminal hairs; tetrasporangia pyriform, sessile and often seriate on the upper side of the ramelli, spermatangial clusters tufted, in the same position; cystocarps single or a few together in the forking of a lateral branch or lateral on a main filament, depressed-spherical, not markedly lobed.

Bermuda, Florida, Bahamas, Guadeloupe, Netherlands Antilles.

REFERENCES: Phyc. Bor.-Am. 698, Collins 1906, Collins and Hervey 1917, Howe 1920, Taylor 1928, 1942; *C. ellipticum* v. *major*, Mazé and Schramm 1870–77, Murray 1889.

Callithamnion uruguayense Taylor Pl. 41, figs. 5–7

Plants rose-red or darker, bushy, to 2.5–3.5 cm. tall, the lower branches entangled, the upper free and complanate with percurrent main axes; chief branching irregularly alternate, the upper branches ascending; branchlets distichous, alternate, stiff, gradually tapering to acute tips, ascending above, below divergent and recurved; base with several long rhizoidal attachments but without cortication; main axes about 115 μ diam., the cells about 100 μ long; in the central parts of the plant filaments are 50–56 μ diam., the cells 200–220 μ long; the branchlets 320–630 μ long, their cells 33–37 μ diam., 55–60 μ long; tetrasporangia 1–5 together, seriate on the upper side of the branchlets, sessile, oval, 50–55 μ long, 42–45 μ diam.; cystocarpic plants somewhat irregularly branched, the cystocarps lateral, bilobed, each portion oval, 175–210 μ long, 95–100 μ diam.

Cuba, Brazil, Uruguay. Plants growing on rocks in moderately shallow water.

REFERENCES: Taylor 1939a, Joly 1957.

Callithamnion byssoides Arnott *in* Hooker

Plants forming very soft globose tufts 3–8(–12) cm. high, usually light pink; filaments extremely delicate, the lower branching obscured by their slenderness and the involved habit, though but slightly corticated; axes repeatedly pinnately branched, the segments to 40(–140?) μ diam., the outer secondary branches freer and flexuous, pinnate with many completely pinnately compound smaller branches, the ultimate branchlets very slender, at the base about 10–20 μ diam., at the terminal cells 5–7 μ diam., the cells 6–10 (or more?) diameters long; tetrasporangia sessile on the upper side of the branchlets, oblique-obovate or subglobose, 38–52 μ long, 28–40 μ diam.

V. **jamaicensis** Collins. Plants compact, the branches shorter and stouter than in the type, densely set, the terminal branchlets often arranged more like *C. corymbosum*.

Bermuda, Jamaica, Virgin Isls., Guadeloupe, Martinique.

REFERENCES: Vickers 1905, Børgesen 1913–20, Taylor 1928, Hamel and Hamel-Joukov 1931. For the v. *jamaicensis*, Collins 1901, Phyc. Bor.-Am. 443, Collins and Hervey 1917; *?C. hypneae,* Mazé and Schramm 1870–77, Murray 1889.

Callithamnion cordatum Børgesen

Plants 2–4 cm. tall, bushy, attached by several rhizoidal outgrowths from the lower cells, the main axis straight below, flexuous above, sparingly branched below, abundantly above; hardly corticated, the lower basal branch cells generally sending down slender rhizoidal filaments along the axis, though they do not envelop it; axis 200 μ diam. very near the base and the cells very short and thick-walled, above about 160 μ diam., the cells about 300 μ long and slightly enlarged at the nodes; the branchlets 8 μ diam. with very long cells; branching alternate, but subdichotomous in the ultimate divisions, the lesser branches and branchlets incurved; hair tips occasionally seen in the youngest parts of the plant; sporangia sessile, solitary on the distal end and upper side of the lesser branch cells, obovate-oblong, about 27 μ diam., 40 μ long; spermatangial clusters small, flat against the branches in the same position; cystocarps binate, the cordate divisions trilobed.

Bermuda, Virgin Isls. Epiphytic on Gracilaria as dredged at a depth of 27 m.; perhaps also in shallow water.

REFERENCES: Børgesen 1913–20, Collins and Hervey 1917, Howe 1920.

Callithamnion corymbosum (J. E. Smith) C. Agardh

Plants forming erect, rounded, soft, bright pink tufts 2–6 cm. tall; below irregularly or pinnately branched, the branches not much entangled, not particularly delicate, 250–450 μ diam., often a little corticate at the base, the cells 4–10 diameters long; above more densely branched, the lesser branchlets dividing dichotomously, forming corymb-like tufts, the tips usually bearing delicate colorless hairs; tetrasporangia 2–3 together on the inner side of the branchlets, sparsely produced.

Bermuda, Jamaica, Brazil. Epiphytic on coarse algae and doubtless on other objects at very moderate depths.

REFERENCES: Collins 1901, Collins and Hervey 1917, Howe 1918a; *Phlebothamnion versicolor*, Martens 1870.

Callithamnion roseum (Roth) Harvey

Plants erect, tufted, to 4–7 cm. tall, rather deep pink or red and relatively firm; lower branches little entangled, or almost free, repeatedly alternately radially branched below, the segments 125–350 μ diam., 3–4 diameters long, eventually a little corticated; above bilaterally pinnate, and subcorymbose near the tips; the branchlets 38–42 μ diam., little tapered, cells 2–4 diameters long; tetrasporangia few, generally paired, nearly spherical, mostly unilateral on the inner side of the branchlets, 45–70 μ diam., 60–85 μ long.

Bermuda.

REFERENCE: Collins and Hervey 1917.

Callithamnion herveyi Howe

Plants dull purple, very soft, 2–4 cm. tall, repeatedly irregularly branched or obscurely branched in 4 rows, the branchlets more or less distichous or subdichotomous and complanate; main axes chiefly and below heavily corticated by rhizoids, tomentose, 0.30–0.35 mm. diam. below; cells 40–80 μ diam., 1.5–2.0 diameters long and thick-walled; lesser branches cylindric-plumose in habit, 0.8–1.5 mm. long, the ramelli spreading or erect, over-all 0.3–0.6 mm. diam., cells of the ultimate branchlets 1.5–2.0(–3.0) diameters long, the tip cells 8–12 μ diam., sometimes piliferous; tetrasporangia mostly solitary or occasionally subsecund, obovoid or subglobose, 38–40 μ diam.; monosporangia terminal, scattered or irregularly clustered, sometimes concatenate in twos or threes, obovoid to subglobose, 36–65 μ long; spermatangial clusters hemispherical to subglobose, 30–50 μ diam.; cystocarps often geminate, subglobose, 100–220 μ diam.

Bermuda.

REFERENCES: Howe 1918a; *C. hookeri*, Collins and Hervey 1917.

Callithamnion felipponei Howe

Plants to 9.5 cm. tall, virgate and subsimple or with several virgate branches from near the base; main axes corticated, bearing few to many radially disposed short branches 3–7 mm. long with more or less dense, tufted or alopecuroid, erect or somewhat spreading fascicles of ecorticate ramuli, these carrying rigid alternate-distichous, irregularly spiral, or subdivaricately dichotomous incurved or sometimes recurved, acuminate (or in age, obtuse) ramelli; ramelli 75–160 μ diam. below, to 13 μ diam. at the tips, the cells 1.0–2.5 diameters long; tetrasporangia 40–50 μ diam., 50–60 μ long.

Brazil, Uruguay. Epiphytic on Gelidium and probably on other algae.

REFERENCES: Howe 1931, Taylor 1939a, Joly 1957.

UNCERTAIN RECORDS[1]

Callithamnion amentaceum Crouan

Guadeloupe, the type locality. De Toni 4:1341; Mazé and Schramm 1870–77, Murray 1889.

Callithamnion apiculatum Crouan

Guadeloupe, the type locality. De Toni 4:1341; Mazé and Schramm 1870–77, Murray 1889.

Callithamnion corniculifructum Crouan

Guadeloupe, the type locality. De Toni 4:1341; Mazé and Schramm 1870–77, Murray 1889.

Callithamnion dasytrichum (Montagne) Montagne

Florida, Brazil. Type from Chile. De Toni 4:1362; Phyc. Bor.-Am. 697, Taylor 1928, 1931.

Callithamnion granulatum (Ducluzeau) C. Agardh

Barbados. Type from western Europe. De Toni 4:1331; *C. spongiosum*, Vickers 1905.

Callithamnion lherminieri Crouan

Guadeloupe, the type locality. De Toni 4:1341; Mazé and Schramm 1870–77, Murray 1889.

Callithamnion polyspermum C. Agardh

North Carolina, South Carolina, Florida. Type from western Europe. De Toni 4:1315; Harvey 1853, Murray 1889, Hoyt 1920.

[1] The names here ascribed to Crouan are all essentially *nomina nuda*.

Seirospora Harvey, 1846

Plants tufted, delicate, alternately branched below, subdichotomously branched above, the ultimate branches capillary, the lower somewhat corticated; cells cylindrical, uninucleate, with many small chromatophores; sporangia on the upper branches, lateral at the nodes, tetrapartite or tetrahedral; on the asexual plants also occur seirospores, produced in radiating clusters at the ends of the branches, the outermost largest, grading to the point of origin; carpogenic branches three- or four-celled, with two auxiliaries; cystocarps with the gonimoblasts loose, entirely converted into carposporangia.

KEY TO SPECIES

1. Ecorticate .... S. occidentalis
1. Corticated below .... S. purpurea

Seirospora occidentalis Børgesen

Plants 1–2 cm. tall, attached by short rhizoids from the lower axis cells, these at the base thick-walled, about 200 μ diam., and about as long; general branching alternate or occasionally opposite, the ecorticate axes about 85 μ diam. and the cells 3–5 diameters long; final 1–2 stages of branching subdichotomous, the ultimate cells about 8–11 μ diam. and 2–10 diameters long or more; sporangia sessile on the upper side and distal end of branchlet cells, generally tetrahedrally divided, seirospores in the same position, as simple or branched rows of 3–5 short, dense, subspherical cells 18–20 μ diam., and spermatangial clusters likewise in the same position; cystocarps of 2 groups of rather elongated gonimoblasts attached to lower cells of a lesser branch system, the carpospores about 40–42 μ diam.

Florida, Bahamas, Virgin Isls., Guadeloupe. Epiphytic on Gracilaria and other algae, dredged from depths to 27 m.

REFERENCES: Børgesen 1913–20, Howe 1920, Taylor 1928, Feldmann and Lami 1937.

Seirospora purpurea Howe

Plants purplish, tufted but entangled below, 1.0–2.5 cm. tall, repeatedly alternately branched, toward the tips subdichotomous; main axes 150–250 μ diam. below, with rhizoidal cortication for about half their length; largest uncorticated cells 40–95 μ diam., subcylindrical or a little enlarged at the nodes, 1.5–2.5 diameters long, the walls 8–20 μ thick; cells of the ultimate branchlets 2–4 diameters long, at the tips 6–13 μ diam., obtuse; sporangia scattered, solitary at the nodes, obovoid or ellipsoid, tetrahedral, 50–65 μ diam.; spermatangial clusters more or less ovoid, at the nodes, about 26–40 μ diam., 48–65 μ long; cystocarps hemispherical, 300–400 μ diam., of free, erect-spreading moniliform gonimoblast filaments, the carpospores 35–40 μ long.

Bermuda. Growing on rocks in caves and on the shaded sides of walls or cliffs somewhat below low-tide line.

REFERENCE: Howe 1918a; *Callithamnion byssoideum* v. *jamaicense,* Phyc. Bor.-Amer. 2045.

Compsothamnion Nägeli, 1861

Plants bushy, the ecorticate filaments alternately branched, complanate in the lesser divisions; tetrasporangia borne on the lateral branchlets; spermatangia clustered on the upper side of the branches, or terminal; carpogonia developed near the tips of lateral branchlets but without a distinctive short axial cell below the central cell bearing the carpogenic branch; cystocarps not involucrate.

Compsothamnion thujoides (Smith) Schmitz

Plants epiphytic, small, in tufts, seldom 2 cm. tall; branching alternately 3–4-pinnate, distally markedly complanate; axes somewhat flexuous, to 60 μ diam., the cells 2–4 diameters long; branchlets slender but not much tapered, about 10 μ diam.; sporangia tetrahedral or irregular in division, about 30–35 μ diam., 50–60 μ long, terminal on little branchlets.

Bermuda. Dredged from 90 m. depth.

REFERENCE: Bernatowicz *in herb.*

Mesothamnion Børgesen, 1917

Plants bushy, attached by numerous branched, multicellular rhizoids, above delicately filamentous, freely branched; the tetrasporangia tetrahedral, borne on the upper side of the branches; spermatangial clusters subcylindrical, replacing divisions of the forks in the lower branch systems; cystocarps with four or five groups of gonimoblasts, terminal on short branches, the branches and branchlets nearby curving over and enveloping the cystocarp as an involucre.

KEY TO SPECIES

1. Spermatangial clusters sessile .... M. boergeseni
1. Spermatangial clusters stalked .... M. caribaeum

Mesothamnion boergeseni Joly

Plants tufted, 2–3 cm. tall; branching alternate below, but the distal branchlet segments subdichotomous, incurved; tetrasporangia and polysporangia subspherical, to 53 μ diam., sessile and seriate on the upper side toward the base of each fertile branchlet; spermatangial clusters sessile, subcylindrical, about 35 μ diam., 100 μ long, similarly placed.

Brazil. Epiphytic on Bryothamnion in shallow water.

REFERENCE: Joly 1957.

Mesothamnion caribaeum Børgesen

Plants tufted, the main axes stout, about 230 μ diam., the cells about 450 μ long, uncorticated; branches repeatedly alternate, of decreasing diameter, the lesser branches slightly incurved, 3–5 times pseudodichotomous and in the terminal portions about 25 μ diam., the cells to 80 μ or more in length; tetrasporangia 45 μ diam., secund and single on the branchlets at the nodes; spermatangial clusters subcylindrical, stalked, about 40 μ diam., 70 μ long.

Bahamas, Virgin Isls. Epiphytic on Gelidiella and other coarse algae in shallow water and dredged from a depth of 30 m.

REFERENCES: Børgesen 1913–20, Howe 1920.

Haloplegma Montagne, 1842

Plants forming flat spongy fronds, irregularly lobed, formed of monosiphonous filaments joined in a network, without a definite midrib system; surface meshes bearing very short erect or incurved free filaments; sporangia on the free filaments, tetrahedral; spermatangial clusters small, subcylindrical, also on the surface filaments; cystocarps in swellings projecting from the surface, with incurving involucral filaments.

Haloplegma duperreyi Montagne

Plants small, 4–5 cm. tall, dark red, partly encrusting, distally flabellate, irregularly lobed, the lobes 4–10 mm. broad, somewhat spatulate with rounded tips, often minutely fimbriate, below increasingly spongy, the surface meshes bearing short, divaricate-branched, blunt-tipped, free filaments at the nodes.

Subsp. **spinulosum** Howe. Free filaments rigid, subspinescent, in groups of 2–3, of 2–11 cells and 25–140 μ long, the cells 1.5–3.0 diameters long.

Bahamas, Jamaica, Puerto Rico, Guadeloupe, Martinique, Barbados, Grenada, Brazil. Growing on the bases of large and coarse algae, under overhanging rocks, and perhaps on other objects in shallow water where sheltered from excessive light.

REFERENCES: Mazé and Schramm 1870–77, Dickie 1874a, Piccone 1886, Murray 1889, Collins 1901, Vickers 1905, Howe 1915. For the subspecies *spinulosum*, Howe 1920.

Corynospora J. Agardh, 1851

UNCERTAIN RECORDS

(Corynospora) Monospora herpestica Vickers

Barbados, the type locality. De Toni 6:468; Vickers 1905.

(Corynospora) Monospora ?belangeri (Montagne) De Toni

Martinique, the type locality. De Toni 4:1301.

Griffithsia C. Agardh, 1817

Plants erect, bushy, of notably large-celled, falsely dichotomous or laterally dividing branches; cells multinucleate, with many small chromatophores and a large central vacuole; cells generally bearing delicate colorless repeatedly tri- to polychotomously branched hairs near their upper margins; tetrahedral sporangia whorled at the fertile nodes, partly covered by involucral cells; plants dioecious, the spermatangia very small, in large caplike sori which are usually on the distal ends of the outer cells of the fertile branches; each procarp formed upon an apical cell, which is displaced laterally by growth of the cell below; mature procarp consisting of a broad central cell, certain sterile cells, and the four-celled carpogenic branch, and after fertilization of an auxiliary in contact with the carpogonium; mature cystocarps seemingly lateral on the branches, partly covered by involucral cells.

KEY TO SPECIES

1. Plants with the cells subcylindrical throughout, or a little tumid in the young parts 2
1. Plants with most of the cells swollen, subspherical, pyriform to clavate 4

2. Tetrasporangia borne singly on the distal ends of the basal cells of the whorled trichoblasts; cystocarps borne on one-celled stalks .... G. barbata, p. 515
2. Tetrasporangia grouped about the nodes on short pedicels 3
2. Tetrasporangial sori terminating stout one-celled branchlets G. caribaea, p. 515

3. Creeping plants, without very conspicuous free portions G. radicans, p. 515
3. Erect plants with inconspicuous creeping filaments; spermatangial clusters terminal on 1–3-celled pedicels ..... G. tenuis, p. 516

4. Spermatangia whorled at the nodes and partly covered by involucral cells .... G. schousboei, p. 516
4. Spermatangia distributed over the terminal cells of the filaments as a cap, and forming a band about the distal end when occurring on subterminal cells, but without any involucre .... G. globulifera, p. 517

Griffithsia barbata (Smith) C. Agardh

Plants tufted, 1–6(–12) cm. tall; filaments dichotomously branched, 200–400 μ diam., the cells 6–8 diameters long, the branching most dense above, where the filaments are somewhat tapered; branched verticillate trichoblasts abundant about the upper ends of the younger cells; tetrasporangia borne singly on the upper ends of branchlet cells which are basal to the trichoblasts; cystocarps terminal on short one-celled branchlets, surrounded by several large, incurved, sausage-shaped, involucral cells.

Virgin Isls. In shallow water, and dredged to 30 m. depth.

REFERENCE: Børgesen 1913–20.

Griffithsia caribaea G. Feldmann

Plants tufted, erect, rose-red, attached by rhizoids and to 2.5 cm. tall; subdichotomous to alternately branched, the cells of the middle part of the plant subfusiform to subclavate, 200–250 μ diam., 5–6 diameters long; tetrasporangial branches lateral, of one coenocyte rather broader than others and reaching 300 μ diam., bearing on the end many tetrasporangia surrounded by about 8 incurved involucral cells, most of which do not reach the height of the mature group of sporangia, but on the outer side one, sometimes 2, become much larger and may reach a length of 1 mm.

Guadeloupe, Barbados. Apparently growing on other small algae in shallow water, but protected from intense light, as in a grotto.

REFERENCE: G. Feldmann [-Mazoyer] 1947.

Griffithsia radicans Kützing

Plants small, 2–3 cm. long, creeping below, attached by rhizoids with scutate ends, the filaments in part assurgent, irregularly branched; cells nearly cylindrical, about 3 times as long as broad, the terminal ones obtuse, the upper cells 180–210 μ diam., 350–500 μ long, the older ones in the middle of the plant 180–250 μ diam., 750–900 μ long; tetrasporangia 80–85 μ diam., irregularly grouped or verticillate about the upper ends of fertile cells, without any involucre, the stalk of each sporangium pyriform, in length about 0.7–1.5 times the diameter of the sporangium.

Brazil.

REFERENCES: Martens 1870, Schmidt 1924, Taylor 1930a.

Griffithsia tenuis C. Agardh

Plants bushy, soft and delicate; basal filaments creeping and attached by rhizoids, erect and fastigiate above, 2–7 cm. tall, the filaments vaguely alternately branched, long and short branches intermixed, attached near or below the middle of the supporting cell; cells little constricted at the nodes, 120–300 μ diam., about 3–6 diameters long; young cells short, bearing a whorl of trichoblasts; tetrasporangia verticillate, appearing on as many as 6 successive nodes, each terminal on a pyriform stalk cell, without involucral cells, the sporangia 50–100 μ diam.; spermatangial clusters terminal on 1–3-celled pedicels; procarps developed on the subaxial cell of a branch, later displaced laterally, the cystocarp at maturity on a prominent one-celled stalk, surrounded by 6–8 large, incurving involucral cells.

Bermuda, North Carolina, Florida, Hispaniola, Virgin Isls., Barbados.

REFERENCES: Collins and Hervey 1917, Howe 1918a, 1920, Børgesen 1913–20, Taylor and Arndt 1929, Blomquist and Humm 1946, Taylor 1954a, Joly 1956b; *G. thyrsigera*, Vickers 1905.

Griffithsia schousboei Montagne

Plants tufted, 2–6 cm. tall; filaments dichotomously branched, fastigiate and somewhat flabellate, 0.5–1.0 mm. diam., moniliform above; below the cells pyriform, to 2–3 diameters long; tetrasporangia spherical, crowded about the nodes and with a few incurved involucral cells about them; spermatangia similarly arranged and also with involucral cells; cystocarps solitary, unilateral, nearly enclosed by involucral cells.

Bermuda, Jamaica, Guadeloupe.

REFERENCES: Mazé and Schramm 1870–77, Murray 1889, Collins and Hervey 1917, Howe 1918a.

Griffithsia globulifera Harvey

Plants tufted, the holdfast fibrous, the height 4.0–6.5 cm., color bright rose, texture delicate, easily crushed; branching erect and fastigiate, occasionally proliferous, the branches generally moniliform, of coenocytic segments which vary from somewhat clavate near the base of the plant to pyriform and nearly spherical at the apex, in length 0.6–3.2 mm., diameter 0.2–0.7(–1.5) mm., not corticated, except the few lowest segments by descending attaching rhizoids; in sporangial and carposporic plants the filaments somewhat tapering at the apex, but in the male plants, the cells increasingly large and spherical toward the tip; spherical tetrahedral sporangia about 8–20 at a fertile node, their diameter 40–50 μ, attached by short single-celled supports, which may bear secondary lateral sporangia, and enveloped by incurved sausage-shaped involucral cells; spermatangia in sori, generally forming a cap over the outer end of the terminal cell of the fertile branch, and sometimes over the outer end of 1–2 cells below the tip.

Bermuda, Florida, Bahamas, Caicos Isls., Cuba, Virgin Isls., Guadeloupe, Barbados, Venezuela. On stones and other objects in shallow water. Sterile specimens, probably of this species, were dredged from a depth of 33 m.

REFERENCES: Howe 1918a, 1920, Sanchez A. 1930; *G. globifera*, Murray 1889, Vickers 1905, Børgesen 1913–20, Taylor 1928, 1942, Feldmann and Lami 1937; *G. corallina* v. *globifera*, Harvey 1853, Mazé and Schramm 1870–77; *G. opuntioides*, Mazé and Schramm 1870–77, Murray 1889.

UNCERTAIN RECORDS

Griffithsia corallina (Lightfoot) C. Agardh

Florida, St. Eustatius. Type from Great Britain. De Toni 4:1279; Murray 1889, Sluiter 1908.

Griffithsia monilis Harvey

Bermuda. Type from Australia. De Toni 4:1283; Collins and Hervey 1917.

Griffithsia opuntioides J. Agardh

Barbados. Type from the Mediterranean. De Toni 4:1272; Vickers 1905.

Griffithsia flosculosa (Ellis) Batters

Guadeloupe, Grenada. Type from Great Britain. *G. setacea,* De Toni 4:1274; Mazé and Schramm 1870–77, Murray 1889.

Spermothamnion Areschoug, 1847

Plants tufted, of uniseriate, branched, uncorticated filaments, the basal part stoloniferous, attaching by unicellular, somewhat elongated holdfasts each ending in a lobed disk; the erect filaments oppositely or unilaterally branched, the cells uninucleate and with many small chromatophores; sporangia tetrahedral or polysporic, forming oval or cylindrical clusters on the upper side of the branchlets, or terminal; carpogonia developed near the tips of lateral branchlets, these tips having three small cells, on the middle one of which, with pericentral and sterile cells, is borne the four-celled carpogenic branch; two auxiliary cells dividing to produce gonimoblasts of which only the outermost cells produce carposporangia.

KEY TO SPECIES

1. Plants forming large tufts or cushions commonly over 1 cm. tall .. 2
1. Plants smaller, usually felted, in low cushions, or entangled among other algae 4

2. Sporangia generally tetrahedral 3
2. Sporangia polysporic .... S. nonatoi, p. 520

3. Erect filaments commonly oppositely branched; tetrasporangia 45–65 μ diam.; cystocarps surrounded by small involucral branchlets .... S. turneri, p. 519
3. Erect filaments irregularly branched; tetrasporangia 60–80 μ diam.; cystocarps naked .... S. gymnocarpum, p. 518

4. Erect filaments mostly 10–20 μ diam. .... S. investiens, p. 520
4. Erect filaments mostly 20–40 μ diam. 5
4. Erect filaments 40–65 μ diam.; cystocarps naked S. macromeres, p. 521

5. Chiefly repent, mostly growing on Codium .... S. gorgoneum, p. 521
5. Not strictly creeping, growing on rocks and coarse Rhodophyceae in very shaded situations .... S. speluncarum, p. 521

Spermothamnion gymnocarpum Howe

Plants in purplish cushions 1–3 cm. thick, or growing among other algae; primary filaments creeping, 50–130 μ diam., the cells 130–

400 μ long, with walls 13–50 μ thick; the erect branches arising near the ends of the cells, 50–115 μ diam., rather freely, widely, and irregularly branched, subdichotomous, alternate, or subsecund, rarely with 3–4 branches at a node; the lower cells cylindrical, 300–750 μ long or 3–7 diameters, with walls 5–40 μ thick, the obtuse terminal cells 40–65 μ diam. or only 12–15 μ diam. in very bushy specimens; tetrasporangia solitary or 2–5 together on unicellular pedicels at the nodes in the distal parts of the main and lateral branches, subglobose, 60–80 μ diam.; monoecious, the spermatangial clusters ovoid to subcylindrical, 26–40 μ diam., 40–80 μ long; cystocarps moriform or hemispherical, 80–150 μ diam., solitary or several together, without any involucre.

Bahamas. Growing with other small algae in shallow water.

REFERENCE: Howe 1920.

Spermothamnion turneri (Mertens) Areschoug

Plants with creeping primary filaments which closely invest the host and support a thick, bushy tuft of erect filaments about 2–5 cm. tall, the purplish rose clumps often crowded together, spongy in texture; creeping filaments 30–45 μ diam., the cells 3–6 diameters long, attached by haptera; erect filaments oppositely pinnate, the long branches spreading, slender, the main filaments 20–80 μ diam., the cells 3–8 diameters long, and the ultimate branchlets 20–45 μ diam.; tetrasporangia spherical, 45–65 μ diam., borne in series on the upper sides of the branchlets near the base on stalks one cell long, or the stalks sometimes forked, sometimes absent; procarps subterminal on the main filaments or the main branches, often of tetrasporic plants, the cystocarps becoming surrounded by several small branchlets.

V. **variabile** (C. Agardh) Ardissone. Branches and branchlets alternate or unilateral instead of opposite; found with the typical form.

North Carolina, Florida, Jamaica. Usually growing on various coarse algae in shallow water or at moderate depths. These southern records seem more doubtful as the years pass without modern confirmation.

REFERENCES: Harvey 1853, Taylor 1928; *Callithamnion turneri,* Murray 1889, Williams 1949a. For the v. *variabile,* Collins 1901, Taylor 1928.

Spermothamnion nonatoi Joly

Plants tufted, rosy, to 3 cm. tall; filaments decumbent below, attached by rhizoids with terminal haptera, of thick-walled cells 90 μ diam. and to 470 μ long; erect axes sparingly alternately branched below, oppositely branched above, the cells to 108 μ diam., 370–400 μ long; polysporangia fasciculate, on the upper branchlets, the fascicles to 108 μ diam.

Brazil. Growing on rocks in shallow water.

REFERENCE: Joly 1957.

Spermothamnion investiens (Crouan) Vickers

Plants forming dense, woolly, rose tufts, closely enveloping the host plant; primary filaments creeping, attached to the substratum at intervals by unicellular haptera; secondary filaments erect, numerous, 1–3 mm. tall, 10–20 μ diam., sparingly alternately branched, sometimes simple; branches usually simple, segments 30–100 μ, usually 55–70 μ long; tetrasporangia opposite or secund, terminating short, usually one-celled lateral branchlets, ellipsoid or slightly obovate, sometimes almost globose, 30–40 μ wide, 37–45 μ long; spermatangia oblong-ovate, borne singly at the apices of more or less prolonged lateral branches or of the main filaments; cystocarps similarly situated.

V. **cidaricola** Børgesen. Plants to 1 cm. tall, the basal filaments 30 μ diam., with thick walls; erect filaments 16–25 μ diam., the cells 3–5 diameters long; tetrasporangia on small one-celled pedicels, attached a little below the distal ends of branch cells, single or sometimes two on a branched pedicel, rounded-ovate, 44–46 μ diam., 46–52 μ long.

North Carolina, Bermuda, Bahamas, Virgin Isls., Guadeloupe, Barbados. Growing on various algae and shells in water of moderate depth, the variety on Eucidaris at 27 m. depth.

REFERENCES: Vickers 1905, Collins and Hervey 1917, Howe 1920, Hoyt 1920; *Callithamnion investiens,* Mazé and Schramm 1870–77, Murray 1889. For the v. *cidaricola,* Børgesen 1913–20.

Spermothamnion speluncarum (Collins and Hervey) Howe

Plants forming a velvety growth, the prostrate filaments 30–40 μ diam., the cells 2–4 diameters long, attached by haptera emerging near the middle of the originating cell; erect filaments cylindrical or slightly tapering, issuing from the dorsal side near the middle of the supporting cell and usually opposite a hapteron, simple or with a few alternate or secund branches, 22–39 μ diam., the lower cell of each branch 2.0–2.5 diameters long but the others 3–4 diameters; tip cells rounded; tetrasporangia solitary or opposite, sessile or on 1–2-celled pedicels attached to the lower parts of the erect filaments, 52–65 μ diam.; spermatangial clusters terminal, short-cylindrical or ovoid-ellipsoid, 20–28 μ diam., 40 μ long.

Bermuda, Bahamas. Growing under shelving rocks, in a cave, and on algae, near low-water line.

REFERENCES: Howe 1920; *Rhodochorton speluncarum,* Collins and Hervey 1917.

Spermothamnion gorgoneum (Montagne) Bornet Pl. 65, fig. 2

Plants forming a soft purple felt, the lower filaments 3–5 mm. long, intricate, in part rhizomatous; free filaments erect, alternately branched, 20–50 μ diam., the cells 2–7 diameters long; cystocarps lateral, on one-celled stalks, spherical, 180 μ diam.

Bermuda, Florida, Mexico, Bahamas, Jamaica, Hispaniola, Guadeloupe, Barbados, Tobago. Epiphytic on Codium and probably on other algae, in shallow water.

REFERENCES: Collins 1901, Vickers 1905, Collins and Hervey 1917, Howe 1918a, 1920, Taylor 1928, 1942, 1943, Humm 1952b; *Callithamnion gorgoneum,* Mazé and Schramm 1870–77, Murray 1889.

Spermothamnion macromeres Collins and Hervey

Plants matted, at least below, the prostrate filaments about 65 μ diam., the cells 3–5 diameters long, attached by short rhizoids; erect filaments 5–8 mm. long, 40–65 μ diam. at the base, 30 μ diam. above, the cylindrical or slightly clavate cells 5–20 diameters long; branching distant, apparently dichotomous, erect; tetrasporangia borne terminally on short ditrichotomous branchlets which arise singly or oppositely at fertile nodes, the sporangia nearly spherical, 50–55 μ diam.; spermatangial clusters on the inner sides of the branches, ovoid to cylindrical, 50–60 μ diam., 125–150 μ long; cystocarps borne terminally on short clavate stalk cells which replace one side of a fork.

Bermuda, Bahamas, Netherlands Antilles. Growing on rocks in the intertidal zone, infiltrated with sand.

REFERENCES: Collins and Hervey 1917, Howe 1918a, 1920, Taylor 1942.

UNCERTAIN RECORDS

(Spermothamnion?) Callithamnion pellucidum Farlow

Florida, probably the type locality, but the name source not found. Perhaps a *nomen nudum,* but at any rate a later homonym of Harvey's Tasmanian species. De Toni 4: 1267; *Callithamnion pellucidum,* Algae Exsic. Am. Bor. no. 90, Murray 1889, J. Agardh 1892.

Spermothamnion roseolum (C. Agardh) Pringsheim

Cuba. Type from northern Europe. De Toni 4: 1261, 6: 456; *Callithamnion repens* and v. *tenellum,* Montagne 1863.

Gymnothamnion J. Agardh, 1892

Plants filamentous, monosiphonous, uncorticated, the primary filaments decumbent and attaching by rhizoids, forming on the upper side plumose fronds which have a percurrent axis and opposite pinnate or bipinnate branching; sporangia pedicellate, tetrahedral, on the upper nodes of the branchlets; spermatangia corymbosely branched on the lower nodes of the fertile pinnules; cystocarps oblong, not involucrate, crowded on the apices of the main axes and adjacent branchlets.

Gymnothamnion elegans (Schousboe) J. Agardh Pl. 66, figs. 1–4

Plants to 1–4 cm. long, or intertwined to form a thin mat over the substratum, generally showing rhizomatous axes bearing erect pinnate branches 0.8–5.0 mm. tall, with descending rhizoids opposite these; the main erect axes 20–25 μ diam., the cells 2–5 diameters long, subcylindrical or somewhat clavate, bearing the branchlets at the forward end; branchlets opposite in pairs, 10–15 μ diam., sometimes reduced or absent, sometimes alternating 90° in position in succeeding nodes; on well-developed or erect branches regularly distichously pinnate and plumose, the pinnae somewhat ascending, about 60–210 μ long, often regularly again pinnate on the upper side only, or secund, varying greatly from the plumose to very irregular forms on the same plant; sporangia tetrahedral, spherical, or somewhat ovoid, 30–35 μ diam.

Bermuda, Florida, Bahamas, Cuba. Especially common on rocks and old corals in the intertidal zone where exposed to the surf, but protected from drying, as under overhanging rocks or in caves. Small and inconspicuous, and often entangled among other algae.

REFERENCES: Howe 1920, Taylor 1928, 1954a; *G. bipinnatum*, Collins and Hervey 1917; *Ptilothamnion bipinnatum*, Howe 1918a.

Ceramium Roth, 1797

Plants erect and bushy, or sometimes partly matted; branching dichotomous or more seldom alternate; branches segmented, the axis uniseriate, of relatively large cells, corticated at the nodes by a zone of smaller cells, in some species this zone spreading so as to cover the axial internodes completely; sporangia spherical, generally tetrahedral, sessile, borne at the nodes or, in the completely corticated species, often between them, immersed, projecting or largely naked; spermatangia formed of minute colorless cells in a layer on the corticated portions; one or two four-celled carpogenic branches borne on lateral nodal cells, which also give rise to an auxiliary cell after fertilization, each auxiliary forming a close mass of carpospores which may become surrounded by a few incurving branchlets.

KEY TO SPECIES

1. Axial cell row in the older parts of the plant completely covered by a corticating layer of small cells 2
1. Axial cell row surrounded at the nodes by bands of small cells, but otherwise partly exposed, at least in the older axes 3

2. Branch tips generally forcipate; internodal cells generally perceptible through the cortication and so the branches somewhat banded and nodose .... C. rubrum, p. 535
2. Branch tips erect; internodal cells hardly discernible through the cortex and the branches smooth and not appearing banded .... C. nitens, p. 535

3. Nodal bands with only 1–2 transverse cell rows 4
3. Nodal bands of the mature axes with more than 2 cell rows, the distribution commonly irregular 6

4. Bands with one complete cell row, sometimes with a few scattered cells above .... C. leptozonum, p. 525
4. Bands with 2 rows, or exceptionally and in very old segments appearing by proliferation to have 3–4 cell rows 5

5. Nodes of the erect filaments about 20–35 μ diam.; plants epiphytic on Codium .... C. codii, p. 526
5. Nodes about 50 μ diam.; plants chiefly erect C. fastigiatum, f. flaccida, p. 527
5. Nodes about 100 μ diam.; plants creeping .... C. comptum, p. 529

6. Chiefly creeping and attached by rhizoids at intervals, though the branches may in part become erect 7
6. Primarily erect and bushy species 8

7. Sporangia tetrapartite; branching dichotomous; the filaments about 200 μ diam. below, 80–100 μ diam. above C. cruciatum, p. 530
7. Sporangia tetrahedral; branching scanty, alternate, unilateral; filaments below 40–65 μ diam., above 40–50 μ diam. C. leutzelburgii, p. 529

8. Branching essentially dichotomous, though sometimes unequally so, and with later proliferations 9
8. Branching notably alternate and complanate 11

9. Lower 1–2 cells of the bands clearly broader than long, in living material often transversely rectangular C. byssoideum, p. 528
9. Lower cells not regularly broader than long 10

10. Cortication in the ultimate branches subconfluent, though limited to the nodes in the older parts of the plant; the corticating cells elongated lengthwise of the axis C. corniculatum, p. 530
10. Cortication not confluent in any region 12

11. Branching close, in the distal parts frondlike; lower nodes not swollen; the cortication extending equally up and down from the nodal septum .... C. floridanum, p. 534
11. Branching loose, not frondlike; lower nodes conspicuously swollen; cortication in the upper part of the plant mostly above the nodal septum .... C. uruguayense, p. 532
C. brasiliense, p. 533

12. Branchlet tips forcipate and externally strongly serrate; gland cells present in the nodal bands .... C. tenuissimum, p. 531
12. Branchlet tips not exceptionally serrate; no gland cells present .. 13

13. Nodal cells becoming irregularly disposed, usually several to be counted lengthwise of the mature node 14
13. Nodal cells in a few fairly distinct transverse series; sporangia where known emergent or naked 16

14. Sporangia strongly exserted, merely surrounded at the base by cortical cells .... C. fastigiatum, p. 526
14. Sporangia immersed or a little emergent 15

15. Proliferous branches numerous; cells of the lower margin of the older nodes usually not averaging over 13 μ diam.
C. diaphanum, p. 532
15. Proliferous branches few or none; cells of the lower margin usually 10–22 μ diam. .... C. strictum, p. 530

16. Nodal cortication of one row of small and one row of large cells, the former only sparingly divided in the oldest filaments .... C. fastigiatum, f. flaccida, p. 527
16. Nodal cortication of more than 2 rows of cells 17

17. Plants to 2.5 cm. tall; internodes to 6 diameters or hardly to 0.75 mm. long .... C. brevizonatum v. caraibica, p. 527
17. Plants to 1.5 dm. tall; internodes to 10 diameters or about 1.25 mm. long .... C. subtile, p. 527

Ceramium leptozonum Howe

Plants delicate, 1.5–3.0 cm. tall; main filaments 40–72 μ diam., the apices slightly forcipate or erect; internodal cells cylindrical below, more turgid above, 1.5–4.0 diameters long; nodal bands narrow, usually of one cell row, the cells elongate lengthwise of the filament and about 4 of them visible across the width of the filament, or with a few smaller cells above; sporangia tetrahedral, naked, solitary at a node, mostly secund along the outer side of the filament, occasionally 2–3 and subverticillate, 50–65 μ diam.

Bermuda. In a sheltered, land-locked salt-water pool, with subterranean communication with the sea.

REFERENCE: Howe 1918a.

Ceramium codii (Richards) Feldmann-Mazoyer

Plants very small, about 1 cm. tall, primarily creeping, the prostrate filaments 30–50 μ diam., without horizontal branches, developing occasional rhizoids at the nodes, and at nearly every node simple erect filaments 0.5–1.5 mm. long, 20–35 μ nodal diam.; erect filaments of very long and slender axial cells, usually with 2 irregular rows of small cells at each node; tetrasporangia secund, single at each node, 20–30 μ diam., 30–45 μ long, about half enveloped from the lower side by branching series of cover cells; spermatangial clusters sheathing the axis well below the apex of each fertile filament, usually covering two nodes and the intervening internode; cystocarps formed just below the filament tips, later exceeded by about 4 involucral branchlets.

Bermuda, Bahamas, Hispaniola, Barbados. Epiphytic on Codium in shallow water.

REFERENCES: Feldmann-Mazoyer 1940; *Ceramothamnion codii,* Richards 1901, Vickers 1905, Collins and Hervey 1917, Howe 1918a, 1920, Børgesen 1924.

Ceramium fastigiatum (Roth) Harvey Pl. 67, figs. 4–6

Plants forming dense tufts about 4–8 cm. tall, bright rose-red, delicately capillary, the basal portions creeping, but the branching regularly dichotomous and fastigiate above; apices erect or slightly incurved; nodes ultimately with 4–6 rows of cells, of which the lowest are of intermediate size, the next higher considerably larger, while those of the upper margin are much smaller and may be a little longitudinally elongate; older nodes in the erect portion about 60–155 μ diam., 55–65 μ long, the internodes 75–150 μ diam., 0.6–1.36 mm. long; tetrasporangia single or to 6 at a node, greatly projecting, somewhat covered on the lower side by the upgrowth of a few cells, 33–65 μ diam., 50–68 μ long; cystocarpic plants sometimes with lateral branchlets, the cystocarps lateral, sometimes 2–3 together, the 2–4 involucral branchlets hardly longer than the cystocarp.

F. **flaccida** H. E. Petersen. Tufts 4–7 cm. tall; nodal cells in 2 series, with the upper cells rather smaller than the lower, or irregular and about 3 cells deep; nodes 50 μ diam., 20–40 μ long; internodal cells 50 μ diam., to 400 μ long; tetrasporangia scattered, very prominent.

Bermuda, North Carolina, Florida, Mexico, Cuba, Jamaica, Hispaniola, Virgin Isls., Guadeloupe, Grenadines. Growing on coarse algae and phanerogams in shallow water, the f. *flaccida* particularly on the roots of mangroves.

REFERENCES: Hemsley 1884, Murray 1889, Collins 1901, Taylor 1928, 1941b, Sanchez A. 1930, Williams 1949a, Humm 1952b. For the f. *flaccida*, Børgesen 1913–20, Taylor 1928 *p. p.*, 1929b, 1954a, Taylor and Arndt 1929, Feldmann and Lami 1936.

Ceramium brevizonatum H. E. Petersen v. **caraibica** Petersen and Børgesen Pl. 67, figs. 7–9

Plants to 2.5 cm. tall, dichotomously branched, the tips forcipate, the nodes well separated; nodes 96–136 μ wide, 40–58 μ and about 3–5 cells long, the cells of the upper margin smaller than in the type and commonly in filaments of 2–3 cells, the cells of the lower margin also relatively somewhat smaller than in the type; gland cells absent; internodes very variable, but to about 6 times as long as broad; polyspores present.

Florida, Hispaniola.

REFERENCES: Børgesen 1924; *C. tenerrimum* v. *tenuissimum* includes this and v. *arachnoideum* (Phyc. Bor.-Am. 346) according to Feldmann-Mazoyer 1940, but my specimen of this number has nodal structure in accord with v. *caraibica*, though not the species type.

Ceramium subtile J. Agardh Pl. 65, figs. 5, 6

Plants forming soft, more or less entangled tufts to 1.5 dm. long; branching more or less dichotomous, the apices strongly incurved;

lower internodes reaching 120 μ diam., 1.25 mm. in length, the nodes to 130 μ diam., and usually shorter than broad; the nodes generally showing 3 series of cells, an upper of small, irregularly placed, rounded cells, a middle of large, irregularly rounded cells, and a lower, which is at first of a single row of cells rather transversely placed but these later divided by vertical or oblique walls to a number of irregular cells; tetrasporangia surrounded below, but protruding from the nodes above, usually single, about 60 μ in diam.; cystocarps at the upper forks, frequently involucrate by a few short branchlets.

Bermuda, Florida, Texas, Mexico, Bahamas, Cuba, Jamaica, Hispaniola, St. Barthélemy, Guadeloupe, Martinique, Grenada, Barbados, Costa Rica, Netherlands Antilles. Common in shallow water on various objects.

REFERENCES: Harvey 1853, Mazé and Schramm 1870–77, Dickie 1874a, Murray 1889, Howe 1920, Taylor 1928, 1933, 1941a, 1942, 1943, 1954a; Taylor and Arndt 1929, Quirós C. 1948; *C. gracillimum*, Collins 1901; *C. gracillimum, C. chilense, C. cornigerum,* and *?C. nodiferum,* Mazé and Schramm 1870–77, Murray 1889.

Ceramium byssoideum Harvey Pl. 67, figs. 1–3

Plants small, tufted, soft, of slender filaments; when young the prostrate filaments attached by numerous unicellular holdfasts, above sparingly and fairly regularly dichotomously branched, the branches 60–90 μ diam., erect to spreading, the lower internodes 1.5–8.0 diameters long; nodal cortication showing occasional gland cells near the upper margin and a notable series of transversely elongated cells in the lower portion of each band; tetrasporangia 1–2 at each node, subsecund, appearing to inflate the lower portion of the cortical band, to 60 μ diam., the wall to 8 μ thick, nearly completely exposed at maturity; cystocarps apparently terminal on short clavate branchlets of 2–5 segments, surrounded by short involucral branchlets.

Bermuda, North Carolina, Florida, Texas, Bahamas, Cuba, Jamaica, Hispaniola, Virgin Isls., Saba Bank, Old Providence I., Colombia, Venezuela, Brazil.

Mazoyer (1940) reduces this species to varietal status under *C. gracillimum* (Griffiths) Harvey, to which it is doubtless related, but the transverse cells of the lower node margins are so very much more distinct in the living state than those she figures, commonly appearing rectangular, that I prefer to retain the old designation for the American plants. The cotype of *C. dawsoni* Joly agrees in nodal structure with this species and differs from his figures.

REFERENCES: Harvey 1853, Murray 1889, Collins 1901, Howe 1918a, 1920, Taylor 1928, 1939b, 1941a, 1942, 1954a, Taylor and Arndt 1929; *C. dawsoni* and *C. gracillimum* v. *byssoideum,* Joly 1957; *C. tenuissimum* v. *pygmaeum* Collins 1901, Collins and Hervey 1917; *C. transversale,* Collins and Hervey 1917, Børgesen 1913–20, 1924, Williams 1951.

Ceramium leutzelburgii Schmidt

Plants small, chiefly creeping, unilaterally and sparingly alternately branched, the rhizoids numerous; nodal cortication 40–55(–65) μ diam., 26–40 μ long; nodes of about 3 cell rows, the upper row of smaller cells, the middle of larger, and the lower of somewhat transversely elongated cells; internodal cells of the creeping filaments generally 40–60 μ diam., 130–170 μ long, of the ascending filaments generally 40–50 μ diam., 78–120 μ long; tetrasporangia single, rarely 2 at a node, almost completely free of the cortex, 80–90 μ diam., the outer wall thick.

Brazil.

REFERENCE: Schmidt 1924.

Ceramium comptum Børgesen

Plants dichotomous, chiefly creeping; the nodes distant, much wider than long, generally of 2 rows of cells, the cells of the upper margin much smaller, those of the lower row rather angular and large, about 35 μ diam., the cortical bands about 100 μ wide, 35–42 μ long; internodes about 300 μ or 3–7 diameters long; tetrasporangia solitary or 2 together, unilaterally developed in the nodal zones.

Hispaniola.

REFERENCE: Børgesen 1924.

Ceramium cruciatum Collins and Hervey

Plants minute and creeping, the prostrate filaments about 200 μ diam., cylindrical or slightly constricted at the nodes, attached by unicellular rhizoids; erect filaments usually developed by assurgent branch tips, less often as branches lateral from prostrate nodes; lesser branches 80–100 μ diam., the apices forcipate; cortication consisting of narrow, uneven bands of cells somewhat elongated lengthwise of the axis, in no definite order; internodes to about 3 diameters long; sporangia tetrapartite, 1–4 at a node, when single at the nodes in a unilateral series, but when more numerous irregular in arrangement, about half the sporangium projecting beyond the cortication, ovoid, about 35 μ diam., 50 μ long, the wall about 6 μ thick.

Bermuda, Bahamas. Growing on various coarse algae in shallow water.

REFERENCES: Collins and Hervey 1917, Howe 1918a, 1920.

Ceramium corniculatum Montagne

Plants with filaments dichotomous below, 90–200 μ diam. above and somewhat alternate with erect, incurved branches; nodal bands not protuberant, the corticating cells with their longer axes lengthwise of the filament; on the lower segments the internodes hyaline, 3–4 diameters long, but in the young portions the nodes adjacent; tetrasporangia verticillate in the central parts of the nodes of clavate or fusiform lateral incurved branchlets.

Florida, Bahamas, Caicos Isls., Guadeloupe, Martinique, Barbados, ?Brazil, Uruguay.

REFERENCES: Mazé and Schramm 1870–77, Murray 1889, Vickers 1905, Howe 1920, Taylor 1928; *?C. rubrum* v. *virgatum*, Martens 1870; *C. strictoides*, Mazé and Schramm 1870–77, Murray 1889, Vickers 1905; *C. vimineum*, De Toni 1903.

Ceramium strictum (Kützing) Harvey

Plants in soft tufts 2–8 cm. tall, dull red, the nodes distinct, the branches rather slender, regularly dichotomous with few lateral proliferations, somewhat spreading or fastigiate above; the tips of the branches forcipate; the nodal diameter in older parts about 210–336 μ, length 85–210 μ, sharply defined at both margins, the cells of somewhat unequal size in the oldest parts; some of the largest cells of the inner series partly exposed across the center of the node, the cells of the lower margin 10–22 μ diam., much smaller than those of the upper margin; internodes nearly colorless, the older to about 185–270 μ diam., the segments 0.6–1.5 mm. long; tetrasporangia immersed in the central part of the nodal band, or somewhat exposed above, 50–60 μ diam., 60–70 μ long; cystocarps lateral on the upper branches, the 4–6 involucral branchlets considerably overtopping them.

North Carolina, Florida, Texas, Virgin Isls., Guadeloupe, Brazil. Occasional in shallow water on various objects.

REFERENCES: Mazé and Schramm 1870–77, Murray 1889, Möbius 1890, Børgesen 1913–20, Hoyt 1920, Taylor 1936, 1941a, b.

Ceramium tenuissimum (Lyngbye) J. Agardh

Plants in large tufts, the filaments slender, repeatedly dichotomous, having scattered adventitious branchlets with forcipate tips and the nodes at the tips projecting along the strongly serrate outer side; lower internodes about 100 μ diam., 3–7 diameters long; nodes 70–110 μ diam., 50–60 μ long; gland cells present in the nodes; tetrasporangia in upper or lateral branchlets, single or verticillate at the nodes on the outer sides; cystocarps terminal on the branches, barely exceeded by 2–3 involucral branchlets.

Bermuda, North Carolina, Florida, Bahamas, Jamaica, Puerto Rico, Guadeloupe, Barbados, Netherlands Antilles, Brazil. A plant of shallow water growing on various objects.

Reports of this species from America are of doubtful value; those from northern waters seem more correctly distributed between *C. fastigiatum* and *C. strictum,* and the same or some similar disposition is probably true of the tropical records. However, for the present a description, modified by data included in some of the more plausible tropical reports, is included. The varieties recorded in our area are even less certainly worthy of credence.

REFERENCES: Harvey 1853, Möbius 1889, Murray 1889, Collins 1901, Vickers 1905, Collins and Hervey 1917, Howe 1918a, 1920,

Hoyt 1920, Schmidt 1924, Taylor 1928, 1954a; *C. nodiferum,* Mazé and Schramm 1870–77. For v. *arachnoideum* (C. Agardh) J. Agardh, De Toni 4:1451; Sluiter 1908, Collins and Hervey 1917, Taylor 1928. For v. *patentissimum* (Harvey) Farlow, Mazé and Schramm 1870–77, Collins and Hervey 1917. For v. *pygmaeum* (?Kützing) Hauck, De Toni 4:1452; Hauck 1888, Vickers 1905.

Ceramium diaphanum (Lightfoot) Roth

Plants bushy, 5–20 cm. tall, dull brownish red, the conspicuous nodes very sharply demarcate from the colorless internodes; branching dichotomous, erect-spreading, commonly with numerous lateral adventitious branches; branch tips forcipate to erect; diameter below 335–450 μ, the nodes broader than long, the cells of unequal size, with larger deep-placed cells partly covered by many very small ones in a single layer; internodes 200–460 μ diam., the cells 0.50–1.25 mm. long; tetrasporangia immersed in the upper part of the nodes, which may become lobulate-swollen, the sporangia to about 50–75 μ diam.; cystocarps on short lateral or sublateral branches, near the ends, little overtopped by the 3–4 involucral branchlets.

F. **strictoides** H. E. Petersen. Close to the typical form, differing in the more strictly dichotomous branching without lateral branchlets, and by having the colorless internodes to about 6 times as long as the nodes.

?North Carolina, ?Florida, ?Cuba, ?Brazil, Uruguay. The early records of this species, referring to the typical form, are very doubtful. The f. *strictoides* was, however, recognized by Petersen in material from Uruguay and Argentina, and for that reason it is included.

REFERENCES: Montagne 1839b, 1863, Harvey 1853, Martens 1870, Zeller 1876, Murray 1889, Taylor 1928, Williams 1949a. For the f. *strictoides,* Taylor 1939a.

Ceramium uruguayense n. sp. Pl. 65, figs. 3, 4

Plants to 4 cm. tall, dark purplish red; the axes subpercurrent, alternately divided, the lesser branches progressively shorter, probably complanate, from alternate becoming deliquescent or subdichotomous in the lesser divisions; small adventitious branchlets often present; branch tips at first incurved, soon divergent, acute, 55–85 μ diam., upper branches about 120 μ diam., the nodes and the internodes cylindrical; axes of the central portion of the plant to 560 μ diam. at the strongly protuberant nodes, the internodes 300 μ diam., but in the lower parts of the plant more slender; nodes in the upper part nearly touching, in the central part separated by more than their width, in the lowest parts the internodes again shorter; cortication at the nodes in the upper parts of the plant, consisting of a row of large deep-placed cells partly covered by smaller irregularly oval ones, which form a fairly straight lower margin, but the cells above the septum a little smaller and less regular; nodes of the central part with more numerous deep rows of cells, the smaller outer cells more elongate; emergent tetrasporangia single in the nodes on the outer side of the small branchlets and also 1–5 in the upper part of each node in the lesser axes, 45–55 μ diam., 55–65 μ long, almost completely covered from below by cells of the nodal cortex; cystocarps small, on the upper branches, exceeded by 3–5 acute, strongly incurved, involucral branchlets; spermatangial plants not seen.

Brazil, Uruguay. Forming tufts on rocks in shallow water.

In the type material of *C. miniatum* Suhr the upper margins of the nodes are quite irregular and extend upward; at a distance of 5 mm. from the branch tips the naked strips across the internodes have almost or quite disappeared, while in *C. uruguayense* they increase in length progressively from the tips.

REFERENCE: Taylor 1939a, p. 151, pl. 4, fig. 2, pl. 7, figs. 2–4 (as *C. miniatum*).

Ceramium brasiliense Joly

Plants tufted, purplish, 5–10 mm. tall; branching alternate, becoming subdichotomous in the lesser divisions, and with numerous proliferations; apices slightly incurved; cortication of a few rows of cells overlapping a deeper layer at the node, the cells of the upper margin somewhat smaller than those below; branchlets about 50 μ diam., older axes to 135 μ, the bare section of the internodes 1.5–2.0 times the diameter; tetrasporangia in swollen branchlets, often secund on the outer side, single or 2–3 at a node, immersed but very strongly projecting; spermatangia partly covering the fertile nodes; cystocarps surrounded, and in part exceeded, by involucral branchlets; carposporangia few, large.

Brazil. Epiphytic on segmented coralline algae.

Additional characters have been added to the scanty original description after a close examination of cotype material. It seems possible that this plant represents a depauperate form of *C. uruguayense,* and if this can be confirmed Joly's name will have precedence.

REFERENCE: Joly 1957.

Ceramium floridanum J. Agardh

Plants small, to about 7 cm. in height, bright rose; at first irregularly alternately branched to about 2 degrees, the diameter of the main divisions reaching about 0.3 mm., the segments 330–390 μ in length; above more regularly branched, the complanate lateral divisions at first deliquescent and of rather even length, but at about the third or fourth degree after reaching a length of about 0.6 mm. a definite subpercurrent axis is quickly defined in each division, these axes in turn bearing branch systems of temporarily limited growth, forming thereby flat frondlike divisions which at first are broad and vague, but ultimately become ovate to lanceolate, about 2–3 mm. broad, and to 10 mm. long; adventitious branchlets absent, or, especially on tetrasporic plants, more or less numerous; ultimate branchlets 30–60 μ diam., the tips abruptly conical, divergent; corticated nodes not swollen, 180–240 μ long, the cortication composed of 1–3 rows of large deep-placed cells partly covered by irregularly placed small cells in a layer which extends about equally up and down the internodal cells, the exposed internodal portions pale, short in the younger branches, to 150 μ long, or about 0.2–0.5 the width of the segment in the older divisions; tetrasporangia whorled in the smaller branches, spherical, to about 75 μ outside diam., greatly distending the upper portion of the nodes, and slightly exposed on their upper sides; cystocarps matured 1–2 forks below the tips of the branches, slightly exceeded by a few involucral branchlets.

Florida, Venezuela, Trinidad. Growing on coarser algae, particularly on segmented corallines.

REFERENCES: Taylor 1928, 1929b, 1942.

Ceramium rubrum (Hudson) C. Agardh

Plants forming large bushy masses, from disklike bases rising to 1–4 dm. tall, color red, of various shades depending on location, obscurely banded; coarse, to 1 mm. diam., and somewhat firm, at least below; branching dichotomous, tapering to the upper divisions, with, in some forms, irregular habit and more or less abundant lateral branches; branch tips forcipate to erect; nodes little swollen, and below indeed more often contracted; nodal rings almost contiguous from the apex, shortly behind which the intervals between the nodal bands may be shown by narrow lines, later obscured; mature axial internodal cells entirely covered by spreading corticating growth from the node, the cortex of small angular to rounded cells, those at the surface not very different about the septa; tetrasporangia 55–80 μ diam., in rows about the nodes, later arising from the spreading cortication, immersed, or causing slight swelling; spermatangia in crowded tufts on younger portions of the plant; cystocarps on lateral branches, invested by involucres of 3–5 branchlets which curve up and over them.

North and South Carolina. Also reported from Bermuda, Florida, Cuba, Virgin Isls., Grenada, and Brazil, but these records are all very doubtful. The species does appear again in Argentina, and may extend a little farther north or south of its present known range. A plant of shallow but not stagnant or brackish water, growing on almost any firm object.

REFERENCES: Montagne 1863, Martens 1870, Hemsley 1884, Murray 1889, Hoyt 1920, Taylor 1928.

Ceramium nitens (C. Agardh) J. Agardh Pl. 66, fig. 14

Plants bright rose, in spreading tufts but entangled, 5–10 cm. tall, the texture of the filaments rather firm and almost cartilaginous on drying; filaments 130–430 μ diam., branching dichotomous, the branches spreading, or above secund and incurved, the tips erect and subulate; axial cells 2–3 times as long as broad, completely and densely corticated by small rounded cells; tetrasporangia randomly distributed in the branchlets, immersed in the cortex, tetrapartite; cystocarps lateral on the upper branches, more or less exceeded by about 3 involucral branchlets.

Bermuda, Florida, Bahamas, Caicos Isls., Cuba, Jamaica, Hispaniola, Puerto Rico, Virgin Isls., Guadeloupe, Martinique, British Honduras, Colombia. Occasional on rocks a little below low-tide line in rather sheltered places, but since most commonly found washed ashore, probably more frequent in somewhat deeper water.

REFERENCES: Harvey 1853, Mazé and Schramm 1870–77, Rein 1873, Hemsley 1884, Murray 1889, Collins 1901, Collins and Hervey 1917, Børgesen 1913–20, Howe 1918a, 1920, Taylor 1928, 1933, 1935, 1941b, 1943, 1954a; Taylor and Arndt 1929, Sanchez A. 1930; *C. lanciferum,* Murray 1889.

UNCERTAIN RECORDS

Ceramium ciliatum (Ellis) Ducluzeau

Brazil. Type from England. De Toni 4:1473; *Echinoceras ciliatum,* Martens 1870.

Ceramium deslongchampii Chauvin

Florida. Type from France. De Toni 4:1467; Taylor 1928; *C. deslongchampii* v. *viminiarum,* Murray 1889.

Ceramium miniatum Suhr

Guadeloupe. Type from Peru. Mazé and Schramm 1870–77, Murray 1889.

Ceramium obsoletum C. Agardh

Brazil. Type from South Africa. De Toni 4:1481; Martens 1870, Taylor 1931; *Sphaerococcus micrococcus,* Martius 1835.

Reinboldiella De Toni, 1895

Plants small, filiform, repent, terete or compressed, irregularly pinnately branched, complanate; axis of one conspicuous series of cells completley corticated; tetrasporangia verticillate in enlarged, stichidial branchlets; spermatangia in cushionlike bands around slender branchlets; cystocarps involucrate, sometimes grouped, terminating short, stout branchlets.

Reinboldiella repens (Taylor) Feldmann and Mazoyer

Plants epiphytic, small; creeping main axes to 5–10 mm. long, attached at numerous points by branched, unicellular holdfasts; alternately pinnately branched, the lateral axes to 5 mm. long, axes of both degrees flattened, 165–250 μ broad, bearing close-placed determinate branches 0.5–2.0 mm. long, which are contracted at the base, flattened and coarsely serrate, or the teeth developing into spinelike branchlets 150–200 μ long; branches of all orders with an axial cell row 80–160 μ diam., the cells 0.5–1.2 diameters long, with a subcortex of cells of intermediate size particularly broad in the horizontal plane, forming the expanded sides of the blade, covered by a sharply delimited cortex of one layer of small cells 5.5–13.0 μ diam. in surface view; tetrapartite sporangia forming bands in swollen, stalked, stichidial branchlets; spermatangia in tiny tufts, forming a swollen belt around subcylindrical branchlets; cystocarps at the ends of small, swollen branchlets, partly involucrate.

Trinidad. Epiphytic on coarse red algae.

REFERENCES: Feldmann and Mazoyer 1937; *Carpoblepharis repens*, Taylor 1929b.

Centroceras Kützing, 1842

Plants matted or bushy, filamentous, dichotomously branched but often laterally proliferated; the axis internodes completely corticated by regular longitudinal rows of rectangular cells, the nodes commonly spinulose; tetrasporangia verticillate in the terminal segments of axillary torulose proliferations; cystocarps lateral, bilobed, partly surrounded by short involucral branchlets.

Centroceras clavulatum (C. Agardh) Montagne

Plants of very variable form, stiff and matted, streaming in the current and 1–2 dm. tall, or forming large disengaged, entangled masses; usually dark purplish red in color, or in exposed places decolorate; filaments 50–200 μ diam., the internodes short above, to 0.75 mm. long below; apices forcipate; nodes spiny, the spines verticillate, generally of 2 cells and most prominent in the younger branches; tetrasporangia mostly exposed, generally formed in verticils in the terminal segments of short, often axillary branchlets.

Bermuda, Florida, Texas, Mexico, Bahamas, Caicos Isls., Cuba, Cayman Isls., Jamaica, Hispaniola, Puerto Rico, Virgin Isls., St. Barthélemy, Guadeloupe, Dominica, Martinique, Barbados, Grenada, British Honduras, Old Providence I., Costa Rica, Panama, Colombia, Netherlands Antilles, Venezuela, Tobago, French Guiana, Brazil. Exceedingly common in shallow water throughout almost the entire range; dredged to a depth of 20 m. Sometimes forming a short turf infiltrated with sand, sometimes entangled among other plants, often plumose or free, characteristically growing in shallow water, and seldom fruiting. So variable, as in the degree to which the spines are developed, that many names have been applied to aspects of the one species, but these have no particular value.

REFERENCES: Harvey 1853, 1861, Mazé and Schramm 1870–77, Dickie 1874b, Hemsley 1884, Piccone 1886, 1889, Murray 1889, Möbius 1890, Vickers 1905, Sluiter 1908, Børgesen 1913–20, 1924, Jennings 1917, Howe 1918a, 1920, 1928, Schmidt 1923, Taylor 1928, 1929b, 1930b, 1933, 1935, 1936, 1939b, 1940, 1941a, b, 1942, 1943, 1954a, Taylor and Arndt 1929, Sanchez A. 1930, Questel 1942, Joly 1951, 1957, Humm 1952b, De Mattos 1952; *C. clavulatum* v. *brachyacanthum*, v. *hyalacanthum*, v. *leptacanthum*, and v. *oxyacanthum*, Mazé and Schramm 1870–77, Murray 1889; *C. clavulatum* v. *crispulum*, Murray 1889; *C. clavulatum* v. *micracanthum*, Piccone 1886, Murray 1889; *C. cryptacanthum*, Dickie 1874b, Hemsley 1884; *C. hyalacanthum*, Hohenacker Meeresalgen no. 537; *C. leptacanthum* and *C. macracanthum*, Martens 1870; *C. micracanthum*, Martens 1870, 1871, Zeller 1876; ?*C. rhizophorum*, Montagne 1850; *Ceramium clavulatum*, Greville 1833, Hooper 1850, Hauck 1888, Möbius 1889, Collins 1901, Grieve 1909, Collins and Hervey 1917, Schmidt 1924; *C. clavulatum* v. *crispulum*, Montagne 1863.

Spyridia Harvey, 1833

Plants forming erect bushy masses; much alternately branched, the branches corticated by transverse series of longitudinally elongate chromatophore-bearing cells, these later subdivided and covered by rhizoidal downgrowths on the larger axes; ultimate branchlets of limited growth, deciduous, consisting of a uniseriate axis of large cells bearing rings of small chromatophore-bearing cells at the nodes, the elongated internodes naked and translucent, the nodes often, and the tips generally, armed with short indurated spine cells; tetrahedral sporangia seriate at the nodes on the upper sides of the branchlets; spermatangia originating on the nodes of the branchlets; cystocarps near the tips of small branches, often surrounded by slender involucral branchlets.

KEY TO SPECIES

1. Smaller lateral branches in part swollen toward the straight tips, which have few branchlets; branching distichous ..... S. clavata, p. 541
1. Branches not club-shaped 2
2. Branchlets with no more than a single terminal spine; branching radial without pronounced main axes .... S. filamentosa, p. 539
2. Branchlets when young generally with recurved spines on the terminal or 2 distal nodes 3
3. Older short branches stout, tapering, and spinelike, becoming denuded; distichous, with axes compressed above ..... S. complanata, p. 540
3. Older short branches not distinctive; main axes percurrent; branching radial or distichous; hamate branch tips common; recurved spines sometimes obsolete .... S. aculeata, p. 541

Spyridia filamentosa (Wulfen) Harvey Pl. 66, fig. 15

Plants bushy, from a rhizoidal disk holdfast, to 1.5–2.0 dm. tall, coarse below, alternately branching, the branches bearing numerous short slender deciduous branchlets, or denuded below, the general color dull rose to often bleached, straw-colored, or dull brownish, the texture often spongy; main axes corticated by a layer of small cells surrounding the large ones of the axial row, these formed as alternating transverse nodal and internodal series of longitudinally elongate cells which eventually become irregular; diameter of the main axes to 1–2 mm.; branchlets radially inserted, about 0.5–1.5 mm. long, 20–45 μ diam., the segments about 2–4 diameters long, with several cells at each node, at first in a single ring, later some cells dividing transversely and obliquely, the branchlet apex with a terminal spine cell; tetrasporangia spherical, sessile, 40–70 μ diam., single or whorled on the nodes of the branchlets; spermatangia about the nodes of the branchlets and extending over the internodes; cystocarps terminal on short branches, with an involucre of incurved branchlets.

V. **refracta** (Wulfen) Harvey. Branching subdichotomous, the branches spreading, the lesser divisions flexuous, curved to hooked, but not crozier-like, and often entangled.

Bermuda, North Carolina, South Carolina, Florida, Mexico, Bahamas, Caicos Isls., Cuba, Cayman Isls., Jamaica, Hispaniola, Puerto Rico, Virgin Isls., St. Barthélemy, St. Eustatius, Guadeloupe, Martinique, Barbados, Grenada, British Honduras, Costa Rica, Panama, Colombia, Netherlands Antilles, I. las Aves, Brazil. Frequent in warm, shallow water and quiet situations, but dredged to a depth of 20 m.

REFERENCES: Harvey 1853, 1861, Mazé and Schramm 1870–77, Rein 1873, Dickie 1874a, Hemsley 1884, Hauck 1888, Murray 1889, Collins 1901, Vickers 1905, Sluiter 1908, Børgesen 1913–20, 1924, Collins and Hervey 1917, Howe 1918a, 1920, Hoyt 1920, Schmidt 1924, Taylor 1928, 1930a, 1933, 1935, 1936, 1941b, 1942, 1943, Taylor and Arndt 1929, Sanchez A. 1930, Hamel and Hamel-Joukov 1931, Joly 1957; *S. arcuata,* Möbius 1890; *S. filamentosa* v. *cuspidata,* v. *friabilis,* v. *villosa,* and *S. insignis,* Mazé and Schramm 1870–77, Murray 1889; *Ceramium filamentosum,* Montagne 1863.

Spyridia complanata J. Agardh

Plants bushy, 6–9 cm. tall, the axes compressed, alternately and distichously pinnate, plumose in the younger parts, the divisions 5–12 mm. broad, rather stiff in the older parts and the persistent short branchlets sharply tapering and spinelike; branches corticated throughout; branchlets with a terminal spine and 2–3 recurved aculei at the tips.

Bermuda, Guadeloupe, Brazil. Generally growing on exposed rocks in the lower intertidal zone. Doubtfully distinct from *S. aculeata* v. *disticha.*

REFERENCES: Mazé and Schramm 1870–77, Murray 1889, Collins and Hervey 1917, Howe 1918a.

Spyridia clavata Kützing

Plants slender to robust, 8–20 cm. long, in color pale pink; the main branches somewhat firm; axes terete below, somewhat compressed and 1–2 mm. wide above, distichously alternately, sometimes oppositely branched, larger and smaller branches intermixed, corticated to the tips; smaller branches about 2.0–4.5 mm. long, at least in part clavate or spindle-shaped, markedly tapering to the base; younger branches distichously clothed with short, determinate, somewhat curved branchlets, except that these are reduced on or absent from the swollen branch tips; branchlets about 600 μ long, the nodes about 55 μ diam., the internodes about 60 μ long; tetrasporangia borne on the lower nodes of the branchlets, tetrahedral, about 50 μ diam.

North Carolina, Florida, Jamaica, Hispaniola, Virgin Isls., Guadeloupe, Barbados, Martinique, Venezuela, Tobago. Plants growing on reefs and rocks in shallow water, and dredged to a depth of at least 12 m.

REFERENCES: Murray 1889, Vickers 1905, Børgesen 1913–20, Hoyt 1920, Taylor 1933, 1942, Hamel and Hamel-Joukov 1931; *S. clavifera* and *S. montagneana*, Mazé and Schramm 1870–77.

Spyridia aculeata (Schimper) Kützing

Pl. 66, fig. 16; pl. 71, fig. 5

Plants large, generally rose-red, becoming dull with age, to 2.0–2.5 dm. tall, densely, repeatedly alternately branched in all directions, or somewhat distichous, with long leading axes quite prominent in determining the aspect of the plants; the branches throughout corticated with equal numbers of longitudinally elongated cells in shorter and longer alternating transverse bands established over each segment; determinate branchlets radially disposed, about 1 mm. long, 60 μ diam. below, the axial cells 100 μ long, but above 37 μ diam. and of about the same length, corticated only at the nodes by 8 small cells; tips of the branchlets with a terminal spine and in addition one or more lateral uncinate spines on the first and second nodes; tetrasporangia naked, sessile, 60–75 μ diam.

V. **berkeleyana** (Montagne) J. Agardh. Uncinate spines generally absent.

V. **disticha** Børgesen. Branches plumose and obviously distichous.

F. **inermis** Børgesen. Determinate branchlets also distichously arranged and uncinate spines lacking.

V. **hypneoides** J. Agardh. Branch tips commonly free of branchlets and strongly hamate.

Bermuda, North Carolina, Florida, Bahamas, Cuba, Jamaica, Hispaniola, Puerto Rico, Virgin Isls., Guadeloupe, Grenada, Colombia, Netherlands Antilles, Venezuela, Brazil. Common in shallow water, especially in rather exposed places.

The varieties are not at all well marked. The forms exposed to wave action and thereby somewhat stunted tend to be more distichous than the lax ones; crozier-shaped tips are commonly lacking in forms of either extreme. Though it is generally difficult to demonstrate the recurved spines, they should be sought on the young branchlets near the branch tips.

REFERENCES: Harvey 1853, Mazé and Schramm 1870–77, Hemsley 1884, Collins 1901, Børgesen 1913–20, 1924, Collins and Hervey 1917, Howe 1918a, 1920, Taylor 1930a, 1942, Blomquist and Humm 1946 (vars.), Humm 1952b. For the v. *berkeleyana,* Collins and Hervey 1917. For the v. *disticha,* with its forma *inermis,* Børgesen 1913–20. For the v. *hypneoides,* Collins and Hervey 1917, Taylor 1928, 1941b; *S. spinella,* Mazé and Schramm 1870–77, Murray 1889.

UNCERTAIN RECORD

Spyridia ceramioides J. Agardh

Florida, the type locality. De Toni 4:1429.

DELESSERIACEAE

Plants usually foliaceous, sometimes the branches so slender as to appear filamentous, simple or bushy, alternately or infrequently dichotomously branched, the branches membranous or infrequently only compressed; growth from an apical cell producing an axial row which originates connected lateral cell rows of several degrees to produce a membrane and sometimes also a cortex, or the initial cells with age obscured and growth diffusely marginal, or, rarely, the apical cells replaced by hairs; sporangia tetrahedral, usually in superficial sori; spermatangia in sori; procarps borne on supporting cells which form one or two four-celled carpogenic branches, groups of sterile cells, and after fertilization auxiliaries; each cystocarp with a basal fusion cell and branched gonimoblasts, the outer cells of which are carposporangial, the whole invested by an inflated, ostiolate pericarp.

KEY TO GENERA

1. Very slender, seemingly filamentous species 2
1. Strap-shaped to foliaceous species 3

2. Small creeping plants with short erect branches and with filiform flat branchlets having terminal hairs which replace the apical cells .... TAENIOMA, p. 548
2. Bushy filamentous plants 2 cm. or more in height, the filaments more or less flattened, the hairs unilateral behind the apical cells .... COTTONIELLA, p. 549

3. Foliaceous species with an entire blade .... GRINNELLIA, p. 547
3. Species with lobed or branched, foliaceous to strap-shaped blades 4
3. Species with lobed or branched blades alternately continuous or netlike in successive zones .... MARTENSIA, p. 555

4. Blades without veins of any sort 8
4. Blades without evident midribs 5
4. Blades with evident midribs 6

5. Tetrasporangial sori solitary below the tips of ordinary branches .... ACROSORIUM, p. 552
5. Tetrasporangial sori marginal or in small marginal proliferations .... CRYPTOPLEURA, p. 554

6. Midribs generally present below only, but veinlets present in all parts of the blade .... HYMENENA, p. 553
6. Midribs, but no lateral veins or veinlets, present 7

7. Small, plum-purple, mostly creeping intertidal species; tetrasporangia in oblique intersecting rows .... CALOGLOSSA, p. 544
7. Larger rose-colored or greenish species; tetrasporangia irregularly arranged in sori beside the midrib .... HYPOGLOSSUM, p. 545

8. Blades one cell thick except at the base; tetrasporangia in scattered sori .... NITOPHYLLUM, p. 551

8. Blades with cortical cell layers; tetrasporangia in marginal stichidia .... CYCLOSPORA, p. 546

Caloglossa (Harvey) J. Agardh, 1876

Plants in the form of flat, dichotomously forking, locally constricted blades; each developing from a prominent apical cell a midrib consisting of a broad axial row of large cells covered by a cortex of elongated cells, this midrib bordered by a monostromatic lateral membrane consisting of subhexagonal cells running in oblique series from costa to margin; secondary branching proliferous from the midrib; tetrasporangia spherical, developed in oblique series from cells of the lateral part of the blade, mostly near the upper end; cystocarps sessile on the midrib, each enveloped in a thin pericarp.

Caloglossa leprieurii (Montagne) J. Agardh Pl. 68, fig. 1

Plants spreading or somewhat erect, to 4–5 cm. across, color violet; the blades to 2 mm. broad, constricted at the forkings and elsewhere, the individual segments lanceolate, 4–6 mm. long, sometimes linear-attenuate, more rarely ovate, often forming rhizoids at the constrictions; secondary segments or blades formed here or proliferously from the midribs of the blades.

V. **hookeri** (Harvey) Post. Branching very irregular, plants developing conspicuous clusters of erect segments at the nodes.

Bermuda, South Carolina, Georgia, Florida, Alabama, Mississippi, Bahamas, Jamaica, Hispaniola, Puerto Rico, Virgin Isls., Guadeloupe, Martinique, Barbados, British Honduras, Canal Zone, Colombia, Surinam, French Guiana, Brazil. Widespread and often very abundant in the intertidal zone, in rather sheltered coves, growing on rocks, woodwork, and especially on the roots of mangroves (Rhizophora); able to grow in areas of very low salinity, and north beyond the tropics as far as Connecticut.

REFERENCES: Hauck 1888, Murray 1889, Cox 1901, Collins 1901, Vickers 1905, Børgesen 1913–20, Collins and Hervey 1917, Howe 1918a, 1920, Taylor 1928, 1935, Taylor and Arndt 1929, Feldmann and Lami 1936, Joly 1951, 1957, de Mattos 1952; *Delesseria leprieurii,* Montagne 1850, Harvey 1853, Mazé and Schramm 1870–77. For the v. *hookeri,* Post 1955a.

Hypoglossum Kützing, 1843

Plants generally bushy, the branches slender, flat, with a midrib but no lateral veins, the blades elsewhere of one cell layer; branching from the midribs of the blades; growth from prominent apical cells at the tip of each blade, the apical cells of each segment-increment reaching the margins even to those of the third degree; midribs with a distinctive central cortical layer of cells; tetrasporangia and spermatangia in interrupted sori upon or along the midrib; pericarps sessile on the midribs, more prominent and with a produced pore on one side.

KEY TO SPECIES

1. Plants larger, blades rather flat-spreading, the segments tapering, but the tips rather obtuse or retuse .... H. tenuifolium
1. Plants smaller and rounded, blades short and rather curled, the tips acute, involute .... H. involvens

Hypoglossum involvens (Harvey) J. Agardh Pl. 68, fig. 4

Plants very delicate, small, tufted, 5–9 cm. diam., when living nearly colorless or pale greenish, but at least when dried becoming pale rose-colored; blades 1.0–2.5 cm. long, generally about 1.0–1.5 mm. wide, narrowly linear-lanceolate, the base obtuse but the apex attenuate and involute, the margin often undulate; costa of large cells; branching from the costa, usually unilaterally, at intervals of 3–10 mm.; tetrasporangia in sori beside the costa.

Florida, ?St. Eustatius. Only collected adrift, so apparently a deep-water plant.

REFERENCES: *Delesseria involvens,* Hooper 1850, Harvey 1853, Murray 1889, Taylor 1928; ?*Zellera boeckii,* Sluiter 1908.

Hypoglossum tenuifolium (Harvey) J. Agardh Pl. 68, fig. 2

Plants to about 10 cm. tall, and spreading; color light pink to greenish or nearly colorless, darkening little on drying; main blades to 3–5 cm. long and 2.5–3.5 mm. broad, oblong to oblanceolate, the apex subacute to broadly obtuse, the midrib hardly visible to the naked eye except in basal blades; proliferous branching from the midribs, usually regular, frequently closely overlapping; blades one cell thick in the wings, the cells in oblique rows to near the margin, which is bordered by about 2 rows of smaller cells; cells on each side of the midrib distinctively larger, in 2 superimposed series; midrib originally 3 cells in thickness, the central row remaining undivided but the outer rows divided and extending irregularly lengthwise of the axis to form a strengthening cortex; reproduction by tetrahedral sporangia formed irregularly in sori toward the tip of the blade, bordering the midrib, giving the naked-eye appearance of a bright red fleck on the pale green blade.

V. **carolinianum** Williams. Plants rarely more than 1 cm. tall.

Bermuda, North Carolina, South Carolina, Georgia, Florida, Bahamas, Caicos Isls., Jamaica, Virgin Isls., Guadeloupe, Netherlands Antilles, Venezuela. Apparently typically a plant of deep water, as it was secured when washed ashore and when dredged to depths of over 58–90 m., though the variety was described from shallow water.

REFERENCES: Howe 1920, Taylor 1942, Williams 1949a; *H. tenuifolium* f. *schoonhoveni*, Sluiter 1908; *H. hypoglossoides*, Collins and Hervey 1917; *Delesseria hypoglossum*, Harvey 1853, Murray 1889, Taylor 1928; *Delesseria tenuifolia*, Harvey 1853, Mazé and Schramm 1870–77, Murray 1889, Børgesen 1913–20. For v. *carolinianum*, Williams 1951.

Cyclospora J. Agardh, 1892

Plants bushy, the branches compressed or terete, below much and repeatedly distichously branched from near the rounded margin, the upper axis more nearly bare and subcaudate; without costa or veinlets, the blades showing a central tissue of large cells and a subcortex and a cortex of smaller ones; sporangia tetrahedral, numerous, in oblique series in marginal lanceolate stichidia.

Cyclospora curtissiae J. Agardh

Plants exceeding 7 cm. in height, below irregularly branched from near the margins of the compressed axes, above more nearly bare and the apices of the axes generally caudate; branches bearing numerous marginal branchlets which are contracted at the base; minor branchlets acting as stichidia, but slightly modified.

Florida.

REFERENCE: Taylor 1928.

Grinnellia Harvey, 1853

Plants of large, usually simple pink blades; a prominent midrib composed of elongate inner cells covered by rounded outer ones; lateral part of the blade one cell in thickness and lateral veins absent; growth from an apical cell; sporangia in scattered, elongate, unilaterally projecting nodular sori, more or less elevated, several cells thick, the tetrahedral sporangia formed from cells below the surface; spermatangia in scattered, often confluent sori; procarps individually developed on special islets of cells usually in the plane of the main blade and surrounded by normal blade cells; cystocarps scattered, in each the carposporangia seriate, with the superficial mature carposporangial cells external and smaller less-developed ones within, attaching to a large fusion cell, the whole covered by a thin-walled hemispherical pericarp, which discharges through an apical pore.

Grinnellia americana (C. Agardh) Harvey Pl. 69, fig. 5

Plants more or less gregarious, of large, erect, simple, translucent pink blades which are occasionally proliferous from the base, and consist of a very short stalk bearing a lanceolate or ovate-oblong blade 1–5(–10) dm. long, 4–10 cm. or more in width, the strong midrib extending nearly the length of the blade; tetrasporangia in elongate sori about 0.3–1.0 mm. wide and 0.5–2.0 mm. long scattered over the surface of the blade, the sporangia about 50–65 μ diam. and occasionally germinating in place; cystocarps scattered over similar blades, the pericarps about 0.3–0.6 mm. diam.; spermatangia in separate or confluent sori on much smaller blades which are only about 1–3 cm. long.

V. **caribaea** Taylor. Plants with a prominent midrib only in the lower part of the blade, which is broadly ovate to cordate; tetrasporangia in small round sori 0.20–0.37 mm. diam.; cystocarps similarly scattered, less numerous and much larger, strongly swollen, the pericarps to 1.0–1.5 mm. diam.

North Carolina, South Carolina, Netherlands Antilles, Venezuela. Growing in moderately shallow water; in southern waters generally dredged from a few meters depth. The variety dredged from 43 m. depth.

REFERENCES: Hoyt 1920, Williams 1949a. For the v. *caribaea*, Taylor 1942, Williams 1949a.

Taenioma J. Agardh, 1863

Plants creeping, abundantly alternately branched, the erect branches compressed; individual branchlets ending in determinate membranous blades showing a definite midrib, the axial cells with four pericentral cells of equal length, the lateral ones on the flat branchlets each bearing two marginal cells of half their length; branchlet tips with evident transversely dividing apical cells, soon replaced by conspicuous hairs with basal growth; tetrasporangia in two rows beside the midrib; spermatangia on both surfaces of the fertile branchlets; cystocarps urceolate.

KEY TO SPECIES

1. Branchlets of 8–20 segments, tipped with 2 hairs .... T. macrourum
1. Branchlets of 20–30 segments, tipped with 3 hairs ... T. perpusillum

Taenioma macrourum Thuret

Plants about 1 mm. tall, creeping, the rhizomatous axes terete, 60–75 μ diam., with upturned tips and rhizoids 30 μ diam. on the lower side; axes with 4 uncorticated pericentral cells; short branches and indeterminate branches formed alternately along the upper-lateral sides on about each fourth to sixth segment; flat determinate branchlets stipitate, linear, 60–75 μ wide, 8–20 segments or 240–400 μ long, the apex of each branchlet showing a large apical cell or 2 long colorless hairs; stichidial branchlets 80–90 μ wide, borne on stalks of 1–2 segments; tetrasporangia formed in a row on each side of the midrib.

Bermuda, Bahamas, Caicos Isls., Virgin Isls. Growing on calcareous reefs in very shallow water.

REFERENCES: Thompson 1910 *p. p.*, Tseng 1944; *T. perpusillum*, Børgesen 1913–20, Collins and Hervey 1917, Howe 1920.

Taenioma perpusillum (J. Agardh) J. Agardh

Plants about 2–3 mm. tall, creeping, the rhizomatous axes terete, 90–110 μ diam., attached by rhizoids which reach 75 μ diam.; erect branches terete, bearing alternate fastigiate branchlets; determinate branchlets stipitate, the stalks 3–4 segments in length, but flattened above and with a midrib, to 70 μ diam., 20–30 segments and 0.6–1.0 mm. long, typically bearing 3 hairs at the summit; stichidial branchlets to 110 μ broad but otherwise similar; sporangia tetrahedral, in 2 rows of generally 8–14 each.

Puerto Rico, Barbados. Growing on old wood or rocks often in exposed situations toward low-water mark.

REFERENCES: Vickers 1905, Tseng 1944; *T. macrourum,* Thompson 1910 *p. p.*

Cottoniella Børgesen, 1919

Plants filamentous, attached by rhizoids and terminal haptera from decumbent portions, then assurgent, becoming erect and alternately branched, subsecund, becoming complanate; branch growth initiated by a prominent apical cell, transversely divided; branches at first arcuate, later straight, in the young state developing one central and four or five pericentral cells, later becoming corticated, compressed, dorsiventral; below denuded, but in the upper part abundantly clothed with determinate monosiphonous ramelli, one to each segment, on the convex sides of the curved branch tips.

KEY TO SPECIES

1. Pericentral cells eventually 5 . . . . C. sanguinea, p. 549
1. Pericentral cells persistently 4 2

2. Lateral pericentral cells each bordered by 2 short alar cells; main axes 250–900 μ broad . . . . C. filamentosa, p. 550
2. Lateral pericentral cells without an alar border; main axes to 250 μ diam. . . . . C. arcuata, p. 550

Cottoniella sanguinea Howe

Plants gregarious, purplish red, very soft and lubricous, 2–3 cm. tall; more or less creeping and attached by rhizoids near the base, pseudodichotomous, the lower parts of the axes strongly corticated,

150–280 μ diam., subterete or compressed; upper portions not corticated, the original 4 pericentral cells generally increased to 5, 25–60 μ broad, tapering to the apex, but scarcely arcuate, bearing secund monosiphonous ramelli 0.3–0.6 mm. long, of 12–20 cells 1–3 diameters long.

Brazil. Growing on Sargassum, probably in shallow water.

REFERENCE: Howe 1928.

Cottoniella filamentosa (Howe) Børgesen

Plants 4–16 cm. long, very delicate and bushy, pseudochotomous below and the branches corticated for 0.5–0.8 of their length, the tips slightly curved; secondary branches originating along the upper midline of the axes; axes subterete at first, becoming flat and dorsiventral, 250–900 μ broad; above narrowly linear, the width increased by a row of 2 short alar siphons outside each of the 2 original large lateral pericentral cells; determinate ramelli in a single secund row along the upper side of the branchlets, in their upper parts sometimes apparently alternately laterally deflected, 0.2–1.0 mm. long, of 7–20 cells which are 2–6 diameters long.

Bermuda, Florida, Cuba, Brazil. Epiphytic on Sargassum and other algae in shallow water, and to a depth of 92 m.

REFERENCES: Børgesen 1913–20, Howe 1928; *Sarcomenia filamentosa,* Howe 1905b, 1909, 1918b.

Cottoniella arcuata Børgesen

Plants about 8 cm. tall, the main axes to 200–250 μ broad, the tips strongly arcuate, the 4 original pericentral cells persistent, with light cortication in the lower parts of the plant; branches originating laterally, the monosiphonous ramelli alternating in 2 rows along the convex side of each branchlet, reaching a length of 1 mm., of cells 7–20 μ diam., 175 μ long.

Virgin Isls. Once dredged from deep water, at 36 m.

REFERENCE: Børgesen 1913–20.

Platysiphonia Børgesen, 1931

UNCERTAIN RECORD

Platysiphonia miniata (C. Agardh) Børgesen

Brazil. Type from Spain. *Sarcomenia miniata,* De Toni 4:735; Murray 1889 *p. p.*

Chondrophyllum Kylin, 1924

UNCERTAIN RECORD

Chondrophyllum monanthos (J. Agardh) Kylin

Brazil. Type from Australia. *Nitophyllum monanthos,* De Toni 4:637; Möbius 1890.

Nitophyllum Greville, 1830

Plants foliaceous with wide segments, or subpalmately or dichotomously branched; branch tips lacking a distinctive apical cell; veins and veinlets absent; blades of one cell layer except near the base; tetrasporangia in scattered sori; gonimoblasts with one, seldom two, terminal carposporangia.

KEY TO SPECIES

1. Margin entire, essentially plane .... N. punctatum
1. Margin aculeate-dentate or undulate .... N. wilkinsoniae

Nitophyllum punctatum (Stackhouse) Greville Pl. 69, fig. 2

Plants attached by a minute disk, delicate, rose red, subsessile, 3–8, occasionally to 20, cm. tall, sometimes subpalmate below, distally irregularly dichotomous, the lobes somewhat narrow, 3–10 (–20) mm. broad; tetrasporangial sori oblong or rounded, to 2 mm. wide, 6.0–6.5 mm. long, scattered over the frond; cystocarps about 1 mm. diam., likewise scattered.

V. **ocellatum** (Lamouroux) J. Agardh. Segments linear, 3–7 mm. broad, the tips forked, the sporangial sori round.

?North Carolina, Florida, ?Netherlands Antilles, Venezuela.

REFERENCES: Murray 1889, Taylor 1942. For v. *ocellatum,* Harvey 1853, Taylor 1928.

Nitophyllum wilkinsoniae Collins and Hervey Pl. 69, fig. 1

Plants densely tufted, to 10 cm. tall, di- polychotomously divided, the segments strap-shaped or cuneate, 2–10 mm. broad, the apices obtuse to spatulate, the margins usually undulate, sometimes sparingly erose-dentate or rarely locally minutely aculeate-dentate; monostromatic, 15 μ thick in younger parts, 60 μ in the older ones, the cells 30–80 μ diam. in surface view; tetrasporangial sori small, scattered, slightly projecting, the sporangia to 60 μ diam.

Bermuda. Occasional, forming dense masses on rocks a little below low-water mark.

REFERENCES: Collins and Hervey 1917, Howe 1918a.

UNCERTAIN RECORD

Nitophyllum lenormandii (Derbès and Solier) Rodriguez

Barbados. Type from the Mediterranean. De Toni 4:625; Vickers 1905.

Acrosorium Zanardini, 1869

Plants freely subdichotomously, or irregularly to alternately branched, the branches commonly with dentate margins, without macroscopic veins but with microscopic veinlets lacking rhizoidal cortication; branch tips without a distinctive apical cell; tetrasporangia in single, large, rounded sori below the tips of the branches; gonimoblasts with terminal carposporangia.

KEY TO SPECIES

1. Branches broad; marginal teeth minute and rather close together .... A. odontophorum
1. Branches narrow; marginal teeth large and scattered; branch tips attenuate and sometimes hooked .... A. uncinatum

Acrosorium uncinatum (Turner) Kylin Pl. 58, fig. 2

Plants 2–5 cm. long, tufted, without any definite stalk; alternately branched, the linear segments irregularly 2.0–4.5 mm. broad, the tips often hooked; the margins entire except for large scattered teeth which intergrade in form with the smaller branchlets or the hooked tips; veinlets irregularly longitudinal and somewhat anastomosing; fertile branches generally small, lateral.

Brazil. Apparently only collected by dredging in our area.

REFERENCE: Taylor 1930a.

Acrosorium odontophorum Howe and Taylor Pl. 69, fig. 3

Blades thin, nearly sessile, 7–10 cm. (or more?) broad, subradiately dichotomous, here and there subpinnate, deep brownish-purple, ecostate, with very obscure veinlets (scarcely visible under a hand-lens); the margins rather regularly denticulate, the teeth deltoid-acute or subacuminate, the older 150–225 μ long; blades 25–42 μ thick in unistratose or bistratose parts, in other parts 5–7 cells and to 250 μ thick; the cells angular in surface view, 25–80 μ in maximum diameter, those of the obscure anastomosing veinlets slightly narrower; apical cells of the younger teeth and lobes distinct.

Brazil. Dredged from deep water.

REFERENCES: Howe and Taylor 1931; *Nitophyllum odontophorum*, Taylor 1930 (*nomen nudum*).

Hymenena Greville, 1830

Plants bushy, marginally subdichotomously or subalternately branched, the segments strap-shaped, or subcuneate; branch apices without transversely dividing apical cells; midribs present or absent, but small veinlets always present and generally anastomosing; pericarps and tetrasporangial sori scattered over the blades, the gonimoblasts with terminal carposporangia.

Hymenena media (Hoyt) n. comb.

Plants erect, light pink to rose colored, 5–22 cm. tall, with a short stipe, above repeatedly and very irregularly dichotomous, the segments strap-shaped, 4–19 mm. broad, the margins undulate and often minutely proliferous; microscopic anastomosing veinlets numerous, usually one cell wide, sometimes visible without a lens, coalescent below into the stipe; thallus of one layer of cells except at the veinlets, where 3–6 cells thick, sometimes with a border along the older veins also 3 cells thick; tetrasporangial sori scattered over the thallus, in locally thickened areas, usually between the veinlets, projecting on both surfaces of the blade.

North Carolina, South Carolina. Occasional in shallow water, but apparently chiefly a species of deeper water; dredged to a depth of 25 m.

REFERENCE: *Nitophyllum medium*, Hoyt 1920, p. 494.

Cryptopleura Kützing, 1843

Plants bushy; in the lower parts often becoming somewhat stalk-like, dichotomously or subpalmately or laterally branched, the margins often proliferous; growth from obscure apical cells, developing a thallus of one cell layer except in the numerous small anastomosing veins and veinlets; tetrasporangia and spermatangia in elongate marginal sori or in marginal proliferations; pericarps somewhat projecting, the end cells of the gonimoblasts forming carposporangia.

KEY TO SPECIES

1. Sori principally in the main blades .... C. lacerata
1. Sori principally in marginal leaflets .... C. fimbriata

Cryptopleura fimbriata (Greville) Kützing — Pl. 58, fig. 1

Plants to 7–12 cm. tall, deep red, rather thick in texture and veined above; blades of irregular width, substipitate, broadening and repeatedly polychotomously divided above, the ultimate segments oblong, obtuse; sori in the marginal lobes of pinnately lobed branches, or conspicuous and solitary in marginal leaflets.

Brazil, Uruguay.

REFERENCES: Martens 1870, Taylor 1939a; *Nitophyllum calophylloides*, J. Agardh 1872–90; *N. fimbriatum*, Greville 1833.

Cryptopleura lacerata (Gmelin) Kützing — Pl. 69, fig. 4

Plants bushy, to 6–15 cm. tall, often several blades arising from a single small holdfast; blades thin, subcuneate to palmatifid, pinnatifid or subdichotomous, the segments often redividing, to 2.5 cm. broad; apices rounded, the margins undulate to subfimbriate; veins macroscopic below, more delicate and inconspicuous above; tetrasporangia and spermatangia in more or less confluent sori, particularly toward the margins, and the sporangia sometimes in marginal leaflets; cystocarps scattered toward the margins of the blades.

Brazil, ?Uruguay.

REFERENCES: Martens 1870, Taylor 1939a, Joly 1957.

Martensia Hering, 1841

Plants foliaceous, lobed or dichotomously or irregularly branched with broad, strap-shaped, plane, undulate or strongly crisped segments, the basal part becoming stalklike and much thickened in some species; adult growth from a marginal row of initial cells producing a simple membrane one cell in thickness, without veins, or growth periodically intercalary, whereby gridlike zones are developed with the thin ribs of the grid considerably extended in a plane at right angles to that of the simple membrane; tetrasporangia in sori over the upper part of the thallus, especially over the grid bands; cystocarps on the margins of the grid bands, projecting.

Martensia pavonia (J. Agardh) J. Agardh Pl. 68, fig. 3

Plants to 3–4 cm. high, sessile, lobed or branched, the segments to 2–10 mm. wide, frequently altered from the simple membrane to a grid of parallel or anastomosed bands, and often rather irregular.

Florida, Mexico, Jamaica, Virgin Isls., Guadeloupe. From shallow-water stations and dredged to a depth of 30 m., often entangled among other algae.

REFERENCES: Mazé and Schramm 1870–77, Murray 1889, Børgesen 1913–20, Taylor 1928.

DASYACEAE

Plants bushy or with long, terete primary axes; branches radially or dorsiventrally bearing monosiphonous branched filaments of limited growth which may be free or united into a network; growth not continuous from a persisting apical cell, but the successive segments before cortication each producing laterally a new growing point, thereby displacing the preceding apex, which develops into a lateral tuft of filaments; axial cell row in some genera becoming surrounded by a circle of pericentral cells, and in many cases also corticated by the development of rhizoidal downgrowths from the bases of the lateral filaments, rarely the polysiphonous origin disappearing in the development of an undifferentiated parenchymatous axis; forking lateral filaments in some cases becoming polysiphonous in the lower segments, but monosiphonous above and often ending in colorless filiform extensions; tetrasporangia produced in distinctive stichidia; colorless spermatangia borne on lateral branchlets; procarps developed near the bases of the lateral tufts of filaments, a fertile pericentral cell producing sterile cells, the four-celled carpogenic branch, and the auxiliary; gonimoblasts monopodial in branching; cystocarp enveloped by an ample ostiolate pericarp.

KEY TO GENERA

1. Branchlets free 2
1. Branchlets regularly anastomosing to form a network 4
2. Branching radial; tetrasporangia substantially exposed at maturity 3
2. Branching dorsiventral; tetrasporangia nearly completely enclosed by the cover cells .... HETEROSIPHONIA, p. 565
3. Axis surrounded by evident pericentral cells, often with rhizoids in addition; tetrasporangia with 2 cover cells .... DASYA, p. 556
3. Axis without pericentral cells, directly corticated by rhizoids; tetrasporangia with 3 cover cells .... DASYOPSIS, p. 563
4. Network supported by a distinct, often branched, axis system 5
4. Network not differentiated into a coarse axis and a delicate mesh .... HALODICTYON, p. 566
5. Thallus network superficial, subterete or prismatic in form; axis showing 4 regular pericentral cells in each segment, eventually in the older parts covered by rhizoids .... DICTYURUS, p. 567
5. Thallus network spongy, the whole laterally branched, with terete divisions; axis without pericentral cells, in the oldest parts becoming corticated by rhizoids .... THURETIA, p. 568

Dasya C. Agardh, 1824

Plants erect, more or less bushy, with stout main branches covered with filiform branchlets; structurally the main branches with five pericentral cells surrounding the axial row, in the older portions of most species the whole corticated by rhizoidal filaments; branchlets crowded on the axes, whorled, polysiphonous about the base or not at all, above dividing into monosiphonous, pseudodichotomously branched, chromatophore-bearing filaments (ramelli); tetrasporangia developed in distinctive siliquose, stalked stichidia, each sporangium with two cover cells; spermatangial clusters lanceolate to subcylindrical on ramelli and often hair-tipped; procarps developed on a fertile segment near the base of a ramellar tuft; the four-celled carpogenic branch associated with sterile cells; after fertilization the auxiliary cell is cut off from the same supporting cell as the carpogonium; carposporangia in moniliform series of about four cells around a large fusion cell; pericarp with a large apical pore, developed immediately after fertilization.

KEY TO SPECIES

1. Ramelli spirally disposed 2
1. Ramelli randomly distributed 6
1. Ramelli subverticillate and sometimes lightly corticated at the base 10

2. Axes ecorticate, except occasionally at the extreme base D. rigidula, p. 558
2. Axes extensively corticated 3

3. Ramelli not specialized as below; nodes not conspicuous 4
3. Ramellar tips indurate-aculeate or hyaline-attenuate, the basal cells thick-walled; nodal cross-walls of the axial cell row refractive, visible through the young cortex D. sertularioides, p. 560

4. Ramelli not ocellate .... D. corymbifera, p. 559
4. Ramelli in conspicuous ocellate clusters at the branchlet tips ... 5

5. Tall, showy plants several centimeters in height, alternately branched .... D. caraibica, p. 560
5. Small, densely bushy, subdichotomously branched plants hardly 3 cm. tall .... D. collinsiana, p. 558

6. Sparingly branched, the ramelli very much crowded to form terminal "ocelli" .... D. ocellata, p. 559
6. More copiously branched, hardly ocellate 7

7. Plants subdichotomously branched below, the divisions rather elongate above .... D. crouaniana, p. 561
7. Plants clearly alternately branched 8

8. Branchlets regularly becoming naked below, only clothed with ramelli near the tips .... D. ramosissima, p. 561
8. Branchlets usually ramellate throughout 9
9. Primary branching sparse, main axis dominant, habit virgate D. pedicellata, p. 562
9. Primary branching abundant, habit broad, main axis not particularly dominant .... D. harveyi, p. 561
10. Ramelli often becoming slightly corticated at the base, less than 2 mm. long, their upper cells about 10 diameters long D. mollis, p. 562
10. Ramelli not corticated at the base, 2.0–4.4 mm. long, their upper cells about 3–4 diameters long .... D. punicea, p. 563

Dasya rigidula (Kützing) Ardissone Pl. 72, fig. 4

Plants small, purplish, to about 2 cm. tall, freely alternately branched, relatively stiff; axes except at the extreme base altogether ecorticate, about 300–500 μ diam.; ramelli spirally arranged, widely dichotomous, incurved; cells about 2 diameters long, swollen between the septa; stichidia replacing lower forks of the branchlets, sessile, ovoid-oblong, subapiculate, 2–3 diameters long.

Bermuda, Florida, Mexico, Bahamas, Cayman Isls., Jamaica, Venezuela. A small and easily overlooked species commonly mixed with larger things. It is very similar in aspect to *Heterosiphonia wurdemanni,* which has 6 (or 4) pericentral cells instead of 5. It is a plant of shallow water.

REFERENCES: Howe 1920, Taylor 1935; *Dasya arbuscula,* Vickers 1905 (? *p.p.*).

Dasya collinsiana Howe

Plants tufted, 1–3 cm. tall, dull red, drying yellowish; the branching 5–10 times subdichotomous, the main divisions corymbose or somewhat fastigiate, the aspect of the last divisions brushlike, 1.0–1.5 mm. diam.; main axes 0.5–0.75 mm. diam. near the base, corticated up to the last 2–3 forkings; the ends of the pericentral cells laterally overlapping at the nodes, short, usually only 2–3 diameters long; ramelli in an evident coma at the tips, below spreading to subsquarrose, 0.5–0.9 mm. long, 4–5 times divaricately dichotomous, firm below, soft in the outer divisions and often with short terminal hairs; short basal cells 100–130 μ diam., subterminal and terminal cells 45–55 μ diam., about 2 diameters long; stichidia acuminate-apiculate, 90–130 μ diam., 300–500 μ long, subsessile or on a one-celled stalk.

Bermuda, Florida, Bahamas, Jamaica. Growing in shallow water on rocks and coarse algae, and dredged to 14 m. depth.

REFERENCES: Howe 1918a, 1920, Taylor 1928.

Dasya ocellata (Grateloup) Harvey

Plants bushy, reddish purple, about 2 cm. tall, little to densely branched; sometimes more elongate and sparingly alternately branched and then to 6–7 cm. tall; axes becoming heavily corticated, the segments 1.5 diameters long below, near the tips 0.5–1.0 diameter; ramelli irregularly disposed over the cortex, fasciculate, dichotomously branched, 2–4 mm. long, 15 μ diam., toward their tips attenuate and very soft, the cells about 2 diameters long below, 6 diameters long toward the tips; stichidia on 1–2-celled stalks on the lower divisions of the ramelli, conical, or when old becoming linear-lanceolate and 7–10 diameters long, the tips often sterile and acute; pericarps sessile on short branchlets, ovate-globose, the tip not produced.

Bermuda, Virgin Isls. Plants of shallow water, growing on mangrove roots, Sargassum, and similar objects.

REFERENCES: Børgesen 1913–20, Collins and Hervey 1917.

Dasya corymbifera J. Agardh

Plants 5–12 cm. tall, pyramidal, bright red; repeatedly alternately branched, the main axes somewhat denuded below, about 0.75 mm. diam., above covered with ramelli; axial segments 1.5–2.0 diameters long below, substantially corticated by rhizoids; ramelli spirally disposed, dichotomously branched, the lower cells 50–70 μ diam., 70–90 μ long, tapering to the upper segments which are exceedingly delicate, about 10 μ diam., 100–120 μ long and strongly incurved; stichidia usually on 1–3-celled stalks, lanceolate, 5-6 diameters long; pericarps sessile on short branches, ovate-conical, sometimes with a short produced apex.

Bermuda, Florida, Jamaica, Hispaniola, Virgin Isls., Guadeloupe, Grenada. Growing in shallow water, especially under overhanging rocks, but also dredged to a depth of 30 m.

REFERENCES: Mazé and Schramm 1870–77, Murray 1889, Børgesen 1913–20, Collins and Hervey 1917, Howe 1918a, Taylor 1933; *D. arbuscula,* Collins 1901, Collins and Hervey 1917, Taylor 1928; *D. arbuscula* f. *subarticulata,* Phyc. Bor.-Am. 493.

Dasya caraibica Børgesen

Plants bushy, subpyramidal, to 20 cm. tall, rose-purple in color; the filiform main axes soon naked below, with few primary but moderately numerous secondary divisions alternately and radially disposed; axial segments about 400–500 μ long, becoming covered by rhizoidal cortication, and about 1 mm. diam.; ramelli spirally disposed, borne on a conspicuous basal cell about 50 μ diam., 60–70 μ long, lying between the pericentral cells of the supporting axis, 2–4 mm. long, ocellate at the branch tips, distally dichotomously divided at narrow angles, the segments 24 μ diam. below, 200 μ or more in length, reduced to 8 μ diam. at the apices.

Florida, Virgin Isls. Dredged from 30–50 m. depth.

REFERENCE: Børgesen 1913–20.

Dasya sertularioides Howe and Taylor

Plants 6–20 cm. tall; primary branching irregular, the secondary subpinnate, the ultimate dichotomous ramelli irregularly quadrifarious or irregularly spiral, falsely distichous on drying; basal parts of stem and main branches denudate for 1–10 cm., strongly corticated, 0.8–1.5 mm. diam., tapering to 100–160 μ diam. near the apex, the cortications thinner but persisting nearly to the tips; pericentral cells somewhat interlocking at the nodes, the axial siphon easily visible distally, the cells 90–160 μ long with somewhat thickened end walls; ramelli 3–6 times dichotomous, 0.75–1.0 mm. long, patent, often crowded and ocellate at the apices, the basal cells 46–54 μ diam., 2.0–2.5 diameters long, with very thick walls, the apical cells acuminate or aculeate, or when young with terminal hairs 1–2 μ diam.

Brazil. Apparently a deep-water species; known only from a dredged collection.

REFERENCES: Taylor 1930a, Howe and Taylor 1931.

Dasya crouaniana J. Agardh Pl. 71, fig. 1

Plants to 6.0–6.5 cm. tall; branches cylindrical, with radiating ramelli, widely subflabellately dichotomous, the axes corticated to the apex; for some distance below penicillate-villous, the very soft ramelli arising from the cortex, branching dichotomously, tapering from the base to the acute apices, the segments 4 times as long as broad or longer; stichidia on the lower forks, oval-lanceolate, the stalks monosiphonous.

Florida, Bahamas, Guadeloupe.

REFERENCES: Howe 1920, Taylor 1928; *D. dichotomo-flabellata*, Mazé and Schramm 1870–77, Murray 1889.

Dasya ramosissima Harvey

Plants bushy, brownish red, to 15–30 cm. tall, the lower axes repeatedly widely alternately branched, the branches spreading, naked, except near the tips; pericentral cells corticated in mature branches; branches of the last order slender, about 2.5 cm. long, naked at the base, above densely ramulate; ramelli issuing from the cortex at random, their lower forks patent, the distal divisions erect, the cells below about 1.5 diameters long, those above 4–6 diameters long; stichidia on the lower divisions of the ramelli, linear-oblong and subacute; pericarps sessile on short branchlets, ovate-globose, not produced at the apex.

Bermuda, Florida, Bahamas, Cuba, Grenada. Growing on overhanging rocks in shallow water and dredged to 55 m. depth.

REFERENCES: Harvey 1853, Murray 1889, Collins and Hervey 1917, Howe 1918b, 1920, Taylor 1928.

Dasya harveyi Ashmead

Plants bushy, rose-red, 20–25 cm. tall; freely alternately or secundly branched, these branches firm, densely corticated, the surface cells considerably elongate; branches of the last orders closely beset with slender, flaccid, alternately dividing branchlets 1.0–2.5 cm. long; dichotomous ramelli 2–5 mm. long, with cells several times longer than broad; stichidia from the lower forks of the ramelli, slender and tapering to each end, with 2 rows of tetraspores.

Florida.

REFERENCES: Harvey 1858, Murray 1889, Taylor 1928.

Dasya pedicellata (C. Agardh) C. Agardh

Plants from a small disklike holdfast, erect, to 2–7 dm. tall, light to deep red-purple; sparingly to freely alternately branched, the lateral branches infrequently redivided, 2–6 mm. diam., sometimes denuded below, but above densely covered with slender ramelli 4–7 (–14) mm. long, the cells near the base 15–40 μ diam., at the tips 7–12 μ diam.; the stichidia lanceolate to linear-lanceolate, acute, 80–120 μ diam., 0.2–1.25 mm. long, the sporangia tetrahedral, 40–50 μ diam.; sexual plants dioecious, the spermatangial clusters lanceolate to linear-lanceolate, acute and usually filament-tipped, 60–75 μ diam., 250–550 μ long; the stalked pericarps urceolate, below transversely oval, to 0.75–1.00 mm. diam., above with a narrow ostiolate neck about 200 μ diam.

Bermuda, North Carolina, South Carolina, Florida, Texas, Bahamas, Cuba, Virgin Isls., Netherlands Antilles, Venezuela. Growing in shallow protected situations on shells, stones, coarse algae, etc., and dredged to a depth of 110 m. In the tropics individuals of this species are much less robust and less deeply colored than in northern waters.

REFERENCES: Børgesen 1913–20, Collins and Hervey 1917, Howe 1918a, 1920, Hoyt 1920, Taylor 1928, 1941a, b, 1942, 1954a; *D. elegans,* Harvey 1853, Rein 1873, Hemsley 1884, Murray 1889; *D. elegans* v. *scotiochroa,* Murray 1889; *Eupogodon grande* and *E. mazei,* Mazé and Schramm 1870–77, Murray 1889.

Dasya mollis Harvey

Plants bushy, to 3–4(–15) cm. tall; irregularly branched, without a well-defined main axis, the chief divisions well corticated; ramelli in ill-defined whorls most distinct near the branch tips, about 50 μ diam. at the base where often a little corticated, the lowest cell of each ramellus 1–2 diameters long; toward the tips sharply tapering, the upper cells about 7 μ diam., 10 diameters long; stichidia attached near the bases of the ramelli, about 135 μ diam.

Bermuda, Florida, Bahamas, Cuba, Virgin Isls. Apparently a moderately deep-water form: known only as dredged from 8–33 m. depth.

REFERENCES: Harvey 1853, Murray 1889, Børgesen 1913–20, Taylor 1928.

Dasya punicea Meneghini

Plants 5–10 cm. tall, the main axes rather slender, densely pyramidally branched in all directions, long and short branches intermixed; main divisions naked at the base or with only degenerating bases of the branchlets, above brushlike with very soft ramelli vaguely whorled, in part issuing from the axis and in part from the cortex; pericentral cells visible in sections of the young axis but less so in the older parts, the axes corticated nearly to the tips; ramelli 2.0–4.4 mm. long, somewhat fasciculate, dichotomously branched at acute angles, the lower cells 1.5–2.0 diameters long, the upper ones 3–4 diameters long; stichidia on 1–2-celled stalks, becoming lanceolate-ovate; pericarps on short stalks, ovate, the apex attenuate.

Bermuda.

REFERENCE: Collins and Hervey 1917.

UNCERTAIN RECORDS

Dasya arbuscula (Dillwyn) C. Agardh

Guadeloupe, Barbados. Type from Ireland. De Toni 4:1205; Mazé and Schramm 1870–77, Vickers 1905.

Dasya hussoniana Montagne

Guadeloupe. Type from the Red Sea. De Toni 4:1190; Mazé and Schramm 1870–77, Murray 1889.

Dasyopsis Zanardini, 1843

Plants erect, bushy, the branches clothed with ramelli; axis and branches without pericentral cells, directly corticated by rhizoids and of a parenchymatous aspect in transverse section; ramelli spirally disposed, deciduous, at the base polysiphonous and this part persistent, aculeiform; stichidia on a short monosiphonous stalk or this with rhizoidal cortication, with several tetrasporangia in each segment, each partly covered by three cortical cells.

KEY TO SPECIES

1. Large and coarse; ramelli softer and with longer cells; stichidia fusiform at maturity .... D. antillarum
1. Small; ramelli somewhat stiffer; stichidia cylindrical at maturity .... D. spinuligera

Dasyopsis spinuligera (Collins and Howe) Howe

Plants small, 2–3 cm. tall, dark red; with few elongate branches, the corticated axes bare below except for the short, acute, occasionally persisting spinous bases of the ramelli; ramelli otherwise deciduous, but very dense on the younger parts of the plant, dichotomous, erect, the cells 2–3 diameters long below, but subequal distally and not much tapered; stichidia conical or cylindrical, arising near the bases of the fascicles of ramelli, with 2–4 tetrasporangia in a whorl.

Bermuda.

REFERENCES: Howe 1920; *Dasya spinuligera,* Collins and Hervey 1917, Howe 1918a.

Dasyopsis antillarum Howe

Plants 4–10 cm. tall, brownish-purple or faded; below irregularly branched, subdichotomous to more usually copiously alternate; main axes subterete or compressed, 1.0–2.5 mm. diam., bearing subspinescent branchlets 1–3 mm. long; the tips and the branchlets, sometimes most of the axes, clothed with dichotomous, deciduous ramelli; ramellar segments 15–26 μ diam., 3–6 diameters long; stichidia fusiform, 75–150 μ diam., 300–450 μ long, borne close to the axes on one-celled pedicels, or apparently sessile, occasionally with hair tips.

Florida, Bahamas, Caicos Isls., Cuba. Growing in shallow water.

REFERENCES: Howe 1920, Taylor 1928.

Heterosiphonia Montagne, 1842

Plants usually erect, sometimes decumbent with dorsiventral organization, main stems often flattened, generally corticated, chief divisions usually sympodial, secondary divisions frequently pinnate, bearing ultimate ramelli which are alternately branched, monosiphonous or, toward the base, polysiphonous; axes with four, six, or more pericentral cells, which remain undivided, or divide on the lower side of the axis, and are corticated by rhizoids from the bases of the branches, from which cortex secondary branches may arise; tetrasporangia in well-developed stichidia, replacing a branch of a ramulus; spermatangial clusters similarly placed, with a polysiphonous base, generally pointed; procarps generally developed near the bases of the ramelli, which may furnish stalks to the pericarps.

KEY TO SPECIES

1. Plants subrepent, 1–6 cm. long; main axes uncorticated, mostly 65–130 μ diam. . . . . H. wurdemanni
1. Plants 1–2 dm. tall; main axes strongly corticated, mostly 0.45–1.1 mm. diam. . . . . H. gibbesii

Heterosiphonia wurdemanni (Bailey *ex* Harvey) Falkenberg

Pl. 72, fig. 9

Plants small, about 1–6 cm. long, repent or tangled among other algae; dichotomously or irregularly branched, the main axis without cortication, to about 200 μ diam., with 6 pericentral cells; somewhat dorsiventral, with incurved monosiphonous ramelli which are attached to alternate segments and are dichotomously or in part alternately 2–4-times branched, about 100 μ diam. at the base to 50 μ diam. in the ultimate divisions, which are sharply tapering at the apex, the cells firm-walled, shorter than broad or subequal, a little turgid.

V. **laxa** Børgesen. Plants lax, sparingly branched, with 4 pericentral cells in the main axes; the ramelli attenuate, the tips filiform, the cells of the ultimate divisions cylindrical, 40 μ diam. and 70 μ long near the base, 120 μ long near the tips.

Bermuda, Florida, Bahamas, Caicos Isls., Jamaica, Hispaniola, Virgin Isls., Guadeloupe, Barbados, British Honduras, Colombia,

Venezuela. Growing entangled among other algae, especially Sargassum, and dredged to a depth of 7 m. in this area and 20 m. in the Mediterranean.

REFERENCES: Collins 1901, Børgesen 1913–20, Collins and Hervey 1917, Howe 1918a, 1920, Taylor 1928, 1933, 1935, 1942, 1943, 1954a; *Dasya wurdemanni*, Bailey 1848 (*nomen nudum*), Harvey 1853, Mazé and Schramm 1870–77, Murray 1889; *Heterodasya wurdemanni*, Vickers 1905. For the v. *laxa*, Børgesen 1913–20, Taylor and Arndt 1929.

Heterosiphonia gibbesii (Harvey) Falkenberg

Pl. 72, fig. 7; pl. 73, fig. 5

Plants large, the tufts reaching a height of 1–2 dm., the main axes sparsely forking, the long flexuous secondary axes with alternate pinnate or bipinnate corticated subdeterminate branches, the brownish diaphragms of the central axial cells visible through the cortex; the main branches complanate and chiefly denuded near the base, but above bearing densely placed ramelli which are alternately branched and polysiphonous below, but branch dichotomously above and are there monosiphonous.

Bermuda, Florida, Bahamas, Cuba, Jamaica, Grenada, Venezuela. Growing on stones or old corals in open and sunny locations, not exposed to heavy surf.

REFERENCES: Howe 1920, Taylor 1928, 1942, 1954a, 1955a; *Dasya gibbesii*, Harvey 1853, Murray 1889, Collins 1901; *Polysiphonia gibbesii*, Hooper 1850.

Halodictyon Zanardini, 1843

Plants in the form of a network of irregular mesh in three dimensions, without any distinct axis, the dichotomous filaments monosiphonous, branches from one filament attaching to another at points of cell juncture; reproductive organs on short polysiphonous branchlets with four pericentral cells, the tetrasporangia in two rows in short, compressed stichidia; spermatangial clusters flattened, oval, with a marginal band of sterile cells; procarps from the second pericentral cell of the branchlet producing an urceolate pericarp with a thin wall.

Halodictyon mirabile Zanardini Pl. 72, fig. 5

Plants small, seldom to 6–8 cm. tall, rosy; the filaments furcate-dichotomous above, forming a soft net, the meshes irregularly polygonal, of cells 3–6 diameters long; stichidia in the distal parts of the frond, single or paired, generally sessile; pericarps formed in the lower parts of the frond, projecting from the reticulum.

Bermuda, Florida, Bahamas, Jamaica, Barbados. Occasionally washed ashore attached to other algae, and dredged from a depth of 92 m.

REFERENCES: Collins 1901, Vickers 1905, Howe 1920, Taylor 1928.

Dictyurus Bory, 1836

Plants erect, radially branched, the main axes bearing lateral determinate polysiphonous branchlets alternately from each second node whose ultimate monosiphonous ramelli anastomose and collectively form a superficial, more or less prismatic network; pericentral cells four in the main branches, soon becoming heavily secondarily corticated; tetrasporangia in stichidia borne on ramelli with thickened polysiphonous bases.

Dictyurus occidentalis J. Agardh Pl. 70, figs. 1, 2

Plants in large tufts or colonies 4–12 cm. tall, dull reddish-purple; erect from a substantial base, the main stems flexuous, sparingly branched, above covered with the characteristic ridged, somewhat locally constricted network or quadrangular veil with concave sides, 4–7 mm. diam., composed of anastomosed filaments, extending for a distance of 2–5(–9) cm.; lower axes denuded, showing distichously placed remnants of the determinate branchlets which bore the network.

Florida, Mexico, Jamaica, Virgin Isls., Guadeloupe, Martinique, Barbados. Found in shallow water, and also dredged to a depth of 30 m.

REFERENCES: J. Agardh 1847, Mazé and Schramm 1870–77, Dickie 1874a, Murray 1889, Collins 1901, Vickers 1905, Howe 1909, Børgesen 1913–20, Hamel and Hamel-Joukov 1931, Humm 1952b.

Thuretia Decaisne, 1843

Plants repeatedly alternately branched, complanate, the axes becoming stalklike and denuded below; without pericentral cells, but with rhizoidal cortication in the oldest portions; above with anastomosing monosiphonous ramelli, rising alternately from successive nodes, which invest the axes with a spongy net; tetrasporangia in superficial conical stichidia at the bases of the ramelli; spermatangial clusters ovoid to globose, the pericarps broadly urceolate with a projecting neck, both formed within the meshes of the network.

Thuretia borneti Vickers — Pl. 70, figs. 6, 7

Plants bushy, to about 3–5 cm. tall, dull red in color and of a spongy texture; branching freely from near the base, the divisions subsimple or 1–3 times irregularly branched at intervals of 3–12 mm.; individual reticulum-covered segments terete, 1–2 mm. diam., the terminal ones obtuse; true axes about 130 μ diam., obscured by the reticulum; areolae of the network polygonal, about 65–100 μ diam., the constituent filaments 18–32 μ diam., the cells 1.0–1.5(–2.0) diameters long.

Bermuda, Jamaica, Barbados. Growing near low-tide level, sheltered by Sargassum and overhanging rocks.

REFERENCE: Vickers 1905.

RHODOMELACEAE

Plants usually bushy, sometimes sparingly branched; branches often delicate, usually terete, occasionally flat; growth from persisting apical cells producing an axial cell row, the principal branches of the thallus developed from successive segments; branched colorless hairs (trichoblasts) often present; axial cells generally surrounded, at least in the fruiting portions, by a series of pericentral cells cut off from them by longitudinal walls, producing a typically polysiphonous structure, and sometimes further corticated either by subsequent divisions of these to several degrees or by appressed rhizoidal downgrowths; tetrahedral sporangia formed from internal segments of the pericentral cells, the branchlets bearing them usually little modified, but in extreme cases like stichidia; spermatangial clusters developed from trichoblast rudiments in the form of colorless tufts, cones, or plates of spermatangia; procarps developed from polysiphonous basal trichoblast segments, the fertile pericentral cell as a supporting cell producing sterile cells and the four-celled carpogenic branch; from the supporting cell beside the carpogonium an auxiliary is cut off after fertilization, from which, by means of a fusion cell, the sympodially branched gonimoblasts are produced; outer cells of the gonimoblasts alone forming carpospores; cystocarps becoming enclosed by an ostiolate pericarp.

KEY TO GENERA

1. Branches flat, or at least the ultimate ones compressed 18
1. Branches terete, the tips erect or in a few cases arcuate 2

2. Ultimate branchlets, to be designated ramelli, polysiphonous 3
2. Ultimate branchlets, at least at the tips monosiphonous, but more or less chromatophorous, thus differing from trichoblasts .. 4

3. Thallus radially organized 8
3. Thallus erect, and branchlets strikingly distichous in arrangement .... PTEROSIPHONIA, p. 593
3. Thallus dorsiventral, more or less creeping 16

4. Ramelli soft, very early deciduous .... WRIGHTIELLA, p. 591
4. Ramelli persisting and characteristically clothing the upper branches 5

5. Radially organized 6
5. Characteristically dorsiventral in organization; pericentral cells subdivided by secondary walls .... BOSTRYCHIA, p. 594

6. Ramelli very lax and delicate; stalks of the stichidia monosiphonous .... LOPHOCLADIA, p. 590
6. Ramelli shorter and relatively stiff; stalks of the stichidia polysiphonous 7

7. Pericentral cells 5–7; sporangia single in each segment, in spiral series in small, more or less specialized branchlets BRONGNIARTELLA, p. 589
7. Pericentral cells 4; distinctive stichidia with 4 spores at a node MURRAYELLA, p. 592

8. Plants filamentous, with 3 uncorticated pericentral cells FALKENBERGIA, p. 571
8. Pericentral cells more numerous; plants filamentous or somewhat fleshy 9

9. Apical cells sunken in pits at the tips of the rather fleshy branches 10
9. Apical cells not in pits 11

10. Pericentral cells not forming a definite circle in section, usually not discernible in mature axes; tetrasporangia not developed from pericentral cells, but lying immediately below the surface layer .... LAURENCIA, p. 621
10. Pericentral cells evident; tetrasporangia developed from them CHONDRIA, p. 610

11. Axes throughout parenchymatously corticated, and more or less fleshy 12
11. Axes filamentous, or in part corticated by rhizoids, seldom in parenchymatous fashion 15

12. Branchlets short and crowded, stiffly filamentous, very thinly corticated .... DIGENIA, p. 588
12. Branchlets stouter, heavily corticated, at least at the base 13

13. Branchlets in 2–4 ranks .... BRYOTHAMNION, p. 586
13. Branchlets primarily radial 14

14. Axes with short spinelike branchlets, but these chiefly on short spur branches .... ACANTHOPHORA, p. 619
14. Branchlets generally contracted at the base, not at all spinelike CHONDRIA, p. 610

15. Branchlets not distinct from the lesser branches POLYSIPHONIA, p. 572
15. Branchlets distinctively short, more or less closely investing the axes .... BRYOCLADIA, p. 585

16. Determinate branchlets of one type 17
16. Showing a simple determinate branchlet followed by a pinnate one .... DIPTEROSIPHONIA, p. 600

17. Erect axes of limited growth, but sometimes sparingly branched, especially on fertile plants, with trichoblasts forming a dorsal crest on the convex tips; branches endogenous and lateral in origin, but not regularly placed LOPHOSIPHONIA, p. 605
17. Erect axes sharply determinate, unbranched, the trichoblasts spirally arranged; such axes and the lateral indeterminate branches formed in regular sequence .... HERPOSIPHONIA, p. 601

18. Branch apices erect 19
18. Branch apices inrolled 20

19. Plants somewhat fleshy, ecostate, alternately pinnately lobed or branched .... LAURENCIA, p. 621
19. Membranous, with a faint costa, irregularly laterally branched, the branches substipitate .... CLADHYMENIA, p. 618

20. Branching alternate 21
20. Branching opposite, but respecting indeterminate axes commonly irregular 22

21. Blades ecorticate, the margins serrated by small involute branchlets .... AMANSIA, p. 608
21. Blades corticated, the margins aculeate-serrate ..... VIDALIA, p. 609

22. Pericentral cells 5; axes narrowly alate by development from two lateral pericentral cells; the margins aculeate-serrate ENANTIOCLADIA, p. 609
22. Pericentral cells 6; axes compressed, not alate; the margins with small curved branchlets at intervals .. PROTOKUETZINGIA, p. 607

Falkenbergia Schmitz, 1897

Plants small, filamentous, attaching at various points by branched multicellular holdfasts; apical cells evident, transversely divided, the segments in turn dividing lengthwise to form three pericentral cells, alternating sixty degrees in position in adjacent sections; uncorticated and without trichoblasts; branching irregular, alternate, the branches each developed by the outgrowth of the midportion of a pericentral cell into a papilla, which is then cut off to become the apical cell of a branch; tetrahedral sporangia rare, formed from the pericentral cells.

J. and G. Feldmann (1946) have interpreted *F. rufolanosa* as the tetrasporophyte phase of *Asparagopsis armata*. It is quite probable that the same relation exists between *F. hillebrandii* and *A. taxiformis*, but this has not yet been experimentally confirmed, and so Falkenbergia is retained here in its traditional place as a convenience.

Falkenbergia hillebrandii (Bornet) Falkenberg Pl. 72, fig. 8

Plants hardly 2 cm. tall, in part of creeping filaments attached by branched holdfasts, in part of more or less erect and entangled filaments, these irregularly alternately branched, without a conspicuous main axis, 30–80 μ diam., the segments of irregular length, 0.5–1.2

diameters long, the cells with pointed ends and alternating in position from one segment to the next; cells smaller and more irregular near the base of a branch.

Bermuda, Florida, Mexico, Bahamas, Cayman Isls., Virgin Isls., Guadeloupe, Martinique, Barbados, Brazil. Epiphytic on various algae in shallow water and dredged to a depth of 14 m.

REFERENCES: Vickers 1905, Børgesen 1913–20, Collins and Hervey 1917, Howe 1918a, 1920, Taylor 1928, Hamel and Hamel-Joukov 1931, Humm 1952b; *Sarcomenia miniata,* Mazé and Schramm 1870–77, Murray 1889 *p. p.*

UNCERTAIN RECORD

Falkenbergia rufolanosa (Harvey) Schmitz

Bermuda, Guadeloupe, Barbados. Type from Australia. De Toni 4:865; *Polysiphonia rufolanosa,* Mazé and Schramm 1870–77, Murray 1889.

Polysiphonia Greville, 1824

Plants entirely erect, or with decumbent basal filaments giving rise to the erect portions; usually abundantly dichotomously or laterally branched, the branches filamentous, terete; all branches polysiphonous, with an axial cell series surrounded by several pericentral cells; the main axes and branches, but not the ultimate branchlets, in some species covered by a rhizoidal but often pseudoparenchymatous cortex; delicate colorless branched hairs (trichoblasts) present in many species; tetrahedral sporangia in the slightly thickened upper branches, solitary in the segments, often seriate; spermatangial clusters colorless, ovoid to subcylindrical, on short stalks, developed from trichoblast rudiments, bearing numerous spermatangia on the pericentral cells of the fertile portion; procarps developed from trichoblast rudiments, the upper pericentral cell acting as a supporting cell for the four-celled carpogenic branch and the sterile cells; subsequent to fertilization the support also cuts off an auxiliary cell, which after fusion develops short gonimoblasts in branched series, the terminal cells becoming the carposporangia, surrounded by a large ostiolate pericarp.

Commonly a difficult genus; in the American tropics species of Polysiphonia are extraordinarily hard to identify. There are numerous species validly recorded; at least as many more are inadequately supported, and probably a substantial proportion of the published identifications are erroneous in any case. Plants grow to reproductive maturity very rapidly and often without attaining all of the characteristic vegetative features potentially present. I noted this rapid growth in Florida (1928, p. 184), where a boat freshly coated with antifouling paint in less than two months developed a heavy turf of fully reproductive *Polysiphonia havanensis* f. *mucosa*. The whole development of these plants probably occurred within a month, since the toxicity of the paint for algal germlings would first have to be dissipated. Consequently, it will always be necessary to remember that specimens of small stature are not necessarily small species. This is an ever-present problem in all latitudes, but doubly serious in the tropics.

Another difficulty lies in the inadequacy of the descriptions of many species. While in many genera it is entirely possible to redescribe the plant from type material, Polysiphonias are so fragile that such details as branch origin cannot well be established from the original, often ill-prepared specimens, and as a consequence their exact nature and relation to other species will remain in doubt. If an attempt is made to describe these features from secondary materials, as is often necessary, a concept of the species may grow up which is actually at variance with the original materials.

KEY TO SPECIES

1. Pericentral cells 4 2
1. Pericentral cells 6, rarely 5 to 8 .... P. denudata, p. 580
1. Pericentral cells 8 to 12 9
1. Pericentral cells 12 or more 15

2. Main branches strongly corticated and clothed with setaceous branchlets .... P. ramentacea, p. 580
2. Main branches ecorticate, or with a very little cortication near the base 3

3. Trichoblasts usually absent, but when present not obscuring the prominent apical cell; color dull purplish; plants not adhering well to paper on drying .... P. subtilissima, p. 575
3. Trichoblasts evident 4

4. Branching clearly alternate in the upper parts of the plants 5

4. Branching subdichotomous through most of the plant, this being most evident below, but sometimes the indeterminate branchlets alternately placed 10

5. Branches arising in the axils of trichoblasts 6
5. Branches replacing trichoblasts 7

6. Main axes mostly 50–90 μ diam.; plants soft and mucous, reddish-purple; the base creeping, the upper branching fastigiate .... P. havanensis, p. 577
6. Main axes mostly 100–300 μ diam.; plants firmer, brownish-purple; branching irregular or near the tips almost distichous .... P. binneyi, p. 577

7. Segments below shorter than broad, or in branchlets 1.0–1.5 diameters long; large and bushy species, the axes to 200–300 μ diam. .... P. ferulacea, p. 578
7. Segments as long or, above, much longer than broad 8

8. Rosy, adhering well to paper when dried; segments about 100–150 μ diam., generally 1.5–2.0 diameters long; tetrasporangia seriate in the branch tips ...... P. macrocarpa, p. 578
8. Dull purple, hardly adhering to paper; segments about 90 μ diam.; generally 3–8 diameters long, though sometimes subequal in the creeping parts or young branchlets; the tetrasporangia seriate in branches near the central portions of the plant .... P. subtilissima, p. 575

9. Plants small, creeping, with short erect branches, the segments shorter than broad 14
9. Plants to 8 cm. tall, the branches arising at the bases of trichoblasts .... P. tepida, p. 581
9. Plants to 15 cm. tall, the branches not arising at the bases of trichoblasts; the segments of the main branches 1.0–1.3 diameters long .... P. foetidissima, p. 581

10. Main axes without a conspicuous abundance of adventitious branchlets; plants to 2.5 cm. tall 11
10. Main axes with an abundance of short, often spinelike branchlets; plants becoming larger 12

11. Base a discoid holdfast; epiphytic on sea grasses, larger algae, corals, and gorgonians .... P. gorgoniae, p. 576
11. Base of creeping filaments; a reef plant growing in exposed places .... P. sphaerocarpa, p. 576

12. Filaments lightly corticated, at least below 13
12. Filaments ecorticate .... P. fracta, p. 576

13. Plants with divaricate branching throughout; coarse below, the close-placed spinelike determinate branchlets about 1 mm. long .... P. echinata, p. 579
13. Plants more delicate, the upper branching fastigiate, the more scattered branchlets about 2 mm. long ... P. hapalacantha, p. 579

14. Pericentral cells 8–10; trichoblasts few or inconspicuous or absent; erect branches to 2.5 cm. tall .... P. exilis, p. 581
14. Pericentral cells 10–16; trichoblasts evident at the stem tips; erect branches shorter .... P. howei, p. 582

15. Plants forming soft cushions, coarse in branching, seldom 1 dm. tall .... P. opaca, p. 583
15. Plants not particularly soft, bushy and progressively finely branched, often 1–3 dm. tall .... P. nigrescens, p. 582

Polysiphonia subtilissima Montagne

Plants to 15 cm. tall, color blackish-purple, texture rather soft; arising from a creeping base; erect filaments subdichotomously branched below, alternately branched above, sometimes with numerous subsimple coarse rhizoids, the segments to 90 μ diam. below, 1.5–2.3(–8.0) diameters long, with 4 pericentral cells, ecorticate; ordinary branching erect, wandlike to clustered; adventitious branchlets not numerous; above the branchlets 33–45 μ diam., the segments to 130 μ, usually about 1.5 diameters, long; apical cells conspicuous; simple or sparingly forked trichoblasts sometimes present, rather persistent, lateral branches replacing certain of these in development from the apex; tetrasporangia seriate in the branches near the central portions of the plant.

Bermuda, North Carolina, Florida, Mississippi, Louisiana, Texas, Bahamas, Cuba, Cayman Isls., Jamaica, Virgin Isls., Guadeloupe, Barbados, Grenada, French Guiana, Brazil. Common on muddy bottoms of shallow lagoons and streams, and on various objects scattered over the bottom. Able to grow well in water of somewhat reduced salinity. Probably much more generally distributed than the list of localities indicates, but this rather unprepossessing plant would not attract the attention of a casual collector.

REFERENCES: Montagne 1850, Mazé and Schramm 1870–77, Dickie 1874b, Hemsley 1884, Murray 1889, Möbius 1890, Vickers 1905, Howe 1920, Taylor 1928, 1933, 1936, 1954a.

Polysiphonia gorgoniae Harvey

Plants somewhat bushy, or more rarely in light brown tufts; basal holdfasts discoid, bearing one or a few primary axes 0.5–3.0 cm. tall, subsetaceous below, attenuate upwards, dichotomously branched, widely divergent near the base and branching 2–3 times above; repeatedly irregularly branched toward the tips, the branchlets spreading fanlike in the water; segments shorter than broad below, in the lesser divisions 1.5–2.0 diameters long, but in the ultimate branchlets very short; pericarps sessile or short-stalked on the small branchlets, oblate-spherical.

Florida, Bahamas. Growing on various "corals," probably in shallow water.

REFERENCES: Harvey 1853, Murray 1889, Howe 1920, Taylor 1928.

Polysiphonia sphaerocarpa Børgesen

Plants small, in tufts 1.0–1.5 cm. tall; base of decumbent filaments attached to the substratum by numerous rhizoids; these filaments 60–200 μ diam., the segments 100–125 μ long; assurgent filaments form the free parts of the tuft, the branches sometimes connected by haptera, the cylindrical segments 100 μ diam. and 150 μ long below, but 60 μ diam., 175 μ long and turgid in the upper divisions; branching pseudodichotomous to alternate, very erect, the branches apparently replacing trichoblasts which are shed very early; pericentral cells 4, uncorticated; tetrasporangia 60 μ diam., in a spiral line in the lesser branches and branchlets; pericarps nearly spherical, about 80 μ diam., on short, thick pedicels.

Hispaniola, Virgin Isls. Growing on reef rocks in exposed situations.

REFERENCES: Børgesen 1913–20, 1924.

Polysiphonia fracta Harvey

Plants small, laxly intricate and spreading below, tufted above, dull brownish-purple, the rather stiff filaments capillary, irregularly flexed and curved, here and there with rhizoidal haptera; trichoblasts present, deciduous; main branching subdichotomous, the shorter branches spreading to erect, sometimes secund, sometimes divaricate, bearing numerous scattered, spreading thornlike branchlets about 2 mm. long; segments ecorticate, shorter than broad, with 4 pericentral cells.

Florida, Grenada.

REFERENCES: Harvey 1853, Murray 1889, Taylor 1928.

Polysiphonia havanensis Montagne

Plants erect, usually rather small, reddish-purple, very soft, to 4–9 cm. tall, subdichotomously branched below, alternately above, the branching fastigiate; apices with numerous trichoblasts, the branches originating in their axils; the main axes 50–90 μ diam., tapering considerably in the branches; the segments with 4 pericentral cells, uncorticated, 1.5 diameters long below, 2–3 diameters long in the middle branches; tetrasporangia few, often 2–3, unilaterally prominent in the fertile branchlets; pericarps small, ovate, sessile and suberect on the branchlets.

V. **mucosa** J. Agardh. Even more delicate and soft than the type of the species, the segments to 4 diameters long, the trichoblasts very conspicuous; pericarps globose below, broadly urceolate above.

Bermuda, North Carolina, Florida, Mexico, Bahamas, Caicos Isls., Cuba, Jamaica, Hispaniola, Puerto Rico, Virgin Isls., Guadeloupe, Dominica, Martinique, Barbados, British Honduras, Guatemala. Probably common on algae, stones, and other objects in shallow water through most of our range.

REFERENCES: Harvey 1853, Montagne 1863, Mazé and Schramm 1870–77, Hauck 1888, Murray 1889, Collins 1901, Vickers 1905, Børgesen 1913–20, 1924, Howe 1918a, 1920, Taylor 1928, Taylor and Arndt 1929, Hamel and Hamel-Joukov 1931, Feldmann and Lami 1936, Williams 1949a; *P. violacea p. p.*, Mazé and Schramm 1870–77, Murray 1889.

Polysiphonia binneyi Harvey

Plants 2–4(–15) cm. tall, yellowish to purplish, somewhat stiff; branching frequent, alternate, the habit irregular and not fastigiate, or the branches sometimes short and pseudodistichous; tips with numerous trichoblasts, the branches arising in their axils; main axes 100–350 μ diam., branchlets to 150 μ diam.; pericentral cells 4, the segments uncorticated, considerably shorter than broad, or subequal.

Bermuda, Florida, Texas, Bahamas, Caicos Isls., Jamaica, Guadeloupe, Netherlands Antilles, Venezuela. Locally common in shallow water on rocks and old corals, and dredged to a depth of 9 m.

REFERENCES: Harvey 1853, Howe 1920, Taylor 1928, 1941a, 1942; *P. havanensis* v. *binneyi*, Mazé and Schramm 1870–77, Murray 1889, Collins 1901.

Polysiphonia macrocarpa Harvey

Plants densely tufted, decumbent below, above erect, to about 2 cm. tall, very soft, rosy, adhering well to paper when dried; decumbent filaments 100–150 μ diam., the segments about 150 μ long; erect filaments subfastigiately alternately branched above, to 100 μ diam. below, 20–25 μ diam. above, the segments with 4 pericentral cells, uncorticated, 1.5–2.0 diameters long, seldom to 4 diameters; trichoblasts present on the older plants, but branches not originating in their axils; tetrasporangia in long rows in the simple or sometimes forked tips of the branchlets; spermatangial clusters cylindrical, obtuse, without sterile tips, 100–110 μ diam., 400–500 μ long; pericarps urceolate, about 150 μ diam., 250 μ long.

Bermuda, North Carolina, Florida, Virgin Isls., Barbados. Growing on mangrove roots and the like in shallow water.

REFERENCES: Vickers 1905, Børgesen 1913–20, Collins and Hervey 1917, Taylor 1928, Williams 1949a.

Polysiphonia ferulacea Suhr

Plants bushy, from a fibrous base, abundantly branched, to 5–15 cm. tall, pseudodichotomous below, subfastigiate to virgate above, the axils, at least above, commonly acute; segments to 200–300 μ diam., subequal or shorter than broad, with 4 pericentral cells, uncorticated, the outer cell walls in old stems to 12–40 μ thick; branches arising without obvious relation to trichoblasts; tetrasporangia unilateral in swollen fastigiate branchlets; spermatangial clusters stout, subcylindrical, about 60 μ diam., 200 μ long, with 1–2 conspicuous sterile cells at the tip; pericarps ovate-globose, sessile, borne below the tips of the branchlets.

Bermuda, Florida, Texas, Mexico, Bahamas, Caicos Isls., Cayman Isls., Jamaica, Hispaniola, Virgin Isls., Guadeloupe, Barbados, Costa Rica, Venezuela. Probably generally found near low-water mark, but also dredged to a depth of 36 m.

REFERENCES: Mazé and Schramm 1870–77, Dickie 1874a, Murray 1889, Collins 1901, Vickers 1905, Børgesen 1913–20, Collins and Hervey 1917, Howe 1918a, 1920, Taylor 1928, 1933, 1941a, 1943; *P. breviarticulata*, ?Hooper 1850, Harvey 1853, Mazé and Schramm 1870–77; *P. fibrillosa*, Hemsley 1884; *P. mollis, P. utricularis,* and *P. violacea p. p.*, Mazé and Schramm 1870–77, Murray 1889; *P. nigrescens p. p.*, Rein 1873, Hemsley 1884.

Polysiphonia echinata Harvey

Plants 6–10 cm. tall, stiff, often hardly adhering to paper; subdichotomously branched, the filaments stout, the branches widely divergent, 1.0–2.5 cm. distant, in the upper portions 2.0–2.5 cm. long, at least above closely beset with simple or bifid spreading branchlets 1 mm. long and about as distant; nodose and dark in color; pericentral cells 4, with a little cortication from the bases of the older adventitious branchlets; segments shorter than broad; membranes thick.

Florida.

REFERENCES: Harvey 1853, Murray 1889, Taylor 1928.

Polysiphonia hapalacantha Harvey

Plants densely tufted, to 4–15 cm. tall, the filaments thick below, capillary above, below subdichotomous and the branching strongly divaricate, above fastigiate and less regular, all divisions bearing many very slender, usually simple quadrifarious branchlets 2(–4) mm. long; pericentral cells 4, very lightly corticated in the larger branches; the segments below about 1.5 diameters long, but in the lesser branches subequal or shorter than broad; trichoblasts numerous, caducous.

Florida, Bahamas. A plant of shallow water.

REFERENCES: Harvey 1853, Murray 1889, Howe 1920, Taylor 1928.

Polysiphonia ramentacea Harvey

Plants tufted, 3–10 cm. tall, brownish-purple; primary filaments moderately coarse, laterally branched, the terminal divisions virgate; axis and branches clothed with simple to generally forked setaceous determinate branchlets 2–5 mm. long, rather uniform in length and decreasing toward the apex on any particular branch; trichoblasts sometimes present; segments with 4 pericentral cells, the principal branches corticated; tetrasporangia few, in upper, somewhat contorted determinate branchlets; spermatangial clusters ovate-lanceolate, similarly clustered at the tips of branchlets; pericarps few, single or grouped, rounded to ovoid.

Florida, Bahamas.

REFERENCES: Harvey 1853, Howe 1920, Taylor 1928, 1936, 1954a.

Polysiphonia denudata (Dillwyn) Kützing

Plants to 25 cm. tall, the base disciform, the lower branches strongly divergent (90°–120°) when young, the upper branches more irregular and erect; color dark reddish-purple in the upper branches and branchlets, where very soft in texture, but below the branches slender though stiff and nearly colorless, or sometimes brownish; lower portions devoid of branchlets, 340–750 μ diam., the segments 250–500 μ long, nearly or quite ecorticate; secondary branches 100–170 μ diam., the segments 400–900 μ long; branchlets 33–45 μ diam., the segments 45–80 μ long; throughout with 6 pericentral cells; trichoblasts absent, rudimentary, or moderately developed, maturing well below the apex and not obscuring the prominent apical cells; branchlets formed laterally at the bases of trichoblasts if these are present; tetrasporangia oval to spherical, 60–100 μ diam.; spermatangial clusters long-conical to lanceolate, mostly near the tips of the branchlets, 240–280 μ long, 50–55 μ diam.; pericarps sessile or short-stalked, subglobose to transversely oval, 330–540 μ diam.

Bermuda, North Carolina, South Carolina, Virgin Isls., Guadeloupe, Barbados, Netherlands Antilles. Growing on rocks, woodwork, shells, etc., in rather shallow and quiet water. It is able to do well in somewhat brackish and befouled localities, but is even more luxuriant in very protected areas with clear, well-circulated water.

REFERENCES: Hoyt 1920; *P. variegata,* Harvey 1853, Mazé and Schramm 1870–77, Murray 1889, Vickers 1905, Sluiter 1908, Børgesen 1913–20.

Polysiphonia foetidissima Cocks

Plants densely tufted and generally entangled, dark red, soft and slippery, readily decomposing, from a creeping base to 3–8(–15) cm. tall; the main filaments 80–160 μ diam. below, alternately or subdichotomously branched in all directions; trichoblasts absent; the segments subequal or one-third longer than broad, generally with 8–10 pericentral cells; tetrasporangial branches torulose-inflated; pericarps with conical tips.

Bermuda. Growing on stones and timbers in shallow water.

REFERENCES: Collins and Hervey 1917, Howe 1918a.

Polysiphonia tepida Hollenberg

Plants forming dark reddish-brown mats of considerable thickness; prostrate filaments to 120 μ diam.; erect filaments to 8 cm. tall or more, 80–130 μ diam., ecorticate; segments 1–2 diameters long in the middle branches, with 7–8 pericentral cells; branches each originating as one member of the basal fork of a trichoblast; branching erect, moderately abundant, mostly alternately distichous, at intervals of 6–8(–30) segments; tetrasporangia scattered, or in long series in the penultimate branches, single in each segment, 30–35 μ diam.

North Carolina, Texas. Growing on stones at low-tide level.

REFERENCES: Hollenberg 1958; *P. taylori,* Williams 1949a (*nomen nudum*).

Polysiphonia exilis Harvey

Plants densely caespitose, dull purplish-brown, the substance firm; basal filaments intricate, with lateral radicular branchlets and erect branches to about 24 mm. long; segments about half as long as broad, with 8–10 pericentral cells; branches slender, beset with numerous branchlets, the apex generally bare and trichoblasts few or inconspicuous; tetrasporangia in distorted, tangled branchlets.

Bermuda, Florida, Bahamas. On rocks and gorgonians near and a little below low-water line.

REFERENCES: Harvey 1853, Dickie 1874b, Farlow 1876, Hemsley 1884, Murray 1889, Howe 1920, Taylor 1928.

Polysiphonia howei Hollenberg

Plants small, creeping, densely matted and forming large colonies; basal filaments entangled, attached by unicellular rhizoids, the segments 100–170 μ diam., subequal or shorter than broad; pericentral cells 10–16; branches arising exogenously on all sides, 2–4 segments apart, some remaining prostrate or undeveloped, others giving rise to erect branches 2–5 mm. tall, 70–150 μ diam., which are at first arched toward the apex of the supporting prostrate filament, and are simple or very sparingly branched; trichoblasts abundant near the tips, caducous, leaving scar cells; tetrasporangia single in each segment, 40–55 μ diam., in rather long somewhat spiral series; sexual organs unknown.

Bermuda, North Carolina, Florida, Mexico, Bahamas, Caicos Isls., Jamaica, Hispaniola, Virgin Isls., Guadeloupe, Grenada, Brazil. Very common over most of the northern part of the range, growing on rocks, algae, corals, and the like near low-tide line. Perhaps even more widespread than the list suggests, as it is not an attractive species and might well be neglected by a non-phycological collector.

REFERENCES: Hollenberg *in* Taylor 1945c, Williams 1949a, Joly 1957; *Lophosiphonia obscura,* Collins 1901, Børgesen 1913–20, Howe 1918a, 1920, Taylor 1928, 1933, 1943, Hamel and Hamel-Joukov 1931; *Polysiphonia obscura,* Mazé and Schramm 1870–77, Murray 1889.

Polysiphonia nigrescens (Hudson) Greville

Plants erect, to 10–30 cm. tall, with a distinct main axis and abundant lateral branching; color dark purple to black, texture firm to coarse; main branches often rather naked below, above more abundantly laterally branched, near the tips becoming brushlike, corymbose to bilateral, the ordinary branches produced directly from the subapical segments, independent of the many deciduous trichoblasts; adventitious branchlets often present; branches with

16 (8–20) pericentral cells, below to 520–850 μ diam., the segments 300–720 μ long, nearly or quite ecorticate; in the middle portions of the plant the segments 0.5–3.0 diameters long, while the branchlets are 60–85 μ diam., the segments 100–150 μ long; tetrasporangia 60–100 μ diam., seriate in somewhat moniliform, distorted, often forked branchlets, these becoming 100–140 μ diam.; antheridia lanceolate, grouped near the tips of the branchlets; pericarps broadly ovoid, short-stalked, 380–420 μ diam.

?Bermuda, North Carolina, South Carolina. Usually growing in rather shallow water on rocks.

REFERENCES: Murray 1889, Hoyt 1920, Williams 1949.

Polysiphonia opaca (C. Agardh) Moris and De Notaris

Plants in large cushions, to 10 cm. in height, dull purple, drying brownish-black and hardly adherent to paper; branching subdichotomous below, alternate above, the branches beset with short, widely divergent, redividing branchlets 1–4 mm. long; trichoblasts numerous; segments with 12–24 pericentral cells, subequal to 1.5 diameters long; pericarps subglobose, sessile on the branches.

Bermuda, Florida, Bahamas, Martinique. Growing on submerged rocks.

REFERENCES: Collins and Hervey 1917, Howe 1918a, 1920, Taylor 1928, Hamel and Hamel-Joukov 1931; *P. nigrescens p. p.*, Rein 1873, Hemsley 1884.

UNCERTAIN RECORDS

Polysiphonia bostrychoides Crouan

Guadeloupe, the type locality. A *nomen nudum.* De Toni 4:961; Mazé and Schramm 1870–77.

Polysiphonia bostrychoides Montagne

French Guiana, the type locality. Montagne 1895.

Polysiphonia callithamnioides Crouan

Guadeloupe, the type locality. A *nomen nudum.* De Toni 4:961; Mazé and Schramm 1870–77, Murray 1889.

Polysiphonia capucina Crouan

Guadeloupe, the type locality. A *nomen nudum.* De Toni 4:961; Mazé and Schramm 1870–77, Murray 1889.

Polysiphonia dasyoeformis Crouan

Guadeloupe, the type locality. Schramm and Mazé 1865; *Bostrychia dasyaeformis,* Mazé and Schramm 1870–77.

Polysiphonia dichotoma Kützing

Brazil. Type from the Adriatic Sea. Martens 1870.

Polysiphonia elongata (Hudson) Harvey

Bermuda, Brazil. Type from England. De Toni 4:903; *P. elongata* C. Agardh, Martens 1870, Zeller 1876; *P. elongata* Greville, Murray 1889.

Polysiphonia fibrillosa Greville

Bermuda. Type from Great Britain. De Toni 4:919, 6:919; Murray 1889.

Polysiphonia foeniculacea (Draparnaud) J. Agardh

Florida. Type from the Mediterranean. De Toni 4:914; *P. hirta* J. Agardh, Murray 1889.

Polysiphonia funebris De Notaris

Guadeloupe. Type from the Mediterranean. De Toni 4:1066; Mazé and Schramm 1870–77, Murray 1889.

Polysiphonia furcellata (C. Agardh) Harvey

Guadeloupe. Type from France. De Toni 4:930; Mazé and Schramm 1870–77, Murray 1889.

Polysiphonia incompta Harvey

Guadeloupe. Type from South Africa. De Toni 4:872; Mazé and Schramm 1870–77, Murray 1889.

Polysiphonia lanosa (Linnaeus) Tandy

Bermuda. Type from Europe. De Toni 4:945 (as *P. fastigiata*). Recently collected attached to Ascophyllum washed ashore, and surely not of local growth.

Polysiphonia mollis Hooker *f.* and Harvey

Grenada. Type from Tasmania. De Toni 4:877; Murray 1889.

Polysiphonia pulvinata (Roth) J. Agardh

Guadeloupe, Barbados. Type from Europe. De Toni 4:895; Mazé and Schramm 1870–77, Murray 1889.

Polysiphonia subulata (Ducluzeau) J. Agardh

Jamaica. Type from France. De Toni 4:901 (as *P. violacea* v. *subulata*); Collins 1901.

Polysiphonia utricularis Zanardini

Grenada. Type from the Red Sea. De Toni 4:891; Murray 1889 *p. p.*

Polysiphonia violacea (Roth) Greville

Brazil. Type from Europe. De Toni 4:900; Murray 1889 *p. p.*, Möbius 1890; *P. violacea* C. Agardh, Martens 1870.

Bryocladia Schmitz, 1897

Plants with a creeping base, the terete main filaments erect, alternately laterally branched; axes with several pericentral cells, ecorticate; lateral branchlets spirally disposed, short, stiff, erect or recurved, simple or branched, sometimes with deciduous trichoblasts; tetrasporangia numerous in linear series on the outer side of the fertile branchlets, single in each segment and covered by two equal cells; urceolate pericarps numerous near the growing tips of stout side branches, between the branchlets.

KEY TO SPECIES

1. Elongate axes directly clothed with short, stout, simple or subsimple branchlets . . . . B. cuspidata
1. Elongate axes with short, pinnate branchlets . . . . B. thyrsigera

Bryocladia thyrsigera (J. Agardh) Schmitz

Plants from a creeping base, 4–8 cm. tall, black when dried; main axes erect, somewhat entangled below, sparingly divided, with 10 pericentral cells, ecorticate, the segments about one-half as long as broad; above with numerous shorter branches as much as 2.5 cm. long, bearing pinnate branchlets, the slender tapering ramelli in sterile plants rather spreading, although more crowded toward the branch tips; the fertile branchlets short and crowded, stiff, sometimes recurved, the segments about as long as broad; tetrasporangia rather scarce, in but slightly modified branchlets; pericarps pedicellate.

Florida, Mexico, Guadeloupe, Grenada, Venezuela, Brazil. Growing on rocks in shallow water, somewhat exposed to wave action.

REFERENCES: Taylor 1928, 1930b, 1942, Joly 1951, 1957, De Mattos 1952, Humm 1952a; *Polysiphonia thyrsigera,* J. Agardh 1847, Harvey 1853, Mazé and Schramm 1870–77, Murray 1889.

Bryocladia cuspidata (J. Agardh) De Toni Pl. 71, fig. 2

Plants 2–8 cm. tall, purplish black, the erect axes sparingly divided; pericentral cells 8, the segments about half as long as broad, ecorticate; branches hardly 1 mm. long, simple, or commonly bearing one smaller subdivision near the base, alternate, spreading or recurved, stiff, tapering strongly from base to apex, the segments about as long as broad.

Florida, Texas, Mexico, Cuba, Jamaica, Puerto Rico, Guadeloupe, Barbados, Venezuela, Trinidad, Brazil. Growing on rocks or other massive objects in shallow water, somewhat exposed to wave action.

REFERENCES: Taylor 1928, 1929b, 1936, 1941a, 1942, 1954a, Humm 1952a; *Polysiphonia cuspidata,* J. Agardh 1847, Hauck 1888, Collins 1901, Vickers 1905; *P. verticillata,* Mazé and Schramm 1870–77, Murray 1889.

UNCERTAIN RECORD

Bryocladia dictyurus (J. Agardh) Taylor

Guadeloupe. Type from the Pacific coast of Mexico. *Polysiphonia dictyurus,* De Toni 4: 931; Mazé and Schramm 1870–77, Murray 1889.

Bryothamnion Kützing, 1843

Plants erect and bushy, rarely cartilaginous, the main branches terete, compressed, or angular in section, alternately divided, the branches bearing short determinate branchlets; axes with several pericentral cells, extensively corticated; stichidia axillary, rough or with short branchlets, the tetrasporangia spirally placed, single in each segment; pericarps short-stalked on transformed branchlets, subglobose, thick-walled.

KEY TO SPECIES

1. Branches terete or compressed in section; the branchlets radially or bilaterally disposed . . . . B. seaforthii
1. Branches triangular in section; the branchlets greatly reduced in length . . . . B. triquetrum

Bryothamnion seaforthii (Turner) Kützing Pl. 73, fig. 3

Plants erect and bushy, reddish-purple, soon fading when dried, in texture very firm, membranous when young, subcartilaginous in the oldest parts, 8–20 cm. tall; the axes compressed and pinnately branched, sparingly below, subfastigiately above; pericentral cells 8–9, heavily corticated; determinate corticated branchlets developed in the axils of the very small deciduous trichoblasts, distichous, marginal or sometimes in 3–4 rows, short, obliquely erect, sharply tapering, bilaterally bearing smaller 1–4-forked spinelike ramelli; stichidia short-stalked, axillary, simple or sparingly branched, torulose; pericarps subglobose, subterminal on the pinnate branchlets.

F. **imbricata** J. Agardh. Branchlets radially disposed on a terete axis, densely imbricate.

Florida, Mexico, Caicos Isls., Cuba, Jamaica, Hispaniola, Puerto Rico, Virgin Isls., Guadeloupe, Barbados, Grenada, British Honduras, Costa Rica, Colombia, Trinidad, Tobago, French Guiana, Brazil. Dredged from a depth of 30 m.; perhaps a deep-water plant, but frequently found cast ashore.

REFERENCES: Martens 1870, Mazé and Schramm 1870–77, Dickie 1874a, b, Piccone 1886, Hauck 1888, Murray 1889, Collins 1901, Vickers 1905, Børgesen 1913–20, Schmidt 1923, 1924, Howe 1928, Taylor 1928, 1929b, 1930a, 1935, 1941b, 1942, 1943, 1954a, Joly 1951 (including f. *disticha*), 1957, De Mattos 1952; *B. hypnoides*, Mazé and Schramm 1870–77, Murray 1889; *Alsidium seaforthii*, Harvey 1861; *Amansia seaforthii*, Greville 1833; *Thamnophora seaforthii*, Montagne 1863. For the f. *imbricata*, Taylor 1928.

Bryothamnion triquetrum (Gmelin) Howe
Pl. 72, fig. 6; pl. 73, fig. 4

Plants to 25 cm. tall, from a discoid base, dull brownish-purple, fleshy-cartilaginous; briefly stalklike and terete below, but above irregularly alternately branched, with 7–9 pericentral cells, heavily corticated, the branches triangular in section, bearing corticated branchlets along the angles in spiral succession in the axils of trichoblasts which are very short, at most once forked; branchlets very short, to about 2 mm. long, closely investing the growing tips, more widely spaced below, their lower portions nearly simple, above trifid, the ramelli stiff and subulate; stichidia stalked, axillary between the ramelli, with short fasciculate branches.

Florida, Mexico, Bahamas, Caicos Isls., Cuba, Jamaica, Hispaniola, Puerto Rico, Virgin Isls., St. Barthélemy, Guadeloupe, Martinique, Grenada, Barbados, Old Providence I., Costa Rica, Panama, Colombia, Netherlands Antilles, Venezuela, Brazil. Principally found growing on rocks in shallow, moderately exposed places, but also dredged to a depth of 20 m.

REFERENCES: Børgesen 1913–20, Howe 1915, 1920, 1928, Schmidt 1923, 1924, Taylor 1928, 1929b, 1930a, 1933, 1939b, 1940, 1941b, 1942, 1943, 1954a, Taylor and Arndt 1929, Sanchez A. 1930, Hamel and Hamel-Joukov 1931; *B. triangulare,* Martens 1870, Mazé and Schramm 1870–77, Dickie 1874a, Hauck 1888, Murray 1889, Collins 1901, Vickers 1905, Sluiter 1908; *Alsidium triangulare,* Harvey 1853, 1861; *A. triangulare* and *A. triquetrum,* Hooper 1850; *Thamnophora triangularis,* Montagne 1850.

UNCERTAIN RECORD

(?*Bryothamnion*) *Alsidium schottii* Harvey

Colombia, the type locality. A *nomen nudum.* Harvey 1861.

Digenia C. Agardh, 1823

Plants erect, bushy, widely dichotomously or irregularly laterally branched, the axes cartilaginous, without a well-defined apical cell or polysiphonous structure, but broadly parenchymatous with larger cells within, smaller without; these branches closely radially beset with short, usually simple branchlets which have an axis originating from an apical cell, which bear near their tips small deciduous trichoblasts, and which develop six to eight pericentral cells covered by a thin parenchymatous cortex; tetrasporangia in the upper, irregularly swollen and chiefly uncorticated parts of the branchlets, each spore covered by three pericentral cells; spermatangia in small ovoid disks clustered at the tips of fertile branchlets; pericarps ovoid, terminal and lateral on the branchlets.

Digenia simplex (Wulfen) C. Agardh

Plants 3–25 cm. tall, dull brownish-red, wiry below, cartilaginous above, the axis and main branches heavily clothed above with slender, stiff branchlets which are 3–5(–15) mm. long, but the branches may be denuded below.

Bermuda, Florida, Texas, Mexico, Bahamas, Turks Isls., Cuba, Jamaica, Hispaniola, Puerto Rico, Virgin Isls., St. Barthélemy, Guadeloupe, Martinique, Grenada, British Honduras, Panama, Brazil. Common and often abundant in the intertidal zone on rather exposed reefs and rocks, frequently much dwarfed and hardly more than 3–5 cm. tall, but in sheltered spots much larger, and dredged to a depth of 20 m. This plant is a favored support for many other algal species and is commonly so heavily epiphytized that a dense colony of the dwarfed form may be completely concealed.

REFERENCES: Hooper 1850, Harvey 1853, Mazé and Schramm 1870–77, Rein 1873, Dickie 1874a, Hemsley 1884, Murray 1889, Collins 1901, Vickers 1905, Børgesen 1913–20, 1924, Collins and Hervey 1917, Jennings 1917, Howe 1918a, b, 1920, 1928, Taylor 1928, 1933, 1935, 1940, 1941a, 1942, 1943, 1954a, Taylor and Arndt 1929, Sanchez A. 1931, Hamel and Hamel-Joukov 1931; *D. wulfeni*, Martens 1870.

Pachychaeta Kützing, 1862

UNCERTAIN RECORD

Pachychaeta griffithsioides Kützing

Antigua, the type locality. Kützing 1862.

Brongniartella Bory, 1822

Plants bushy, alternately branched, the axes polysiphonous, with five or seven pericentral cells which ultimately are corticated by rhizoidal investment; lateral branchlets formed from the basal cells of the monosiphonous ramelli, which are subdichotomously or alternately branched; stichidia with tetrasporangia spirally arranged, single in each segment; pericarps sessile on the branches.

Brongniartella mucronata (Harvey) Schmitz Pl. 66, figs. 11, 12

Plants with a basal disk to 1 cm. diam., producing several axes to 15–20 cm. tall, about 1 mm. diam., naked below, branched and ramellate above; pericentral cells 5, corticated; ramelli monosiphonous, subdichotomously to somewhat alternately branched, the cells short, the apices aculeate, indurated; tetrasporangia in slightly modified branchlets which bear many ramelli.

Bermuda, North Carolina, Florida, Jamaica, Grenada.

REFERENCES: Hoyt 1920, Taylor 1928, 1954a; *Dasya mucronata*, Harvey 1853, Hemsley 1884, Murray 1889, Collins 1901.

Lophocladia Schmitz, 1893

Plants repeatedly dichotomous; denuded below, the axes with four pericentral cells, becoming more or less corticated, but the terminal divisions densely ramellate, the dichotomously or alternately branched monosiphonous ramelli radially disposed; stichidia formed by transformation of the ramelli flexuous and irregular in contour, the tetrasporangia single in the segments, conspicuously spirally disposed.

Lophocladia trichoclados (Mertens *in* C. Agardh) Schmitz

Plants to 7–10 cm. tall, rosy, widely dichotomously branched, slender and soft; the upper portions villous with lax penicillate ramelli about 2 mm. long, dichotomous, the angles acute, the cells 2–3 diameters long near the base of the cluster, 6–8 diameters long in the upper divisions; stichidia single in each tuft, flexuous-distorted, acuminate, with 6–8 spirally placed tetraspores.

Bermuda, Florida, Mexico, Bahamas, Jamaica, Hispaniola, Virgin Isls., St. Barthélemy, Guadeloupe, Barbados, Venezuela, French Guiana. A moderately deep-water species, often washed ashore in abundance; dredged from a depth of 20 m.

REFERENCES: Vickers 1905, Børgesen 1913–20, 1924, Howe 1920, Taylor 1928, 1942, Prat 1935c, Tandy 1936, Feldmann and Lami

1937, Williams 1951; *Dasya lophoclados*, Harvey 1853, *D. trichoclados*, Murray 1889, *D. trichoclados* v. *oerstedi*, Mazé and Schramm 1870–77, Murray 1889; *Conferva trichoclados* Mertens *ms.*; *Griffitsia ?trichoclados*, C. Agardh 1828; *Polysiphonia leptoclada*, Montagne 1850.

UNCERTAIN RECORD

Lophocladia lallemandi (Montagne) Schmitz

Guadeloupe. Type from the Red Sea. De Toni 4:1015; *Dasya lallemandi*, Mazé and Schramm 1870–77, Murray 1889.

Wrightiella Schmitz, 1893

Plants erect, radially branched, throughout beset with short, soft, spur branchlets which in the younger parts bear monosiphonous, chromatophorous ramelli; axes with four pericentral cells, corticated by rhizoids; stichidia on monosiphonous stalks, the tetrasporangia in a spiral row; pericarps on polysiphonous stalks, very broadly urceolate with a wide opening.

KEY TO SPECIES

1. Main branches subpercurrent, beset with numerous tetrastichous spur branchlets 1–5 mm. long................ W. blodgettii
1. Main branches commonly deliquescent, bearing filiform branchlets or a few spur branchlets .... W. tumanowiczi

Wrightiella blodgettii (Harvey) Schmitz

Plants to several decimeters in length, repeatedly branched, rather firm in texture, the axes smooth below, above with 4 rows of simple or once-forked subulate spur branchlets 1–5 mm. long, soft, when young bearing conspicuous alternately branched monosiphonous ramelli which have a distinctive small cell at the base of each branch; tetrasporangia in spiral series in the stichidia, which have a very nodulose aspect and are borne on long, monosiphonous stalks; pericarps short-stalked, subglobose, the ostiole projecting.

Bermuda, Florida, Bahamas, Guadeloupe. Dredged from 7–36 m. depth.

REFERENCES: Collins and Hervey 1917, Howe 1918a, 1920, Taylor 1928; *Alsidium blodgettii*, Harvey 1853, Rein 1872, Hemsley 1884,

Murray 1889; *Schimmelmannia bollei,* Mazé and Schramm 1870–77, Murray 1889.

Wrightiella tumanowiczi (Gatty) Schmitz

Plants bushy, with a discoid base reinforced by a few fibrous branchlets, to 30–75 cm. tall or perhaps more, rather copiously branched below, without a dominant primary axis, but the many long leading branches sparingly alternately subdivided, above bearing few to numerous delicate, lax branchlets about 0.5–3.0 cm. in length; pericentral cells 4, the cortication somewhat delayed, but heavy on all the older branches and branchlets; monosiphonous ramelli not conspicuous; stichidia borne on the outer divisions of the ramelli, often bearing simple lateral ramelli in turn, strongly nodulose, the sporangia covered by the pericentral cells; pericarps short-stalked, ovate to urceolate.

Bermuda, Florida, Texas, Bahamas, Virgin Isls., Guadeloupe, Brazil. Dredged from 18–27 m. depth. While it is drying, a very strong sulphurous odor is emitted by this plant.

REFERENCES: Børgesen 1913–20, Collins and Hervey 1917, Howe 1918a, 1920, Taylor 1928, 1936, 1941a; *Dasya tumanowiczi,* Harvey 1853, Mazé and Schramm 1870–77, Dickie 1875, Murray 1889.

UNCERTAIN RECORD

(Wrightiella) Alsidium helminthochorton (La Tourrette) Kützing

Florida. Type from the Mediterranean. De Toni 4:862; *Gracilaria helminthochorton,* Hooper 1850.

Murrayella Schmitz, 1893

Plants erect, subdichotomously to radially alternately branched; axes with four pericentral cells, ecorticate; naked below, densely ramellate above; ramelli originating as alternate polysiphonous branchlets which quickly divide into the monosiphonous ramelli; stichidia formed in the upper parts of the ramellar clusters, the sporangia in four rows or obliquely verticillate; pericarps with monosiphonous stalks.

Murrayella periclados (C. Agardh) Schmitz

Plants deep purple, 2.0–4.5 cm. tall from a creeping base, subdichotomous below, alternately branched above and towards the tips fasciculate; ramelli tardily deciduous, of determinate growth, on the younger branches alternately radially disposed, alternately branched, monosiphonous throughout, the cells about 28 μ diam., 60 μ long, or polysiphonous below, with the branchlets monosiphonous as above; stichidia borne terminally on the polysiphonous branchlet bases, nearly cylindrical, tapering toward the apex, about 100 μ diam. and 900 μ long, with 4 tetrasporangia in a whorl, each about 55 μ diam.; simple ramelli sometimes borne laterally on the stichidia.

Bermuda, Florida, Bahamas, Cuba, Cayman Isls., Jamaica, Virgin Isls., Guadeloupe, Martinique, Barbados, Grenada, British Honduras, Guatemala, Panama, Venezuela, French Guiana. A characteristic plant of mangrove thickets; also found in caverns or under rocks and in similar places growing in shallow, quiet water.

REFERENCES: Collins 1901, Vickers 1905, Børgesen 1913–20, Collins and Hervey 1917, Howe 1918a, 1920, Taylor 1928, 1935, 1954a, Feldmann and Hamel 1936, Post 1955a; *Bostrychia periclados*, Mazé and Schramm 1870–77, Murray 1889; *B. tuomeyi*, Harvey 1853.

Pterosiphonia Falkenberg, 1889

Plants rhizomatous, the erect branches sparingly divided, alternately pinnately branched, the short branchlets simple or once forked; pericentral cells five to ten, naked or with a parenchymatous cortex; vegetative trichoblasts absent; tetrasporangia single in each fertile segment, lying along the face of the branch axis; spermatangia stalked, in clusters on the fertile branchlet tips; cystocarps short-stalked, ovoid, lateral on the axes of fertile branches.

Pterosiphonia pennata (Roth) Falkenberg

Plants tufted, soft, becoming fragile when dried, blackish purple, to 2–6 cm. tall, distichously pinnate or irregularly partly bipinnate, the main axes with 8–9 pericentral cells, the segments about as long as broad, becoming corticated below; fronds linear, about 2 mm.

broad, the branchlets 1.0–1.5(–3.0) mm. long, simple or sparingly forked, incurved when young, obliquely erect when mature.

North Carolina, Brazil. Growing on stone jetties, etc., somewhat below low-tide level.

REFERENCES: Williams 1949a, Joly 1951, 1957, de Mattos 1952; *Bostrychia rivularis,* Blomquist and Humm 1946.

Bostrychia Montagne, 1838

Plants filiform, black or dull purplish; stoloniferous, rhizoidal and erect branches often distinguishable, even the rhizoidal branches polysiphonous, ordinarily regularly bilaterally branched, the branches near the apex usually incurved; for the most part polysiphonous, several cells of equal length being disposed about the central axis, or these pericentral cells regularly transversely divided, but the ramellar branchlets often monosiphonous at the tips; sporangia whorled in special stichidial branchlets, several tetrahedral sporangia being formed in each segment; pericarps subglobose, terminal on the branchlets.

KEY TO SPECIES

1. Primary haptera consisting of transformed branches formed in the ordinary branch position 2
1. Primary haptera on special flagellar outgrowths from the axes, generally formed on the lower side and at a fork, but not in the ordinary position and succession of branching 5

2. Ecorticate 3
2. Corticated; regularly distichously branched .... B. pilulifera, p. 597

3. Branchlets or ramelli of the last order, except hapteral branches, altogether monosiphonous .... B. moritziana, p. 596
3. Branchlets of the last order chiefly polysiphonous, only the last 1–3 segments monosiphonous 4

4. Plants compact, mosslike .... B. radicans, p. 595
4. Plants loosely branched, mostly free and wide-spreading B. rivularis, p. 595

5. Cortication rhizoidal in nature; conspicuously pinnate, the branchlets simple, polysiphonous .... B. calliptera, p. 597
5. Cortication parenchymatous 6

6. Branch system not differentiated into long and short branches, all having about the same growth potential; not conspicuously pinnate 7
6. Branch system distinctly divided into long axial branches and short 1–3-pinnate lateral branches of rather limited growth, the branchlets less regularly pinnate 8

7. Pericentral cells surrounded by several layers of similar large cells and of smaller ones external to these ... B. montagnei, p. 598
7. Pericentral cells surrounded by 1–2 layers of cells B. scorpioides, p. 597

8. Branchlets or ramelli of the last order altogether monosiphonous B. tenella, p. 599
8. Branchlets of the last order polysiphonous at the base, only the distal segments monosiphonous .... B. binderi, p. 598

Bostrychia radicans Montagne

Plants dull purplish, creeping below and attaching by haptera, above somewhat erect, repeatedly pinnately branched, the branches somewhat clustered and incurved near the tips, but below more widely separated, ecorticate, but polysiphonous throughout except for the tips of the branchlets; stichidia developed below these tips on long pedicels, ovate-oblong, about 2 diameters long, with 3–4 series of tetrasporangia, of which 2 seem most prominent.

F. **moniliforme** Post. The terminal 5–15 segments of the branchlets monosiphonous.

Florida, Mississippi, Louisiana, French Guiana, Brazil. Growing on rocks or other heavy objects in the lower intertidal zone.

REFERENCES: Montagne 1850, Möbius 1890, 1895, Post 1936, Joly 1951, 1954, 1957, De Mattos 1952, Taylor 1954a; *B. guadeloupensis*, Mazé and Schramm 1870–77, Murray 1889; *B. leprieurii*, Montagne 1850. For the f. *moniliforme*, Post 1936; *B. radicans* f. *brasiliana*, Möbius 1889.

Bostrychia rivularis Harvey

Plants dull purplish-violet, diffuse, becoming erect from creeping stolons, attached by holdfasts, which are to 3 cm. long; repeatedly dichotomously branched, the lower branches spreading, forming short branches bilaterally, but the terminal ones erect, with branchlets subcorymbosely incurved; branchlets 1–2 mm. long, attenuate, polysiphonous almost to their tips, the terminal 1–3 segments monosiphonous; segments of the principal branches ecorticate, half as long as broad, with 6–8 pericentral cells which are transversely divided; branchlets with 4 or fewer pericentral cells; stichidia in the middle portions of the ultimate branches, about 15 segments with tetrasporangia; pericarps ovate, formed on older branchlets.

Bermuda, North Carolina, South Carolina, Florida, Bahamas, Cuba, Jamaica, Hispaniola, Puerto Rico, Guadeloupe, Costa Rica, Tobago, Surinam, French Guiana, Brazil. Growing on mud, stones, and other objects in very shallow, often brackish or quite fresh water; to be expected in such places as salt marshes and estuaries. The range of this species extends to the north as far as New Hampshire, far beyond the tropical area.

REFERENCES: Harvey 1853, Mazé and Schramm 1870–77, Murray 1889, Collins and Hervey 1917, Hoyt 1920; *Bostrychia polysiphonioides,* Mazé and Schramm 1870–77, Murray 1889; *Amphibia rivularis,* Taylor 1928.

Bostrychia moritziana (Sonder) J. Agardh

Plants rather erect, 2.0–6.5 cm. tall, repeatedly pinnately branched and beset with lateral branchlets about 2.2 mm. long; branches subfastigiate, becoming denuded below, closely distichously branched above, the axes with 7–8 pericentral cells, remaining ecorticate; determinate branchlets polysiphonous at the base, with segments half as long as broad, incurved, becoming completely monosiphonous in the outer divisions, where the cells are one-half longer than broad or more; stichidia elongate-lanceolate, with 3–4 sporangia in each segment arranged so that they appear as 2 rows in lateral view.

Florida, Texas, Jamaica, Hispaniola, Guadeloupe, St. Lucia, Grenada, Canal Zone, Panama, Colombia, Venezuela, Trinidad, Surinam, French Guiana. Growing on mangrove roots in sheltered places, but also migrating into freshwater streams and there growing on rocks.

REFERENCES: Mazé and Schramm 1870–77, Murray 1889; *B. cornigera* and *B. leptoclada,* Montagne 1850; *B. guadeloupensis p. p.*, Mazé and Schramm 1870–77, Murray 1889; *B. moritziana* v. *intermedia,* Collins 1901; *B. monosiphonia,* Montagne 1850; *Amphibia moritziana,* Taylor and Arndt 1929.

Bostrychia pilulifera Montagne

Stems creeping, 4–7 cm. long, attached by haptera, distichously repeatedly pinnate, corticated, the branchlets alternate, polysiphonous, rarely more than 2 mm. long, deflected, often attaching at the tips; pericarps terminal on the outer branchlets, ovate-globose, 700–750 μ diam.

British Guiana, Surinam, French Guiana.

REFERENCES: Montagne 1850, Post 1936.

Bostrychia calliptera Montagne

Basal portions of the plants more or less decumbent and attached by hapteral disks, the haptera developed as ventral flagellar structures usually at a point of branching; in large part erect, 6–11 cm. tall, freely di- polychotomously branched, the divisions distichously pinnate; main axes with 6 very short pericentral cells, which become corticated by rhizoidal filaments; branchlets simple, borne directly on indeterminate axes, with 4 pericentral cells, polysiphonous to the tips, ecorticate; stichidia stipitate, lanceolate and a little curved, the tips attenuate; with 2 rows of sporangia visible from the side.

French Guiana, Brazil. Growing on the trunks of the red mangrove in shallow, sheltered situations.

REFERENCES: Montagne 1850, Joly 1954, 1957, Post 1955b.

Bostrychia scorpioides (Gmelin) Montagne

Plants dull violet, drying brownish, repeatedly pinnate, the branchlets near the tips incurved; pericentral cells 6–8, corticated except at the growing tips and in the branchlets, usually with 1–2 layers of large cells around the pericentral cells, the external layers of smaller cells not sharply marked; main axes flexuous, subdistichously widely branched at intervals of about 2 mm.; stichidia stalked, developed from lower divisions of the determinate branches, lanceolate, with a sterile apex, subtorulose, about 4 diameters long, with

3–4 sporangia in each segment; pericarps terminal on ultimate branchlets.

Bermuda, Florida, Bahamas, Cuba, Jamaica, British Honduras, French Guiana, Brazil.

REFERENCES: Rein 1873, Murray 1889, Post 1936.

Bostrychia montagnei Harvey Pl. 74, fig. 1

Plants tufted, dull reddish-purple, very coarse, to 7.5 cm. tall, alternately and distichously pinnate and spreading to a breadth of 2.5–5.0 cm.; lateral branches below stiff, becoming denuded of branchlets, but above repeatedly pinnate, softer and incurved; pericentral cells about 6–7, surrounded by 5–7 layers of corticating cells in the main branches, of which 1–3 outer layers are of much smaller and rather irregularly placed cells; uncorticated branchlets polysiphonous except near the tips, where there may be short monosiphonous terminal portions; main axes to 650 μ diam., the segments obscured; uncorticated polysiphonous portions of the branchlets 50–70 μ diam., the segments subequal, soon thereafter obscured; monosiphonous tips when present about 25 μ diam., the cells 0.8–1.2 diameters long; stichidia short-stalked, formed by the nearly total transformation of ultimate branchlets, showing 2 rows of tetrasporangia in side view.

Bermuda, Florida, Louisiana, Bahamas, Cuba, Jamaica, British Honduras, Brazil. Growing almost entirely on red mangrove stilt roots in very sheltered situations.

REFERENCES: Hooper 1850, Harvey 1853, Rein 1873, Hemsley 1884, Murray 1889, Collins and Hervey 1917, Howe 1918a; *B. scorpioides* v. *montagnei*, Post 1936, Joly 1954, 1957; *Amphibia montagnei*, Howe 1920, Taylor 1928, 1935.

Bostrychia binderi Harvey

Plants dull purplish, matted, to 2 cm. tall, tripinnate, the lower branches hardly over 2 mm. long, crowded, alternately simply pinnate, the upper elongate and bipinnate; ramuli spreading, the last divisions rather short and spinuliform, only the tips, of 2–20 cells, frequently monosiphonous; pericentral cells usually 7, parenchymatously corticated with 1–3 series of cells, except sometimes at the tips of the branchlets; stichidia formed from the distal parts of the branchlets, with 4–5 tetrasporangia in each segment, the extreme tips sterile.

Bermuda, Florida, Bahamas, Cuba, Jamaica, Hispaniola, Virgin Isls., Tortola I., Guadeloupe, Martinique, St. Vincent, Barbados, Panama, Venezuela, Trinidad, French Guiana, Brazil. Forming dense mats on rocks in the intertidal zone.

REFERENCES: Post 1936, Feldmann and Lami 1937, Taylor 1942, 1943, 1954a, Joly 1951, 1954, 1957; *B. capillacea*, Mazé and Schramm 1870–77, Murray 1889; *B. mazei*, Mazé and Schramm 1870–77, Murray 1889, Collins 1901; *B. pectinata*, Howe 1920, Taylor 1928; *B. sertularia*, Montagne 1859, Mazé and Schramm 1870–77, Martens 1871, Hemsley 1884, Murray 1889, Möbius 1890, Collins and Hervey 1917, Howe 1918a; *Amphibia sertularia*, Howe 1920, Taylor 1929b.

Bostrychia tenella (Vahl) J. Agardh

Plants densely tufted, 2–5 cm. tall, or in mats 1–2 cm. thick, dull reddish-purple or faded to pale brownish, more or less erect, repeatedly pinnate, denuded below, densely branched above, the tips strongly incurved; main axes with 6–8 pericentral cells and for the most part corticated, with 3–4 layers of parenchymatous cells; determinate branches spreading, alternately distichous, polysiphonous, the lower divisions corticated, the upper divisions ecorticate, the terminal branchlets monosiphonous, 25–35 cells long, about 20 μ diam., the cells 30–45 μ long; stichidia replacing terminal branchlets, linear-lanceolate, with a sterile pedicel and an attenuate tip, showing 4 tetrasporangia in each segment; pericarps on denuded branchlets, rounded-globose.

Bermuda, Florida, Bahamas, Caicos Isls., Cuba, Cayman Isls., Jamaica, Hispaniola, Virgin Isls., Guadeloupe, Martinique, Barbados, Grenada, British Honduras, Costa Rica, Venezuela, Brazil. Widespread and often very abundant on rocks, walls, and other objects in the intertidal zone.

REFERENCES: Mazé and Schramm 1870–77, Murray 1889, Möbius 1890, Collins 1901, Børgesen 1913–20, Collins and Hervey 1917, Howe 1918a, Hamel and Hamel-Joukov 1931, Feldmann and Lami

1937, Joly 1957; *B. calamistrata,* Harvey 1853, Hemsley 1884; *B. elegans, B. muscoides, B. pilifera,* and *Polysiphonia bostrychoides,* Mazé and Schramm 1870–77, Murray 1889; *B. viellardi,* Zeller 1876; *Amphibia tenella,* Howe 1920, Taylor 1928, 1930b, 1933, 1935, 1954a; *Rhodomela calamistrata,* Montagne 1883.

UNCERTAIN RECORD

Bostrychia brasiliana Leuderwaldt

Brazil, the type locality; a *nomen nudum.* Leuderwaldt 1919, Joly 1954.

Streblocladia Schmitz, 1897

UNCERTAIN RECORDS

Streblocladia camptoclada (Montagne) Falkenberg

Guadeloupe. Type from Peru. De Toni 4:1062; *Polysiphonia camptoclada,* Mazé and Schramm 1870–77, Murray 1889.

Streblocladia collabens (C. Agardh) Falkenberg

Guadeloupe. Type from Europe. De Toni 4:1063; *Polysiphonia collabens,* Mazé and Schramm 1870–77, Murray 1889.

Dipterosiphonia Schmitz and Falkenberg, 1897

Plants creeping, the filaments terete or compressed, usually attached by haptera; lateral branching consisting of secondary axes and of determinate, pinnately branched structures which in reduced form simulate the main branch system, and each of which typically lies above a simple, terete, aculeate branchlet in alternate fashion along the axis; trichoblasts sometimes discernible on the branchlets; tetrasporangia numerous in somewhat enlarged branchlets of either type, the sporangia greatly projecting, covered by small cells; spermatangial clusters crowded at the tips of simple branchlets; pericarps ovoid, one or two together on similar branchlets.

KEY TO SPECIES

1. All divisions terete, the generic branch characters obscure except in young portions of well-developed creeping axes ..... D. ringens
1. Determinate lateral branchlets flattened, though in some parts these may be poorly developed .... D. dendritica

Dipterosiphonia ringens (Schousboe) Falkenberg

Plants in minute dull purple tufts, entangled, attached by rhizoids with scutate ends issuing from the creeping primary branches; secondary filaments somewhat erect, of about 60 short segments, 60–80 μ diam., repeatedly laterally branched, bearing both simple and pinnately divided rather stiff branchlets; pericentral cells 4–6, ecorticate; tetrasporangia 2–6 in a longitudinal series in subtorulose, pod-shaped branchlets; spermatangial clusters cylindric-conical, 40–50 μ diam., 160–200 μ long, clustered at the tips of the small branchlets; pericarps subglobose, relatively large and numerous, about 250 μ diam., on short horizontal branchlets.

Bermuda, Hispaniola. Epiphytic on various algae and also growing on limpets in shallow water, and dredged to 6 m. depth.

REFERENCES: Howe 1918a; *Lophosiphonia bermudensis*, Collins and Hervey 1917, Børgesen 1924.

Dipterosiphonia dendritica (C. Agardh) Schmitz

Plants small, creeping, brownish red, forming colonies, attaching by rhizoids with terminal disks from the lower side of the principal branches; secondary branches ascending, branching distichous, successive branchlets dimorphic, the one stout, simple, spinelike, the other compressed and complanate, distichous in turn with its divisions a simplified version of the main axial system, the ramelli with a merely dentate margin; pericentral cells 5; the segments about half as long as broad, or, in very attenuate parts, subequal; tetrasporangia in rows of 3–4 in thickened obtuse ramelli of much-branched fronds; cystocarpic individuals loosely branched, the pericarps on the outer side of simple branchlets below the apex.

Bahamas, Virgin Isls., Barbados, Colombia, Brazil. Epiphytic on larger algae in shallow water.

REFERENCES: Vickers 1905, Børgesen 1913–20, Howe 1920, Schmidt 1924, Taylor 1941b; *Polysiphonia dendritica*, Martens 1870.

Herposiphonia Nägeli, 1846

Plants creeping, the wide-ranging and freely branched rhizomatous filaments attaching by unicellular rhizoids, though some branches or even large portions of the plant often free from the substratum; axes terete or compressed, with several pericentral cells, ecorticate, dorsiventral, bearing erect branchlets on the dorsal side; lateral branches or their rudiments alternating with branchlets in a definite sequence from the nodes, some of which may regularly be naked; erect axes of limited growth, simple, polysiphonous, like the rhizomatous axes in the young state with the tips inrolled; on the dorsal side the erect axes bearing a tuft of deciduous trichoblasts; tetrasporangia in the central portions of the erect branchlets, forming a more or less straight single series along the axis; spermatangial clusters developed from the trichoblast rudiments in groups near the apices of the branchlets; pericarps ovate or globose, single, placed similarly to the spermatangia.

KEY TO SPECIES

1. Main axes with a branchlet, a long branch, or a branch rudiment from each segment 2
1. Main axes with some segments regularly free from branches or branchlets 3
2. Main axes and branches with strongly downcurved apices; branchlets 8–10 segments long, also curved when young; plants often free from the substratum in large part, and several centimeters long . . . . H. pecten-veneris, p. 603
2. Main axes and branches nearly straight or slightly upcurved 4
3. Plants small, about 1–2 cm. diam., often crowded; growing tips sharply upcurved, the main axes persistently creeping, 75–150 μ diam. . . . . H. secunda, p. 604
3. Plants large, to 5 cm. diam., repeatedly pinnate, in part free from the host, the growing tips a little upcurved, the axes 150–250 μ diam. . . . . H. bipinnata, p. 602
4. Plants creeping, dull purple, small; the branchlets 12–45 segments long . . . . H. tenella, p. 604
4. Plants usually in great part free, the main axes often yellow; the branchlets rose-red, with 8–12 segments H. pecten-veneris v. laxa, p. 603

Herposiphonia bipinnata Howe

Plants free or only partly creeping, the main axes 2–5 cm. long, straw-colored or bleached, the branchlets rose; filicoid or floccose,

2–3-pinnate, the branchlets in part pectinate-secund; axes 150–250 μ diam., the tips slightly upturned, the segments 1.5–2.0 diameters long; pericentral cells usually 10; a branchlet usually arising from each of 2 successive nodes, followed by 1–2 naked nodes; branchlets slightly curved, 8–12 segments long, crowned by coarse trichoblasts, the segments 0.5–1.25 diameters long; tetrasporangia somewhat protuberant, the branchlets to 105–135 μ diam.; cystocarps ovoid, subtruncate, 300–450 μ diam., 380–540 μ long.

Bahamas. Growing attached to marine phanerogams in shallow water.

REFERENCE: Howe 1920.

Herposiphonia pecten-veneris (Harvey) Falkenberg

Plants moderately small, 2–10 cm. diam., widely alternately branched, creeping below but erect above, the more remote branches unilaterally pectinate with short secund branchlets arising from each node; main branches flexuous, the tips strongly recurved with the convexity dorsal; segments to 165 μ diam., to 2.0–3.5 diameters long in the branches, with 9–12 pericentral cells; branchlets 8–10 segments long, the segments somewhat longer than broad, to 65–85 μ diam., with 10–12 pericentral cells.

V. **laxa** n. var. Plants attached in the older parts, but mostly free, the branchlets rosy, the axes often yellow; main axes straight or slightly but not distinctively curved; alternately to markedly bilaterally branched; usually with a leading branch developing from each fourth node, or represented by a rudiment, alternating with a series of three branchlets; main axes 120–160 (?–250) μ diam., with 9–10 pericentral cells; branchlets with 8–12 segments, 35–60 μ diam., with 6–7 pericentral cells.

Bermuda, Florida, Bahamas, Caicos Isls., Jamaica. Growing among coarse algae and on phanerogams in shallow water.

REFERENCES: Harvey 1853, Collins and Hervey 1917, Howe 1920, Taylor 1928; *Polysiphonia pecten-veneris*, Murray 1889, Collins 1901. For the v. *laxa, Herposiphonia* sp.?, Taylor 1928, p. 177, pl. 25, fig. 12.

Herposiphonia secunda (C. Agardh) Ambronn Pl. 72, figs. 10, 11

Plants small, usually about 1–2 cm. diam., closely attached to the substratum, sometimes forming close mats, or entangled upon and attached to masses of other algae; color dark reddish-brown throughout, the cell membranes often yellowish-brown; branching of the main axes irregularly alternate, the apices strongly curved toward the dorsal surface, but the older axes nearly straight, 75–150 μ diam., with 7–8 pericentral cells; axes with an indeterminate branch or a rudiment of one usually preceding each branchlet, these originating from every fifth or sixth node, and to 60–70 μ diam., with 7–8 pericentral cells; pericarps solitary near the summits of the branchlets, displacing the apex.

Bermuda, Florida, Bahamas, Caicos Isls., Cuba, Jamaica, Hispaniola, Virgin Isls., Guadeloupe, Barbados, Grenada, British Honduras, Panama, Netherlands Antilles. Growing on coarser algae, rocks, woodwork, and other objects in shallow water, and sometimes covering extensive areas.

REFERENCES: Sluiter 1908, Collins and Hervey 1917, Børgesen 1913–20, Howe 1918a, 1920, Taylor 1928, 1929b, 1935, Taylor and Arndt 1929; *Polysiphonia secunda*, Harvey 1853, Montagne 1863, Mazé and Schramm 1870–77, Hemsley 1884, Murray 1889, Collins 1901.

Herposiphonia tenella (C. Agardh) Ambronn Pl. 72, fig. 12

Plants small, closely attached to the substratum, usually forming close mats; color purplish; branching of the main axes irregularly alternate, infrequent, the apices slightly upcurved, the older portions straight; axes 100–150 μ diam., with either an erect branchlet or an indefinite branch, or a rudiment of one, from each node, 1–3 branchlets alternating with each indefinite branch; branchlets with as many as 35–45 segments and 5–7 mm. tall, 65–90 μ diam. with 12–14 pericentral cells; tetrasporangia to 60 μ diam., in series of 15–20 to each branchlet.

Bermuda, North Carolina, South Carolina, Florida, Mexico, Bahamas, Caicos Isls., Jamaica, Virgin Isls., Martinique, Barbados, Colombia, Brazil. Growing on various algae, rocks, and other objects in shallow, protected situations.

REFERENCES: Vickers 1905, Børgesen 1913–20, Collins and Hervey 1917, Howe 1918a, 1920, Hoyt 1920, Taylor 1928, 1935, 1940, 1941b, Joly 1957.

UNCERTAIN RECORDS

Herposiphonia monocarpa (Montagne) De Toni

Guadeloupe. Type from South Africa. De Toni 4:1061; *Polysiphonia monocarpa*, Mazé and Schramm 1870–77, Murray 1889.

Herposiphonia pectinella (Harvey) Falkenberg

Puerto Rico. Type from Australia. De Toni 4:1055; *Polysiphonia pectinella*, Hauck 1888.

Herposiphonia prorepens (Harvey) Schmitz

Barbados. Type from South Africa. De Toni 4:1057; Vickers 1905.

Polyzonia Suhr, 1834

UNCERTAIN RECORD

Polyzonia divaricata Crouan

Guadeloupe, the type locality; a *nomen nudum*. De Toni 6:417; Mazé and Schramm 1870–77, Murray 1889.

Lophosiphonia Falkenberg, 1897

Plants filamentous, the primary axes creeping and attached by rhizoidal holdfasts, cylindrical, laterally branched, dorsiventral with the apex recurved; erect filaments of limited growth, like the basal ones uncorticated, very sparingly branched, the branch origin generally endogenous, the apices sometimes recurved and bearing on the convex side a series of deciduous trichoblasts; tetrasporangia formed in the upper branches of somewhat more freely divided filaments.

KEY TO SPECIES

1. Rhizoids slender; pericentral cells generally 6 or more 2
1. Rhizoids saccate; pericentral cells 4 .... L. saccorhiza

2. Trichoblasts spirally arranged; pericentral cells 8–10... L. subadunca

2. Trichoblasts in a single dorsally secund series on the hooked tips of the erect branches; pericentral cells 6–12........... L. cristata

Lophosiphonia saccorhiza Collins and Hervey

Plants rosy red, soft, the prostrate filaments 50–70 μ diam., the segments 1–2 diameters long; prostrate filaments attached by saccate rhizoids up to 160 μ diam., 800 μ long; pericentral cells 4 throughout; erect filaments numerous, up to 2 mm. tall, 25–45 μ diam,. contracted abruptly at the base, tapering more gradually toward the apex, the segments 2–3 diameters long below, but much shorter above; trichoblasts well developed, but quickly shed; tetrasporangia in a single series in the upper, much inflated part of the filaments; spermatangial clusters ovoid-conical, also near the tips of the filaments; cystocarps subspherical, to 100 μ diam., with one or more short subtending branchlets near the tips of the filaments.

Bermuda. Plants epiphytic on Codium, the rhizoids lying between the utricles.

REFERENCES: Collins and Hervey 1917, Howe 1918a.

Lophosiphonia subadunca (Kützing) Falkenberg

Plants with a creeping branched rhizome attaching at intervals by rhizoids; erect branchlets alternately subsecund along the dorsal side of the rhizomes, sparingly divaricately branched; segments of the axes a little longer than broad, those of the branchlets rather shorter; pericentral cells 8–10.

Texas, Bahamas.

REFERENCES: Howe 1920, Taylor 1941a.

Lophosiphonia cristata Falkenberg

Plants rhizomatous, creeping, often much entangled, the apices curved downward, the older stems 80–120 μ diam.; the erect branchlets successively developed along the dorsal side of the rhizomes at intervals of several segments, the apices strongly curved toward the forward end of the plant, with a conspicuous crest of pseudodichotomous trichoblasts; pericentral cells (6–)9–10, occasionally to 12 in the erect branchlets; tetrasporangia spirally placed near the tips of the erect branchlets.

Bermuda, Bahamas, Jamaica, Virgin Isls. On rocks and reefs in shallow water.

REFERENCES: Børgesen 1913–20, Howe 1920, Hollenberg 1958.

UNCERTAIN RECORD

Lophosiphonia neglecta (Harvey) De Toni

Brazil. Type from Australia. De Toni 4:1071; *Polysiphonia neglecta*, Zeller 1876.

Protokuetzingia Falkenberg, 1897

Plants erect, more or less complanate, the divisions compressed, the branch tips incurved, the axis polysiphonous with five or six pericentral cells, which become corticated, the subcortical layer of large cells, the surface layer of small ones; axes producing secondary branches laterally at regular intervals; the ultimate branchlets formed similarly, often in clusters; tetrasporangia in two rows in strongly incurved branchlets; spermatangia stalked, on the backs of the branchlets; pericarps broadly sessile, in a row on the back of each incurved branchlet.

Protokuetzingia schottii Taylor

Plants bushy, nearly black when dry, to 18 cm. tall or more, arising from a rather thick disk 3–4 mm. diam.; erect axes flexuous, about 1.7–2.0 mm. diam., sparingly alternately divided below, more closely above, the branches similar to the axes, when young slightly incurved, 0.5 mm. diam.; determinate branchlets opposite, distichous, at intervals of 1–3 mm., commonly with accessory branchlets originating at the bases of those first formed, the branchlets occasionally forked, strongly incurved and tapered to base and apex, 0.08–0.16 mm. diam., 1.0–1.15 mm. long, the apices with deciduous branched trichoblasts; spermatangial plants more symmetrical than the carpogonial, with longer, simpler, and more erect branchlets; spermatangial clusters conic-oval to subcylindrical, 75–115 μ diam., 150–225 μ long; carpogonial plants irregularly spreading, the fertile branchlets 0.16–0.24 mm. diam., to 3 mm. long, the crowded dorsal row of pericarps prominent; pericarps rounded-conical to ovate, the apex sometimes slightly pointed, generally 45–80 μ diam. and tall, but these perhaps immature, for others were seen to 230–260 μ diam.

Colombia.

REFERENCE: Taylor 1941b.

Halopithys Kützing, 1843

UNCERTAIN RECORD

Halopithys pinastroides (Gmelin) Kützing

Brazil. Type from England. De Toni 4:1081; Schmidt 1924.

Amansia Lamouroux, 1809

Plants erect, foliaceous, pinnately branched, the narrow divisions membranous, with inrolled tips; costate, the costa showing five pericentral cells, often becoming corticated; the lateral membrane transversely zonate; stichidia incurved, developed from marginal teeth or from superficial pinnate processes formed on the frond, with the tetrasporangia in two rows; pericarps rounded-ovate, in similar positions.

Amansia multifida Lamouroux Pl. 70, fig. 5

Plants to 15 cm. tall, reddish-purple, very bushy, attached by a small discoid holdfast, the stipe terete, the blades repeatedly and rather regularly pinnately divided, proliferating from the stipe and the costa; the divisions strap-shaped, to 2–5 mm. wide, or sometimes attenuate, serrate, of 2 cell layers except at the midrib; the stichidia developed from elongated marginal teeth.

Florida, Bahamas, Cuba, Jamaica, Hispaniola, Puerto Rico, Guadeloupe, Martinique, Panama, Venezuela, Trinidad, Tobago, Brazil.

REFERENCES: Hooper 1850, Harvey 1853, Mazé and Schramm 1870–77, Martens 1870, 1871, Piccone 1886, 1889, Hauck 1888, Murray 1889, Collins 1901, Howe 1920, 1928, Schmidt 1923, 1924, Taylor 1928, 1929b, 1942, 1943, 1954a, Feldmann and Lami 1937; *Epineurum multifidum*, Martens 1870.

Vidalia Lamouroux, 1824

Plants erect, fleshy-membranous, in the branches flat and ligulate, sometimes with a midrib evident on the lower side, sometimes the blade twisted; axes polysiphonous with five primary pericentral cells, laterally expanded into the membrane, which has a medulla of large cells covered by a small-celled cortex, the cells arranged in oblique rows; branch apices inrolled, margins undulate to serrate or pinnatifid, minute veinlets connecting the teeth with the costa; tetrasporangia in two rows in stichidium-like projections from the pinnate tips of the teeth; spermatangial clusters and pericarps near the apices of similar teeth.

Vidalia obtusiloba (Mertens) J. Agardh Pl. 70, figs. 3, 4

Plants briefly stalked from a basal disk, 10–25 cm. tall, bright red or blackish when dried, denuded below or quickly expanded to the flat blades, 2–3 times pinnate, the divisions alternate, 4–9 mm. broad when dried or to about 12 mm. soaked, a little narrowed toward the base, broadly rounded at the tips, the margins undulate to alternately serrate, the median line with a costa, very inconspicuous distally, more evident below, connected with the marginal spines by very faint veinlets; the medulla distally of 2 layers of large cells, below of 4 layers, covered throughout by a small-celled cortex; marginal teeth in the young parts becoming conspicuous, deltoid to linear, incurved, in the older parts somewhat intergrading with small leaflets; pericarps usually subsessile, single, projecting from the curved side of marginal teeth.

Florida, Mexico, Puerto Rico, Guadeloupe, Grenada, Colombia, Venezuela, Brazil.

REFERENCES: Harvey 1861, Mazé and Schramm 1870–77, Piccone 1886, 1889, Möbius 1889, Murray 1889, Schmidt 1923, 1924, Taylor 1928, 1940, 1941b, 1942, Howe 1928, Sanchez A. 1930; *Rytiphloea hilariana*, Greville 1833, Martens 1870; *R. obtusiloba*, Martens 1870.

Enantiocladia Falkenberg, 1889

Plants erect, somewhat fleshy, complanate, oppositely branched, the primary branches elongate, with five pericentral cells and a substantial cortex of small cells forming a midrib; laterally alate with a membrane of large medullary cells covered with a small-celled cortex and, in the older portions, with a subcortex as well, the margins oppositely aculeate-dentate; slender branches hardly alate, often bearing small branchlets on one margin only, or these reduced to mere teeth; stichidia crowded on the warty, often proliferous upper portions of the branchlets; pericarps subglobose, the wall rather thick.

Enantiocladia duperreyi (C. Agardh) Falkenberg

Plants reddish-purple, drying black, to 10–18 cm. tall, the base small and discoid, the short terete stalk expanding to the complanate, 1–3-pinnate blade; branches irregularly opposite, the tips involute, the bases often attenuate and subterete, but above similar to the primary axes, flat or a little canaliculate, 2.0–4.5 mm. wide, the margins oppositely serrate, the teeth subulate, incurved, often forked, or in fertile plants near the tips fimbriate; stichidia pinnate, the divisions linear, strongly incurved, the sporangia in 2 rows on the outer side.

Florida, Cuba, Jamaica, Puerto Rico, Guadeloupe, Martinique, Brazil. A plant of moderate to quite deep water, growing on the shells of living Strombus and doubtless other objects.

REFERENCES: *Amansia duperreyi,* Mazé and Schramm 1870–77, Dickie 1874b, Hauck 1888, Murray 1889.

Chondria C. Agardh, 1817

Plants bushy, much and alternately branched, the branches terete, those of the upper divisions usually markedly constricted at the base and the ultimate ones spindle- or club-shaped, the branchlets tipped by clusters of trichoblasts; growth from an apical cell, which is often depressed to the bottom of a pit by rapid growth and division of the lateral segments; axis with five pericentral cells surrounded by a loose cortex of branching cell series which terminate in small cells closely approximated as an epidermal layer; tetrasporangia usually single at a node, developed from the pericentral cells in ultimate or penultimate branchlets which are little altered; spermatangial clusters of various shapes, commonly in the form of flat or twisted colorless plates borne near the apices of the branchlets; procarps developed from the second segment of a trichoblast rudiment; immediately after fertilization the active pericentral cell cuts off an auxiliary beside the carpogonium; carposporangia borne on a large fusion cell; pericarp in its lower portion becoming several cells thick.

KEY TO SPECIES

1. Plants mostly repent, with many multicellular haptera .... C. polyrhiza, p. 617
1. Plants erect or sprawling 2

2. Tips of the branchlets acute, the growing points exposed 3
2. Tips of the branchlets obtuse or truncate, the growing points not projecting 6

3. Plants large, light brownish-purple, the branches markedly more slender than the main axes .... C. littoralis, p. 612
3. Plants with a rather regular gradation between the main axes and the branchlets 4

4. Plants dark purple when living, black when dried, irregularly branched and sometimes very bushy .... C. atropurpurea, p. 613
4. Plants light reddish-purple or decolorate, not blackening when dried 5

5. Coarse, growing on stones or shells; branchlets spindle-shaped; tetrasporangia chiefly formed toward the distal ends of the fertile branchlets .... C. tenuissima, p. 613
5. Slender, generally growing on other plants; branchlets little tapered distally, the ends blunt; tetrasporangia formed in the thickened middle portion of even the oldest fertile branchlets .... C. baileyana, p. 614

6. Branchlets generally contracted at the base 7
6. Branchlets not contracted at the base; without conspicuous trichoblasts .... C. floridana, p. 616

7. Pericentral cells ordinarily without local wall thickenings visible through the cortex 8
7. Pericentral cells in living plants with conspicuous thickenings of their anterior cell walls, which are visible through the cortex 11

8. Plants bushy, the axes about 1.0–1.5 mm. diam., the branches spreading, the branchlets about 0.3–0.7 mm. diam., with a

conspicuous tuft of trichoblasts at each apex; strongly staining the paper brown in drying .... C. dasyphylla, p. 616

8. Plants not strongly staining the paper 9

9. Habit erect and bushy 10

9. Plants sprawling, the rather angularly flexuous main axes 0.75–1.25 mm. diam. .... C. cnicophylla, p. 614

10. Plants erect and slender, the axes 0.22–0.32 mm. diam., the principal branches erect, the branchlets narrowly clavate, about 0.1–0.2 mm. diam. .... C. leptacremon, p. 615

10. Plants pyramidal, bushy, the straight axes 1.0–2.5 mm. diam. C. sedifolia, p. 615

11. Plants tufted, the cushions 1–3 cm. high; the main axes mostly 0.22–0.35 mm. diam.; apical tufts of trichoblasts inconspicuous .... C. curvilineata, p. 617

11. Plants solitary or gregarious, mostly 3–8 cm. tall; main axes 0.4–0.75 mm. diam.; apical tufts of trichoblasts often 0.75 mm. long or more, becoming brownish on drying C. collinsiana, p. 617

Chondria littoralis Harvey

Plants tufted, pale or somewhat purplish, 1–3 dm. tall or somewhat more; main axes 1–2 mm. diam., becoming denuded below, virgate above, where more or less densely and rather narrowly paniculately branched, the ultimate indefinite branches beset with elongate tapered branchlets with contracted bases, about 0.5 mm. diam., 1–15(–25) mm. long, or these rebranched; branchlets terminating in prominent tufts of trichoblasts; tetrasporangia in the branchlets, 100–120 μ diam., appearing in the older branchlets chiefly near the outer end; spermatangial clusters flat, about 450 μ long, borne around the trichoblast tufts and laterally on the branchlets; pericarps single on a branchlet, lateral, urceolate, with a very wide mouth, about 800 μ diam., 1.5 mm. long.

Bermuda, North Carolina, Florida, Mexico, Bahamas, Caicos Isls., Cuba, Hispaniola, Virgin Isls., St. Barthélemy, Guadeloupe, Colombia. Commonly growing on rocks and old corals in shallow, sheltered waters, and dredged to a depth of 9 m.

REFERENCES: Harvey 1853, Hemsley 1884, Børgesen 1913–20, 1924, Howe 1920, Hoyt 1920, Taylor 1928, 1933, 1942, 1954a;

Chondriopsis capensis and *C. littoralis,* Mazé and Schramm 1870–77, Murray 1889.

Chondria atropurpurea Harvey

Plants bushy, dark purple, firm in texture, subpyramidal, to 2 dm. tall, the branches rather elongate, often naked of branchlets below, above at intervals bearing single or fasciculate linear-fusiform branchlets which may become 2–3 cm. long, or continue growth and branch again; young branchlets abruptly narrowed below, gradually tapering above; tetrasporangia in fusiform branchlets, which become somewhat rough of surface; cystocarps on densely crowded branchlets, the conspicuous pericarps to 1.5 mm. diam., subspherical to broadly urceolate.

Bermuda, North Carolina, South Carolina, Florida, Texas, Bahamas, Cuba, Hispaniola, Virgin Isls., Guadeloupe, Colombia, Venezuela, Brazil.

REFERENCES: Harvey 1853, Vickers 1905, Børgesen 1913–20, Collins and Hervey 1917, Howe 1920, Hoyt 1920, Taylor 1928, 1933, 1941a, b, 1942, 1954a, 1955a; *Chondriopsis atropurpurea,* Dickie 1874b, Murray 1889; *C. harveyana p. p.,* Mazé and Schramm 1870–77, Murray 1889.

Chondria tenuissima (Goodenough and Woodward) C. Agardh

Plants bushy, 10–20 cm. tall, pale straw-colored or dull purple; main branches coarse, firm, the lesser segments rather soft, the main axis simple or with a few similar main branches, 1.0–2.5 mm. diam.; primary axes bearing numerous long, widely divergent, secondary axes 5–8 cm. long; ultimate branchlets usually rather distantly placed, spindle-shaped, 2–7(–15) mm. long, 0.25–0.5 mm. diam., tapering to both ends, the apices pointed, bearing evident tufts of trichoblasts; tetrasporangia 40–60 μ diam., scattered in the upper portion of the fruiting branchlets; pericarps short-stalked or subsessile, scattered along the branchlets.

North Carolina, Florida, Bahamas, Cuba, Jamaica, St. Barthélemy, Guadeloupe, Venezuela. Plants of shallow water in rather sheltered situations, chiefly growing on stones or shells.

REFERENCES: Collins 1901, Howe 1920, Taylor 1928, 1933, 1936, 1942, 1954a, Williams 1949a; *Chondriopsis tenuissima* and f. *crassa*, Mazé and Schramm 1870–77, Murray 1889.

Chondria baileyana (Montagne) Harvey

Plants gregarious, 7–10(–25) cm. tall, slender in aspect and soft in texture, pale straw-colored or pale purple; the main axis subsimple or with a few long similar branches, so habit generally narrowly pyramidal, less commonly broader; main axis 0.3–1.0 mm. diam.; branchlets not congested, to about 5 mm. long, 80–200 μ diam., elongate and club-shaped, the distal ends little narrowed, ultimately obtuse; tetrasporangial branchlets, with the sporangia 55–105 μ diam., in a circumscribed band well below the apex, and in the oldest branchlets rather near the center.

North Carolina, Florida, Jamaica, Dominica, Barbados. Plants of shallow water in rather sheltered situations, growing on shells, stones, or coarser plants. Primarily a northern species, and these records are open to suspicion.

REFERENCES: Collins 1901, Vickers 1905; *C. tenuissima* v. *baileyana*, Hoyt 1920.

Chondria cnicophylla (Melvill) De Toni Pl. 74, fig. 3

Plants entangled, at least to 2 dm. diam., dark purplish, but with age decolorate, arising from a discoid holdfast, the main axes indistinct, 0.75–1.25 mm. diam., flexuous, somewhat complanate and repeatedly alternately branched nearly at right angles and at considerable intervals, so that the habit becomes broad and sprawling; branchlets 2–3(–5) mm. long, very numerous along the main and secondary branches, radially disposed, 1–2 together, or on old branches to 6 in a cluster, set obliquely or at right angles, abruptly contracted at the base, somewhat tapered toward the tip but blunt rather than acute; tetrasporangia distributed throughout the fertile branchlets.

Florida, Cuba. Apparently a plant of sheltered locations, often associated with sea grasses in shallow water. This species seems anomalous respecting the position of the growing point. There is no pronounced terminal pit, but the branchlets are certainly obtuse and the young trichoblasts conceal the growing point.

REFERENCES: Taylor 1954a; *Chondriopsis cnicophylla,* Murray 1889.

Chondria leptacremon (Melvill) De Toni

Plants erect, probably several shoots arising from a common base, pale reddish in color, drying and staining the paper brown; 1.5 dm. tall, main axes rather sparingly divided, very slender, 0.22–0.32 mm. diam., the branches erect, abundantly radially alternating, with numerous short subdeterminate branches each bearing about 2–5 branchlets, which in turn may fork once or twice; ultimate branchlets very narrowly clavate, tapering from the blunt apex to the base.

Florida, Bahamas, Cuba.

REFERENCES: Howe 1920, Taylor 1928; *Chondriopsis leptacremon,* Murray 1889.

Chondria sedifolia Harvey

Plants densely bushy, 10–15 cm. tall, dull reddish-purple or somewhat faded, coarse and rather firm in texture; main axis 1.0–2.5 mm. diam., with several large erect-spreading branches, in habit broadly pyramidal; the branchlets single or clustered, much contracted to the base, short pyriform to more elongate and club-shaped, diameter 0.3–0.6 mm., length 1–2(–5) mm., the distal end truncate, with a tuft of trichoblasts; tetrasporangia clustered near the ends of the fertile branchlets; pericarps ovate, subsessile, 1–few on a branchlet or lesser branch.

North Carolina, Florida, Guadeloupe, Barbados. Plants of shallow water growing on rocks and shells in moderately open situations. Primarily a northern species, and these southern records are open to suspicion.

REFERENCES: Harvey 1853, Vickers 1905, Hoyt 1920, Taylor 1928; *Chondriopsis dasyphylla* v. *sedifolia,* Murray 1889; *Laurencia chondriopsides* and *L. thujoides,* Mazé and Schramm 1870–77, Murray 1889.

Chondria dasyphylla (Woodward) C. Agardh

Plants bushy, 10–20 cm. tall, pale straw-colored or light brownish-purple, showing a considerable tendency to stain the paper when mounted and dried; the main axis erect, 1.0–1.5 mm. diam., with several long similar branches, the habit of each broadly pyramidal; branchlets single or clustered, club-shaped, contracted at the base, obtuse and finally retuse at the apex, with a central apiculus bearing a particularly conspicuous tuft of trichoblasts, the older branchlets generally of uneven, somewhat torulose contour, length 2–3(–10) mm., 200–500 μ diam; tetrasporangia formed in the distal portion of the fertile branchlets; spermatangial clusters flat, transversely oval, developed from a portion of a normal trichoblast; pericarps lateral on the branchlets, with a very short stalk, single or 2–3 on one branchlet.

Bermuda, North Carolina, Florida, Bahamas, Jamaica, Hispaniola, Puerto Rico, Virgin Isls., Guadeloupe, Barbados, Grenada, Uruguay. Plants of shallow water in moderately open situations. Essentially a southern plant in America, and northern reports are open to suspicion.

REFERENCES: Harvey 1853, Hauck 1888, Collins 1901, Vickers 1905, Børgesen 1913–20, 1924, Collins and Hervey 1917, Howe 1920, Hoyt 1920, Taylor 1928, 1939a; *Chondriopsis dasyphylla* with f. *gracilis* and *C. harveyana p. p.*, Mazé and Schramm 1870–77, Murray 1889.

Chondria floridana (Collins) Howe Pl. 74, fig. 4

Plants to 1–2 dm. tall, pinkish or yellowish, the color intensified by drying; bushy, the habit pyramidal, the branching regularly and progressively alternate to 4–5 degrees, the main axes 2–3 mm. diam., the ultimate branchlets straight or curved, cylindrical or slightly clavate to the blunt tips, but not contracted at their bases, 0.3–0.5 mm. diam., 5–10 mm. long, without conspicuous trichoblasts; tetrasporangia forming a band about 1–2 mm. below the tips of the fertile branchlets, with a few scattered below the band.

Florida, Jamaica, St. Barthélemy, Guadeloupe, Venezuela, Brazil. Apparently a plant of moderately deep water, as dredged from depths of 2–36 m.

REFERENCES: Taylor 1928, 1930a, 1942, 1954a; *C. dasyphylla* f. *floridana,* Collins 1906; *Laurencia flexuosa, L. virgata* and ?v. *denudata,* Mazé and Schramm 1870–77, Murray 1889.

Chondria polyrhiza Collins and Hervey

Plants very small, slender, color pale red, substance rather firm; dense fascicles of short, unicellular rhizoids issuing from all portions of the plant; the main axes seldom reaching 0.5 mm. diam., with more or less numerous alternate, somewhat more slender, rather flexuous spreading branches and 1–2 orders of smaller and shorter branchlets, with more or less contracted bases, tipped with tufts of short fugacious trichoblasts; segments of pericentral cells indistinctly visible in the youngest parts only; cortical cells linear, 3–4 times as long as broad; tetrasporangia in the ultimate branchlets, the fertile portion swollen and subdentate in outline.

Bermuda, Florida, Bahamas, Virgin Isls. From shallow water, and also dredged from a depth of 18 m.

REFERENCES: Collins and Hervey 1917, Børgesen 1913–20, Howe 1918a, 1920, Taylor 1928.

Chondria curvilineata Collins and Hervey

Plants dull red or yellowish, soft in texture, 1–3 cm. tall; very slender throughout, the main axes 0.22–0.35(–0.5) mm. diam., bearing a few alternate branches of two orders, seldom more, which are cylindrical or slightly contracted at base and apex; growing point at the bottom of an apical pit, with a tuft of short trichoblasts; swollen ends of the pericentral cells showing through the cortication as lines curved toward the apex of the branch; cortical cells about twice as long as broad; pericarps sessile on the branches.

Bermuda, Bahamas, Hispaniola.

REFERENCES: Collins and Hervey 1917, Howe 1918a, 1920, Børgesen 1924.

Chondria collinsiana Howe

Plants solitary or gregarious, erect, 3–8 cm. tall, yellowish or pinkish, soft in texture; branching virgate or somewhat paniculate, the well-defined main axes 0.4–0.75 mm. diam.; subcorticated, and pericentral cells with conspicuous curved thickenings of their anterior cell walls visible through the surface layer; surface cells 26–40 μ broad, 65–160 μ long; ultimate branchlets solitary and rather distant, fusiform-obovoid to nearly cylindrical, moderately attenuate at the base and obtuse or subtruncate at the apex with conspicuous trichoblasts which become brown on drying, the growing point concealed in the apical pit; branchlets 2–5 diameters long, about 0.28–0.45 mm. diam., 0.75–4.5 mm. long; tetrasporangia distributed generally, or confined to the apical portion of the fruiting branchlets.

Florida, Bahamas. Growing on various coarse algae and phanerogams in shallow water.

REFERENCE: Howe 1920.

UNCERTAIN RECORDS

Chondria pumila Vickers

Barbados, the type locality. De Toni 6:383; Vickers 1905.

Chondria tenera (Crouan) De Toni

Guadeloupe, the type locality; essentially a *nomen nudum*. De Toni 6: 386; *Chondriopsis tenera*, Mazé and Schramm 1870–77, Murray 1889.

Cladhymenia Hooker and Harvey, 1845

Plants forming flat blades which branch from the margins, frequently in a subflagelliform manner; the tips not inrolled, the blades with an obscure costa within which the nominally five pericentral cells are generally obscured by the growth of slender filaments between them; tetrasporangia in slender, terete, marginal pinnules; pericarps lateral on marginal pinnules, sessile, subovate.

Cladhymenia? lanceifolia Taylor

Plants to 10 cm. tall, perhaps more, dark brownish-red when dried, from a sparingly branched, coarsely fibrous holdfast, which gives rise to one or several erect axes; primary axes terete and firm below, forming a stalk 2–5 mm. long, or this obsolescent, above gradually expanded to a lanceolate blade which may reach 8 mm. in width,

5–7 cm. in length, the margin vaguely undulate, the midrib barely distinguishable, the texture membranous; branching from the lower part of the axes flagelliform, the flagellar extensions expanded at the tip and reinforcing the primary holdfast, but branching higher on the blades marginal or submarginal, irregularly disposed, these branches linear to linear-lanceolate, or approaching the size and shape of the primary blades.

Colombia. Dredged from a bottom of lithothamnia at 14–16 m. depth.

REFERENCE: Taylor 1942.

Acanthophora Lamouroux, 1813

Plants erect, bushy, coarse and stiff; alternately branched, the terete branches of indefinite growth simulating the primary axis, or, short and of limited growth, more or less closely beset with short, spirally placed, spinous branchlets; seemingly parenchymatous, but derived from an axis which is polysiphonous near the tips, with five pericentral cells; trichoblasts principally about the branch apices and the reproductive organs; tetrasporangia in the short lateral branches; spermatangial clusters disciform, stalked; pericarps sessile, in the axils or at the bases of spinous branchlets.

KEY TO SPECIES

1. Plants relatively loosely branched, the terminal divisions often long; determinate branches with numerous short spines, but these absent from the indeterminate shoots .... A. spicifera
1. Plants relatively closely and irregularly branched, the terminal divisions crowded; short spines present on the indeterminate branches .... A. muscoides

Acanthophora muscoides (Linnaeus) Bory — Pl. 72, fig. 3

Plants densely bushy, to 6–16 cm. tall or somewhat more, irregularly branched, the main axes seldom to 2.5 mm. diam.; both main and spur branches bearing spirally disposed short spines about 1 mm. long; tetrasporangia in small, very spiny branches; pericarps urceolate, lateral on thick spinelike branchlets, the aperture broad.

Florida, Mexico, Bahamas, Cuba, Jamaica, Hispaniola, Virgin Isls., Guadeloupe, Martinique, Barbados, British Honduras, Costa Rica, Panama, Colombia, Tobago, Brazil. Growing in shallow water, and dredged to a depth of 18 m.; not very common.

REFERENCES: Grunow 1867, 1870, Martens 1870, 1871, Mazé and Schramm 1870–77, Dickie 1874a, Zeller 1876, Murray 1889, Möbius 1890, Vickers 1905, Børgesen 1913–20, Howe 1920, 1928, Taylor 1928, 1929b, 1930a, 1933, 1935, 1941b, 1943; *A. delilei,* Harvey 1853, Murray 1889; *Chondria muscoides,* Montagne 1863.

Acanthophora spicifera (Vahl) Børgesen

Pl. 71, fig. 3; pl. 72, figs. 1, 2

Plants erect, sparingly branched to somewhat bushy, to 25 cm. tall, the ultimate divisions often long, arcuate, except near the base usually abundantly beset with short determinate branches which are markedly spinose, but spines absent elsewhere; diameter of the main axes 2–3 mm.; trichoblasts fugacious, borne on the branch tips and branchlets; tetrasporangia in swollen, very spiny, short branches; spermatangial clusters platelike, each developed on trichoblasts of which the basal cell persists as the stalk of the disk; pericarps on a short, stout stalk, subtended by a small spine, urceolate, the apex rather produced.

Bermuda, Florida, Mexico, Bahamas, Caicos Isls., Cuba, Jamaica, Hispaniola, Puerto Rico, Virgin Isls., St. Barthélemy, Guadeloupe, Dominica, Martinique, Barbados, Grenada, Guatemala, Old Providence I., Costa Rica, Panama, Colombia, Netherlands Antilles, I. las Aves, Venezuela, Trinidad, Tobago, Brazil. Widespread and often very common in shallow water, either in exposed situations or sheltered places with strong currents; frequently heavily epiphytized. Dredged to a depth of 3 m. only.

REFERENCES: Børgesen 1913–20, Collins and Hervey 1917, Howe 1918a, b, 1920, Schmidt 1924, Taylor 1928, 1936, 1939b, 1940, 1941b, 1942, 1943, 1954a, Taylor and Arndt 1929, Sanchez A. 1930, Questel 1942, Quiros C. 1948, Joly 1957; *A. antillarum,* Harvey 1861; *A. intermedia* and *A. thierii* f. *gracilis,* Mazé and Schramm 1870–77, Murray 1889; *A. thierii,* Harvey 1853, Martens 1870, Mazé and Schramm 1870–77, Dickie 1874a, b, Hemsley 1884, Zeller 1876,

Hauck 1888, Murray 1889, Collins 1901, Vickers 1905, Sluiter 1908, Grieve 1909, Schmidt 1923, 1924.

Laurencia Lamouroux, 1813

Plants erect and bushy, of a more or less fleshy consistency and even at times subcartilaginous, with a scutate or fibrous base, the axes terete or compressed, radially or bilaterally branched; growth from an apical cell, which is located in a terminal pit and surrounded by rudimentary trichoblasts; segmentation at first on the plan of an axial siphon surrounded by pericentral cells, but this early obscured and in the mature axes quite unrecognizable; adult axis structure a parenchyma of large colorless cells covered usually with a single layer of small cortical cells with chromatophores; terminal branchlets if sterile usually clavate, with the base hardly constricted, though gradually enlarged upward and the apex locally swollen about the terminal pit; fertile branchlets similar, or reduced and clustered; tetrasporangia usually in a diffuse zone below the tips of the branchlets, lying just under the surface cells, often causing a superficial roughness, in origin not directly derived from the pericentral cells; spermatangial clusters small, ovoid or barrel-shaped, in the expanded apical pit of the fertile branchlets; pericarps partly immersed, the carpostome projecting.

KEY TO SPECIES

1. Thallus broad and flat, the pinnate branches wide .. L. lata, p. 628
1. Branches for the most part terete; where compressed much less than 3 mm. wide 2
2. Branching subcorymbose, chiefly dichotomous below, though in part often alternate above 3
2. Branching chiefly alternate and radial, usually with fairly evident main axes at least in the youngest parts of the plant 4
3. Plants 4–12 cm. tall, subcartilaginous; branches not concrescent, 0.75–2.0 mm. diam. .... L. corallopsis, p. 623
3. Plants 1–2 cm. tall, fragile; the branches commonly concrescent, 0.15–0.45 mm. diam. .... L. nana, p. 622
4. Walls of the axial cells of the main branches with refractive lenticular thickenings; branching narrowly pyramidal or even virgate, particularly near the tips of the main branches; branchlets 0.20–0.45 mm. diam. .. L. microcladia, p. 627

4. Walls of the axial cells not thickened on the ends; branchlets larger 5
5. Plants dull greenish-purple; branchlets short, more or less tubercle-like, closely set on the ultimate branches L. papillosa, p. 623
5. Branchlets of sterile plants not notably crowded 6
6. Plants dull greenish-purple, drying blackish, almost wiry, the base frequently with flagelliform branches, the erect main branches sparingly divided and nearly naked below, but above freely alternately branched, the lateral divisions virgate, with rather short erect branchlets; fertile branchlets in small groups, shortened and crowded .... L. scoparia, p. 625
6. Plants not notably dark and the branches spreading, or if dark generally with compressed branches 7
7. Cortex surface essentially smooth 8
7. Surface cells in the apical parts mammilliform, or occasionally acutely projecting; plants rather divergently branched and sprawling .... L. gemmifera, p. 624
8. Plants soft, repeatedly branched, rather pink in the branch tips, but more often green or yellow in the main axes; ultimate branchlets obovoid to subclavate 9
8. Plants stiff, rather irregularly and sparingly branched below, the main axes evident in the distal portions, generally compressed above; the ultimate branchlets wart- or peglike, on the flattened marginal portions .... L. poitei, p. 625
9. Plants erect, with well-defined leading axes, repeatedly branched, paniculate .... L. obtusa, p. 626
9. Plants decumbent or entangled, without leading axes continuously evident throughout the plant 10
10. Branchlets subcylindrical or a little clavate, long; the main axes fairly discernible in the upper parts of well-developed plants; diameter fairly regular, to about 0.75 mm. in the main branches .... L. intricata, p. 626
10. Branchlets at first cylindrical, becoming fusiform; the axes rather irregular in diameter, but about 0.35 mm. diam. L. chondrioides, p. 627

Laurencia nana Howe

Plants densely caespitose and intricate, rose-red to garnet-brown, mostly 1–2 cm. high, fragile, often innovating from broken ends;

branching dichotomous or subdichotomous, the branches more or less concrescent or coherent, often showing rhizoidal haptera; main segments 0.15–0.45 mm. diam., the ultimate branchlets cylindric or subcylindric, mostly 0.1–0.35 mm. diam., broad at the truncate apex; surface of the cortex smooth or the cells somewhat swollen, firm-walled, rounded-hexagonal and mostly 25–45 μ broad; tetrasporic branches slightly enlarged.

Bahamas, Jamaica. Growing near low-water mark on corals, lithothamnia, etc.

REFERENCE: Howe 1920.

Laurencia corallopsis (Montagne) Howe

Plants 4–12 cm. tall, dull reddish or brownish-purple, fleshy to subcartilaginous, bushy, with subdichotomous branches arising from near the holdfast, above irregularly alternately branched, the branches sometimes arising close together, sometimes more scattered, subcorymbose in aspect, the terminal divisions rather short, erect; branches rather uniformly 0.75–2.0 mm. diam., except for slight swellings at the tips.

Bermuda, Florida, Bahamas, Cuba, Jamaica, Virgin Isls., Guadeloupe, Martinique, Barbados, Venezuela. Probably rather frequent on the reefs in a dwarf form, but more characteristic plants come from pools and deeper water; the species has been dredged from 14 m.

REFERENCES: Howe 1918a, b, 1920, Taylor 1928, 1941b; *L. cervicornis,* Harvey 1853, Mazé and Schramm 1870–77, Murray 1889, Collins 1901, Vickers 1905, Børgesen 1913–20, Collins and Hervey 1917; *Corallopsis sagraeana,* Mazé and Schramm 1870–77, Murray 1889; *Sphaerococcus corallopsis,* Montagne 1863.

Laurencia papillosa (Forsskål) Greville Pl. 74, fig. 2

Plants densely clustered, to 5–8(–16) cm. tall, sometimes more, olive-green or greenish-purple, subcartilaginous, not adherent to paper; the primary axes alternately branched, all leading divisions subsimple, 1–2 mm. diam., or above with rather short, spreading, secondary divisions; naked or nearly so below, but the upper portions and especially the tip and secondary divisions sparingly to very densely covered with short, clavate, truncate branchlets; cortical cells in section radially twice as long as broad or more; tetrasporangia in irregular, lobed branchlets; cystocarps nearly spherical, usually clustered about the fertile branchlets.

Bermuda, Florida, Mexico, Bahamas, Caicos Isls., Cuba, Cayman Isls., Jamaica, Hispaniola, Puerto Rico, Virgin Isls., St. Barthélemy, Guadeloupe, Barbados, Grenada, British Honduras, Costa Rica, Panama, Colombia, Netherlands Antilles, Venezuela, Trinidad, Brazil. Common and widespread throughout most of the range, growing in the intertidal zone, seldom a little deeper, on reefs and rocks exposed to the waves. In extremely exposed situations these plants become greatly dwarfed and simplified in form.

REFERENCES: Greville 1833, Harvey 1853, 1861, Martens 1870, Mazé and Schramm 1870–77, Dickie 1874a, Hemsley 1884, Piccone 1886, 1889, Hauck 1888, Murray 1889, Collins 1901, Sluiter 1908, Børgesen 1913–20, Collins and Hervey 1917, Howe 1918a, 1920, 1928, Taylor 1928, 1929b, 1930a, 1933, 1935, 1940, 1941a, b, 1942, 1943, 1954a, 1955a, Taylor and Arndt 1929, Sanchez A. 1930, Joly 1951; *L. canariensis p. p.* and *L. hybrida p. p.*, Mazé and Schramm 1870–77, Murray 1889; *L. papillosa* v. *subsecunda,* Murray 1889; *L. papillosa* f. *thyrsoidea,* Schmidt 1923, 1924; *L. thyrsoidea,* Martens 1870; *Chondria thyrsoidea,* Montagne 1863.

Laurencia gemmifera Harvey

Plants 10–15 cm. tall, cartilaginous, repeatedly alternately branched, above sometimes slightly compressed and distichous, the last divisions bearing branchlets reduced to tubercles about 2 mm. long; tetrasporangia in much reduced branchlets; surface cells of the main axes about 40–130 μ long, nearly square in section, in the apical regions protuberant, mammillate, or even acute.

Bermuda, North Carolina, Florida, Texas, Mexico, Bahamas, Caicos Isls., Cuba, Jamaica, Hispaniola, Puerto Rico, Virgin Isls., Barbados, Guadeloupe, Panama, Colombia, Trinidad, Brazil. From rocks and reefs in shallow water, and dredged from a depth of 20 m.

REFERENCES: Harvey 1853, 1861, Mazé and Schramm 1870–77, Rein 1873, Dickie 1874b, Hemsley 1884, Howe 1920, Hoyt 1920,

Børgesen 1924, Taylor 1929a, 1935, 1941a, b, 1942, 1954a; *L. tuberculosa* v. *gemmifera,* Murray 1889, Collins 1901.

Laurencia poitei (Lamouroux) Howe

Plants to about 1 dm. tall, sparse to bushy, in color pale buff to pinkish, alternately branched from the main axis, the branches often approximating the primary axis in length; ultimate branchlets short, truncate, not closely packed, opposite or alternate on the last indeterminate branches, or radially placed or marginal along flattened branches, though frequently both conditions on the same plant; branches 1.0–1.5 mm. diam., where flattened reaching a width of 2 mm.; cortical cells neither radially elongate nor projecting when viewed in section.

Bermuda, North Carolina, Florida, Texas, Mexico, Bahamas, Caicos Isls., Cuba, Jamaica, Hispaniola, Puerto Rico, Virgin Isls., Guadeloupe, Dominica, Barbados, Grenada, Colombia, Netherlands Antilles, Venezuela, Brazil. On reefs and rocks in shallow water and not infrequently dredged from depths to 16 m.

REFERENCES: Børgesen 1913–20, Collins and Hervey 1917, Howe 1918a, b, 1920, Schmidt 1923, 1924, Taylor 1928, 1935, 1936, 1941a, b, 1942, 1955a; *L. brasiliana,* Martens 1871, Zeller 1876; *L. mexicana,* Kützing 1865, Hauck 1888; *L. pinnatifida* and *L. tuberculosa,* Harvey 1853, Mazé and Schramm 1870–77, Murray 1889 *p. p.,* Grieve 1909, Hoyt 1920; *Gracilaria poitei,* Harvey 1853, Dickie 1884, Hauck 1888, Murray 1889 *p. p.?,* Vickers 1905.

Laurencia scoparia J. Agardh

Plants arising from a coarsely fibrous base, densely tufted, 4–12 cm. tall, greenish-purple, drying blackish, firmly cartilaginous, not adherent to paper; primary axes numerous, terete, the branches erect throughout, subdichotomous below, above alternate to opposite, paniculate or somewhat virgate on the long upper divisions; surface cells regularly arranged, columnar or, when older, nearly square in section; tetrasporiferous branchlets short, blunt, and densely fasciculate.

?Bermuda, ?North Carolina, Jamaica, Guadeloupe, Barbados, Grenada, Costa Rica, Netherlands Antilles, Venezuela, Trinidad, Brazil.

REFERENCES: Harvey 1853, Mazé and Schramm 1870–77, Martens 1871, Dickie 1874a, Piccone 1886, Murray 1889, Vickers 1905, Schmidt 1923, Taylor 1933, 1942, Williams 1951; *L. scoparia* f. *minor*, Mazé and Schramm 1870–77, Murray 1889.

Laurencia obtusa (Hudson) Lamouroux

Plants to 1.5–2.5 dm. tall, bushy, in color with green or yellow axes and rose branchlets; below showing long main stems which are sparingly alternately branched, 0.75–1.50 mm. diam., but above are increasingly closely paniculately branched and spreading, the smallest branches and the short, truncate, ultimate branchlets opposite or subverticillate, 0.5–0.75 mm. diam.; tetrasporangia in a band below the apex of the hardly-modified branchlets.

Bermuda, Florida, Mexico, Bahamas, Caicos Isls., Cuba, Cayman Isls., Jamaica, Hispaniola, Virgin Isls., St. Barthélemy, Guadeloupe, Barbados, British Honduras, Panama, Colombia, Netherlands Antilles, I. las Aves, Tobago, Brazil. Very widespread and common in sheltered situations throughout most of the range, growing on various objects in shallow water.

REFERENCES: Greville 1833, Hooper 1850, Harvey 1853, Martens 1870, Mazé and Schramm 1870–77, Dickie 1874a, b, Hemsley 1884, Möbius 1889, 1890, Murray 1889, Collins 1901, Børgesen 1913–20, 1924, Collins and Hervey 1917, Howe 1918a, b, 1920, Schmidt 1923, 1924, Taylor 1928, 1935, 1936, 1939b, 1940, 1941b, 1942, 1943, 1954a, Taylor and Arndt 1929, Sanchez A. 1930; *L. caespitosa p. p.*, *L. canariensis* f. *major*, *L. crassifrons* and f. *dendroidea*, *L. dendroidea* vars. *corymbifera*, *denudata*, and *tenuifolia*, and *L. divaricata*, Mazé and Schramm 1870–77, Murray 1889; *L. obtusa* v. *gracilis*, Martens 1870, Collins and Hervey 1917; *L. obtusa* v. *racemosa*, Mazé and Schramm 1870–77, Murray 1889; *L. paniculata*, Collins 1902, 1917 *p. p.*

Laurencia intricata Lamouroux

Plants to 5(–10) cm. tall, matted or somewhat tufted, rose-pink, crisply fleshy; branching alternate, quite irregular below, without recognizable leading axes, but above for the last 2–3 orders of branching the main axes usually distinguishable and in exceptional cases the general habit somewhat paniculate; main branches 0.5–0.75 mm. diam., the rather elongate terminal divisions hardly less, a little clavate in shape.

Bermuda, Florida, Bahamas, Cuba, Cayman Isls., Jamaica, Hispaniola, Virgin Isls., Guadeloupe, Grenada, Costa Rica, Panama. Common, intricate, and irregular in shallow water and exposed situations, more ample and regular in deep water, where this species has been dredged to a depth of 36 m.

REFERENCES: Mazé and Schramm 1870–77, Murray 1889, Howe 1918a, b, 1920, Taylor 1928, 1933, 1936, 1954a, Taylor and Arndt 1929, Quirós C. 1948; *L. implicata,* Harvey 1853, Mazé and Schramm 1870–77, Murray 1889, Collins 1901, Børgesen 1913–20; *Chondria intricata,* Montagne 1842, 1863.

Laurencia chondrioides Børgesen

Plants in tufts to 6 cm. tall, rosy red in color; branching repeatedly radially alternate at very irregular intervals, long and short branches intermixed, the main branches terete, slender, rather irregular in thickness, about 350 μ diam.; lesser branches and branchlets subfusiform, tapering to both ends but especially downward, the summit remaining obtuse; surface cells about 35 μ broad, 140 μ long in the older portions, the axis and pericentral cells visible in young branchlets only; tetrasporangia near the summit of fertile branchlets, rather few, about 80 μ diam.

Virgin Isls. Dredged from 30 m. depth.

REFERENCE: Børgesen 1913–20.

Laurencia microcladia Kützing

Plants to about 5–10 cm. tall, texture firm, almost wiry, often adhering imperfectly to paper; in color generally with greenish pigment in the main stem and secondary axes, the ultimate branchlets more or less rose-pink; usually narrowly pyramidal, densely branched, diameter of the main branches 0.25–1.50 mm., that of the ultimate branchlets 0.20–0.45 mm.; walls of the inner cells of the main axes in fresh material showing numerous refringent rounded thickenings.

Bermuda, Florida, Bahamas, Caicos Isls., Cuba, Cayman Isls., Jamaica, Hispaniola, Virgin Isls., St. Barthélemy, Barbuda, Guadeloupe, Aves I., Costa Rica, Panama, Netherlands Antilles, Venezuela. Frequent in the intertidal zone, particularly on rocks in exposed situations.

REFERENCES: Howe 1918a, 1920, Taylor 1928, 1929b, 1933, 1942, 1954a; *L. caespitosa p. p.* and *L. canariensis p. p.*, Mazé and Schramm 1870–77, Murray 1889; *L. obtusa* v. *gelatinosa*, Mazé and Schramm 1870–77, Murray 1889, Børgesen 1913–20, Collins and Hervey 1917; *L. obtusa* v. *microcladia*, Murray 1889.

Laurencia lata Howe and Taylor

Plants stipitate, complanate, to 15 cm. long, to 4–10 mm. wide, color deep reddish; distichously bi- tripinnate, the pinnae alternate or occasionally subopposite, spreading, rather sparingly redivided, the ultimate pinnules mostly elliptic-oblong, oblong, or subspatulate, 2–15 mm. long, 1.5–3.5 mm. wide, obtuse or emarginate-obtuse, 7–13 mm. distant; frond 175–380 μ thick near the median line, 115–135 μ thick at the margins; superficial cells in mature parts mostly 40–90 μ long.

Brazil. Known from dredged specimens only.

REFERENCES: Taylor 1930a, Howe and Taylor 1931.

UNCERTAIN RECORDS

Laurencia brongniartii J. Agardh

Martinique, the type locality. De Toni 4:805.

Laurencia concinna Montagne

Guadeloupe, Grenada. Type from Torres Strait. De Toni 4:806; Mazé and Schramm 1870–77. Referred by Yamada (1931) to the above species.

Laurencia corymbosa J. Agardh

Guadeloupe. Type from South Africa. De Toni 4:783; Mazé and Schramm 1870–77, Murray 1889.

Laurencia divaricata J. Agardh

Barbados. Type from the Red Sea. De Toni 4:786; Vickers 1905.

Laurencia heteroclada Harvey

Guadeloupe. Type from Australia. De Toni 4:782; ?*L. arbuscula,* Mazé and Schramm 1870–77, Murray 1889.

Laurencia hybrida (De Candolle) Lenormand

Brazil. Type from France. De Toni 4:796; Martens 1870, Zeller 1876, Murray 1889.

Laurencia paniculata J. Agardh

Bermuda, Florida, Guadeloupe, Colombia. Type from the Mediterranean. De Toni 4:788; Harvey 1861, Mazé and Schramm 1870–77, Murray 1889, Collins and Hervey 1917, Taylor 1928.

Laurencia pinnatifida (Gmelin) Lamouroux

Jamaica, Brazil. Type from Europe. De Toni 4:798; Greville 1833, Martens 1870, Murray 1889 *p. p.*

Laurencia vaga Kützing

Guadeloupe. Type from New Caledonia. De Toni 6:374; Mazé and Schramm 1870–77.

DESCRIPTIONS OF NEW TAXA

Caulerpa floridana sp. n. Pl. 11, figs. 5, 6; pl. 18, fig. 10

Plantae stolones parce furcatos, late divergentes, ad 7 dm. long. vel plus, plerumque 2.0–2.5 mm. diam. habentes; stolones ramos crassos descendentes rhizoidea-ferentes intervallis 2–4 cm., et intervallis paululum maioribus ramos foliares ascendentes promentes; rami foliares e stipitibus 1–3 cm. long., in laminas pinnatas ovatas ad oblongas 3–13 cm. long., 1.7–2.7 cm. lat. desinentes constantes, pinnulae confertae, patentes ad modice ascendentes, in basi constrictae atque teretes, supra compressae atque sensim latae factae, plerumque 1.5–2.0 mm. latae, versus cacumen rotundatum atque acute apiculatum non inflatae.

Plantae typicae ex aqua ca. 21 metrorum profunda subductae prope locum White Shoal dictum in I. Dry Tortugas, Florida, legit W. R. Taylor *num. 361*, 17 VII 1924, in herbariis Taylorii et Horti Botanici Neo-Eboracensis (isotypus).

Dictyota ciliolata Kütz. v. **bermudensis** var. n. Pl. 59, fig. 1

Plantae ubique minores, plerumque 5–8 cm. altitudine, arcte subdichotome ad manifeste alterne ramosae, infra obscurae colore, segmentis 1.0–2.0 mm. lat., 3–10 mm. long., marginibus parce minute dentatis.

Plantae typicae in loco dicto Turnup Breaker, I. St. Davids, colonia Bermuda, legit W. R. Taylor et A. J. Bernatowicz *num. 49-753*, 12 IV 1949, in herbariis Taylorii et Universitatis Michiganensis (isotypus).

Dictyota jamaicensis sp. n. Pl. 32, figs. 4, 5

Plantae fruticosissimae, ex haptero parvo fibrato ad 12 cm. altitudine; axes infra paululum stuposi, saepe manifeste torti, intervallis 4–15 mm., plerumque 7 mm., dichotome ramosi, ramis infra latissime patentibus sed in divisionibus summis angulos angustos habentibus; segmenta 1–2 mm. lata, marginibus eroso-dentatis intervallis irregularibus 0.5–5.0 mm. plerumque circa 1.0 mm.; tetrasporangia ad circa 80 μ diam., sparsa aut ad centrum segmentorum parce aggregata; oogonia in soris ad circa 500 μ long., 250 μ lat., per segmenta fertilia sparsa.

Plantae typicae in loco dicto Christofers Cove, Drax Estates, in Paroecia St. Ann, in colonia Jamaicae, legit W. R. Taylor, *num. 56-227*, 20 III 1956, in herbariis Taylorii et Universitatis Michiganensis (isotypus).

Dictyopteris jamaicensis sp. n. Pl. 32, fig. 2

Plantae altitudine usque ad 15 cm. vel plus, axibus inferioribus denudatis atque stirpideis; primum subdichotomae, mox semel ad ter intervallis 1–2 cm. alterne ramosae, laminis ligulatis dilute brunneis, 7–11 mm. lat., marginibus integris ad minute eroso-dentatis, aliquantulum undulatis; membrana in partibus iuvenibus ex uno strato cellularum, in partibus vetustioribus 40–60 μ crass. e duobus stratis constans, cellulis a superficie visis quasi rectangularibus circa 24–33×33–42 μ; costae conspicuae, venis lateralibus marginalibusque absentibus; sori pilorum per laminas sparse; organa reproductiva non visa.

Plantae typicae ex aqua 40 metrorum profunda subductae prope locum Hellshire Hills dictum in Paroecia St. Catherine, in colonia Jamaicae, per W. R. Taylor, *num. 56-227;* 20 III 1956, in herbariis Taylorii, Universitatis Michiganensis (isotypus), necnon Musei Scientiarum (isotypus), Institutum Jamaicensis, in loco Kingston, Jamaica dicto.

Dictyopteris hoytii sp. n. Pl. 32, fig. 1

Plantae fruticosae luteo-brunneae, ad 30 cm. altitudine, ex haptero parvo pulvinato, axibus inferioribus simplicibus ad saepe irregulariter ramosos, per 0.5–6.0 cm. denudatis, stuposis; laminae supra plerumque a bis ad sexies dichotome furcatae, sinibus angustis, apicibus obtusis ad retusos, ligulatae, 1.4–3.0 cm. lat., 90–150 μ crass., undulatae, marginibus rectis aut obscure crenatis, manifeste aculeato-dentatis, dentibus 0.5–1.0 mm. long., intervallis 0.75–2.0 mm.; in structura 2–4 strata medullaria cellularum magnarum atque stratum corticeum cellularum multo minorum utraque in superficie praebentes; costae conspicuae, ad apices ramorum extendentes;

venae laterales delicatae sed manifestae, alternatae aut oppositae intervallis circa 2–3 mm., ad margines oblique extendentes; pili super laminas irregulariter aggregati sparsi; tetrasporangia in soris parvis in lineis ad venas parallelis relata; oogonia 65–130 μ diam., zona angusta marginali excepta, super laminas dispersa.

Plantae typicae ex aqua 25 metrorum profunda subductae prope loco dicto Beaufort, North Carolina, legit Lewis Radcliffe 1914, in herbario Horti Botanici Neo-Eboracensis.

Padina haitiensis Thivy sp. n. Pl. 75, fig. 1

Plantae stipitatae, ex haptero coacto enascentes, 5–6 cm. alt., ad 8 cm. lat., in lobos latos etiam ad stirpem plerumque fissae, versus basim ferrugineae ac stuposae, infra brunneae, in superficie inferiore vix calcifactae lineis pilorum manifestis, in superficie superiore paululum densius calcifactae, zonis inter lineas pilorum 1.5–2.0 mm. lat.; laminae e duobus stratis cellulatum a vicinitate marginalis apicalis quasi ad stirpem constantes, 65 μ crass. in triente basali, 105 μ autem, proxime super stirpem; cellulae 25–38 μ lat., celluli strati inferioris circa 0.66 minus profundae quam strati superioris, sori indusiati in parte media vel superiore cuiusque zonae fertilis siti, 0.4–0.6 mm. lat., sporangia 80–110 μ diam.

Planta typica in loco dicto I. Tortuga, Hispaniola, legit W. L. Schmitt et G. R. Lunz *num. 17,* 21 III 1937, in herbario Taylorii.

Padina perindusiata Thivy sp. n. Pl. 75, fig. 2

Plantae altitudine maiores 10 cm., utrinque paulum calcifactae, lineae piliferae in superficiebus oppositis alternantes, zonis latitudine irregularibus alterne fertilibus 1.5–3.0 mm. lat. et sterilibus 0.75–2.0 mm. lat.; plantae bistratosae, infra 105 μ crass., media in parte 90 μ cellulis 30–50 μ lat., cellularum strato inferiore 0.5–0.33 minus profundo quam strato superiore; sori spongariorum 0.50–0.75 mm. lat. in una linea continua aut in 2–3 lineis interruptis media in parte cuiusque zonae fertilis, per indusium perspicuum praetexti; sporangia 170 μ diam.

Planta typica in loco dicto Southwest Channel, I. Dry Tortugas, Florida, legit W. R. Taylor *num. 1356,* 20 VI 1926, in herbario Taylorii.

Galaxaura obtusata v. **major,** var. n. Pl. 44, fig. 5

Plantae ad 17 cm. altae, rami minus crebri, 3–4 mm. diam., interdum ad furcas articulati, segmentis maturis 3–9 plo longioribus quam latis, infra modice calcifactis, supra plerumque fere sine calce, plantis viventibus rosaceis translucentibus.

Plantae typicae in loco dicto Tuckers Town Bay, I. Hamilton, colonia Bermuda, legit W. R. Taylor et A. J. Bernatowicz *num. 49-288,* 15 III 1949. In herbariis Taylorii et Universitatis Michiganensis (isotypus).

WURDEMANNIACEAE n. fam.

Plantae gregariae tenues rigidaeque, ramis tenuibus; in structura matura filamentosae multiaxialesque, proprietate filamentosa aetate obscuranta, membranis cellularum nisi in centro medullae crassis; reproductio per tetrasporangia zonata in ramis admodum immutatis peripheraliter nata. Genus typicum Wurdemannia Harvey a. 1853, p. 245.

Cryptonemia bengryi n. sp. Pl. 80, fig. 1

Plantae ad 14 cm. alt., fruticosae, rubidae, axis unus vel plures ex haptero parvo lobato enascens, infra teres stirpideus 1.5 cm. et alterne radiatim semel ad ter ramosus, divisionibus minoribus confertis factis, ferentibus laminas stipitatas ecostatas 2–8 cm. long., 10–15 mm. lat., 100–195 μ crass., firme membranaceas, irregulariter lanceolatas ad oblanceolatas, 1–5 lobos alternatos marginibus integris aut prope cacumina paululum erososubdentatis plerumque habentes; cortex e strato cellularum uno constans, cellulae a superficie visae subangulares, 5.0–8.5 μ diam., parietibus satis crassis, in sectione rotundo-rectangulares, circa 8.0–8.5 μ lat., 10.0–11.5 μ alt., subcortex e duobus stratis cellularum transverse ovatarum, strato interiore discontinuo, constans; medulla modice laxa, e filamentis 2.5–6.6 μ diam. constans; ganglia non multa, circa 8.5 μ diam., 6–7 radios habentia, aut filamenta refractiva iunctiones 3–5 filorum sine amplificationibus gangliaribus praebentia; organa reproductiva non visa.

Plantae typicae in loco dicto Alligator Pond Bay, Paroecia Manchester, colonia Jamaicae, legit R. P. Bengry *num. A. 2645,* 3 XI

1957, in herbariis Museum Scientiarum, Institutum Jamaicensis et Taylorii (isotypus).

Rhodymenia pseudopalmata v. **caroliniana** var. n.

Plantae maiores quam typus speciei factae, stirpes brevissimae, laminae patentes complanatae longitudinem usque ad 12 cm. latitudinem usque ad 15 cm. attingentes; usque ad septies dichotome ramosae, segmentis latitudine infra ad 6 mm., supra ad 3.5 mm., vix attenuatis, longitudine inter furcas 8–60 mm., cacuminibus obtusis retusisve; sori tetrasporangiales transverse ovati, proximi cacuminibus segmentorum terminalium paululum spatulatorum.

Plantae typi collectionis ex loco I. Pawley, So. Carolina dicto, anno 1873 per L. R. Gibbes acquisitae, et in herbario Horti Botanici Neo-Eboracensis. *Cfr.* Hoyt 1920, p. 487, pl. CI, figs. 3, 4 (sub nom. *R. palmetta*).

Ceramium uruguayense sp. n. Pl. 65, figs. 3, 4

Plantae ad 4 cm. altitudine, axibus supercurrentibus alterne divisis, ramuli minores progredienter breviores, probabiliter complanati, primum alternantes in divisionibus minoribus deliquescunt; ramelli parvi adventicii saepe adsunt; cacumina ramulorum divergentia acuta 55–58 μ diam.; axes parte in centrali plantae ad 560 μ diam. ad nodos valde protuberantes 300 μ diam., tenuioribus, autem, partibus in inferioribus plantae; nodi parte in centrali per spatium maius quam diametron remoti; corticatio ad nodos parte in superiore plantae plerumque super septum nodale, constans ex ordine cellularum magnarum infra sitarum per cellulas minores irregulariter ovatas partim obtectarum, margine inferiore corticationis subrecto, superiore, autem, minus regulari; tetrasporangia emergentia in ramellis parvis adventiciis necnon parte in superiore nodorum axiliam frequentia; cystocarpi in ramulis superioribus.

Plantae typicae in loco dicto Puerto la Paloma, Uruguay, legit W. L. Schmitt *num. 277*, 6 XII 1925, in herbario Taylorii. *Cfr.* Taylor 1939a, p. 151, pl. 4, fig. 2, pl. 7, figs. 2–4 (sub nom. *C. miniatum*).

Herposiphonia pecten-veneris v. **laxa** n. var.

Plantae partibus in venustioribus affixae, plurim, autem, discretae, rami rosei, axes saepe flavi; axes principales recti aut paululum curvati, alterne ad valde bilateraliter ramosi, ramo primario, e omnibus nodis quartis evolvente, atque serie trium ramulorum plerumque alternantibus, ramuli 35–60 μ diam., 6–7 cellulas pericentrales habentes.

Planta typica in loco Bird Key Reef, Dry Tortugas I., Florida dicto, legit W. R. Taylor *num. 9223*, 16 VI 1924, in herbario Taylorii. *Cfr.* Taylor 1928, p. 177, pl. 25, fig. 12.

BIBLIOGRAPHY[1]

AGARDH, C. A. 1821, 1828. Species algarum rite cognitae, cum synonymis, differentiis specificis et descriptionibus succinctis. I (Fucoideae, Florideae, Ulvoideae): i + 531. 1821. II (Lemanieae, Ectocarpeae, Ceramieae): lxxviii + 189. 1828. Greifswald.

AGARDH, J. G. 1841. In historiam algarum symbolae. Linnaea, 15:1–50, 443–457.

—— 1847. Nya alger från Mexico. Öfvers. af Kongl. Vetensk.-Akad. Förhandl., 4(1): 5–17.

—— 1848–1901. Species, genera et ordines algarum, seu descriptiones succinctae specierum, generum et ordinum, quibus algarum regnum constituitur. I. Species, genera et ordines Fucoidearum . . . , algas fucoideas complectens, 1: i–viii + 1–363. 1848. II. Species, genera et ordines Floridearum . . . , 2(1): i–xii + 1–351, 1851; *ibid.*, 2(2): 337–720, 1852; *ibid.*, 2(3): i–iv + 701–1291, 1863. III[i]. Epicrisis systematis Floridearum, 3(1): i–vii + 1–724, 1876. III[ii]. Morphologia Floridearum, 3(2): i–iii + 1–301, 1880. III[iii]. De dispositione Delesseriearum, mantissa algologica, 3(3): i–vi + 1–239, 1898. III[iv]. De Florideis mantissa collectanea, tum de speciebus novis aut aliter interpretandis commentaria, tum indices sistens specierum antea seorsim descriptarum, 3(4): i–iv + 1–149. 1901. Lund.

—— 1872–90. Till algernes systematik, nya bidrag. Första afd. (I. Caulerpa, II. Zonaria, III. Sargassum), Lunds Univ. Årsskr., 9(8): 1–71, 1872; Andra afd. (IV. Chordarieae, V. Dictyoteae), *ibid.*, 17(4): 1–134+2, 3 pls., 1881 (1882); Tredje afd. (VI. Ulvaceae), *ibid.*, 19(2): 1–177+4, 4 pls., 1883; Fjerde afd. (VII. Florideae), *ibid.*, 21(8): 1–117+3, 1 pl., 1885; Femte afd. (VIII. Siphoneae), *ibid.*, 23(2): 1–174+6, 3 pls., 1887; Sjette afd. (IX. Sporochnoideae, X. Fucaceae, XI. Florideae), *ibid.*, 26(3): 1–125, 3 pls., 1890.

—— 1879. Florideernes morphologi. K. Svenska Vetensk.-Akad. Handl., 15(6): 1–199, 33 pls.

[1] Although publications which came to the attention of the author after June 1957 have, where possible, been included in this bibliography, it has generally not been possible to refer to them throughout the systematic portion of the text.

AGARDH, J. G. 1892–99. Analecta algologica: observationes de speciebus minus cognitis earumque dispositione. Lunds Univ. Årsskr., 28(6): 1–182, pls. 1–3, 1892; Continuatio I, *ibid.*, 29(9): 1–144, pls. 1, 2, 1894a; Continuatio II, *ibid.*, 30(7): 1–98, pl. 1, 1894b; Continuatio III, *ibid.*, 32(2): 1–140, pl. 1, 1896; Continuatio IV, *ibid.*, 33(9): 1–106, pls. 1, 2, 1897; Continuatio V, *ibid.*, 35(II,4): 1–160, pls. 1–3, 1899.

ANON. 1885. St. Thomas, Virgin Islands. Rep. Sci. Res. Expl. Voyage of H.M.S. "Challenger." Narrative of the Cruise. 1(1): 126–130.

ARESCHOUG, J. E. 1855. Phyceae novae et minus cognitae in maribus extra-Europaeis collectae, quas descriptionibus observationibusque illustravit. Nova Acta Reg. Soc. Sci. Upsaliensis, iii, 1: 329–372, 1854.

ASHMEAD, S. 1857, 1858. Marine algae collected by Mr. S. Ashmead at Key West, Florida . . . Proc. Acad. Nat. Sci. Phila., 8: Donations . . . V–VI, 1857 (1856); Marine algae from Key West, *ibid.*, 9: 74, 1858 (1857).

BAILEY, J. W. 1847–48. Notes on the algae of the United States. Amer. Journ. Sci. and Arts, ii, 3: 80–85, 399–403, 1847; *ibid.*, 6: 37–42, 1848.

BARTON, E. S. 1891. A systematic and structural account of the genus *Turbinaria* Lamouroux. Trans. Linn. Soc. London, ii, 3: 215–226, pls. 54, 55.

—— 1901. The genus Halimeda. Siboga Expeditie. Uitkomsten op zoologisch, botanisch, oceanographisch en geologisch gebied verzameld in Nederlandsch Oost-Indië aan boord H. M. "Siboga" onder commando van Luitenant ter Zee 1[e] kl. G. F. Tydeman, 60: 1–32, pls. 1–4. Leiden.

BIGELOW, H. B. 1905. The shoal-water deposits of the Bermuda Banks. Proc. Amer. Acad. Arts and Sci., 40: 559–592.

BLINKS, L. R. 1927. On Valonia and Halicystis in eastern America. Science, 65: 429–430.

—— AND BLINKS, A. H. 1931. Two genera of algae new to Bermuda. Bull. Torrey Bot. Club, 57: 389–396, fig. 18, pls. 22, 23 (1930).

BLOMQUIST, H. L. 1954. A new species of *Myriotrichia* Harvey from the coast of North Carolina. Journ. Elisha Mitchell Sci. Soc., 70: 37–41, pl. 6.

BLOMQUIST, H. L. 1955. *Acinetospora* Bornet new to North America. *Ibid.*, 71(1): 46–49, 10 figs.

—— 1958a. *Myriotrichia scutata* Blomq. conspecific with *Ectocarpus subcorymbosus* Holden in Collins. *Ibid.*, 74(1): 24.

—— 1958b. The taxonomy and chromatophores of *Pylaiella antillarum* (Grun.) De T. *Ibid.*, 74(4): 25–30, 16 figs.

—— AND HUMM, H. L. 1946. Some marine algae new to Beaufort, North Carolina. *Ibid.*, 62(1): 1–8, pls. 1–3.

—— AND PYRON, J. H. 1943. Drifting "seaweed" at Beaufort, North Carolina. Amer. Journ. Bot., 30(1): 28–32, 2 figs.

BØRGESEN, F. 1900a, b. A contribution to the knowledge of the marine alga vegetation on the coasts of the Danish West-Indian Islands. Bot. Tidsskr., 23: 49–57, 4 figs. En Bidrag til kundskaben om Algevegetation ved Kysterne af Dansk Vestindien. *Ibid.*, 23: 58–60.

—— 1905. Contributions à la connaissance du genre *Siphonocladus* Schmitz. Overs. Kongel. Danske Vidensk. Selsk. Forhandl., 1905 (3): 259–291, figs. 1–13.

—— 1907. An ecological and systematic account of the Caulerpas of the Danish West Indies. Kongel. Danske Vidensk. Selsk. Skrifter, vii, Naturvid. og Math. Afd., 4(5): 337–392, 31 figs.

—— 1908a. The Dasycladaceae of the Danish West Indies. Bot. Tidsskr., 28: 271–283, 8 figs.

—— 1908b. The species of Avrainvillea hitherto found on the shores of the Danish West Indies. Vidensk. Medd. Naturh. Foren. København, 1908: 27–44, 8 figs., pl. 3.

—— 1909a, 1910. Some new or little known West Indian Florideae. Bot. Tidsskr., 30: 1–19, 1909. *Id.*, II, *ibid.*, 117–207, 1910.

—— 1909b. Notes on the shore vegetation of the Danish West Indian Islands. *Ibid.*, 29: 201–259, 40 figs., pls. 3–6.

—— 1909c. Vegetationen i Dansk Vestindien. Atlanten, 1909: 601–632, figs. 277–300.

—— 1911a. The algal vegetation of the lagoons in the Danish West Indies. Biol. Arbejd. tilegnede E. Warming, pp. 1–56, figs. 1–9.

Børgesen, F. 1911b, 1912a. Some Chlorophyceae from the Danish West Indies. Bot. Tidsskr., 31: 127–152, figs. 1–13, 1911. *Id.*, II, *ibid.*, 32: 241–273, figs. 1–17, 1912.

—— 1912b. Two crustaceous brown algae from the Danish West Indies. Nuova Not., 23: 123–129, 3 figs.

—— 1913–1920. The marine algae of the Danish West Indies. I. Chlorophyceae. Dansk Bot. Arkiv, 1(4): 1–158 + 2, 126 figs., map, 1913. *Id.*, II. Phaeophyceae, *ibid.*, 2(2): 1–66 + 2, figs. 1–44, 1914a. *Id.*, III, Rhodophyceae, a, *ibid.*, 3: 1–80, figs. 1–86, 1915. *Id.*, b, *ibid.*, 3: 81–144, figs. 87–148, 1916. *Id.*, c, *ibid.*, 3: 145–240, figs. 149–230, 1917. *Id.*, d, *ibid.*, 3: 241–304, figs. 231–307, 1918. *Id.*, e, *ibid.*, 3: 305–368, figs. 308–360, 1919. *Id.*, f, *ibid.*, 3: 369–504, figs. 361–435, 1920. *Reprinted as* Vol. 1(1), Chlorophyceae, pp. 4 + 1–158, figs. 1–126; vol. 1(2), Phaeophyceae, pp. 6 + 159–228, figs. 127–170, 1913–1914; vol. 2, Rhodophyceae, pp. 2 + 1–504, figs. 1–435 + map, 1915–1920. Copenhagen.

—— 1914b. The species of Sargassum found along the coasts of the Danish West Indies, with remarks upon the floating forms of the Sargasso Sea. Mindeskr. i Anledning af Hundredaaret af Japetus Steenstrups Fødsel., 2(32): 1–20, 8 figs.

—— 1924. Marine algae, pp. 14–35, 17 text figs. *In* Ostenfeld, C. H., Botanical results of the Dana-Expedition, 1. Plants from Beata Island, St. Domingo, collected by C. H. Ostenfeld. Dansk Bot. Arkiv, 4(7): 1–36.

—— 1925–1930. Marine algae from the Canary Islands, especially from Teneriffe and Gran Canaria. I. Chlorophyceae. Kongel. Danske Vidensk. Selsk., Biol. Medd., 5(3): 1–123, 49 figs., 1925. *Id.*, II. Phaeophyceae, *ibid.*, 6(2): 1–112, 37 figs., 1926. *Id.*, III. Rhodophyceae 1, Bangiales and Nemalionales, *ibid.*, 6(6): 1–97, 47 figs., 1927. *Id.* 2, Cryptonemiales, Gigartinales and Rhodymeniales (Mélobésiées by Mme P. Lemoine), *ibid.*, 8(1): 1–97, 31 figs., 4 pls., 1929. *Id.* 3, Ceramiales, *ibid.*, 9(1): 1–159, 60 figs., 1930.

—— 1932. A revision of Forsskål's algae mentioned in *Flora Aegyptico-Arabica* and found in his herbarium in the Botanical Museum of the University of Copenhagen. Dansk Bot. Arkiv, 8(2): 1–15, 4 text figs., 1 pl.

—— 1949. On the genus *Titanophora* (J. Ag.) Feldm. and description of new species. Dansk Bot. Arkiv, 13(4): 1–8, 3 figs., 2 pls.

Børgesen, F., and Paulsen, O. 1900. La végétation des Antilles Danoises. Rev. Gén. de Bot., 12: 99–107, figs. 7–12; 138–153, figs. 13–44; 224–245, figs. 48–93; 289–297, figs. 94–105; 344–354; 434–446, figs. 152–154; 480–508, figs. 166–208, pls. 4–14.

Bory de St. Vincent, J. B. 1827. Botanique. Agamie. Hydrophites, pp. 62–177, pls. 1–24. *In* Duperry, L. I., Voyage autour du monde sur "La Coquille" pendant les années 1822, 1823, 1824 et 1825. 1827–1829. Paris.

Bowman, H. H. M. 1918. Botanical ecology of the Dry Tortugas. Publ. no. 252; Papers, Dept. Marine Biol., Carnegie Inst. Washington, 12(5): 1–138, pls. 1–6.

Brieger, F. G. 1946. O desenvolvimento de novas "Rhodophyceae" do Brasil. Anais da Primera Reunião Sul-Americana de Botânica, Rio de Janeiro 1938, 2: 11–17, pls. A–E.

Browne, P. 1756, 1789. Algae, pp. 71–74. *In* The civil and natural history of Jamaica. Part I, An accurate description of that island . . . , 2 maps, pp. [4+] i–viii, 1–27[+1]. Part II, pp. [4+] xxix–xxxiv; Book I, An account of the fossils. . . . , pp. 35–66; Book II, History of the vegetable productions . . . , pp. lxviii–lxx, 71–374, pls. 1–38; Book III, Account of such animals . . . , pp. ccclxxiv–ccclxxix[+1], 381–503[+1], pls. 39–49. [Part III not published.] 1756, London. *Id.*, 2nd edition, paged as before, *with:* Four additional indexes. . . . , 46 unnumbered pages. 1789. London.

Butler, E. 1902. Botanizing in Jamaica. Postelsia, 1: 87–131, pls. 10–13.

Chapman, V. J. 1946. Algal zonation in the West Indies. Ecology, 27(1): 91–93.

Chou, R. C.-Y. 1945, 1947. Pacific species of Galaxaura, I. Asexual types. Papers Michigan Acad. Sci., Arts and Lett., 30: 35–56, 2 figs., 11 pls. (1944). *Id.*, II. Sexual types, *ibid.*, 31: 3–24, 3 figs., 13 pls. (1945).

Collins, F. S. 1901. The algae of Jamaica. Proc. Amer. Acad. Arts and Sci., 37: 229–270.

—— 1902. Marine algae, pp. 12–14. *In* Northrop, A. R., Flora of New Providence and Andros . . . Mem. Torrey Bot. Club, 12: 1–98, map, 19 pls.

—— 1906. New species in the Phycotheca. Rhodora, 8: 104–113.

Collins, F. S. 1909b–1918a. The green algae of North America. Tufts Coll. Stud., 2(3): 79–480, pls. 1–18, 1909. First Suppl., *ibid.*, 3(2): 91–109, pls. 1, 2, 1912. Second Suppl., *ibid.*, 4(7): 1–106, pls. 1–3, 1918.

—— 1911. Notes on algae, X. Rhodora, 13: 184–187.

—— and Hervey, A. B. 1917. The algae of Bermuda. Proc. Amer. Acad. Arts and Sci., 53: 1–195, pls. 1–6.

—— Holden, I., and Setchell, W. A. 1895–1919. Phycotheca Boreali-Americana. (Exsiccata). Fascicles 1–46 + A–E. Malden, Mass.

—— and Howe, M. A. 1916. Notes on species of Halymenia. Bull. Torrey Bot. Club, 43: 169–182.

Cox, J. D. 1901. Algae, pp. 142–148. *In* Mohr, C., Plant life of Alabama. Contr. U. S. Nat. Herb., 6: 1–921.

Curtis, M. A. 1867. Algae, pp. 155–156. *In:* Geological and natural history survey of North Carolina, III, Botany; containing a catalogue of the indigenous and naturalized plants of the state. Raleigh, N. C.

Curtiss, A. H. 1899. Mrs. Floretta A. Curtiss, a biographical sketch by her son. 14 + iv pp., 3 figs., 4 pls. Jacksonville, Fla.

Dawson, E. Y. 1953. On the occurrence of Gracilariopsis in the Atlantic and Caribbean. Bull. Torrey Bot. Club, 80(4): 314–316.

Decaisne, J. 1842. Mémoire sur les corallines ou polypiers calcifères. Ann. Sci. Nat., Bot., ii, 18: 96–128.

De Toni, G. B. 1889–1924. Sylloge algarum omnium hucusque cognitarum. 1 (Sylloge Chlorophycearum): 1–12 + i–cxxxix + 1–1315, 1889. 3 (Sylloge Fucoidearum): i–xvi + 1–638, 1895. 4 (Sylloge Floridearum) (1): i–xx + i–lxi + 1–388, 1897; *Id.*, (2): 387–776, 1900; *Id.*, (3): 775–1525, 1903; *Id.*, (4): 1523–1973, 1905. 6 (Sylloge Floridearum 5, Additimenta): i–xi + 1–767, 1924. Padua.

Dickie, G. 1874a. On the marine algae of Barbados. Journ. Linn. Soc., Bot., 14: 146–152, pl. 11. (1875).

—— 1874b. I. On the marine algae of St. Thomas and the Bermudas, and on *Halophila baillonis* Asch., pp. 311–317; X. Enumeration of the algae collected at St. Paul's Rocks by H. N. Moseley, M.A., Naturalist to H.M.S. "Challenger," pp. 355–359; XII. Enumeration of algae from Fernando de Noronha . . . , pp. 363–365; XIV.

DICKIE, G. 1874b. (*Cont.*)
Enumeration of algae from 30 fathoms at Barra Grande, near Pernambuco, Brazil . . . , pp. 375, 376; XV. Enumeration of algae from Bahia . . . , p. 377. *In* Hooker, J. D., Contributions to the botany of the expedition of H.M.S. "Challenger." *Ibid.*, 14:311–390. (1875).

—— 1877. Supplemental notes on algae collected by H. N. Moseley, M.A., of H.M.S. "Challenger," from various localities. *Ibid.*, 15: 486–489.

DOLLEY, C. S. 1889. The botany of the Bahamas. Proc. Acad. Nat. Sci. Phila., 1889:130–135. (1890).

EATON, D. C. 1875. List of the marine algae collected by Dr. Edward Palmer on the coast of Florida and at Nassau, Bahama Islands, March to August, 1874. 6 pp. New Haven, Conn.

EDWALL, G. 1876. Algae (*det.* J. J. Puiggari), pp. 185–190. *In* Indice das plantas do herbario da Comissão Geographica e Geologica de São Paulo. Boletim da Comm. Geogr. e Geol. São Paulo, 11:49–215. (Not seen).

ELLIS, J., AND SOLANDER, D. 1786. The natural history of many curious and uncommon zoophytes collected from various parts of the globe. xii + 208 pp., 63 pls. London.

FARLOW, W. G. 1871. Cuban seaweeds. Amer. Nat., 5:201–209, figs. 46–59.

—— 1876. List of the marine algae of the United States. Rep. U. S. Comm. Fish and Fisheries, Rep. of the Commissioner for 1873–74 and 1874–75:691–718.

—— 1881. The marine algae of New England. Rep. U. S. Comm. Fish and Fisheries for 1879, Appendix A-1:1–210, 1882. [Separate copies of the independently circulated appendix carry the title "Marine Algae of New England and Adjacent Coast" and the date 1881 on the title page, but the completed volume when issued was dated 1882. The work was re-issued with the shorter name on the title page in 1891, with the same pagination.]

—— ANDERSON, C. L., AND EATON, D. C. 1877–1889. Algae exsiccatae Americae-Borealis. Fascicles 1–5. Boston.

FELDMANN, G. 1947. Contribution à l'étude des Céramiacées. Bull. Soc. Bot. de France, 94(5, 6): 176–179, 2 figs.

FELDMANN, J. 1937–1942. Les algues marines de la côte des Albères. I–III, Cyanophycées, Chlorophycées, Phéophycées, Rev. Algol., 9: 141–335, figs. 1–67, pls. 1–10, 1937. IV, Rhodophycées, *ibid.*, 11: 247–330, figs. 1–25, 1939a; *ibid.*, 12: 77–100, figs. 26–34, 1941; *ibid.*, 13 (Trav. Algol., 1.): 29–113, figs. 35–62, pl. 1, 1942.

—— 1939b. Une Nemalionale à carpotetraspores: *Helminthocladia hudsoni* (C. Ag.) J. Ag. Bull. Soc. Hist. Nat. Afrique du Nord, 30: 87–97, figs. 1–4.

—— 1948. La végétation marine des Antilles Françaises. Assoc. Franç. Avancem. Sci., Congr. de la Victoire, 3: 585–586. (1945).

—— 1955. Les plastes des Caulerpa et leur valeur systématique. Rev. Gen. Bot., 62: 422–431, figs. 1, 2.

—— AND FELDMANN (-MAZOYER), G. 1946. A propos d'un recent travail du Prof. H. Kylin sur l'alternance de générations du *Bonnemaisonia asparagoides*. J. Ag. Bull. Soc. Hist. Nat. Afrique du Nord, 37: 35–38.

—— AND LAMI, R. 1935. La végétation marine, pp. 251–260. *In* Stehlé, H. Flore de la Guadeloupe et Dependances, I., Essai d'ecologie et geographie botanique. xiii + 284 pp., illus., map. Basse-Terre.

—— 1936. Sur la végétation de la mangrove à la Guadeloupe. Comptes Rendus Acad. Sci. (Paris), 203(17): 883–885.

—— 1937. Sur la végétation marine de la Guadeloupe. *Ibid.*, 204(3): 186–188.

—— AND MAZOYER, G. 1937. Sur la structure et les affinités du *Ceramium poeppingianum* Grunow (*Reinboldiella poeppingianum* comb. nov.), Rhodophycées. Bull. Soc. Hist. Nat. Afrique du Nord, 28: 213–223, 4 figs.

FELDMANN-MAZOYER, G. 1940. Recherches sur les Céramiacées de la Méditerranée occidentale. pp. 510 + 3, 191 text figs., 4 pls. Algiers.

FOSLIE, M. 1897. On some Lithothamnia. Kongel. Norske Vidensk. Selsk. Skr., 1897(1): 1–20.

—— 1898. Some new or critical Lithothamnia. *Ibid.*, 1898(6): 1–19.

—— 1900a. New or critical calcareous algae. *Ibid.*, 1899(5): 1–34.

Foslie, M. 1900b. Remarks on Melobesieae in Herbarium Crouan. *Ibid.*, 1899(7): 1–16.

—— 1901. New Melobesieae. *Ibid.*, 1900(6): 1–24.

—— 1904. Algologiske notiser I. *Ibid.*, 1904(2): 1–9.

—— 1905a. New Lithothamnieae and systematical remarks. *Ibid.*, 1905(5): 1–9.

—— 1905b. Den botaniske samling. Kongel. Norske Vidensk Selsk. Skr., Åarsber., 1904: 15–18.

—— 1906. Algologiske notiser II. Kongel. Norske Vidensk Selsk. Skr., 1906(2): 1–28.

—— 1907a. Algologiske notiser III. *Ibid.*, 1906(8): 1–34.

—— 1907b. Algologiske notiser IV. *Ibid.*, 1907(6): 1–30.

—— 1908a. Algologiske notiser V. *Ibid.*, 1908(7): 1–20.

—— 1908b. Pliostroma, a new subgenus of Melobesia. *Ibid.*, 1908(11): 1–7.

—— 1908c. Nye kalkalger. *Ibid.*, 1908(12): 1–9.

—— 1909. Algologiska notiser VI. *Ibid.*, 1909(2): 1–63.

——† 1929. Contributions to a monograph of the Lithothamnia. H. Printz, editor. 60 pp., 75 pls. Quarto. Trondheim.

—— and Howe, M. A. 1906a. New American coralline algae. Bull. New York Bot. Gard., 4(13): 128–136, pls. 80–93.

—— 1906b. Two new coralline algae from Culebra, Porto Rico. Bull. Torrey Bot. Club, 33: 577–580, pls. 23–26.

Funk, G. 1923. Über einige Ceramiaceen aus dem Golf von Neapel. Beih. zum Bot. Centralbl., 39(II,2): 223–247, pl. 5.

—— 1927. Die Algenvegetation des Golfes von Neapel. Pubb. Staz. Zool. Napoli, 7(Supplement): 1–507, 50 figs., 20 pls.

Gardiner, T., Brace, L. J. K., and Dolley, C. S. 1890. Provisional list of the plants of the Bahama Islands. Proc. Phila. Acad. Nat. Sci., 1889: 349–407.

Gepp, A., and Gepp, E. S. 1905a. Notes on Penicillus and Rhipocephalus. Journ. Bot., Brit. and For., 43: 1–5, pl. 468.

GEPP, A., AND GEPP, E. S. 1905b. Atlantic algae of the "Scotia." *Ibid.*, 43: 109, 110.

—— 1911. The Codiaceae of the Siboga Expedition, including a monograph of the Flabellarieae and the Udoteae. Siboga Expeditie . . . , 62: 1–150, 22 pls.

—— 1912. Marine algae of the Scottish National Antarctic Expedition. Rep. Sci. Res., Voyage of S.Y. "Scotia" during the years 1902, 1903 and 1904. 3: 73–88, pls. 1, 2.

GÓMEZ DE LA MAZA, MANUEL 1887. Flora de Cuba. Doctoral thesis. 53 + 2 pp. Havana.

—— 1889. Ensayo de farmacofitologica Cubana. 7 pp. Havana.

—— 1893. Nociones de botánica sistemática. 96 pp. Havana.

—— 1906. Formas interresantes del reino vegetal. Revista de la Faculdad de Letras y Ciencias, Universidad de Habana, 1: 128–138. (1905).

GRAY, J. E. 1866. On Anadyomene and Microdictyon, with the description of three allied genera, discovered by Menzies in the Gulf of Mexico. Journ. Bot., Brit. and For., 4: 41–51; 65–72, pl. 44.

GREVILLE, R. K. 1833a. Algae, vol. 2, pp. 423, 424, 436, 447–450. *In* A. de Saint Hilaire, Voyage dans le district des diamans et sur le littoral du Brésil . . . Vol. 1, pp. xx + 398 + 1; vol. 2, pp. 1–456. Paris.

—— 1833b. Algae, pp. 448–450. *In* Dunal, F., *Review of* A. de Saint Hilaire, Voyage dans le district des diamans et sur le littoral du Brésil. Arch. de Bot., 2: 444–456.

GRIEVE, S. 1909. Note on some sea-weeds from the Island of Dominica, British West Indies. Trans. Bot. Soc. Edinburgh, 24(1): 7–12.

GRUNOW, A. 1867. Algae. Reise der Österreichischen Fregatte "Novara" um die Erde in den Jahren 1857–1858–1859. Bot. Theil, 1: 1–104. Vienna.

—— 1915, 1916. Additamenta ad cognitionem Sargassorum. Verhandl. d. K. K. Zool.-Bot. Gesellsch., 65: 329–448, 1915. *Id., ibid.*, 66: 1–48, 136–185, 1916.

HAMEL, G. 1929. Contribution à la flore algologique des Antilles. Ann. Crypt. Exotique, 2: 53–58, 9 figs.

HAMEL, G. 1931–39. Phaeophycées de France: pp. 1–80 (Ectocarpacées), 1931; pp. 81–176 (Myrionematacées-Spermatochnacées), 1935; pp. 177–240 (Spermatochnacées-Sphacelariacées), 1937; pp. 241–336 (Sphacelariacées-Dictyotacées), 1938; pp. 337–432, I–XLVII (Dictyotacées-Sargassacées), 1939. Paris.

—— AND HAMEL-JOUKOV, A. 1931. Algues des Antilles Françaises. (Exsiccata). Fasc. 1–3. Paris.

——†, AND LEMOINE, MME P. 1953. Corallinacées de France et d'Afrique du Nord. Arch. du Mus. Nat. d'Hist. Nat., vii, 1: 15–136, 83 figs., frontisp., 23 pls.

HARVEY, W. H. 1852–1858a. Nereis Boreali-Americana. I, Melanospermae. Smithsonian Contrib. to Knowledge, 3(4): 1–150, pls. 1–12, 1852; II, Rhodospermae, *ibid.*, 5(5): 1–258, pls. 13–36, 1853. III, Chlorospermae, including supplements, *ibid.*, 10: ii + 1–140, pls. 37–50, 1958a.

—— 1861. Algae; with notes by Arthur Schott. Report of the Secretary of War communicating Lieutenant Michlin's report of his survey for an inter-oceanic ship canal near the Isthmus of Darien. 36th Congress, 2nd Session, Senate (U. S. A.) Exec. Doc. 9, IX, Appendix B, Botany, pp. 175–178.

HAUCK, F. 1888. Meeresalgen von Puerto-Rico. Engler's Bot. Jahrb., 9: 457–470.

—— AND RICHTER, P. 1886–1889. Phykotheka universalis. (Exsiccata). Fasc. 1–3. Trieste and Leipzig.

HEILPRIN, A. 1890. The corals and coral reefs of the western waters of the Gulf of Mexico. Proc. Acad. Nat. Sci. Phila., 1890: 303–316, maps, pls. 6, 7.

HEMSLEY, W. B. 1884. Algae, i, ii, pp. 104–128. *In* Report on the botany of the Bermudas and various other islands of the Atlantic and southern oceans. Rep. Sci. Res. Exploring Voyage of H.M.S. "Challenger," 1873–76. Botany I, 2: 1–135, 13 pls.; *ibid.*, 3: 1–299, 53 pls. (1885).

HOHENACKER, R. T. 1852–1862. Algae marinae exsiccatae: Eine Sammlung europäischer und ausländischer Meeresalgen . . . Fasc. 1–12. Esslingen n. Kirchheim.

HOLLENBERG, G. J. 1958. Phycological notes II. Bull. Torrey Bot. Club, 85(1): 63–69, 2 figs.

Hooper, J. 1850. Introduction to algology, with a catalogue of American algae, or seaweeds, according to the latest classification of Prof. Harvey. 34 pp. Brooklyn.

Howe, M. A. 1903a. Report on a trip to Florida. Journ. New York Bot. Gard., 4:44–49, figs. 2–5.

—— 1903b. The museum exhibit of seaweeds. *Ibid.*, 4:56–63, figs. 9–12.

—— 1903c. Report on a trip to Porto Rico. *Ibid.*, 4:171–176, figs. 18–21.

—— 1904a. Collections of marine algae from Florida and the Bahamas. *Ibid.*, 5:164–166.

—— 1904b. Notes on Bahaman algae. Bull. Torrey Bot. Club, 31:93–100, pl. 6.

—— 1905a. Phycological studies—I. New Chlorophyceae from Florida and the Bahamas. *Ibid.*, 32:241–252, pls. 11–15.

—— 1905b. Phycological studies—II. New Chlorophyceae, new Rhodophyceae, and miscellaneous notes. *Ibid.*, 32:563–586, pls. 23–29.

—— 1907a. Report on a visit to Jamaica for collecting marine algae. Journ. New York Bot. Gard., 8:51–60, figs. 9–14.

—— 1907b. Phycological Studies—III. Further notes on Halimeda and Avrainvillea. Bull. Torrey Bot. Club, 34:491–516, pls. 25–30.

—— 1909a. Report on an expedition to Jamaica, Cuba and the Florida Keys. Journ. New York Bot. Gard., 10:115–118.

—— 1909b. Phycological Studies—IV. The genus Neomeris and notes on other Siphonales. Bull. Torrey Bot. Club, 36:75–104,pls. 1–8.

—— 1910. Report on a botanical visit to the Isthmus of Panama. Journ. New York Bot. Gard., 11:30–44, figs. 7–15.

—— 1915. Report on a visit to Porto Rico for collecting marine algae. *Ibid.*, 16:219–225.

—— 1917. A note on the structural dimorphism of sexual and tetrasporic plants of *Galaxaura obtusata*. Bull. Torrey Bot. Club, 43: 621–624. (1916).

—— 1918a. Algae, pp. 489–540. *In* Britton, N. L., Flora of Bermuda. ix + 585 pp., illus. New York.

—— 1918b. The marine algae and marine spermatophytes of the Tomas Barrera Expedition to Cuba. Smithsonian Misc. Coll., 68(11): 1–13.

Howe, M. A. 1918c. Further notes on the structural dimorphism of sexual and tetrasporic plants in the genus Galaxaura. Brooklyn Bot. Gard. Mem., 1: 191–197, figs. 1–4, pls. 3, 4.

—— 1919. On some fossil and recent Lithothamnieae of the Panama Canal Zone. Bull. United States Nat. Mus., 103: 1–13, pls. 1–8.

—— 1920. Algae, pp. 553–618. *In* Britton, N. L., and Millspaugh, C. F., The Bahama flora, vii + 695 pp. New York.

—— 1924. Notes on algae of Bermuda and the Bahamas. Bull. Torrey Bot. Club, 51: 351–359, 14 figs.

—— 1928. Notes on some marine algae from Brazil and Barbados. Journ. Washington Acad. Sci., 18(7): 186–194, 2 figs.

—— 1931. Notes on the algae of Uruguay. Bull. Torrey Bot. Club, 57: 605–610. 1 pl.

—— and Taylor, Wm. Randolph. 1931. Notes on new or little known marine algae from Brazil. Brittonia, 1: 7–33, 16 figs., pls. 1, 2.

Hoyt, W. D. 1920. Marine algae of Beaufort, North Carolina. Bull. Bur. Fisheries [U. S.], 36: 367–556, 3 maps, 47 figs., pls. 84–109. (1921).

—— 1927. The periodic fruiting of Dictyota and its relation to the environment. Amer. Journ. Bot., 14: 592–619.

Humm, H. J. 1952a. Notes on the marine algae of Florida. 1. The intertidal rocks at Marineland. Florida State Univ. Stud., 7: 17–23.

—— 1952b. Marine algae from Campeche Banks. *Ibid.*, 7: 27.

—— and Caylor, R. L. 1957. The summer marine flora of Mississippi Sound. Publ. Inst. Marine Sci., 4(2): 228–264, 1 fig., 9 pls.

Ives, J. E. 1890. Echinoderms from the northern coast of Yucatan and the harbor of Vera Cruz. Proc. Acad. Nat. Sci. Phila., 1890: 317–340.

Jennings, O. E. 1917. A contribution to the botany of the Isle of Pines, Cuba, . . . Ann. Carnegie Mus., 11: 19–290, pls. 5–28.

Joly, A. B. 1950. Resultados científicos do cruziero do "Bapendi" e do "Vega" à Ilha da Trinidade. Bol. Inst. Paulista de Oceanogr., 1(2): 73–75.

—— 1951. Contribuição para o conhecimento da flora algológica marinha do estado do Paraná. *Ibid.*, 2(1): 125–138, 3 figs.

Joly, A. B. 1953a. Re-discovery of *Mesogloia brasiliensis* Montagne. Bol. Inst. Oceanogr., 3(1,2): 39–47. (1952).

—— 1953b. An approach to the bibliography of Brazilian algae. *Ibid.*, 3(1,2): 101–113.

—— 1953c. Scientific results of the "Baependi" and "Vega" cruise to the Trinidade Island. *Ibid.*, 4(1,2): 147–156.

—— 1953d. Considerações sôbre a flora algológica marinha da Ilha da Trinidade. Anais IV Cong. Nacional Soc. Bot. Brasil, 1953: 41–43.

—— 1954. The genus *Bostrychia* Montagne, 1838, in northern Brazil. Taxonomic and ecological data. Bol. Fac. Filos., Ciências e Letras da Univ. de São Paulo, 173(Bot. 11): 53–74, 4 pls.

—— 1956a. Additions to the marine flora of Brazil. I. *Ibid.*, 209(Bot. 13): 1–24, 3 pls.

—— 1956b. The sexual female plants of *Griffithsia tenuis* C. Agardh. *Ibid.*, 209(Bot. 13): 25–31 + 1, 1 pl.

—— 1957. Contribuição ao conhecimento da flore ficológica marinha da Baía de Santos e Arredores. Thesis. 196 pp., map, 19 pls. São Paulo.

Kemp, A. F. 1857. Notes on the Bermudas and their natural history, with special reference to their marine algae. Canadian Nat. and Geol., 2: 145–156.

Kützing, F. T. 1845–1871. Tabulae phycologicae, oder Abbildungen der Tange. 1: vi + 54 pp., 100 pls., 1845–1849. 2: i + 37 pp., 100 pls., 1850–1852. 3: 28 pp., 100 pls., 1853. 4: xvi + 23 pp., 100 pls., 1854. 5: ii + 30 pp., 100 pls., 1855. 6: iv + 35 pp., 100 pls., 1856. 7: ii + 40 pp., 100 pls., 1857. 8: ii + 48 pp., 100 pls., 1858. 9: viii + 42 pp., 100 pls., 1859. 10: iv + 39 pp., 100 pls., 1860. 11: ii + 32 pp., 100 pls., 1861. 12: iv + 30 pp., 100 pls., 1862. 13: i + 31 pp., 100 pls., 1863. 14: i + 35 pp., 100 pls., 1864. 15: i + 36 pp., 100 pls., 1865. 16: i + 35 pp., 100 pls., 1866. 17: i + 30 pp., 100 pls., 1867. 18: i + 35 pp., 100 pls., 1868. 19: iv + 36 pp., 100 pls., 1869. 20: i + 57 pp. (Index). 1871. Nordhausen.

—— 1847. Diagnosen und Bermerkungen zu neuen oder Kritischer Algen. Bot. Zeit., 5: 161–167.

—— 1849. Species algarum. vi + 922 pp. Leipzig.

KYLIN, H. 1931. Die Florideenordnung Rhodymeniales. Lunds Univ. Årsskr., N.F., Avd. 2, 27(11): 1–48, 20 pls.

—— 1932. Die Florideenordnung Gigartinales. *Ibid.*, 28(8): 1–88, 28 pls.

—— 1949. Die Chlorophyceen der schwedischen Westküste. *Ibid.*, 45(4): 1–79.

LEMOINE, MME P. 1911. Structure anatomique des Melobesiées. Application à la classification. Ann. Inst. Oceanogr. [Monaco], 2(2): 1–213, 105 figs., 5 pls.

—— 1915. Calcareous algae. Rep. of the Danish Oceanogr. Exped. 1908–10 to the Mediterranean and adjacent seas, 2(Biol.)K–1: 1–30, 6 figs., 1 pl., 4 maps.

—— 1917. Corallinaceae, Melobesieae, pp. 147–182. *In* Børgesen, F., 1913–20, *q. v.*

—— 1924. Melobesieae, p. 36. *In* Ostenfeld, C. H., Botanical results of the Dana Expedition, I. Plants from Beata Island, St. Domingo, collected by C. H. Ostenfeld. Dansk Bot. Arkiv, 4(7): 1–36.

LEWIS, I. F., AND BRYAN, H. F. 1941. A new protophyte from the Dry Tortugas. Amer. Journ. Bot., 28: 343–348, 58 + 5 figs.

LUNAN, J. 1814. Hortus Jamaicensis . . . Seaweeds, vol. 2, pp. 157–160. Vol. 1, viii + 538 pp.; vol. 2, 402 + 2 pp. St. Jago de la Vega [Spanish Town].

MANZA, A. V. 1940. A revision of the genera of articulated corallines. Philippine Journ. Sci., 71: 239–316, 20 pls.

MARTENS, G. 1870. Conspectus algarum Brasiliae haectenus detectarum. Vidensk. Medd. Natur-hist. Foren. Kjøbenh., iii, 2(18–20): 297–314.

—— 1871. Algae brasilienses circa Rio de Janeiro a cl. A. Glaziou, horti publici directore, botanico indefesso, annis 1869 et 1870 collectae. *Ibid.*, iii, 3(8–10): 144–148.

MARTIUS, K.-F. P. VON 1828–1834. Icones selectae . . . , Algae, pp. 5–8, pls. 1–5. *In* Icones plantarum Cryptogamicarum quas in itinere per Brasiliam annis 1817–1820 . . . collegit et descripsit. Pp. ii + 138, 76 pls. Folio. Munich. [Differences in the typesetting, the legends on the plates and other details indicate that two issues of the algal portion of this work exist.]

Martius, K.-F. P. von, Eschweiler, F. G., and Nees von Esenbeck, C. C. 1833. Algae, Lichenes, Hepaticae exposueront, pp. 1–50. *In* Martius, K.-F. P. von, Flora Brasiliensis seu enumeratio plantarum in Brasilia tam sua sponte quam accedente cultura provenientium . . . , Pars prior. iv + 390 pp. Folio. Stuttgart.

Mason, L. R. 1953. The crustaceous coralline algae of the Pacific coast of the United States, Canada and Alaska. Univ. Calif. Publ. Bot., 26(4): 313–390, pls. 27–46.

Mattos, A. de 1952. Notas sôbre algae do litoral Paranaense. Arquivos do Museu Paranaense, 9: 245–260, pls. 16–36, map.

M., H. [=H. Mazé] 1868. Hydrophytes de la Guyane Française. Nomenclature générale des plantes cellulaires de la Guyane recueillies par divers de 1835 à 1868. iii + 41 pp. Basse-Terre, Guadeloupe.

Mazé, H., and Schramm, A. 1870–77. Essai de classification des algues de la Guadeloupe. [Designated the second edition, but actually the third edition of this work, originally issued with A. Schramm as the senior author.] xix + 283 + iii pp. Basse-Terre.

Melvill, J. C. 1875. Notes on the marine algae of South Carolina and Florida. Journ. Bot., Brit. and For., 4: 258–265.

Mitchell, M. O. 1893. On the structure of *Hydroclathrus* Bory. Phycol. Mem., 2: 53–57, pls. 14, 15.

Möbius, M. 1889. Bearbeitung der von H. Schenck in Brasilien gesammelten Algen. Hedwigia, 28: 309–347, pls. 10, 11. (1889–1890).

—— 1890. Algae Brasilienses a cl. Dr. Glaziou collectae. Notarisia, 5: 1065–1090.

—— 1892. Über einige brasilianische Algen. Ber. d. d. Bot. Gesellsch., 10: 17–26, pl. 1.

—— 1895. Über einige brasilianische Algen. Hedwigia, 34: 173–180, pl. 2.

Montagne, J. F. C. 1834. Description de plusieurs nouvelles espèces de Cryptogames découvertes par M. Gaudichaud dans l'Amérique Méridionale. Ann. Sci. Nat., Bot., ii, 2: 73–79, pl. 4.

—— 1837. Centurie de plantes cellulaires exotiques nouvelles. *Ibid.*, ii, 8: 345–370.

MONTAGNE, J. F. C. 1839a. Cryptogamae Brasilienses seu plantae cellulares quas in itinere per Brasiliam à celeb. Auguste de Saint Hilaire collectas recensuit observationibusque nonnulis illustravit. *Ibid.*, ii, 12: 42–44, pl. 1.

—— 1839b, 1847 (Atlas). Botanique, Cryptogamie. II, Florula Boliviensis stirpes novae et minus cognitae. Algae, pp. 13–39, pls. 1–7. *In* d'Orbigny, A., Voyage dans l'Amérique Méridionale. 7(2): 1–119.

—— 1840. Seconde centurie de plantes cellulaires exotiques nouvelles. Décades I, II. Ann. Sci. Nat., Bot., ii, 13: 193–207.

—— 1842. Algae, pp. 1–104, pls. 1–5. *In* Ramon de la Sagra, Histoire physique, politique et naturelle de l'Ile de Cuba. Botanique—plantes cellulaires. x + 549 pp. (1838–1842). Paris. [The 12 octavo volumes or fascicles of text and 10 folio fascicles of plates may be found variously bound together, and only the title pages of the first two bear volume numbers.]

—— 1843. Quatrième centurie de plantes cellulaires exotiques nouvelles. Décade VII. Ann. Sci. Nat., Bot., ii, 20: 249–306.

—— 1846. Cryptogames cellulaires, pp. 1–112, pls. 141–145. *In* Gaudichaud, C. B., Histoire naturelle, botanique. Voyage autour du monde . . . sur la corvette "La Bonité." xi + 335 pp. Paris.

—— 1849. Sixième centurie de plantes cellulaires nouvelles, tant indigènes qu'exotiques. Décades III à VI. Ann. Sci. Nat., Bot., iii, 11: 33–66.

—— 1850. Cryptogamia Guayanensis, seu plantarum cellularium in Guayana Gallica annis 1835–1849 a cl. Leprieur collectarum enumeratio universalis. Note sur la station insolité de quelques floridées dans les eaux douces et courantes des ruisseux des montagnes de la Guayane. *Ibid.*, iii, 14: 283–309.

—— 1853. Criptogamia ó plantas celulares, Algae. Vol. 1, pp. 7–69 and Atlas, vol. 5, pls. 1–5. *In* Ramon de la Sagra, Flora Cubana, ó descripcion botanica usos y aplicaciones de la plantas reunidas en la Isla de Cuba. (Algae also listed in vol. 1, Introduccion, p. 46, separately paged). 4 vols. Folio. Paris.

—— 1857–1859. Huitième centurie de plantes cellulaires nouvelles, tant indigènes qu'exotiques. Décades IV, V. Ann. Sci. Nat., Bot., iv, 7: 134–153, 1857. *Id.*, Décades VI, VII, *ibid.*, 9: 53–68, 1858. *Id.*, Décade VIII, *ibid.*, 9: 142–163, 1858. *Id.*, Décades IX, X, *ibid.*, 12: 167–192, 1859.

MONTAGNE, J. F. C. 1860. Neuvième centurie de plantes cellulaires nouvelles, tant indigènes qu'exotiques. Décades I, II. *Ibid.*, iv, 14: 167–185, pls. 10, 11.

—— 1863. Algae, pp. 46, 62, 63 and index. *In* Ramon de la Sagra, Icones plantarum in flora Cubana descriptarum ex historia physica, politica et naturali. ii + 64 pp., pls. 1–20 + 1–89 (actually 102). Paris.

MURRAY, G. 1887. [*Valonia ovalis*]. Journ. Bot., Brit. and For., 25: 379–380.

—— 1888, 1889. Catalogue of the marine algae of the West Indian region. *Ibid.*, 26: 193–196, 1888; *Id.*, *ibid.*, 27: 237–242, 257–262, 298–305, 1889. Repaged, bound, and reissued, 46 pp., pls. 284, 288. 1889. London. References in this book derive from the reissue.

—— 1891a. Algae, pp. 75–80. *In* Ridley, H. N., Notes on the botany of Fernando Noronha. Journ. Linn. Soc., Bot., 27: 1–95, pls. 1–4.

—— 1891b. On a new species of Caulerpa, with observations on the position of that genus. Trans. Linn. Soc. London, Bot., ii, 3: 207–213, pls. 52, 53.

—— 1891c. The distribution of marine algae in space and time. Proc. and Trans. Liverpool Biol. Soc., 5: 164–180.

—— 1892. On the structure of *Dictyosphaeria* Decne. Phycol. Mem., 2: 16–20, pl. 6.

—— 1893a. On Halicystis and Valonia. *Ibid.*, 2: 47–52, pl. 13.

—— 1893b. A comparison of the marine floras of the warm Atlantic, Indian Ocean, and the Cape of Good Hope. *Ibid.*, 2: 65–70.

OLIVEIRA, L. P. H. DE 1948. Distribução geográfica da fauna e flora da Baía de Guanabara. Mem. Inst. Oswaldo Cruz, 45: 709–734, 1 pl.

"O. S." [OTTYS SANDERS]. 1931. Texas marine algae. Bio-Log (The Southwestern Biological Supply Co., Dallas, Texas), 1: 26.

PAPENFUSS, G. F. 1943. Notes on algal nomenclature. II, *Gymnosorus* J. Agardh. Amer. Journ. Bot., 30: 463–468, 14 figs.

—— 1945a. Review of the Acrochaetium-Rhodochorton complex of the red algae. Univ. Calif. Publ. Bot., 18(14): 299–334.

—— 1945b. Further contributions toward an understanding of the Acrochaetium-Rhodochorton complex of the red algae. *Ibid.*, 18(19): 433–447.

Papenfuss, G. F. 1955. Classification of the algae, pp. 115–224. *In* A century of progress in the natural sciences, 1853–1953. 807 pp. California Academy of Sciences, San Francisco.

Parr, A. E. 1939. Quantitative observations on the pelagic Sargassum vegetation of the western North Atlantic. Bull. Bingham Oceanogr. Coll., 6(7): 1–94. 33 figs.

Pearse, A. S., and Williams, L. G. 1951. The biota of the reefs off the Carolinas. Journ. Elisha Mitchell Sci. Soc., 67(1): 133–161, 5 figs.

Petersen, H. E. 1918. Algae (excluding calcareous algae). Rep. of the Danish Oceanogr. Exped. 1908–10 to the Mediterranean and adjacent seas. 2(Biol.) K—3: 1–20, 10 figs.

—— *See* Børgesen, F., 1915–20, 1924.

Phycotheca Boreali-Americana. *See* Collins, Holden, and Setchell.

Piccone, A. 1885. Notizie preliminari intorno alle alghe della "Vettor Pisani" raccolte dal Sig. C. Maracci. Nuovo Giorn. Bot. Ital., 17(3): 185–188.

—— 1886. Alghe del viaggio di circumnavigazione della "Vettor Pisani." 97 pp., 2 pls. Genoa.

—— 1889a. Nuove alghe del viaggio di circumnavigazione della "Vettor Pisani." Atti R. Accad. Lincei, Mem. Cl. Sci. Fisiche, Mat. e Nat., iv, 6: 10–63.

—— 1889b. Alcune specie di alghe del Mar di Sargasso. *Ibid.*, iv, 6: 79–86.

Post, E. 1936. Systematische und pflanzengeographische Notizen zur Bostrychia-Caloglossa-Assoziation. Rev. Algol., 9: 1–84, 4 figs.

—— 1937–1955b. Weitere Daten zur Verbreitung des Bostrychietum. Hedwigia, 77: 11–14. 1937. *Id.*, II, *ibid.*, 78: 202–215, 1938. *Id.*, III. Arch. Protistenk., 93: 6–37, pls. 1, 2, 1939. *Id.*, IV, *ibid.*, 100: 351–377, pls. 11–15, 1955a. *Id.*, V. Ber. d. d. Bot. Gesellsch., 68(5, 6): 205–216, 3 figs, 1955b.

Prat, H. 1934. Comparison bionomique entre les rivages de l'estuaire du Saint-Laurent et ceux des îles Bermudes. Proc. and Trans. Roy. Soc. Canada, iii, 28(5): 25–28.

—— 1935a. Remarques sur la faune et la flore associées aux Sargasses flottantes. Le Nat. Canad., 62: 120–129, 3 figs.

PRAT, H. 1935b. Les Bermudes, bases d'explorations dans les jardins de corail. La Terre et la Vie, 5: 17–26, 6 figs.

—— 1935c. Notes botanique sur l'archipel des Bermudes. Bull. Soc. Bot. de France, 82: 162–168, 2 figs.

—— 1936. Notes sur les atolls et les récifs coralliens des Bermudes. Comptes Rendus Somm. Séances Soc. Biogeogr., 3: 13–14.

PREDA, A. 1908, 1909. Florideae. *In* Flora Italica Cryptogama. Pars II: Alghe, vol. 1(2): 1–358, figs. 1–111, 1908; *id.*, 1(3): 359–462 + 1, figs. 112–130, 1909. Rocca S. Casciano, Italy.

PRINTZ, H. 1927. Chlorophyceae (nebst Conjugatae, Heterocontae und Charophyta). *In* Engler, A., and Prantl, K., Die natürlichen Pflanzenfamilien, 2te Aufl., Bd. 3, iv + 463 pp., 366 figs. Leipzig.

QUESTEL, A. 1942. Marine Algae, p. 69. *In* The flora of the Island of St. Bartholomew and its origin. vii + 224 pp., frontisp., maps. (1941). Basse-Terre, Guadeloupe.

—— 1951. Algae, pp. 193–208. *In* La flore de la Guadeloupe. Géographie générale de la Guadeloupe et dépendences, II et III. Géographie biologique, I.—La Flore. 327 pp., 8 pls., 117 figs., 2 maps. Basse-Terre.

QUIRÓS CALVO, M. 1948. Flora de Costa Rica: Algas Marinas. Farmacia, 2(1): 14.

REED, C. T. 1941. Marine life in Texas waters. Texas Acad. Publ. Nat. Hist., xii + 88 pp., 22 figs., diagr. and maps.

REIN, J. J. 1873. Über die Vegetations-Verhältnisse der Bermuda-Inseln. Bericht ü. d. Senkenb. Naturf. Gesellsch. in Frankfurt am Main, 1872–1873: 131–153.

RICHARDS, H. M. 1890. Notes on *Zonaria variegata*, Lam'x. Proc. Amer. Acad. Arts and Sci., 25: 83–92, 10 figs.

—— 1901. *Ceramothamnion codii,* a new rhodophyceous alga. Bull. Torrey Bot. Club, 28(5): 257–265, pls. 21, 22.

RUSSELL, J. L. 1871. The seaweeds at home and abroad. Amer. Nat., 4: 274–297, figs. 69–75. (1870).

SANCHEZ Y ALFONSO, M. 1930. Las algas de la Habana. Mem. Trab. Real. Inst. Nac. Invest. Cient. y Mus. Hist. Nat. Habana, 1: 35–44.

SAUVAGEAU, C. 1900–1914. Remarques sur les Sphacelariacées. Journ. de Bot., 14:213–234, 247–259, 304–322, 1900. *Ibid.*, 15:22–36, 50–62, 94–116, 137–149, 222–255, 368–380, 408–418, 1901. *Ibid.*, 16:325–349, 379–416, 1902. *Ibid.*, 17:45–56, 69–95, 332–353, 378–422, 1903. *Ibid.*, 18:88–104, 1904. *Id.*, continuation independently publ., pp. 349–634, 1914. 128 figs. Reprinted and repaged as Fasc. 1: pp. 1–320 from Journ. de Bot., 14–17, 1903; Fasc. 2: 321–480, *ibid.*, 17, 18 *p. p.*, 1904; Fasc. 3: 481–634, 1914. 128 figs.

SCHMIDT, O. C. 1923. Marine Algae, vol. 2, pp. 10, 11, vol. 3, pp. 229–231. *In* Leutzelberg, P. von, Estudo botanico do Nordéste. Ministerium da Viação e Obras Públicus. Inspec. Fed. de Obras Contra as Seccas, Viação e Obras Públicas 57, ser. 1A, 3 vols. Rio de Janeiro. 1922, 1923.

—— 1923a. Beiträge zur Kenntnis der Gattung Codium Stackh. Biblio. Bot., 91: i + 68 pp., 44 text figs.

—— 1924. Meeresalgen der Sammlung von Leutzelburg aus Brasilien. Hedwigia, 65: 85–100.

SCHRAMM, A., AND MAZÉ, H. 1865. Essai de classification des algues de la Guadeloupe. (1st ed., printed). ii + 52 pp. Basse-Terre, Guadeloupe. [New species listed on the Crouans' authority in this and the 1866 edition may be located by the references in the present text to the 1870–1877 edition of Mazé and Schramm].

—— 1866. Essai de classification des algues de la Guadeloupe. (2nd ed., lithographed). iii + 144 pp. Cayenne, French Guiana.

SETCHELL, W. A. 1914. The Scinaia assemblage. Univ. Calif. Publ. Bot., 6(5): 79–152, pls. 10–16.

—— 1925. Notes on Microdictyon. *Ibid.*, 13: 101–107.

—— 1929. The genus Microdictyon. *Ibid.*, 14: 453–588, 105 figs.

—— 1933. Some early algal confusions. II. *Ibid.*, 17(9): 187–254, pls. 26–45.

—— AND MASON, L. R. 1943. Goniolithon and Neogoniolithon. Two genera of crustaceous coralline algae. Proc. Nat. Acad. Sci., 29(3, 4): 87–92.

SILVA, P. C. 1955. The identity of *Hydrodictium marinum* Bory. Rev. Algol., n.s., 1(4): 179–180, 1 fig.

—— 1960. Codium (Chlorophyta) in the tropical western Atlantic. Nova Hedwigia, 1(3, 4): (In press).

SLOANE, H. 1696a. An account of four sorts of strange beans, frequently cast ashore on the Orkney Islands, with some conjectures on the manner of their being brought thither from Jamaica. Philos. Trans. Roy. Soc., 19: 398[=298]-300. 1695–1697.

—— 1696b. Plantae submarinae, pp. 1–7. *In* Catalogus plantarum, quae in insula Jamaica sponte proveniunt vel vulgo coluntur cum earundem synonymis et locis natalibus; adjectis aliis quibusdam, quae in insulis Maderae, Barbados, Nieves, et St. Christophori nascuntur, seu prodromi historiae naturalis Jamaicae. Pp. 232 + 43. London.

—— 1707–1725. Submarine Plants, vol. 1, pp. 49–64. *In* A voyage to the islands Madera, Barbados, Nieves, St. Christophers and Jamaica, with the natural history of the herbs and trees, four-footed beasts, fishes, birds, insects, reptiles, etc., of the last of these islands. Vol. 1, clvi + 264 pp., 156 + 4 pls., 1707. Vol. 2, xviii + 449 pp., 157–274 +11 pls. Folio. London.

SLUITER, C. P. 1908. List of algae collected by the Fishery Inspection at Curaçao. Rec. Trav. Bot. Néerl., 4: 231–241, pl. 8.

SOLMS-LAUBACH, H. VON 1895. Monograph of the Acetabularieae. Trans. Linn. Soc., Bot., ii, 5: 1–39, pls. 1–4.

STEPHENSON, T. A., AND STEPHENSON, A. 1950. Life between tide marks in North America. I. The Florida Keys. Journ. Ecol., 38(2): 354–402, 10 figs., pls. 9–15.

—— 1954. The Bermuda Islands. Endeavor, 13(50): 72–80, 14 figs.

SVEDELIUS, N. 1906. On likheten mellan Västindiens samt Indiska och Stilla Oceanens marina vegetation. Bot. Notiser, 1906: 49–57.

TANDY, G. 1936. Algae, pp. 107–110. *In* Rendle, A. B., Notes on the flora of the Bermudas. Journ. Bot., Brit. and For., 74: 42–50, 65–71, 101–112.

—— AND COLMAN, J. 1931. Superficial structure of coral reefs; animal and plant successions on prepared substrata. Carnegie Inst. Washington, Yearb., 30: 371–378.

TAYLOR, WM. RANDOLPH 1924–1926. Report on the marine algae of the Dry Tortugas. Carnegie Inst. Washington, Yearb., 23: 206–207, 1924; Second report . . . *Ibid.*, 24: 239–240, 1925; Third report . . . *Ibid.*, 25: 255–257, 1926.

TAYLOR, WM. RANDOLPH. 1926. The marine algae of the Dry Tortugas. Rev. Algol., 2: 113–135.

—— 1928. The marine algae of Florida, with special reference to the Dry Tortugas. Carnegie Inst. Wash., Publ. 379; Papers from the Tortugas Lab., 25: v + 219, 3 figs., 37 pls.

—— 1929a. Notes on the marine algae of Florida. Bull. Torrey Bot. Club, 56: 199–210, 2 figs.

—— 1929b. Notes on algae from the tropical Atlantic Ocean [I]. Amer. Journ. Bot., 16: 621–630, 13 figs., pl. 62.

—— 1930a. Algae collected on the Hassler, Albatross and Schmitt Expeditions: I. Marine algae from Brazil. *Ibid.*, 17: 627–634, 1 fig., pl. 39.

—— 1930b. Note on marine algae from São Paulo, Brazil. *Ibid.*, 17: 635, 1 fig.

—— 1931. A synopsis of the marine algae of Brazil. Rev. Algol., 5: 279–313.

—— 1933. Notes on algae from the tropical Atlantic Ocean, II. Papers Mich. Acad. Sci., Arts and Lett., 17: 395–407, pl. 26.

—— 1935. Marine algae from the Yucatan Peninsula. Carnegie Inst. Wash., Publ. 461: 115–124. (1936).

—— 1936. Notes on algae from the tropical Atlantic Ocean, III. Papers Michigan Acad. . . . , 21: 199–207. (1935).

—— 1937. Marine algae of the northeastern coast of North America. Univ. Michigan Stud., Sci. Ser. 13. vii + 427 pp., 60 pls. Ann Arbor.

—— 1939a. Algae collected by the Hassler, Albatross and Schmitt expeditions, II. Marine algae from Uruguay, Argentina, the Falkland Islands and the Strait of Magellan. Papers Michigan Acad. . . . , 24(1): 127–164, 7 pls. (1938).

—— 1939b. Algae collected on the Presidential Cruise of 1938. Smithsonian Misc. Coll., 98(9): 1–18, 14 figs., 2 pls.

—— 1940. Marine algae of the Smithsonian-Hartford Expedition to the West Indies, 1937. Contr. U. S. Nat. Herb., 28: 549–562, pl. 20.

—— 1941a. Notes on the marine algae of Texas. Papers Michigan Acad. . . . , 26: 69–79. (1940).

TAYLOR, WM. RANDOLPH 1941b. Tropical marine algae of the Arthur Schott Herbarium. Field Mus. Nat. Hist., Publ. 509, Bot. Ser., 20(4): 87–104, 2 pls.

—— 1942. Caribbean marine algae of the Allan Hancock Expedition, 1939. Rep. Allan Hancock Atlantic Exped., 2. 193 pp., 20 pls. Los Angeles.

—— 1943. Marine algae from Haiti collected by H. H. Bartlett in 1941. Papers Michigan Acad. . . . , 28: 143–163, 4 pls. (1942).

—— 1944, 1945a. The collecting of seaweeds and freshwater algae. Instructions to naturalists in the armed forces for botanical field work. Suppl. to Company D Newsletter, I. 17 pp. 1944. *Id.*, 2nd ed., *ibid.*, 18 pp. 1945.

—— 1945b. William Gilson Farlow, promoter of phycological research in America, 1844–1919. Farlowia, 2(1): 53–70.

—— 1945c. Pacific marine algae of the Allan Hancock Expeditions to the Galapagos Islands. Allan Hancock Pacific Exped., 12. iv + 528 pp., 3 figs., 100 pls. Los Angeles.

—— 1950a. Plants of Bikini and other northern Marshall Islands. Univ. Michigan Stud., Sci. Ser. 18. xv + 227 pp., frontisp., 79 pls. Ann Arbor.

—— 1950b. Field preservation and shipping of biological specimens. Turtox News (General Biological Supply Co., Chicago), 28(2): 42.

—— 1950c. Reproduction of *Dudresnaya crassa* Howe. Biol. Bull., 99(2): 272–284, 52 figs.

—— 1951a. Structure and taxonomic status of *Trichogloea herveyi.* Hydrobiol., 3(2): 113–121, figs. A, B, 1–10.

—— 1951b. Structure and reproduction of *Chrysophaeum lewisii. Ibid.*, 3(2): 122–130, 26 figs.

—— 1952a. The algal genus Chrysophaeum. Bull. Torrey Bot. Club, 79(1): 79.

—— 1952b. Reproduction of *Acrosymphyton caribaeum.* Papers Michigan Acad. . . . , 36: 31–37, 3 pls. (1950).

—— 1952c. Survey of the marine algae of Bermuda. Year Book Amer. Philos. Soc., 1951: 167–171.

TAYLOR, WM. RANDOLPH 1954a. Distribution of marine algae in the Gulf of Mexico. Papers Michigan Acad. . . . , 39: 85–109. (1953).

—— 1954b. Marine algal flora of the Caribbean and its extension in neighboring seas. Huitième Congr. Internat. Bot., Paris 1954, Rapp. et Comm., Sect. 17: 149–150.

—— 1954c. Character of the marine algal vegetation of the Gulf of Mexico, pp. 177–192, figs. 48–50. *In* Galtsoff, P. S., editor, Gulf of Mexico, its origin, waters and marine life. U. S. Dept. Inter., Fish and Wildlife Serv., Fishery Bull., 55(89): xiv + 604 pp., illus.

—— 1955a. Notes on algae from the tropical Atlantic Ocean, IV. Papers Michigan Acad. . . . , 40: 67–76, 8 figs., 5 pls. (1954).

—— 1955b. Marine algal flora of the Caribbean and its extension into neighboring seas, pp. 259–270, 8 figs. *In* Essays in the natural sciences in honor of Captain Allan Hancock . . . xii + 345 pp., Univ. of So. Calif. Press, Los Angeles.

—— 1957. Marine algae of the northeastern coast of North America. Univ. Michigan Stud., Sci. Ser. 13 (2nd ed.). ix + 509 pp., 60 pls. Ann Arbor.

—— AND ARNDT, C. H. 1929. The marine algae of the southeastern peninsula of Hispaniola. Amer. Journ. Bot., 16: 651–662, 10 figs.

—— AND BERNATOWICZ, A. J. 1952. Bermudian marine Vaucherias of the section Piloboloideae. Papers Michigan Acad. . . . , 37: 75–85, 3 pls. (1951).

—— 1953. Marine species of Vaucheria at Bermuda. Bull. Mar. Sci., Gulf and Carib., 2(2): 405–413, 2 pls.

—— JOLY, A. B., AND BERNATOWICZ, A. J. 1953. The relation of *Dichotomosiphon pusillus* to the algal genus Boodleopsis. Papers Michigan Acad. . . . , 38: 97–107, 3 pls. (1952).

THOMPSON, E. I. 1910. The morphology of Taenioma. Bull. Torrey Bot. Club, 37: 97–106, pls. 9, 10.

TSENG, C. K. 1944. Notes on the algal genus Taenioma. Madroño, 7(7): 215–226, 1 fig., pl. 25.

TURNER, D. 1808–1819. Fuci, sive plantarum fucorum generi a botanicis ascriptarum icones descriptiones et historia. Vol. 1, pp. 164 + 2, 71 pls., 1808; vol. 2, pp. 162 + 2, 63 pls., 1809; vol. 3, pp. 148 + 2, 62 pls., 1811; vol. 4, pp. 153 + 7, 62 pls., 1819.

VAHL, M. 1802. Endeel kryptogamiske planter fra St. Croix. Skrifter af Naturhistorie-Selskabet [Copenhagen], 5(2): 29–47.

VICKERS, A. 1905. Liste des algues marines de la Barbade. Ann. Sci. Nat., Bot., ix, 1: 45–66.

——†. 1908. Phycologia Barbadensis. Iconographie des algues marines récoltées à l'île Barbade (Antilles): (Chlorophycées et Phéophycées). Text by M. H. Shaw, i–ix. Part I, Chlorophyceae, pp. 1–30, 53 pls. + 14b, 43b, 44b, 44c. *Id.*, II, Phaeophyceae, pp. 31–44, 34 pls. + 6b, 24b. Paris.

WEBER-VAN BOSSE, A. 1898. Monographie des Caulerpes. Ann. Jard. Bot. Buitenzorg, 15: 243–401, pls. 20–34.

—— 1899. Note sur quelques algues rapportées par le yacht "Chalazie." Journ. de Bot., 13: 133–135.

—— 1913–1923. Liste des algues du Siboga. I. Myxophyceae, Chlorophyceae, Phaeophyceae [with Th. Reinbold], Siboga Expeditie . . . , 59a: 1–186, figs. 1–52, pls. 1–5, 1913; *Id.*, II. Rhodophyceae i, *ibid.*, 59b: ii + 187–310, figs. 53–109, pls. 6–8, 1921; *Id.*, ii, *ibid.*, 59c: 311–392, figs. 110–142, pls. 9, 10, 1923. Leiden.

—— AND FOSLIE, M. 1904. The Corallinaceae of the Siboga-Expedition. Siboga Expeditie . . . , 61: 1–110, 34 figs., pls. 1–16. Leiden.

—— 1916–1917. Rhizophyllidaceae, Squamariaceae, pp. 128–146. *In* Børgesen, F. 1913–20, *q. v.*

WILLIAMS, L. G. 1948. The genus Codium in North Carolina. Journ. Elisha Mitchell Sci. Soc., 64(1): 107–116, 3 figs., pl. 9.

—— 1949a. Seasonal alternation of marine floras at Cape Lookout, North Carolina. Amer. Journ. Bot., 35(10): 683–695, 20 figs.

—— 1949b. Marine algal ecology at Cape Lookout, North Carolina. Furman Stud., Bull. Furman Univ., 31(5): 1–21, 1 fig.

—— 1951. Algae of the Black Rocks, pp. 149–159. *In* Pearse, A. S., and Williams, L. G., The biota of the reefs off the Carolinas. Journ. Elisha Mitchell Sci. Soc., 67(1): 133–161, 5 figs.

—— AND BLOMQUIST, H. L. 1947. A collection of marine algae from Brazil. Bull. Torrey Bot. Club, 74(5): 383–397, 3 figs.

WINGE, Ö. 1923. The Sargasso Sea, its boundaries and vegetation. Rep. of the Danish Oceanogr. Exped. 1908–10 to the Mediterranean and adjacent seas, 3 (Misc.): 1–34, 17 figs.

YAMADA, Y. 1931. Notes on Laurencia, with special reference to the Japanese species. Univ. Calif. Publ. Bot., 16(7) : 185–310, 20 figs., pls. 1–20.

ZEH, W. 1912. Neue Arten der Gattung Liagora. Notizbl. des K. bot. Gart. u. Mus. zu Berlin, 5: 268–273.

ZELLER, G. 1876. Algae Brazilienses circa Rio de Janeiro a Dr. Glaziou, horti publici directore, collectae. *In* Warming, E., editor. Symbolae ad floram Brasiliae centralis cognoscendam, XXII. Vidensk. Medd. Naturh. Foren. København, 1875: 426–432.

PLATES AND DESCRIPTIONS

Many of these plates were drawn by Olivia Embrey, Janet Roemhild, Barbara Gagnon, Wm. L. Brudon, or E. A. Salgado; some of these they have signed. Others were drawn by the author, either for publication in earlier works or particularly for this one.

PLATE 1

PAGE

FIG. 1. *Cystodictyon pavonium:* individual plant to show lacunae and part of the venation, × 1.4 124

FIG. 2. *Ernodesmis verticillata:* portion of a plant, × 1.7 113

FIG. 3. *Enteromorpha lingulata:* habit of two plants, × 3 60

FIG. 4. *Ulva fasciata:* a single young plant showing short and broad blades as well as long strap-shaped ones, × 0.75 66

For additional figures of *Cystodictyon* see Plate 8; for *Ernodesmis,* Plate 6; for *Ulva,* Plate 8.

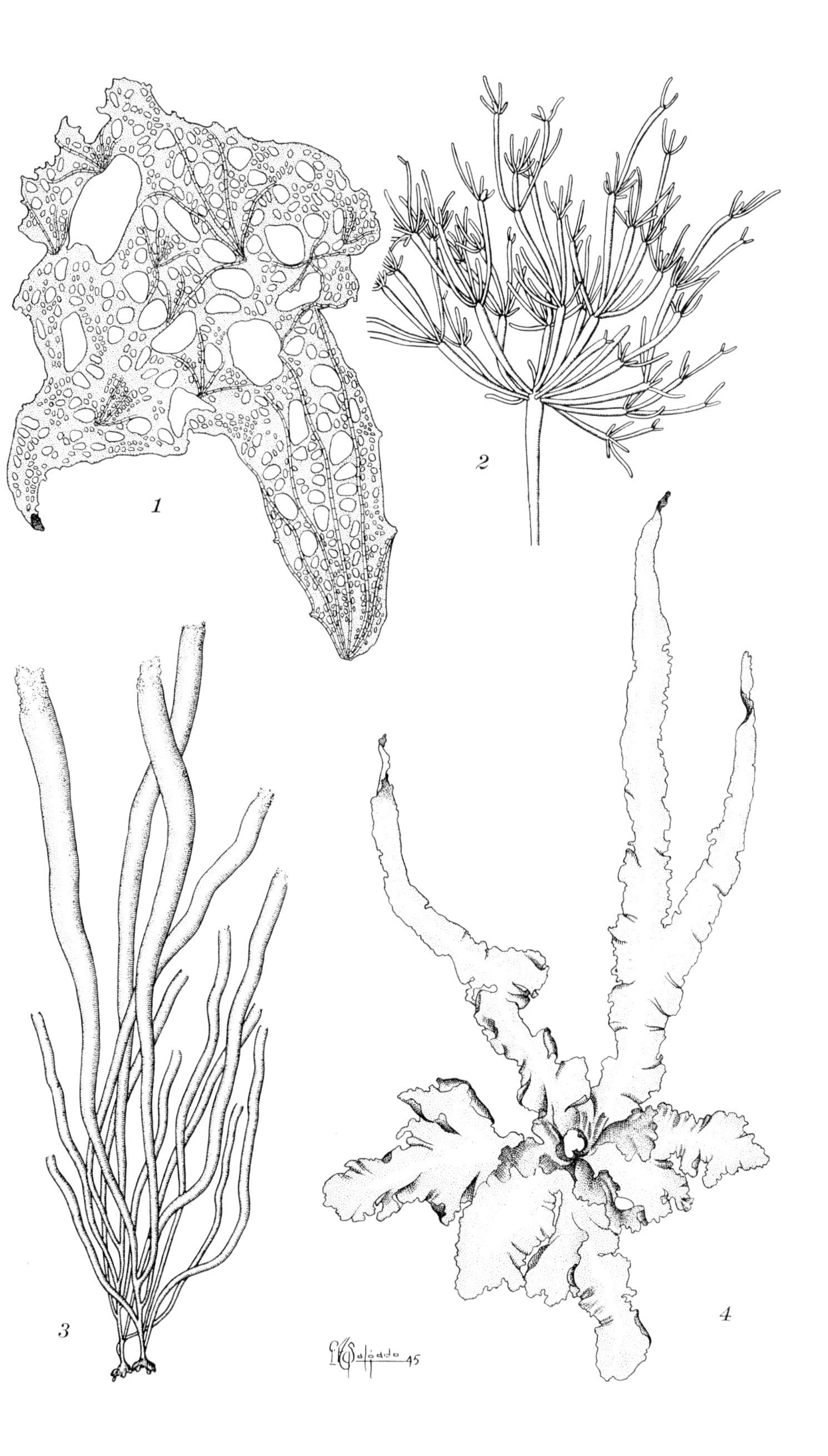
1
2
3
4

PLATE 2

PAGE

FIG. 1. *Cladophoropsis membranacea:* upper portion of a filament showing typical branching, × 10 117

FIG. 2. *Cladophoropsis macromeres:* upper portion of an unusually evenly branched filament, × 3 118

FIG. 3. *Cladophora fuliginosa:* upper portion of a filament showing the branching, × 3 83

FIG. 4. *Phaeophila dendroides:* portions of two filaments showing the twisted setae, × 250 48

FIG. 5. *Rhizoclonium hookeri:* portions of two filaments showing the characteristic branching, × 53 77

FIG. 6. *Valonia macrophysa:* three individuals from a clump showing characteristic branching, × 1.25 110

FIG. 7. *Ulvella lens:* portion of a thallus to show cell arrangement, × 475 52

FIG. 8. *Chaetomorpha linum:* two portions of filaments showing cell proportions, × 30 71

FIG. 9. *Chaetomorpha brachygona:* two portions of filaments showing cell proportions, × 30 70

For additional figures of *Cladophora* see Plate 3; for *Cladophoropsis*, Plate 3; for *Valonia*, Plates 7, 9. These figures are from Taylor, *Marine Algae of Florida* (1928).

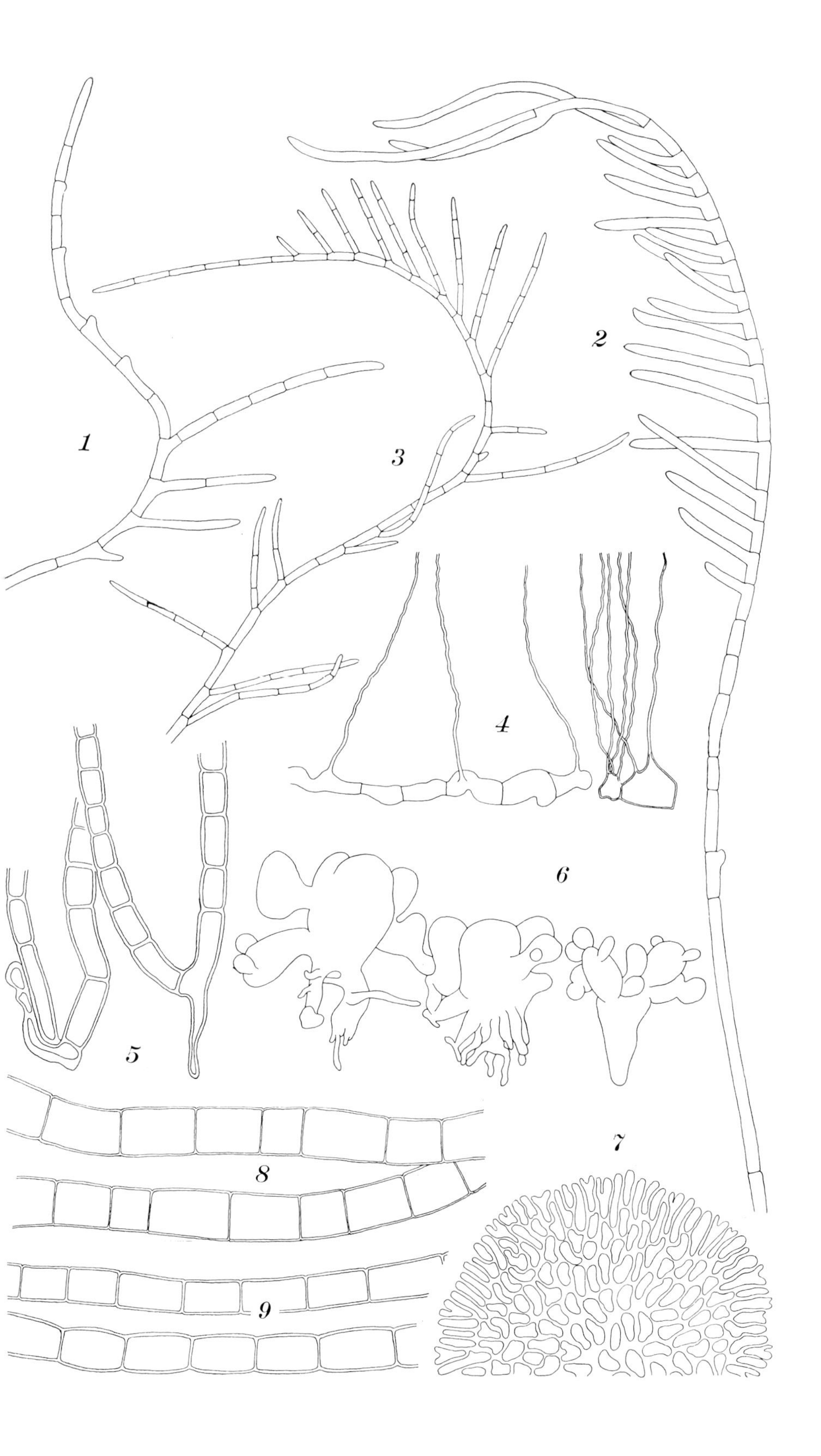
1
2
3
4
5
6
7
8
9

PLATE 3

PAGE

For additional figures of *Cladophora* and *Cladophoropsis,* see Plate 2.

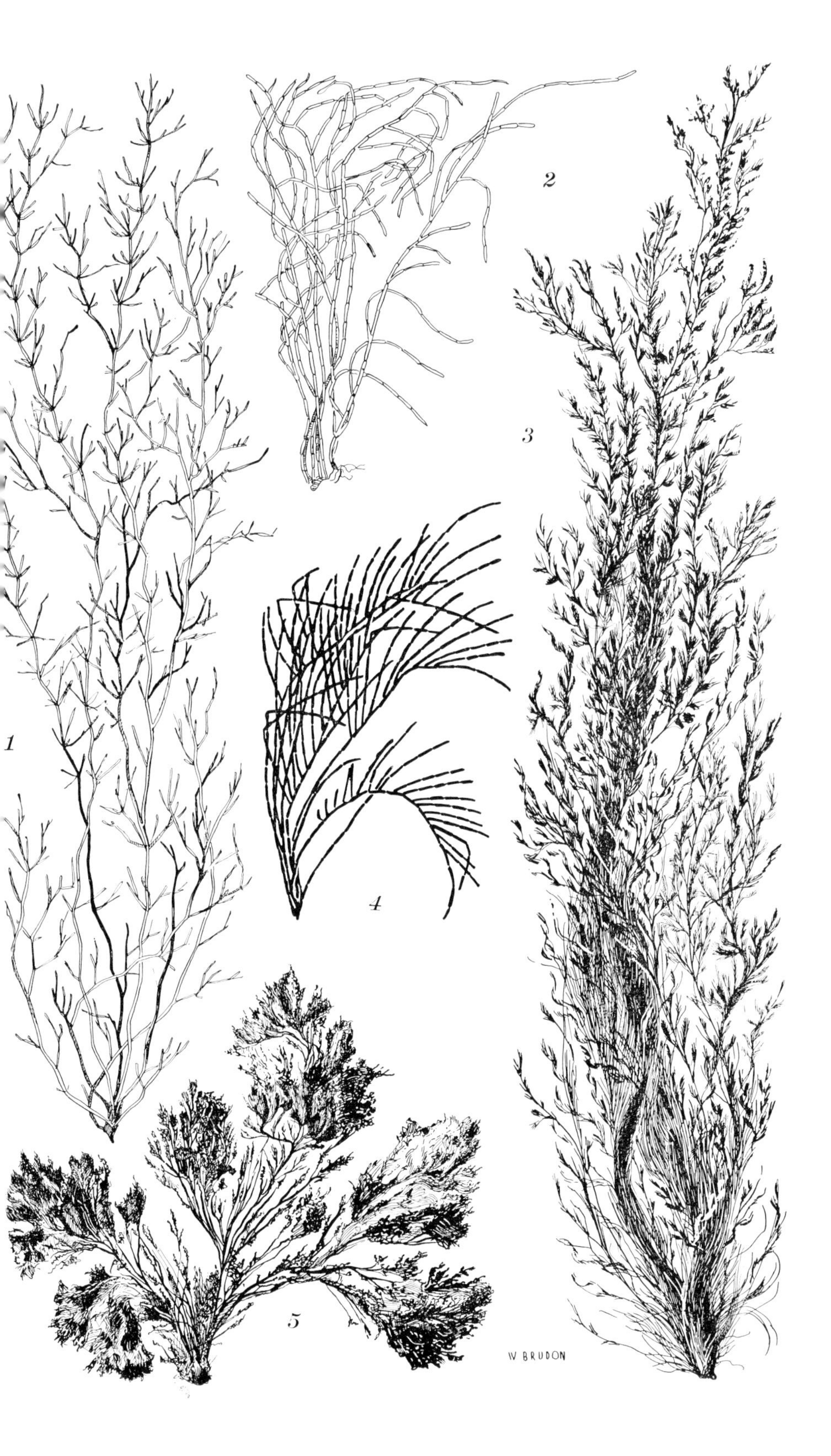
1
2
3
4
5
W BRUDON

PLATE 4

PAGE

Fig. 1. *Cymopolia barbata:* portion of a plant to show segmentation and terminal filament tufts, × 1.0 102

Fig. 2. *Dasycladus vermicularis:* a cluster of several plants, × 2 .. 99

Fig. 3. *Batophora oerstedi:* a plant in the vegetative state and part of another with young sporangia, × 1.5 98

Fig. 4. *Batophora oerstedi* v. *occidentalis:* vegetative and fertile plants showing the shorter more compact branchlets, × 3 98

Fig. 5. *Acetabularia crenulata:* a cluster of several plants showing the concave disks, × 1.5 105

For additional figures of *Acetabularia* see Plate 6; for *Batophora,* Plates 5, 6; for *Cymopolia,* Plate 6; for *Dasycladus,* Plate 6.

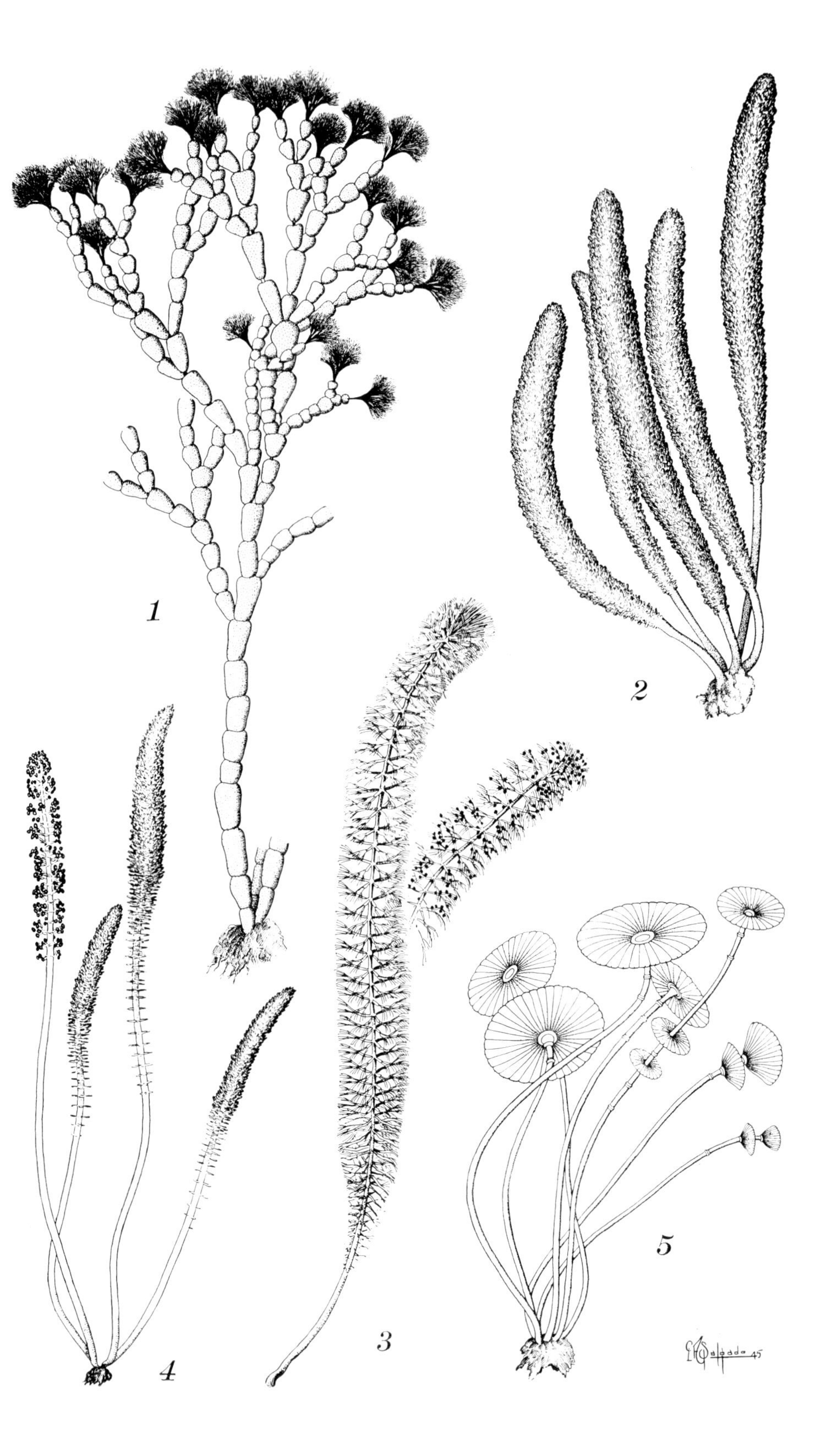
1
2
3
4
5

PLATE 5

PAGE

FIG. 1. *Struvea anastomosans* v. *caracasana:* portions of filaments showing the terminal netlike frond, × 5 122

FIG. 2. *Chamaedoris peniculum:* several plants showing annulate stipes and terminal brushes of filaments, × 1.2 115

FIG. 3. *Neomeris cokeri:* general aspect of a plant showing scattered sporangia, × 5.5 100

FIG. 4. *Batophora oerstedi:* portion of a fruiting axis showing sporangia. The delicate branchlet tips have fallen off. × 8 98

FIG. 5. *Neomeris annulata:* two plants showing sporangial zones and flaking off of the limy coat below, × 10 101

For additional figures of *Batophora* see Plates 4, 6; for *Dasycladus,* Plates 4, 6; for *Neomeris,* Plate 6; for *Struvea,* Plate 9.

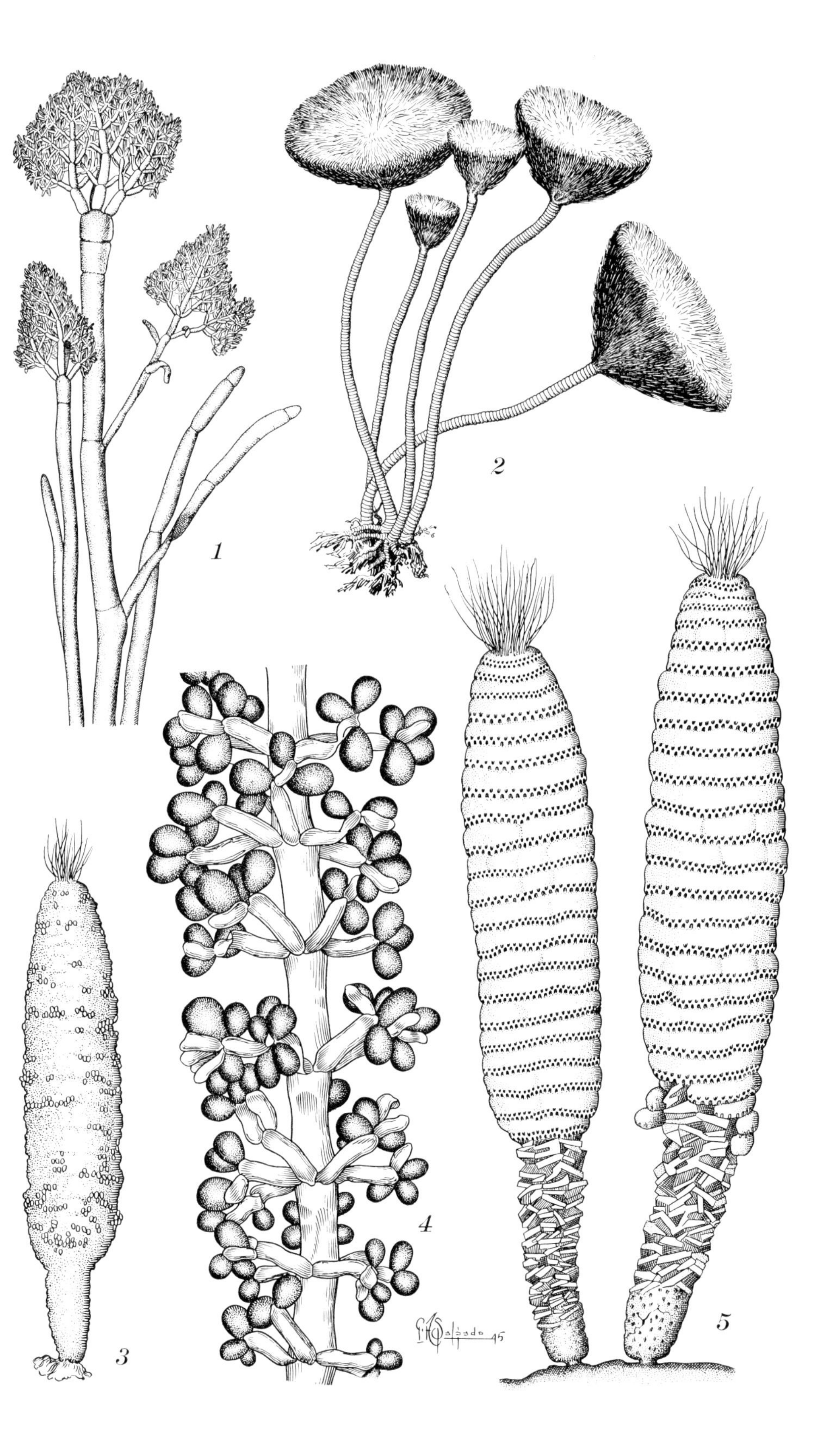

1
2
3
4
5

PLATE 6

PAGE

FIG. 1. *Cymopolia barbata:* three branchlets bearing clavate assimilators and sporangia, × 53 102

FIG. 2. *Dasycladus vermicularis:* branchlet with a sporangium on the the basal cell, the delicate terminal filaments having fallen from the surrounding cells, ×22 99

FIG. 3. *Batophora oerstedi:* branchlet with sporangia, after loss of the delicate terminal filaments, × 22 98

FIGS. 4–6. *Neomeris annulata:* branchlets with a sporangium on each basal cell subtended by two assimilators, and in Fig. 5 very young and retaining the delicate terminal filaments, × 63 101

FIG. 7. *Siphonocladus rigidus:* portions of two small plants to show branching, × 3 114

FIG. 8. *Dasycladus vermicularis:* a lateral branchlet in the vegetative state, × 22 99

FIG. 9. *Batophora oerstedi:* a lateral branchlet in the vegetative state, × 22 98

FIG. 10. *Ernodesmis verticillata:* part of a plant showing branching, × 1.9 113

FIG. 11. *Acicularia schenckii:* tips of a few disk rays with aplanospores; portions of the corona superior and corona inferior, × 53 .. 107

FIG. 12. *Acetabularia crenulata:* tips of two disk rays; portions of the corona inferior and corona superior, × 53 105

FIG. 13. *Acetabularia pusilla:* two plants showing rays both separate, and forming a disk, × 12 104

For additional figures of *Acetabularia* see Plate 4; for *Batophora,* Plates 4, 5; for *Cymopolia,* Plate 4; for *Dasycladus,* Plate 4; for *Ernodesmis,* Plate 1; for *Siphonocladus,* Plate 7. These figures are from Taylor, *Marine Algae of Florida* (1928).

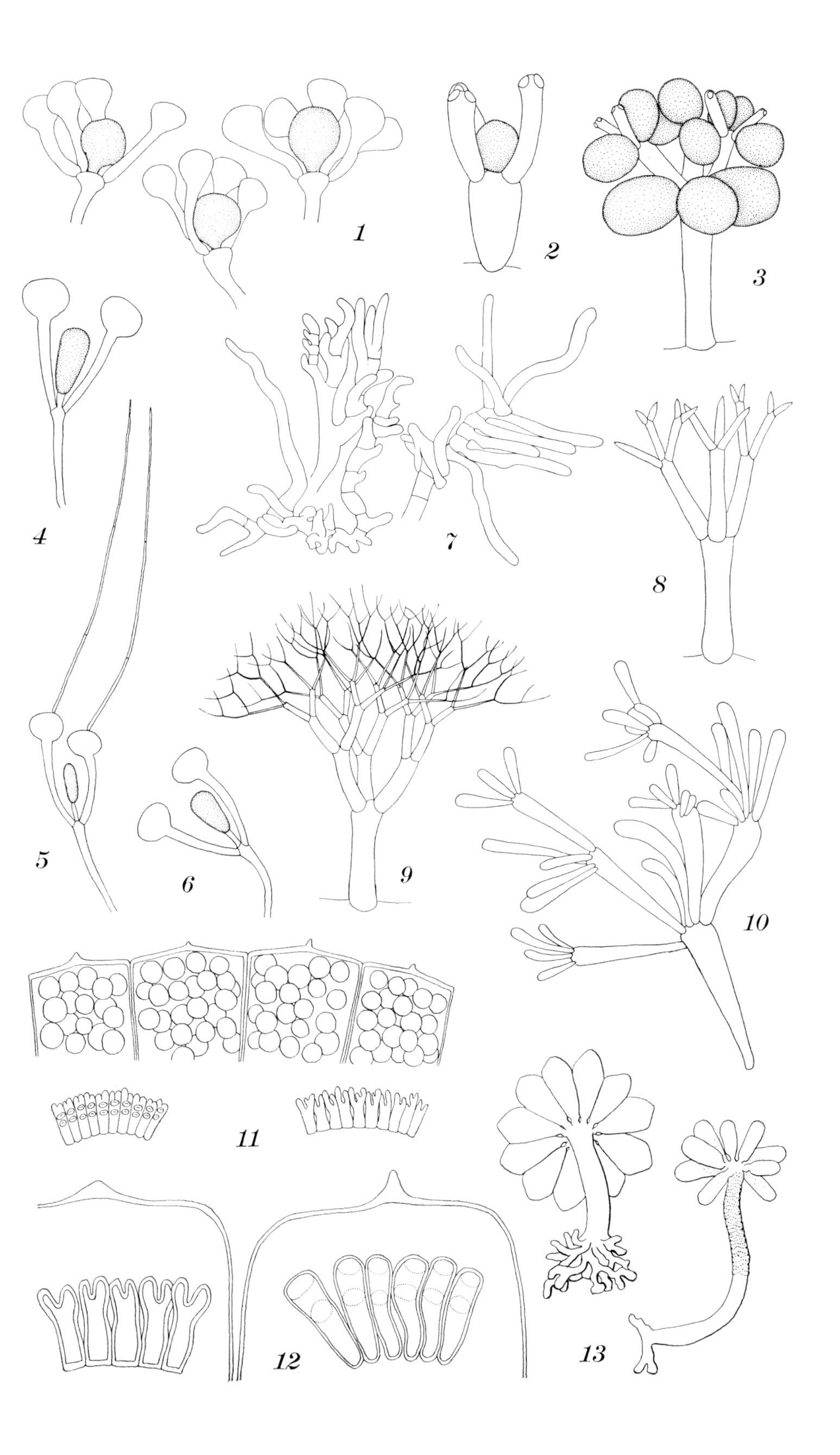
1
2
3
4
5
6
7
8
9
10
11
12
13

PLATE 7

For additional figures of *Anadyomene* see Plate 8; for *Siphonocladus,* Plate 6; for *Valonia,* Plates 2, 9.

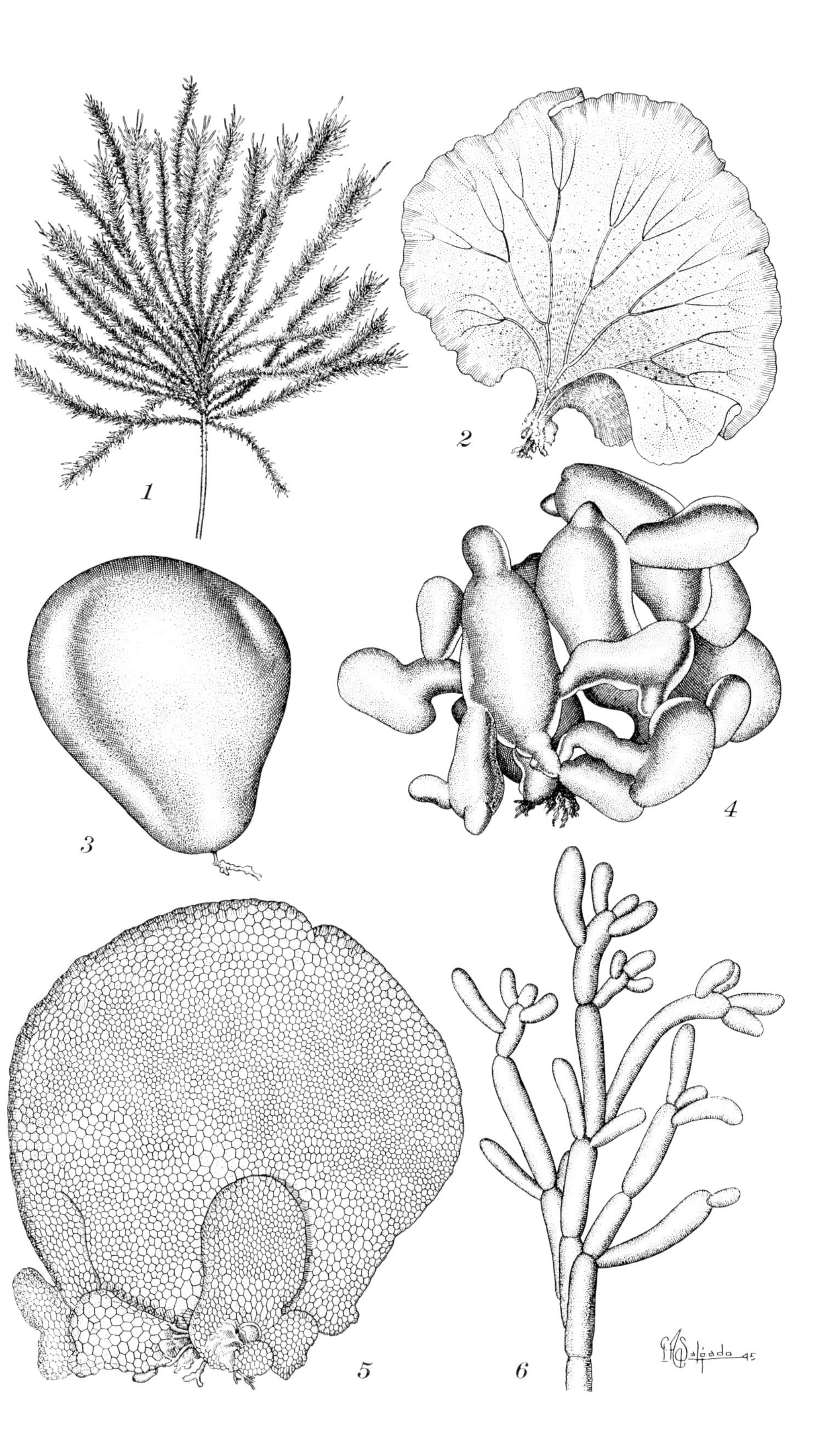
1
2
3
4
5
6

PLATE 8

PAGE

FIG. 1. *Microdictyon boergesenii:* details of part of a blade showing the venation and the cells making up the meshes of the net, × 5.. 120

FIG. 2. *Anadyomene stellata:* details of the marginal portion of a blade showing the venation and the arrangement of the intercalary cells, × 15.......... 125

FIG. 3. *Ulva profunda:* the marginal portion of a blade showing the characteristic perforations, × 1.0.......... 67

FIG. 4. *Cystodictyon pavonium:* details of part of a blade showing the ultimate venation, the lacunae and the intercalary cells, × 7.. 124

For additional figures of *Anadyomene* see Plate 7; for *Cystodictyon,* Plate 1; for *Ulva,* Plate 1.

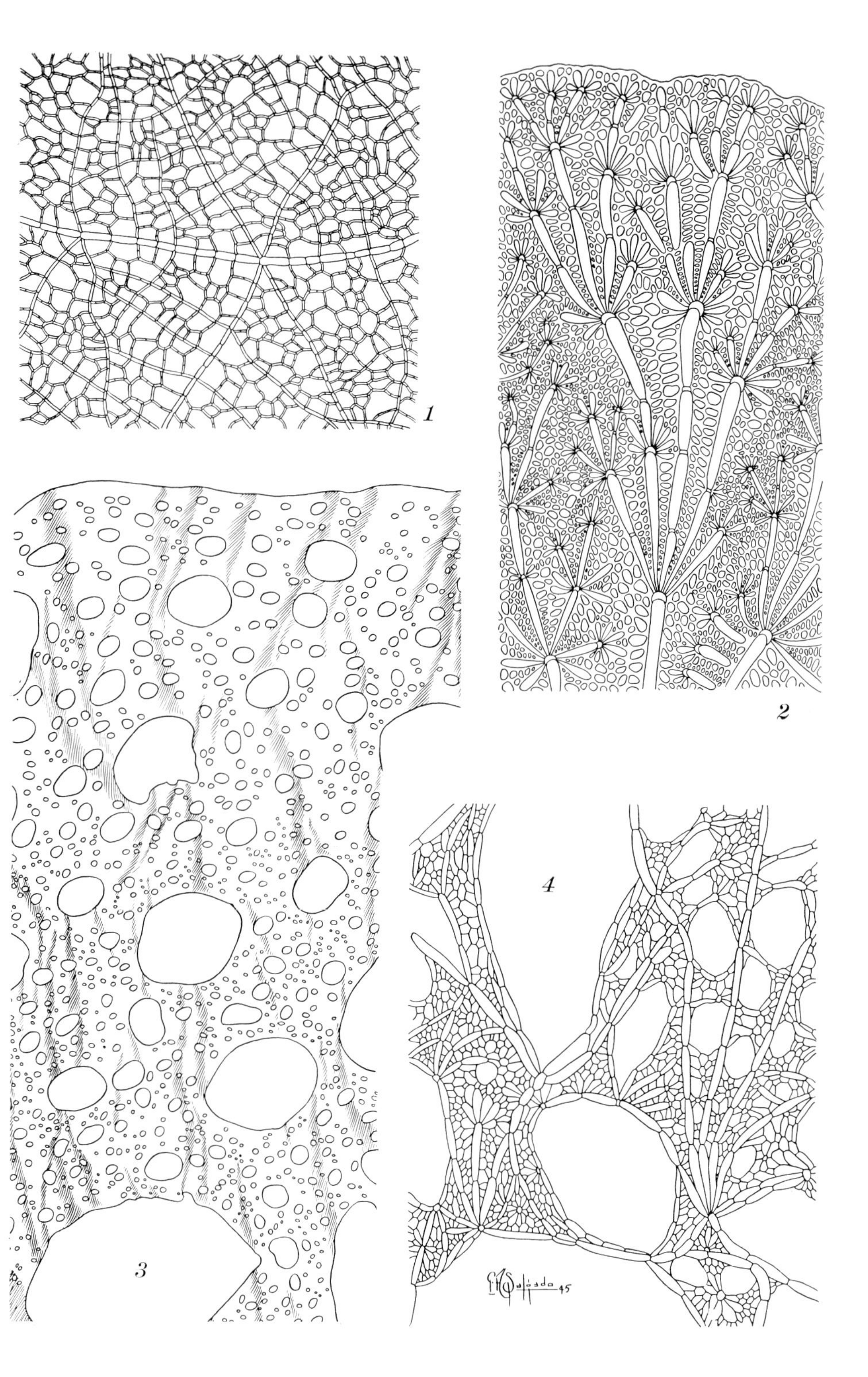
1
2
3
4
ECSalgado 45

PLATE 9

For additional figures of *Struvea* see Plate 5; for *Valonia* see Plates 2, 7. These figures are in part from Taylor, *Marine Algae of Florida* (1928).

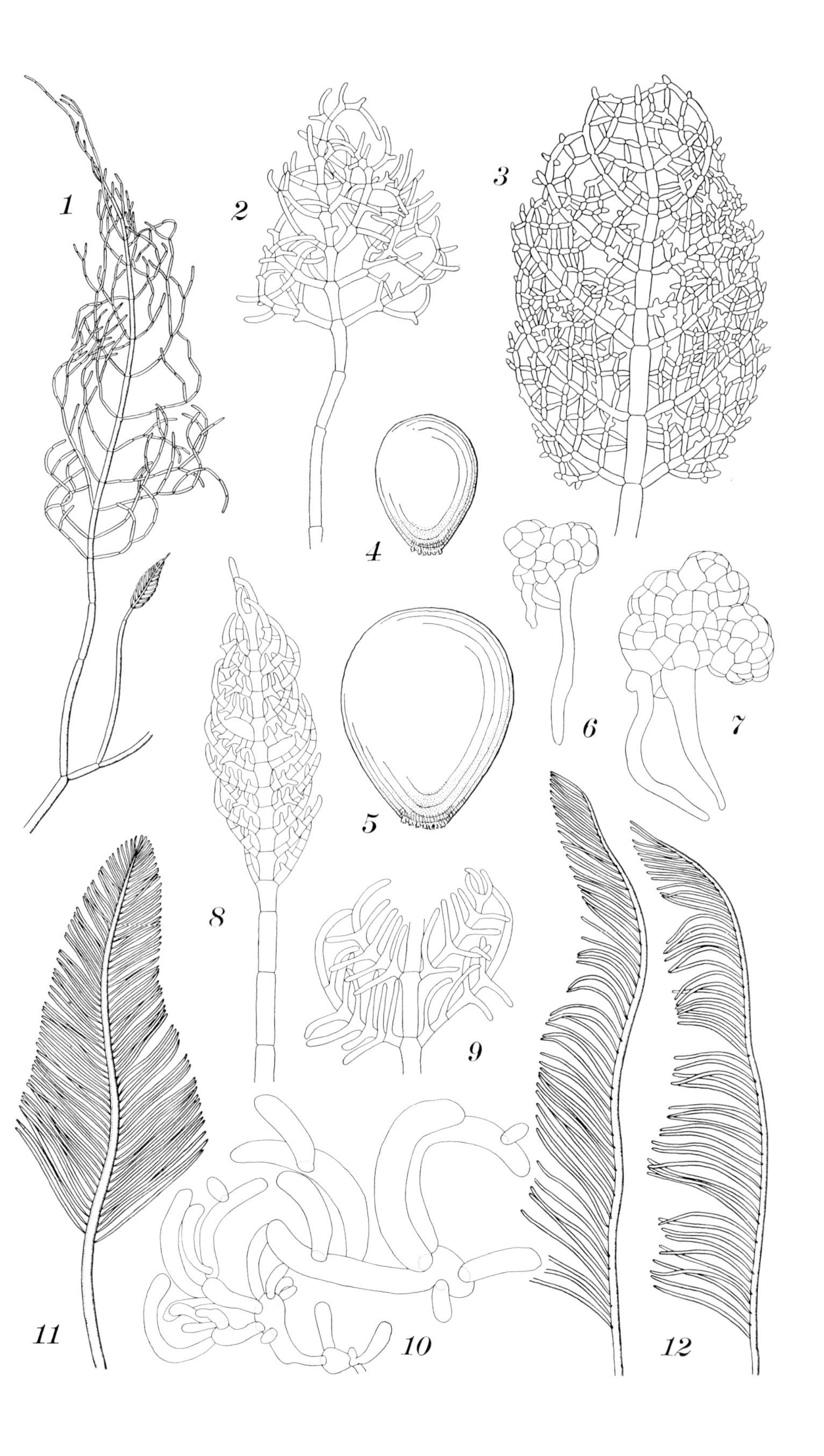

1
2
3
4
5
6
7
8
9
10
11
12

PLATE 10

PAGE

FIGS. 1–2. *Caulerpa verticillata:* portion of a plant with verticils and tips of a branchlet from a verticil, × 1.5, × 30. 138

FIG. 3. *Caulerpa vickersiae* v. *luxurians:* portion of a plant showing pennate branchlets, × 10. 137

FIGS. 4–6. *Caulerpa vickersiae:* habit of portions of three plants, × 2.5 137

FIG. 7. *Caulerpa vickersiae* v. *furcifolia:* details of a portion of a plant showing radially placed, often forked, branchlets, × 8. 137

FIGS. 8–9. *Caulerpa vickersiae* v. *luxurians:* habit of portions of two plants, × 2.5. 137

FIG. 10. *Caulerpa webbiana:* detail of branchlets on a portion of a branch, × 50. 139

FIG. 11. *Caulerpa pusilla:* portion of a tomentose rhizome with an erect branch, × 7. 138

FIG. 12. *Caulerpa fastigiata:* habit of a plant with stoloniferous and erect portions, × 2. 136

For additional figures of *Caulerpa* see Plates 11–18.

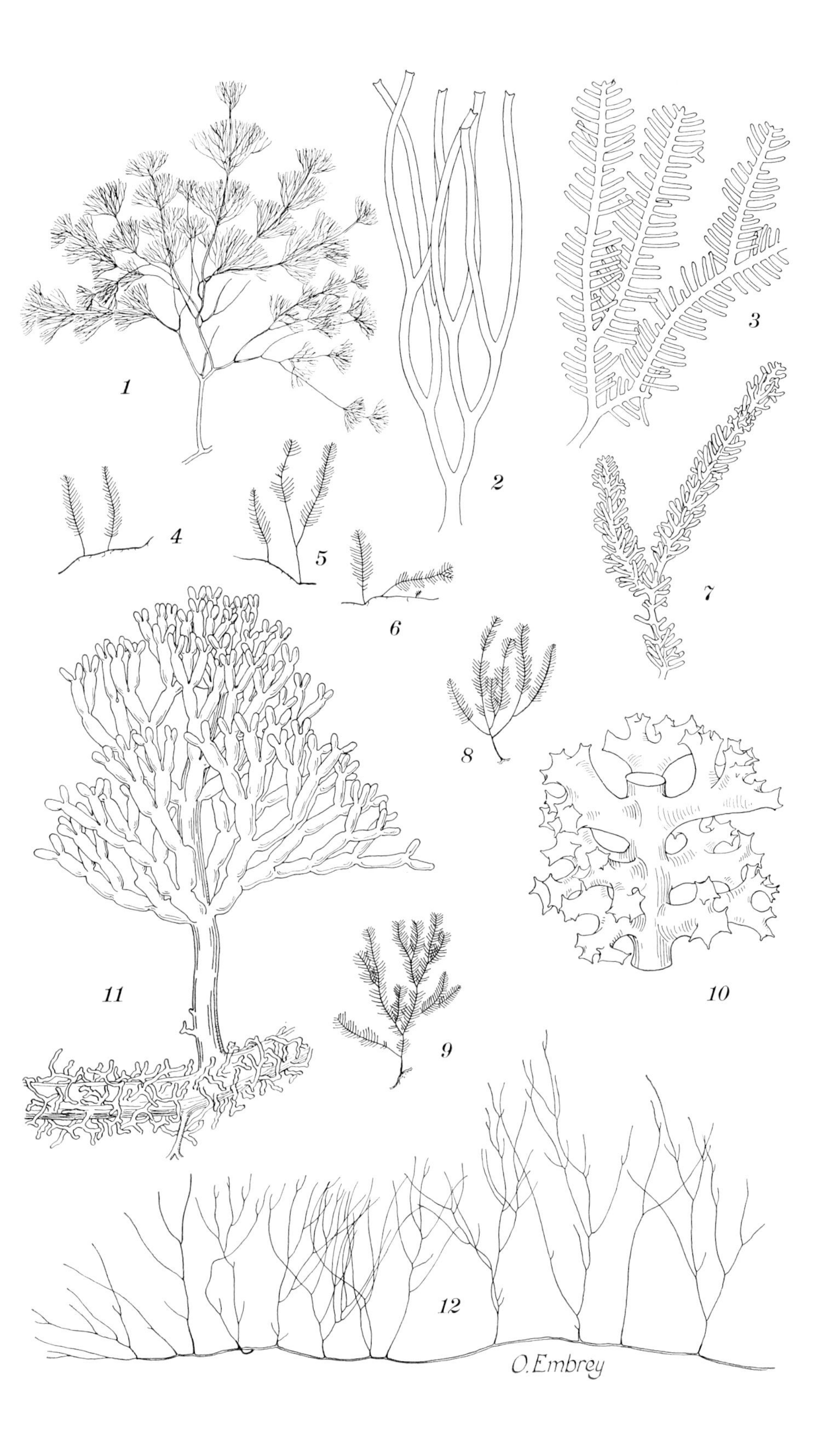
1
2
3
4
5
6
7
8
9
10
11
12
O. Embrey

PLATE 11

For additional figures of *Caulerpa* see Plates 10, 12–18.

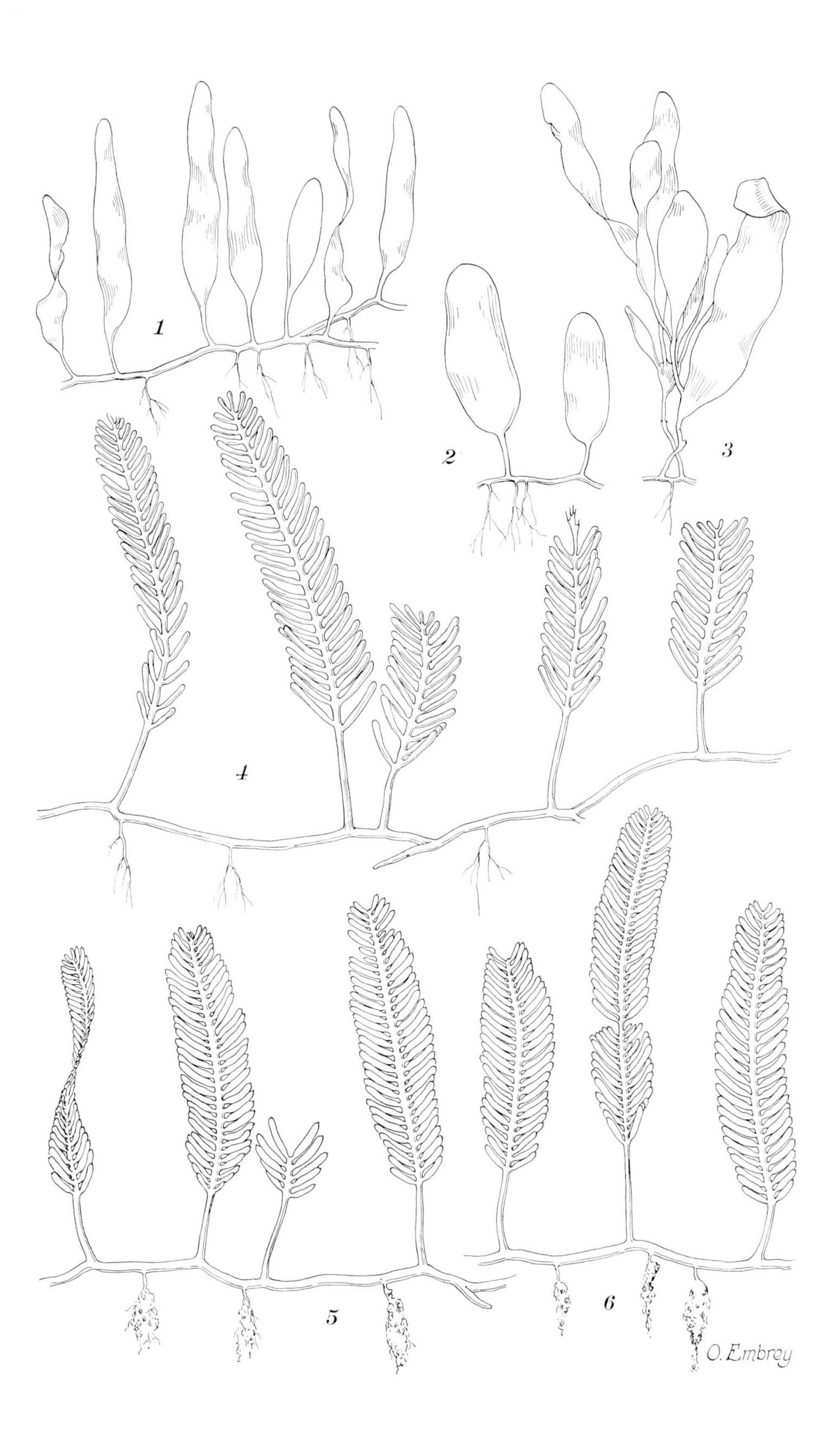
1
2
3
4
5
6
O. Embrey

PLATE 12

PAGE

FIG. 1. *Caulerpa taxifolia:* habit of part of a plant, × 0.6 142

FIGS. 2–3. *Caulerpa mexicana* f. *laxior:* portions of two plants to show the habit, × 0.5 141

FIGS. 4–5. *Caulerpa mexicana:* habit of portions of two plants to show variation in the blades, × 0.5 141

For additional figures of *Caulerpa* see Plates 10, 11, 13–18.

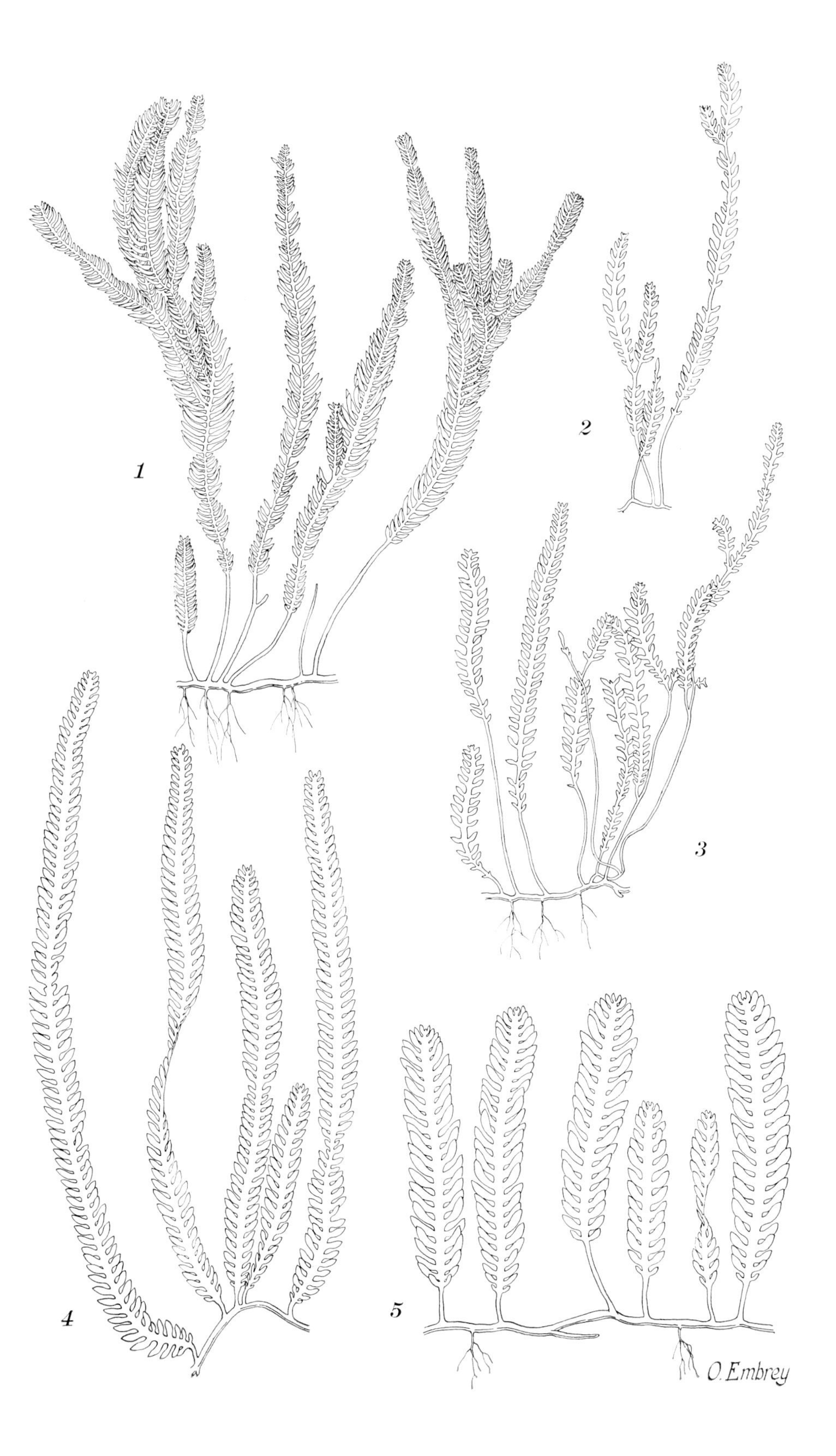
1
2
3
4
5
O. Embrey

PLATE 13

PAGE

FIG. 1. *Caulerpa sertularioides* f. *longiseta:* habit of part of an exceptionally freely branched specimen, × 1.1 144

FIGS. 2–3. *Caulerpa sertularioides* f. *brevipes:* habits of parts of two characteristic specimens, × 0.55 144

FIGS. 4–5. *Caulerpa sertularioides* f. *farlowii:* habit of part of a characteristic specimen, and details of one erect branch, × 0.55, × 1.6 144

FIGS. 6–7. *Caulerpa sertularioides* f. *longiseta:* detail of part of an erect blade, and habit of part of characteristic plant, × 1.1, × 0.55 144

For additional figures of *Caulerpa* see Plates 10–12, 14–18.

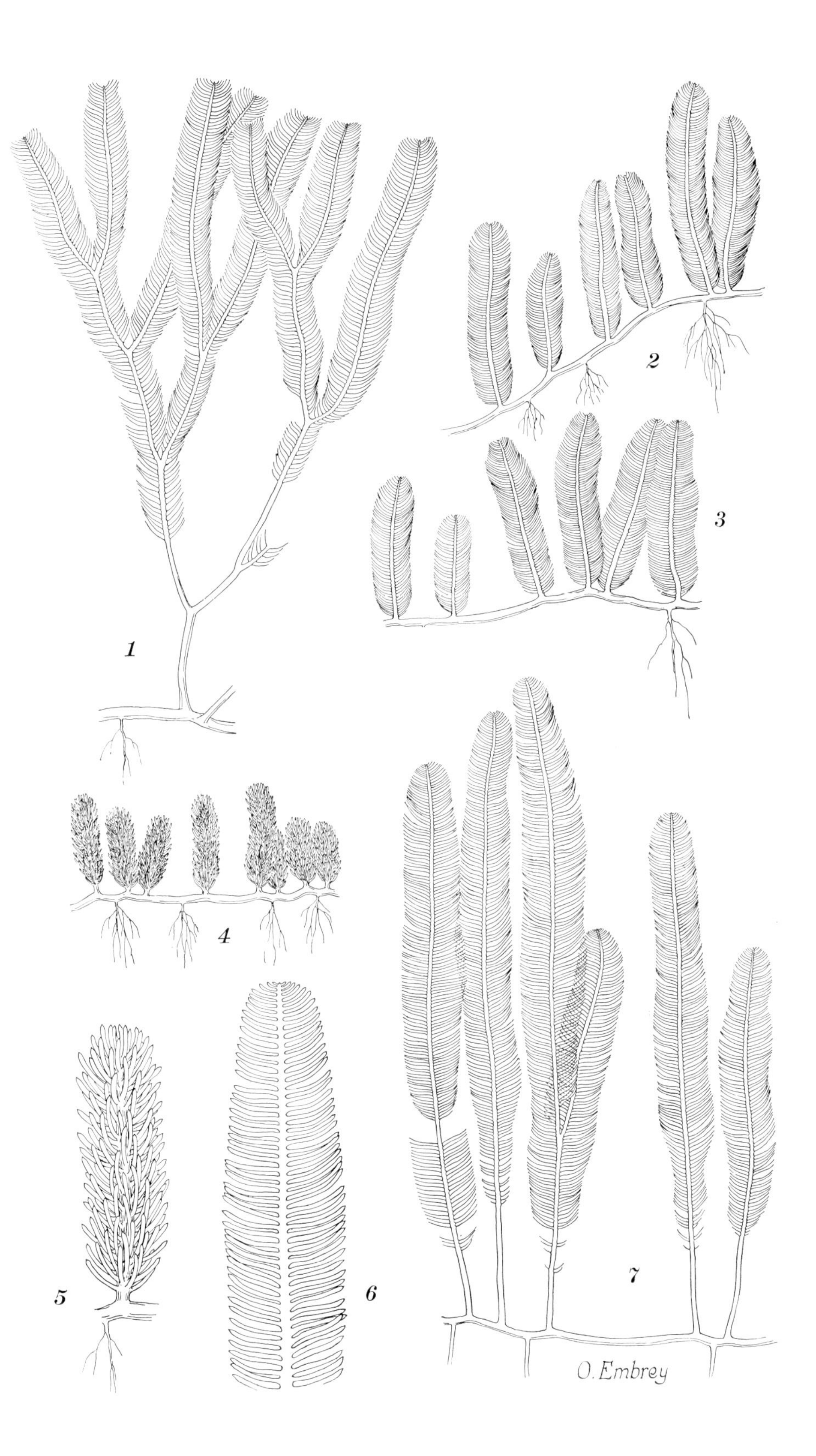
1
2
3
4
5
6
7
O. Embrey

PLATE 14

PAGE

FIGS. 1–2. *Caulerpa lanuginosa:* habit of parts of two plants, × 0.5. 145

FIG. 3. *Caulerpa cupressoides* v. *lycopodium:* detail of an erect branch, × 1.2.. 147

FIGS. 4, 6. *Caulerpa cupressoides* v. *flabellata:* habit of two erect branches, × 1.3.. 147

FIG. 5. *Caulerpa serrulata:* habit of part of a plant, × 3.......... 145

For additional figures of *Caulerpa* see Plates 10–13, 15–18.

1
2
3
4
5
6
O. Embrey

PLATE 15

PAGE

FIG. 1. *Caulerpa cupressoides:* habit of part of a typical plant, × 1.0. 146

FIGS. 2–3. *Caulerpa cupressoides* v. *lycopodium* f. *elegans:* erect fronds, characteristic except for a rather broad axis, × 1.5.... 148

FIG. 4. *Caulerpa cupressoides* v. *mamillosa:* habit of a portion of a plant, × 1.0.. 148

For additional figures of *Caulerpa,* see Plates 10–14, 16–18.

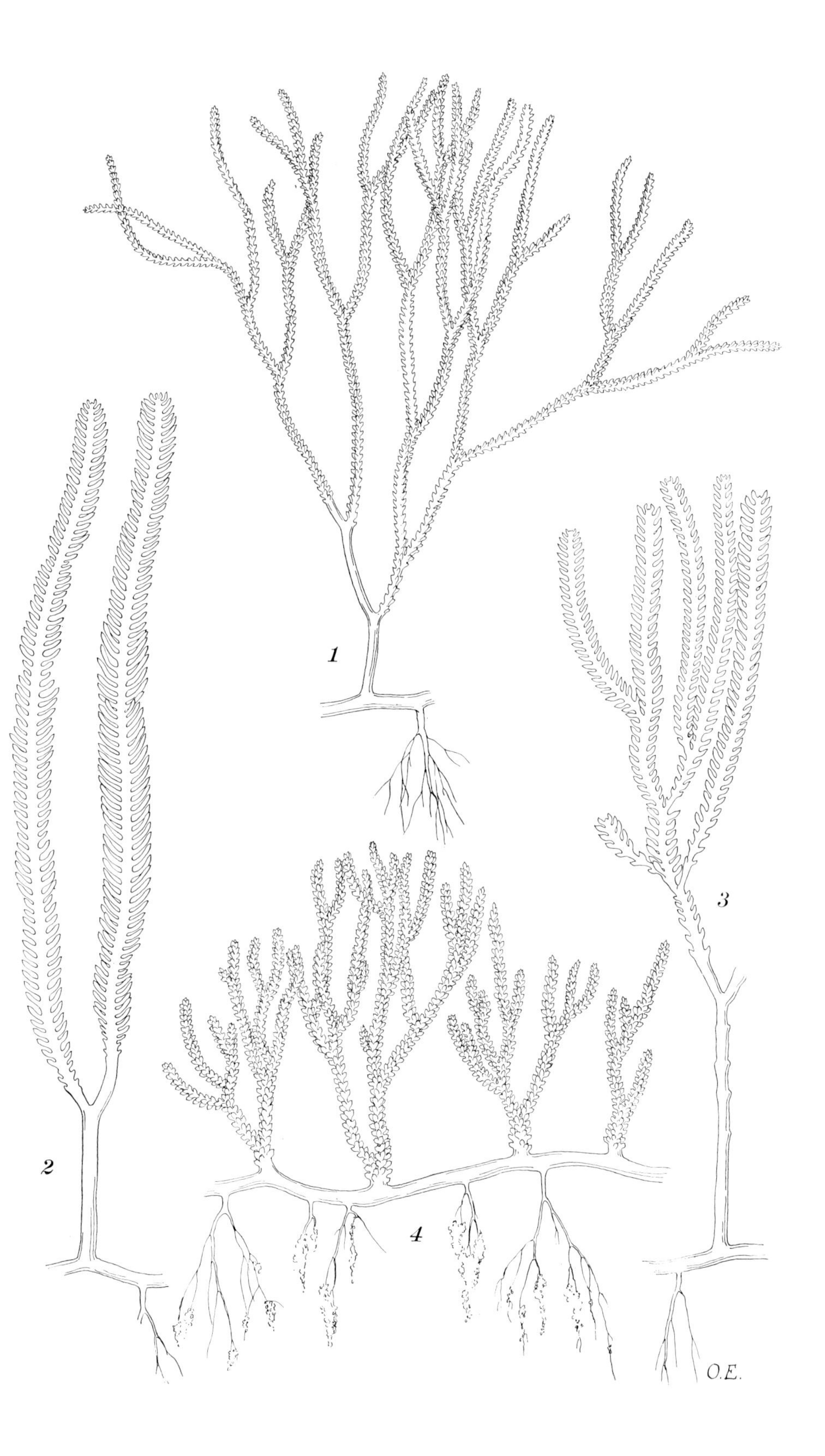
1
2
3
4
O.E.

PLATE 16

For additional figures of *Caulerpa* see Plates 10–15, 17, 18.

1
2
4
3
O. Embrey

PLATE 17

PAGE

FIG. 1. *Caulerpa racemosa* v. *macrophysa:* habit of part of a plant, × 0.55 153

FIG. 2. *Caulerpa peltata:* habit of part of a plant, × 1.6.......... 155

FIG. 3. *Caulerpa racemosa* v. *uvifera:* habit of part of a plant, × 0.55 153

FIG. 4. *Caulerpa racemosa* v. *laetevirens:* habit of part of a plant, × 0.55 153

FIG. 5. *Caulerpa microphysa:* habit of part of a plant, × 1.0..... 155

FIG. 6. *Caulerpa racemosa* v. *occidentalis:* habit of part of a plant, × 0.55 153

FIG. 7. *Caulerpa racemosa* v. *clavifera:* habit of part of a plant, × 0.55 152

For additional figures of *Caulerpa* see Plates 10–16, 18.

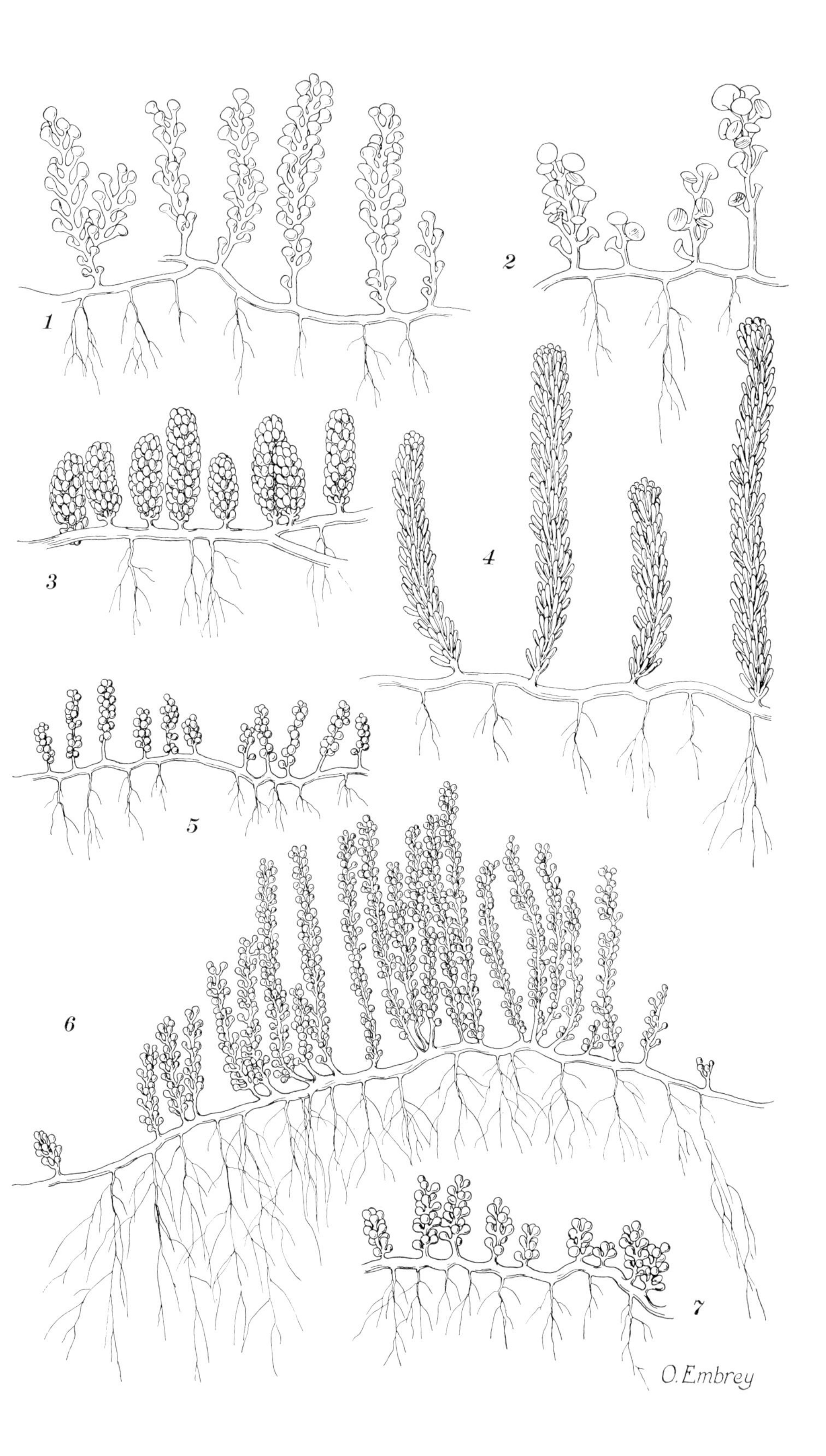
1
2
3
4
5
6
7
O. Embrey

PLATE 18

For additional figures of *Caulerpa* see Plates 10–17.

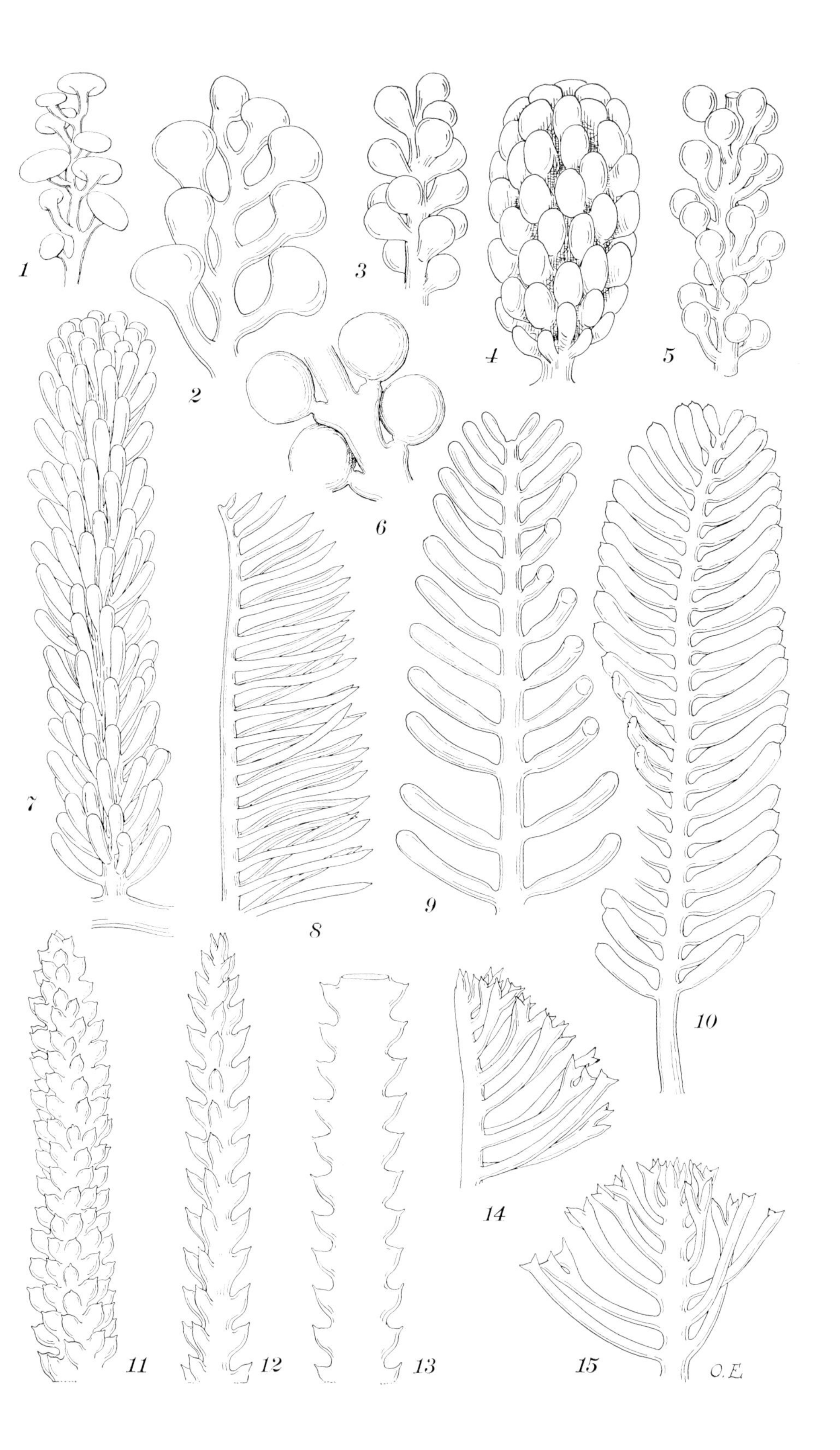

1
2
3
4
5
6
7
8
9
10
11
12
13
14
15
O. E.

PLATE 19

For additional figures of *Avrainvillea* see Plate 25.

1
2
W BRUDON
3

PLATE 20

PAGE

FIG. 1. *Cladocephalus scoparius:* habit of two plants, × 0.8...... 164

FIG. 2. *Udotea spinulosa:* plant habit, × 0.8..................... 167

FIG. 3. *Udotea conglutinata:* plant habit, × 0.55................ 165

FIG. 4. *Udotea flabellum:* habit of a very large and old plant with many marginally proliferated blades, × 0.35.................... 168

FIG. 5. *Udotea flabellum:* habit of a plant with a narrowly lacerate blade, × 0.55.. 168

For additional figures of *Cladocephalus* see Plate 25; for *Udotea,* Plates 22, 25.

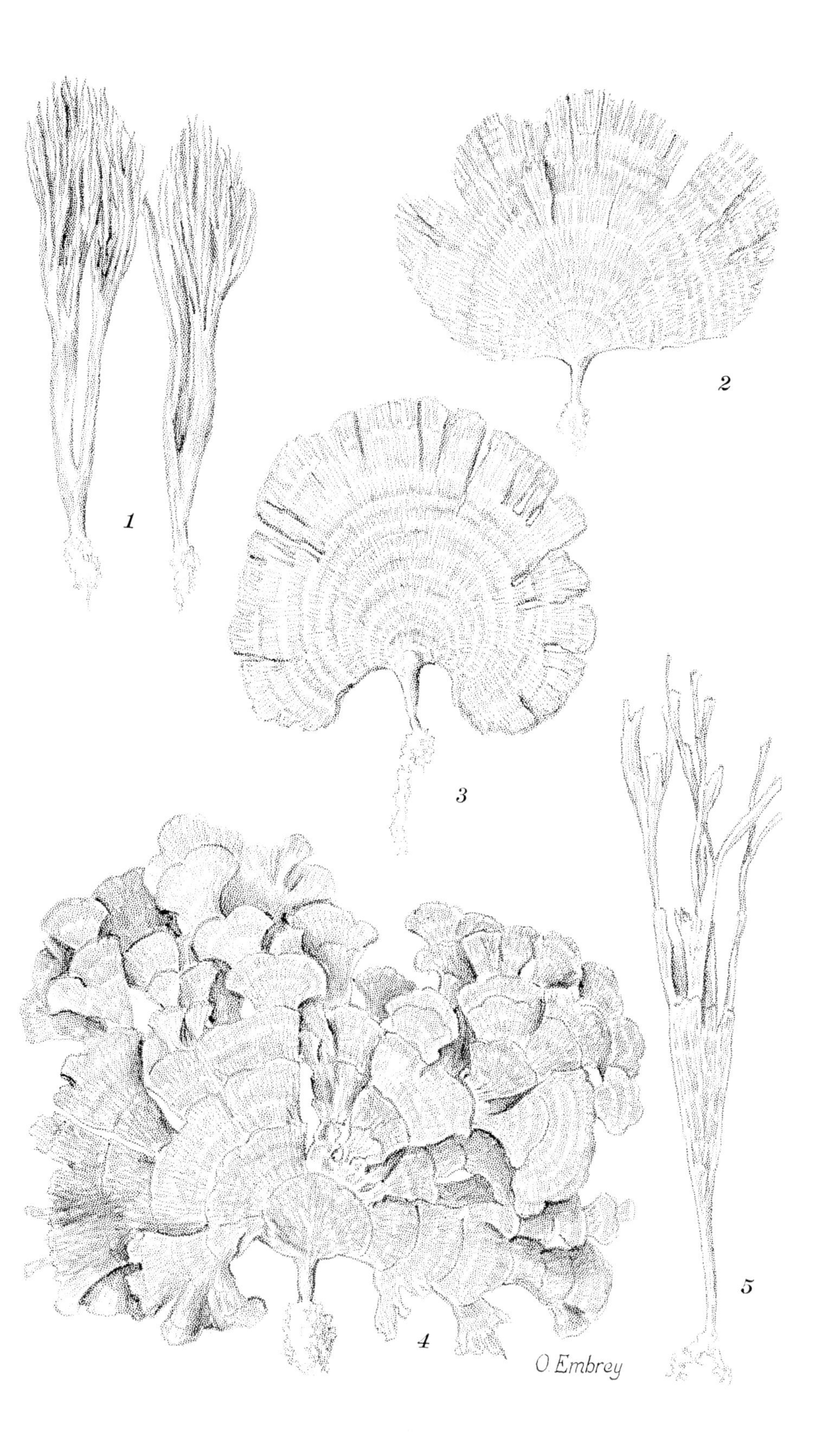
1
2
3
4
5
O. Embrey

PLATE 21

For additional figures of *Penicillus* see Plate 25.

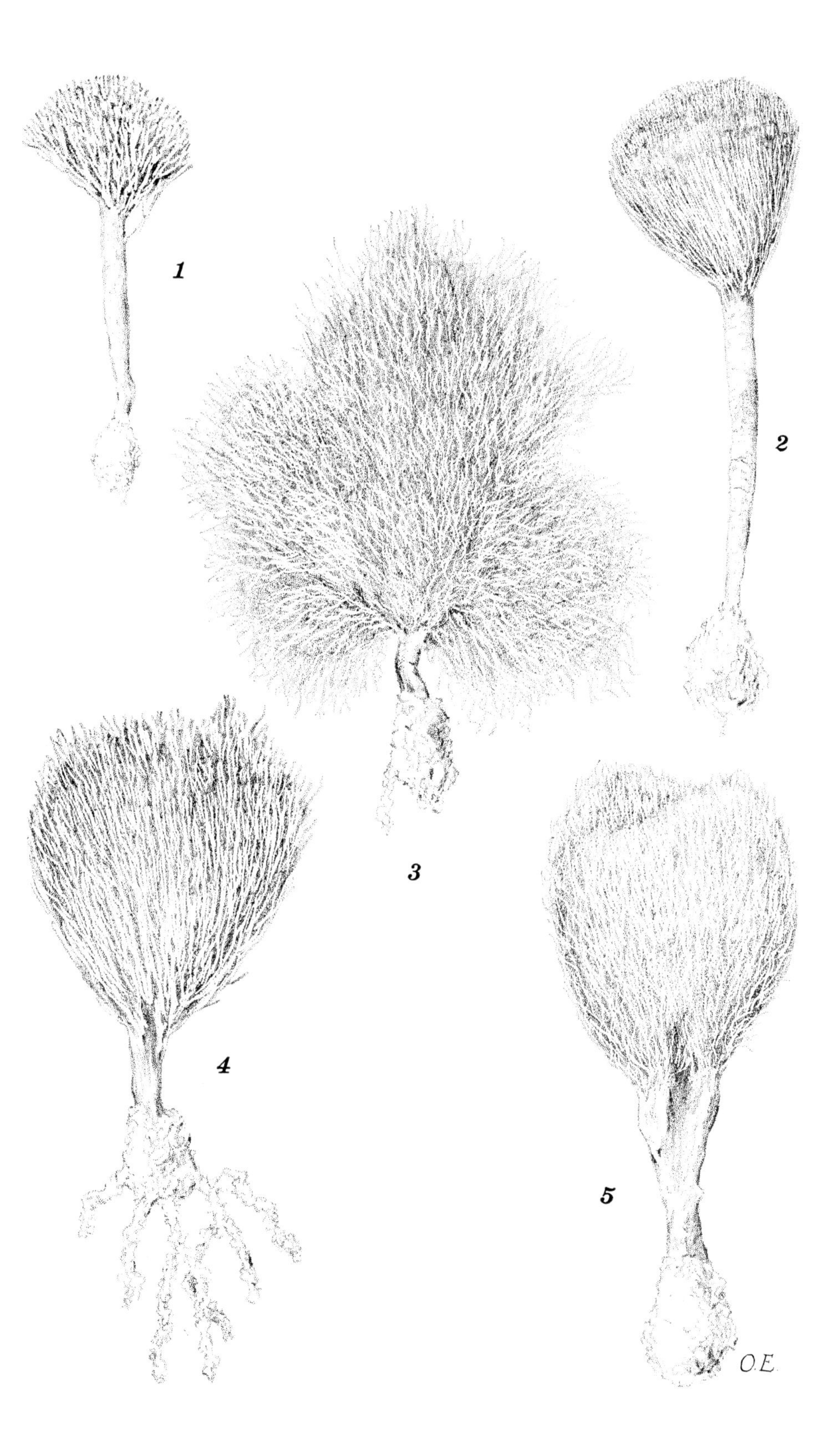
1
2
3
4
5
O.E.

PLATE 22

PAGE

Fig. 1. *Rhipocephalus oblongus:* habit of two plants, × 1.2........ 173

Fig. 2. *Rhipocephalus phoenix* f. *longifolius:* plant habit, × 1.2.... 174

Fig. 3. *Rhipilia tomentosa:* plant habit, × 1.0.................. 162

Fig. 4. *Udotea cyathiformis:* plant habit, × 0.8.................. 166

Fig. 5. *Rhipocephalus phoenix* f. *brevifolius:* plant habit, × 1.2.... 174

Fig. 6. *Udotea sublittoralis:* plant habit, × 1.2.................. 165

For additional figures of *Rhipocephalus* see Plate 25; for *Udotea* see Plates 20, 25.

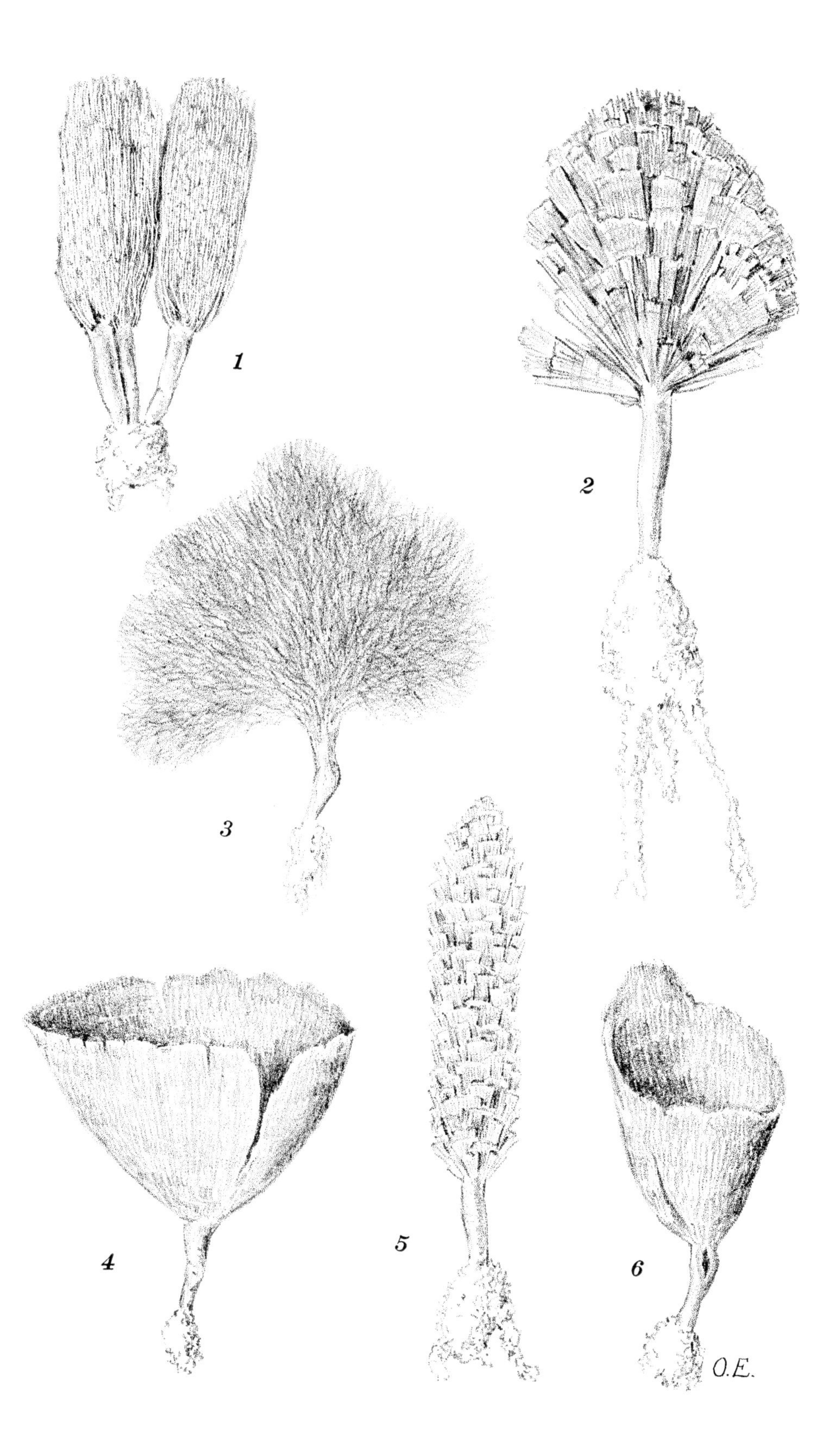
1
2
3
4
5
6
O.E.

PLATE 23

For additional figures of *Halimeda* see Plates 24, 25.

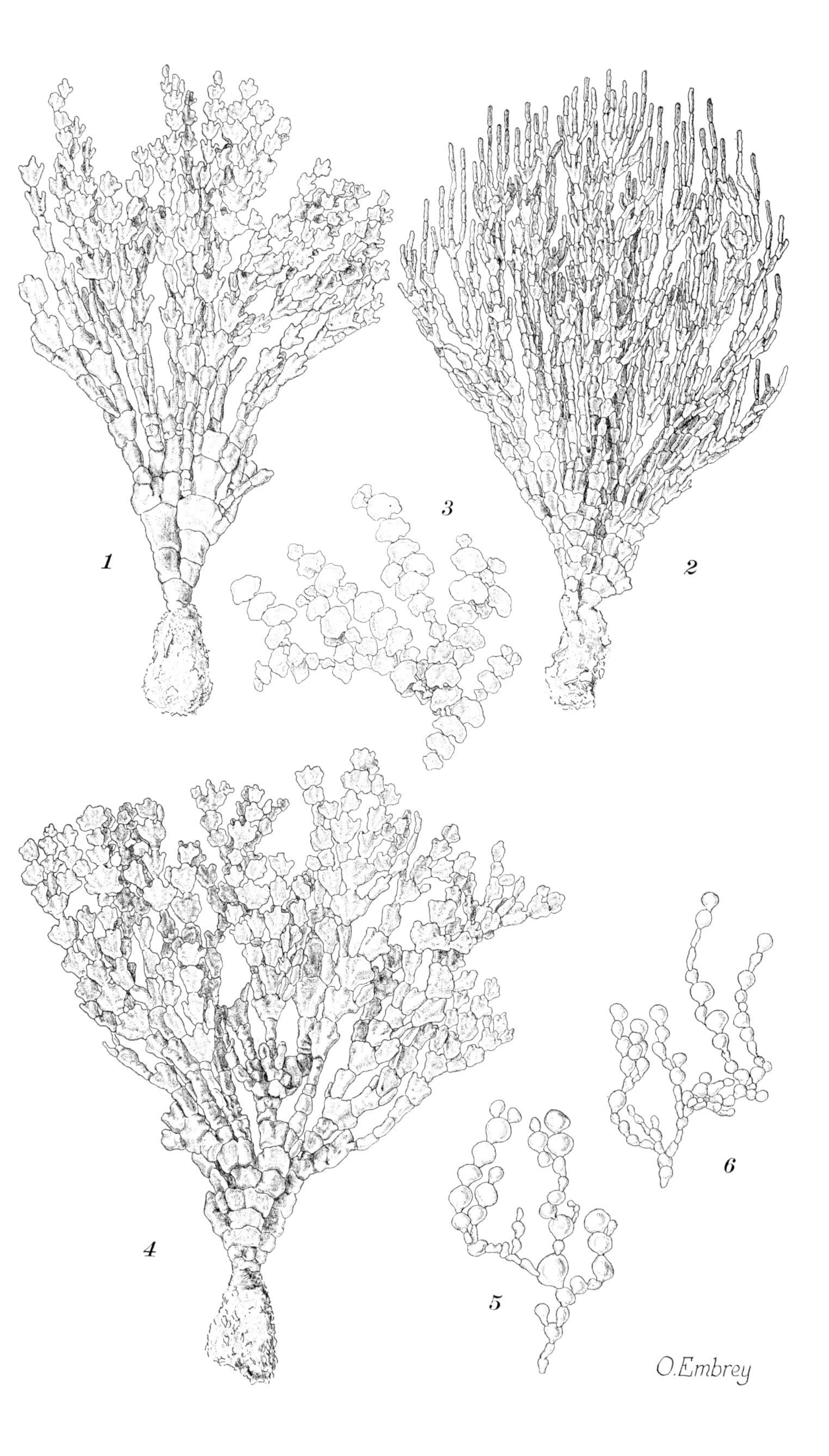
1
2
3
4
5
6
O.Embrey

PLATE 24

PAGE

FIG. 1. *Halimeda opuntia* f. *triloba:* habit of a portion of a plant, × 0.4 176

FIG. 2. *Halimeda discoidea:* plant habit, × 0.55........ 179

FIG. 3. *Halimeda favulosa:* plant habit, × 0.5........ 183

FIG. 4. *Halimeda simulans:* plant habit, × 0.6........ 180

FIG. 5. *Halimeda tuna:* plant habit, × 0.55........ 178

For additional figures of *Halimeda* see Plates 23, 25.

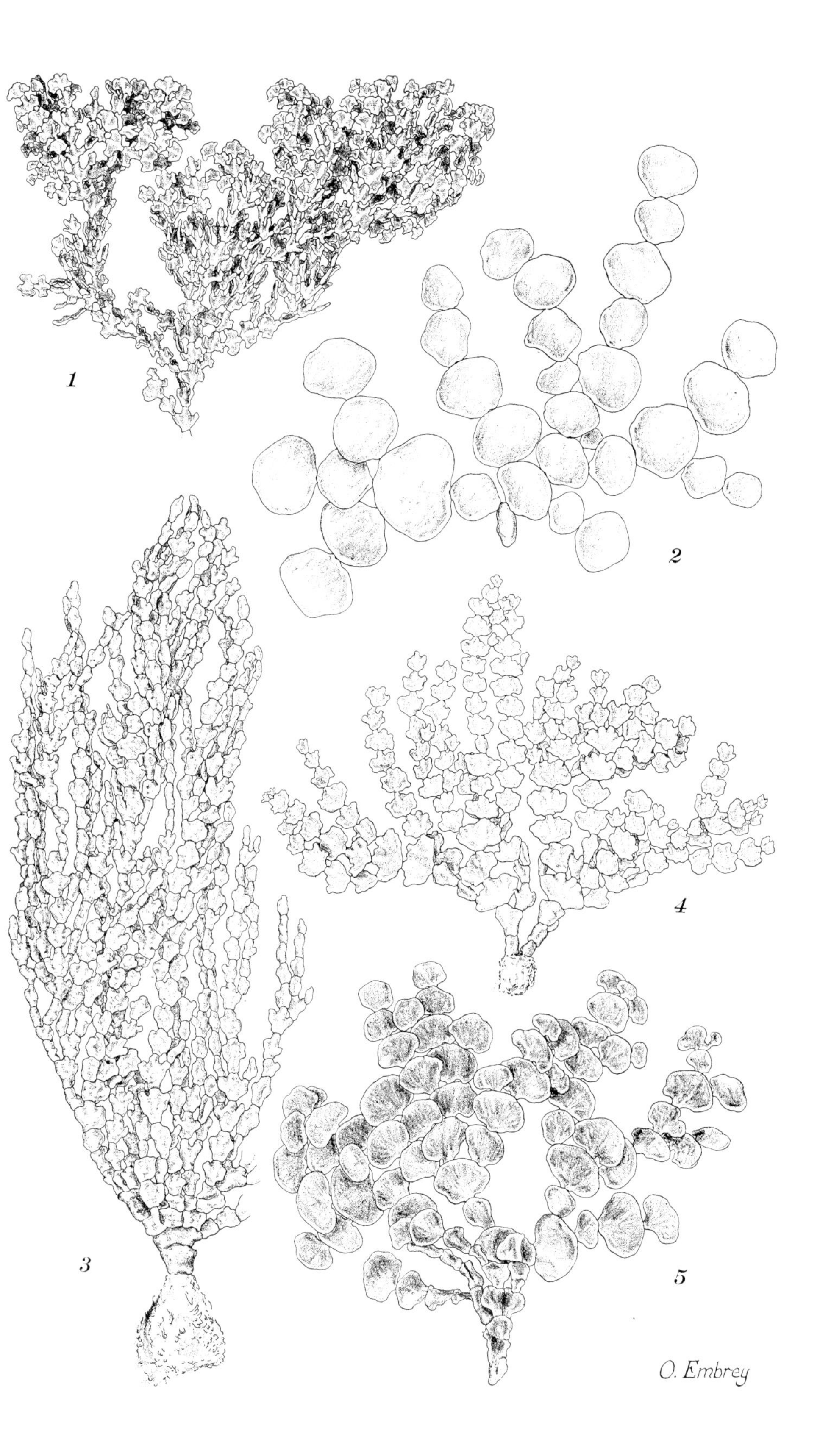
1
2
3
4
5
O. Embrey

PLATE 25

PAGE

FIG. 1. *Penicillus pyriformis:* tips of a cortical filament from the stalk, × 100 170

FIG. 2. *Penicillus lamourouxii:* terminal portion of a cortical filament from the stalk, × 100 172

FIG. 3. *Udotea flabellum:* tips of a cortical filament from the stalk, × 100 168

FIG. 4. *Penicillus capitatus:* tips of two cortical filaments from the stalk, × 100 171

FIG. 5. *Udotea conglutinata:* tips of a cortical filament from the stalk, × 105 165

FIG. 6. *Udotea spinulosa:* tips of a cortical filament from the stalk, × 100 167

FIG. 7. *Rhipocephalus oblongus:* tips of a cortical filament from the stalk, × 100 173

FIGS. 8–9. *Cladocephalus luteofuscus:* three cortical stipe filaments, × 105 163

FIG. 10. *Halimeda scabra:* two clusters of sporangia, × 20 180

FIGS. 11–12. *Avrainvillea nigricans* f. *fulva,* portions of two filaments from the blade, × 55 160

FIG. 13. *Halimeda scabra:* a few cortical filaments showing the acute projections of the surface utricles in lateral view, × 175 180

FIG. 14. *Halimeda scabra:* a portion of the cortex in surface view, × 250 180

FIG. 15. *Penicillus dumetosus:* portions of two filaments from the terminal tuft, × 6.5 172

FIGS. 16–17. *Udotea spinulosa:* portions of two filaments from the blade showing unilateral simple and forked projections, × 100 ... 167

FIG. 18. *Udotea wilsoni:* portion of a filament showing branched projections bilaterally arranged, × 195 167

For additional figures of *Avrainvillea* see Plate 19; for *Cladocephalus,* Plate 20; for *Halimeda,* Plates 23, 24; for *Penicillus,* Plate 21; for *Rhipocephalus,* Plate 22; for *Udotea,* Plates 20, 22. These figures are from Taylor, *Marine Algae of Florida* (1928).

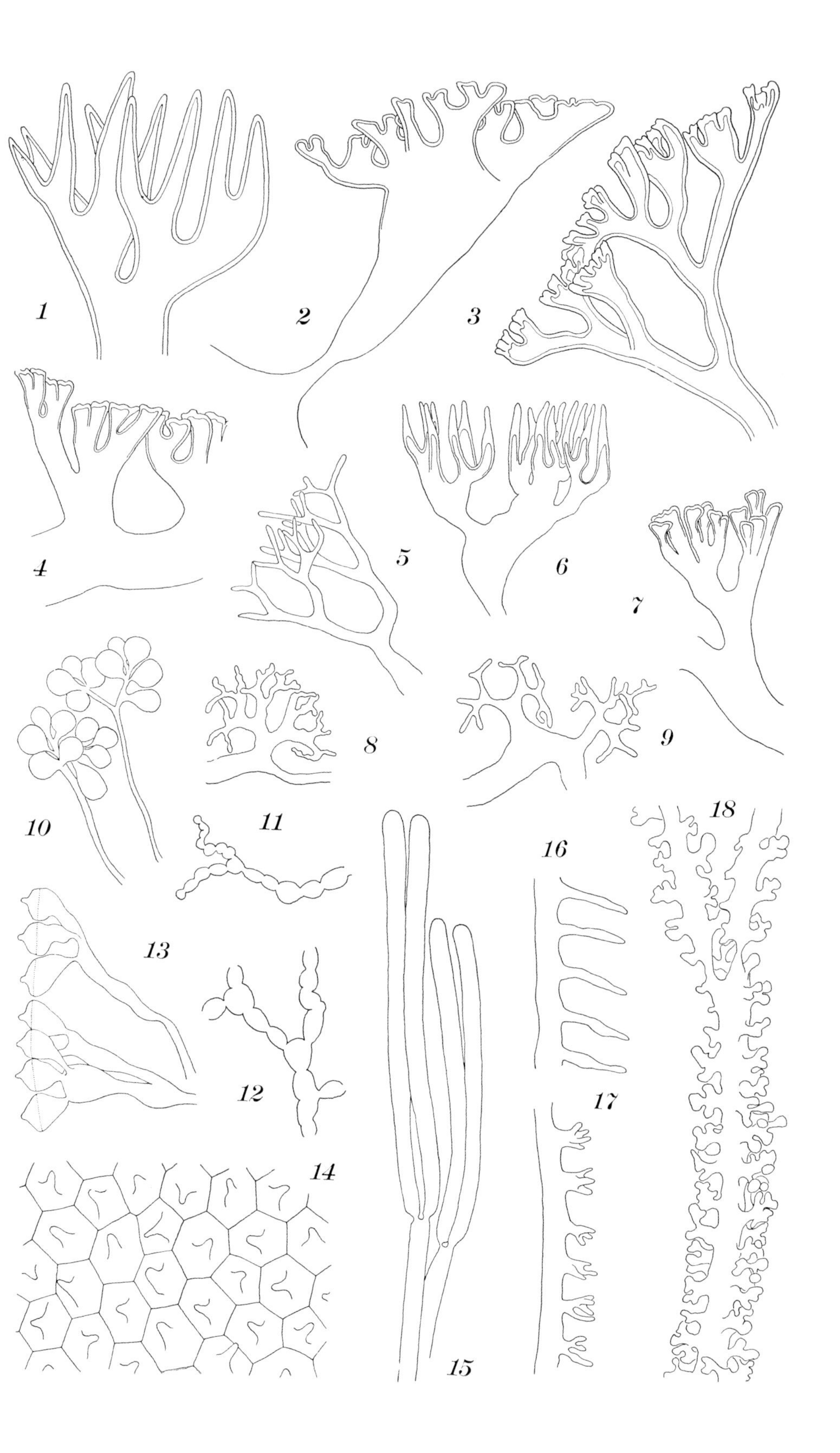
1
2
3
4
5
6
7
8
9
10
11
12
13
14
15
16
17
18

PLATE 26

PAGE

FIG. 1. *Codium decorticatum:* terminal branching in a very bushy specimen, × 0.5 188

FIG. 2. *Codium decorticatum:* habit of a sparsely branched specimen with well-flattened axils, × 0.5 188

FIG. 3. *Codium isthmocladum:* habit of part of a large plant, × 0.75. 186

FIG. 4. *Codium taylori:* habit of part of a plant, × 0.75 188

1
2
3
4

PLATE 27

PAGE

FIGS. 1–2. *Boodleopsis pusilla:* free vegetative filaments showing constrictions; vegetative filaments showing coralloid branching; a filament with a sporangium-like structure, × 145, × 235........ 157

FIGS. 3–5. *Vaucheria bermudensis:* an antheridium and an empty cell subjacent; an oöspore in an oögonium with an antheridium on an adjacent branch; an aplanosporangium, × 155, × 155, × 110 192

FIGS. 6–9. *Vaucheria nasuta:* two antheridia each with an empty cell subjacent, × 365, × 325, and two oögonia each with a beak, × 260 191

FIGS. 10–11. *Vaucheria piloboloides:* an antheridium with an empty cell subjacent; oöspore in an oögonium, with an antheridium on an adjacent branch, × 155................................ 192

These figures are from Taylor and Bernatowicz (1952, 1953) and Taylor, Joly, and Bernatowicz (1953).

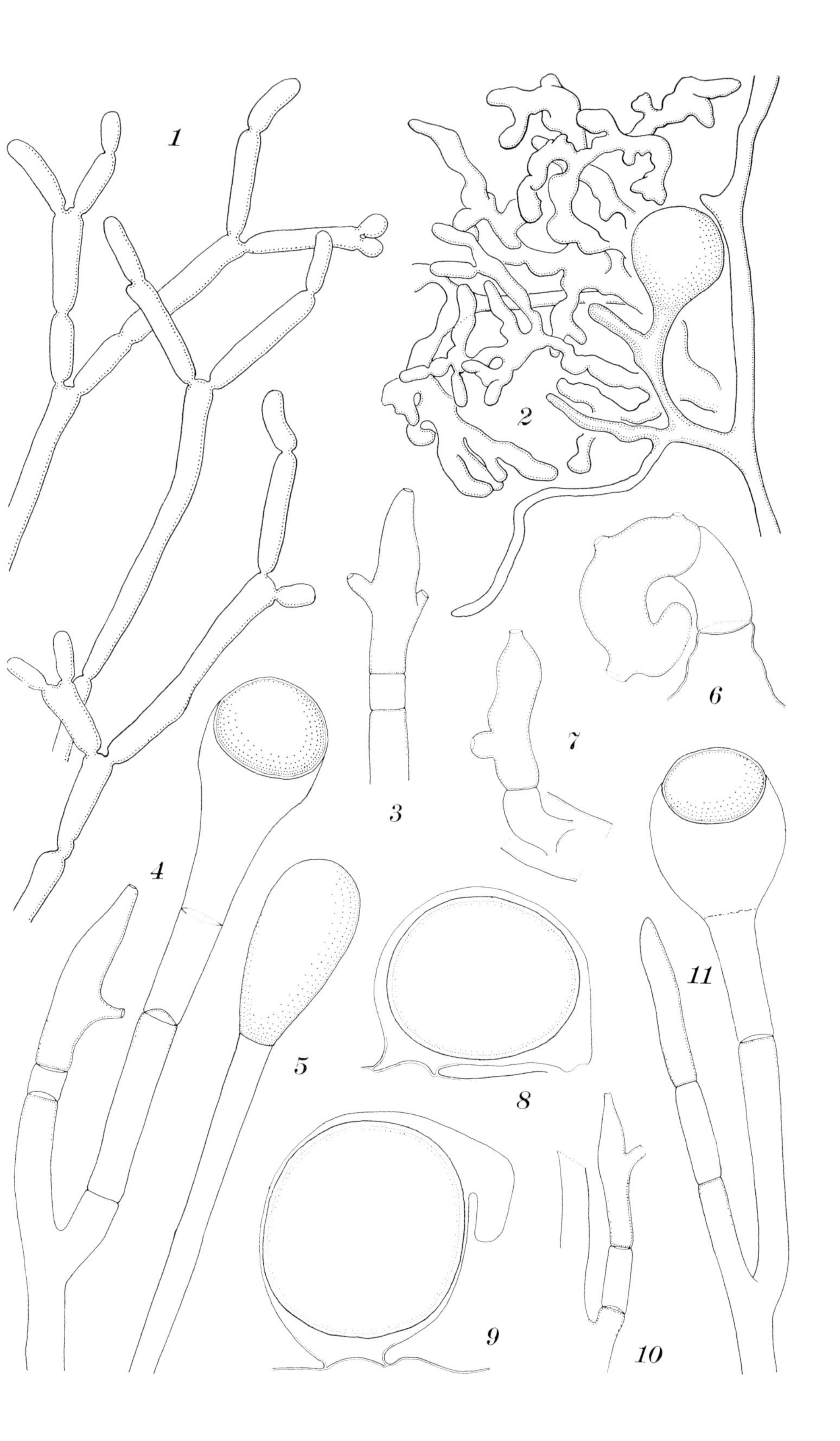
1
2
3
4
5
6
7
8
9
10
11

PLATE 28

PAGE

FIG. 1. *Stypopodium zonale:* plant habit × 0.6 232

FIG. 2. *Chrysonephos lewisii:* habit of part of a branch, × 110 195

FIG. 3. *Chrysonephos lewisii:* fork of a main branch, × 490 195

FIG. 4. *Chrysonephos lewisii:* apex of a branch newly divided, × 490. 195

FIG. 5. *Chrysonephos lewisii:* chromatophores in the cells of a main branch, × 1100 195

FIG. 6. *Chrysonephos lewisii:* apex of a branch during zoöid formation, × 460 195

FIG. 7. *Chrysonephos lewisii:* single zoöid, × 1100 195

FIG. 8. *Sphacelaria novae-hollandiae:* lateral views of three propagulae, × 245 211

For additional figures of *Sphacelaria* see Plate 29. Figures 2–7 are from Taylor (1951b, redrawn).

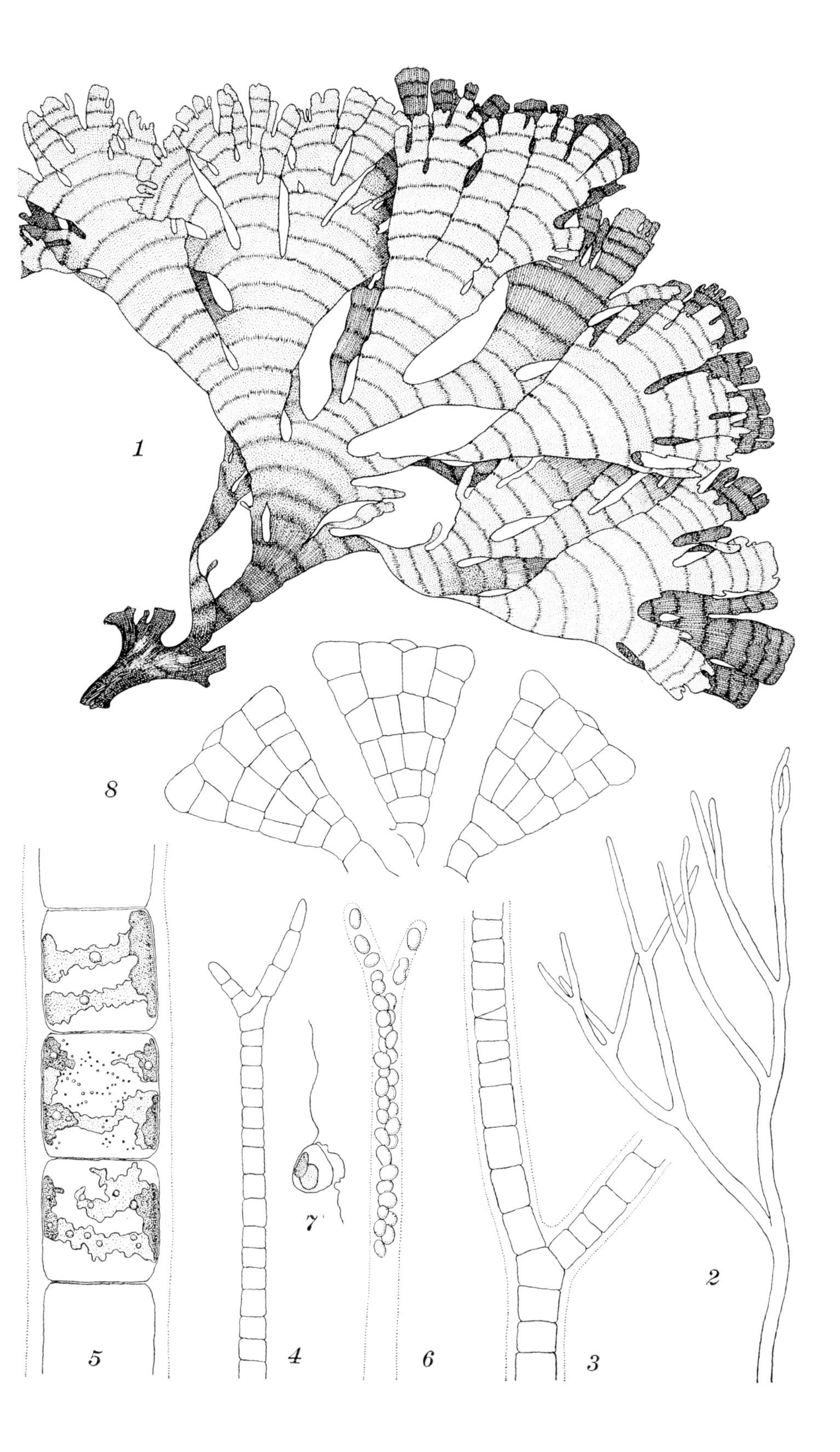
1
8
5
4
7
6
3
2

PLATE 29

PAGE

For additional figures of *Nereia* see Plate 35; for *Rosenvingea*, Plate 36; for *Sphacelaria*, Plate 28. These figures are in part from Taylor, *Marine Algae of Florida* (1928).

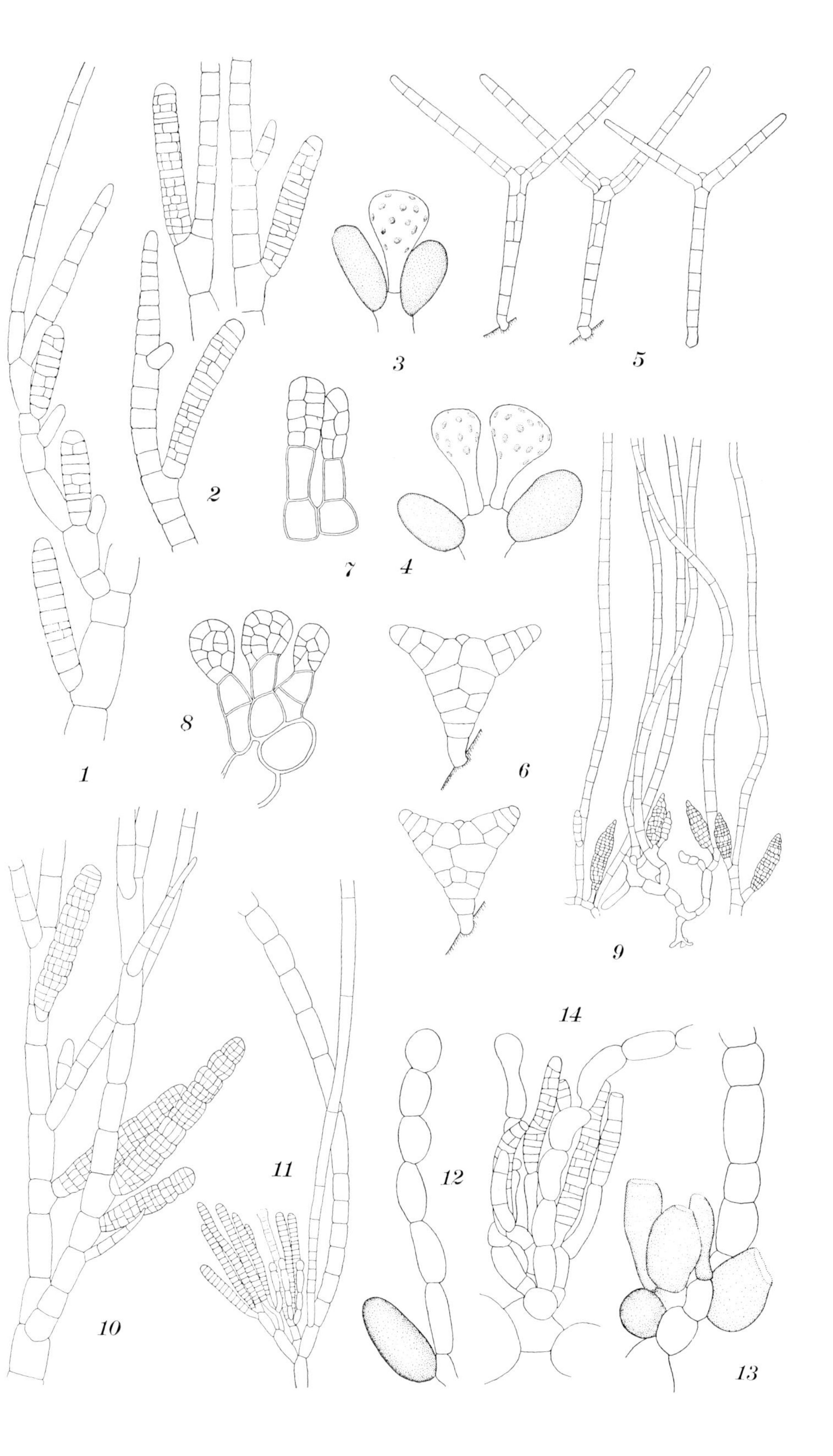
3
5
2
7
4
8
6
1
9
14
11
12
10
13

PLATE 30

PAGE

FIG. 1. *Dilophus guineensis:* a small portion of a plant, × 1.1 216

FIG. 2. *Dictyota bartayresii:* habit of a small, rather plane plant, × 0.55 219

FIG. 3. *Dilophus alternans:* a small portion of a plant, × 1.1 216

FIGS. 4–5. *Dictyota dentata:* portions of two plants to show the acute and the rounded branch tip forms, × 0.55 224

For additional figures of *Dictyota* see Plates 31, 32, 59.

1
2
3
4
5

PLATE 31

For additional figures of *Dictyota* see Plates 30, 32, 59.

1
2
3
4
5
6

PLATE 32

PAGE

FIG. 1. *Dictyopteris hoytii:* habit of part of a plant, × 0.75. Plants are normally more regularly dichotomous 229

FIG. 2. *Dictyopteris jamaicensis:* habit of part of a plant, × 1.1 . . 228

FIG. 3. *Dictyota ciliolata:* habit of part of a plant, × 0.6 223

FIG. 4. *Dictyota jamaicensis:* plant habit, × 0.75 223

FIG. 5. *Dictyota jamaicensis:* detail of part of a branch system to show the character of the marginal teeth, × 1.5 223

For additional figures of *Dictyopteris* see Plate 33; for *Dictyota,* Plates 30, 31, 59.

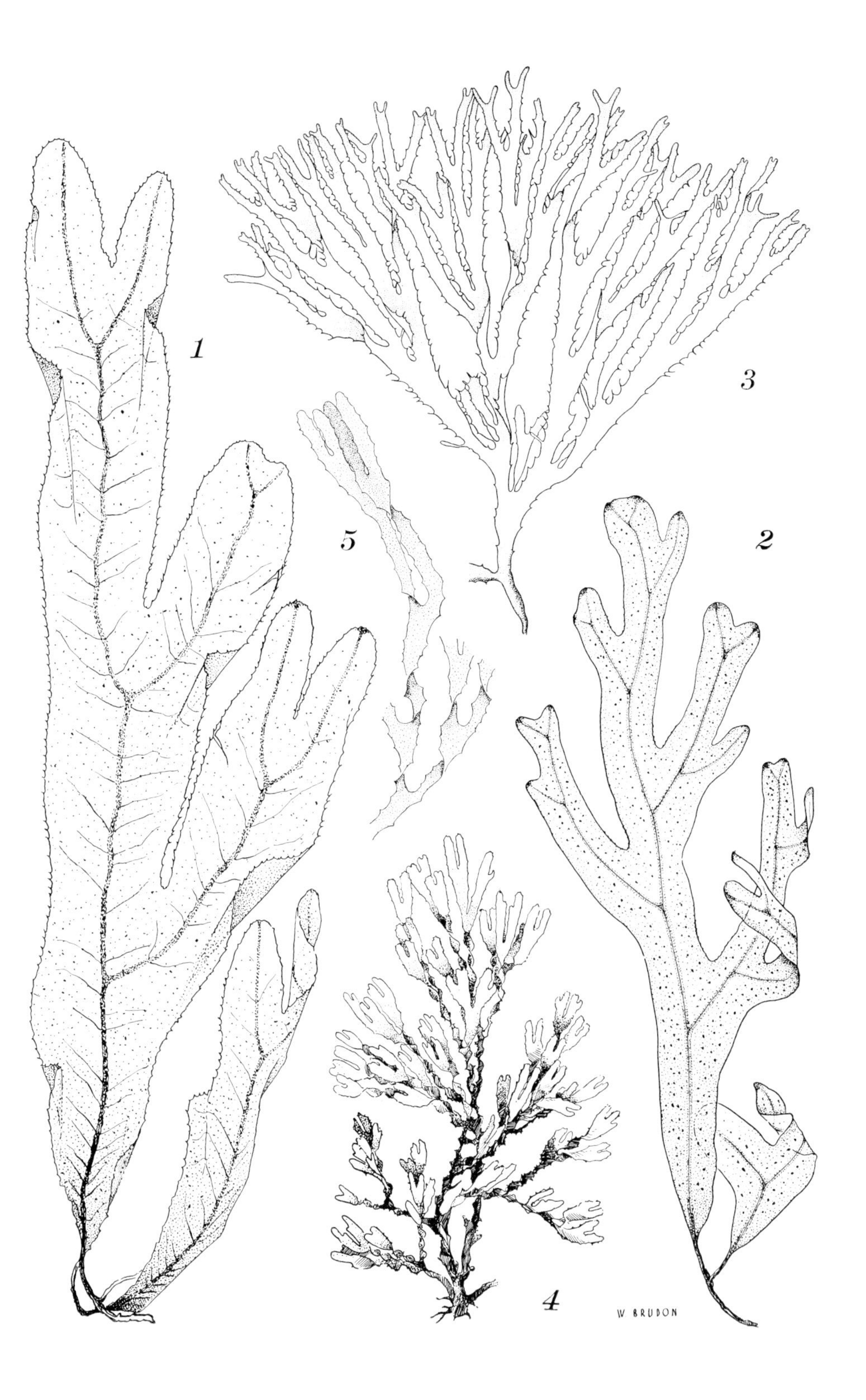
1
2
3
4
5
W BRUDON

PLATE 33

PAGE

FIG. 1. *Dictyopteris justii:* plant habit, × 0.5.................... 226

FIG. 2. *Dictyopteris plagiogramma:* plant habit, and details of venation in a branchlet, × 0.5, 0.85.......................... 229

FIG. 3. *Dictyopteris delicatula:* plant habit, × 1.1............... 227

FIG. 4. *Pocockiella variegata:* plant habit, × 1.1................. 231

FIG. 5. *Spatoglossum schroederi:* plant habit, × 0.85.............. 225

For additional figures of *Dictyopteris* see Plate 32.

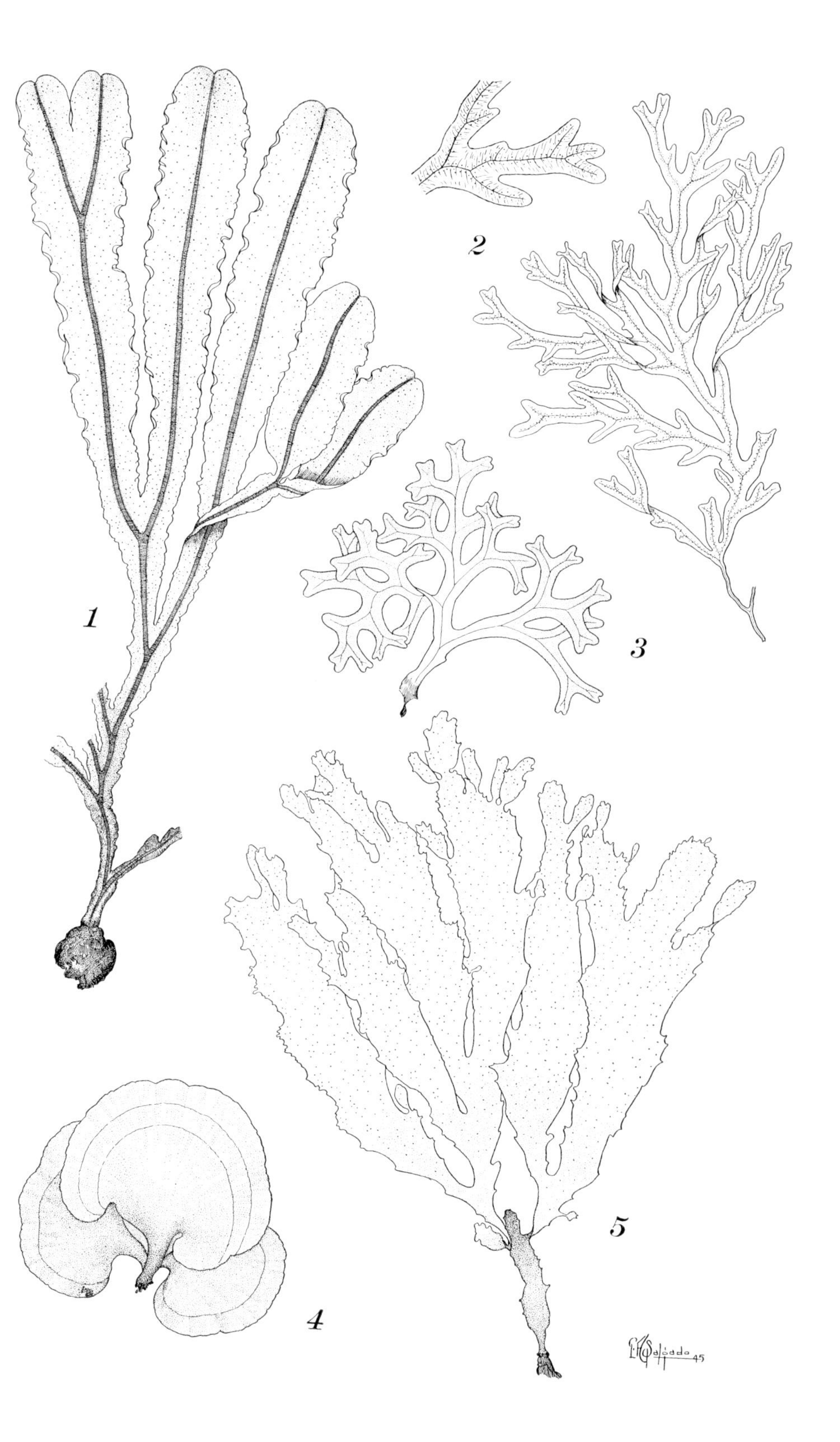

1
2
3
4
5

PLATE 34

PAGE

FIG. 1. *Padina vickersiae:* plant habit, × 1.0 236

FIG. 2. *Padina sanctae-crucis:* plant habit, × 1.0 237

For additional figures of *Padina* see Plate 75.

1
2

PLATE 35

PAGE

Fig. 1. *Nereia tropica:* habit of part of a plant, × 0.75 252

Fig. 2. *Sporochnus bolleanus:* habit of part of a plant, × 0.6 253

Fig. 3. *Sporochnus bolleanus:* details of two fertile branchlets, × 5.0. 253

Fig. 4. *Sporochnus pedunculatus:* habit of part of a plant, × 0.9 ... 253

Fig. 5. *Sporochnus pedunculatus:* details of a fertile branchlet, × 8.0. 253

For additional figures of *Nereia* see Plate 29.

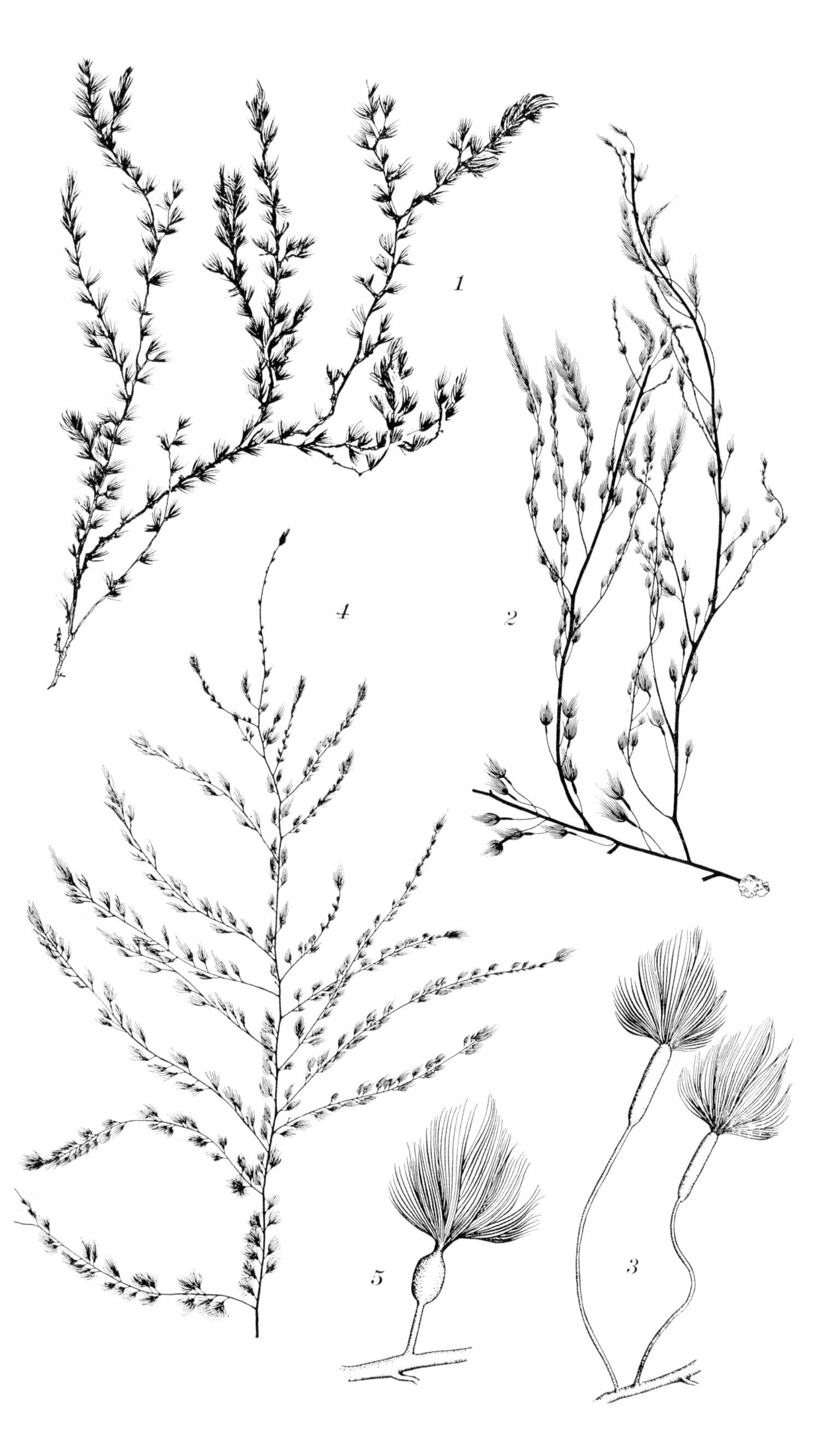
1
4
2
5
3

PLATE 36

PAGE

FIG. 1. *Colpomenia sinuosa:* plant habit, × 0.5.................. 260

FIG. 2. *Rosenvingea intricata:* habit of part of a plant, × 0.75..... 262

FIGS. 3–4. *Chnoospora minima:* habit of a small, tufted plant and of a large loosely branched one, × 0.8......................... 263

FIG. 5. *Hydroclathrus clathratus:* habit of part of a plant, × 0.5... 261

For additional figures of *Rosenvingea* see Plate 29.

1
2
3
4
5

PLATE 37

For additional figures of *Sargassum* see Plates 38–40.

1
2
3

PLATE 38

PAGE

FIG. 1. *Sargassum vulgare:* details of a main branch, × 1.3........ 272

FIG. 2. *Sargassum hystrix* v. *buxifolium:* details of a main branch, × 1.0 279

FIG. 3. *Sargassum platycarpum:* details of two lateral branches, × 0.8 280

FIG. 4. *Sargassum cymosum:* details of a lateral branch, × 1.3.... 278

For additional figures of *Sargassum* see Plates 37, 39, 40.

1
2
3
4

PLATE 39

For additional figures of *Sargassum* see Plates 37, 38, 40.

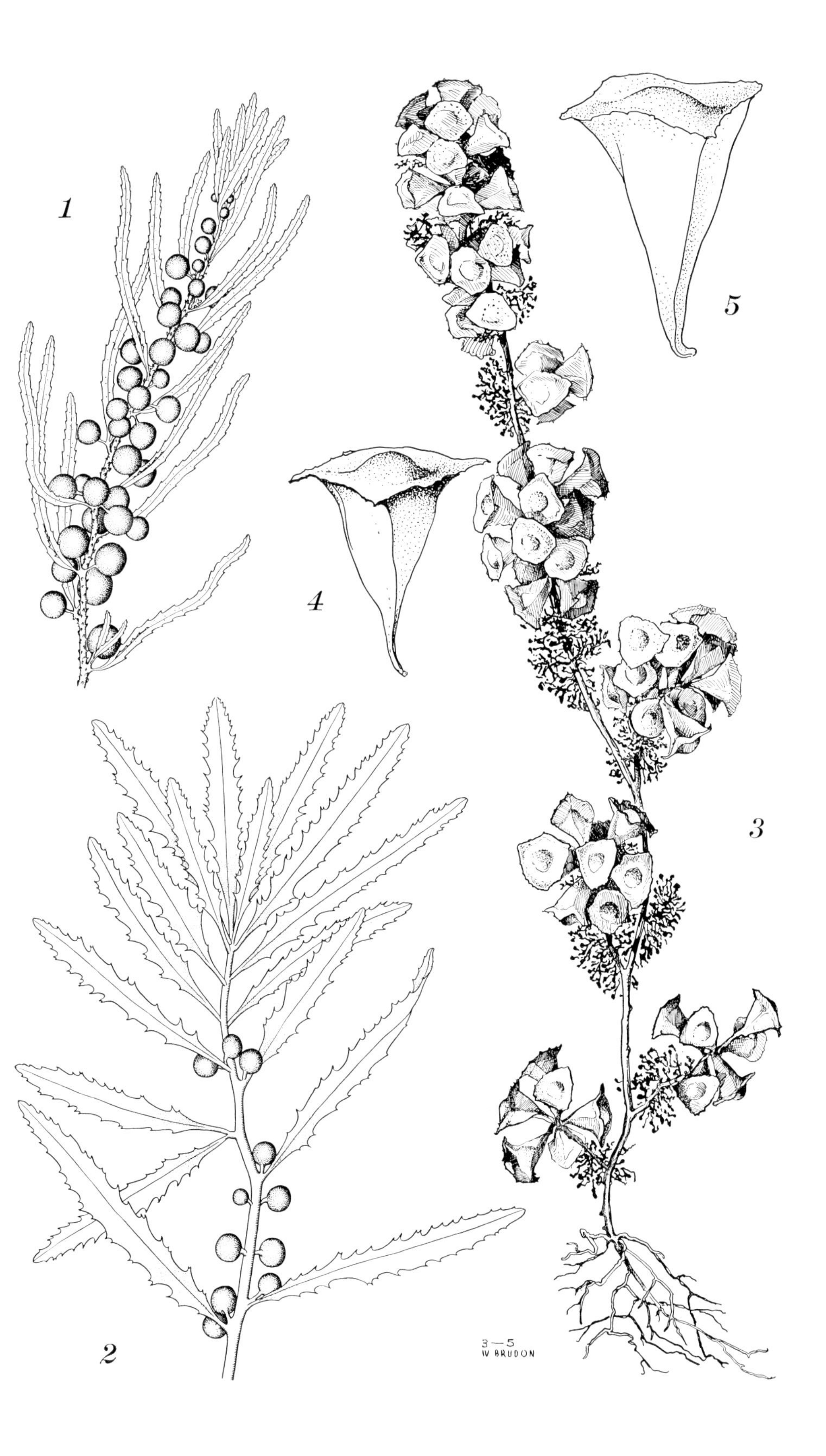
1
2
3
4
5
3—5
W BRUDON

PLATE 40

PAGE

FIG. 1. *Sargassum polyceratium:* leaves, × 1.1 276

FIG. 2. *Sargassum filipendula:* leaves, × 1.1 270

FIG. 3. *Sargassum natans:* tip portion of a branchlet, and one vesicle with a phylloid tip, × 1.1 281

FIG. 4. *Sargassum pteropleuron:* leaves, × 1.1 274

FIG. 5. *Sargassum vulgare:* leaves, × 1.5 272

FIG. 6. *Sargassum hystrix* v. *buxifolium:* leaves, × 1.1 279

FIG. 7. *Sargassum fluitans:* leaves, × 1.1 281

FIG. 8. *Sargassum natans:* leaves, × 1.1 281

FIG. 9. *Sargassum pteropleuron:* tip of a main branch, × 1.1 274

For additional figures of *Sargassum* see Plates 37–39. These figures are in part from Taylor (1928, redrawn).

1
2
3
4
5
6
7
8
9

PLATE 41

PAGE

FIG. 1. *Erythrocladia subintegra:* plant habit, × 555 290

FIGS. 2–3. *Erythrotrichia vexillaris:* habit of a rather young plant and base of an older one with secondary attachment filaments from the lower cells, × 335 292

FIG. 4. *Erythrocladia pinnata:* habit of part of a plant, × 555 290

FIGS. 5–7. *Callithamnion uruguayense:* mature branchlet, branch tip with branchlets and sporangial rudiments, and branchlets with tetrasporangia × 155, × 70, × 155 506

These figures are from Taylor (1939a, 1942).

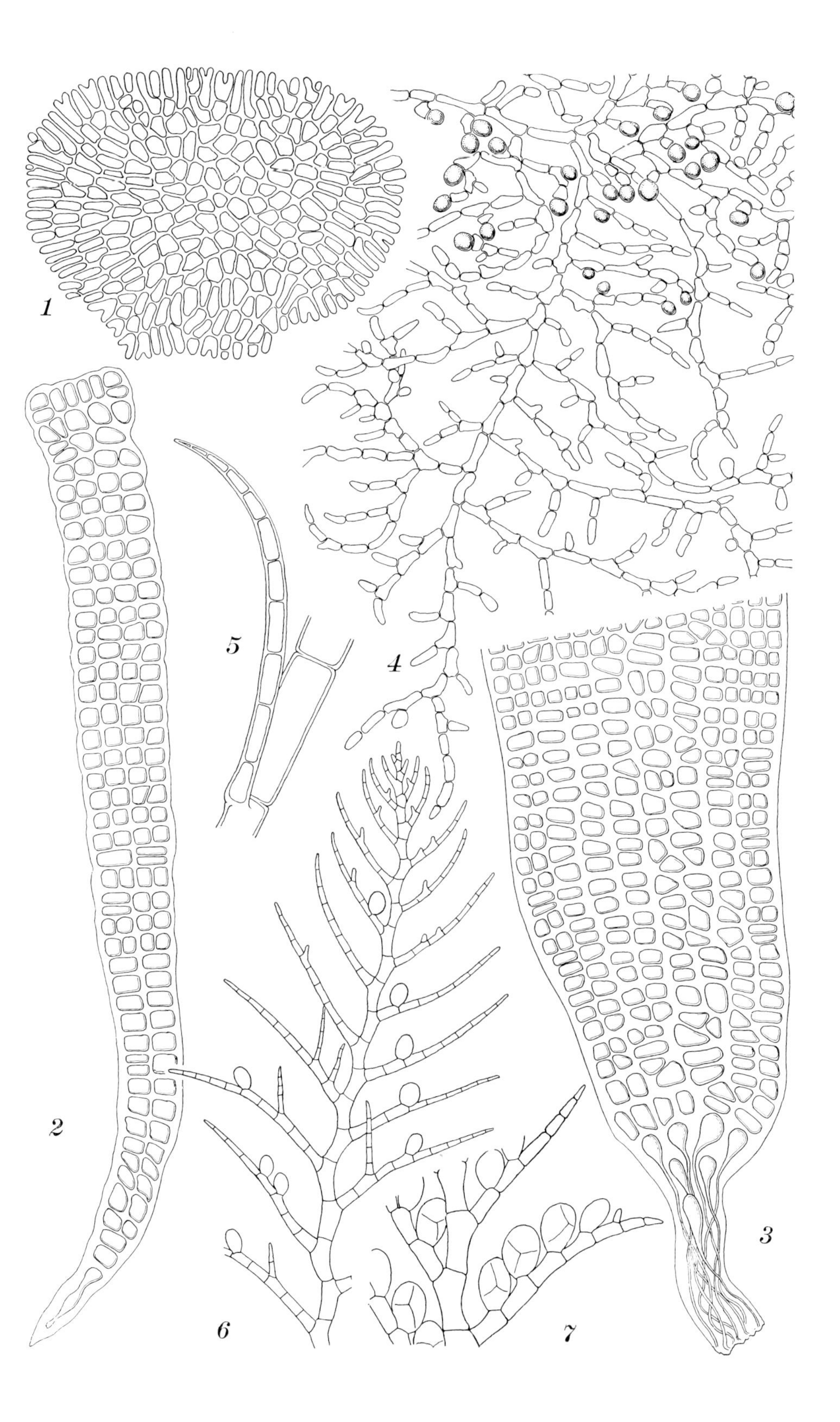

1
2
3
4
5
6
7

PLATE 42

PAGE

1
2
3
4

PLATE 43

PAGE

FIG. 1. *Liagora ceranoides:* habit of part of a plant, × 0.75...... 326

FIG. 2. *Liagora valida:* habit of part of a plant, × 1.0............. 327

FIG. 3. *Liagora farinosa:* plant habit, × 0.75.................... 326

FIG. 4. *Platoma cyclocolpa:* branch of a moderately dissected plant, × 1.0 437

FIG. 5. *Helminthocladia calvadosii:* details of a branch, × 0.75.... 324

FIG. 6. *Dudresnaya crassa:* details of a branch, × 1.0........... 363

For additional figures of *Liagora* see Plate 45.

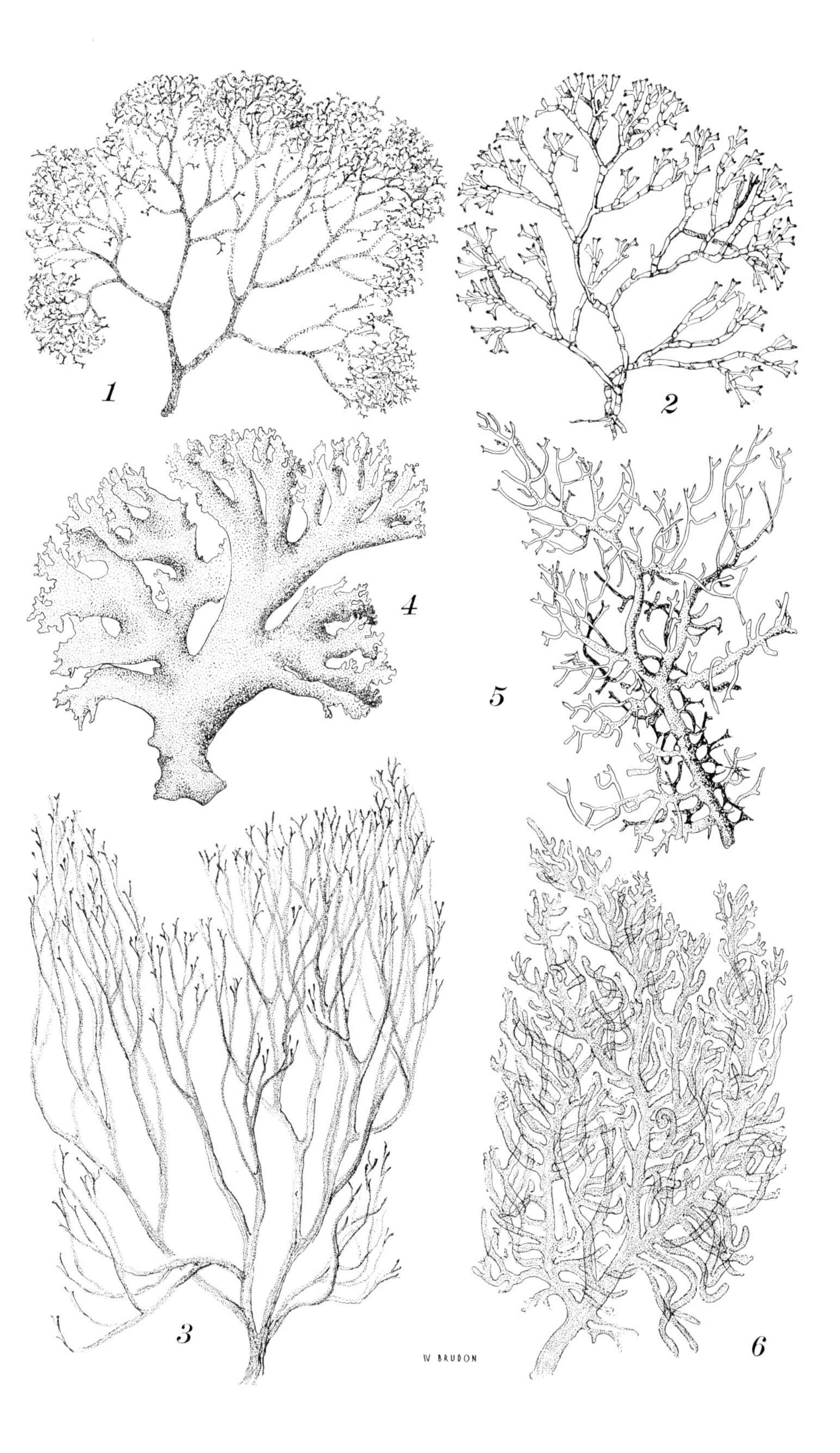
1
2
4
5
3
W. BRUDON
6

PLATE 44

PAGE

FIG. 1. *Galaxaura cylindrica:* a portion of a plant, × 1.0.......... 341

FIG. 2. *Galaxaura marginata:* habit of a small plant, × 0.75........ 343

FIG. 3. *Galaxaura squalida:* a portion of a plant, × 0.5........... 339

FIG. 4. *Galaxaura obtusata:* part of a plant, × 0.5................ 342

FIG. 5. *Galaxaura obtusata* v. *major:* part of a plant, × 0.5....... 342

FIG. 6. *Galaxaura subverticillata:* a portion of a plant, × 0.5...... 339

For additional figures of *Galaxaura* see Plate 45.

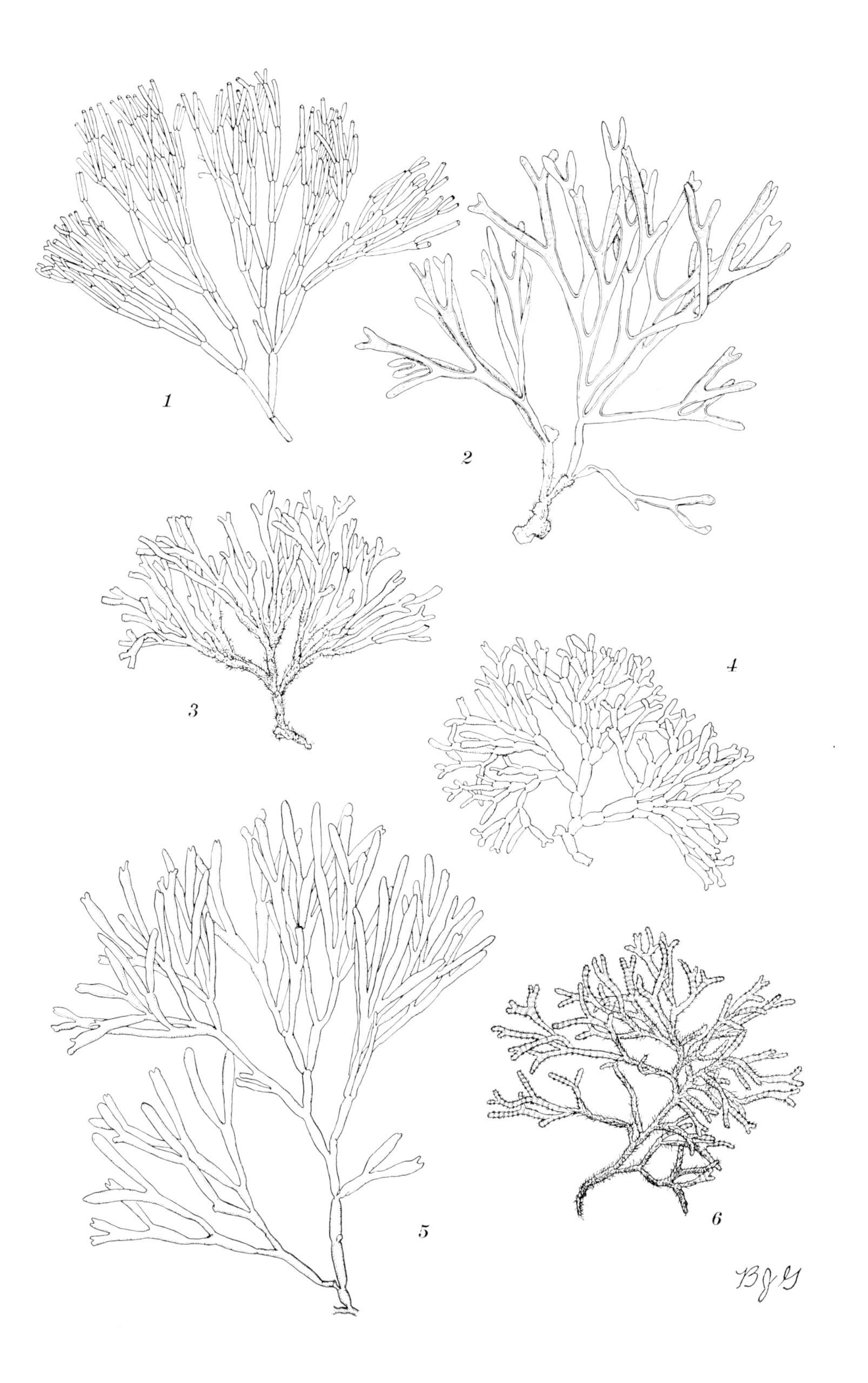
1
2
3
4
5
6

PLATE 45

PAGE

Fig. 1. *Liagora ceranoides:* surface branchlets bearing spermatangia, × 425. 326

Fig. 2. *Liagora farinosa:* stalked cluster of spermatangia, × 425... 326

Fig. 3. *Coelothrix irregularis:* transverse section of the stem with filaments and a gland in the cavity, × 125.......... 488

Fig. 4. *Gelidium pusillum:* habit of two plants, × 4.5.......... 354

Fig. 5. *Galaxaura obtusata:* t.s. of portion of the cortex, × 280.... 342

Fig. 6. *Galaxaura squalida:* t.s. of portion of the cortex from the lower part of a plant, × 280.......... 339

Figs. 7–8. *Galaxaura marginata:* t.s. of portions of the cortex from tetrasporic and sexual plants, × 395, × 280.......... 343

Fig. 9. *Galaxaura subverticillata:* t.s. of a portion of the cortex from a zone with long filaments, × 280.......... 339

Fig. 10. *Gracilaria debilis:* t.s. of part of an axis, × 85.......... 442

Fig. 11. *Eucheuma isiforme:* t.s. of part of an axis, × 19.......... 459

Fig. 12. *Halymenia floresia:* t.s. of part of a blade, × 90........ 418

For additional figures of *Coelothrix* see Plate 46; for *Eucheuma*, Plate 50; for *Galaxaura*, Plate 44; for *Gracilaria*, Plates 54–57, 59; for *Halymenia*, Plates 51–53; for *Liagora*, Plate 43. These figures are from Taylor, *Marine Algae of Florida* (1928, redrawn).

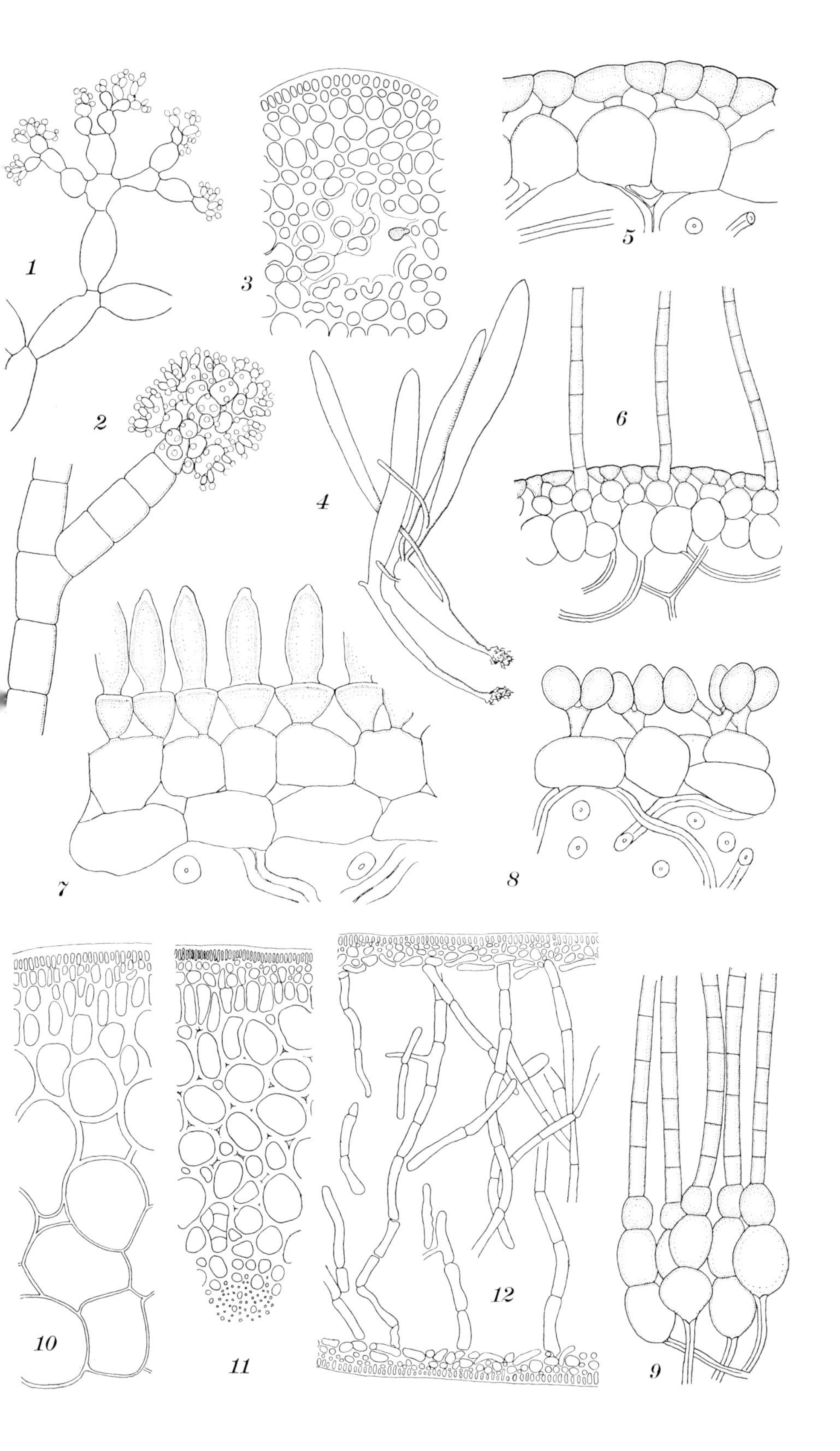
1
2
3
4
5
6
7
8
9
10
11
12

PLATE 46

	PAGE
FIG. 1. *Pterocladia pinnata:* plant habit, × 1.0	361
FIG. 2. *Pterocladia bartlettii:* plant habit, × 2.0	359
FIG. 3. *Pterocladia americana:* plant habit, × 1.0	360
FIG. 4. *Coelothrix irregularis:* plant habit, × 1.5	488
FIG. 5. *Gelidiella acerosa:* plant habit, × 1.0	351

For an additional figure of *Coelothrix* see Plate 45.

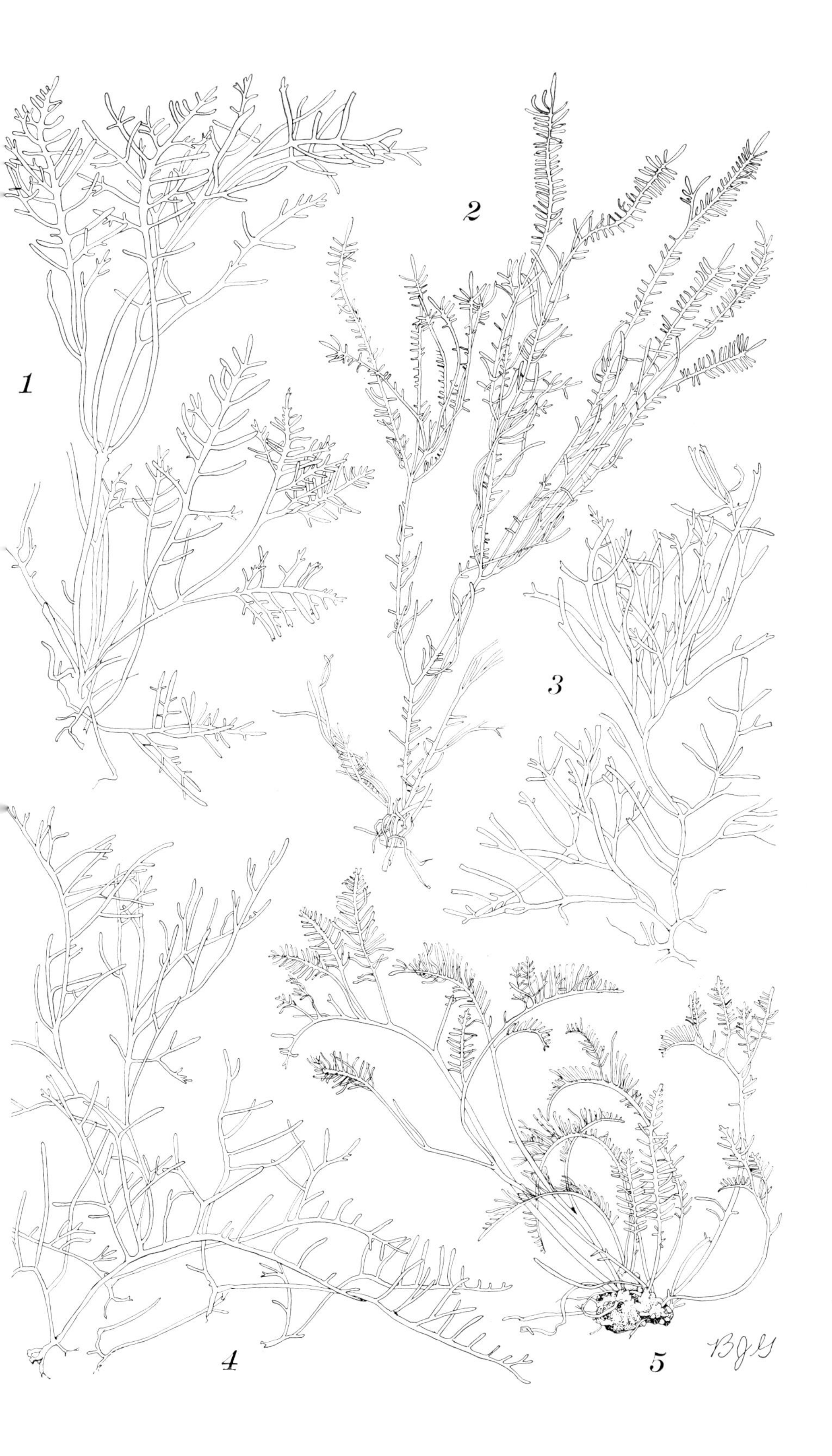
1
2
3
4
5
BJG

PLATE 47

PAGE

FIG. 1. *Amphiroa fragilissima:* habit of the upper part of a plant, × 3.5 403

FIG. 2. *Amphiroa fragilissima:* details of part of a branch with conceptacles and of a fork to show the terminal branch swellings more exactly, × 10 403

FIG. 3. *Amphiroa rigida* v. *antillana:* habit of part of a plant, × 3.5. 404

FIG. 4. *Amphiroa tribulus:* plant habit, most of the branches in the subterete condition, × 1.8 406

FIG. 5. *Amphiroa tribulus:* detail of one branch segment of the flat type, with conceptacles, × 3.0 406

FIGS. 6–7. *Amphiroa hancockii:* habit of portions of a plant with the segments chiefly in the flat phase, × 0.6 406

FIG. 8. *Amphiroa hancockii:* habit of a portion of a plant with the segments in the subterete phase, × 0.8 406

For additional figures of *Amphiroa* see Plate 48.

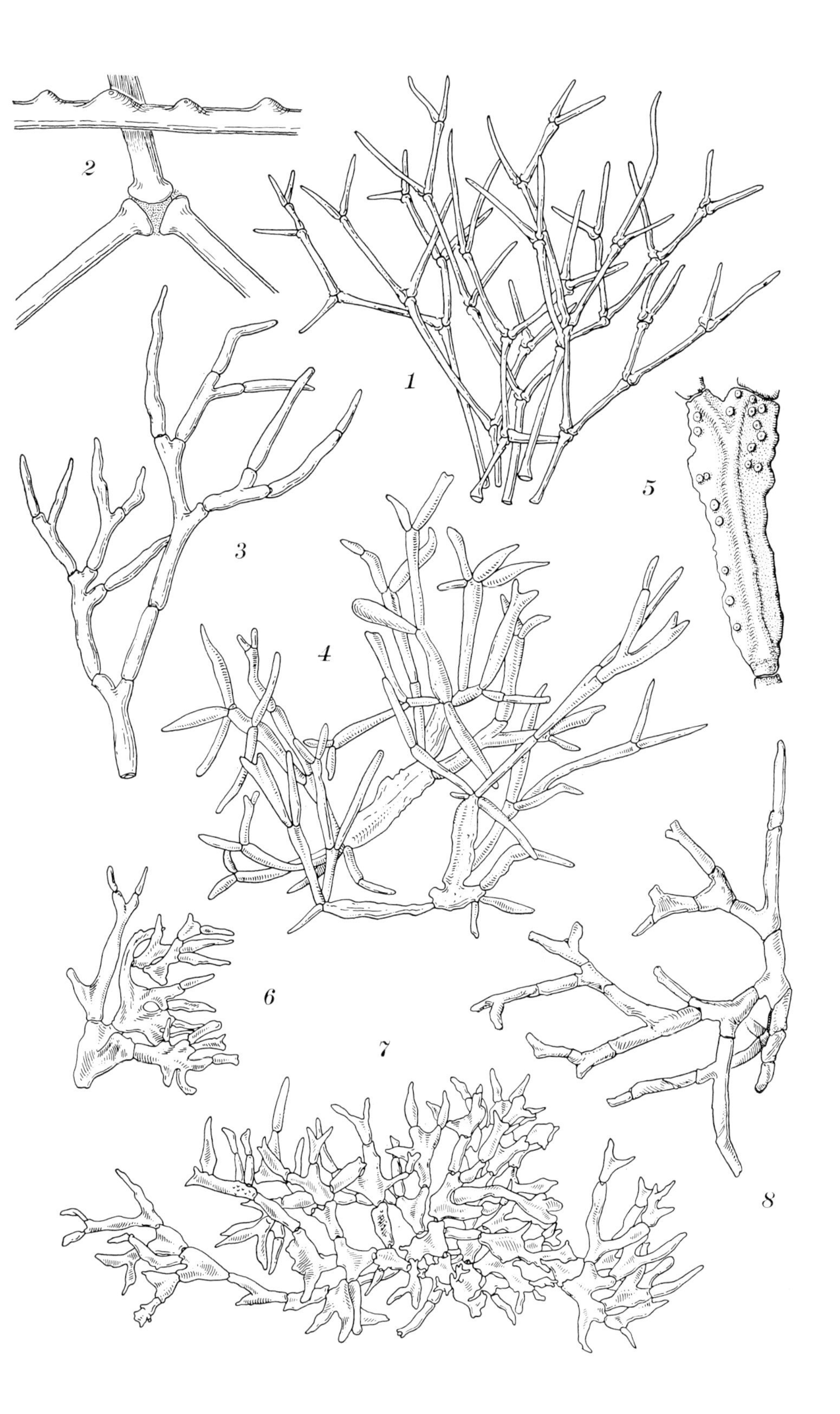
2
1
5
3
4
6
7
8

PLATE 48

PAGE

Fig. 1. *Amphiroa rigida* v. *antillana:* habit of part of a plant, × 1.6. 404

Fig. 2. *Amphiroa brasiliana:* habit of part of a plant with conceptacles, × 2.75 405

Fig. 3. *Cheilosporum sagittatum:* details of two branches, × 2.75.. 408

Fig. 4. *Goniolithon strictum:* portions of two branches, × 1.1..... 399

For additional figures of *Amphiroa* see Plate 47; for *Goniolithon,* Plates 76–78.

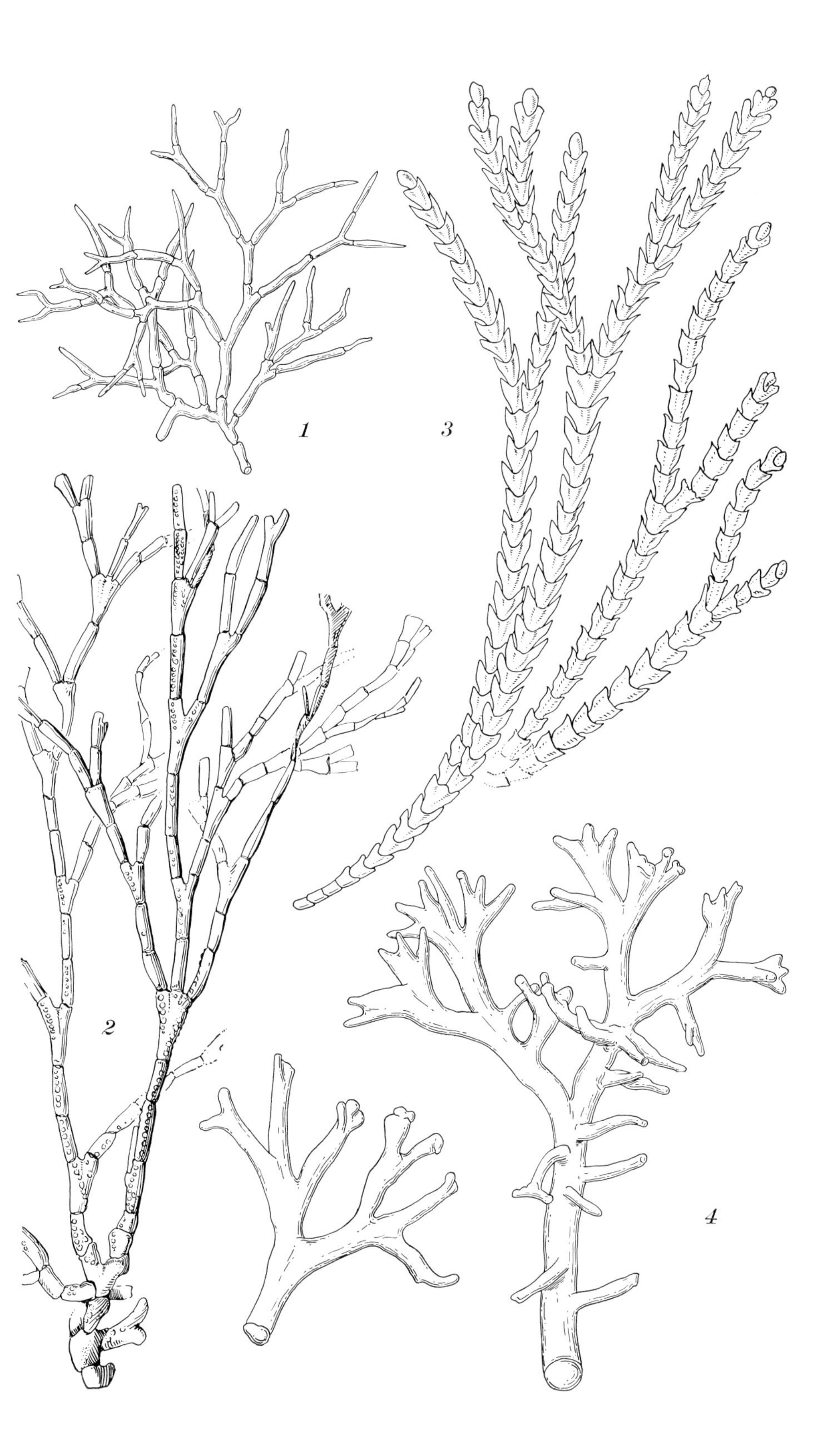
1
3
2
4

PLATE 49

PAGE

Figs. 1–2. *Jania adherens:* habit of a portion of a plant, and a part in more detail, × 10, × 15 413

Fig. 3. *Jania rubens:* habit of part of a plant, × 5 413

Fig. 4. *Jania capillacea:* habit of part of a fertile plant, × 15 412

Fig. 5. *Jania pumila:* habit of a plant group, × 17 414

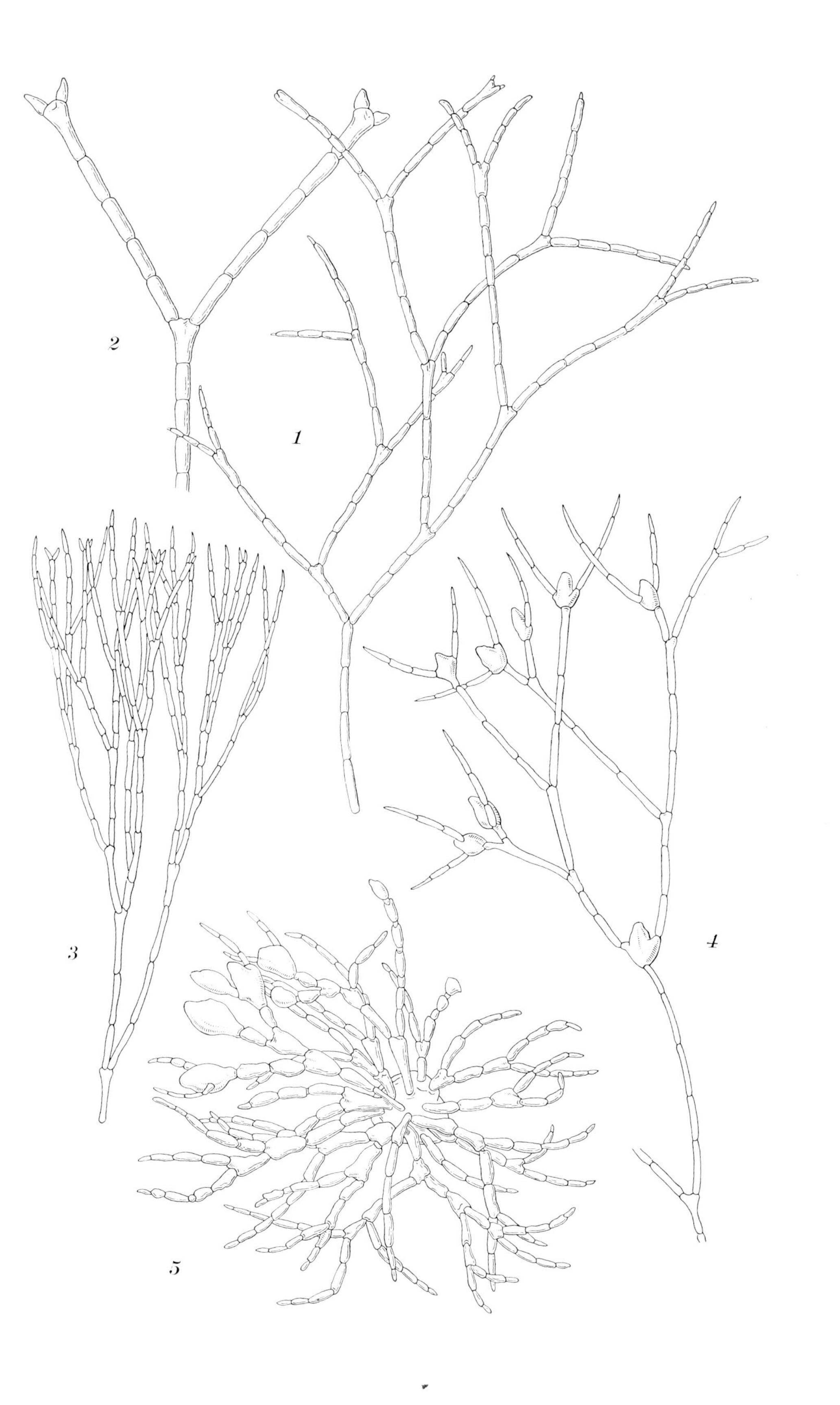

2
1
3
4
5

PLATE 50

PAGE

FIGS. 1–2. *Corallina subulata:* portions of two branches, fig. 1 with conceptacles, × 11 410

FIGS. 3–4. *Corallina cubensis:* portions of two branches, × 9, × 18. 409

FIG. 5. *Eucheuma acanthocladum:* a portion of a branch, × 1.8. 458

FIG. 6–7. *Eucheuma isiforme:* portions of two branches, fig. 7 with cystocarps, × 0.9, × 1.33. 459

For an additional figure of *Eucheuma* see Plate 45.

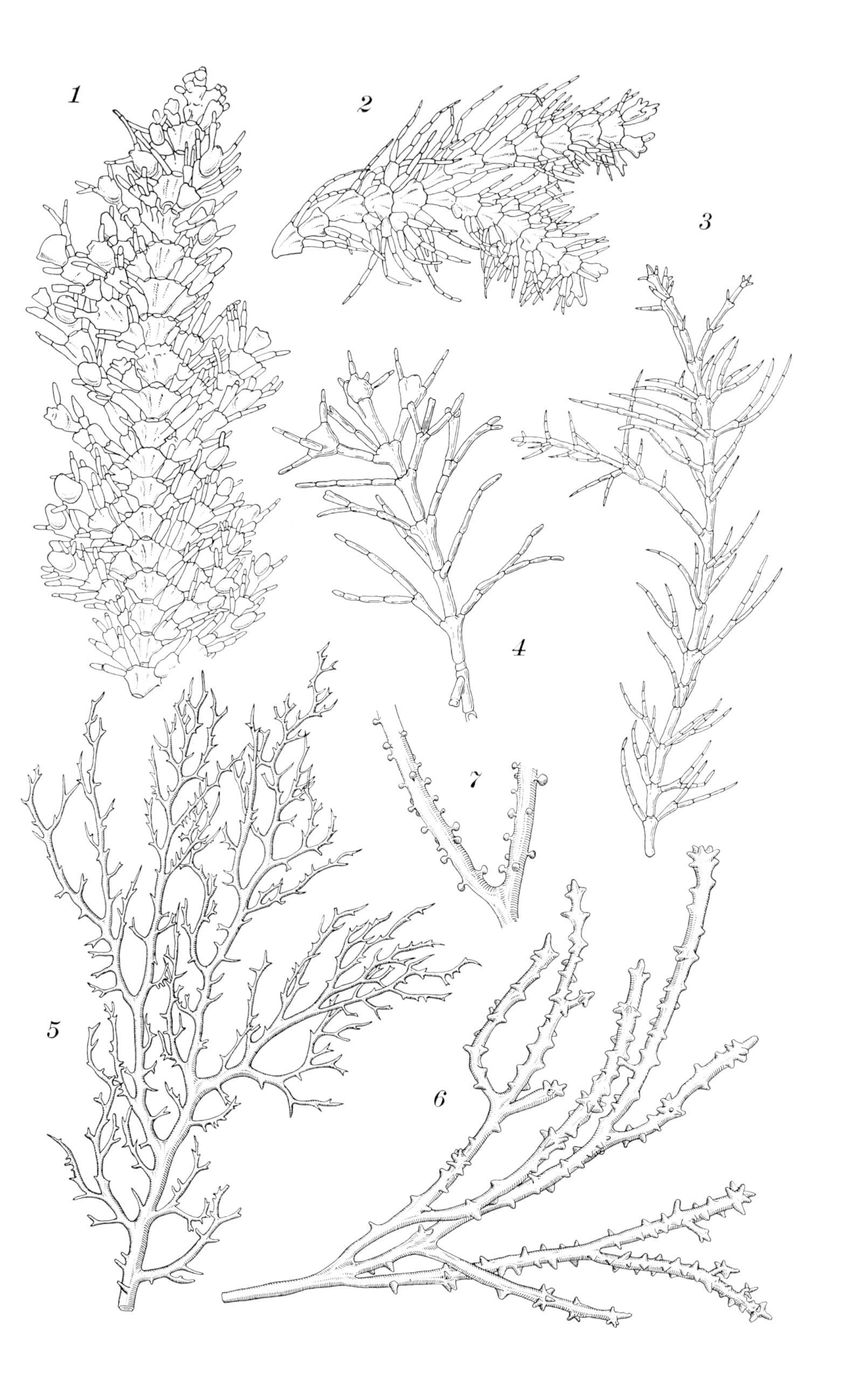
1
2
3
4
7
5
6

PLATE 51

PAGE

FIGS. 1–2. *Halymenia agardhii:* habit of two plants showing more flattened and more terete branches, × 1.0, 0.75 417

FIG. 3. *Halymenia floresia:* a small portion of a plant showing pinnate branching, × 1.0 418

For additional figures of *Halymenia* see Plates 45, 52, 53.

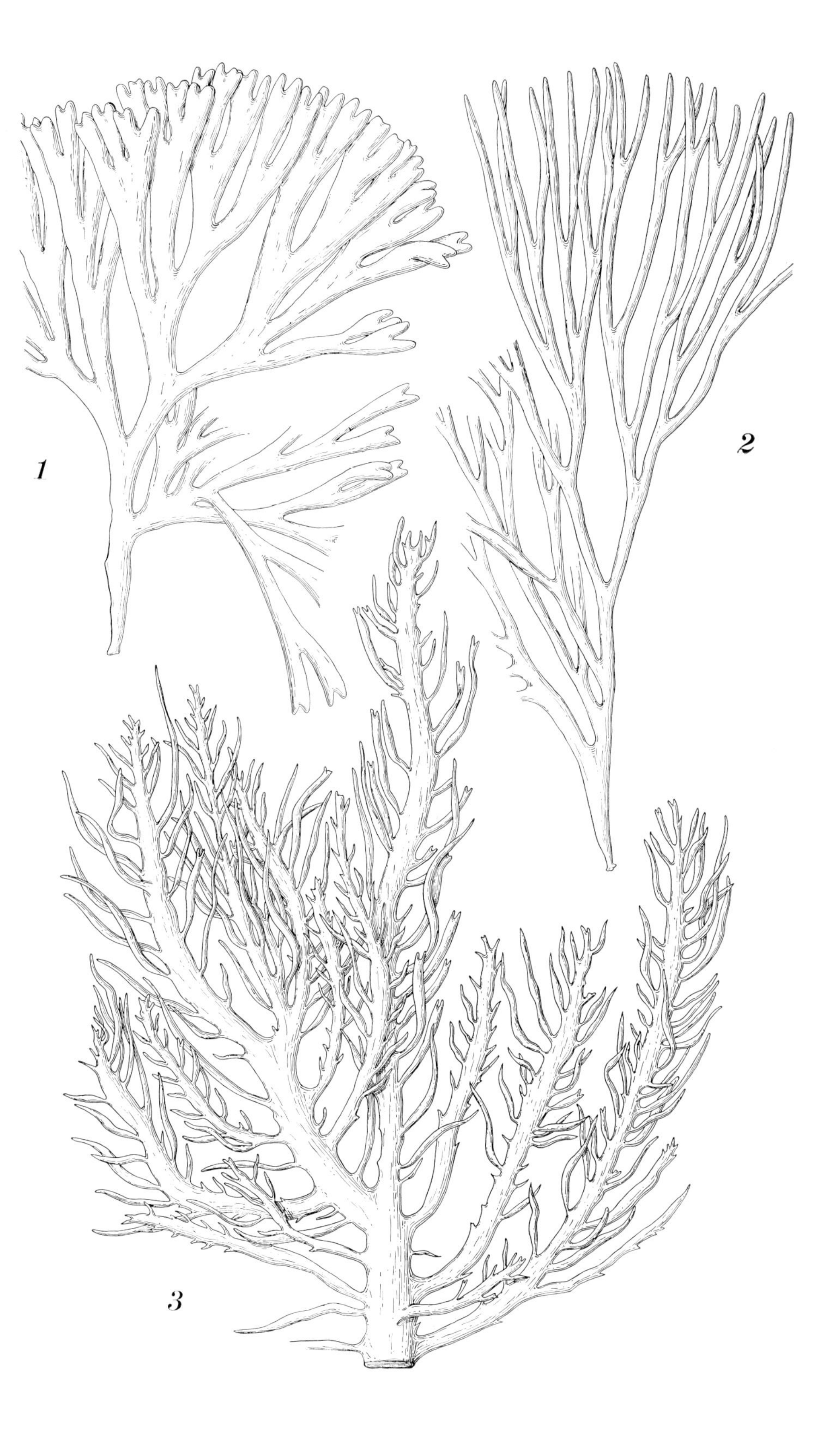
1
2
3

PLATE 52

PAGE

FIG. 1. *Halymenia vinacea:* plant habit, × 0.6 422

FIG. 2. *Halymenia duchassaignii:* plant habit, × 0.9 419

For additional figures of *Halymenia* see Plates 45, 51, 53.

1
2
ROEMHILD

PLATE 53

PAGE

FIG. 1. *Halymenia bermudensis:* plant habit, × 0.95 419

FIG. 2. *Halymenia floridana:* plant habit, × 0.3 420

For additional figures of *Halymenia* see Plates 45, 51, 52.

1
2

PLATE 54

PAGE

FIG. 1. *Gracilaria curtissiae:* habit of a plant with a rather broad base, × 0.5 449

FIGS. 2–3. *Grateloupia filicina:* habit of rather irregularly pinnate specimens, one with a broad, the other with a narrow axis, × 1.0 .. 424

FIG. 4. *Grateloupia cuneifolia:* habit of an unusually branched specimen, × 1.0 425

FIGS. 5–7. *Fauchea peltata:* habit of three plants, × 1.0 477

For additional figures of *Gracilaria* see Plates 45, 55–57, 59.

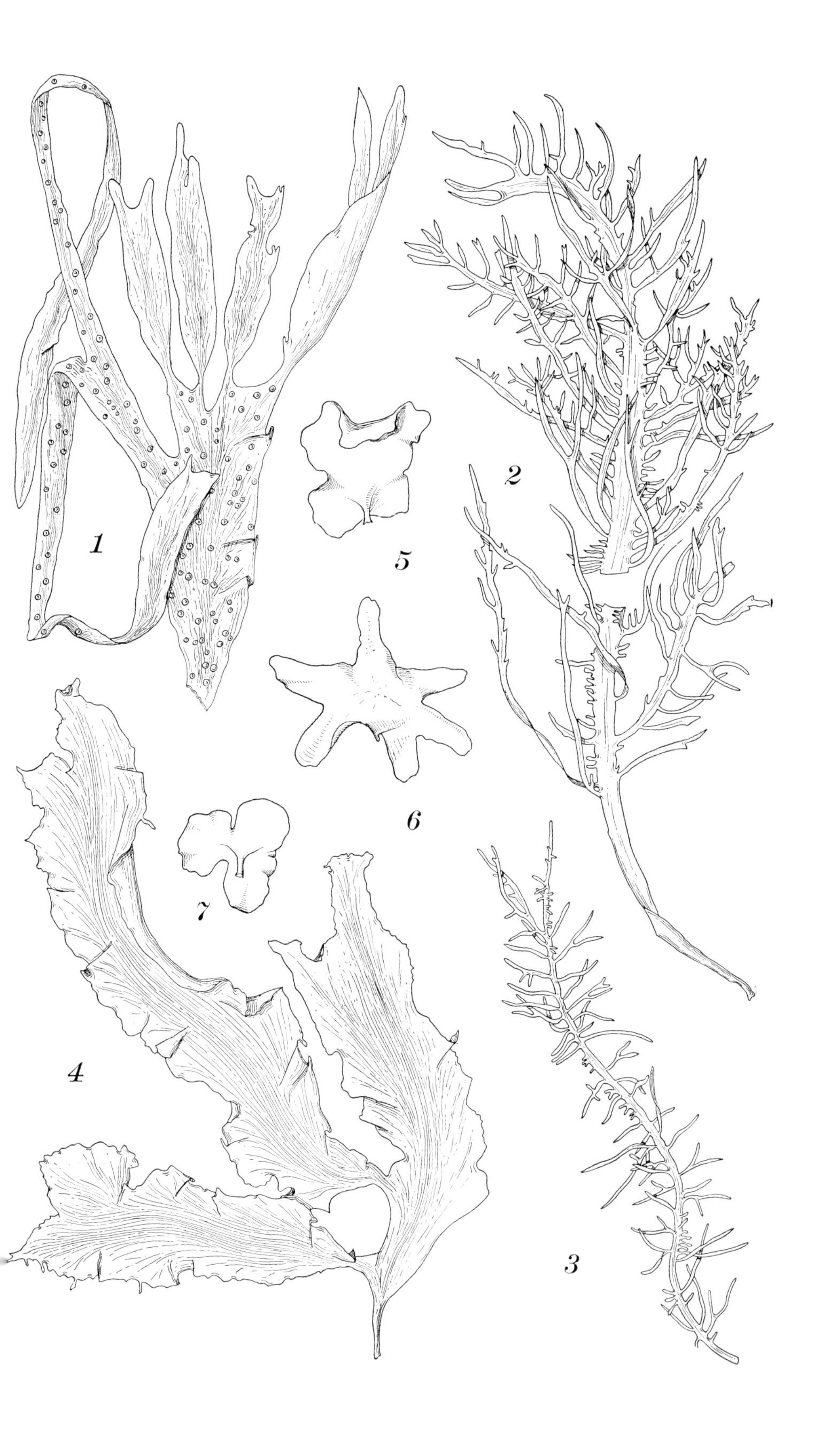
1
2
3
4
5
6
7

PLATE 55

PAGE

FIG. 1. *Gracilaria foliifera:* habit of a narrowly branched plant, × 1.0 446

FIG. 2. *Gracilaria damaecornis:* habit of a rather irregularly branched plant, × 1.0. The flat-topped habit becomes obscured in preparing plants for the herbarium.......... 443

FIG. 3. *Gracilaria venezuelensis:* habit of part of a plant, × 1.0.... 448

FIG. 4. *Gracilaria crassissima:* plant habit, × 1.0.......... 443

For additional figures of *Gracilaria* see Plates 45, 54, 56, 57, 59.

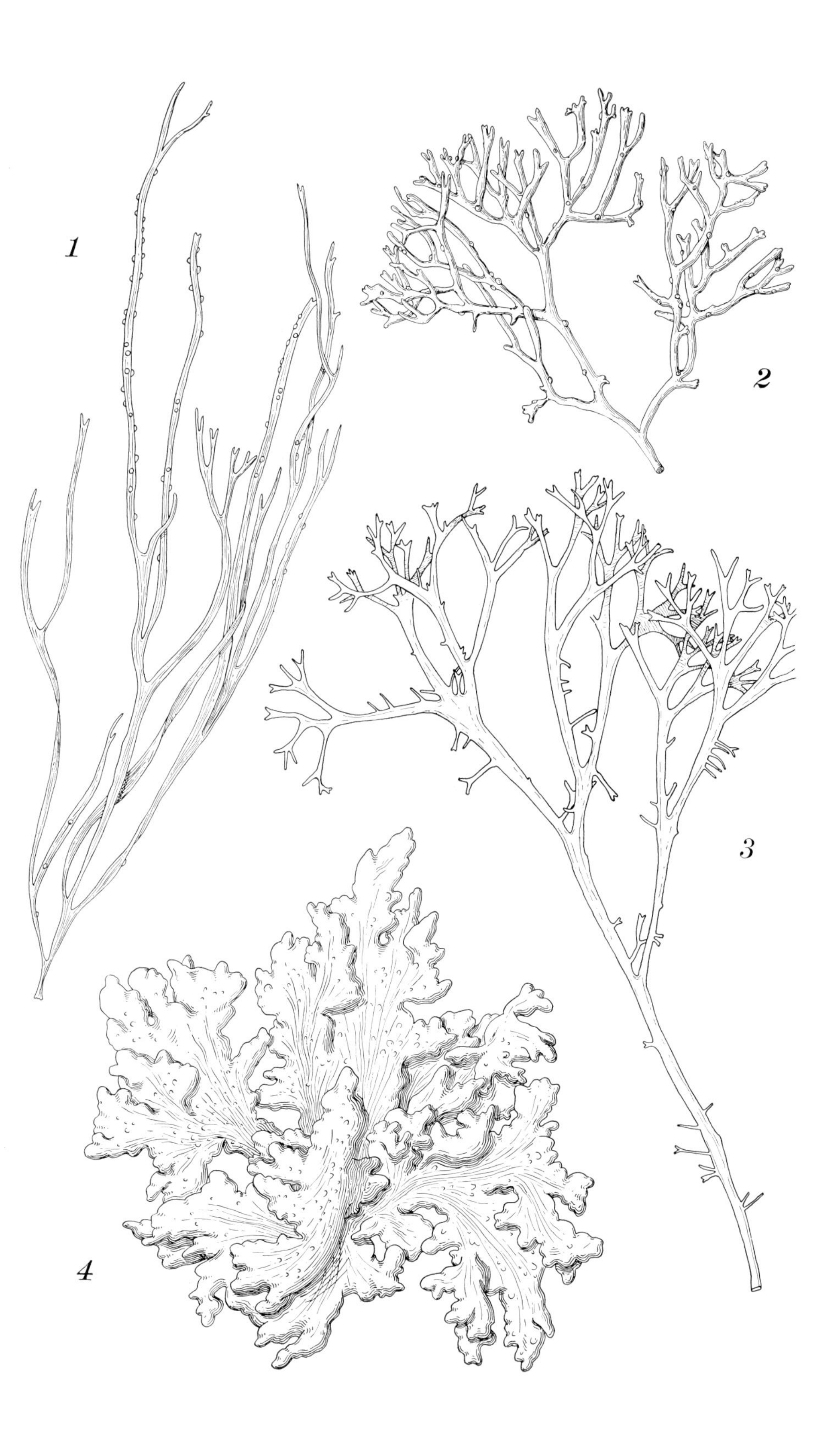
1
2
3
4

PLATE 56

For additional figures of *Gracilaria* see Plates 45, 54, 55, 57, 59.

1
2
3
4

PLATE 57

PAGE

FIGS. 1–2. *Gracilaria domingensis:* a narrow branch from a slender specimen and a portion of the axis with a side branch from a more flattened one, × 0.9, × 1.35 446

FIG. 3. *Gracilaria debilis:* plant habit, × 0.75 442

FIG. 4. *Gracilaria crassissima:* a portion of a specimen with narrow branches, × 0.9 443

For additional figures of *Gracilaria* see Plates 45, 54–56, 59.

1
3
2
4

PLATE 58

PAGE

Fig. 1. *Cryptopleura fimbriata:* part of a fertile plant, × 1.0..... 554

Fig. 2. *Acrosorium uncinatum:* a small plant showing the hooked branchlets, with one enlarged, × 1.0.......................... 552

Fig. 3. *Cryptonemia luxurians:* a branch with proliferating blades, × 1.0 428

Fig. 4. *Cryptonemia crenulata:* plant habit, × 1.0................ 427

Fig. 5. *Agardhiella ramosissima* v. *dilatata:* detail of a few lateral branches, × 1.0.. 457

For additional figures of *Acrosorium* and of *Cryptopleura* see Plate 69, for *Cryptonemia*, Plate 80.

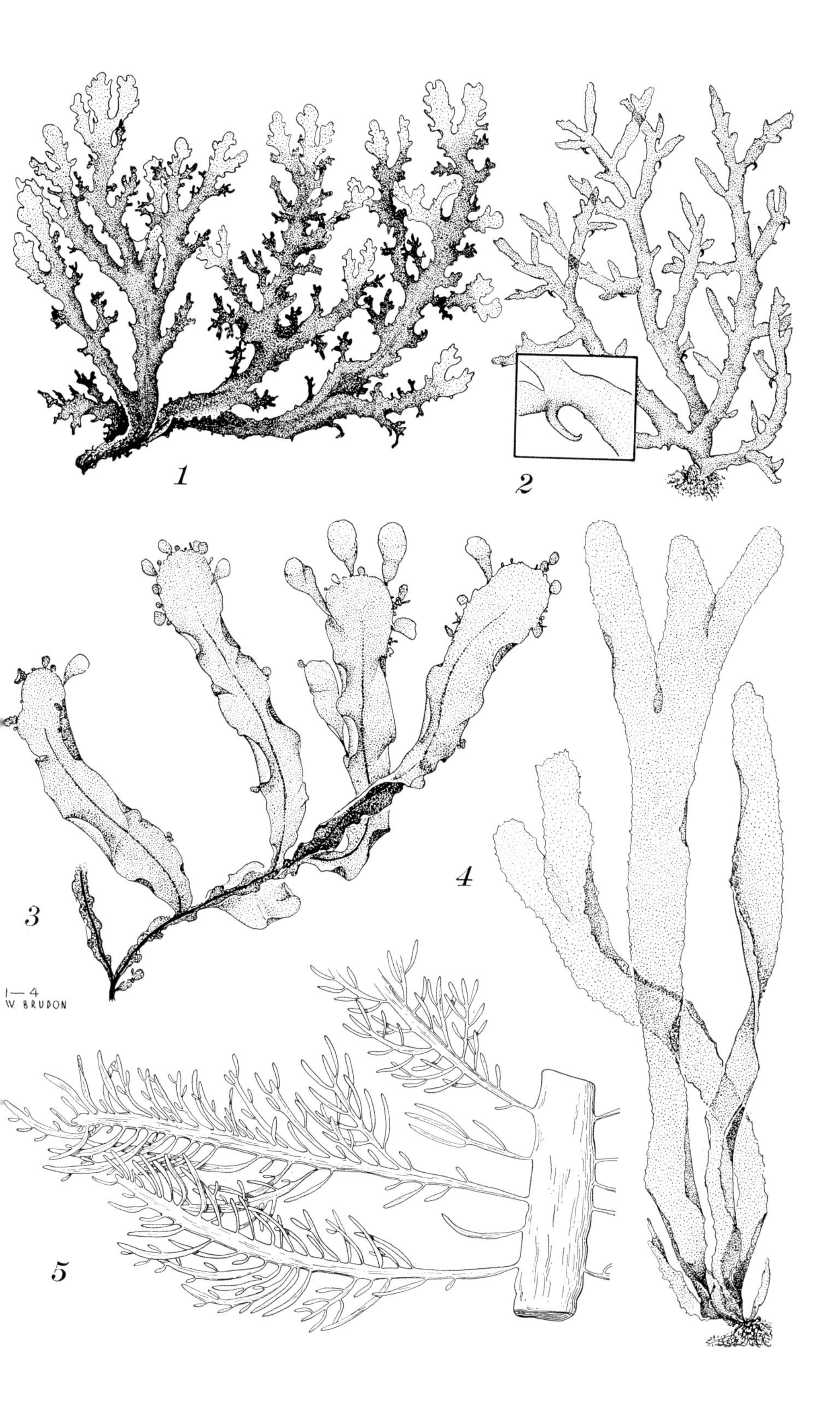
1
2
3
4
5
1—4
W. BRUDON

PLATE 59

PAGE

FIG. 1. *Dictyota ciliolata* v. *bermudensis:* a main branch, × 1.25, and a branch tip, × 3.6 223

FIG. 2. *Plocamium brasiliense:* part of a plant, × 1.6 453

FIG. 3. *Rhodymenia occidentalis:* plant habit, × 0.8 485

FIG. 4. *Gracilaria mammillaris:* plant habit, × 1.1 447

For additional figures of *Dictyota* see Plates 30–32; for *Gracilaria,* Plates 45, 54–57.

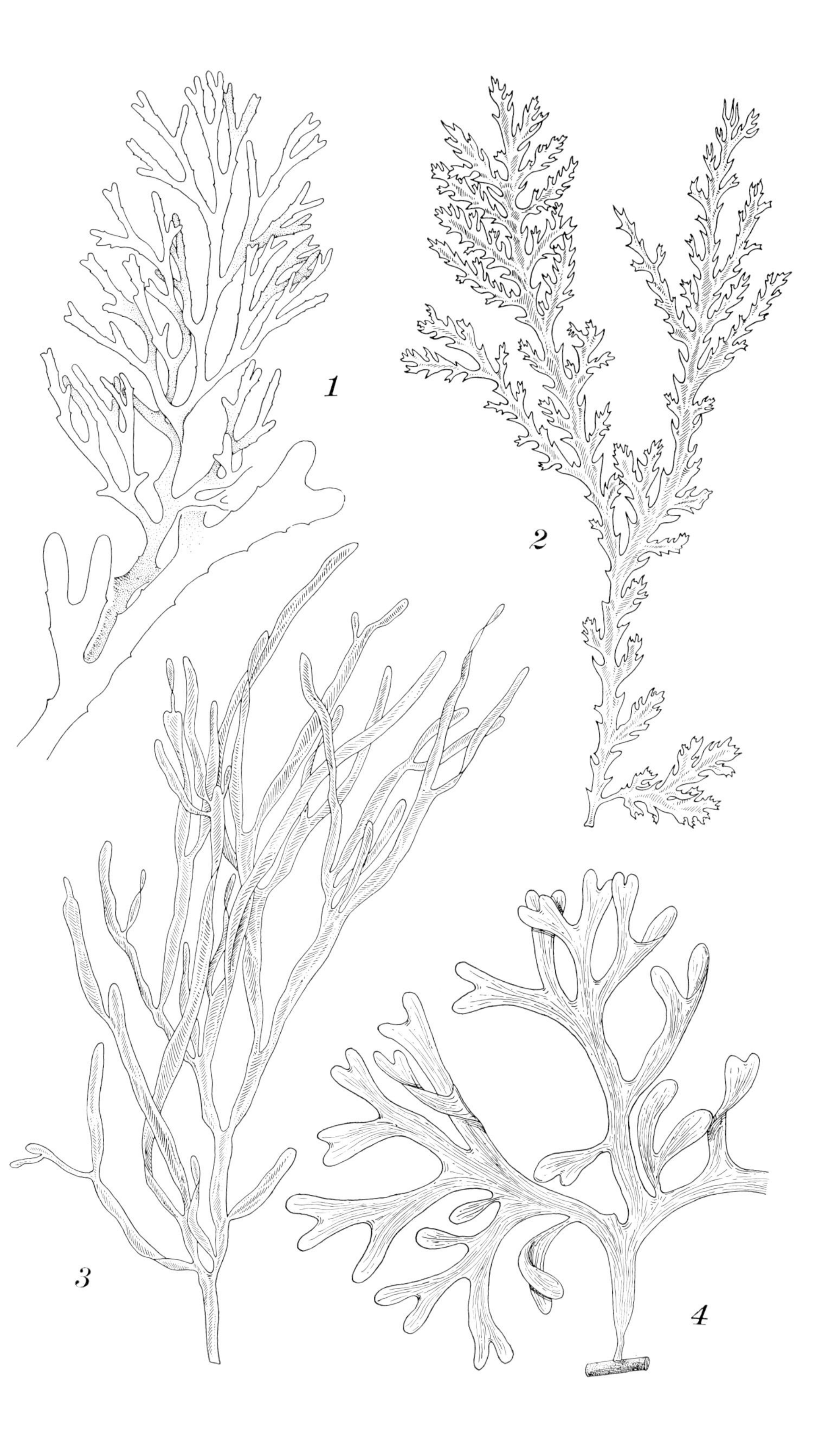
1
2
3
4

PLATE 60

PAGE

For additional figures of *Gigartina* see Plate 61.

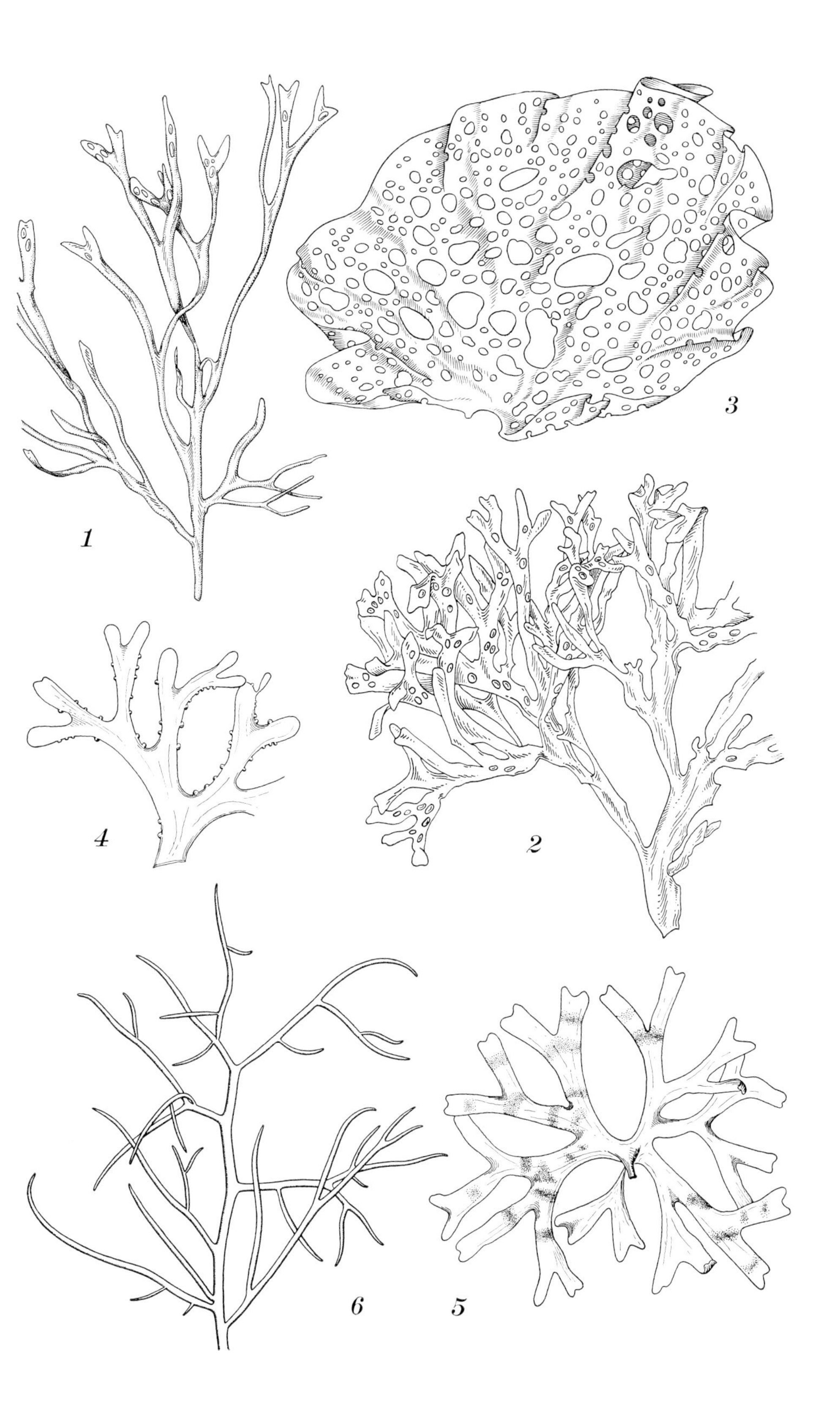

1
2
3
4
5
6

PLATE 61

	PAGE
Fig. 1. *Gigartina elegans:* habit of part of a plant, × 2...........	474
Fig. 2. *Gastroclonium ovatum:* plant habit, × 1.0................	490
Fig. 3. *Gigartina teedii:* habit of part of a plant, × 2.25..........	473
Fig. 4. *Champia parvula:* part of a plant, × 2..................	490
Fig. 5. *Champia salicornoides:* plant habit, × 2................	491
Fig. 6. *Coelarthrum albertisii:* plant habit, × 1.5..............	482

For additional figures of *Gigartina* see Plate 60.

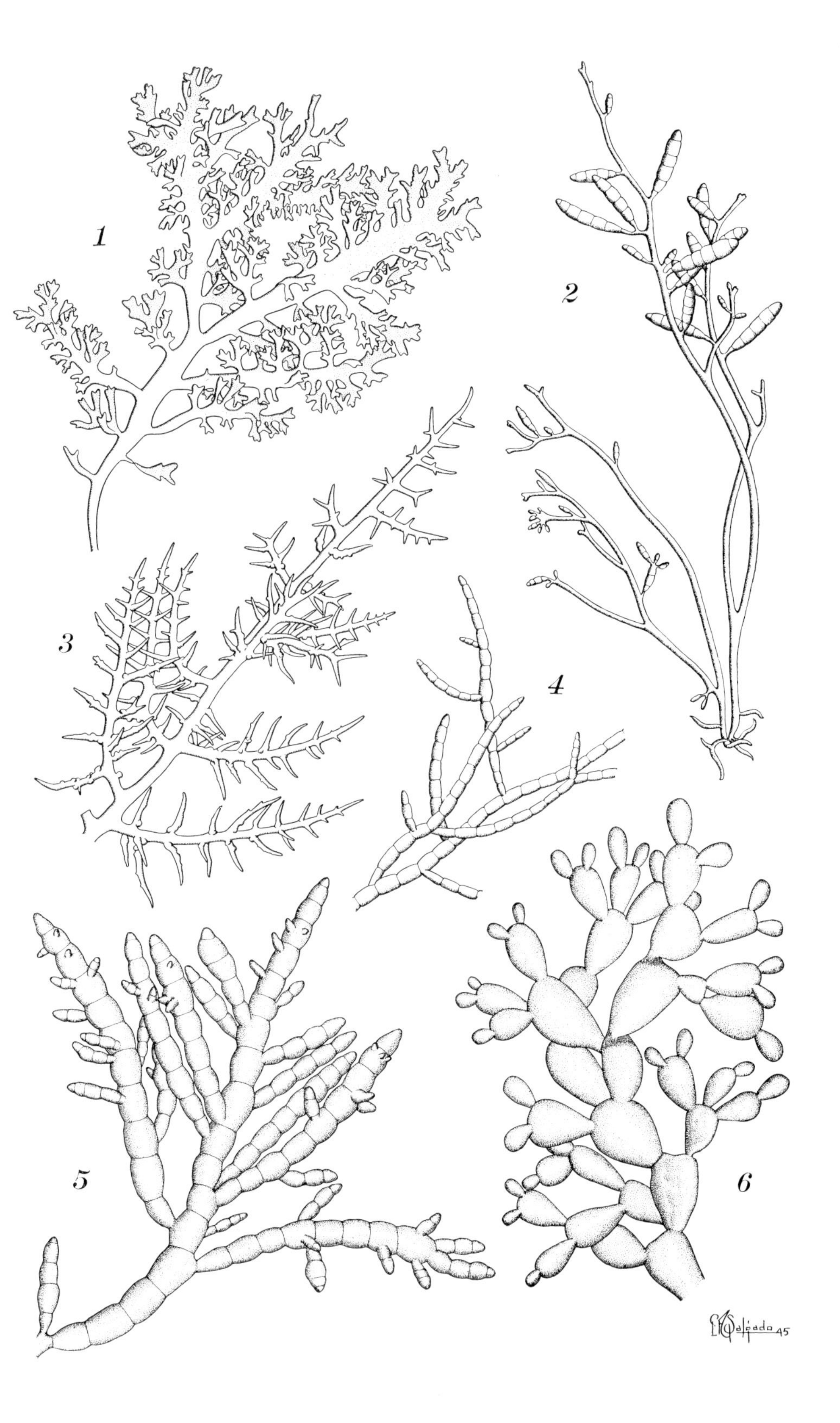
1
2
3
4
5
6
E.Y. Salgado 45

PLATE 62

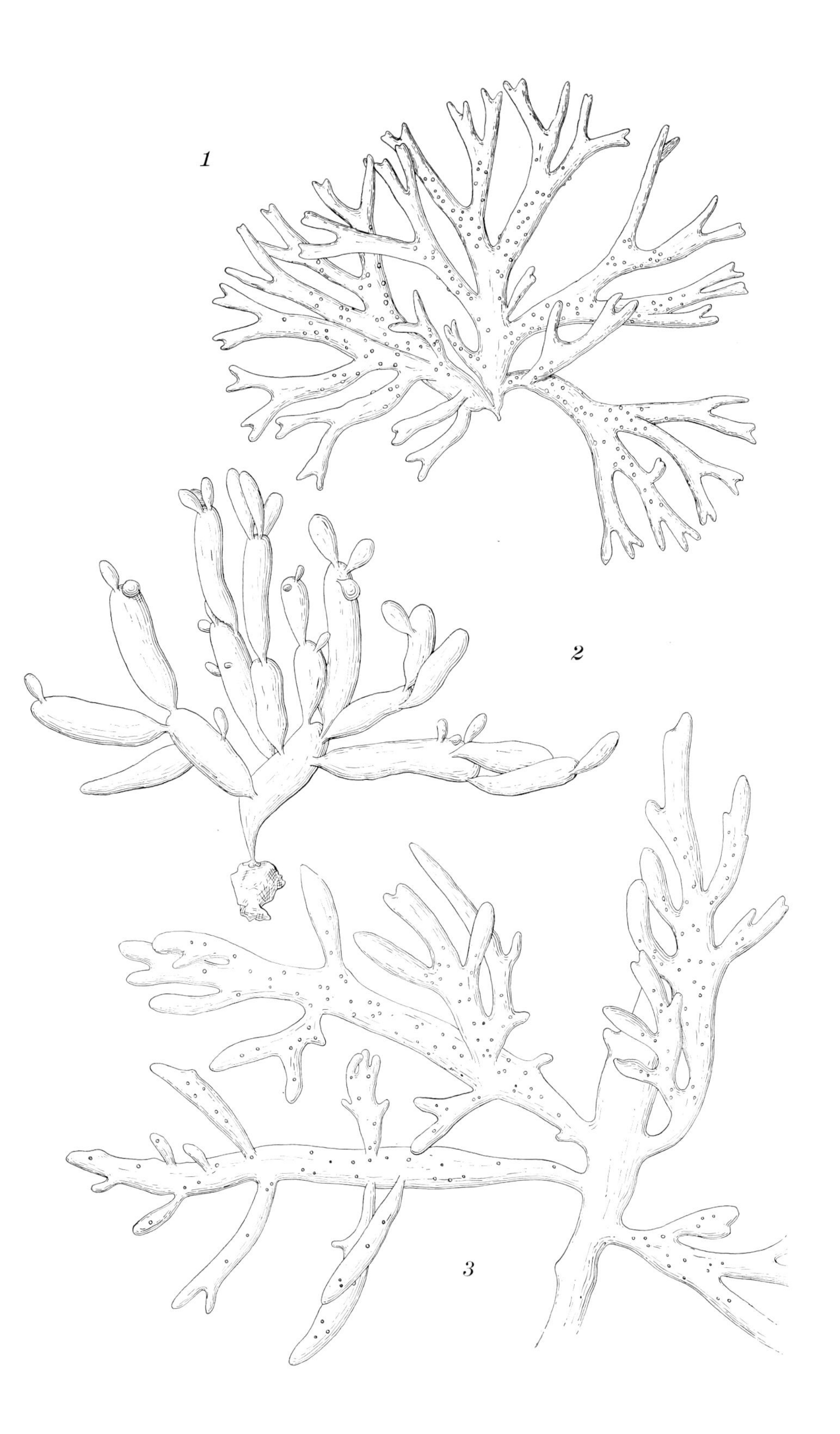
1
2
3

PLATE 63

1
2
ROEMHILD

PLATE 64

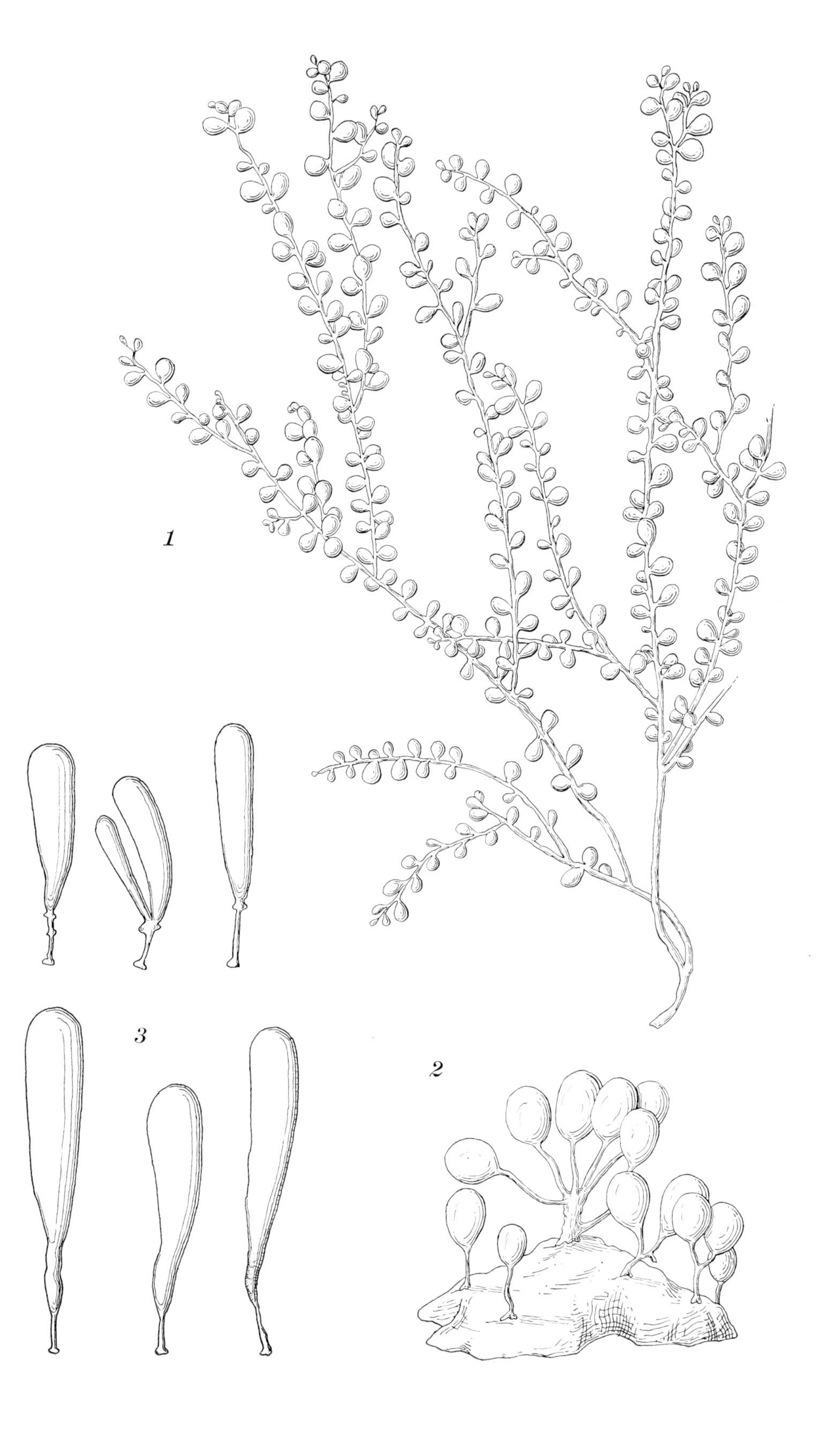
1
3
2

PLATE 65

PAGE

Fig. 1. *Dohrniella antillarum:* part of an axis to show young branching, and details of two mature branchlet tips, × 125, × 310...... 501

Fig. 2. *Spermothamnion gorgoneum:* portion of a plant with polysporangia, × 60... 521

Figs. 3–4. *Ceramium uruguayense:* young and old nodes, × 145.... 532

Figs. 5–6. *Ceramium subtile:* young and old nodes, × 340, × 220.. 527

For additional figures of *Ceramium* see Plates 66, 67. These figures are in part from Taylor and Arndt (1929) and Taylor (1939a, 1942).

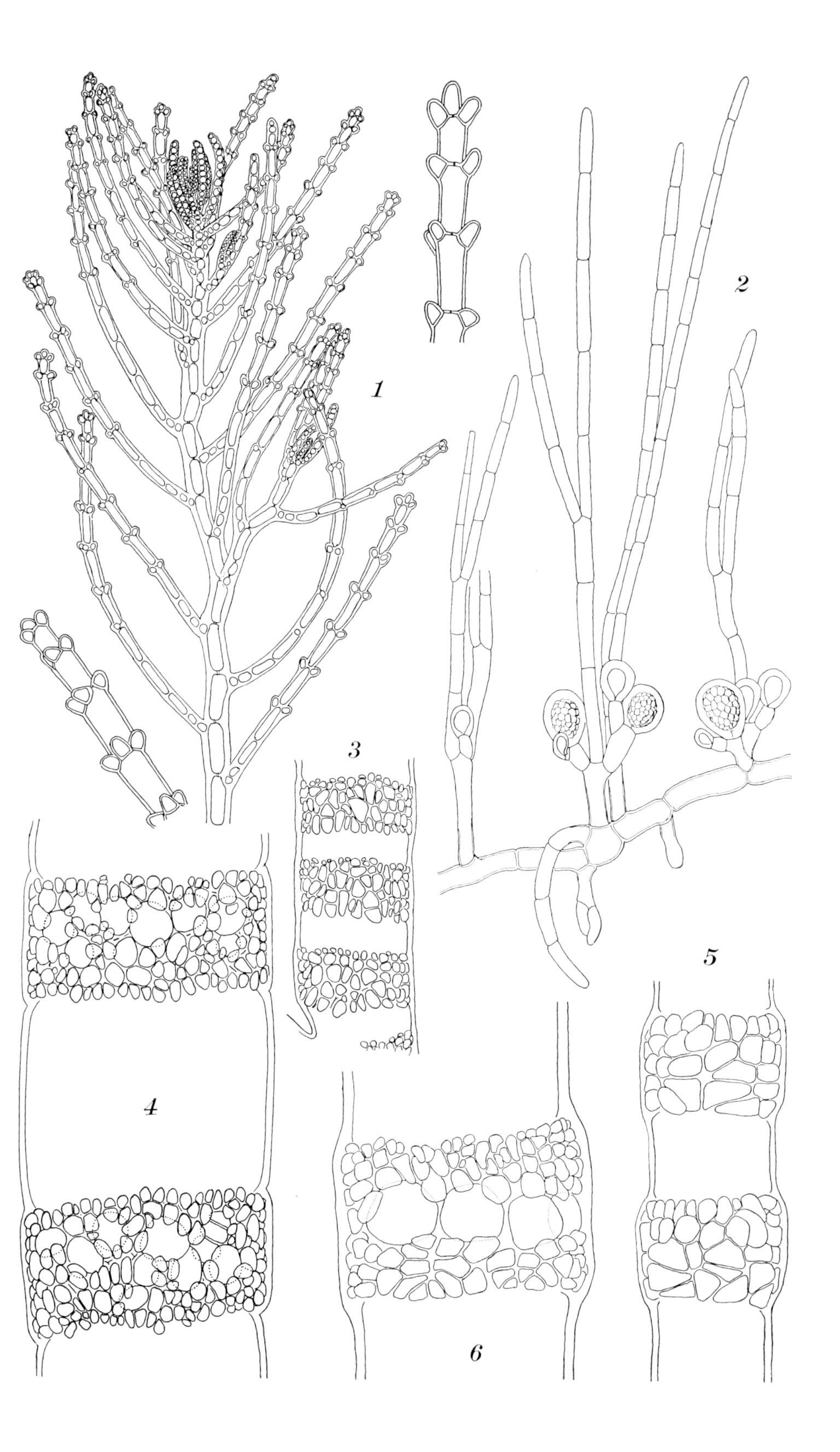
1
2
3
4
5
6

PLATE 66

PAGE

FIGS. 1–4. *Gymnothamnion elegans:* branch tip; portions from the upper, middle and attenuate lower portions of an axis, × 165.... 522

FIGS. 5–6. *Wrangelia penicillata:* branchlet with involucrate spermatangial cluster; lower portion of an axis showing advance of cortication, × 45.. 503

FIGS. 7–8. *Wrangelia argus:* branchlet with involucrate tetrasporangium; lower portion of an axis showing advanced cortication, × 85, × 45.. 502

FIGS. 9–10. *Wrangelia bicuspidata:* two branchlets showing forked tips; lower node showing loose cortication, × 165, × 45........ 503

FIGS. 11–12. *Brongniartella mucronata:* branchlet and an enlarged tip, × 45, × 85.. 590

FIG. 13. *Catenella repens:* habit of part of a plant, × 5............ 462

FIG. 14. *Ceramium nitens:* part of a branch showing relation of cortication to the axial septa, × 220........................ 535

FIG. 15. *Spyridia filamentosa:* tip of a branchlet, × 415........... 539

FIG. 16. *Spyridia aculeata:* tip of a branchlet, × 415.............. 541

For additional figures of *Ceramium* see Plates 65, 67; for *Spyridia,* Plate 71; for *Wrangelia,* Plate 74. These figures are from Taylor, *Marine Algae of Florida* (1928).

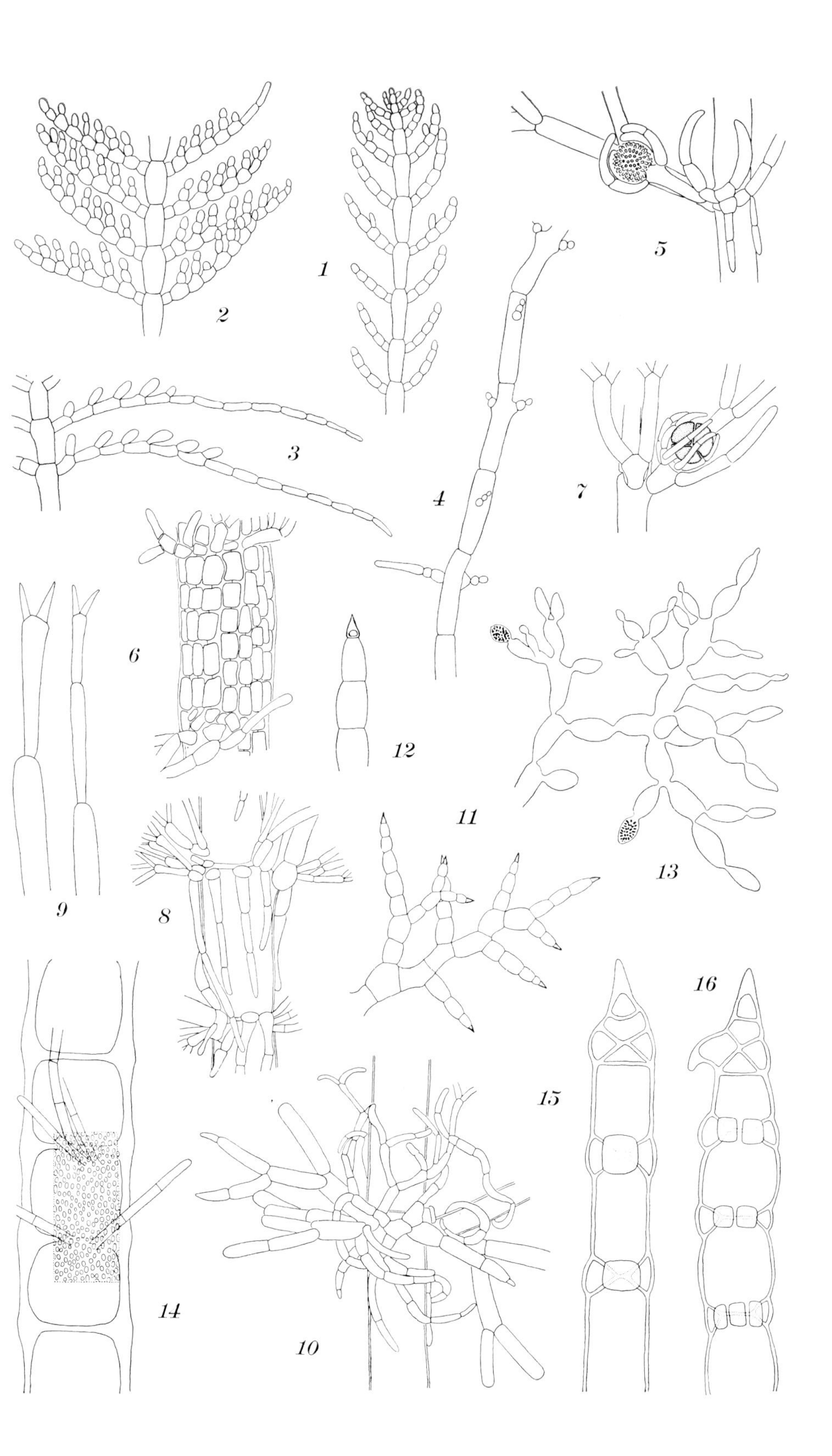
1
2
3
4
5
6
7
8
9
10
11
12
13
14
15
16

PLATE 67

PAGE

FIGS. 1–3. *Ceramium byssoideum:* cortication of the nodes in three stages, × 325, × 325, × 240 528

FIGS. 4–6. *Ceramium fastigiatum* f. *flaccida:* cortication of the nodes in three stages, × 370, × 325, × 240 527

FIGS. 7–9. *Ceramium brevizonatum* v. *caraibica:* cortication of the nodes in three stages, × 370, × 325, × 155 527

For additional figures of *Ceramium* see Plates 65, 66.

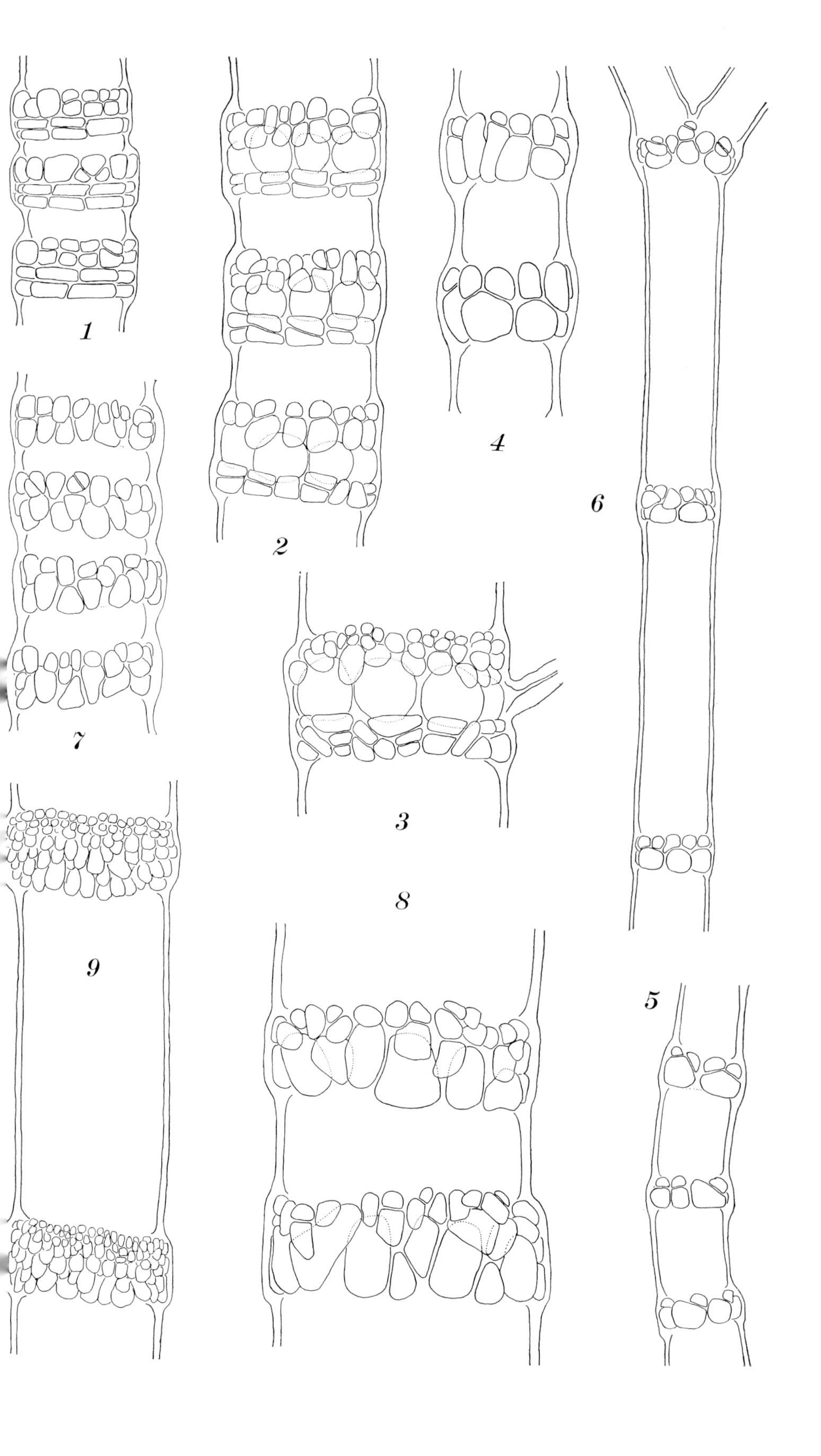
1
2
3
4
5
6
7
8
9

PLATE 68

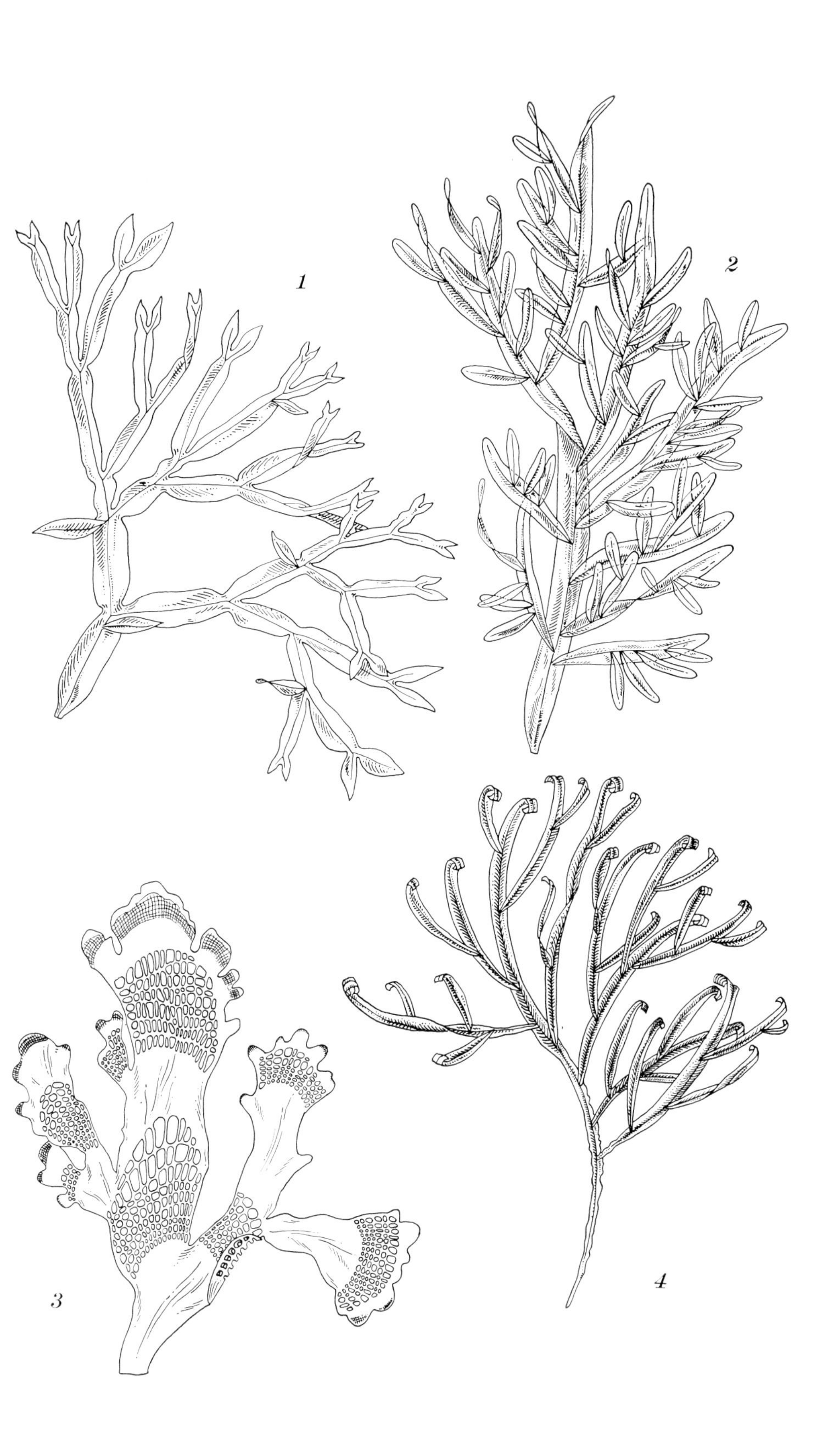
1
2
3
4

PLATE 69

PAGE

FIG. 1. *Nitophyllum wilkinsoniae:* a portion of a plant, × 1.6...... 552

FIG. 2. *Nitophyllum punctatum:* part of a plant, × 1.6........... 551

FIG. 3. *Acrosorium odontophorum:* a portion of a plant, × 1.1..... 553

FIG. 4. *Cryptopleura lacerata:* a small part of a plant, × 1.1...... 554

FIG. 5. *Grinnellia americana* v. *caribaea:* plant habit, × 0.55....... 547

For additional figures of *Acrosorium* and of *Cryptopleura* see Plate 58.

1
2
3
4
5

PLATE 70

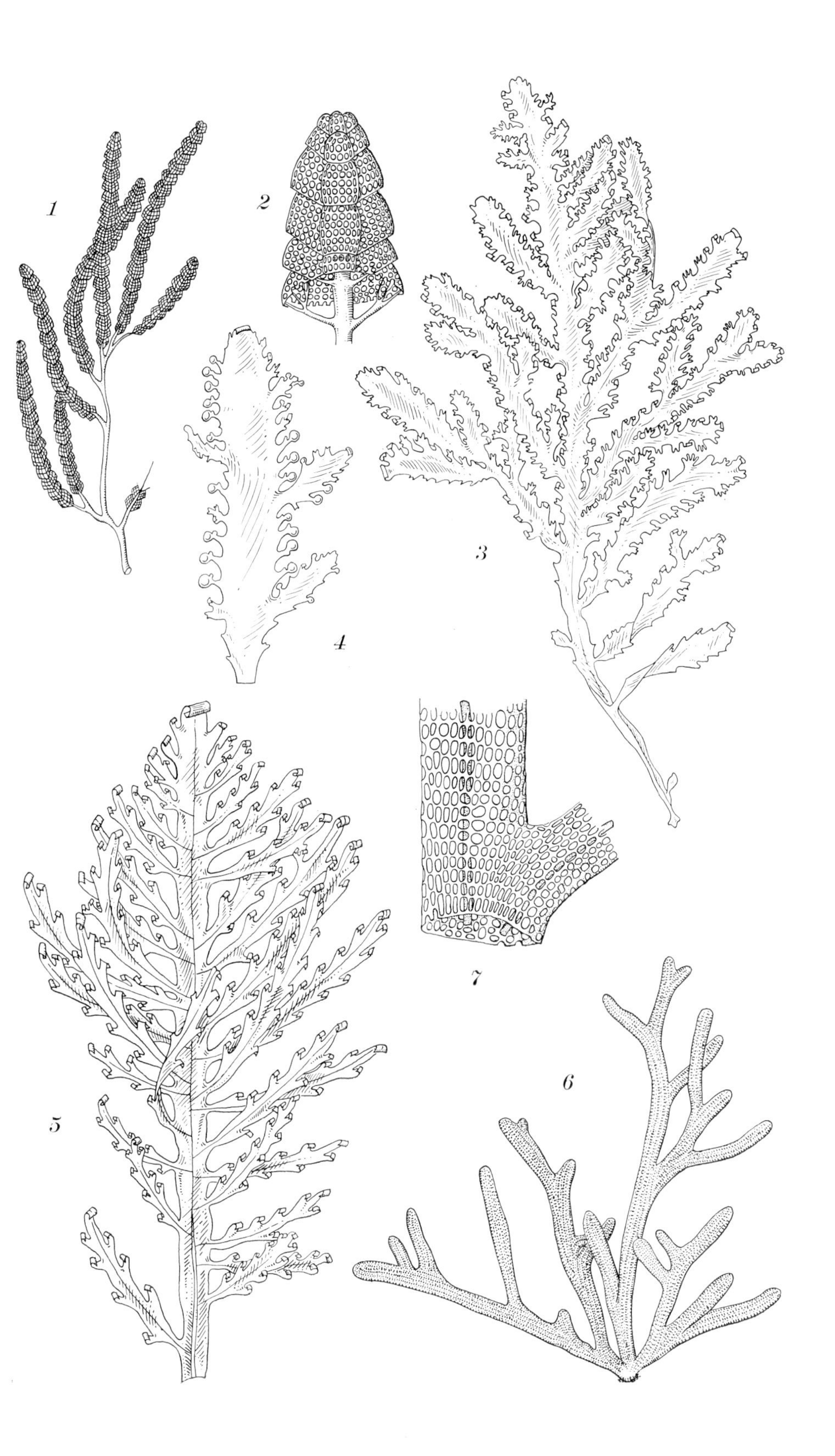
1
2
3
4
5
6
7

PLATE 71

PAGE

Fig. 1. *Dasya crouaniana:* habit of part of a plant, × 1.5.......... 561

Fig. 2. *Bryocladia cuspidata:* habit of a few branches, × 1.5...... 586

Fig. 3. *Acanthophora spicifera:* habit of part of a plant, × 0.9.... 620

Fig. 4. *Asparagopsis taxiformis:* habit of part of a plant, × 0.75... 348

Fig. 5. *Spyridia aculeata* v. *hypneoides:* habit of a branch, × 2.. 541

For additional figures of *Acanthophora* see Plate 72, for *Dasya*, Plate 72, for *Spyridia*, Plate 66.

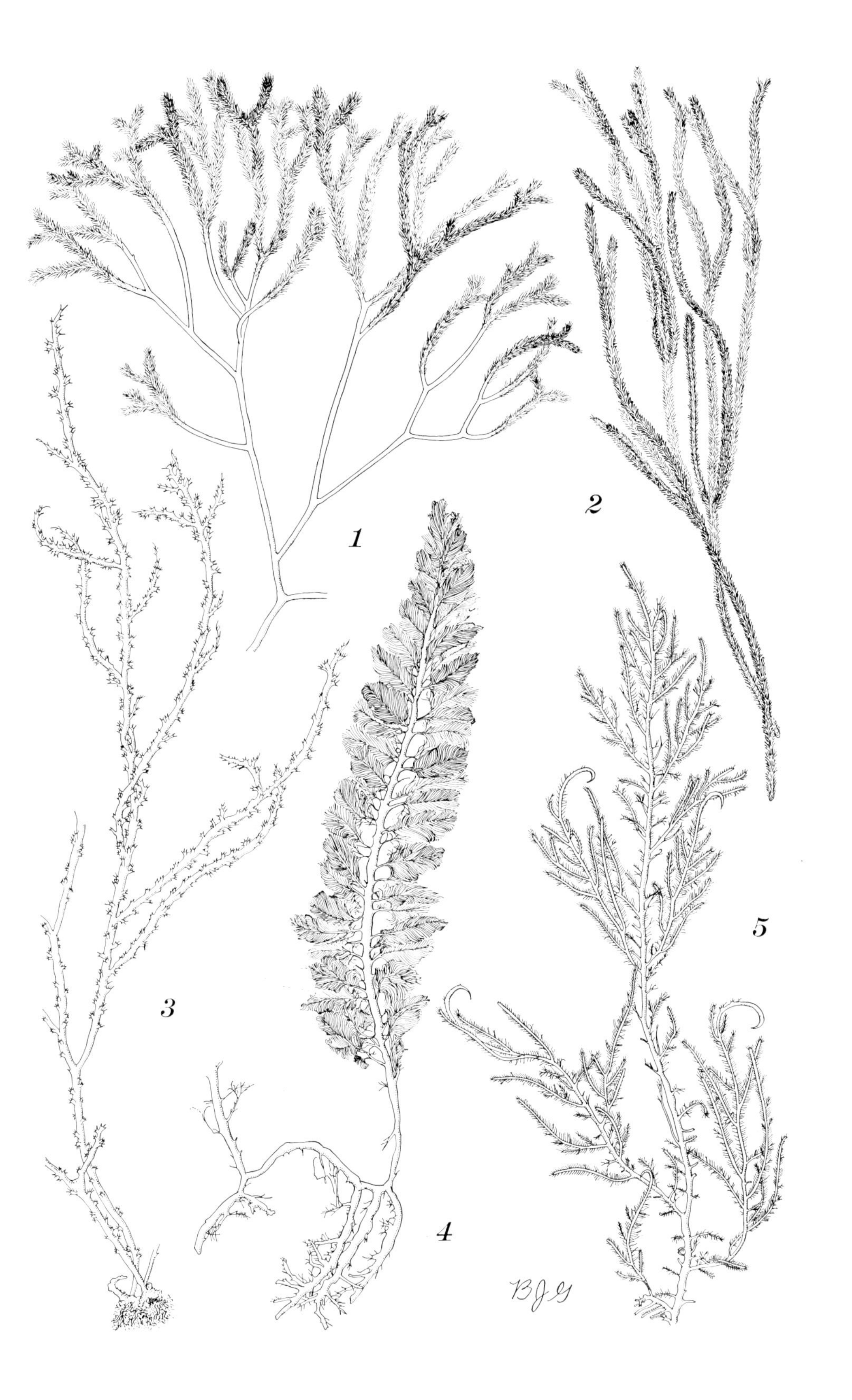
1
2
3
4
5
BJG

PLATE 72

PAGE

FIGS. 1–2. *Acanthophora spicifera:* tip and middle portion of a branch, × 4.5 620

FIG. 3. *Acanthophora muscoides:* tip of a branch showing spines on the main axis as well as on the branchlets, × 4.5 619

FIG. 4. *Dasya rigidula:* portion of an axis with ramellar base, × 92. 558

FIG. 5. *Halodictyon mirabile:* portion of a plant, × 23 567

FIG. 6. *Bryothamnion triquetrum:* terminal fork of an axis, × 46. 587

FIG. 7. *Heterosiphonia gibbesii:* ramulus with basal segments polysiphonous, terminal uniseriate, × 28 566

FIG. 8. *Falkenbergia hillebrandii:* portion of a branch, × 135 571

FIG. 9. *Heterosiphonia wurdemanni:* a ramellar tuft on an axis, × 28. 565

FIGS. 10–11. *Herposiphonia secunda:* tips of a creeping axis, and a mature erect branchlet, × 50 604

FIG. 12. *Herposiphonia tenella:* a portion of a creeping axis with a mature, erect branchlet, × 50 604

For additional figures of *Acanthophora* see Plate 71; for *Bryothamnion,* Plate 73; for *Dasya,* Plate 71; for *Heterosiphonia,* Plate 73. These figures are from Taylor, *Marine Algae of Florida* (1928).

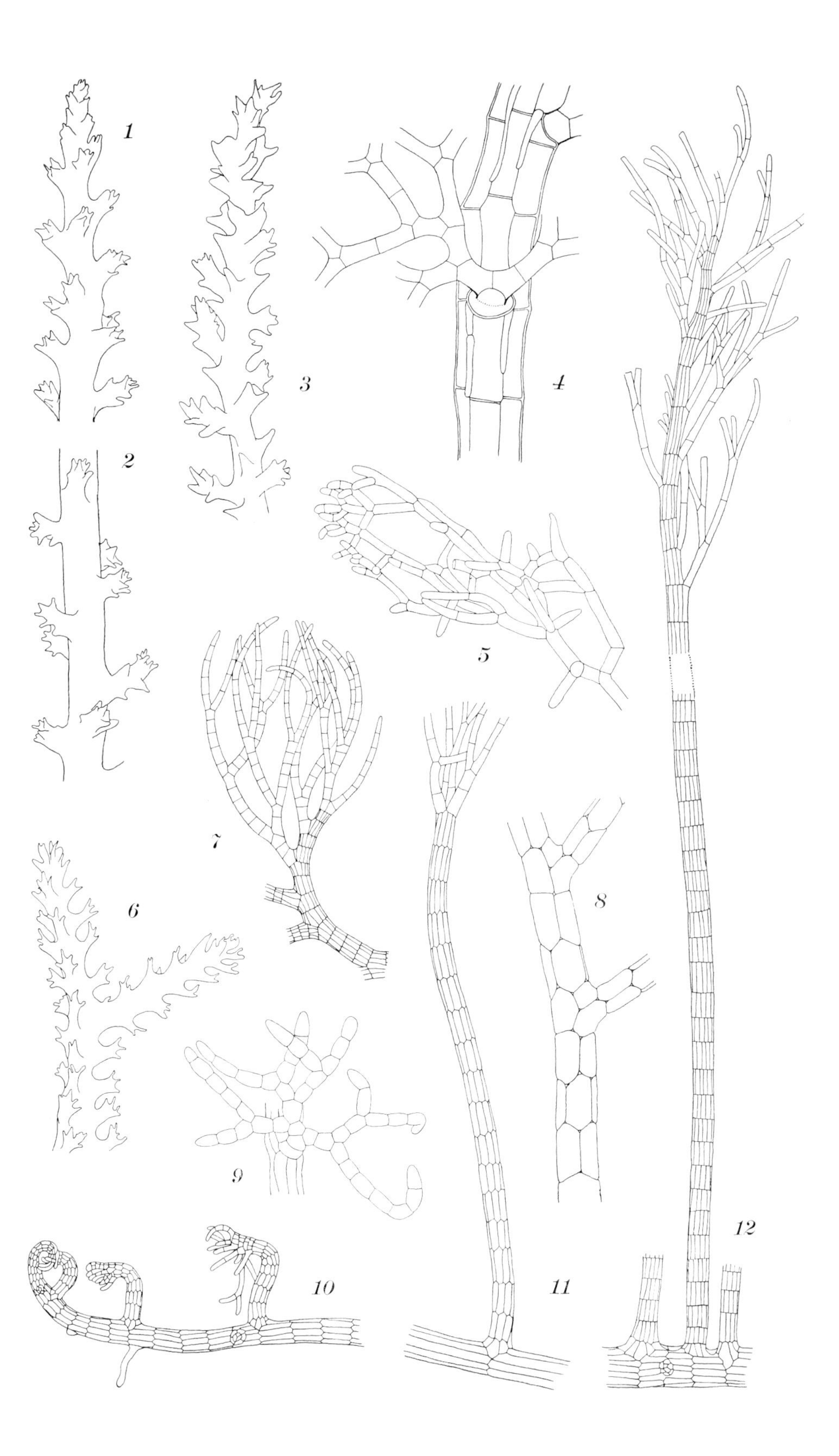
1
2
3
4
5
6
7
8
9
10
11
12

PLATE 73

PAGE

FIG. 1. *Hypnea musciformis:* plant habit, × 1.2 467

FIG. 2. *Hypnea cervicornis:* plant habit, × 1.0 466

FIG. 3. *Bryothamnion seaforthii:* habit of a portion of a plant, × 0.8 587

FIG. 4. *Bryothamnion triquetrum:* habit of a portion of a plant, × 1.0 587

FIG. 5. *Heterosiphonia gibbesii:* plant habit, × 1.0 566

For additional figures of *Bryothamnion* and *Heterosiphonia*, see Plate 72.

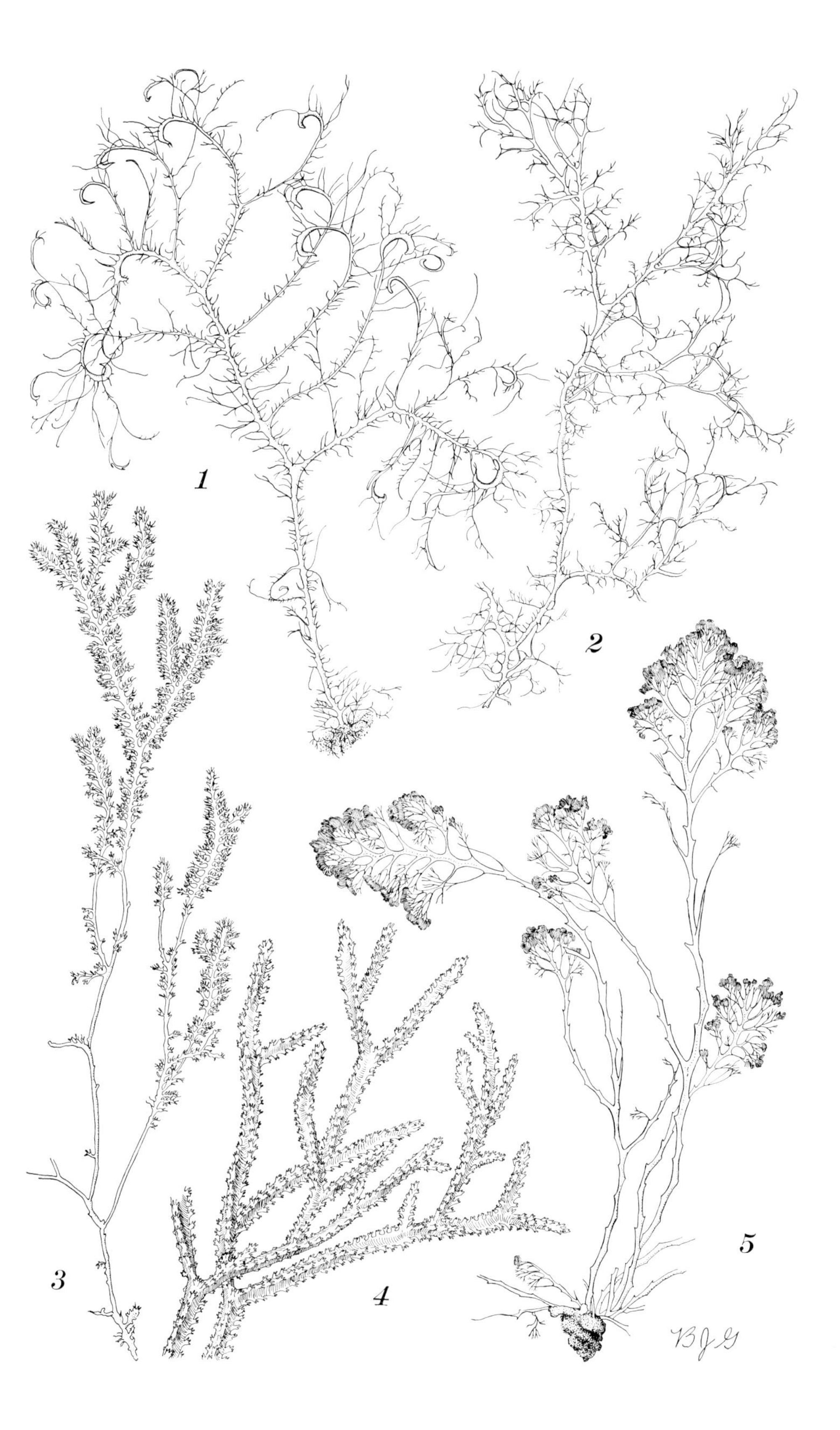
1
2
3
4
5
BJG

PLATE 74

PAGE

Fig. 1. *Bostrychia montagnei:* habit of part of a plant, × 2...... 598

Fig. 2. *Laurencia papillosa:* habit of part of a plant, × 1.0......... 623

Fig. 3. *Chondria cnicophylla:* habit of a lateral branch, × 1.5...... 614

Fig. 4. *Chondria floridana:* habit of a lateral branch, × 1.5........ 616

Fig. 5. *Wrangelia penicillata:* plant habit, × 3................... 503

For additional figures of *Wrangelia* see Plate 66.

1
2
3
4
5
BJG

PLATE 75

PAGE

FIG. 1. *Padina haitiensis:* photograph of the tetrasporic type specimen, × 1.2 235

FIG. 2. *Padina perindusiata:* photograph of the tetrasporic type specimen, × 1.25 235

For additional figures of *Padina* see Plate 34.

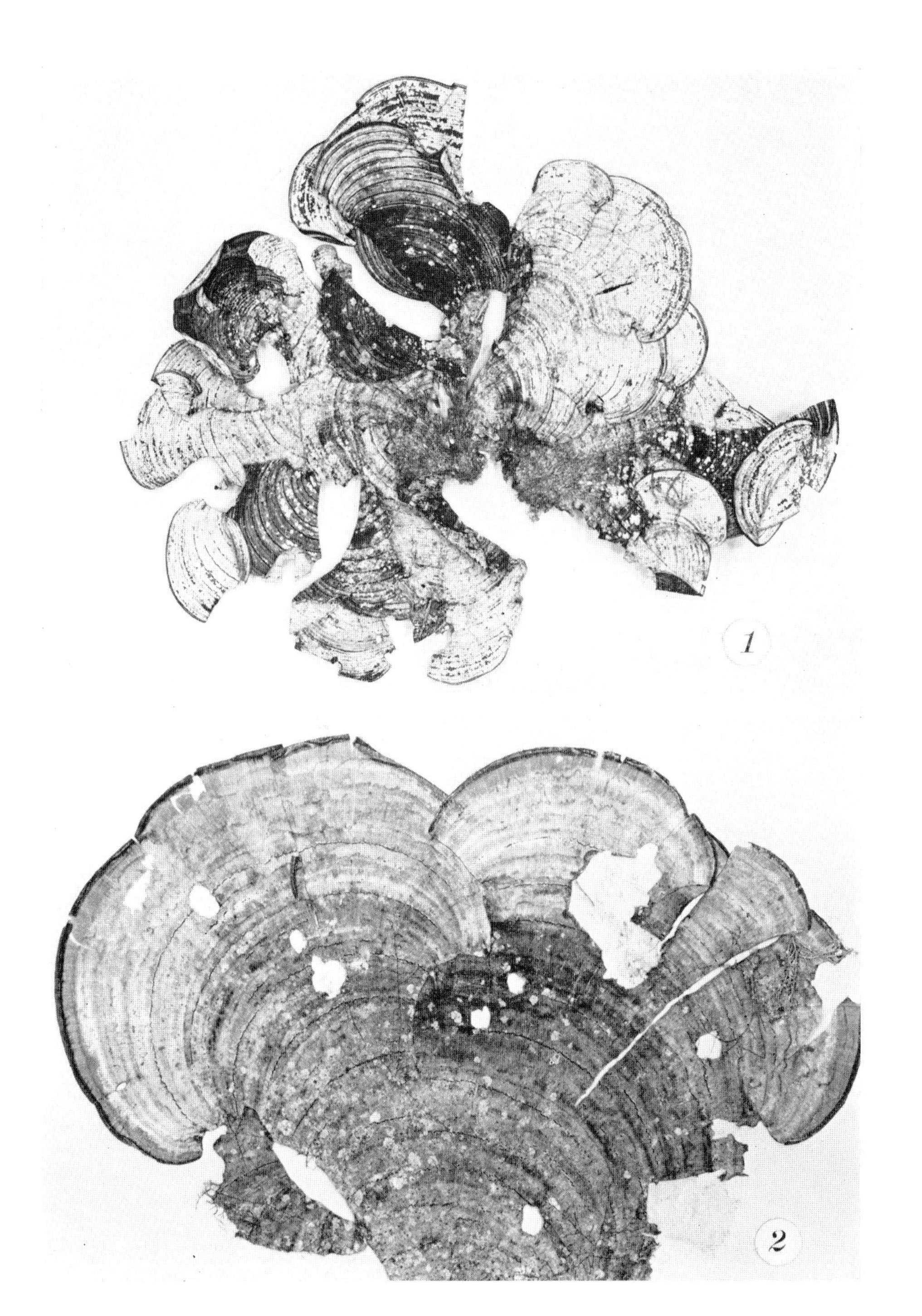

1
2

PLATE 76

PAGE

FIG. 1. *Archaeolithothamnium episporum:* general view of a specimen, × 1.55 378

FIG. 2. *Goniolithon boergesenii:* general view of a specimen with conceptacles, × 1.9 397

FIG. 3. *Goniolithon solubile:* general view of a specimen with conceptacles, × 1.3 396

For additional figures of *Goniolithon* see Plates 48, 77, 78.

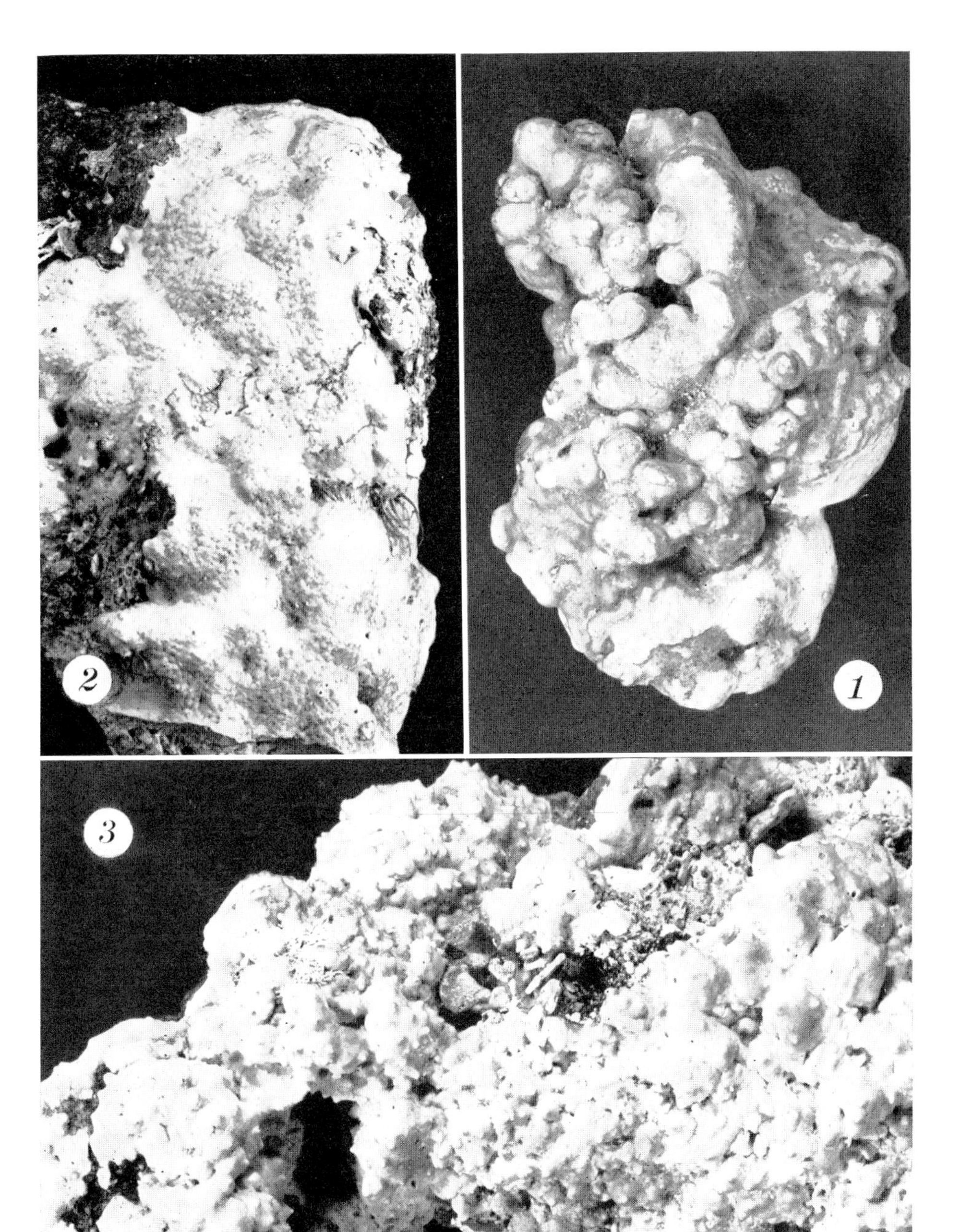

PLATE 77

For additional figures of *Goniolithon* see Plates 48, 76, 78.

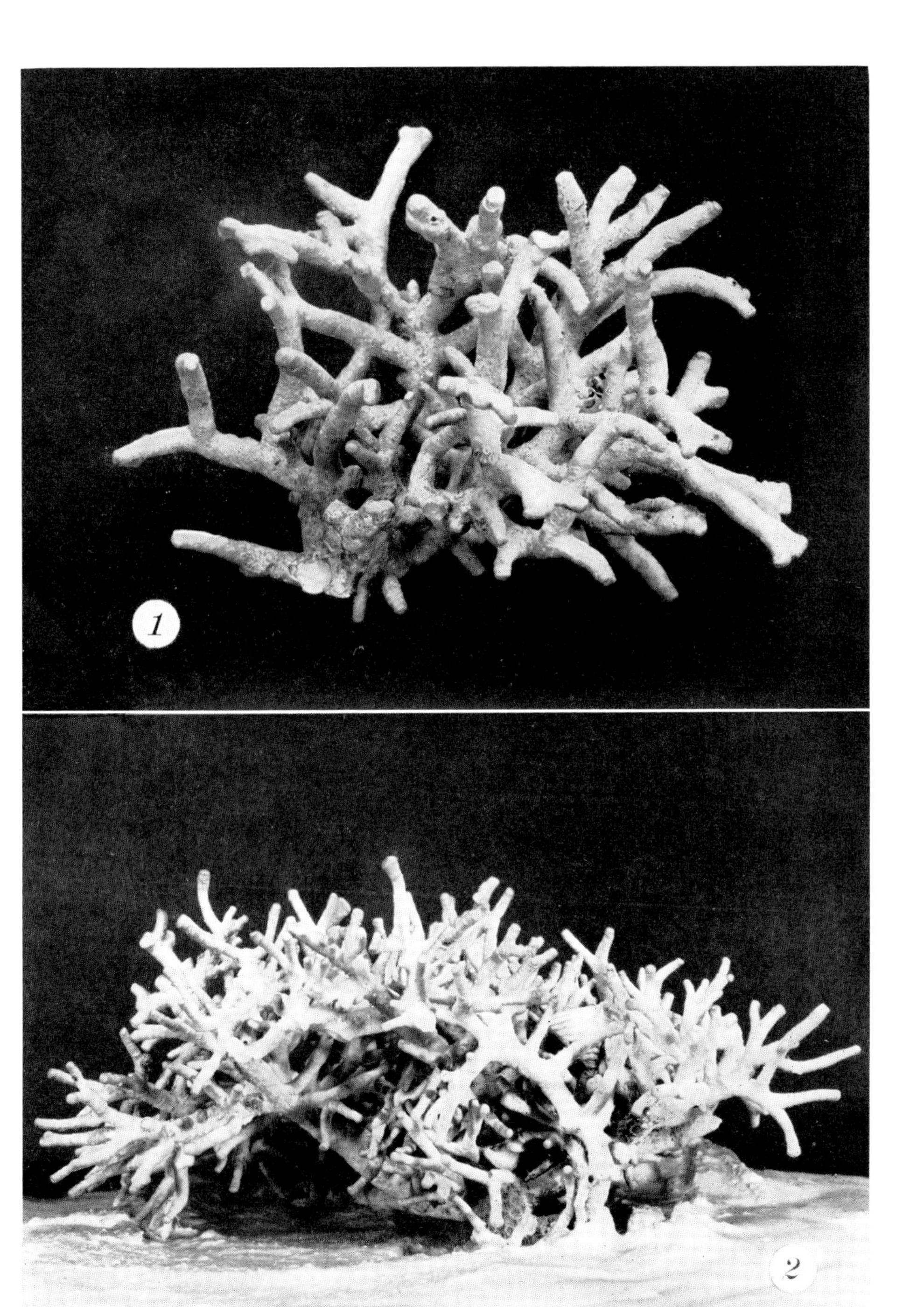
1
2

PLATE 78

PAGE

FIG. 1. *Goniolithon spectabile* f. *nana* (prox.): general view of plants on a massive substrate, × 1.3 399

FIG. 2. *Lithothamnium incertum:* general view of a plant attached to a coral, × 1.2 384

For additional figures of *Goniolithon* see Plates 48, 76, 77; for *Lithothamnium* see Plate 79.

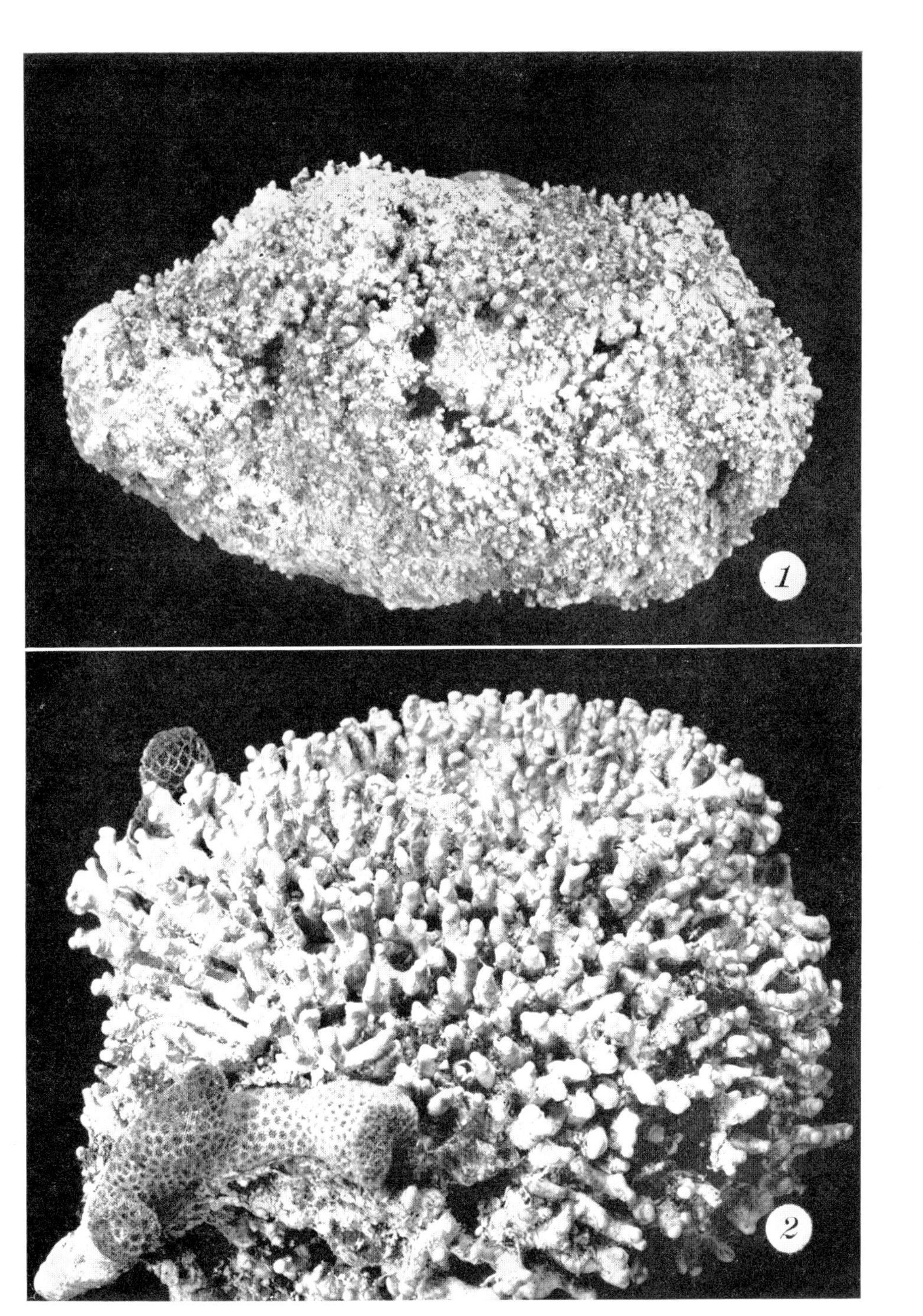
1
2

PLATE 79

For an additional figure of *Lithothamnium* see Plate 78.

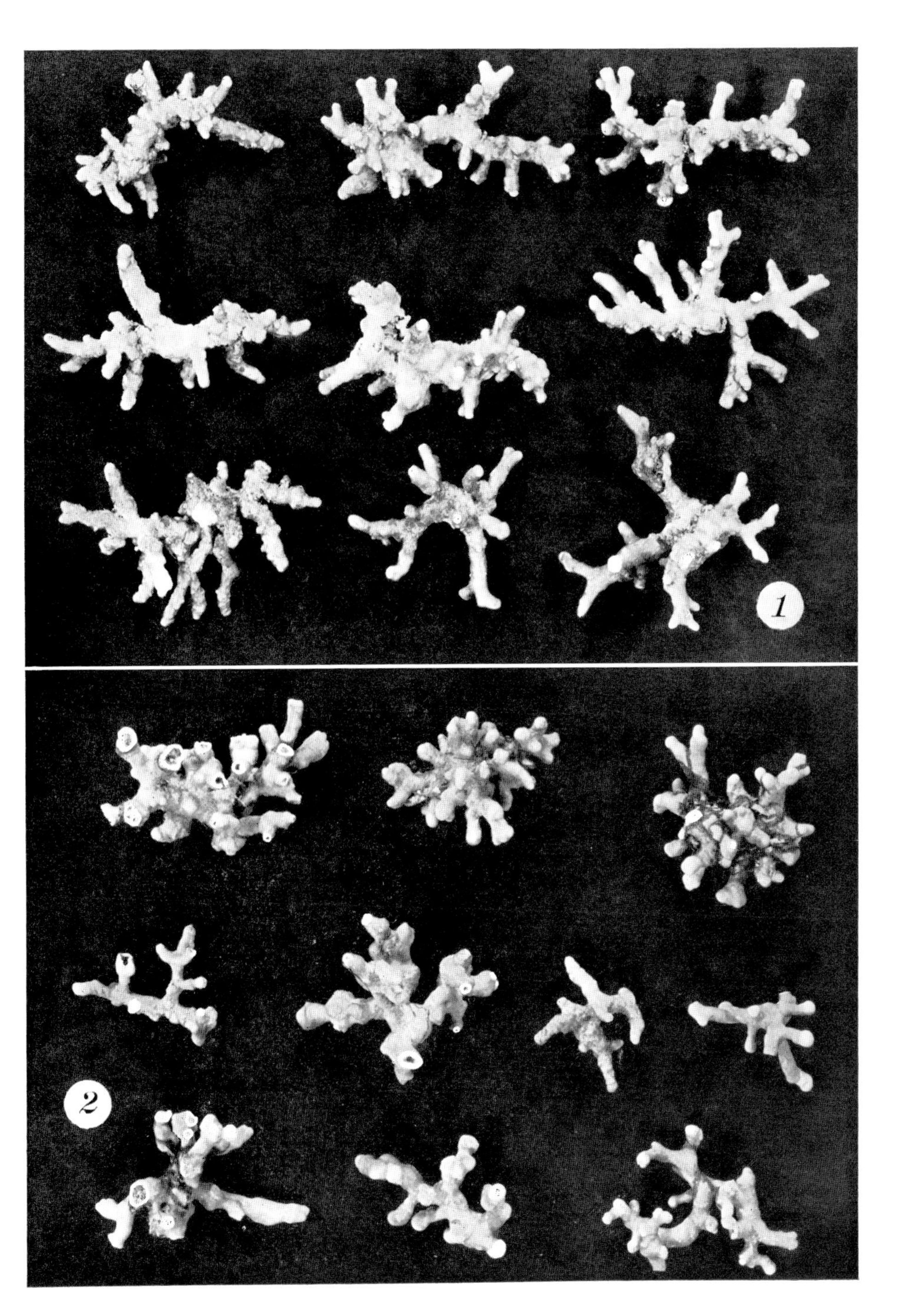
1
2

PLATE 80

PAGE

FIG. 1. *Cryptonemia bengryi:* habit of the type specimen × 0.6.... 427

FIG. 2. *Kallymenia limminghii:* habit of several plants; *a*, a very proliferous specimen, *b*, *c*, two juvenile blades, *d*, seven other plants with strap-shaped blades having erose margins and small proliferations, × 0.95.. 432

For additional figures of *Cryptonemia* see Plate 58.

1
2
a
b
c
d

INDEX

INDEX